AERODYNAMICS OF
TURBINES AND COMPRESSORS

VOLUME X

HIGH SPEED AERODYNAMICS

AND JET PROPULSION

AERODYNAMICS OF TURBINES AND COMPRESSORS

EDITOR: W. R. HAWTHORNE

PRINCETON, NEW JERSEY

PRINCETON UNIVERSITY PRESS

1964

London: OXFORD UNIVERSITY PRESS

L. C. CARD 58-5029

PRINTED IN THE UNITED STATES OF AMERICA BY
THE MAPLE PRESS COMPANY, INC., YORK, PENNA.

FOREWORD

On behalf of the Editorial Board, I would like to make an acknowledgement to those branches of our military establishment whose interest and whose financial support were instrumental in the initiation of this publication program. It is noteworthy that this assistance has included all three branches of our Services. The Department of the Air Force through the Air Research and Development Command, the Department of the Army through the Office of the Chief of Ordnance, and the Department of the Navy through the Bureau of Aeronautics, Bureau of Ships, Bureau of Ordnance, and the Office of Naval Research made significant contributions. In particular, the Power Branch of the Office of Naval Research has carried the burden of responsibilities of the contractual administration and processing of all manuscripts from a security standpoint. The administration, operation, and editorial functions of the program have been centered at Princeton University. In addition, the University has contributed financially to the support of the undertaking. It is appropriate that special appreciation be expressed to Princeton University for its important over-all role in this effort.

The Editorial Board is confident that the present series which this support has made possible will have far-reaching beneficial effects on the further development of the aeronautical sciences.

Theodore von Kármán

PREFACE

Rapid advances made during the past decade on problems associated with high speed flight have brought into ever sharper focus the need for a comprehensive and competent treatment of the fundamental aspects of the aerodynamic and propulsion problems of high speed flight, together with a survey of those aspects of the underlying basic sciences cognate to such problems. The need for a treatment of this type has been long felt in research institutions, universities, and private industry and its potential reflected importance in the advanced training of nascent aeronautical scientists has also been an important motivation in this undertaking.

The entire program is the cumulative work of over one hundred scientists and engineers, representing many different branches of engineering and fields of science both in this country and abroad.

The work consists of twelve volumes treating in sequence elements of the properties of gases, liquids, and solids; combustion processes and chemical kinetics; fundamentals of gas dynamics; viscous phenomena; turbulence; heat transfer; theoretical methods in high speed aerodynamics; applications to wings, bodies and complete aircraft; nonsteady aerodynamics; principles of physical measurements; experimental methods in high speed aerodynamics and combustion; aerodynamic problems of turbo machines; the combination of aerodynamic and combustion principles in combustor design; and finally, problems of complete power plants. The intent has been to emphasize the fundamental aspects of jet propulsion and high speed aerodynamics, to develop the theoretical tools for attack on these problems, and to seek to highlight the directions in which research may be potentially most fruitful.

Preliminary discussions, which ultimately led to the foundation of the present program, were held in 1947 and 1948 and, in large measure, by virtue of the enthusiasm, inspiration, and encouragement of Dr. Theodore von Kármán and later the invaluable assistance of Dr. Hugh L. Dryden and Dean Hugh Taylor as members of the Editorial Board, these discussions ultimately saw their fruition in the formal establishment of the Aeronautics Publication Program at Princeton University in the fall of 1949.

The contributing authors and, in particular, the volume editors, have sacrificed generously of their spare time under present-day emergency conditions where continuing demands on their energies have been great. The program is also indebted to the work of Dr. Martin Summerfield who guided the planning work as General Editor from 1949–1952. The cooperation and assistance of the personnel of Princeton University Press and of the staff of this office has been noteworthy. In particular, Mr. H. S. Bailey, Jr., the Director of the Press, and Mr. R. S. Snedeker, who has supervised the project at the Press have been of great help. The figures were prepared by Mr. Zane Anderson. Special mention is also due Mrs. E. W. Wetterau of this office who has handled the bulk of the detailed editorial work for the program.

Coleman duP. Donaldson
General Editor

PREFACE TO VOLUME X

This volume deals with the problems of the flows in gas turbines and compressors. Such flows are complex, and there is an obvious difficulty in any attempt to describe comprehensively the flow in say, a compressor or a turbine in an exact and rigorous fashion. The choice of topics for the Sections and each author's treatment of any one topic reflect this difficulty. Originally, the application of aerodynamic theory to flow through axial compressors and turbines followed the blade element theory of propellers. Sections B, E, G and K which discuss sub- and supersonic flow and unsteady flow in cascades develop this approach. The next step, that of describing the three-dimensional aspects of the flow, is exemplified in the ideal flow theory developed in Section C. In Sections F and H the attempt is made to describe more completely the flow in axial compressor and turbine stages. With each step towards the final goal the approximations required increase and the unsatisfactory gaps in exact knowledge become more obvious.

The vast effort which has been expended in experimental research, development and testing has warranted the presentation of an account of experimental techniques, Section D. The radial flow turbine and the centrifugal compressor are briefly described in Sections I and J.

As in one or two other volumes in the Series the dates of completion or final revision of the articles have stretched over a period of years. Two manuscripts (Sections K and J) were completed in 1951 and 1952. Sections B, G, I, E and C were completed in their final form between 1956 and 1958, and the remaining three (H, D and F) were finished in 1960. It is regrettable that the publication of the earlier manuscripts has been subject to such vexatious delays. However, although some recent work has inevitably not been included, it is hoped that some of the timeless quality which the authors have aimed at has been achieved.

My sincere thanks go to all the authors for their patience and hard work, to Dr. Coleman duP. Donaldson for his assistance and his friendly but firm pressure, and to the Princeton University Press for their careful work.

W. R. Hawthorne
Volume Editor

CONTENTS

Chapter 2. Real Flow through a Radial Turbine

AERODYNAMICS OF TURBINES AND COMPRESSORS

SECTION A

INTRODUCTION

W. R. HAWTHORNE

A,1. The Gas Turbine. It is now twenty years since the first flight of an aircraft powered by a gas turbine. The importance of this event in the history of aviation may be judged by the astonishing advances which have been made in the last two decades. The stimulus of the new engine, simpler and lighter than the piston engine, has made jet propulsion and high speed flight possible. It has also encouraged the development of other methods of propulsion: the rocket, the ramjet, and the pulse jet. The gas turbine, either in the form of a turbojet or of a turboprop engine, is at present the chief method of aircraft propulsion. The present volume and Vol. XI of this Series are both concerned with the fundamental aspects and components of the gas turbine as an aircraft prime mover. Questions which relate to the performance and design of each different type of engine are dealt with in Vol. XII.

Historically, the trend from the piston engine to the gas turbine developed for the same reasons that led to the replacement of the reciprocating steam engine by the steam turbine. Turbine engines became attractive as larger powers were required, because, for a given size and weight, they can swallow a much larger flow of air than piston engines.

For a long time, however, there were difficulties which prevented the development of turbines using the internal combustion principle. Because internal combustion engines use air and not a condensing fluid, the compression part of the working process requires considerably more energy than that required to drive the feed pump in a steam engine. Therefore, to produce a satisfactory net output of work from an internal combustion engine, the compressor must be efficient and the temperature at the inlet to the turbine much higher than in the external combustion steam engine. The difficulty of making efficient compressors and of producing materials which could withstand high stresses at elevated temperatures led to the failure of early gas turbine schemes. The price of even partial success was illustrated by the gas turbine of the Societé Anonyme des Turbomoteurs, which was built in 1905 according to the designs of Armengaud and Lemale [*1*, p. 1285; *2*, p. 17]. This engine, the first constant pressure gas turbine to run under its own power, had an efficiency of barely 3 per cent. Water injection was used at the entry to the turbine for cooling purposes so that the maximum gas temperature was probably in the neighborhood of 880°F. The multistage centrifugal compressor (consisting of 20 stages or more) had a pressure ratio of 4 and an efficiency of about 60 per cent. (In 1901, an axial flow compressor had been built by Parsons, but its efficiency was no more than 60 per cent. The blades were probably stalled [*3*], for the compressor was built before the first airplane had flown and at a time when aerodynamics was in its infancy.)

It was not until the 1920's that compressors of high enough efficiency to be useful in a gas turbine began to be developed. The improvement in efficiency is attributable to the use of the growing knowledge of aerodynamics of airfoils and airplanes. In the 1930's, materials suitable for high gas turbine temperatures began to appear and the stage was set for the first successful gas turbine.

The history of the development of the aircraft gas turbine is given in XII,A. This volume and Vol. XI consider some of the fundamental aspects of aircraft gas turbines, particularly their components. Vol. XI is devoted to the combustion process, to some aspects of materials and mechanical strength, and to engine performance. The flow in compressors and turbines is the subject of Vol. X.

Compressors, turbines, and combustion chambers are not the only components in a gas turbine. The intake and the propelling nozzle have become increasingly important as speeds have been raised. Ducts between compressors or between the turbine and the propelling nozzle, the diffusers, which are placed after the intake or between the compressor and the combustion chamber, are components which also require careful design if losses are to be avoided. Discussion of these components will be found in VII,E, which deals with the theoretical aspects, and in Vol. XII, where intakes, diffusers, and nozzles are considered in relation to the engine of which each forms a component.

In the development of the gas turbine, major efforts in research have been applied to the axial flow compressor and turbine. The centrifugal compressor and its mate, the inward flow radial turbine, have been the subject of research and development, too, but in spite of its early successful application the centrifugal compressor has gradually given way to the more complex and costly axial compressor whose efficiency, pressure ratio, and mass flow per unit frontal area are potentially (and practically) higher.

A,2. Development of Theories of Flow through Compressor and Turbine Blades. At the beginning of the century, turbines were designed on the assumption of one-dimensional flow through the blade passages. Bernoulli's equation had by then been extended from incompressible to compressible flows. In fact, the equation for the flow of a compressible gas through a nozzle (isentropic flow) was first given by

St. Venant and Wantzel in 1839 [*4*]. The results were unnoticed until 1867 when Kloster [*5*] and Rankine [*6*] rederived the formula for the critical pressure ratio. The fact that the velocity in the throat was equal to the local velocity of sound appears to have been recognized first by Holtzmann in 1861 [*7*], although Reynolds [*8*, Vol. 2, p. 311] and Hugoniot [*9*] clarified the whole question in 1885 and 1886. The Swedish engineer de Laval patented the convergent-divergent nozzle in 1894 and used it in his steam turbines [*10*].

Steam turbines designed by these methods were on the whole successful, largely because, in a flow with decreasing pressure, there is no tendency for the boundary layer to separate. In the axial compressor, on the other hand, the positive pressure gradients in the passages between the blades tend to cause separation. The development of two-dimensional flow theories started in the 1920's when axial compressor blades began to be designed by considering each blade as an isolated airfoil. References given by Keller [*11*], Tyler [*12*], and Howell [*13*] describe the application of airfoil theory to propellers and fans and the development of theories of two-dimensional flow through cascades or lattices of airfoils. Various methods were employed later for determining the effects of neighboring blades. Betz [*14*], for instance, calculated the corrections to be applied to the flow about an isolated airfoil by replacing the neighboring blades by vortices. Weinig [*15*] determined the potential flow through a cascade of flat plates and used the results to compute the correction factors. Today the literature contains descriptions of methods for determining the potential flow through cascades of blades with arbitrary shapes, or for determining the shapes of blades for reasonably specified distributions of surface pressure. For compressible flows, Lin [*16*] has suggested an exact method, and several approximate methods are available. The solution of the mathematical problem still leaves computational difficulties which are considerable, especially for closely pitched blades such as those found in turbines. Numerous papers, many of them describing approximate methods, bear witness to the complexity of the computations. In Sec. B, which discusses the theory of two-dimensional flow through cascades, considerable attention has therefore been paid to approximate methods.

The theory of two-dimensional flow through cascades should be of great value in predicting the onset of separation. However, control of the boundary layer by altering the profile shape has only limited possibilities. In a compressor or turbine, the blades are pitched so closely together that it is possible to consider them as the boundaries of bent, diverging passages. The diffusion is then determined by the flow inlet and outlet angles. Howell showed in his summary of cascade data in 1942 [*17*] that losses began to increase when the diffusion exceeded a certain value dependent on the pitch-chord ratio and the flow outlet angle, and that at low Mach numbers the influence of blade camber and profile was small.

Furthermore, at the normal Reynolds number and turbulence level in compressors and turbines, transition of the boundary layer probably occurs very near the leading edge, so that the blades are not as sensitive to profile shape and Reynolds number effects as they are in the cascade wind tunnel. Nevertheless, potential flow theory and the requisite boundary layer predictions have been useful in the selection of profiles. Profile shape—i.e. blade thickness and camber line—has been found to affect the performance of cascades at high Mach numbers. At supersonic speeds the shape of a compressor blade is particularly important (Sec. G).

The theory of the two-dimensional flow about cascades of compressor and turbine blades is fundamental to the fluid mechanics of turbomachinery because it can be used with fair approximation to represent the flow about the blades of an axial flow compressor or turbine. In fact, the flow in any compressor or turbine is three-dimensional; it is neither axially symmetric nor steady, it is rotational and irreversible, and owing to the presence of many fixed and moving boundaries there are substantial boundary layer effects.

About 1930, a number of workers began to apply aerodynamic theory to the three-dimensional flow in compressors and turbines. The main question was how the blade shape and setting should vary along its length. In 1926, Griffith [*13*] advocated blades with constant circulation, i.e. blades which would shed no vorticity. It was thought that shed vorticity might cause loss of efficiency, just as it creates an induced drag in isolated airfoils. Designers of steam turbine blades were interested in obtaining the same work output from each particle of fluid. In a flow with no losses, both of these requirements lead to the same type of flow, namely, one in which the tangential velocity varies inversely with the radius between each row of blades. In an annulus with inner and outer walls of constant radius, the above requirements lead, for incompressible flow, to streamlines which lie on cylindrical surfaces and to a uniform axial velocity everywhere. When the velocities are high enough to introduce compressibility effects, the change of density from root to tip will not be the same as the flow passes through consecutive blade rows. The streamlines will therefore shift from one radius to another, and there will be radial as well as axial and circumferential velocities [*10*]. Whittle [*18*] assumed that between the blade rows of a turbine the radial velocities and accelerations would quickly vanish, and a condition of "radial equilibrium" would be established. He showed that the axial velocity would then be invariant with radius. Blades designed to suit the above conditions are termed "free vortex" blades; it is assumed in their design that all the radial shift of the flow takes place within the blading.

The free vortex or constant circulation design in the compressor leads to large frontal areas, if high Mach numbers at the tip of the rotor and root of the stator are to be avoided. The rotor blades have to be considerably twisted to obtain the correct radial variation of the blade angles. Cautiously, at first, constant circulation blading was abandoned in favor of designs in which the radial variation of tangential velocity was reduced. Some compressors have been built with at least one row of blades, about which the circulation has varied six-fold between hub and tip. In some turbines, radial displacements of the streamlines have been eliminated or reduced by making the product of local density and axial velocity invariant with the radius. The resulting "constant specific mass flow" design introduces radial variation of circulation and work when the turbine pressure drop is sufficiently large.

In the design of blading with radially varying circulation, it has been frequently assumed that radial equilibrium is established between the blade rows, and the axial velocities there have been calculated for the required distributions of tangential velocity. The angles at inlet and outlet to the blades are often assumed to be those given by the radial equilibrium calculation, i.e. the basic assumption is that all radial shift occurs *within* the blade passages. Other designers have averaged the inlet and outlet axial velocities computed on the assumption of radial equilibrium and have used the average velocity to determine air angles, thereby implying that most of the radial shift occurs *outside* the blades.

To improve design methods, more exact theories have been presented. The most exact tend to lead to numerical solutions which are too complex for use in design. The assumption of axially symmetric flow has accordingly been made (it has been tacitly made in the foregoing account) and is the basis for the theoretical work which forms a useful improvement of the "radial equilibrium" theory. Sec. C contains an account of the theory of axially symmetric flow through rows of axial compressor or turbine blades. From the theory various approximations can be made, many of them based on the actuator disk concept, which enable radial variations in velocities and angles to be more accurately predicted.

A,3. Axial Flow Compressors and Turbines. In the development of compressors and turbines, various experimental techniques have had to be evolved, both for examining the over-all and stage-by-stage performance of multistage units and for determining the behavior of components. Since the principal tool of the turbo-aerodynamicist has been the cascade wind tunnel, Sec. D discusses the layout of an experimental plant, of instruments, and of two-dimensional and annular cascade wind tunnels. The intermediate stage between the cascade and the compressor or turbine is the single-stage compressor and turbine, whose equipment and testing are also described. Some of the problems of testing multistage units and engines are also discussed.

The results obtained from numerous cascade wind tunnel tests have been used to establish design criteria. The theory of two-dimensional flow through cascades has been found to give satisfactory agreement with experiment in the low-loss or high L/D range of incidence. It is also possible to use the theory to select profiles which are satisfactory at high Mach numbers. The theoretical calculations are laborious, however, and it has been found easier and more direct to test profiles in cascade over the required range of incidence and then, guided by theoretical results, to derive semiempirical correlations which simplify the work of the designer. The need for such simplification is more obvious, perhaps, in compressors than in turbines. An eight-stage axial flow compressor requires that at least 51 profiles be designed (at mean radius, hub, and tip on each rotor, stator, and inlet guide vane) for operation at the design point. The closeness to the stall and the effects of compressibility need to be assessed. During the design work, not only must blades be designed to satisfy certain flow requirements, but their performance under different angles of attack and flow Mach numbers should also be known.

Generalizations of cascade data have been devised which, with methods of interpolation and stall prediction, represent valuable interpretations of what would otherwise be an indigestible mass of test results. Sec. E gives the background and some of the results of this work.

In the theories developed in Sec. B and C, the growth of boundary layers on the walls of the annulus at the hub and tip of the blade rows is mentioned. The existence of such boundary layers has been responsible for considerable difficulty in the design of compressors. By reducing the effective area of the annular duct, they increase the axial velocities of the main flow. There is then a reduction of incidence and work compared to the values obtained if the thickness of the wall boundary layers is neglected. There are opportunities for loss of efficiency and work within the wall boundary layers as they sweep over the ends of the blades. In a cascade with solid (i.e. nonporous) walls there is appreciable thickening of the wall boundary layers when the pressure rise is large, and the stall of blades of high stagger is often preceded by an apparent separation of the flow from the walls. When passing through a rotor, on the other hand, the wall boundary layer thickness often appears to diminish.

The behavior of the wall boundary layer as it passes through a row of blades is described in Sec. E and F. Secondary flows—i.e. flows in a direction transverse to the direction of the mainstream—sweep the wall boundary layer into the core of a vortex which trails downstream from the suction side of the blade passage. A stalled region frequently develops in the corner between the suction side of the blade and the end wall. Some analysis of these phenomena has been attempted, but estimates of loss and boundary layer thickness are still empirical and uncertain.

The application of the theories of two- and three-dimensional flow and the results of cascade tests to single-stage and multistage axial compressors is the subject of Sec. F, which describes methods of design and of predicting the behavior of a given design at other conditions of operation. Sec. F also describes phenomena that have not so far been mentioned, such as the effects of tip clearance, stalling, and surging.

Axial compressors are normally built with unshrouded rotor blades so that there is a clearance at the tip of each rotor. Similar clearances may be necessary at the hub of stators unless they are shrouded (thereby changing the type of leakage path). The effect of leakage through clearances and seals has been shown to be considerable.[1]

As the flow through a compressor is throttled (at constant rotative speed) below a certain value, the compressor stalls or surges. The pressure and efficiency suddenly drops and the flow pulsates—sometimes mildly and, sometimes, particularly in centrifugal compressors, noisily with a large amplitude. Stall or surge of the compressor must be avoided under all conditions of operation. Control over the location of the "surge line" on the map of compressor characteristics is highly desirable. It appears to be possible for a number of *stages* of an axial compressor to stall without causing the whole compressor to stall or surge. Changes in the matching of the relative operating ranges of the individual stages in a compressor have been found to affect the position of the surge line.

A remarkable discovery in the development of the axial compressor was the phenomenon of rotating or propagating stall. At the date the Series was first planned, the existence of a propagating stall had not been properly appreciated. The phenomenon was first described in 1953 [*19*]. Because of the possibility that rotating stall cells may excite dangerous blade vibrations, a large amount of work has been, and is being, done on this problem.

Many of the aerodynamic theories and concepts used in the design of axial flow compressors are applicable to turbines, for example, the theories of two- and three-dimensional flow described in Sec. B and C, and the use of cascade data summarized in Sec. E. Sec. H describes the application of these theories and results to axial flow turbine design. In the turbine, the gas velocities are often appreciably higher in Mach number than in the compressor, so that more attention is paid to compressibility effects in Sec. H than in Sec. F. Because of the higher temperatures, mechanical stress considerations also more markedly influence the aerodynamic design in turbines than in compressors, and a balance has to be struck between weight and performance. Mechanical factors such as fatigue and differential expansion problems make thick trailing edges on the blades

[1] As an extreme example, the "contraflow" system proposed by Griffith may be mentioned. This gas turbine scheme consisted of a number of mechanically independent disks, each carrying a row of double-tier blades. Either the inner or outer tier is a row of compressor blades, the other tier being a row of turbine blades. Neighboring disks rotated in opposite directions. A major difficulty with the scheme is the problem of leakage and of disturbances due to leakage between the seals separating the tiers of blades.

desirable. The aerodynamic effects and losses due to thickened trailing edges are also discussed in Sec. H.

Various means of cooling turbine blades have been suggested, either to permit the use of cheaper materials or higher gas temperatures and consequently of more engine power or thrust. Sec. E contains an account of the results of work on heat transfer in cascades of turbine blades. The effects of blade cooling on turbine stage efficiency is not large [*20*], so that it has not been necessary to include a discussion of it in Sec. H.

Except in the example of the phenomenon of propagating stall, it has been assumed that the flows hitherto analyzed have been steady. The assumption of steady flow has been generally accepted in most theoretical work on compressors and turbines. However, owing to the relative motion between successive rows of blades, the flow is far from steady. The implications for performance and efficiency in the assumption of steady flow are uncertain. The unsteady effects due to interactions between the blade rows are discussed in an introductory fashion in Sec. K.

A,4. Radial Flow Turbomachinery. Both the von Ohain He S-3 and the Whittle W-1, the first turbojet engines to fly in Germany and England, respectively, employed centrifugal compressors, and the He S-3 had a radial inflow turbine. Of the two components, the centrifugal compressor has been most commonly used in aircraft power plants. The radial turbine is being used, however, in numerous auxiliary power plants.

As a consequence of their use, a large amount of research and development effort has been spent on radial flow components, with most effort being devoted to the aerodynamics of the centrifugal compressor. As a result, the pressure ratio obtainable from a single-stage compressor has been appreciably increased in the last twenty years, and the efficiency at pressure ratios of about 4:1 has been raised to 80 per cent or more.

The knowledge of the flow in centrifugal compressors has followed a pattern of development similar to that of axial compressors, save in one important respect. It has not been possible to use a wind tunnel to test elements of the centrifugal compressor. The equivalent of the two-dimensional axial flow cascade is the circular cascade with radial throughflow. The theoretical methods for solving the potential flow through such a cascade are known, but the analogy is too different from the impellers or diffusers that are normally used to be worth pursuing. The diffuser with a large number of small blades has been found to give a characteristic with a poorer range of useful operation than the diffuser with a few vanes. More recently, solutions have been obtained by numerical methods for the compressible flow in an impeller. But theoretical solutions to the problem have been hampered by the large boundary layer effects which occur in the impeller and diffuser passages and by the effects of compressibility. Clearance losses, particularly on small machines, have been found to be large.

An account of the aerodynamics of the centrifugal compressor is given in Sec. J. Sec. I gives a companion account of the radial inflow turbine which has reached pressure ratios and efficiencies somewhat greater than those of the compressor.

A,5. Supersonic Compressors. At the inlet to the diffuser vanes of some centrifugal compressors, velocities above that of sound have been obtained. The term "supersonic compressor," however, is generally applied to axial flow compressors in which the inlet velocities relative to one or more rows of blades are supersonic. The shocks in the blade passages produce pressure rises that give the supersonic compressor its main advantage—a high pressure ratio per stage. Unfortunately, the shocks produce steep adverse pressure gradients which result in losses, owing either to wall boundary layer separation or to other effects. The efficiency of supersonic compressors, therefore, has not been high, and their range of mass flow (at constant rotational speed) has been narrow (see Fig. G,1i and G,1j). The frontal area has been no less than that of axial compressors, and often somewhat larger. Consequently, the supersonic compressor has not yet found an application in aircraft gas turbines.[2]

The supersonic compressor, which is the subject of Sec. G, has been included, largely because of its novelty, its interesting principle of shock compression, and its promise, but not because it has shown signs of superseding subsonic axial compressors. Rather, the trend appears to be toward "transonic" compressors, in which only part of the flow relative to the blades (generally that towards the tip region) is supersonic. This type of compressor stage has given very satisfactory results in the first stage or two of compressors, but is one in which the difficulty of detailed analysis of the flow is even greater than in those discussed in this volume.

A,6. Cited References

1. Stodola, A. *Steam and Gas Turbines.* Transl. by Loewenstein, L. C. Peter Smith, New York, 1945.
2. Constant, H. *Gas Turbines and Their Problems.* Todd, London, 1948.
3. Constant, H. *Proc. Inst. Mech. Engrs. London 153,* 411 (1945).
4. St. Venant, B. de and Wantzel. *Compt. rend. 8,* 294 (1839).
5. Kolster, R. *Z. Ver. deut. Ing. 11,* 435 (1867).
6. Rankine, W. J. M. *Engineer 27,* 165 (1869).
7. Holtzmann, Carl. *Lehrbuch Der Theoretischen Mechanik,* J. B. Metzler, Stuttgart, 1861.
8. Reynolds, O. *Scientific Papers.* Cambridge Univ. Press, 1901.
9. Hugoniot, H. *Compt. rend. 103,* 241 (1886).

[2] One turbojet engine with a single-stage supersonic compressor of over 2.5:1 was tested some time ago; although the performance was poor, partly because of the low pressure ratio and efficiency of the compressor, nevertheless the engine could be started and ran successfully.

10. Hawthorne, W. R. *J. Aeronaut. Sci. 24*, 713 (1957).
11. Keller, C. Axialgeblase Von Stundpunkt Der Traflugel Theorie. *Mitt. ETH Inst. Aerodynam., Zurich. 2* (1934).
12. Tyler, R. A. *Natl. Research Council of Canada, Aero Note AN4*, Ottawa, 1949.
13. Howell, A. R. *J. Roy. Aeronaut. Soc. 52*, 329 (1948).
14. Betz, A. *Ing.-Arch. 2*, 359 (1931).
15. Weinig, F. *Die Stromung Um Die Schauffeln Von Turbomaschinen.* Barth, Leipzig, 1935.
16. Lin, C. C. *J. Math. and Phys. 28*, 117 (1949).
17. Howell, A. R. *Brit. Aeronaut. Research Council Repts. and Mem. 2095*, 458 (1942).
18. Whittle, F. *Proc. Inst. Mech. Engrs. London 152*, 419 (1945).
19. Emmons, H. W., Pearson, C. E., and Grant, H. P. *Trans. Am. Soc. Mech. Engrs. 77*, 455 (1955).
20. Hawthorne, W. R. *Trans. Am. Soc. Mech. Engrs. 78*, 1765 (1956).

SECTION B

THEORY OF TWO-DIMENSIONAL FLOW THROUGH CASCADES

F. S. WEINIG

CHAPTER 1. INTRODUCTION

B,1. The Real Flow and Its Simplification. As a final result, the aerodynamics of turbomachines should integrate all the factors that contribute to the flow and the energy transfer inside of them. This means that for given inlet conditions, which are not necessarily uniform and stationary across the inlet, it is desirable to compute this truly three-dimensional and nonstationary flow taking into consideration compressibility, viscosity, and clearance effects, as well as the mutual interference between the stationary and rotating blade rows. This task is so complex that certain short cuts are essential.

First of all, no consideration is normally given to nonuniform inlet conditions due to varied angles of attack of the airplane or to unsymmetric inlet scoops. Also, the influence of the inlet struts and other features of a secondary nature are normally neglected. The remaining problem can then be broken down into three problems:

1. In the first the average flow viewed in a meridional plane is considered, which brings about the so-called axially symmetric flow pattern used later as the "basic flow" of the cascades.
2. The next problem considers the flow as if it occurred between adjacent rotational stream surfaces, as found by the axially symmetric treatment, but which now contains a finite number of blades. By cutting along these rotational surfaces the two-dimensional problem of cascade flow is established.
3. The third problem considers the flow as if it could be observed in cross sections normal to the rotational stream surfaces, that is, in axial turbomachines, practically normal to the axis of rotation. This brings about the problems of the so-called "secondary flow," which compensates for the simplifications inherent in the concepts of axially symmetric and two-dimensional flow. In the first instance, such a correction must extend the axially symmetric treatment to allow for some

exchange of matter and therewith of energy across the axially symmetric surfaces. Furthermore, the assumption of two-dimensional cascade flow must be corrected in a manner similar to that which adjusts the result of the lifting line theory to small aspect ratios in the theory of the airfoil of finite span.

Although this section is concerned with cascade flow, it seems advisable, however, to show how this kind of flow is affected by the axially symmetric flow problem and in some way to take into account the effects of the secondary flow phenomenon.

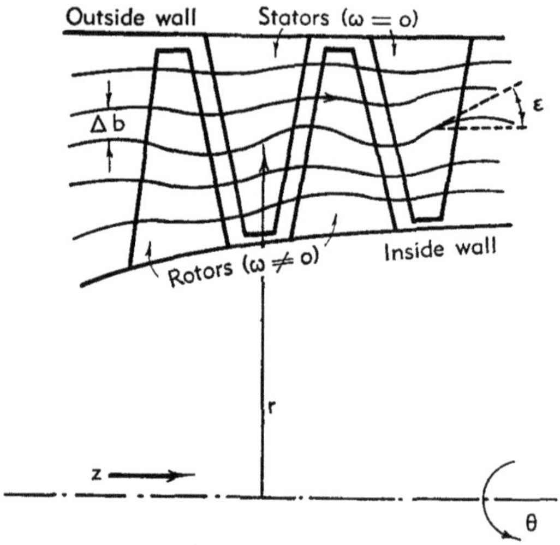

Fig. B,2a. Some stages of a multistage compressor with streamlines of the axially symmetric flow pattern (schematic).

B,2. Transformation of Flow along Rotational Stream Surfaces to Flow in a Plane. Let Fig. B,2a represent the streamline pattern of an axially symmetric flow through some stages of a multistage compressor. Then the corresponding rotational stream surfaces may each be plotted, conformally, into an x^*, y^* plane by

$$y^* = \theta r_m$$
$$x^* = -\int_{\sigma_0}^{\sigma} \frac{r_m}{r}\, d\sigma \tag{2-1}$$

where r_m is a reference radius of the stream surface under consideration, θ is the longitudinal (or circumferential) angle, and σ is the curve length of the meridional streamline, with $d\sigma = dr/\sin \epsilon$, where ϵ is the angle between the direction of the streamline and the axis.

By this transformation the intersections of the axially symmetric stream surfaces with each of the blade rows go over into cascades of profiles. The profiles of each cascade are equally spaced, provided that they are correctly designed and manufactured, along straight lines (x = const), parallel to the cascade axis (Fig. B,2b).

Consider the space between two adjacent axially symmetric stream surfaces. Let Δb be the small distance between them, $\Delta Q = (\bar{\rho}_m/\bar{\rho})\,\Delta Q_m$

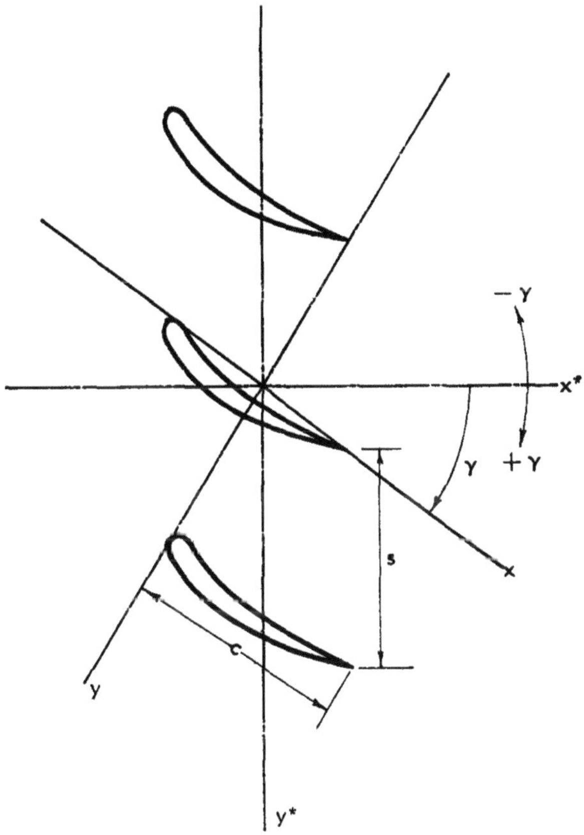

Fig. B,2b. Cascade of profiles obtained by conformal mapping of the axially symmetric stream surfaces into the picture plane x^*, y^*.

the volume flowing between them, and Γ_1 the circulation at the *inlet* of a blade row, where $\bar{\rho}_m$ is the density at the reference radius r_m and $\bar{\rho}$ is the density as found by computation of the axially symmetric flow. Then the components of the velocity relative to the blade row of the axially symmetric flow in the meridional direction and in the direction of the circumference are, respectively:

$$W_{s_{\text{basic}}} = \frac{\Delta Q}{2\pi r\,\Delta b}$$

and

$$W_{\theta_{\text{basic}}} = \frac{\Gamma_1}{2\pi r} + \omega r \tag{2-2}$$

The angular speed of the blade row is ω, which is zero in the case of a stationary blade row or stator.

For each blade row a basic flow may be defined from the axially symmetric flow. This basic flow follows the predetermined axially symmetric stream surfaces, but its circulation does not change when passing through a blade row and, therefore, is considered to be equal to the inlet circulation. It might be convenient to choose, each time, the radii at the exit of the blade row as the reference radius. Hence this basic flow between two adjacent rotational stream surfaces separated by a small distance is considered to be a potential flow with, in the case of a rotating blade row or rotor, a superimposed flow with constant curl or rotation.

Between the velocities along the rotational stream surfaces and in the picture plane (x^*, y^*), the following relations exist generally:

$$\begin{aligned} w_{x^*} &= \frac{r}{r_m} W_z \\ w_{y^*} &= \frac{r}{r_m} W_\theta \end{aligned} \tag{2-3}$$

Thus for the basic flow

$$\begin{aligned} w_{x^*_{\text{basic}}} &= \frac{\bar{\rho}_m}{\bar{\rho}} \frac{\Delta b_m}{\Delta b} \frac{\Delta \tilde{Q}}{2\pi r_m \, \Delta b_m} \\ w_{y^*_{\text{basic}}} &= \frac{\Gamma_1}{2\pi r_m} + \frac{\omega r^2}{r_m} \end{aligned} \tag{2-4}$$

Hence for each blade row the transformed basic flow is a potential flow with, in the case of the rotor, a superimposed vortex flow. In general the curl of the vortex flow is not constant,

$$\text{curl } \mathbf{W}_{\text{basic}} = \frac{\partial w_{x^*}}{\partial x^*} - \frac{\partial w_{y^*}}{\partial y^*} = -2\omega \left(\frac{r}{r_m}\right)^2 \sin \epsilon \tag{2-5}$$

except in the case where

$$\left(\frac{r}{r_m}\right)^2 \sin \epsilon = \text{const}$$

$$\text{curl } \mathbf{W}_{\text{basic}} = \text{const}$$

This case is practically uninteresting, except when $\epsilon = 0$, i.e. in strictly axial flow or when $r = \text{const}$ for which curl $\mathbf{W}_{\text{basic}} = 0$.

By the action of the blades, the basic flow is changed to the flow in

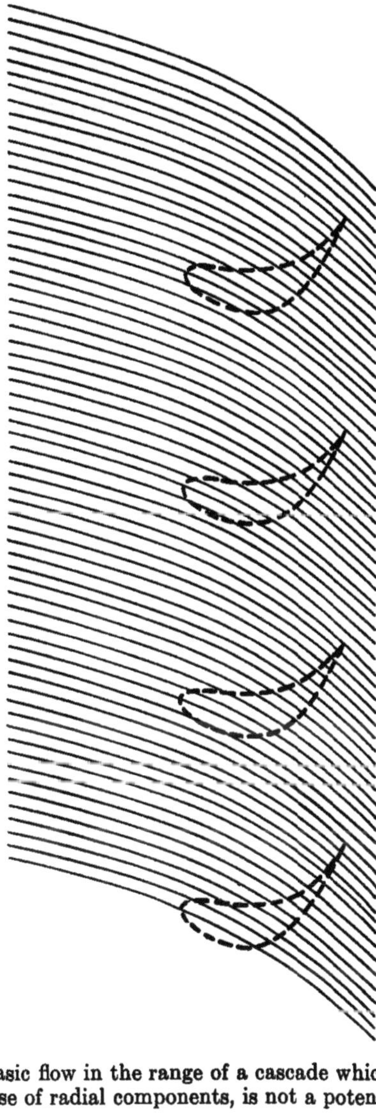

Fig. B,2c. Basic flow in the range of a cascade which, for a rotor, in the case of radial components, is not a potential flow.

which the profiles are streamlines. This *relative cascade flow,* therefore, is obtained by superposition of the *basic flow* and the *vane flow* (Fig. B,2c; B,2d; and B,2e).

The expression vane flow may be chosen since this flow is originated by the vanes or blades, whereas the basic flow exists without the influence of the vanes and corresponds only to the predetermined axially symmetric stream surfaces and to the inflow condition. These are given in the case of each blade row by the flow volume $\Delta\bar{Q}$, by the inlet circu-

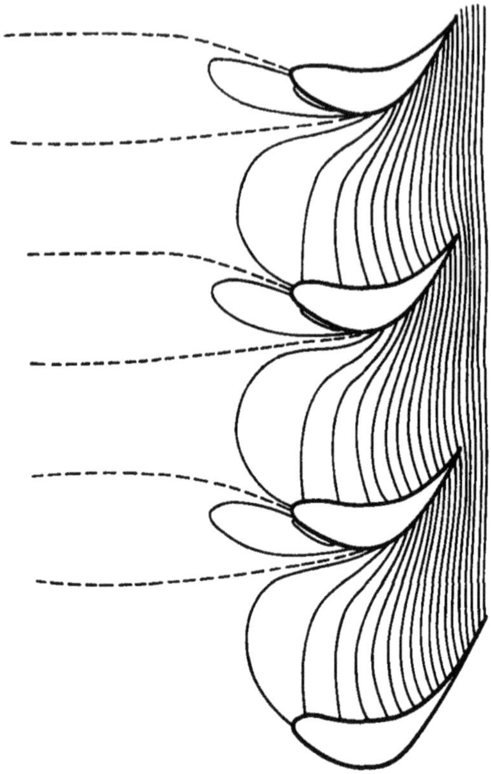

Fig. B,2d. Vane flow caused by action of the blade row which is, in the inviscid case, a potential flow. It has vanishing velocities in front of the cascade and velocities in circumferential direction only in the rear.

lation Γ_1, and, in the case of a rotor, by the angular speed ω. The vane flow is a potential flow if viscosity is neglected.

The substitution of the abbreviations,

$$\hat{\rho} = \frac{\Delta b}{\Delta b_m} \frac{\bar{\rho}}{\bar{\rho}_m}$$

$$\frac{\Delta b}{\Delta b_m} = \frac{\Delta b}{\Delta b_m}(x)$$

$$\frac{\bar{\rho}}{\bar{\rho}_m} = \frac{\bar{\rho}}{\bar{\rho}_m}(x)$$

$$\hat{\rho} = \hat{\rho}(x) \tag{2-6}$$

yields the differential equations for the vane flow

$$\text{curl } \mathbf{W}_{\text{vane}} = \frac{\partial(w_{x^*})_{\text{vane}}}{\partial y^*} - \frac{\partial(w_{y^*})_{\text{vane}}}{\partial x^*} = 0$$

Fig. B,2e. Relative flow through a cascade obtained by superposition of the vane flow on the basic flow. It has profiles as streamlines and is identical upstream with the basic flow.

and

$$\operatorname{div} \hat{\rho}\mathbf{W}_{\text{vane}} = \frac{\partial(\hat{\rho}w_{x^*})_{\text{vane}}}{\partial x^*} + \frac{\partial(\hat{\rho}w_{y^*})_{\text{vane}}}{\partial y^*} = 0 \qquad (2\text{-}7)$$

Here it is assumed that the circumferential changes of the density may be neglected. In the case where these changes must be considered,

$$\hat{\rho}(x^*) = \frac{\Delta b}{\Delta b_m}\frac{\hat{\rho}}{\hat{\rho}_m}$$

is to be replaced by

$$\hat{\rho}(x^*, y^*) = \frac{\Delta b}{\Delta b_m}\frac{\rho}{\rho_m}$$

where ρ is the local density and ρ_m is the corrected average at the reference point. The usefulness of such corrections, however, is somewhat impaired by those effects already neglected, in order to establish the axially symmetric flow pattern.

For the vane flow a potential function φ_{vane} and a stream function ψ_{vane} may be introduced by

$$(w_{x^*})_{\text{vane}} = \frac{\partial \varphi_{\text{vane}}}{\partial x^*} = \frac{1}{\hat{\rho}} \frac{\partial \psi_{\text{vane}}}{\partial y^*}$$
$$(w_{y^*})_{\text{vane}} = \frac{\partial \varphi_{\text{vane}}}{\partial y^*} = -\frac{1}{\hat{\rho}} \frac{\partial \psi_{\text{vane}}}{\partial x^*} \tag{2-8}$$

For the basic flow a stream function also exists:

$$\psi_{\text{basic}} = \frac{\hat{\rho}_m \Delta \tilde{Q}_m}{2\pi r_m \Delta b_m} y^* - \hat{\rho}_m \int \left(\frac{\Gamma_1}{2\pi r_m} + \frac{\omega r^2}{r_m} \right) \hat{\rho} dx^* \tag{2-9}$$

where $r = r(x^*)$ and $\hat{\rho} = \hat{\rho}(x)$. A potential function of the basic flow exists in the case of stator blading only.

Applying Eq. (2-6) to Eq. (2-7) and using the Laplace operator

$$\Delta = \frac{\partial^2}{\partial x^{*2}} + \frac{\partial^2}{\partial y^{*2}}$$

it is found that

$$\Delta \psi_{\text{vane}} = \frac{\partial \psi_{\text{vane}}}{\partial x^*} \frac{d(\ln \hat{\rho})}{dx^*}$$
$$\Delta \varphi_{\text{vane}} = \frac{\partial \varphi_{\text{vane}}}{\partial x^*} \frac{d(\ln \hat{\rho})}{\partial x^*} \tag{2-10}$$

These are the differential equations which should be solved with the boundary condition

$$(w_n)_{\text{vane}} = -(w_n)_{\text{basic}} \tag{2-11}$$

where

$$(w_n)_{\text{basic}} = \frac{1}{\hat{\rho}} \frac{\partial \psi_{\text{basic}}}{\partial s}$$
$$(w_n)_{\text{vane}} = \frac{1}{\hat{\rho}} \frac{\partial \psi_{\text{vane}}}{\partial s} = -\frac{\partial \varphi_{\text{vane}}}{\partial n}$$

with s and n respectively taken along, and normal to, the profile contours.

B,3. Approximate Treatment of Flow Induced by the Blades (Vane Flow). Since the exact solution of these equations is quite complicated, an approximation is usually made. For this purpose let

$$\mathbf{W}_{\text{vane}} = \mathbf{w}_0 + \mathbf{w}_{\text{add}} \tag{3-1}$$

where

$$\operatorname{div}(\hat{\rho}\mathbf{w}_0) = \frac{\partial(\hat{\rho}w_{0x^*})}{\partial x^*} + \frac{\partial(\hat{\rho}w_{0y^*})}{\partial y^*}$$

$$\operatorname{curl}(\hat{\rho}\mathbf{w}_0) = \frac{\partial(\hat{\rho}w_{0x^*})}{\partial y^*} - \frac{\partial(\hat{\rho}w_{0y^*})}{\partial x^*} \tag{3-2}$$

with the boundary condition

$$\hat{\rho}w_{0n} = -\hat{\rho}(w_n)_{\text{basic}} = -\frac{\partial\psi_{\text{basic}}}{\partial s} \tag{3-3}$$

Then

$$\operatorname{curl}\mathbf{w}_0 = \frac{\partial w_{0x^*}}{\partial y^*} - \frac{\partial w_{0y^*}}{\partial x^*} = \frac{1}{\hat{\rho}}\left(\frac{\partial(\hat{\rho}w_{0x^*})}{\partial y^*} - \frac{\partial(\hat{\rho}w_{0y^*})}{\partial x^*}\right) + w_{0x^*}\frac{d(1/\hat{\rho})}{dx^*}$$

$$= -\frac{w_{0x^*}}{\hat{\rho}^2}\frac{d\hat{\rho}}{dx^*}$$

Hence for $\mathbf{w}_{\text{add}}$

$$\operatorname{div}(\hat{\rho}\mathbf{w}_{\text{add}}) = 0$$

$$\operatorname{curl}(\hat{\rho}\mathbf{w}_{\text{add}}) = -\operatorname{curl}\mathbf{w}_0 = \frac{w_{0x^*}}{\hat{\rho}^2}\frac{d\hat{\rho}}{dx^*} \tag{3-4}$$

Furthermore,

$$\psi_{\text{vane}} = \psi_0 + \psi_{\text{add}} \tag{3-5}$$

At the profile contours, $\psi_{\text{add}} = 0$. While the vane flow has a stream function and a velocity potential function, the approximate flow (0) and the additional flow (add) have only a stream function. There exists, however, a potential Φ_0 of the mass flow vector $\mathbf{v}_0 = \hat{\rho}\mathbf{w}_0$.

Hence the approximate flow is equivalent to a plane two-dimensional flow of an incompressible medium with

$$v_{0x^*} = \frac{\partial\Phi_0}{\partial x^*} = \frac{\partial\psi_0}{\partial y^*}$$

$$v_{0y^*} = \frac{\partial\Phi_0}{\partial y^*} = -\frac{\partial\psi_0}{\partial x^*} \tag{3-6}$$

(This flow is considered in Chap. 2.) The error occurring when ψ_0 is taken for ψ_{vane} is described by Eq. 3-4 [*1*]. No consideration is given to this error in the following discussion. It should also be noted that the flow through the blade rows is treated here as though the axial spacing were very large, that is, their mutual and essentially nonstationary interference is disregarded here.

Secondary flow in the blade passages. Having shown how the axially symmetric flow problem affects cascade flow, we now consider how the cascade flow might be modified to adjust it for some effects of the secondary flow phenomenon.

The entering flow is changed by the action of the blades while passing the blade row into the exit flow. At a location sufficiently far downstream of the blade row, the changes of circumferential and radial components are represented by the flow pattern of Fig. B,3a, which may be called the *induced vane flow*. There exist circumferential velocity components that are not at all constant in the circumferential direction. In addition there are also radial velocity components that vary along the circumferential direction and that cannot be covered by the axially symmetric treatment. At the exit of the blade row, this would correspond to

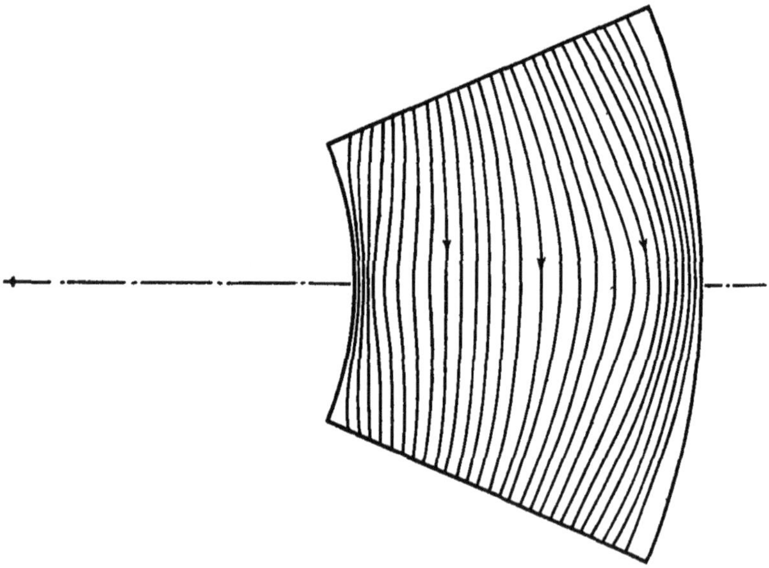

Fig. B,3a. Effect of blading on the flow under consideration. Under the influence of wall boundary layers, for instance, the circulation is not constant along the radius.

the flow pattern shown in Fig. B,3b where the circumferential velocities would not change circumferentially. Let $\Delta(W_\theta)_{\text{vane}}$ be the circumferential components and $\Delta(W_\theta)_p$ the components on the trails of the trailing edges. The flow pattern resulting from the radial velocity components and the difference $\Delta(W_\theta)_p - \Delta W_\theta$ (Fig. B,3c) might be called the *secondary vane flow*. With

$$r\int_1^2 d\theta = s$$

being the spacing of the blades, the circulation around a radial blade element is

$$\Gamma = r\int_1^2 \Delta W_\theta d\theta$$

When expressing this circulation by $\Delta(W_\theta)_p$ an effective spacing may be

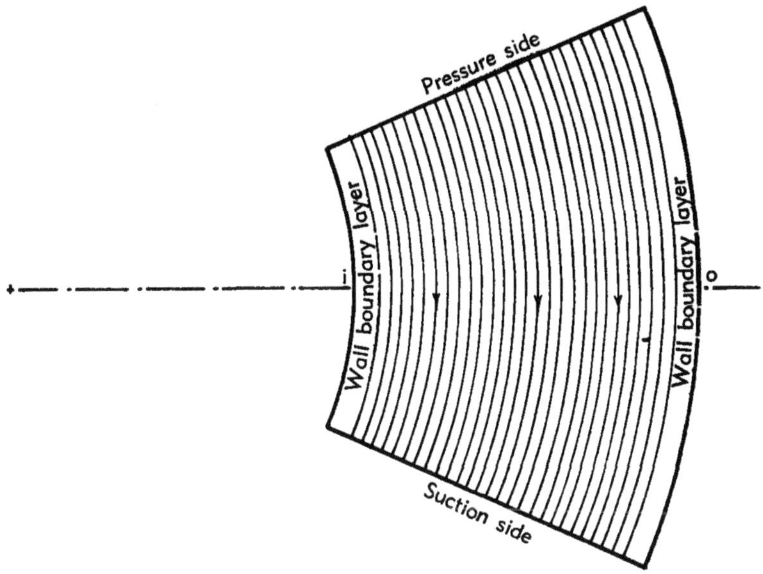

Fig. B,3b. Effect of blading on the flow if flow is treated as axially symmetric, i.e. disregarding circumferential variations.

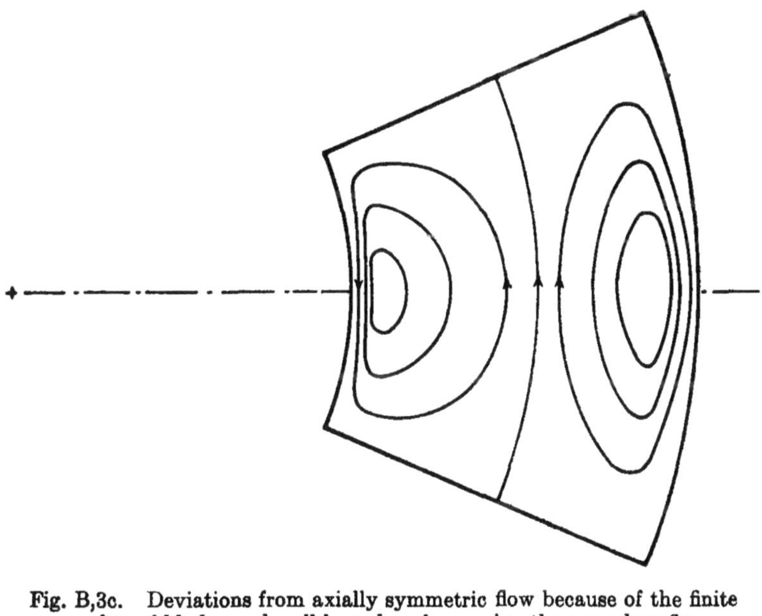

Fig. B,3c. Deviations from axially symmetric flow because of the finite number of blades and wall boundary layers, i.e. the *secondary* flow.

defined by [2]

$$s_{\text{eff}} = \frac{\Gamma}{\Delta(W_\theta)_p}$$

In the case of a single wing of finite span, the spacing s would be infinite, but the effective spacing s_{eff} would be finite. This suggests the treatment of the flow around an element of a wing of finite span rather as a flow through an unstaggered cascade of a spacing ratio s_{eff} than as a profile in infinitely extended flow, thus modifying Prandtl's lifting line theory, especially in the case of small aspect ratios. The application of this modification to the wing of finite span gives quite satisfying results. If the induced vane flow is established, it may be advisable to extend the idea to the flow through blade rows and to use blade interference corrections for s_{eff}/c rather than s/c, with the stagger of the cascade profiles remaining unchanged. Secondary flow effects are also considered in C,1; E,3; F,5; and H,8.

CHAPTER 2. GENERAL TREATMENT OF INCOMPRESSIBLE INVISCID TWO-DIMENSIONAL CASCADE FLOW

B,4. The Forces Acting on a Blade in Cascade. Let us consider the plane two-dimensional flow of an incompressible fluid through a cascade of profiles (Fig. B,4). At a distance sufficiently far upstream and downstream the pressures are p_1 and p_2 respectively. With ρ as the density, Bernoulli's equation, neglecting losses, is [3,4,5]

$$p_1 = p^0 - \tfrac{1}{2}\rho V_1^2$$

$$p_2 = p^0 - \tfrac{1}{2}\rho V_2^2 \tag{4-1}$$

where V_1 and V_2 are the velocities upstream and downstream and p^0 is the stagnation pressure.

The pressure change between the upstream and downstream locations

$$-\Delta p = p_1 - p_2 = \tfrac{1}{2}\rho(V_2^2 - V_1^2) \tag{4-2}$$

or with

$$V_1^2 = V_z^2 + V_{\theta_1}^2, \qquad V_2^2 = V_z^2 + V_{\theta_2}^2$$

also

$$-\Delta p = \tfrac{1}{2}\rho(V_{\theta_1}^2 - V_{\theta_2}^2) \tag{4-3}$$

where $V_{\theta_{1,2}}$ and V_z are the tangential and the meridional or axial velocity components at the inlet and the exit, respectively.

Consider a strip of the cascade flow corresponding to the spacing between the adjacent profiles. The streamlines on the sides of this strip are

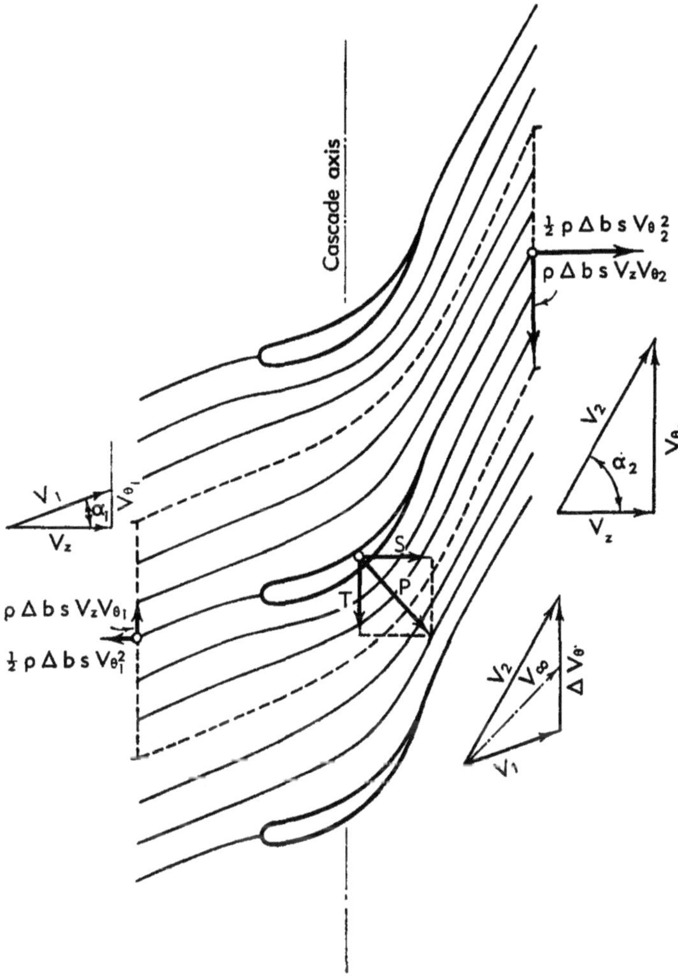

Fig. B,4. Origin of the flow forces acting on the profiles of a cascade and the velocity diagram.

congruent to each other and have equal pressures in corresponding points. Hence no force results from the pressure along the edges of the strip. Let Δb be the spanwise extension of the profiles; then the thrust normal to the cascade axis acting on the profile included in this strip

$$S = \Delta b s \tfrac{1}{2}\rho(V_{\theta_2}^2 - V_{\theta_1}^2) \tag{4-4}$$

Besides this thrust there is also a tangential force T. Let the mass flow through the strip

$$m = \rho \Delta b s V_z \tag{4-5}$$

Then the tangential force

$$T = \rho \Delta b s V_z (V_{\theta_2} - V_{\theta_1}) \tag{4-6}$$

Hence

$$\frac{S}{T} = \frac{V_{\theta_1} + V_{\theta_2}}{2V_z} \tag{4-7}$$

This means that the resultant force $P = \sqrt{S^2 + T^2}$ is normal to the direction of a mean flow velocity

$$\mathbf{V}_\infty = \frac{\mathbf{V}_1 + \mathbf{V}_2}{2} \tag{4-8}$$

of which the tangential component

$$V_{\theta\infty} = \frac{V_{\theta_1} + V_{\theta_2}}{2} \tag{4-9}$$

The resultant force is found to be

$$P = \rho \Delta b s (V_{\theta_2} - V_{\theta_1}) V_\infty \tag{4-10}$$

The quantity [4]

$$s(V_{\theta_2} - V_{\theta_1}) = \Gamma \tag{4-11}$$

is the circulation around each profile. Substitution of Γ yields

$$P = \rho \Delta b \Gamma V_\infty \tag{4-12}$$

with P perpendicular to $\mathbf{V}_\infty$. By this the Kutta-Joukowski theorem for a single profile is extended to the profile in a cascade. For a single profile $\mathbf{V}_1 = \mathbf{V}_2 = \mathbf{V}_\infty$ as $s \to \infty$. As in the case of single profiles, cascade profiles have the force P which is called lift normal to $\mathbf{V}_\infty$, the equivalent free flow velocity. Usually a lift coefficient is defined by

$$C_L = \frac{L}{\frac{1}{2}\rho V_\infty^2 \Delta b c} = \frac{2\Gamma}{V_\infty c} \tag{4-13}$$

Therefore for cascade profiles

$$C_L = \frac{s}{c}\frac{2(V_{\theta_2} - V_{\theta_1})}{V_\infty} \tag{4-14}$$

B,5. Conformal Mapping of the Exterior of a Cascade of Straight-Line Profiles into the Exterior of a Circle. The simplest flow through a cascade of straight-line profiles is that which is parallel to these straight lines. Let the cascade be located in the z plane, where

$$z = x + iy$$

Let $s = 2\pi$ be the spacing, γ the stagger, and c the chord of the profiles. Let $z = iy^* e^{-i\gamma}$ be the cascade axis. Let the leading and trailing edges be located at

$$x = \mp \frac{c}{2} + 2\pi n \sin \gamma \tag{5-1}$$

$$y = 0 \pm 2\pi n \cos \gamma \tag{5-2}$$

$$n = 0, \pm 1, \pm 2, \pm 3, \ldots \tag{5-3}$$

or

$$z_{1,2} = \mp \frac{c}{2} + i2\pi n e^{-i\gamma} \tag{5-4}$$

The complex potential of the parallel flow of unit velocity $\mathbf{V}_\infty = 1$ through this cascade is (Fig. B,5a)

$$\chi_0(z) = \varphi_0 + i\psi_0 = z \tag{5-5}$$

Consider the points

$$z_n = z_0 + i2\pi n e^{-i\gamma} \tag{5-6}$$

to be corresponding points in the congruent strips of this cascade flow. The complex potential in such corresponding points is then

$$\chi_0 = z_0 + i2\pi n e^{-i\gamma} \tag{5-7}$$

Hence the potential and stream functions are

$$\begin{aligned} \varphi_0 &= x_0 + 2\pi n \sin \gamma \\ \psi_0 &= y_0 + 2\pi n \cos \gamma \end{aligned} \tag{5-8}$$

From strip to strip, i.e. from n to $n + 1$, the potential increases by

$$\Gamma_0 = s \sin \gamma = 2\pi \sin \gamma \tag{5-9}$$

and the stream function increases by

$$E_0 = s \cos \gamma = 2\pi \cos \gamma \tag{5-10}$$

Any strip of the cascade plane z can be mapped conformally into the inside or the outside of a circle in a ζ plane. The origin and the goal of the cascade flow are then transformed into two different singular points S_1 and S_2, respectively, of the ζ plane.

When circling a singular point, it is found that the potential and the stream functions of the transformed image of the undisturbed cascade flow change by Γ_0 and E_0, respectively. Hence the transformed image of the origin of the cascade flow must contain a vortex source

$$\mathbf{E}_0 = E_0 - i\Gamma_0 = s(\cos \gamma - i \sin \gamma) = 2\pi e^{-i\gamma} \tag{5-11}$$

and the transformed image of the goal of the cascade flow must contain a vortex sink

$$-\mathbf{E}_0 = -E_0 + i\Gamma_0 = -s(\cos \gamma - i \sin \gamma) = -2\pi e^{-i\gamma} \tag{5-12}$$

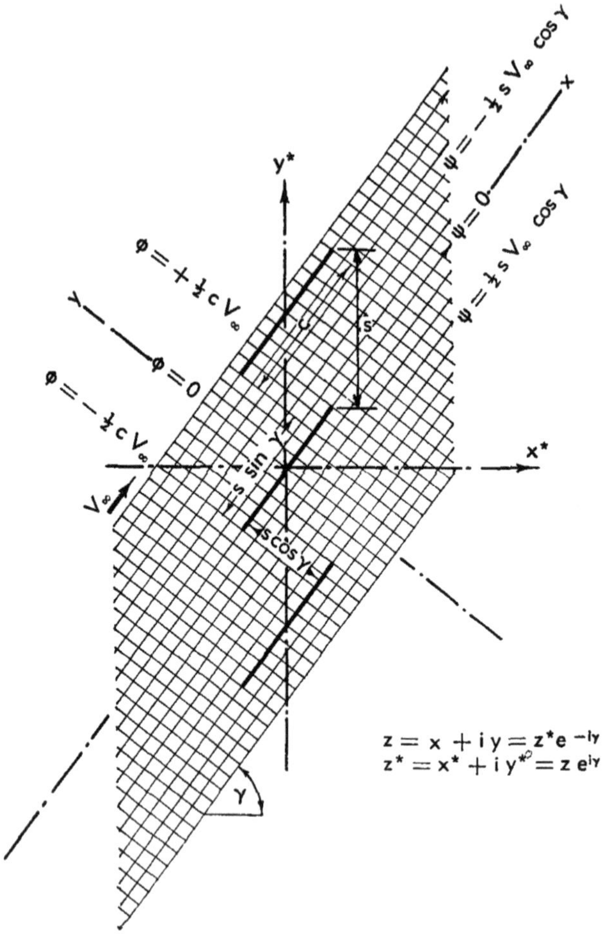

Fig. B,5a. Relations on the straight-line profile cascade in undisturbed parallel flow.

The radius of the picture circle may be prescribed, as well as the location of S_1 and S_2. From a multitude of choices, some deserve preference for the following reasons:

1. The symmetric arrangement of the singularities, especially when the transformed image of the profile is the unit circle $\zeta_o = e^{i\alpha}$ with the singularities placed symmetrically on the ξ axis: $S_1 = -R$, $S_2 = +R$, is very helpful in understanding the problem and provides certain symmetric properties of the ensuing equations [*5*; *6*, p. 91; *7*].
2. Putting one of the singularities at infinity, especially when the transformed image of the profile is the unit circle, $\zeta_o = e^{i\alpha}$, with the other singularity on the real axis, $S_1 = -a$, $S_2 = \infty$, simplifies the equa-

tions, although this is done at the cost of symmetric properties. In some cases the appropriate functions are of a simpler nature and easy to evaluate [7,8,9].

3. The transformed image of the profile may be made a straight line, especially the real axis, $\zeta_s = \xi$, with the singularities located in the lower half-plane relative to the image $\zeta_L = 0$ of the leading edge at

$$S_1 = i(1 - q)e^{i\gamma} - (1 - q^2)\sin\gamma$$
$$S_2 = i(1 + q)e^{i\gamma} - (1 - q^2)\sin\gamma$$

and the trailing edge at infinity [10]. Alternately the singularities may be located relative to the transformed image of the stagnation point $\zeta_{L_s} = 0$, with the image of the trailing edge at ζ_T

$$S_1 = \zeta_T - (1 - q^2)\sin\gamma + i(1 + q)e^{i\gamma}$$
$$S_2 = \zeta_T - (1 - q^2)\sin\gamma - i(1 + q)e^{i\gamma}$$

and the leading edge at infinity.

The latter arrangement results when the hodograph of the flow through a straight-line profile cascade is used for the derivation of the conformal mapping function [11]. The first or symmetric arrangement is used in the following [5].

With R real, the points

$$\zeta = -R \qquad \text{and} \qquad \zeta = +R$$

are chosen as the transformed images of the origin and the goal. Since the unit circle in the transformed flow is a streamline, the vortex source $\mathbf{E}_0$ and the vortex sink $-\mathbf{E}_0$ must be reflected on the unit circle by their counterparts $\bar{\mathbf{E}}_0$ at $-1/R$ and $-\bar{\mathbf{E}}_0$ at $+1/R$. Therefore the complex potential of the transformed flow is given by Fig. B,5b

$$\chi_0^*(\zeta) = e^{-i\gamma}[\ln(R + \zeta) - \ln(R - \zeta)] + e^{+i\gamma}\left[\ln\left(\zeta + \frac{1}{R}\right) - \ln\left(\zeta - \frac{1}{R}\right)\right]$$

Hence with

$$\chi_0(z) = \chi_0^*(\zeta)$$

the conformal mapping function is

$$z^* = e^{-i\gamma}\ln\frac{R + \zeta}{R - \zeta} + e^{+i\gamma}\ln\frac{\zeta + \frac{1}{R}}{\zeta - \frac{1}{R}} \tag{5-13}$$

where $z^* = ze^{+i\gamma}$, or $x^* = x\cos\gamma - iy\sin\gamma$, $y^* = x\sin\gamma + iy\sin\gamma$.

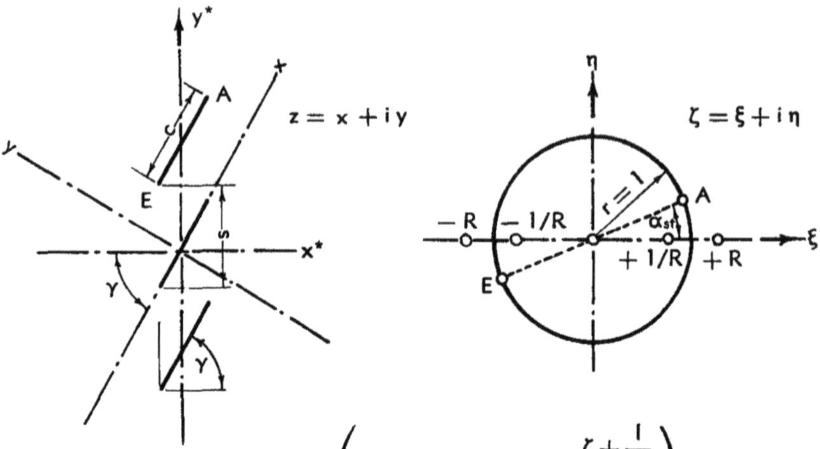

$$Z^* = \frac{s}{2\pi}\left(e^{-i\gamma} \ln \frac{\zeta + R}{\zeta - R} + e^{i\gamma} \ln \frac{\zeta + \frac{1}{R}}{\zeta - \frac{1}{R}} \right)$$

Fig. B,5b. Conformal mapping of a straight-line profile cascade on the unit circle with symmetrically located singularities.

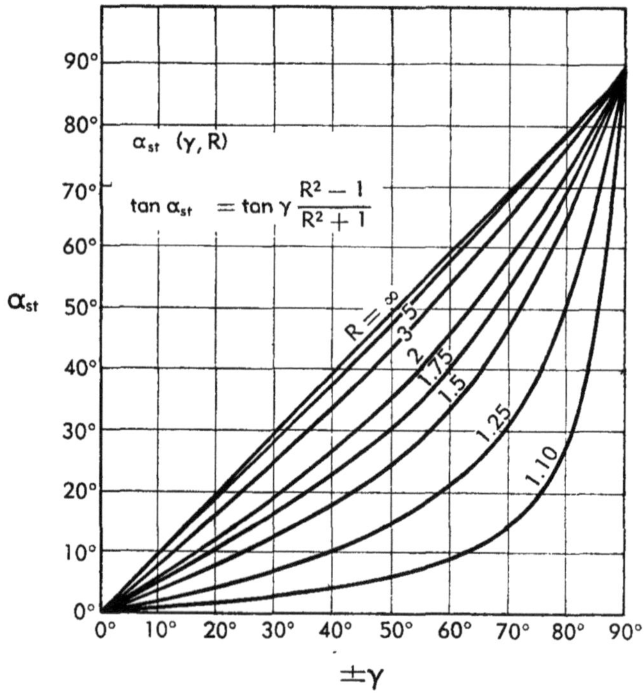

Fig. B,5c. Relations between α_{st}, R, and γ.

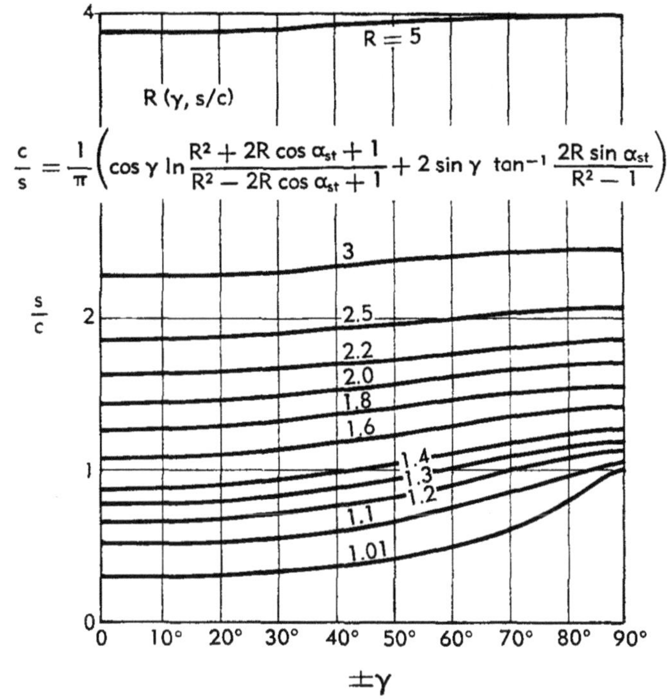

Fig. B,5d. Relations between R, γ, and s/c.

The leading and trailing edges of the straight-line profiles are transformed into the branch points $\zeta_{st_{1,2}} = \mp e^{i\alpha_{st}}$ on the circle. Their locations are determined by

$$\frac{dz^*}{d\zeta} = 0 = e^{-i\gamma}\left(\frac{1}{\zeta_{st} + R} - \frac{1}{\zeta_{st} - R}\right) + e^{+i\gamma}\left(\frac{1}{\zeta_{st} + \frac{1}{R}} - \frac{1}{\zeta_{st} - \frac{1}{R}}\right)$$

The solution yields (Fig. B,5c)

$$\tan\alpha_{st} = \tan\gamma\,\frac{R^2 - 1}{R^2 + 1} \tag{5-14}$$

Substitution of $\zeta_{st_{1,2}} = \mp e^{i\alpha_{st}}$ in the mapping function yields $z_{st_{1,2}} = \mp c/2$ and with $s = 2\pi$ (Fig. B,5d)

$$\frac{c}{s} = \frac{1}{\pi}\left\{\cos\gamma \ln \frac{R^2 + 2R\cos\alpha_{st} + 1}{R^2 - 2R\cos\alpha_{st} + 1} + 2\sin\gamma \tan^{-1}\frac{2R\sin\alpha_{st}}{R^2 - 1}\right\} \tag{5-15}$$

Several interesting limiting cases are as follows:

For $\gamma = 0$, $\alpha_{st} = 0$,

$$\left(\frac{c}{s}\right)_{\gamma\to 0} = \frac{1}{\pi}\ln\left(\frac{R+1}{R-1}\right)^2 = \frac{4}{\pi}\coth^{-1} R \tag{5-16}$$

For $\gamma = 90°$, $\alpha_{st} = 90°$,

$$\left(\frac{c}{s}\right)_{\gamma\to 90°} = \frac{2}{\pi}\tan^{-1}\frac{2R}{R^2 - 1} \tag{5-17}$$

For wide spacing $c/s \to 0$, $\alpha_{st} \to \gamma$,

$$\left(\frac{s}{c}\right)_{R\to\infty} = \frac{\pi}{4}R \tag{5-18}$$

It may be remarked that straight-line cascades may be conformally mapped into other straight-line cascades of the same spacing if, and only if, their mapping into the unit circle results in the same value of R. R therefore plays the same role for cascades as does the modulus of the elliptic integral in the case of biplane problems.

B,6. Effect of Spacing on the Lift Coefficient of Straight-Line Profiles in Cascade. Consider the flow through the cascade with an angle of attack δ. The angle of attack is the angle between the direction of the free flow velocity $\mathbf{V}_\infty$ and the straight-line profiles. Then the complex potential may be considered as the superposition of two parts, the one a flow $\chi_0 \cos\delta$ parallel to the straight lines as just considered above and the other a flow $\chi_{\pi/2} \sin\delta$ corresponding to a free flow velocity normal to it. The circulation is taken as

$$\Gamma_\delta = \Gamma_{\pi/2}\sin\delta \tag{6-1}$$

and of such a quantity that the trailing edges are confluence points corresponding to the Kutta-Joukowski theorem. The complex potential $\chi^*_{\pi/2}$ in the picture plane is then composed of a vortex source at $\zeta = +R$

$$+i\mathbf{E}_0 = +2\pi i e^{-i\gamma} = s(\sin\gamma + i\cos\gamma)$$

and a vortex sink at $\zeta = -R$

$$-i\mathbf{E}_0 = -2\pi i e^{-i\gamma} = -s(\sin\gamma + i\cos\gamma)$$

and their counterparts $+\overline{i\mathbf{E}_0}$ and $-\overline{i\mathbf{E}_0}$ in the points $\zeta = -1/R$ and $\zeta = +1/R$. In addition, the points $\zeta = -R$ and $\zeta = +R$ contain vortices $-\frac{1}{2}\Gamma_{\pi/2}$ and their counterparts $\zeta = -1/R$ and $\zeta = +1/R$ contain opposite vortices $+\frac{1}{2}\Gamma_{\pi/2}$. Then

$$\begin{aligned}\chi^*_{\pi/2} = {} & -ie^{-i\gamma}[\ln(R+\zeta) - \ln(R-\zeta)] \\ & + ie^{+i\gamma}\left[\ln\left(\zeta + \frac{1}{R}\right) - \ln\left(\zeta - \frac{1}{R}\right)\right] - \frac{i}{4\pi}\Gamma_{\pi/2} \\ & \left[\ln(R+\zeta) + \ln(R-\zeta) - \ln\left(\zeta+\frac{1}{R}\right) - \ln\left(\zeta - \frac{1}{R}\right)\right]\end{aligned} \tag{6-2}$$

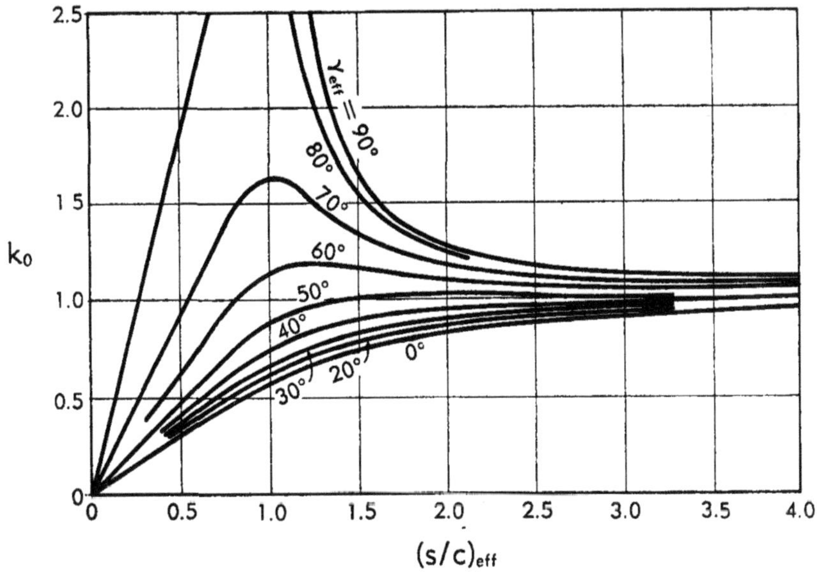

Fig. B,6a. Cascade interference coefficient k_0 for comparison with single-profile theory.

In the transformed image $\zeta = +e^{i\alpha_{st}}$ of the trailing edge

$$\frac{d\chi^*_{\pi/2}}{d\zeta} = 0 \tag{6-3}$$

which determines the circulation $\Gamma_{\pi/2}$. It is found that

$$\Gamma_{\pi/2} = 4s \frac{R}{R^2 + 1} \frac{\cos \alpha_{st}}{\cos \gamma} V_\infty \tag{6-4}$$

In the case of wide spacing,

$$s = c\frac{\pi}{4} R, \quad \alpha_{st} = \gamma, \quad \frac{R^2}{R^2 + 1} \rightarrow 1$$

may be substituted, leading to

$$\Gamma_{\pi/2} = \pi c V_\infty \tag{6-5}$$

which is the theoretical circulation around a single straight-line profile under the Kutta-Joukowski confluence condition (Eq. 6-3) for $\delta = 90°$. The coefficient of mutual interference of the straight-line profiles in a cascade (Fig. B,6a) is then

$$k_0 = \frac{\Gamma_{\text{cascade}}}{\Gamma_{\text{single}}} = \frac{4s}{\pi c} \frac{R}{R^2 + 1} \frac{\cos \alpha_{st}}{\cos \gamma} \tag{6-6}$$

The lift coefficient of a straight-line profile in a cascade therefore is

$$C_L = 2\pi k_0 \sin \delta \tag{6-7}$$

Interesting limiting cases are as follows:

For $\gamma = 0$, $\alpha_{st} = 0$

$$k_0 = \frac{2s}{\pi c} \frac{2R}{R^2 + 1}$$

By the substitution of Eq. 5-16,

$$R = \coth \frac{\pi c}{4s} \quad \text{or} \quad \frac{2R}{R^2 + 1} = \tanh \frac{\pi c}{2s}$$

$$k_0 = \frac{2s}{\pi c} \tanh \frac{\pi c}{2s} \tag{6-8}$$

For $\gamma = 90°$, $\alpha_{st} = 90°$ (Eq. 5-14)

$$k_0 = \frac{2s}{\pi c} \frac{2R}{R^2 + 1} \frac{\sin \gamma}{\sin \alpha_{st}} \frac{\tan \gamma}{\tan \alpha_{st}} = \frac{2s}{\pi c} \frac{2R}{R^2 - 1}$$

or (Eq. 5-17)

$$k_0 = \frac{2s}{\pi c} \tan \frac{\pi c}{2s} \tag{6-9}$$

For wide spacing $k_0 \to 1$, corresponding to the definition. For infinitely narrow spacing k is zero. For finite but narrow spacing, k is proportional to s/c. Hence for small s/c

$$k_0 = \frac{s}{c} \frac{\partial k_0}{\partial (s/c)} \tag{6-10}$$

Let the deflection coefficient q be defined by (Fig. B,6b)

$$q = \frac{k_0}{\frac{2s}{\pi c} \frac{1}{\cos \gamma}} = \frac{2R}{R^2 + 1} \cos \alpha_{st} \tag{6-11}$$

For narrow spacing $s/c \to 0$, $R \to 1$, $2R/(R^2 + 1) \to 1$ and (Eq. 5-14) $\tan \alpha_{st} \to 0$, $\cos \alpha_{st} \to 1$. Hence

$$\lim_{s/c \to 0} q = 1 \tag{6-12}$$

and (Eq. 6-11)

$$\lim_{s/c \to 0} k_0 = \frac{2s}{\pi c} \frac{1}{\cos \gamma} \tag{6-13}$$

As Fig. B,6b shows, this is a good approximation up to $s/c \cong 0.7$. Let β_1, β_∞, and β_2 be the angles between the upstream, free stream, and downstream velocity directions, respectively, and the normal to the cascade axis; then the angle of attack

$$\delta = \gamma - \beta_\infty \tag{6-14}$$

and (Eq. 4-14)

$$2\frac{s}{c}\frac{V_{s_2} - V_{s_1}}{V_\infty} = 2\pi k_0\left(\gamma, \frac{s}{c}\right)\sin\delta \tag{6-15}$$

or

$$\frac{s}{c}\cos\beta_\infty(\tan\beta_2 - \tan\beta_1) = \pi k_0\left(\gamma, \frac{s}{c}\right)\sin(\gamma - \beta_\infty) \tag{6-16}$$

where (Eq. 4-8)

$$\tan\beta_\infty = \tfrac{1}{2}(\tan\beta_1 + \tan\beta_2) \tag{6-17}$$

For given values of β_1, β_2, and s/c the stagger angle γ of the cascade may be determined by combining Eq. 6-17 with the results of Eq. 6-6 as they

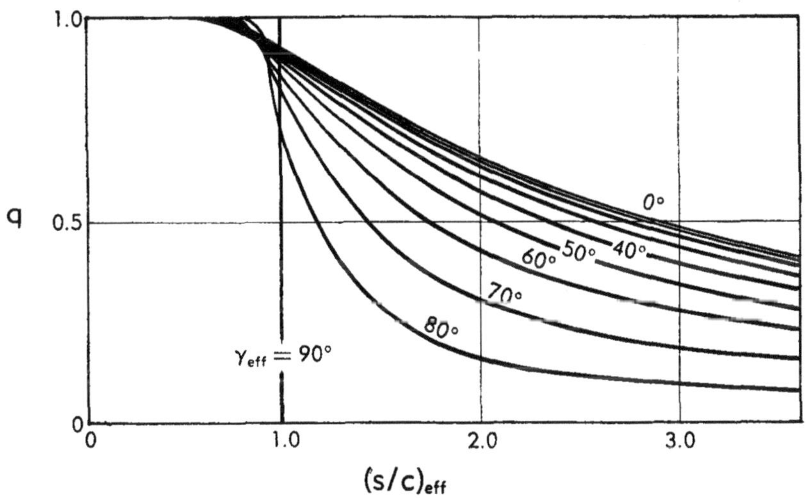

Fig. B,6b. Cascade deflection coefficient q for comparison with stream filament theory.

are plotted in Fig. B,6a. Generally the angles β_1 and β_2 follow from the design problem. The spacing ratio s/c has to be chosen considering the applicability of the theory to the real flow.

For narrow spacing (Eq. 6-13, 6-16, and 6-17)

$$\begin{aligned}\frac{s}{c}\cos\beta_\infty(\tan\beta_2 - \tan\beta_1) &= 2\frac{s}{c}\cos\beta_\infty(\tan\beta_2 - \tan\beta_\infty)\\ &= \pi\frac{2s}{\pi c}\frac{1}{\cos\gamma}\sin(\gamma - \beta_\infty)\end{aligned}$$

Hence

$$\tan\beta_2 = \tan\gamma \tag{6-18}$$

This means, with ideal flow, that for any profile cascade of small spacing ($s/c < 0.7$) the exit angle β_2 is independent of the approach angle β_1 and is equal to the stagger γ of the equivalent straight-line profile cascade. For less close spacing one may substitute Eq. 6-7 and 6-11 and

$$V_\infty = \frac{V_z}{\cos \beta_\infty} = \frac{V_z}{\cos (\gamma - \delta)} \tag{6-19}$$

in Eq. 4-14. Then

$$\frac{V_{\theta_2} - V_{\theta_1}}{V_z} = 2q \frac{\sin \delta}{\cos \gamma \cos (\gamma - \delta)} \tag{6-20}$$

By some transformation, one finds

$$\frac{\sin \delta}{\cos \gamma \cos (\gamma - \delta)} \equiv \tan \gamma - \tan (\gamma - \delta) \tag{6-21}$$

Substitution of

$$\tan \gamma = \frac{V_\theta^*}{V_z}, \quad \tan (\gamma - \delta) = \frac{V_{\theta\infty}}{V_z} = \frac{1}{2} \frac{V_{\theta_1} + V_{\theta_2}}{V_z} \tag{6-22}$$

yields

$$V_{\theta_2} - V_{\theta_1} = 2q\left(V_\theta^* - \frac{V_{\theta_1} + V_{\theta_2}}{2}\right) \tag{6-23}$$

or

$$\frac{V_{\theta_2} - V_{\theta_1}}{V_\theta^* - V_{\theta_1}} = \frac{2q}{1 + q} \tag{6-24}$$

If $V_\theta^* = V_z \tan \gamma$ and $V_{\theta_1} = V_z \tan \beta_1$ are given by the stagger angle of the straight-line profile cascade and the inlet condition, then $V_{\theta_2} = V_z \tan \beta_2$, that is, the exit condition can easily be determined by Eq. 6-24.

The coefficient q, which may be called "the deflection coefficient," therefore makes the comparison of cascade theory with stream filament theory possible.

B,7. Treatment of More General Profiles by Conformal Mapping.

Basic relations leading to conformal mapping functions. If the complex potential

$$\chi = \chi(z) = \varphi + i\psi \tag{7-1}$$

then

$$\frac{d\chi}{dz} = \overline{w} = \overline{w}(z) = u - iv = we^{-i\nu} \tag{7-2}$$

and

$$\ln \frac{d\chi/dz}{\overline{w}_{\text{ref}}} = L = L(z) = \ln \frac{w}{w_{\text{ref}}} - i(\nu - \nu_{\text{ref}}) \tag{7-3}$$

Consider $\chi = \chi(z)$ and $L = L(z)$ to be transformed by $z(\zeta)$ from the

z plane into a picture plane ζ. Then

$$\chi = \chi^*(\zeta) \tag{7-4}$$

$$L = L^*(\zeta) \tag{7-5}$$

From this $z(\zeta)$ can be found

$$z = \int \frac{d\chi}{d\chi/dz} = \int \frac{d\chi^*/d\zeta}{e^{L^*(\zeta)}}\, d\zeta \tag{7-6}$$

Influence of angle of attack. Let the complex potential of the flow through a cascade at zero lift conditions be

$$\chi_0 = \chi_0(z) \tag{7-7}$$

and at an angle of attack δ

$$\chi_\delta = \chi_\delta(z) \tag{7-8}$$

Let

$$\ln \frac{d\chi_0/dz}{\overline{W}_{\text{ref}_0}} = L_0 = L_0(z) \tag{7-9}$$

and

$$\ln \frac{d\chi_\delta/dz}{\overline{W}_{\text{ref}_\delta}} = L_\delta = L_\delta(z) \tag{7-10}$$

Then

$$\Delta L = L_\delta - L_0 = \ln \frac{w_\delta}{w_0} - i(\nu_\delta - \nu_0) - \ln \frac{w_{\text{ref}_\delta}}{w_{\text{ref}_0}} + i(\nu_{\text{ref}_\delta} - \nu_{\text{ref}_0}) \tag{7-11}$$

Along the contours of the profiles,

$$(\nu_\delta - \nu_0) - (\nu_{\text{ref}_\delta} - \nu_{\text{ref}_0}) = \begin{cases} 0 \\ \pm\pi = \pi \operatorname{sign} \delta \end{cases} \tag{7-12}$$

i.e. the velocity direction is the same for both flows, except at that part over which the stagnation point has migrated with the change of the angle of attack. There it is changed to the opposite direction.

When the profile of the z plane is mapped into the unit circle of the ζ plane, the stagnation point may change its location from $E_0 = e^{-i\alpha_{st}}$ to $E_\delta = e^{-i(\alpha_{st}+2\delta^*)}$. Then [5]

$$\Delta L^*(\zeta) = L^*_\delta(\zeta) - L^*_0(\zeta) = \ln \frac{\zeta - e^{-i(\alpha_{st}+2\delta^*)}}{\zeta - e^{-i\alpha_{st}}} - \ln \frac{\overline{W}_{\text{ref}_\delta}}{\overline{W}_{\text{ref}_0}} \tag{7-13}$$

and for points on the profile, $\zeta = e^{i\alpha}$,

$$\Delta L^*(\zeta) = L^*_\delta(\zeta) - L^*_0(\zeta) = \ln \frac{e^{i\alpha} - e^{-i(\alpha_{st}+2\delta^*)}}{e^{i\alpha} - e^{-i\alpha_{st}}} - \ln \frac{\overline{W}_{\text{ref}_\delta}}{\overline{W}_{\text{ref}_0}}$$

$$= \ln \frac{\sin\left(\dfrac{\alpha + \alpha_{st}}{2} + \delta^*\right)}{\sin \dfrac{\alpha + \alpha_{st}}{2}} - i\delta^* - \ln \frac{W_{\text{ref}_\delta}}{W_{\text{ref}_0}} \tag{7-14}$$

Profiles for prescribed velocity distribution. On this may be based the solution of the indirect problem of finding a cascade which for a given value of R, α_{st}, and δ^* has a prescribed but compatible velocity distribution. If, except for a constant, $L_\delta^*(\zeta)$ is given along the unit circle, $L_0^*(\zeta)$ can also be determined. $L_\delta^*(\zeta)$ may be established by determining the potential along the picture circle and by prescribing $\ln (w/w_{ref}(\varphi))$.

Generally a power series may be used to represent $L_0^*(\zeta)$:

$$L_0^*(\zeta) = C_0 + \frac{C_1}{\zeta} + \frac{C_2}{\zeta^2} + \cdots \tag{7-15}$$

where the coefficients C_0, C_1, and C_2, etc., may be found by harmonic analysis. However, the condition

$$L_0(+R) = L_0(-R) \tag{7-16}$$

must be fulfilled to obtain an $L_\delta^*(\zeta)$ which is compatible. Therefore it is necessary [*5*] that

$$C_1 + \frac{C_3}{R^2} + \frac{C_5}{R^4} + \cdots = 0 \tag{7-17}$$

If this is not the case, $L_\delta^*(\zeta)$ or $L_0^*(\zeta)$ has to be adjusted.

Profiles by the inverse method. Another use of the relations of Eq. 7-3 is the inverse method of the production of mapping functions for profile cascades [*5,12,13,14*]. Let

$$L_0^*(\zeta) = \sum \frac{\kappa_n}{\pi} \ln (\zeta - C_n) \tag{7-18}$$

where

$$\sum \kappa_n = 0 \tag{7-19}$$

$$\sum \frac{\kappa_n}{\pi} \ln \frac{C_n - R}{C_n + R} = 0 \tag{7-20}$$

and

$$|C_n| \leqq 1 \tag{7-21}$$

If

$$\kappa_1 = \kappa_T \qquad C_1 = C_T = +e^{i\alpha_{st}}$$

$$\kappa_2 = \kappa_L - \pi \qquad C_2 = C_{L_\delta} = -e^{i(\alpha_{st}+2\delta^*)}$$

$$\kappa_3 = \pi \qquad C_3 = -e^{i\alpha_{st}}$$

then the resulting profiles will have wedge-shaped trailing and leading edges, the wedge angles being κ_T and κ_L. The other values of κ_n and C_n have then to be chosen so as to fulfill the previously stated compatibility conditions (Eq. 7-19, 7-20, and 7-21). Since (Eq. 5-12)

$$\frac{d\chi_0^*(\zeta)}{d\zeta} = e^{-i\gamma}\left(\frac{1}{\zeta_{st} + R} - \frac{1}{\zeta_{st} - R}\right) + e^{+i\gamma}\left(\frac{1}{\zeta_{st} + \frac{1}{R}} - \frac{1}{\zeta_{st} - \frac{1}{R}}\right)$$

it is found by integration that

$$z = z(\zeta) = \int \frac{d\chi_0^*(\zeta)/d\zeta}{e^{L_0^*(\zeta)}}\, dz \tag{7-22}$$

In the case of blunt or cusped leading and trailing edges, the integration can easily be carried through analytically. For generally wedge-shaped trailing or leading edges numerical methods have to be used, and since these profiles are derived from prescribed velocity distributions, it is easy to analyze them.

Mapping of arbitrary profiles. Although this method leads only in a sort of inverse way to cascades of given profile, it is possible by a proper choice of the C_n at least to approximate any given profile having rounded nose and tail with a relatively small number of terms. By such an approximation, the real profile would be transformed into a circle-like figure in the picture plane. By the usual numerical methods, such a near-circle can be transformed easily into a true circle [*15,16*].

Thin, slightly cambered profiles. If the real and imaginary parts of a function of a complex variable are interpreted as the velocity potential and the stream function, then the real (R.P.) and imaginary (I.P.) parts of the kth derivatives are the $(k-1)$st derivatives of the velocity components in the direction of the real and negative imaginary axes. For slightly cambered profiles, which deviate little from the real axis and are described by a polynomial of nth order, the imaginary part of the nth derivative of the complex potential is constant in first order approximation [*17*]. It is possible to establish such functions of the complex variable in the case of single as well as cascade profiles. Integration then yields the intended results. While the method is not needed for single profiles, it is useful for the cascades of profiles. It may also be applied to circular arc profiles. For these profiles at smooth flow conditions one finds

$$\begin{aligned} \frac{d^2\chi}{dz^2} &= \chi^{*(2)}(\zeta) = \frac{ie^{i\gamma}}{2} C \int_{R^2}^{\zeta^2} \frac{d\zeta^2}{(\zeta^2 - E_{st}^2)^2} \\ &= -\frac{ie^{i\gamma}}{2} C \left(\frac{1}{\zeta^2 - E_{st}^2} - \frac{1}{R^2 - E_{st}^2} \right) = \frac{ie^{i\gamma}C}{2(R^2 - E_{st}^2)} \frac{\zeta^2 - R^2}{\zeta^2 - E_{st}^2} \end{aligned} \tag{7-23}$$

$$\begin{aligned} \frac{d\chi}{dz} &= \chi^{*(1)}(\zeta) = -\frac{ie^{i\gamma}C[(e^{i\gamma}/R) + e^{-i\gamma}R]}{R^2 - E_{st}^2} \int_\infty^\zeta \frac{d\zeta}{\zeta^2 - \frac{1}{R^2}} \\ &= ie^{i\gamma}C \frac{R}{2} \frac{(e^{i\gamma}/R) - e^{-i\gamma}R}{R^2 - E_{st}^2} \ln \frac{\zeta + \frac{1}{R}}{\zeta - \frac{1}{R}} \end{aligned} \tag{7-24}$$

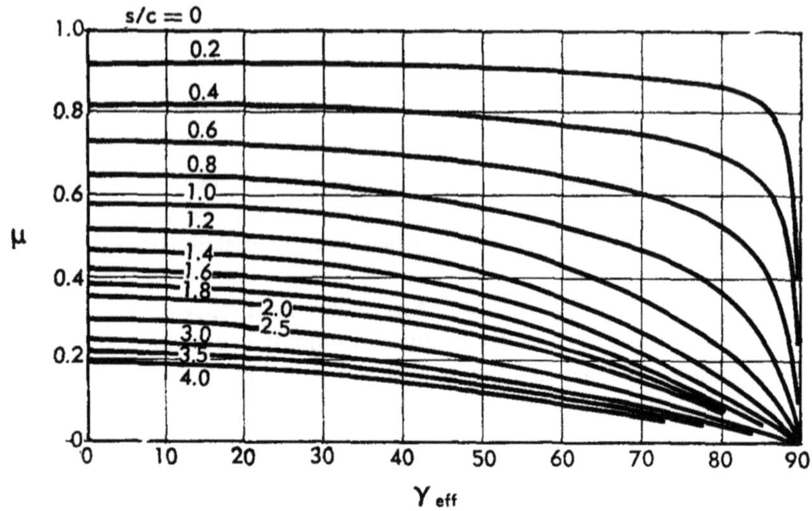

Fig. B,7a. Coefficient μ for computation of the angle deviation for smooth flow through a circular arc profile cascade.

For $\zeta = e^{i\alpha}$,

$$\text{R.P.}\left\{e^{i\gamma}\ln\frac{\zeta+\frac{1}{R}}{\zeta-\frac{1}{R}}\right\} = \frac{x}{2} \tag{7-25}$$

Let

$$C = 8\nu_2\frac{R^2 - E_{st}^2}{(e^{i\gamma}/R) + e^{-i\gamma}R}\frac{1}{Rc} \tag{7-26}$$

Then

$$y' = -\text{I.P.}[\chi^{*1}(\zeta)] = -\nu_2\frac{x}{c/2} \qquad \left(-\frac{c}{2} \leq x \leq +\frac{c}{2}\right)$$

$$y = \int_0^x y'dx = -\nu_2\int_0^x \frac{x}{c/2}\,dx = -\nu_2\frac{c}{4}\left(\frac{x}{c/2}\right)^2 \tag{7-27}$$

For $x = \mp\frac{c}{2}$,

$$y' = \pm\nu_2$$

$$y = -\nu_2\frac{c}{4}$$

For $x = 0$,

$$y' = 0$$

$$y = 0$$

The resulting cascade profiles are of the second order, i.e. parabolic or, in the first order, circular arcs of center angle $\theta = 2\nu_2$.

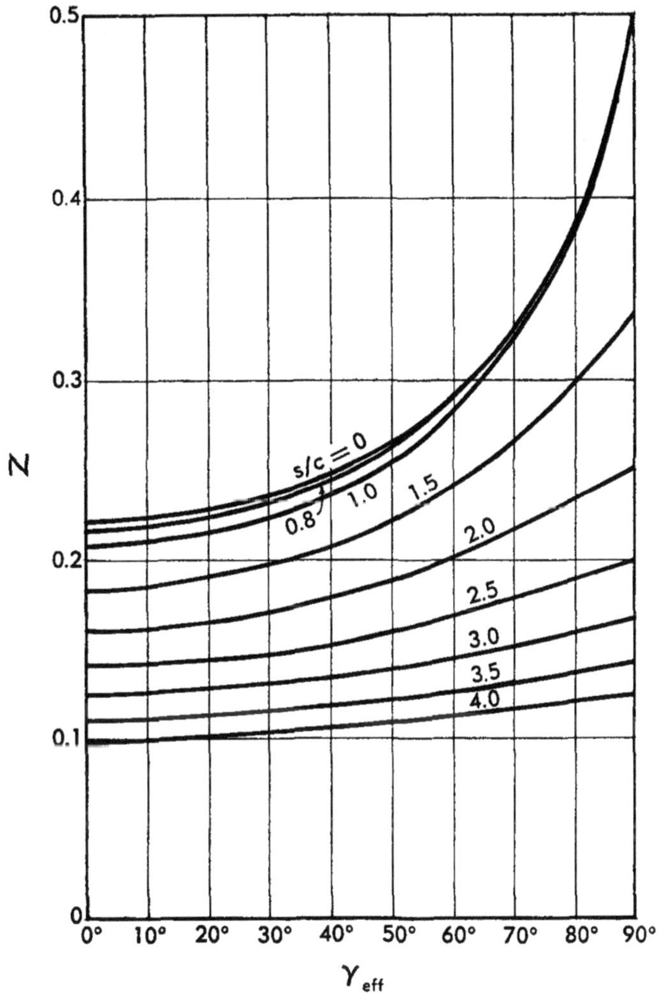

Fig. B,7b. Coefficient N for computation of angle deviation from the radius of curvature of the profile camber line.

For $\zeta = \mp R$, $x = \mp \infty$, $\frac{1}{c} = \frac{1}{2\pi}\frac{s}{c}$, then

$$y'_{-R,+R} = \pm \nu_2 \frac{s/c}{\pi/2} \cos \gamma \ln \frac{R^2 + 1}{R^2 - 1} \tag{7-28}$$

The difference between these directions and those of the leading and trailing edges is called angle exaggeration or deviation (Fig. B,7a), and

$$m = \frac{1}{\mu} = \frac{1}{\frac{s/c}{\pi/2} \cos \gamma \ln \frac{R^2 + 1}{R^2 - 1}} \tag{7-29}$$

may be called the coefficient of angle exaggeration or deviation [*4*,*5*].

The distribution of the x components of the velocity and therefore a first order approximation to the tangential velocities on the profile is given by:

$$\frac{u}{u_0} = 1 - \nu_2 \frac{s/c}{\pi/2}\left\{\sin\gamma \text{ R.P.}\left[\ln\frac{e^{i\alpha}+\frac{1}{R}}{e^{i\alpha}-\frac{1}{R}}\right] + \cos\gamma \text{ I.P.}\left[\ln\frac{e^{i\alpha}+\frac{1}{R}}{e^{i\alpha}-\frac{1}{R}}\right]\right\}$$

$$= 1 - \nu_2 \frac{s/c}{\pi/2}\frac{1}{\cos\gamma}\left[\frac{x}{2}\sin\gamma + \text{I.P.}\left\{\ln\frac{e^{i\alpha}+\frac{1}{R}}{e^{i\alpha}-\frac{1}{R}}\right\}\right]$$

$$= 1 - \nu_2 \frac{s/c}{\pi/2}\frac{1}{\sin\gamma}\left\{\text{R.P.}\left[\ln\frac{e^{i\alpha}+\frac{1}{R}}{e^{i\alpha}-\frac{1}{R}}\right] - \frac{x}{2}\cos\gamma\right\} \qquad (7\text{-}30)$$

The coefficient μ also gives the ratio of deflection θ under a smooth flow condition to the center angle θ_∞ of the circular arc:

$$\mu = \frac{\theta}{\theta_\infty} = \left(1 - 2\frac{s}{c}N\right) \qquad (7\text{-}31)$$

By the use of N (Fig. B,7b) one finds the distance Δl of the point on the profile from the leading or trailing edges, respectively, where the profile has the same direction as the approach or the exit flow:

$$\Delta l = Nt \qquad (7\text{-}32)$$

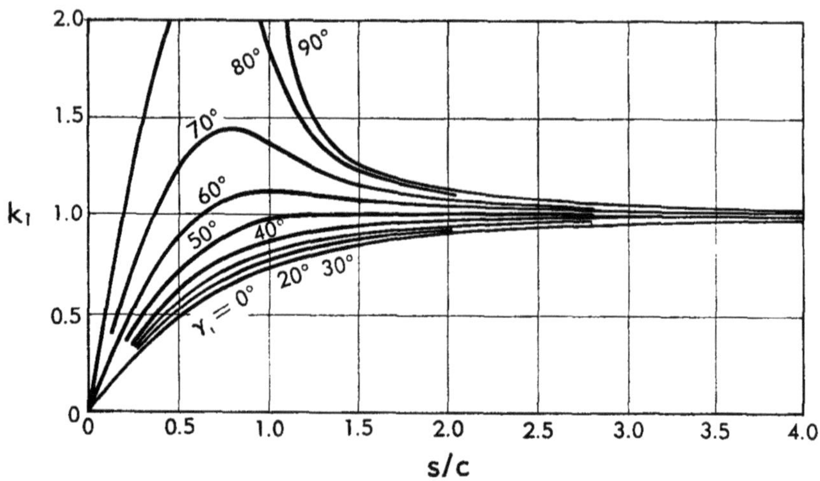

Fig. B,7c. Cascade interference coefficient k_1 for circular arc profiles with smooth inflow conditions.

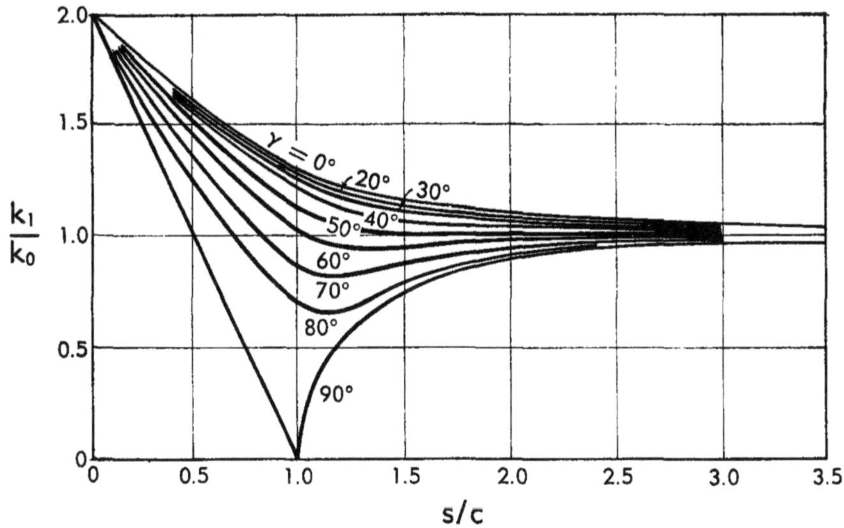

Fig. B,7d. Ratio k_1/k_0 for computation of zero lift condition of circular arc cascades.

With some degree of approximation, this holds also if the profiles are not strictly circular arcs, at least if there is sufficiently high solidity [18]. The lift coefficient c_L for smooth flow conditions of the circular arc profile in the cascade compared to the single circular arc profile follows from Fig. B,7c:

$$k_1 = \frac{C_L}{C_{L_{\text{single}}}} = \frac{4}{\pi}\frac{s}{c}\frac{1}{\cos\gamma}\left(1 - 2\frac{s}{c}N\right) \tag{7-33}$$

The angle of attack of circular arc profiles measured from the chord is zero for smooth flow. Hence the angle of attack for zero lift is

$$\delta_0 = -\frac{k_1}{k_0}\frac{\theta_\infty}{4} \tag{7-34}$$

The quotient k_1/k_0 is shown in Fig. B,7d [5,19].

B,8. Representation of Profiles by Rows of Singularities.

Theoretical basis. Instead of treating a profile cascade by conformal mapping of a cascade strip into the exterior or interior of a circle, there also exists the possibility of making a more direct use of the periodicity of the flow. The complex potential of a row of vortices of equal circulation Γ along a straight line $z = |z|e^{i\beta}$ with equal spacing π is [20]

$$\chi(z) = \frac{i\Gamma}{2\pi}\ln\sinh e^{-i\beta}z \tag{8-1}$$

If a profile is considered to be a fluid line separating the fluid flowing *outside* of the profile from a fluid at rest in its *inside*, then this fluid line is a vortex sheet and each element ds carries the vortex

$$d\Gamma = \pm w(z_s)ds \tag{8-2}$$

the positive or negative sign denoting whether w is in a direction of ds or opposite to it. Hence the complex potential of the cascade flow to be superimposed upon the undisturbed parallel flow becomes

$$\chi(z) = \frac{i}{2\pi} \oint w(z_s) \ln \sinh e^{-i\beta}(z - z_s)ds \tag{8-3}$$

Thin slightly cambered profiles. In the case of infinitely thin cambered profiles, Eq. 8-3 transforms [*21,22,23,24,25,26,27*] to

$$\chi_0(z) = \frac{i}{2\pi} \int_{\mathrm{L}}^{\mathrm{T}} (w_s - w_p) \ln \sinh e^{-i\beta}(z - z_s)ds \tag{8-4}$$

In the case of small camber, the profile almost coincides with the chord, which is considered to be in the direction of the x axis. Different assumptions for the vortex or lift distribution $\gamma = (w_s - w_p) = \gamma(s)$ then lead to different profiles.

If $\chi_{\mathrm{L}}/(c/2) = -1$ and $\chi_{\mathrm{T}}/(c/2) = +1$, and if $\cos \alpha = \chi/(c/2)$ is substituted, the distributions

$$\gamma_0 = \tan \frac{\alpha}{2}$$

$$\gamma_1 = \sin \alpha$$

$$\gamma_2 = \sin 2\alpha$$

$$\cdots$$

etc., have been investigated. The resulting component profile cascades have then been superimposed so as to yield cascades of straight-line profiles under an angle of attack, cascades of circular arc profiles, or S-shaped profiles under smooth flow conditions.

The thickness of the profiles may be treated in the following way [*28,29*]. In zero order approximation in smooth flow and at zero camber, the velocity along the profiles may be considered constant $W_s = W_p = u$. Then the vorticity along the suction side $_s$ and along the pressure side $_p$ may be replaced by dipoles having the moment $\lambda(x) = t(x) \cdot u$ of this vorticity, where $t(x)$ is the local thickness. Hence the contribution of the profile thickness of a slightly cambered profile to the complex potential is

$$\chi_t(z) = \frac{1}{2\pi} \int_{\mathrm{L}}^{\mathrm{T}} \frac{d\lambda(z_s)}{dz_s} \ln \sinh e^{-i\beta}(z - z_s)\, dz_s$$

$$= \frac{1}{2\pi} \int \lambda(z_s) \coth e^{-i\beta}(z - z_s)\, dz_0 \tag{8-5}$$

and the complex potential of the cascade flow becomes

$$\chi = \chi_s + \chi_t \tag{8-6}$$

In this way, the cascade problem may be linearized [*22*]. Since

$$\ln \sinh e^{-i\beta}(z - z_s) \equiv [\ln \sinh e^{-i\beta}(z - z_s) - \ln (z - z_s)] + \ln (z - z_s) \tag{8-7}$$

$$e^{i\beta} \coth e^{-i\beta}(z - z_s) \equiv \left[e^{i\beta} \coth e^{-i\beta}(z - z_s) - \frac{1}{z - z_s}\right] + \frac{1}{z - z_s} \tag{8-8}$$

it is also possible to treat separately the properties of the single profile and the influence of stagger and spacing. This linearization, however, is not always sufficient, and then the problem must be defined by an integral equation.

Exact treatment by integral equation [*30,31*]. Consider that the undisturbed parallel flow t has normal components on the profile surfaces: The perturbance a by the profiles must have opposite and equal normal components $w_n = \partial\varphi_a/\partial\varphi_n = -\partial\varphi_t/\partial\varphi_n$ in order to obtain the profiles as streamlines by superposition. Besides the normal components, the perturbance flow also has tangential components. Then for the potential of the perturbance

$$\varphi_a(x, y) = \frac{1}{\pi} \oint \frac{\partial\varphi_a(z_s)}{\partial n} \text{ I.P.}\{\ln \sinh e^{-i\beta}(z - z_s)\} ds + \frac{1}{\pi} \oint \frac{\partial\varphi_a(z_s)}{\partial s} \text{ R.P.}\{\ln \sinh e^{-i\beta}(z - z_s)\} ds \tag{8-9}$$

and on the profile $(x, y \to x_s, y_s)$

$$\varphi(x_\sigma, y_\sigma) = \frac{1}{\pi} \oint \frac{\partial\varphi_a(s)}{\partial n} \text{ I.P.}\{\ln \sinh \varphi^{-i\beta}(z_\sigma - z_s)\} ds + \frac{1}{\pi} \oint \frac{\partial\varphi_a(s)}{\partial s} \text{ R.P.}\{\ln \sinh \varphi^{-i\beta}(z_\sigma - z_s)\} ds \tag{8-10}$$

With $\partial\varphi_a(s)/\partial n = -\partial\varphi_t(s)/\partial n$ given by the undisturbed parallel flow the last equation is an integral equation for the determination of $\varphi_a(s)$. Since the corresponding homogeneous problem has a solution which is the pure circulating flow, one may fulfill the Kutta-Joukowski confluence condition by prescribing $\partial\varphi/\partial s$ to be zero at the trailing edge [*15*].

Another possibility of reducing the problem of the flow through a given cascade to an integral equation is by the use of Eq. 7-3 [*32,33*]:

$$L(z) = \ln \frac{d\chi/dz}{w_{\text{ref}}} = \ln \frac{w}{w_{\text{ref}}} - i\nu \tag{8-11}$$

If the front and rear stagnation points are also prescribed, the imaginary part ν = I.P.$\{L\}$ is known and the real part $\ln w/w_{\text{ref}}$ = R.P.$\{L\}$ has to be determined. For a cascade, if constants are properly chosen

$$L(z) = \oint L(z_s) \coth [e^{-i\beta}(z - z_s)]dz_s \tag{8-12}$$

Hence

$$\ln w/w_{ref} (\sigma) = \oint \ln \frac{w}{w_{ref}} (s)\text{R.P.}\{\coth [e^{i\beta}(z - z_s)]\}ds$$
$$- \oint \nu(s)\text{I.P.}\{\coth [e^{-i\beta}(z - z_s)]\}ds \tag{8-13}$$

B,9. Methods of Flux Plotting. The differential equation

$$\Delta\varphi = 0$$

occurs in various fields and when a potential flow problem must be solved, one may solve instead an analogous problem in another field [*34*]. Two of these analogies are especially important: the analogy of the electric field where the voltage is analogous to the velocity potential or to the stream function of the ideal flow and the analogy of the highly viscous fluid where the static pressure is analogous to the velocity potential or to the stream function of the ideal flow. In the use of electric analogy, either liquid or solid electrolytes may be used, the latter preferably in the form of conducting paper. The analogy of highly viscous fluid uses the flow between parallel plates placed so close together that the velocities become so small by the effect of the viscosity that their squares are negligible compared to the pressure differences.

A disadvantage of these analogies is the impossibility of directly imitating circulatory flow. This can be done only if the stream function is made analogous to the electric potential of the electric field or to the static pressure of the viscous fluid, respectively. Other analogies have also been attempted, e.g. in the magnetic field, but the obtainable accuracies have been insufficient.

There are advantages in the graphical or mechanical methods [*34*] of flux plotting. They give expert, satisfying results, after some practical training and experience, especially if an accurate solution for similar boundaries is available. In this case, one has to apply the flux plotting only to the difference. A preferable way to obtain useful flux plots consists first in the determination of an approximate flux plot. This flux plot is then evaluated for its isoclines, i.e. for the lines of constant flow direction. The isotaches $\ln w/w_{ref}$, i.e. the lines of constant magnitude of velocity, are drawn and the whole plot improved in order to yield a field of small curvilinear squares (Eq. 7-3). The velocity distribution found in this way is then used along any of the previously approximated potential lines to establish an improved distribution of the stream function along these potential lines [*5*]. In this way, an improved flux plot is found which may be further improved in a similar way. The compatible flow directions upstream and downstream are found from the differences of stream function and potential at the front (1) and trailing (2) edge stagnation points of adjacent profiles n and $n - 1$, respectively,

$$\tan \beta_{1,2} = \frac{\varphi_{1,2(n+1)} - \varphi_{1,2(n)}}{\psi_{1,2(n+1)} - \psi_{1,2(n)}} \tag{9-1}$$

This graphical method may also be used for the improvement of flux plots found by the analogies previously mentioned. The method of graphical flux plotting may also be used to find the circulatory flow around a near-circle in order to establish the conformal mapping into a true circle [*5*].

B,10. Hodograph of the Cascade Flow. Since the complex potential $\chi = \chi(z)$ and its derivative, the conjugate velocity $\bar{w}(z) = d\chi/dz$ (Eq.

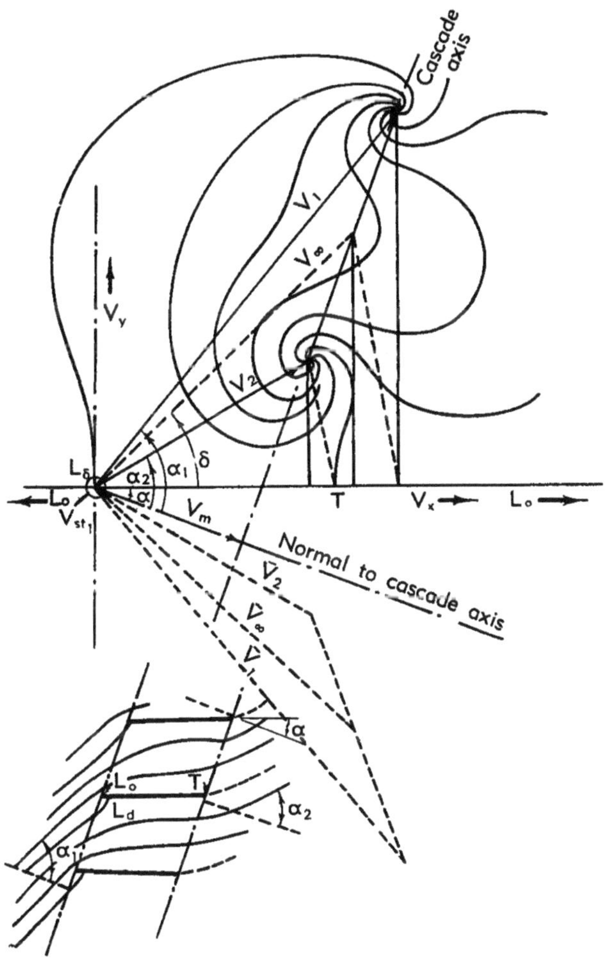

Fig. B,10a. Hodograph of flow through a straight-line profile cascade (schematic).

7-2), are functions of the same complex variable z, the elimination of z yields

$$\chi = \chi_1(\bar{w}) \tag{10-1}$$

and the conformal map of the complex potential into the $\bar{w}$ plane. This map is called the conjugate hodograph or, briefly, the hodograph of the

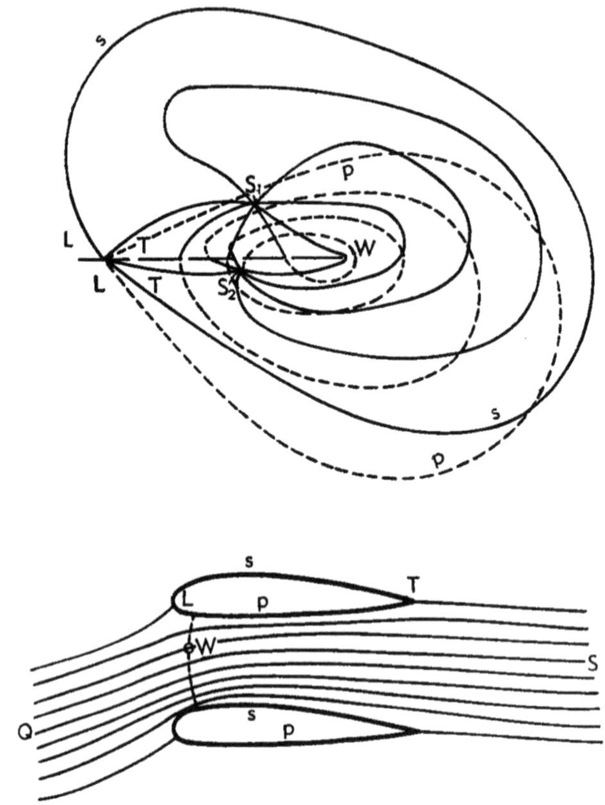

Fig. B,10b. Hodograph of a cascade flow, which in general is not single-valued.

flow. A streamline of this hodograph is the geometric locus of the free end points of the conjugate velocity vectors along this streamline.

Since the logarithm $L(z) = \ln\,(d\chi/dz)/\bar{w}_{\text{ref}}$ of the conjugate velocity (Eq. 7-3) is a function of the complex variable z, the elimination of z yields

$$\chi = \chi_2(L) \tag{10-2}$$

and the conformal map of the complex potential into the L plane. This map is called the logarithmic hodograph. Fig. B,10a shows the hodograph $\chi_1(\bar{w})$ of the flow through a straight-line profile cascade at an angle of attack [*11*].

In general, the hodograph of the flow past a single profile or a cascade is not "*schlicht*" or single-valued as Fig. B,10b shows [*35*]. Profiles exist, however, for which, at least under certain angles of attack, the hodograph becomes "*schlicht,*" as Fig. B,10c shows. The images of the origin and

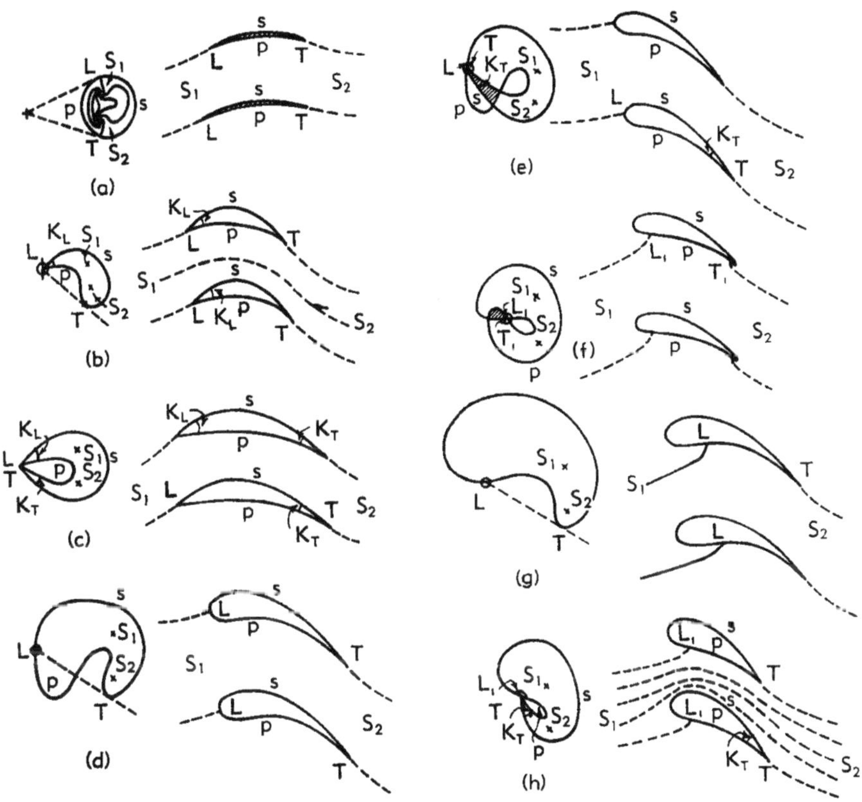

Fig. B,10c. Examples of single-valued and other simple hodographs of cascade flows.
(a) Profile having cusped leading and trailing edges.
(b) Profile having wedged leading and cusped trailing edges.
(c) Profile having wedged leading and trailing edges.
(d) Profile having blunt leading and cusped trailing edges.
(e) Profile having blunt leading and wedged trailing edges.
(f) Profile having blunt leading and trailing edges.
(g) Profile having blunt leading and cusped trailing edges and stagnation point in the point of reversed curvature.
(h) Profile having blunt leading and wedged trailing edges and stagnation point in the point of reversed curvature.

goal of the flow are given by the inlet and exit velocities and contain the corresponding vortex source and sink, thus corresponding to the condition that the profiles must be closed contours. A necessary but not sufficient condition that these contours do not intersect themselves, or are Jordanians, is that the stagnation points of the hodograph flow coincide with $\bar{w} = 0$, as in the case of blunt or wedge-shaped leading or trailing edges, or that they are situated where the velocity vector of extreme direction is tangent to the edge of the hodograph, as in the case of cusped edges.

B,11. Mapping a General Cascade into a Cascade of Straight-Line Profiles. In the case of a single arbitrary profile having a defined trailing edge as a confluence point of the flow, there always exists an angle of attack at which the lift is zero. Let the angle of attack δ be measured

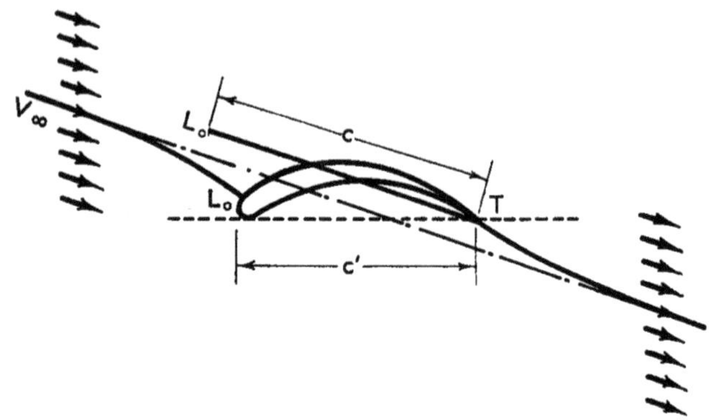

Fig. B,11a. The concept of a straight-line profile, equivalent to a given profile with respect to chord length and direction.

from the direction of zero lift. One streamline of the zero lift flow meets the profile in a branch point L_0, is divided there into upper (s_u) and lower (s_l) sides of the profile, and joins again in a confluence point T, the trailing edge (Fig. B,11a). When computing the potential difference between these points along the upper side s_u and the lower side s_l

$$(\Phi_T - \Phi_{L_0})_{u,l} = \int_{L_0}^{T} v ds_{u,l}$$

the potential differences are found to be equal on both sides. If they were different by an amount Γ, this would be the circulation, which for zero lift, as assumed, must be zero.

A straight-line profile of the length

$$c = \frac{\Phi_T - \Phi_{L_0}}{V_\infty}$$

and in the direction of V_∞ would also experience no lift and has the same potential difference between the leading and trailing edges. The outside of the profile under consideration and the outside of the straight-line

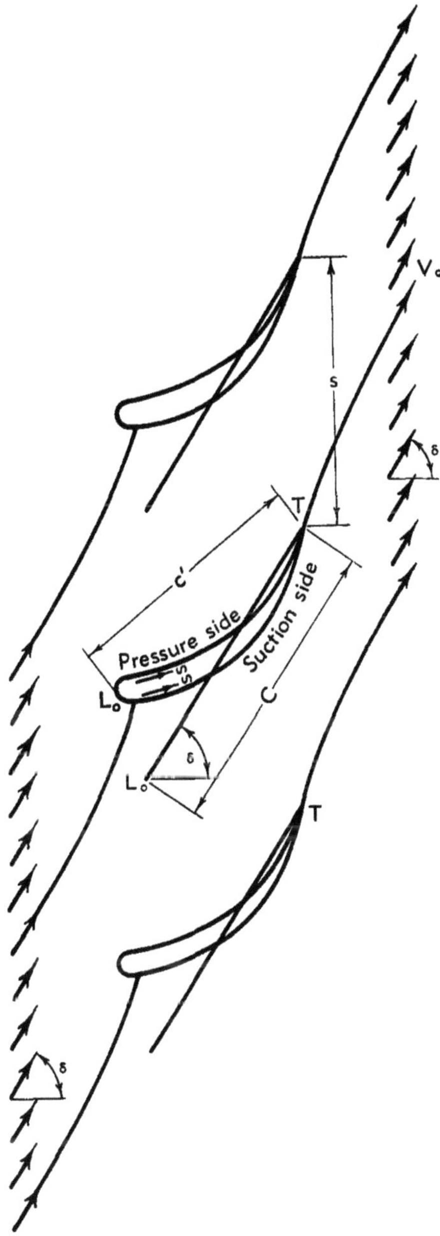

Fig. B,11b. The concept of a cascade of straight-line profiles, equivalent to a cascade of given profiles with respect to chord length and stagger at equal spacing.

profile can be uniquely transformed conformally into one another. The leading (L_0) and trailing (T) edges would correspond to each other, and at infinity the fields would be congruent. At corresponding points there would be equal values of the potential and the stream functions of flow of any equal angle of attack.

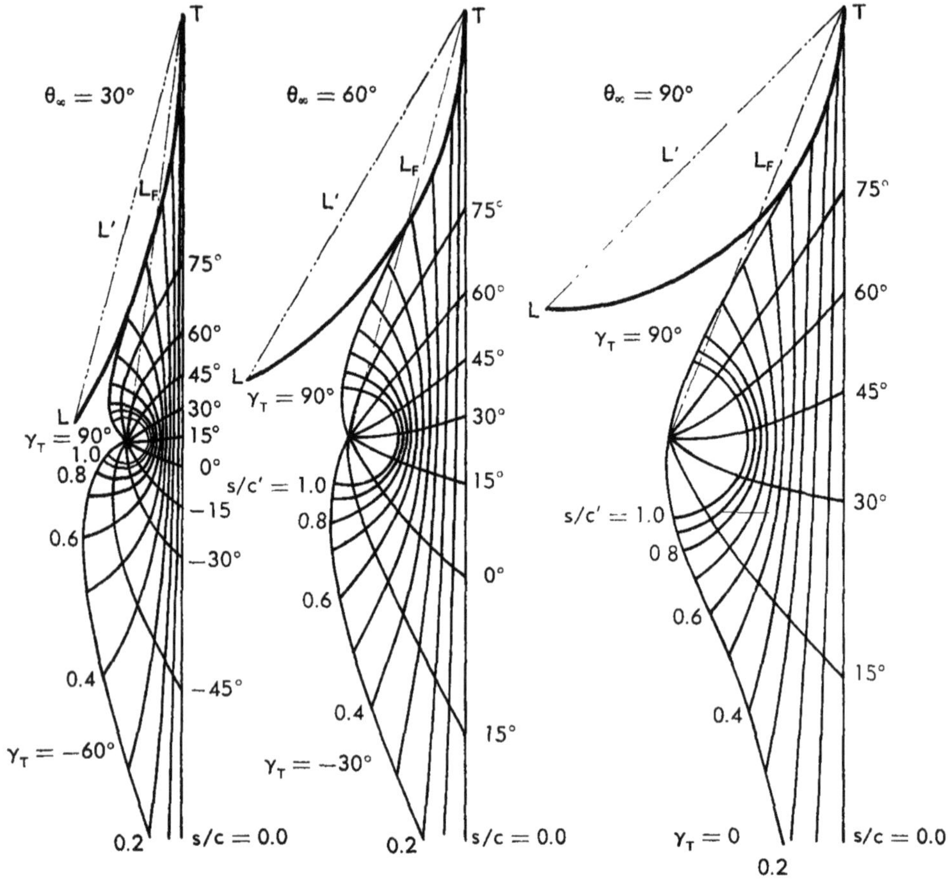

Fig. B,11c. Equivalent straight-line profile cascades of circular arc profile cascades for camber angles of θ_∞ = 30°, 60°, and 90°.

Under any equal angle of attack, therefore, both profiles would experience equal circulation and equal lift. The straight-line profile would consequently be equivalent to the given profile.

Also, for any cascade of profiles a cascade of equivalent straight-line profiles can be found. Again, consider the zero lift flow through the profile cascade. In this case the cascade does not deflect the flow, $V_1 = V_2 = \mathbf{V}_\infty$. The equivalent straight-line profiles then have the direction $\mathbf{V}_\infty$ and their length is

$$c = \frac{\Phi_T - \Phi_{L_0}}{V_\infty}$$

Since the amount of flow between two adjacent straight-line profiles must be the same as between the profiles of the cascade under consideration, the spacing and the cascade axis are the same. If one considers the trailing edges to be coinciding, the comparison of the given profile cascade and the equivalent straight-line profile cascade can be found as shown in Fig. B,11b [*5*].

As previously done, the field of the given cascade may be uniquely transformed conformally into the field of the straight-line profile cascade.

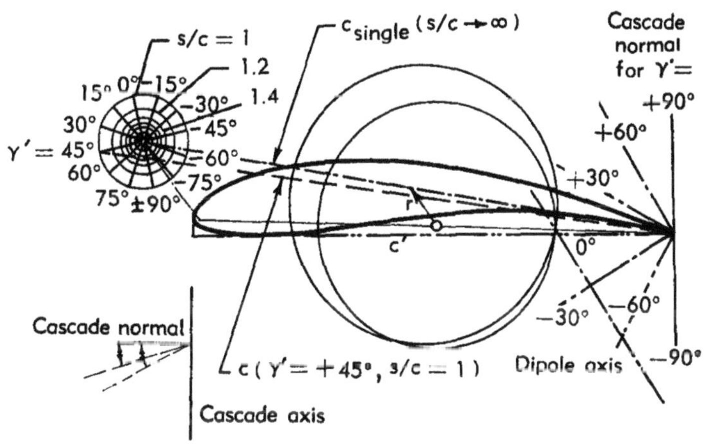

Fig. B,11d. Equivalent straight-line profile cascades of cascades consisting of Joukowski profiles.

When changing the inflow conditions equally for both cascades, equal exit conditions are also obtained. Hence the deflection, circulation, and lift are equal under any equal inlet condition for both cascades. With respect to deflection, circulation, and lift, therefore, the designation of equivalent cascades seems to be justified.

Fig. B,11c shows the straight-line profile cascades which are equivalent to circular arc profile cascades for some finite camber angles depending on spacing and stagger. Fig. B,11d shows the same for a cascade consisting of Joukowski profiles depending on stagger for moderate spacing.

It can be recognized that for a given profile the orientation and length of the equivalent straight-line profile depend on stagger and spacing. Thus there is a cascade interference effect on the zero lift angle of attack as well as on the slope of the lift coefficient, the latter in excess of the interference already encountered by the straight-line profiles themselves.

The knowledge of the straight-line profile cascade which is equivalent to a cascade of given profiles is a great help since, except for the velocity distribution, the performance of a given profile cascade may be derived from the straight-line profile cascade.

CHAPTER 3. METHODS OF TREATING COMPRESSIBILITY, HIGH SOLIDITY, AND VISCOUS EFFECTS IN TWO-DIMENSIONAL CASCADES

B,12. Consideration of Compressibility for Subsonic Flows.

Small perturbation theory. With few exceptions the blades of multistage compressors work in the subsonic range. For the high subsonic, transonic, and supersonic ranges, the theory is still in its beginnings [*36,37,38*] and is not treated here. (See Sec. G for treatment of supersonic range.) In the lower subsonic range, small perturbation theory may be applied.

Consider again the plane two-dimensional cascade flow. Upstream and downstream the flow deviates but little from a parallel flow and, in the case of smooth inflow and exit flow, only slight deviation from upstream or downstream parallel flows will occur near the leading or trailing edges respectively. For the treatment of the inlet flow consider the x axis to be parallel to the upstream velocity and, for the treatment of the exit flow, parallel to the downstream velocity. Then, for the relative flow which deviates only slightly from a parallel flow one obtains, considering compressibility,

$$(1 - M^2)\frac{\partial^2\varphi}{\partial x^2} + \frac{\partial^2\varphi}{\partial y^2} = 0 \tag{12-1}$$

where $M = V_x/a$ is the Mach number of the undisturbed upstream or downstream parallel flow [*39,40*].

Consider the potential φ to be composed of the potential of the undisturbed parallel flow φ_0 and of a small, additional flow φ_a. φ_0 automatically fulfills Eq. 12-1. Hence for φ_a

$$(1 - M^2)\frac{\partial^2\varphi_a}{\partial x^2} + \frac{\partial^2\varphi_a}{\partial y^2} = 0 \tag{12-2}$$

Consider this x, y plane to be mapped into a ξ, η plane by

$$\xi = x, \qquad \eta = \sqrt{1 - M^2}\, y \tag{12-3}$$

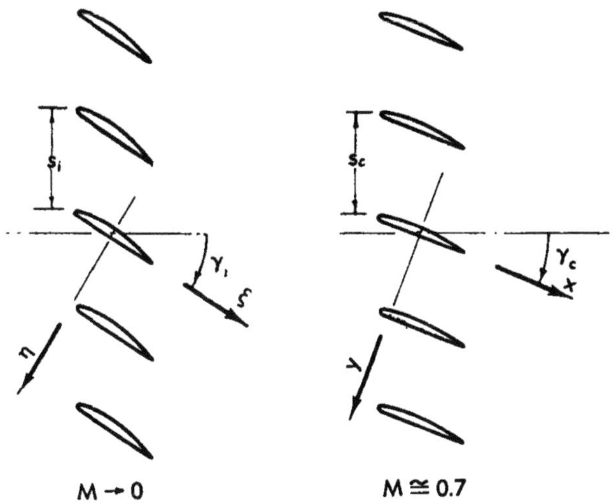

Fig. B,12a. Cascades at different Mach numbers corresponding to each other.

and consider a corresponding absolute flow of an incompressible medium in this plane, having the potential

$$\Phi_a = \sqrt{1 - M^2}\,\varphi_a \tag{12-4}$$

Also $\Phi_0 = \varphi_0$. Then, any solution Φ_a of

$$\frac{\partial^2 \Phi_a}{\partial \xi^2} + \frac{\partial^2 \Phi_a}{\partial y^2} = 0 \tag{12-5}$$

corresponds to a solution φ_a of Eq. 12-2. The velocities of these corresponding flows are given by

$$V_x = V_0 + \frac{\partial \varphi_a}{\partial x} = V_0 + \frac{1}{\sqrt{1 - M^2}} \frac{\partial \Phi_a}{\partial \xi}; \quad V'_\xi = V'_0 + \frac{\partial \Phi_a}{\partial \xi} \tag{12-6}$$

$$V_x \cong V_0 = V'_0 \cong V'_\xi \tag{12-7}$$

$$V_y = \frac{\partial \varphi_a}{\partial y} = \sqrt{1 - M^2}\,\frac{\partial \Phi_a}{\partial y} \frac{1}{\sqrt{1 - M^2}} = \frac{\partial \Phi_a}{\partial \eta} = V'_\eta \tag{12-8}$$

The directions of the flow in both planes at corresponding points are therefore equal:

$$\tan \nu = \frac{V_y}{V_x} \cong \frac{V'_\eta}{V'_\xi} \cong \nu \tag{12-9}$$

By the transformation (Eq. 12-3 and Fig. B,12a) all the streamlines, and therefore the profiles, are unchanged.

Compressibility corrections for angles. Only the distance between the streamlines is changed and with it the stagger angle of the cascade

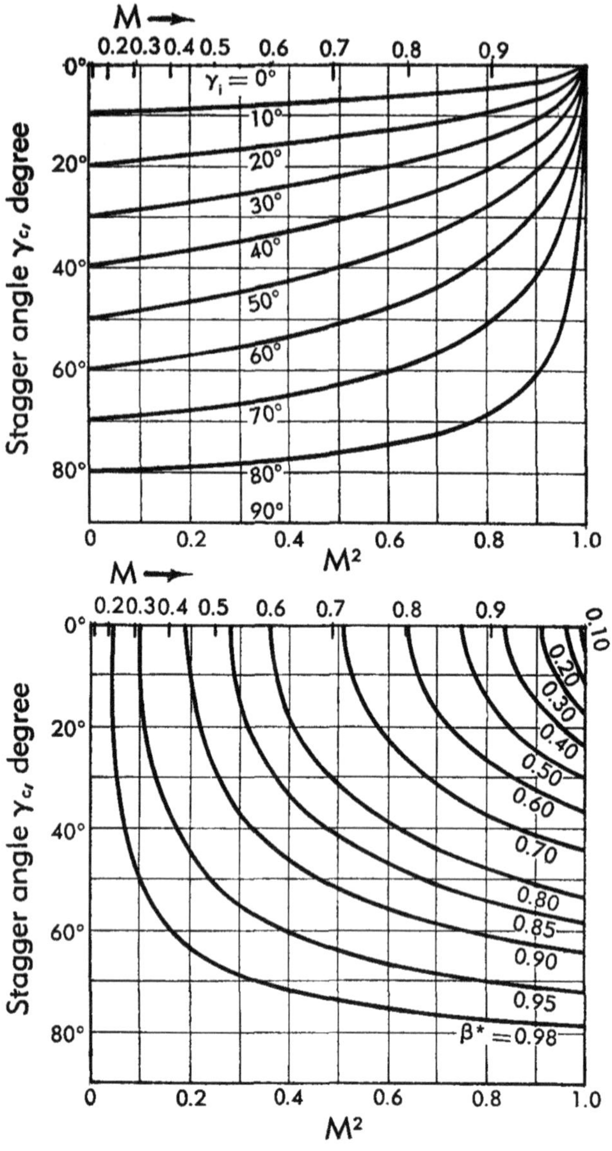

Fig. B,12b. Stagger and spacing of corresponding cascades dependent on Mach number ($\beta^* = \sqrt{1 - \cos \gamma_c^2 M^2}$).

and the spacing ratio [*41*]. (See Fig. B,12b.) Thus

$$\tan \gamma_i = \frac{\tan \gamma_c}{\sqrt{1 - M^2}} \tag{12-10a}$$

$$\sin \gamma_i = \frac{\sin \gamma_c}{\sqrt{1 - \cos^2 \gamma_c M^2}} = \frac{\sin \gamma_c}{\beta^*} \tag{12-10b}$$

$$\left(\frac{s}{c}\right)_i = \left(\frac{s}{c}\right)_c \sqrt{1 - \cos^2 \gamma_c M^2} = \left(\frac{s}{c}\right)_c \beta^* \tag{12-11}$$

where the subscripts $_i$ and $_c$ denote incompressible and compressible terms, respectively. For the cascade flow it is advisable, when computing the flow angle at the inlet and exit, to treat the inlet and exit flows separately and to determine (Eq. 7-17)

$$\left(\frac{\nu_{1,2}}{s/R_{1,2}}\right)_c = \left(\frac{\nu_{1,2}}{s/R_{1,2}}\right)_i = \left[\frac{1}{2}(1 - \mu_{1,2})\frac{c}{s}\right]_i = N\left(\gamma_{1,2}, \frac{s}{c}\right)_i \tag{12-12}$$

with

$$(\nu_{1,2})_i = (\nu_{1,2})_c$$

$$(R_{1,2})_i = (R_{1,2})_c$$

$$(\tan \gamma_{1,2})_i = \frac{(\tan \gamma_{1,2})_c}{\sqrt{1 - M^2}}$$

$$\left(\frac{s}{c}\right)_i = \left(\frac{s}{c}\right)_c \sqrt{1 - (\cos^2 \gamma_{1,2})_c M_{1,2}^2}$$

where $N(\gamma, s/c)$ may be taken from Fig. B,6d.

Compressibility corrections for pressure distribution by the hodograph method. Let λ be a parameter for which [*42,43*]

$$d\lambda = \sqrt{1 - M^2}\,\frac{dW}{W} = \sqrt{1 - M^2}\, d \ln W \tag{12-13}$$

and $\lambda = 0$ for $M = M^* = 1$.

$$M = \frac{W}{a} = \left(\frac{\frac{2}{\gamma + 1} M^{*2}}{1 + \frac{\gamma - 1}{\gamma + 1} M^{*2}}\right)^{\frac{1}{2}} \tag{12-14}$$

$$M^* = \frac{W}{a^*} = \left(\frac{\frac{\gamma + 1}{2} M^2}{1 + \frac{\gamma - 1}{2} M^2}\right)^{\frac{1}{2}} \tag{12-15}$$

are the Mach number and the velocity ratio referred to the critical velocity a^* respectively, and γ here is the ratio of specific heats. The λ, ν plane may be called the quasi-logarithmic hodograph plane.

Let

$$l(\lambda) = \frac{\rho_0}{\rho}\sqrt{1 - M^2} \tag{12-16}$$

be a parameter. Then the differential equations for the potential φ and the stream function ψ are

$$\frac{\partial\varphi}{\partial\lambda} = -l(\lambda)\frac{\partial\psi}{\partial\nu}$$

$$\frac{\partial\varphi}{\partial\nu} = +l(\lambda)\frac{\partial\psi}{\partial\lambda}$$

or with

$$N(\lambda) = \frac{l'(\lambda)}{l(\lambda)} = \frac{d\ln l}{d\lambda}$$

$$\frac{\partial^2\varphi}{\partial\nu^2} + \frac{\partial^2\varphi}{\partial\lambda^2} - N(\lambda)\frac{\partial\varphi}{\partial\lambda} = 0 \tag{12-17}$$

$$\frac{\partial^2\psi}{\partial\nu^2} + \frac{\partial^2\psi}{\partial\lambda^2} + N(\lambda)\frac{\partial\psi}{\partial\lambda} = 0 \tag{12-18}$$

In the case of the incompressible flow ($a^* = \infty$) in a layer of constant thickness:

$$N \equiv 0, \qquad l \equiv 1, \qquad \lambda - \lambda_\infty = \ln\frac{W}{W_\infty}$$

In the ν, λ plane, the picture of the subsonic flow of the compressible medium in a layer of constant thickness is analogous to the flow of an incompressible medium in a layer of the thickness $l = l(\lambda)$. For a constant λ_∞ let

$$\Delta\lambda = \lambda - \lambda_\infty$$

and be small. Then $N(\lambda)$ has only a very small influence on the solution of the differential equation. Thus with the unchanged contour of the quasi-logarithmic hodograph and with the unchanged singularities inside of this contour, when λ_∞ or the Mach number M_∞ is changed, the distribution of the potential on this hodograph contour is changed only slightly.

The assumption of small $\Delta\lambda = \lambda - \lambda_\infty$ is equivalent to the assumption of small excess velocities $\Delta W = W - W_\infty$ in the physical plane x, y. Thus, with changed λ_∞ or Mach number M_∞, the distribution of the potential along the contour of the profile is only slightly changed.

The contour of a profile may be considered to be determined by

$$\nu = \nu(x)$$

By the flow along the profile contour in the physical plane x, y

$$\varphi = \varphi_p(x) \tag{12-19}$$

is given. By the flow picture in the quasi-logarithmic hodograph plane λ, ν

$$\varphi = \varphi_h(\nu) \tag{12-20}$$

is given along the contour of the quasi-logarithmic hodograph, which corresponds to the profile contour.

From the assumption of small $\Delta\lambda$, it follows that $\varphi = \varphi_p(x)$ and $\varphi = \varphi_h(\lambda)$ are only slightly dependent on λ_∞ or the Mach number M_∞. Considering the contour of the quasi-logarithmic hodograph and the singularities contained in it to be given, the corresponding profile contour will change only slightly with change of λ_∞ or the Mach number M_∞ and at corresponding points, having the same ν, equal $\Delta\lambda$ will prevail.

For a first approximation of the effects of compressibility in subsonic flow, it is possible to neglect the changes in the profile contour, when in the quasi-logarithmic hodograph plane the picture of the contour is considered to be unchanged [44].

The effect of the compressibility at the profile contour or in the field may be described by the influence on the pressure coefficient

$$\frac{p - p_\infty}{\frac{1}{2}\rho W_\infty^2} = c_p \tag{12-21}$$

which for incompressible flow is

$$\left(\frac{p - p_\infty}{\frac{1}{2}\rho W_\infty^2}\right)_i = c_{p_i}$$

Consider corresponding values to be

$$\Delta\lambda = \lambda - \lambda_\infty = (\lambda - \lambda_\infty)_i$$

Then

$$C = \frac{c_p}{c_{p_i}} \tag{12-22}$$

may be determined.

For incompressible flow

$$(\lambda - \lambda_\infty)_i = \left(\ln \frac{W}{W_\infty}\right)_i$$

Hence

$$c_{p_i} = \left(\frac{p - p_\infty}{\frac{1}{2}\rho W_\infty^2}\right)_i = 1 - \left(\frac{W}{W_\infty}\right)_i^2 = 1 - e^{2(\lambda-\lambda_\infty)_i} = 1 - e^{2\Delta\lambda}$$

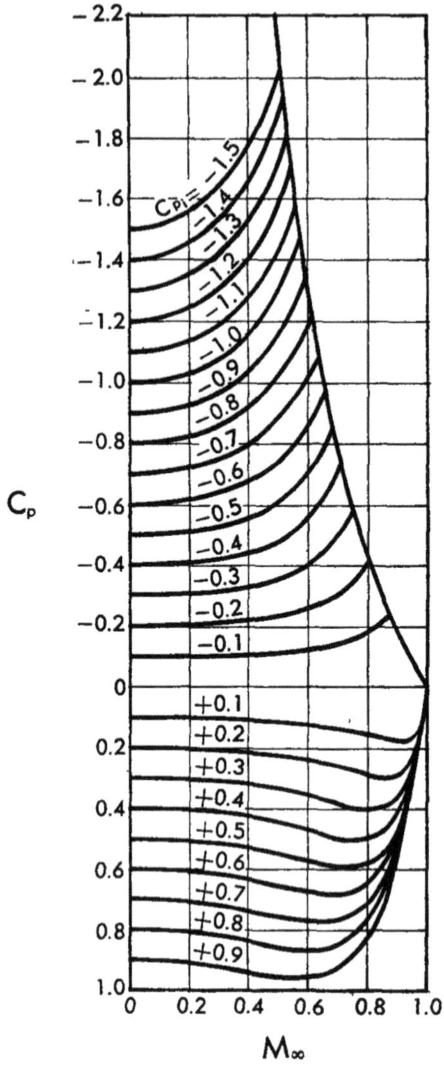

Fig. B,12c. Relation between pressure coefficient and Mach number.

For compressible flow,

$$\frac{p}{p^0} = \left(1 + \frac{\gamma - 1}{2} M^2\right)^{-\frac{\gamma}{\gamma - 1}}$$

$$\frac{p}{p_\infty} = \left(\frac{1 + \frac{\gamma - 1}{2} M^2}{1 + \frac{\gamma - 1}{2} M_\infty^2}\right)^{-\frac{\gamma}{\gamma - 1}}$$

Further,

$$\frac{\frac{1}{2}\rho W_\infty^2}{p_\infty} = \frac{\gamma}{2} M^2$$

Hence with

$$M^2 = M_\infty^2 + \Delta M^2 \tag{12-23}$$

$$c_p = \frac{2}{\gamma M_\infty^2}\left[\left(1 + \frac{\frac{\gamma - 1}{2}\Delta M^2}{1 + \frac{\gamma - 1}{2} M_\infty^2}\right)^{-\frac{\gamma}{\gamma-1}} - 1\right] \tag{12-24}$$

and

$$e^{2\Delta\lambda} = e^{2(\lambda-\lambda_\infty)} = \frac{\frac{1-(1-M^2)^{\frac{1}{2}}}{1+(1-M^2)^{\frac{1}{2}}}\left[\frac{1+h(1-M^2)^{\frac{1}{2}}}{1-h(1-M^2)^{\frac{1}{2}}}\right]^{1/h}}{\frac{1-(1-M_\infty^2)^{\frac{1}{2}}}{1+(1-M_\infty^2)^{\frac{1}{2}}}\left[\frac{1+h(1-M_\infty^2)^{\frac{1}{2}}}{1-h(1-M_\infty^2)^{\frac{1}{2}}}\right]^{1/h}} \tag{12-25}$$

where

$$h = \left(\frac{\gamma - 1}{\gamma + 1}\right)^{\frac{1}{2}} \tag{12-26}$$

After some transformation, neglecting terms of the higher order of $\Delta\lambda$, one finally obtains (Fig. B,12c)

$$C = \frac{c_p}{c_{p_i}} = \frac{1}{\sqrt{1 - M_\infty^2} + \frac{c_{p_i}}{2} M_\infty^2\left(\frac{1}{1 + \sqrt{1 - M_\infty^2}} + \frac{\gamma + 1}{4}\frac{M_\infty^2}{1 - M_\infty^2}\right)} \tag{12-27}$$

In order to apply this result to pressure distributions on cascades, the pressure distribution of the identical cascade in incompressible flow must be known.

B,13. High Solidity Cascades.

Modification of the filament theory. Despite the fact that the cascade theory covers the entire range between the single-profile theory and the stream filament theory, there still seems to be a need for some simplified treatment of cascades of high solidity, especially as they occur in turbines.

In filament theory a force field represents the blade action and the streamlines obtained are all geometrically similar. The theory would give approximately correct solutions only if the spacing were very small, if the thickness of the profiles were very small as compared with this spacing, and if the curvature radius of the profile camber line were everywhere very large compared to the spacing.

By adaptation of the filament theory to finite spacing, finite profile thickness and finite curvature of the camber line can be considered. This improved filament theory may suffice practically up to spacing ratios of $s/c = 1$ and slightly higher; it is directly applicable, however, only for operating conditions of smooth inflow.

This adaptation of the filament theory may be based on the following assumptions [*45*]:

1. The deviation $\nu = \Delta l/R$ of the smooth inflow and the exit flow directions from the camber line tangents at the leading and trailing edges where R is the curvature radius of the camber line near the leading or the trailing edge, respectively, can be estimated (Eq. 7-32 or 12-12) with or without consideration of compressibility.
2. Between the blades the velocity along any parallel to the cascade axis is approximately parallel to the camber line at the corresponding point and its inverse is linearily distributed along such parallels. This assumption is not confined to incompressible flow and not primarily confined to subsonic flow.

From

$$N_{1,2} = \frac{\nu_{1,2}}{s/R_{1,2}} \tag{13-1}$$

the points are found where the camber-line tangents have the direction of inlet and exit velocity by

$$\Delta l_{1,2} = \nu_{1,2}R_{1,2} = sN_{1,2} \tag{13-2}$$

Let $\Delta l_{1,2}$ be the length of the head and tail part of the profile camber line. The remainder $l_{\text{eff}} = l - (\Delta l_1 + \Delta l_2)$ is, in the sense of stream filament theory, the length of the representative part.

Computation of the velocity distribution for incompressible flows. Using the designation of Fig. B,13a, the mean velocity $\tilde{V}$ in the passages may be found from

$$\tilde{V} = \frac{V_m}{\cos\gamma}\frac{s}{s - \dfrac{t}{\cos\gamma}} = \frac{V_m}{\cos\gamma}\frac{s}{s - \Delta s} \tag{13-3}$$

where γ is the directional angle of the camber line of the profile measured from the normal to the cascade axis, s is the spacing, t is the profile thickness as measured normal to the camber line, and V_m is the axial component of the flow far upstream or downstream of the cascade. If V_{θ_1} is the circumferential component of the inlet velocity, and V_{θ_2} of the exit velocity, then

$$\Delta V_\theta = V_{\theta_2} - V_{\theta_1} \tag{13-4}$$

is the change of the circumferential component. Hence the circulation around every blade is

$$\Delta\Gamma = \Gamma_p = \Gamma_2 - \Gamma_1 = s\Delta V_\theta = s(V_{\theta_2} - V_{\theta_1}) \tag{13-5}$$

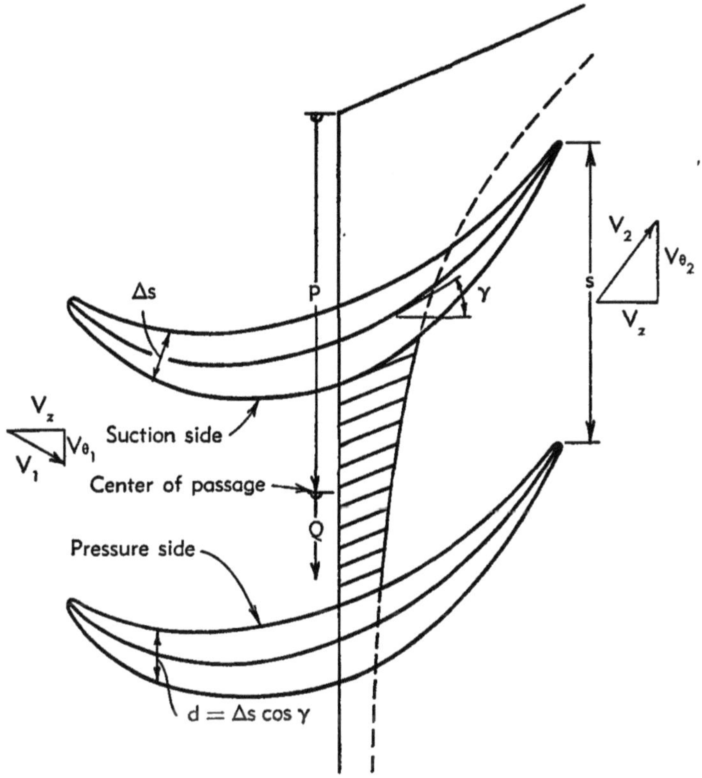

Fig. B,13a. Velocity distribution across a passage parallel to the cascade axis.

The mean circumferential component across the passages parallel to the cascade axis is

$$\bar{V}_\theta = \bar{V} \sin \gamma \tag{13-6}$$

or by substitution of Eq. 13-3,

$$\bar{V}_\theta = V_m \tan \gamma \frac{s}{s - \Delta s} \tag{13-7}$$

Hence the change of the mean circumferential component along the passage is

$$\frac{d\bar{V}_\theta}{dl} = V_m \frac{d}{dl}\left(\tan \gamma \frac{s}{s - \Delta s}\right) \tag{13-8}$$

The contribution of any passage to the circulation around the turbine axis is

$$\Gamma = (s - \Delta s)\bar{V}_\theta = V_m s \tan \gamma \tag{13-9}$$

and the rate of change of this circulation along the camber line of the blade is

$$\frac{d\Gamma}{dl} = sV_m \frac{d \tan \gamma}{dl} = \frac{s}{l_e} V_m \frac{d \tan \gamma}{d\lambda} \qquad \left(0 < \lambda = \frac{l}{l_e} < 1\right) \tag{13-10}$$

This may be called the intensity of the vortex distribution along the camber line of the profile.

Now, if the velocity is denoted on the pressure side by V_p and on the suction side by V_s, then the intensity of the vortex distribution is given by

$$d\Gamma = (V_s - V_p)dl \tag{13-11}$$

Thus

$$V_s - V_p = \frac{d\Gamma}{dl} = \frac{s}{l_e} V_m \frac{d \tan \gamma}{d\lambda} \tag{13-12}$$

where l_e is the total length of the camber line, l is the length of the camber line measured from the trailing edge, and

$$\lambda = \frac{l}{l_e} \tag{13-13}$$

With the assumption that the velocities are to vary inversely as the distance $P + Q$ across the passage (Fig. B,4a),

$$V = \tilde{V} \frac{P}{P + Q} \tag{13-14}$$

where Q is the distance from the center of the passage measured in the direction of the cascade axis, P is an effective mean curvature radius and approximately of the order of the curvature radius of the profile camber line, and $\tilde{V}$ is the velocity in the middle of the passage. Then

$$V_s = \tilde{V} \frac{P}{P - \dfrac{s - \Delta s}{2}} \quad \text{or} \quad \frac{1}{V_s} = \frac{1}{\tilde{V}}\left(1 - \frac{s - \Delta s}{2P}\right) \tag{13-15a}$$

$$V_p = \tilde{V} \frac{P}{P + \dfrac{s - \Delta s}{2}} \quad \text{or} \quad \frac{1}{V_p} = \frac{1}{\tilde{V}}\left(1 + \frac{s - \Delta s}{2P}\right) \tag{13-15b}$$

Hence

$$\frac{1}{2}\left(\frac{1}{V_p} - \frac{1}{V_s}\right) = \frac{1}{\tilde{V}} \frac{s - \Delta s}{2P} \quad \text{or} \quad \frac{2P}{s - \Delta s} = \frac{2V_sV_p}{V_s - V_p} \tag{13-16a}$$

$$\frac{1}{2}\left(\frac{1}{V_p} + \frac{1}{V_s}\right) = \frac{1}{\tilde{V}} \quad \text{or} \quad \tilde{V} = \frac{2V_sV_p}{V_s - V_p} \tag{13-16b}$$

The mean flow velocity

$$\bar{V} = \frac{1}{s - \Delta s} \int_{-\frac{s-\Delta s}{2}}^{+\frac{s+\Delta s}{2}} V dQ = \frac{\bar{V}}{2} \int_{-1}^{+1} \frac{P}{P + Q} \frac{dQ}{\frac{s - \Delta s}{2}}$$

$$= \frac{\bar{V}}{2} \frac{2P}{s - \Delta s} \ln \frac{P + \frac{s - \Delta s}{2}}{P - \frac{s - \Delta s}{2}}$$

or

$$\bar{V} = \frac{V_s V_p}{V_s - V_p} \ln \frac{V_s}{V_p} \tag{13-17}$$

Let

$$V_s + V_p = S \qquad V_s = \frac{S + D}{2} \tag{13-18a}$$

$$V_s - V_p = D \qquad V_s = \frac{S - D}{2} \tag{13-18b}$$

then

$$\bar{V} = \frac{1}{4} \frac{S^2 - D^2}{D} \ln \frac{S + D}{S - D} \cong \frac{S}{2}\left(\frac{S}{D} - \frac{D}{S}\right)\left[\frac{D}{S} + \frac{1}{3}\left(\frac{D}{S}\right)^3 + \cdots\right]$$

or

$$\frac{\bar{V}}{D} = \frac{1}{4}\left[\left(\frac{S}{D}\right)^2 - 1\right] \ln \frac{\frac{S}{D} + 1}{\frac{S}{D} - 1} \tag{13-19}$$

For a given $\bar{V}/D$ the corresponding S/D may be found by the use of Fig. B,13b which evaluates Eq. 4-19. The value of P may be found from

$$\frac{S}{D} = \frac{V_s + V_p}{V_s - V_p} = \frac{2P}{s - \Delta s} \tag{13-20}$$

whence

$$P = \frac{s - \Delta s}{2} \frac{S}{D} \tag{13-21}$$

Then

$$\frac{V_s}{\bar{V}} = \frac{1}{2} \frac{D}{\bar{V}} \left(\frac{S}{D} + 1\right) \tag{13-22a}$$

$$\frac{V_p}{\bar{V}} = \frac{1}{2} \frac{D}{\bar{V}} \left(\frac{S}{D} - 1\right) \tag{13-22b}$$

can be found, after

$$D = V_s - V_p = \frac{s}{l_c} V_m \frac{d \tan \gamma}{d\lambda} \tag{13-23}$$

and

$$\bar{V} = \frac{V_m}{\cos\gamma}\frac{s}{s - \Delta s} \tag{13-24}$$

have been determined.

The velocity distribution found in this way, however, has to be adapted to the finite spacing. The velocity found for the representative part corresponds to the circulation, but head and tail parts also contribute to the circulation. Therefore the circulation of the representative part has to be reduced by their amount and the calculated velocities have

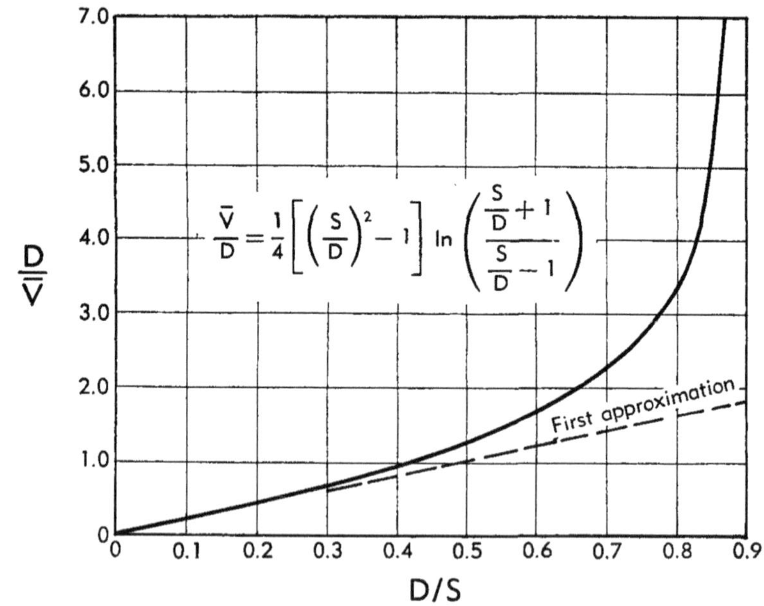

Fig. B,13b. Relationship between difference D and sum S of velocities on suction and pressure sides and the mean axial approach velocity.

to be adjusted correspondingly. This may be done graphically with some judgment as shown in Fig. B,13c, where a first correction is made using elongations ending at the inlet and exit velocities and where a second correction is made to meet the stagnation point conditions and to adjust the difference in the length of the profile contour on the suction and pressure sides. In general a velocity distribution found in this way is sufficient to determine the potential distribution on the suction and on the pressure sides. This can be used for improvement of the analysis, e.g. by flux plotting.

Extension to compressible flows. Under the assumption of velocity varying inversely as the distance $P + Q = r$ along a parallel to the cascade axis, the distribution of velocity and therefore of density is the same as in compressible single vortex flow.

Consider, at a radius r^*, the velocity of this vortex flow to be equal to the sound velocity a^*. Then

$$M^* = \frac{V}{a^*} = \frac{r^*}{r} \tag{13-25}$$

and the distribution of density becomes

$$\frac{\rho}{\rho^*} = \left(\frac{\gamma - 1}{2}\right)^{\frac{1}{\gamma-1}} \left[1 - \frac{\gamma - 1}{\gamma + 1}\left(\frac{r^*}{r}\right)^2\right]^{\frac{1}{\gamma-1}} \tag{13-26}$$

The stream function is then found nondimensionally from

$$\Psi = \int_{r/r^*}^{1} \frac{\rho}{\rho^*}\frac{r^*}{r}\, d\frac{r}{r^*} \tag{13-27}$$

with $\Psi = 0$ for $r/r^* = 1$. ρ/ρ^* and Ψ depend only slightly on $\gamma = c_p/c_v$ and are shown in Fig. B,13d for $\gamma = 1.33$.

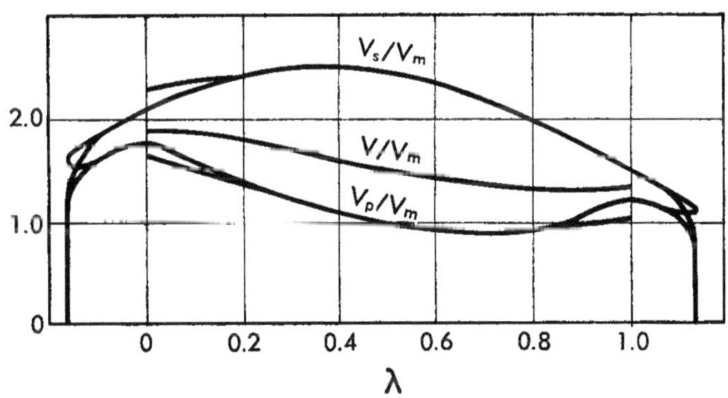

Fig. B,13c. Velocity distributions along the pressure and suction sides as found by approximation and their adjustment for the tail and head parts of the profile.

Let $\hat{V}_m$ be the axial velocity in a blade-free annulus at the location under consideration and V_m the axial velocity in this annulus if the density equals the upstream density. Let ρ_m be the mean density in the annulus under consideration and ρ_1 the density upstream. Let

$$f = \frac{\rho_1}{\rho_m} \tag{13-28}$$

Then

$$\hat{V}_m = fV_m \tag{13-29}$$

The factor f may be determined from the compressible vortex flow by

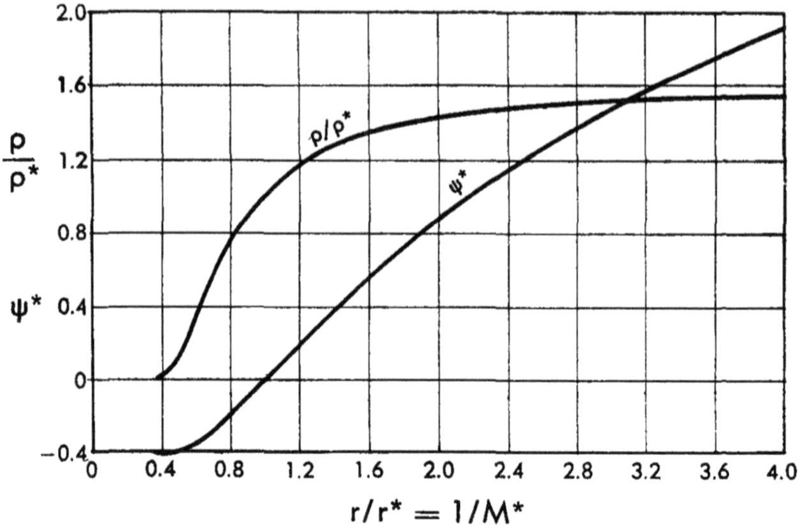

Fig. B,13d. Distribution of density ρ and stream function ψ of a vortex for $\gamma = c_p/c_v = 1.33$.

an iterative procedure as described later. By the use of this factor, the equations of incompressible flow may easily be modified. It is found that when the ^ values designate the compressible case,

$$\hat{\bar{V}} = \frac{\hat{V}_m s}{\cos\gamma(s - \Delta s)} = f\frac{V_m s}{\cos\gamma(s - \Delta s)} = f\bar{V} \tag{13-30}$$

$$\hat{\bar{V}}_\theta = \hat{\bar{V}}\sin\gamma = fV_m\tan\gamma\frac{s}{s - \Delta s} = f\bar{V}_\theta \tag{13-31}$$

$$\hat{\Gamma} = (s - \Delta s)\hat{\bar{V}}_\theta = fV_m s\tan\gamma \tag{13-32}$$

$$\frac{d\hat{\Gamma}}{dl} = fsV_m\frac{d\tan\gamma}{dl} = f\frac{s}{l_c}V_m\frac{d\tan\gamma}{d\lambda} \tag{13-33}$$

$$\hat{V}_s - \hat{V}_p = \frac{d\hat{\Gamma}}{dl} = f(V_s - V_p) = f\frac{s}{l_c}V_m\frac{d\tan\gamma}{d\lambda} \tag{13-34}$$

Hence

$$\frac{\hat{D}}{\hat{\bar{V}}} = \frac{\hat{V}_s - \hat{V}_p}{\hat{\bar{V}}} = \frac{f(V_s - V_p)}{f\bar{V}} = \frac{D}{\bar{V}} \tag{13-35}$$

and

$$\frac{\hat{S}}{\hat{D}} = \frac{S}{D} \tag{13-36}$$

Therefore for given $\hat{V}/\hat{D} = V/D$ the corresponding $\hat{S}/\hat{D} = S/D$ may again be found by the use of Fig. B,13b.

Since Eq. 13-30 gives

$$\hat{\bar{V}} = f\bar{V}$$

it follows that

$$\hat{D} = fD \qquad \text{and} \qquad \hat{S} = fS \tag{13-37}$$

Hence

$$\hat{V}_s = \frac{\hat{S} + \hat{D}}{2} = f\frac{S + D}{2} = fV_s \tag{13-38a}$$

$$\hat{V}_p = \frac{\hat{S} - \hat{D}}{2} = f\frac{S - D}{2} = fV_p \tag{13-38b}$$

and

$$\frac{\hat{V}_s}{\hat{\bar{V}}} = \frac{1}{2}\frac{\hat{D}}{\hat{\bar{V}}}\left(\frac{\hat{S}}{\hat{D}} + 1\right) \tag{13-39a}$$

$$\frac{\hat{V}_p}{\hat{\bar{V}}} = \frac{1}{2}\frac{\hat{D}}{\hat{\bar{V}}}\left(\frac{\hat{S}}{\hat{D}} - 1\right) \tag{13-39b}$$

may be found after

$$\hat{D} = \hat{V}_s - \hat{V}_p = f\frac{s}{l_c}V_m\frac{d\tan\gamma}{d\lambda} \tag{13-40}$$

and

$$\hat{\bar{V}} = f\frac{V_m}{\cos\gamma}\frac{s}{s - \Delta s} \tag{13-41}$$

have been determined.

It may be mentioned that, except for these changes, the other equations found for the incompressible case also hold true for the compressible case if the quantities are denoted by ^.

The factor f can be determined in the following way: The mass flowing through a passage is computed from upstream conditions

$$m = s\rho_1 V_{m_1} = s\rho_1 V_1 \cos\beta_1$$

For any station

$$m = (s - \Delta s)\rho_m \bar{V}\cos\beta$$

Hence to fulfill the mass flow condition

$$\frac{\bar{V}}{V_1} = \frac{s}{s - \Delta s}\frac{\rho_1}{\rho_m}\frac{\cos\beta_1}{\cos\gamma} \tag{13-42}$$

Since

$$m = \psi_s - \psi_p = \rho^* a^* s^*(\Psi_s - \Psi_p) = s\rho_1 V_{m_1}$$

$$\Psi_s - \Psi_p = \frac{\rho_1}{\rho^*}\frac{V_{m_1}}{a^*}\frac{s}{r^*} = \Psi_s\left(\frac{V_s}{a^*}\right) - \Psi_p\left(\frac{V_p}{a^*}\right) \tag{13-43}$$

In the case of incompressible flow, let

$$\frac{V_{s(0)}}{V_{m_1}} = \frac{V_s}{V_{m_1}}$$

and

$$\frac{V_{p(0)}}{V_{m_1}} = \frac{V_p}{V_{m_1}}$$

Then for a given V_{m_1}/a^*, $V_{s(0)}/a^*$ and $V_{p(0)}/a^*$ are also known. From this $\psi_{s(0)}$ and $\Psi_{p(0)}$ can be obtained. Then

$$\frac{\Psi_s - \Psi_p}{\Psi_{s(0)} - \Psi_{p(0)}} = \frac{\rho_{1(0)}}{\rho_m} = f_{(0)} \tag{13-44}$$

is found. In the case of a given profile, this is an approximation because of the change of mean density due to compressibility.

An improved value of $\bar{V}/V_{m_1}$ is then found and from this improved values of $\hat{V}_\theta$, $\hat{\Gamma}$, $d\hat{\Gamma}/dl$ are obtained, yielding $V_{s(1)}/a^*$ and $V_{p(1)}/a^*$ and improved values, $\psi_{s(1)}$ and $\psi_{p(1)}$. Then with

$$\frac{\Psi_s - \Psi_p}{\Psi_{s(1)} - \Psi_{p(1)}} = \frac{\rho_{1(0)}}{\rho_m} = f_{(1)}$$

this procedure may be iterated. If the procedure converges, the result is the desired one.

If there is no convergence, it may be concluded that the smooth shock-free flow assumed for the given conditions does not exist, and shocks may be expected.

For a somewhat higher mass flow, therefore, choking may be expected.

There are also other methods [*46,47*] for the high solidity cascade, which are based mainly on the change of velocity normal to a streamline depending on the curvature radius R_ψ of the streamline

$$\frac{d \ln (V/V_{\text{ref}})}{dn} = \frac{1}{R_\psi}$$

Although this is an exact relation, the application is somewhat difficult and, when short cuts are applied, it does not offer more accurate results than the procedure described above, where for each section along the cascade axis, R_ψ is taken as equal to the radius of the camber line.

B,14. Boundary Layers and Wakes.

Theoretical treatment of boundary layer. In the case of a single profile having a relatively small lift, the boundary layer has so small an influence on the pressure distribution that it is generally disregarded. In the case of a cascade of higher solidity the boundary layer has a larger influence because its displacement of the external flow cannot be neglected when compared with the spacing. This is especially true in the range of larger lift coefficients. Before true boundary layer investigations are attempted, the pressure distribution should already be known. The following is an approximate method based on potential flow, which gives the pressure distribution and a boundary layer model [*48*].

In boundary layer theory, it is considered that the boundary layer extends only a small distance δ from the profile contour, while outside of this range the flow ψ_0 is supposed to be that of an ideal fluid (Fig.

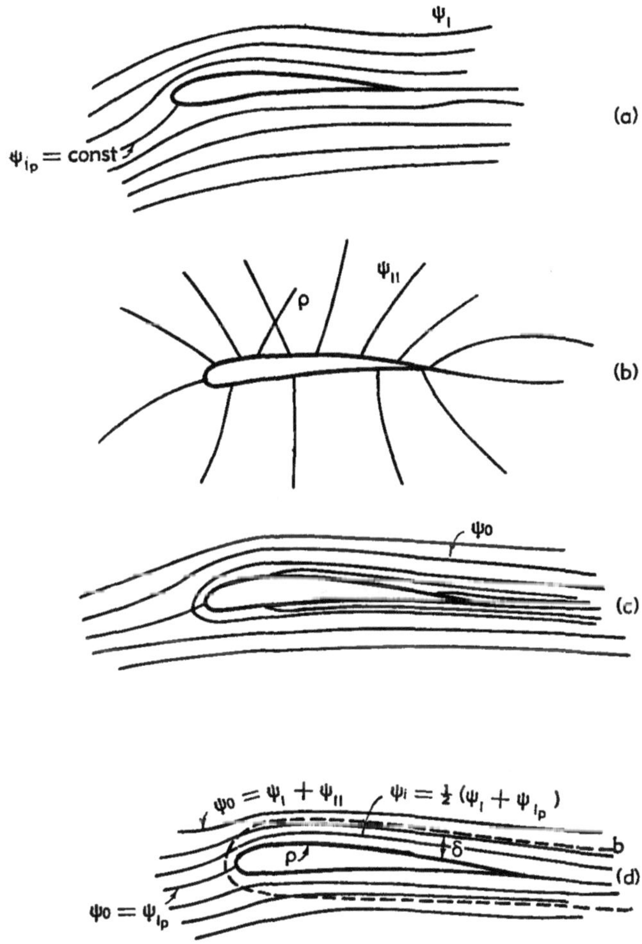

Fig. B,14a. Imbedding of boundary layer in potential flow: (a) ideal potential flow ψ_{I}, (b) displacement flow ψ_{II}, (c) displaced flow $\psi_0 = \psi_{\mathrm{I}} + \psi_{\mathrm{II}}$, and (d) flow with boundary layer (outside $\psi_0 = \psi_{\mathrm{I}} + \psi_{\mathrm{II}}$; inside $\psi_i \cong \frac{1}{2}(\psi_{\mathrm{I}} + \psi_{\mathrm{I}_p})$.

B,14a). Consider this flow ψ_0 to originate by superposition of an ideal flow ψ_{I} without a boundary layer and a flow ψ_{II} due to sources on the profile contour, corresponding to the increase of displacement thickness δ^* of the boundary layer, thus

$$\psi_0 = \psi_{\mathrm{I}} + \psi_{\mathrm{II}} \tag{14-1}$$

For a given momentum thickness θ of the boundary layer, the displacement thickness δ^* is a minimum if the velocity in the range of the boundary layer is constant and one half of the velocity outside the boundary layer. In this case the displacement thickness δ^* and the momentum thickness θ are in the ratio

$$H = \frac{\delta^*}{\theta} = 2 \tag{14-2}$$

This value of the boundary layer form factor H is representative of values near the trailing edge of profiles.

Therefore the boundary layer and the wake may be substituted for by a flow ψ_i having one half the velocity of the ideal flow ψ_I which has no boundary layer. Thus

$$\psi_i = \tfrac{1}{2}(\psi_I - \psi_{I_p}) + \psi_{I_p} = \tfrac{1}{2}(\psi_I + \psi_{I_p}) \tag{14-3}$$

where $\psi_{I_p} = \text{const}$ is the stream function of ψ on the contour of the profiles. The edge of this boundary layer is then determined by

$$\psi_0 = \psi_i \tag{14-4}$$

In this way no new streamlines originate on the profile or in the boundary layer. The replacement flow contains one half of its vorticity on the edge (b) of the boundary layer and one half on the profile contour (p), whereas in the real flow the vorticity is more evenly distributed across the boundary layer. Consider the total pressure across the boundary layer to be constant, and the static pressure along the edge to be equal on the outside and inside. The resulting static pressure change across the boundary layer compensates for the curvature of the streamlines in the boundary layer.

Now consider the circulation of the flow ψ_I to be such that in the confluence point the static pressure becomes equal on both sides of the confluence streamline. The confluence point then does not necessarily coincide exactly with the trailing edge. Generally the circulation will be less than that following from the Kutta-Joukowski theorem.

In first order approximation, the displacement sources might be considered to be distributed proportionately to the squares of the velocities of the flow ψ_I and in the total to correspond to the drag.

In this way, at least approximately, a relation between the drag and lift deficiency might be established along with a lift maximum. Furthermore the established pressure distribution might be an improved starting point for an application of the boundary layer theory itself.

Separated flow when wake and downstream pressure are equal. Along the profile contour generally, and depending on the pressure distribution and curvature, only a certain pressure recovery is possible. With an increased angle of attack the flow separates from the profile and then resembles a potential flow with free streamlines separating from the profile. The pressure along these free streamlines must equal the downstream pressure so that such a flow can exist (Fig. B,14b).

The case of a cascade of straight-line profiles has been treated by Betz and Peterson [*49,50,51*]. Downstream velocity and direction are found to depend on upstream velocity and direction and on stagger and spacing of the cascade.

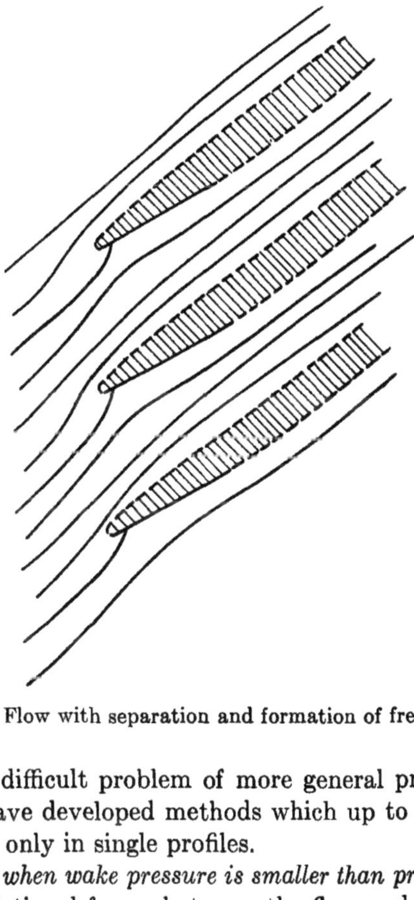

Fig. B,14b. Flow with separation and formation of free streamlines.

For the more difficult problem of more general profiles, Levi-Civita and Schmieden have developed methods which up to now seem to have found application only in single profiles.

Separated flow when wake pressure is smaller than pressure downstream. Because of the frictional forces between the flow and the wake, at least in the case of low solidity, the pressure in the wake is lower near the profile than the pressure farther downstream [*52*].

In the case of symmetric profiles in symmetric flow, Riabouchinsky has suggested treating the flow upstream and, to some extent, downstream by means of potential flow in the following way. (Fig. B,14c):

Consider downstream a straight line normal to the free stream velocity on which the upstream contour is reflected. Consider the streamlines to separate on the upstream contour and to attach to the reflected contour correspondingly.

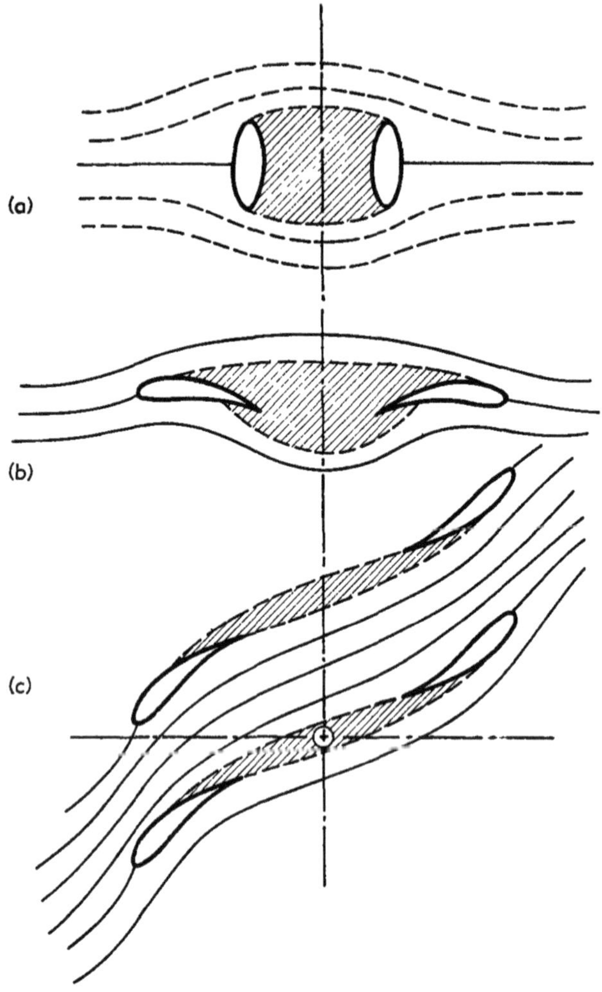

Fig. B,14c. Treatment of low pressure wake problems: (a) symmetric body under symmetric conditions (Riabouchinski [52]), (b) single airfoil profile treated by reflection (Schmieden [42]), and (c) single airfoil or cascade treated central symmetrically.

Determine this line of reflection so that the wake pressure or the drag of the body equals the observed one. Schmieden proposed that this method also be used for single airfoils. In order to apply this idea to profiles of cascades, instead of a reflection on a straight line a central symmetric arrangement might be used as shown in Fig. B,14c. It is not known to the author whether cascade flow has actually been treated in this way.

Effect of wake dissipation on downstream flow. Along a parallel to the line through the trailing edges, the downstream flow has neither constant direction nor constant velocity, mainly as a result of the profile wake. The conditions become more uniform as the distance downstream increases.

The conditions on two such parallels may be distinguished by the superscripts $'$ and $''$. Let p be the pressure, ρ the density, and V the

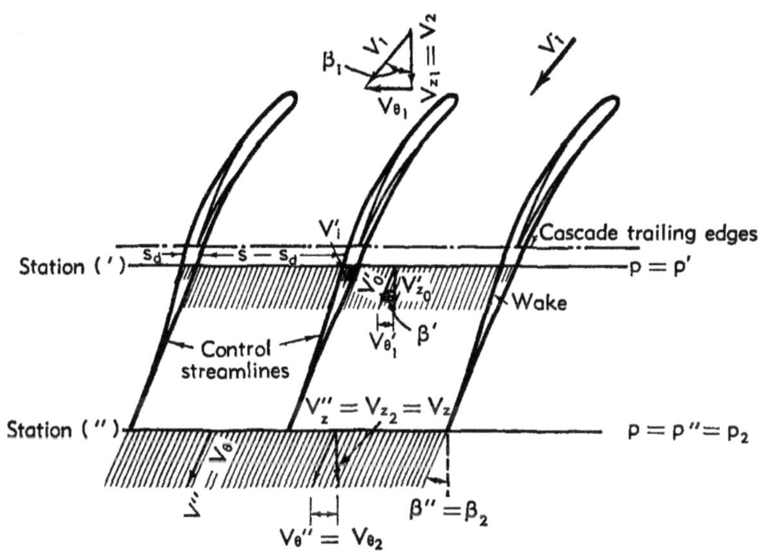

Fig. B,14d. Dissipation of the wakes behind cascade profiles.

velocity, having the components V_θ in the tangential direction and V_z normal to it.

There are certain relations between the conditions at these parallels, and in order to establish them the equations of continuity and momentum may be applied to a control surface bounded by two streamlines having the spacing s of the cascade and two parallels to the cascade trailing edges.

No resulting force component acts in the tangential direction. Hence, according to the law of momentum, when two-dimensional flow is considered (Fig. B,14d),

$$\int_0^s \rho' V_z' V_\theta' dy = \int_0^s \rho'' V_z'' V_\theta'' dy \tag{14-5}$$

The pressure does not contribute to the resulting tangential forces, because in corresponding points at the two bounding streamlines the pressure is the same.

Also, in the axial direction, no resulting force component acts. Therefore, corresponding to the law of momentum,

$$\int_0^s \rho' V_z'^2 dy + \int_0^s p' dy = \int_0^s \rho'' V_z''^2 dy + \int_0^s p'' dy \tag{14-6}$$

and, corresponding to the law of continuity,

$$\int_0^s \rho' V_z' dy = \int_0^s \rho'' V_z'' dy \tag{14-7}$$

Consider now the parallel (′) to be just far enough from the cascade so that the pressure may be considered to be constant, but the velocity to be not yet equalized, though of constant direction. Consider the other parallel (″) to be so far downstream that not only the pressure but, as a result of mixing, the velocity may also be considered to be uniform and of constant direction. Consider the density to be unchanged. Then

$$\int_0^s V_z' V_\theta' dy = V_z'' V_\theta'' s \tag{14-8}$$

$$\int_0^s V_z'^2 dy + \frac{p'}{\rho} s = V_z''^2 s = \frac{p''}{\rho} s \tag{14-9}$$

$$\int_0^s V_z' dy = V_z'' s \tag{14-10}$$

In order to simplify these integrals the velocities (′) along the parallel near the cascade outside of the wake ($s_d < y < s$) are assumed to be

$$V_{z_0}' = \text{const} \qquad V_{\theta_0}' = \text{const}$$

and in the range of the wake ($0 < y < s_d$)

$$V_{z_1}' = \text{const} \qquad V_{\theta_1}' = \text{const}$$

If the flow angles are everywhere equal,

$$\frac{V_{z_1}'}{V_{z_0}'} = \frac{V_{\theta_1}'}{V_{\theta_0}'} = \epsilon \tag{14-11}$$

Then the following relation between stations ′ and ″ are found:

$$V_z'' = V_{z_0}' \left[1 - (1 - \epsilon) \frac{s_d}{s} \right] \tag{14-12}$$

$$V_\theta'' = V_{\theta_0}' \frac{1 - (1 - \epsilon^2) \dfrac{s_d}{s}}{1 - (1 - \epsilon) \dfrac{s_d}{s}} \tag{14-13}$$

$$\frac{p''}{\rho} = \frac{p'}{\rho} + V_{z_0}'^2 \frac{s_d}{s} \left(1 - \frac{s_d}{s} \right) (1 - \epsilon)^2 \tag{14-14}$$

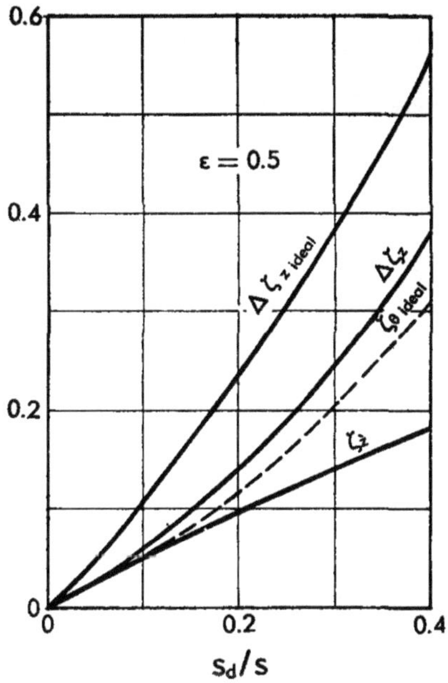

Fig. B,14e. Coefficients ζ_z, $\zeta_{z_{\text{ideal}}}$, $\zeta_{\theta_{\text{ideal}}}$ and $\Delta\zeta_z$ for $\epsilon = \frac{1}{2}$.

or

$$V'_{z_0} = \frac{V''_z}{1 - (1 - \epsilon)\dfrac{s_d}{s}} \tag{14-15}$$

$$V'_{\theta_0} = V''_\theta \frac{1 - (1 - \epsilon)\dfrac{s_d}{s}}{1 - (1 - \epsilon^2)\dfrac{s_d}{s}} \tag{14-16}$$

$$\frac{p'}{\rho} = \frac{p''}{\rho} - V''^2_z \frac{\dfrac{s_d}{s}\left(1 - \dfrac{s_d}{s}\right)(1 - \epsilon)^2}{\left[1 - (1 - \epsilon)\dfrac{s_d}{s}\right]^2} \tag{14-17}$$

The pressure change between stations (′) and (″) is

$$\Delta p = \zeta_z \tfrac{1}{2}\rho V''^2_z \tag{14-18}$$

where (Fig. B,14e)

$$\zeta_z = \frac{\dfrac{s_d}{s}\left(1 - \dfrac{s_d}{s}\right)(1 - \epsilon)^2}{\left[1 - (1 - \epsilon)\dfrac{s_d}{s}\right]^2} = 2\,\frac{\dfrac{s_d}{s}}{1 - \dfrac{s_d}{s}}\left[1 - \frac{\epsilon}{1 - (1 - \epsilon)\dfrac{s_d}{s}}\right]^2 \tag{14-19}$$

The pressure p'' farther downstream is always larger than the pressure p' near the cascade. If there were no losses the pressure change

$$\Delta p_{\text{ideal}} = \tfrac{1}{2}\rho(V_0'^2 - V''^2) \tag{14-20}$$

or

$$\Delta p_{\text{ideal}} = \tfrac{1}{2}\rho V_z''^2 \zeta_{z\text{ideal}} + \tfrac{1}{2}\rho V_\theta''^2 \zeta_{\theta\text{ideal}} \tag{14-21}$$

where (Fig. B,14e)

$$\zeta_{z\text{ideal}} = \frac{1}{\left[1 - (1 - \epsilon)\dfrac{s_d}{s}\right]^2} - 1 \tag{14-22}$$

$$\zeta_{\theta\text{ideal}} = \frac{\left[1 - (1 - \epsilon)\dfrac{s_d}{s}\right]^2}{\left[1 - (1 - \epsilon^2)\dfrac{s_d}{s}\right]^2} - 1 \tag{14-23}$$

The total pressure loss is

$$\Delta p_{\text{loss}} = \Delta p_{\text{ideal}} - \Delta p \tag{14-24}$$

or

$$\Delta p_{\text{loss}} = \tfrac{1}{2}\rho V_z''^2(\zeta_{z\text{ideal}} - \zeta_z) + \tfrac{1}{2}\rho V_\theta'' \zeta_{\theta\text{ideal}} \tag{14-25}$$

The substitution of

$$\tan \beta_2 = \frac{V_\theta''}{V_z''} \tag{14-26}$$

yields

$$\Delta p_{\text{loss}} = \frac{\tfrac{1}{2}\rho V_z^2}{\cos^2 \beta_2} [\cos^2 \beta_2(\zeta_{z\text{ideal}} - \zeta_z) + \sin^2 \beta_2 \zeta_{z\text{ideal}}] \tag{14-27}$$

Hence the pressure loss coefficient related to $V_z = V_z''$

$$\Delta C_z^* = \frac{\Delta p_{\text{loss}}}{\tfrac{1}{2}\rho V_z''^2} = \frac{1}{\cos^2 \beta_2} [\cos^2 \beta_2(\zeta_{z\text{ideal}} - \zeta_z) + \sin^2 \beta_2 \zeta_{z\text{ideal}}] \tag{14-28}$$

The components $\Delta\zeta_z = \zeta_{z\text{ideal}} - \zeta_{\theta\text{ideal}}$ and $\Delta\zeta_\theta - \zeta_{\theta\text{ideal}}$ of the pressure loss coefficient ΔC_z^* are represented in Fig. B,14e.

As long as more exact data are not known, $\epsilon = \frac{1}{2}$ may be considered to be a useful approximation for a station (′) near the trailing edges of the cascade profiles.

For $\epsilon = \frac{1}{2}$

$$\cos^2 \beta_2 \Delta C_z^* = \frac{\Delta p_{\text{loss}}}{\tfrac{1}{2}\rho V_z^2} \cos^2 \beta_2 = \frac{\Delta p_{\text{loss}}}{\tfrac{1}{2}\rho V_2^2} \tag{14-29}$$

as shown in Fig. B,14f. Further,

$$\frac{V_{z0}'}{V_z''} = \frac{1}{1 - (1 - \epsilon)\dfrac{s_d}{s}}$$

and

$$\frac{V'_{\theta_0}}{V''_{\theta}} = \frac{1 - (1 - \epsilon)\frac{s_d}{s}}{1 - (1 - \epsilon^2)\frac{s_d}{s}}$$

Hence

$$\tan \beta' = \frac{V'_{\theta_0}}{V'_{z_0}} = \frac{V'_{\theta_i}}{V'_{z_i}} = \frac{\left[1 - (1 - \epsilon)\frac{s_d}{s}\right]^2}{1 - (1 - \epsilon^2)\frac{s_d}{s}} \tan \beta'' \qquad (14\text{-}30)$$

The angle between the downstream flow direction and the axial direction is always larger far downstream than it is near the cascade.

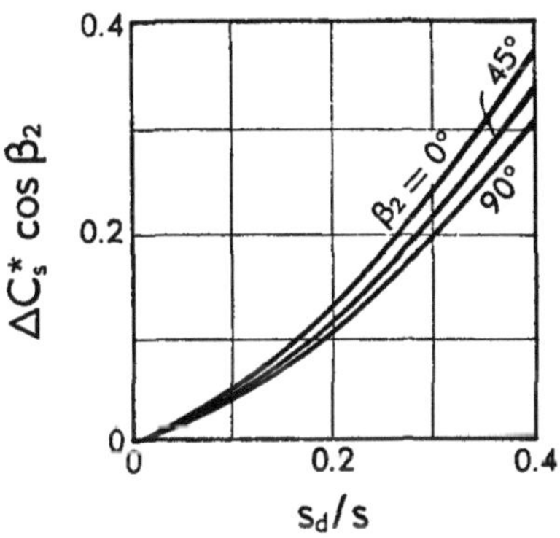

Fig. B,14f. Pressure loss coefficient $\Delta C_s^* = \Delta p_{loss}/\frac{1}{2}\rho V_2^2$ for $\epsilon = \frac{1}{2}$.

B,15. Cited References and Bibliography

Cited References

1. Betz, A., and Fluegge-Lotz, I. Berechnung der Schaufeln von Kreiselraedern. *Ing.-Arch. 11,* 486–501 (1938).
2. Weinig, F. S. Beitrag zur Theorie des Tragfluegels endlicher, imbesondere kleiner Spannweite. *Luftfahrtforschung 13,* 405 (1936).
3. Betz, A. Tragfluegel und hydraulische Machinen. *Handbuch d. Physik Geiger und Scheel* 7, 214. (Springer, Berlin, 1927.)
4. Foettinger, H. Ueber die physikalischen Grundlagen der Turbinen-und Propellerwirkung. *Z. Flugtech. Motorluftschiffahrt 13,* 233 (1922).
5. Weinig, F. S. *Die Stroemung um die Schaufeln von Turbomachinen.* Barth, Leipzig, 1935.
6. Durand, W. F., ed. *Aerodynamic Theory,* Vol. II. Springer, Berlin, 1934–1936.
7. Deich, M. E. Flow of gas through turbine lattices. *NACA Tech. Mem. 1393,* 1956.
8. Busemann, A. Das Foerderhoehenverhaeltnis radialer Kreiselpumpen mit logarithmisch spiraligen Schaufeln. *Z. angew. Math. u. Mech. 8,* 373 (1938).
9. Manea, G. *Ermittlung des Geswindigskeits-und Druckverlaufs laengs beliebig geformter Schaufeln eines Zellenlosen Turbinenrades (Determination of Velocity and*

Pressure along Arbitrary Shaped Blading of Turbomachinery). *Dissertation*, Tech. Hochschule, Berlin, 1932.
10. Koenig, E. Potentialstroemung durch Gitter. *Z. angew. Math. u. Mech. 2*, 422 (1922).
11. Weinig, F. S. Properties of straight line profile cascades and the hodograph of the flow through them. (Unpublished.)
12. Costello, G. R. Method of designing cascade blades with prescribed velocity distributions in compressible potential flows. *NACA Rept. 978*, 1950. Also *NACA Tech. Note 2281*, 1951.
13. Goldstein, A. W., and Jerison, M. Isolated and cascade airfoils with prescribed velocity distribution. *NACA Rept. 869*, 1947.
14. Hansen, A. G., and Yohner, P. L. A numerical procedure for designing cascade blades with prescribed velocity distributions in incompressible potential flow. *NACA Tech. Note 2101*, 1950.
15. Garrick, I. E. On the plane potential flow past a lattice of arbitrary airfoils. *NACA Rept. 788*, 1944.
16. Hargest, T. J. The theoretical pressure distributions around some related airfoils in cascade. *Brit. Aeronaut. Research Council Repts. and Mem. 2765*, 1950.
17. Weinig, F. S. A new approach to the theory of thin slightly cambered profiles and cascades. (To be published in *J. Appl. Mech.*)
18. Diesendruck, L. Iterative interference methods in the design of thin cascade blades. *NACA Tech. Note 1254*, 1947.
19. Klingemann, G. Verfahren zur Berechmung der theoretischen Kennlinien von Turbomachinen. *Ing.-Arch. 9*, 151 (1940). Transl. *Gt. Brit. RTP 1377*.
20. Pistolesi, E. On the calculation of flow past an infinite screen of thin airfoils. *NACA Tech. Mem. 968*, 1941.
21. Katzoff, S., Finn, R. S., and Laurence, J. C. Interference method for obtaining the potential flow past an arbitrary cascade of airfoils. *NACA Tech. Note 1252*, 1947.
22. Lieblein, V. Zur Berechnung der Auftriebscharacteristik eines Profiles im Gitterverband (On the calculation of the lift characteristic of a profile in cascade arrangement). Transl. in *Brit. Ministry Air Production LFA Rept., Volkenrode Rept. and Transl. 422*, 1944.
23. Funaioli, E. Sul Calcolo di Schieri alari di Profili sottili di Curvatura non trascurabile (Upon the calculation of cascades of airfoils, the camber of which is not negligibly small). *Acta Pontif. Acad. Sci. 14, 5*, 45 (1950).
24. Schilhansl, M. Naeherungsweise Berechnung von Auftrieb and Druckverteilung in Fluegelgittern. *Jahrbuch Wiss. Ges. Luftfahrtforschung*, 151 (1927).
25. Staufer, F. Verfahren zur Bestimmung der Schaufelform gerader Schaufelgitter. *Die Wasserwirtsch, 27*, 421, 456, 407 (1933).
26. Howell, A. R., and Carter, A. D. S. Fluid flow through cascades of airfoils. *Sixth Intern. Congress Appl. Mech. 11*, 173 (1946).
27. Carter, A. D. S., and Hughes, H. P. A theoretical investigation into the performance of airfoils in cascades. *Brit. Aeronaut. Research Council Repts. and Mem. 2384*, 1950.
28. Schlichting, H. Berechnung der reibungslosen Stroemung fuer ein vorgegebenes ebenes Schaufelgitter in inkompressibler Stroemung (Calculations of the nonviscous flow for a two-dimensional cascade in incompressible flow). *Fluid Mech. Inst., Braunschweig, Rept. 52/18*, 1952.
29. Scholz, N. Ein Einfaches Singularitaetenverfahren zur Erzeugung von Schaufelgittern (A simple singularity method for producing cascades). *Z. angew. Math. u. Mech. 30*, 262–263 (1950). Also *J. Aeronaut. Sci. 18*, 67, 637 (1951).
30. Weinig, F. S. Beschreibung eines Verfahrens Tragfluegelprofile vorgeschriebener Druckverteilung zu ermitteln und eines Verfahrens die Druckverteilung an einem beliebigen Tragfluegelprofil zu berechnen. *Z. W. B. Forschungbericht FB632*, 1936.
31. Mutterperl, W. A solution of the direct and inverse potential problems for arbitrary cascades of airfoils. *NACA Wartime Rept. L81*, 1954.

32. Lighthill, M. J. A mathematical method of cascade design. *Brit. Aeronaut. Research Council Repts. and Mem. 2104*, 1945.
33. Weinig, F. S. Influence of spacing and stagger on a profile in a cascade. *Proc. Fourth Midwestern Congress Fluid Mech.*, Purdue, 1956.
34. Poritsky, H., Sells, B. E., and Danforth, C. E. Graphical, mechanical, and electrical aids for compressible fluid flow. *J. Appl. Mech. 17*, 37–46, 1950.
35. Weinig, F. S. Der Hodograph der Gitterstroemung als Weg zur Ermittlung guenstiger Turbinen und Propeller Profile. *Verhandl. Third Intern. Congress Appl. Mech., Stockholm, 1*, 437 (1930).
36. Kraft, H. *Proc. Eighth Intern. Congress Appl. Mech.*, Istanbul, 1952.
37. Wu, C. H., and Brown, C. A. A theory of the direct and inverse problems of compressible flow past cascade of arbitrary airfoils. *J. Aeronaut. Sci. 19*, 183–196 (1952).
38. Yeh, H. *Design of Cascades in Compressible Flow. Sci. D. Thesis*, Mass. Inst. Technol., 1950.
39. Prandtl, L. Ueber Stroemmugen deren Geschwindigkeiten mit der Schallgeschwindigkeit vergelichbar sind. *J. Univ. Tokio Aeronaut. Research Inst. 6*, 14 (1930).
40. Glauert, H. The effect of compressibility on the lift of airfoils. *Proc. Roy. Soc. London A118*, 113 (1927).
41. Rao, G. V. R. Compressible flow past a cascade of thin airfoils. *J. Aeronaut. Sci. 20*, 217–218 (1953).
42. von Kármán, Th. Compressibility effects in aerodynamics. *J. Aeronaut. Sci. 8*, 377 (1941).
43. Tsien, H. S. Two-dimensional subsonic flow of compressible fluids. *J. Aeronaut. Sci. 6*, 399 (1939).
44. Weinig, F. S. Subsonic influence of compressibility on the pressure distribution of a profile. *J. Aeronaut. Sci. 23*, 279 (1956).
45. Weinig, F. S. Analysis of profile lattices introducing a large deflection into an airstream. *Wright Field Tech. Rept. F-TR-2165-ND, ASTIA-ATI22838.*
46. Alpers, S. Design method for two-dimensional channels for compressible flow with application to high solidity cascades. *NACA Tech. Note 1931*, 1949.
47. Stanitz, J. D. Approximate design method for high-solidity blade elements in compressors and turbines. *NACA Tech. Note 248*, 1926; *2593*, 1952.
48. Weinig, F. S. Consideration of the influence of the boundary layer, when computing pressure distribution on single profiles or profile cascades. (Unpublished.)
49. Betz, A., and Peterson, E. Anwendung der Theorie der freien Strahlen. *Ing.-Arch. 2*, 190 (1931).
50. Cornell, W. G. The stall performance of cascades. *Proc. Second U. S. Natl. Congress Appl. Mech.*, Ann Arbor, Mich., 1954.
51. Schmieden, C. Unstetige Stroemungen durch Gitter. *Ing.-Arch. 3*, 130 (1932).
52. Riabouchinski, D. On steady fluid motion with free surfaces. *Proc. London Math. Soc. 19*, 206 (1921); *Ser. 2, Vol. 25, Part. 3*, 185 (1936).

Bibliography

Collar, A. R. The flow of a perfect fluid through cascades of aerofoils. *Roy. Aeronaut. Soc. 45*, 183–213 (1941). Also *Brit. Aeronaut. Research Council Repts. and Mem. 1885*, 1940; *4589*, 1940.

Katzoff, S., Bogdonoff, H. E., and Boyet, H. Comparisons of theoretical and experimental lift and pressure distributions on airfoils in cascades. *NACA Tech. Note 1376*, 1947.

Katzoff, S., and Hannah, M. E. Further comparisons of theoretical and experimental lift and pressure distributions on airfoils in cascades at low subsonic speed. *NACA Tech. Note 2391*, 1951.

Merchant, W., and Hansford, J. T. Flow of an ideal fluid through a cascade of blades. *Metropolitan-Vickers Summary Rept.*, 1945. Also *Brit. Aeronaut. Research Council Repts. and Mem. 1890, 1893*, 1940.

Numachi, F. Aerofoil theory of propeller turbines and propeller pumps. *Proc. World Eng. Congress*, Tokio, 1929.

Traupel, W. Die Berechnung der Potentialstroemung durch Schaufelgitter (The calculation of the potential flow past blades in cascades). *Schweiz. Arch. 12*, *Sulzer Tech. Rev. 1*, 25 (1945).

Tyler, R. A. The available theoretical analysis of two-dimensional cascade flow. *Div. Mech. Eng., Can. Natl. Research Council Aeronaut. Note AN-4*, 1949.

Weinig, F. S. Zur Frage der Abrundung und Zuschaerfung umtroemter Kanten. *Z. angew, Math. u. Mech. 13*, 224 (1933).

SECTION C

THREE-DIMENSIONAL FLOW IN TURBOMACHINES

FRANK E. MARBLE

C,1. Introduction. The flow through a turbomachine blade row may be separated conceptually into two parts: that which is due to the gross influence of the blade row and that which is associated with the details of blade shape. The general idea of decomposing the flow field in this manner is a direct extension of classical propeller theory which, in turn, was motivated by the success of the Lanchester-Prandtl wing theory. In wing theory, one first computes the downwash at a given point of the wing induced by the entire wing and there analyzes the detailed flow over the element in question. In turbomachine theory, one computes the flow induced by all blades or blade rows at a given point and then analyzes the detailed flow over the particular blade element in question.

It was Ruden [*1*] who first gave the intuitive description of turbomachine flow in the language of wing theory. In spite of the fact that he did not pursue analytical development of this picture, Ruden did recognize some essential relationships between linearized wing and turbomachine theory. One of these was the fact that, for a linearized theory, the axial velocity distortion in the plane of a blade row is half that induced by the blade row far downstream in much the same manner that the downwash in the plane of a finite wing is half that in the Trefftz plane.

The computational difficulties of the full three-dimensional flow field caused investigators to seek a simpler but less general description of the interesting portion of the flow. This search led many to develop independently the so-called radial equilibrium theory. In the radial equilibrium theory one looks only at the flow far upstream and far downstream of the blade row where radial accelerations have vanished and assumes a completely axially symmetric flow where no trace of the individual blades remains. In fact this description became so widespread by 1942, at least among workers in Germany, Switzerland, United States, and England, and the publication of these elementary results was so casual or nonexistent, that it is impossible to cite accurately any priority in the matter. The work of Traupel [*2*] is one of the earliest complete accounts in which the radial equilibrium idea is employed, but it was used by others at an earlier date. One of particular interest, developed and used extensively in the period of 1940, was due to von Kármán and Rannie although it was not published in detail until 1949 [*3*]. This particular treatment differs from others by being somewhat more than a linear theory.

The drawback of the radial equilibrium theory is that it gives no account of how the changes in velocity distribution take place across a blade row. One has no idea whether the change in profile is complete immediately behind or many blade lengths behind the blade row. To rectify this difficulty, Meyer [*4*] studied the detailed potential flow through a stationary row of a finite number of blades, using a technique which was originally introduced by Ackeret for analysis of flow through cascades. More recently Wu [*5,6,7*] developed a fairly general method of treating turbomachine problems from a purely numerical viewpoint. The difficulty experienced here is the general complexity of results, the time consumed in calculation of a specific example and the relative impossibility of extracting from these theories some essential features that may be extended to apply to technological problems. On the other hand the complete three-dimensional theories do offer the possibility of investigating in detail very particular problems if they arise.

The three-dimensional problem simplifies greatly when it is assumed that the flow is axially symmetric, as was done in the radial equilibrium treatment. Physically this implies that the blade row must consist of an infinite number of infinitely thin blades. To achieve the most realistic representation, these blades may be given their proper chordwise load distribution but averaged circumferentially. A three-dimensional flow theory based on these ideas was introduced by the present author [*8*] and expanded to cover a variety of typical examples by Marble [*9*] and Marble and Michelson [*10*]. The idea of this axially symmetric throughflow is exactly that outlined in the introductory paragraph: to compute an induced flow field in which the individual blades may be considered operating. The assumption of axial symmetry indicates merely that the induced flow in the plane of the blade row will be computed as if the trailing vortex field were circumferentially uniform rather than separated into discrete sheets.

The actual flow pattern over a turbomachine blade row differs from the axially symmetric throughflow because of the finite peripheral spacing of the blades and the detailed geometrical shape. The detailed blade geometry is the main concern of cascade studies. In applying the results of cascade theory and experiments to the three-dimensional flow problem it is usual to approximate the effect of blade geometry, at a given radius, by the local two-dimensional flow about the blade; the approach stream direction and magnitude are given by the axially symmetric throughflow theory. In this manner the blade angle of attack, the blade camber and thickness, the critical Mach number, etc., may be selected or deduced from two-dimensional cascade information, the process being repeated at as many radii as required to define the blade. In this process it is tacitly assumed that the flow about the blade at one radius does not influence appreciably the flow about the blade at another radius.

The calculations involved in axially symmetric throughflow theory are still sufficiently complex that application of the theory is restricted to examples requiring special attention and is not very useful in routine design. In fact the axially symmetric throughflow theory would be of only modest help to the designer were it not for the fact that it leads to a very simple and useful approximation. In his 1948 paper [*8*] the present author introduced the so-called "exponential approximation" to the throughflow. In essence this is a method of describing the transition from the known radial equilibrium patterns far upstream and downstream of the blade row; it employs the ideas of the axially symmetric throughflow theory in obtaining this approximate transition. These ideas were subsequently applied by the author [*9,10*] and later independently by Railly [*11*] and by Horlock [*12,13*] to describe a wide variety of technological problems including mutual interference of neighboring blade rows. This latter problem was also treated in a somewhat less general manner, and by a completely different technique, by Wu and Wolfenstein [*14*]. At present the radial equilibrium theory, the exponential approximation, and its extensions given in the following work constitute a method of sufficient accuracy and simplicity for utilization in the majority of turbomachine problems.

There are effects, sometimes significant ones, that are in no way accounted for in the combination of axially symmetric throughflow theory and two-dimensional cascade theory. One of these is the fact that the vorticity trailing from the actual blade row is not distributed but is concentrated in the blade row wakes. It would be expected that this modification to the induced velocity field is significant for wide blade spacings and for large spanwise gradients of shed vorticity. This problem has not, in fact, been investigated to a significant extent. A second property of the actual flow, which is unaccounted for in the throughflow-cascade theory, is the effect of boundary layers. These phenomena were discussed physically by Weske [*15*]. Of particular importance is the interaction of the individual blades with thick boundary layers on the hub and tip casings of the turbomachine. This problem of "secondary flow" has been analyzed by Hawthorne [*16*] and several others as a source of losses and as an influence upon blade angle setting. Finally there is the rather singular problem of the transonic compressor, that is a blade row in which the relative velocities are supersonic at the tip and subsonic at the root. Here the question is whether cascade theory can be applied locally to determine blade performance and required angles. It is not unreasonable to expect that the flow about a blade element in the transonic region would alter significantly the flow about blade elements at different radii and hence negate the assumption that blade elements at different radii are independent. A detailed analysis by McCune [*17*] has recently indicated that the interference is indeed so large that cascade theory is not applicable in this instance.

Of the various flow processes comprising the complete three-dimensional flow field, only the axially symmetric throughflow and the two-dimensional cascade flow have been developed to the point of usefulness in the design of turbomachine components. Consequently it is with the axially symmetric throughflow that the present section is principally concerned.

C,2. Formulation of Axially Symmetric Throughflow. The throughflow is described by the axially symmetric motion of an ideal

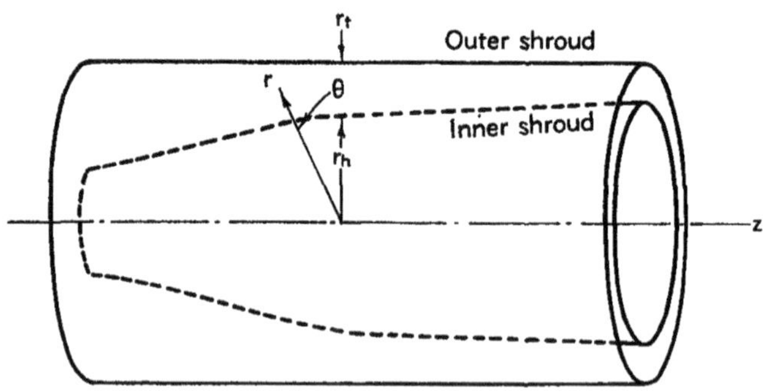

Fig. C,2a. Schematic diagram of turbomachine.

gas passing through a prescribed force field representing the blade row and moving tangentially to hub and tip boundaries, Fig. C,2a, of given shape. There are, in general, two types of problem of interest to turbomachine designers differing according to the information prescribed about the blade row, the direct and inverse problems. The *inverse problem* considers the flow field generated when a prescribed distribution of enthalpy or angular momentum is added by the blade row. Here the aim is to proportion the blades and to choose their angles appropriately to meet conditions at a certain design point. The *direct problem*, on the other hand, considers the flow field induced by blades of given geometric configuration. Here it is desired to determine the blade row performance at operating conditions other than those for which it was designed.

The aerothermodynamic equations. It is most convenient to describe the throughflow in a cylindrical coordinate system, Fig. C,2b, where u, v, and w are the velocity components in the r, θ, and z directions. Denoting the axially symmetric force components corresponding to the blade row by F_r, F_θ, and F_z, the equations of motion are

$$u\frac{\partial u}{\partial r} + w\frac{\partial u}{\partial z} - \frac{v^2}{r} = -\frac{1}{\rho}\frac{\partial p}{\partial r} + F_r \tag{2-1}$$

$$u\frac{\partial(rv)}{\partial r} + w\frac{\partial(rv)}{\partial z} = rF_\theta \tag{2-2}$$

$$u\frac{\partial w}{\partial r} + w\frac{\partial w}{\partial z} = -\frac{1}{\rho}\frac{\partial p}{\partial z} + F_z \tag{2-3}$$

In writing the equations of motion, account has been taken of the general axial symmetry of the flow, and consequently all peripheral

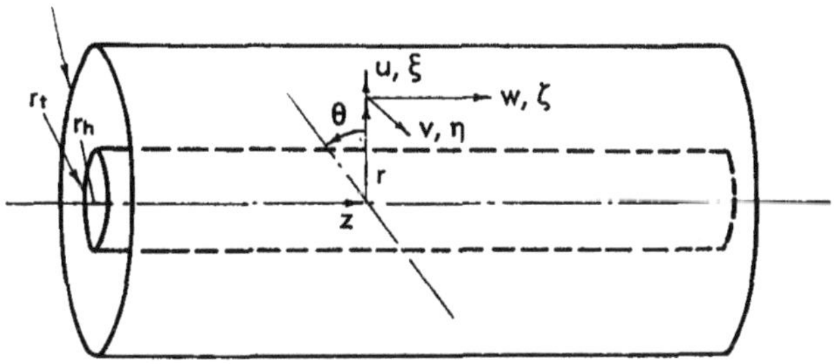

Fig. C,2b. Notation for axes, velocity components, and vorticity components.

derivatives vanish. The continuity equation may be written similarly as

$$\frac{1}{r}\frac{\partial}{\partial r}(r\rho u) + \frac{\partial}{\partial z}(\rho w) = 0 \tag{2-4}$$

It will often be convenient to describe the flow field in terms of the radial, tangential, and axial vorticity components defined respectively as

$$\xi = -\frac{\partial v}{\partial z} \tag{2-5}$$

$$\eta = \frac{\partial u}{\partial z} - \frac{\partial w}{\partial r} \tag{2-6}$$

$$\zeta = \frac{1}{r}\frac{\partial(rv)}{\partial r} \tag{2-7}$$

The equations of motion, Eq. 2-1, 2-2, and 2-3, may be rewritten in terms of vorticity components as

$$w\eta - v\zeta = F_r - \left[\frac{1}{\rho}\frac{\partial p}{\partial r} + \frac{\partial}{\partial r}\left(\frac{q^2}{2}\right)\right] \tag{2-8}$$

$$u\zeta - w\xi = F_\theta \tag{2-9}$$

$$v\xi - u\eta = F_z - \left[\frac{1}{\rho}\frac{\partial p}{\partial z} + \frac{\partial}{\partial z}\left(\frac{q^2}{2}\right)\right] \tag{2-10}$$

where q is the magnitude of the velocity, $q^2 = u^2 + v^2 + w^2$.

For some analysis it will be convenient to write the equations of motion in terms of the enthalpy and entropy gradients rather than the pressure gradient. Denote the general pressure variation δp as

$$\frac{1}{\rho}\,\delta p = \delta h - T\delta s \tag{2-11}$$

where s and h are respectively the entropy and thermodynamic enthalpy of the gas. Then defining the stagnation enthalpy

$$h^0 = h + \frac{q^2}{2} \tag{2-12}$$

equations of motion in the radial and axial direction, Eq. 2-8 and 2-10, mav be rewritten

$$w\eta - v\zeta = F_r - \frac{\partial h^0}{\partial r} + T\frac{\partial s}{\partial r} \tag{2-13}$$

$$v\xi - u\eta = F_z - \frac{\partial h^0}{\partial z} + T\frac{\partial s}{\partial z} \tag{2-14}$$

It is worth noting also, at this point, that the combined first and second laws of thermodynamics, Eq. 2-11, may be written to give the entropy variation of the gas along a stream surface in the form

$$T\left(u\frac{\partial s}{\partial r} + w\frac{\partial s}{\partial z}\right) = u\frac{\partial e}{\partial r} + w\frac{\partial e}{\partial z} + p\left(u\frac{\partial}{\partial r} + w\frac{\partial}{\partial z}\right)\left(\frac{1}{\rho}\right) \tag{2-15}$$

Although the heat transfer to the gas is usually negligible, the losses in the blades make the flow process a nonisentropic one. Consequently it is necessary to calculate the entropy variation explicitly.

Conditions at the blades. In the absence of any shear stresses or losses the local blade force is normal to the local blade surface. When losses such as surface shear stresses are present, however, the complete force system must also have a component tangential to the local blade surface. It is appropriate then to express the blade forces in two parts: one set, $F_r^{(1)}, F_\theta^{(1)}, F_z^{(1)}$, the resultant of which is normal to the blade surface; and a second set, $F_r^{(2)}, F_\theta^{(2)}, F_z^{(2)}$, which is tangential to the blade surface.

The first set of blade forces, normal to the blade surface, bears a simple relation to the velocity vector. For since the relative velocity $u, v - \omega r, w$ is tangential to the blade surface locally, it is likewise normal to the force $F_r^{(1)}, F_\theta^{(1)}, F_z^{(1)}$, that is, the scaler product vanishes,

$$uF_r^{(1)} + (v - \omega r)F_\theta^{(1)} + wF_z^{(1)} = 0 \tag{2-16}$$

The second set of blade forces, representing the blade drag and leading to losses, will be considered of a specialized form. It will be assumed that this component is not only tangential to the blade surface but parallel with the gas velocity vector $(u, v - \omega r, w)$ relative to the blade. This restriction is clearly appropriate when the force $F_r^{(2)}, F_\theta^{(2)}, F_z^{(2)}$ arises only from simple skin friction. It is an oversimplification of fact, however, when there exists a cross flow within the boundary layer, i.e. when the local direction of boundary layer flow deviates from the direction of flow just outside the boundary layer. These values of transverse skin friction are generally of minor importance and there is a question, moreover, whether the effect of such flow may properly be evaluated within the axially symmetric throughflow theory. Under these restrictions this second force system may be written as

$$(F_r^{(2)}, F_\theta^{(2)}, F_z^{(2)}) = -\frac{1}{2} C_D[u^2 + (v - \omega r)^2 + w^2] \frac{(u, v - \omega r, w)}{\sqrt{u^2 + (v - \omega r)^2 + w^2}} \tag{2-17}$$

where C_D is an appropriate local drag coefficient which can be considered a prescribed function $C_D(r, z)$ within the blade row.

There exists an additional restriction [*18*] upon the blade force system $F_r^{(1)}, F_\theta^{(1)}, F_z^{(1)}$ arising from the fact that it is generated by a continuous, generally smooth, blade surface. Suppose the blade surface is defined by a relation,

$$\beta(r, \theta, z) = \text{const} \tag{2-18}$$

so normalized that

$$\left(\frac{\partial \beta}{\partial r}\right)^2 + \left(\frac{1}{r}\frac{\partial \beta}{\partial \theta}\right)^2 + \left(\frac{\partial \beta}{\partial z}\right)^2 = 1$$

Then the unit vector normal to the blade surface has components

$$\frac{\partial \beta}{\partial r}, \frac{\partial \beta}{r \partial \theta}, \frac{\partial \beta}{\partial z}$$

and if the blade loading is given by a function $\Lambda(r, z)$, the blade force vector $(F_r^{(1)}, F_\theta^{(1)}, F_z^{(1)})$ is given by

$$(F_r^{(1)}, F_\theta^{(1)}, F_z^{(1)}) = \Lambda(r, z)\left(\frac{\partial \beta}{\partial r}, \frac{\partial \beta}{r \partial \theta}, \frac{\partial \beta}{\partial z}\right) \tag{2-19}$$

Taking the curl operation of Eq. 2-19 gives curl $\mathbf{F}$ = curl (Λ grad β). If the right side of this relation be expanded and note is taken of the fact that curl grad $\beta \equiv 0$, then

$$\operatorname{curl} \mathbf{F}^{(1)} = \operatorname{grad} \Lambda \times \operatorname{grad} \beta \tag{2-20}$$

Furthermore it is clear from Eq. 2-19 that the force vector $\mathbf{F}^{(1)}$ is parallel to grad β with the consequence that $\mathbf{F}^{(1)} \cdot (\operatorname{grad} \Lambda \times \operatorname{grad} \beta) = 0$. Thus the left-hand side of Eq. 2-20 yields the condition that the force must satisfy the relation

$$\mathbf{F}^{(1)} \cdot \operatorname{curl} \mathbf{F}^{(1)} = 0 \tag{2-21}$$

in order that the flow follow a blade surface that is constructable and continuous.

The condition given by Eq. 2-21 may be written down explicitly for the present special case of axial symmetry, and gives

$$-F_r^{(1)}\frac{\partial F_\theta^{(1)}}{\partial z} + F_\theta^{(1)}\left(\frac{\partial F_r^{(1)}}{\partial z} - \frac{\partial F_z^{(1)}}{\partial r}\right) + F_z^{(1)}\frac{1}{r}\frac{\partial}{\partial r}(rF_\theta^{(1)}) = 0 \qquad (2\text{-}22)$$

This may be written in the more compact form

$$\frac{\partial}{\partial z}\left(\frac{F_r^{(1)}}{rF_\theta^{(1)}}\right) = \frac{\partial}{\partial r}\left(\frac{F_z^{(1)}}{rF_\theta^{(1)}}\right) \qquad (2\text{-}23)$$

The restriction imposed by Eq. 2-23 is of mathematical significance in that it reduces somewhat the freedom of choice in selecting the force field to represent a blade row. From a strictly utilitarian point of view, however, there is some doubt as to how seriously Eq. 2-23 should be taken. In the first place the axially symmetric throughflow itself is an approximation which makes the concept of detailed blade shape rather dubious. Secondly, in the preponderant number of cases concerned in jet propulsion application, the radial force component F_r is negligible. It is, therefore, very defensible to assume $F_r^{(1)} = 0$ and neglect the condition given by Eq. 2-23 or its equivalent. In the present work Eq. 2-23 will be retained in the general theoretical consideration but will be dropped in particular appropriate examples.

Variations in stagnation enthalpy and entropy. The so-called Euler turbine equation is one of the few, very general results of turbomachine theory. In its usual form it relates the stagnation enthalpy rise through a blade row to the change in angular momentum of the fluid imparted by a moving blade row. Since it will be advantageous to follow fluid mass along the streamlines it is appropriate to denote by l the length measured, from an arbitrary origin, along the axially symmetric stream surface in a fixed meridional plane. Then the derivative along the stream surface, moving with the fluid, is

$$V_l\frac{\partial}{\partial l} \equiv u\frac{\partial}{\partial r} + w\frac{\partial}{\partial z} \qquad (2\text{-}24)$$

where $V_l = \sqrt{u^2 + w^2}$ is the meridional velocity along the stream surface. For example, dynamic equilibrium in the tangential direction, Eq. 2-2, becomes

$$V_l\frac{\partial(rv)}{\partial l} = rF_\theta \qquad (2\text{-}25)$$

The product rv is referred to as the angular momentum (per unit mass). Eq. 2-25 states physically that, along a stream surface, the angular momentum changes at a rate proportional to the moment of the applied tangential force.

The Euler turbine equation must follow from the first law of thermodynamics, given by Eq. 2-15. The second term on the right-hand side, representing the work associated with volume changes, may be rewritten as

$$p\left(u\frac{\partial}{\partial r} + w\frac{\partial}{\partial z}\right)\left(\frac{1}{\rho}\right) = \left(u\frac{\partial}{\partial r} + w\frac{\partial}{\partial z}\right)\left(\frac{p}{\rho}\right) - u\frac{1}{\rho}\frac{\partial p}{\partial r} - w\frac{1}{\rho}\frac{\partial p}{\partial z} \qquad (2\text{-}26)$$

The pressure derivatives $(1/\rho)\partial p/\partial r$ and $(1/\rho)\partial p/\partial z$ may be eliminated through use of the first and third equations of motion, Eq. 2-1 and 2-3. Substituting from these into Eq. 2-26 gives, after some simplification,

$$p\left(u\frac{\partial}{\partial r} + w\frac{\partial}{\partial z}\right)\left(\frac{1}{\rho}\right) = \left(u\frac{\partial}{\partial r} + w\frac{\partial}{\partial z}\right)\left(\frac{p}{\rho} + \frac{q^2}{2}\right) - \left(u\frac{\partial}{\partial r} + w\frac{\partial}{\partial z}\right)\left(\frac{v^2}{2}\right)$$
$$- u\frac{v^2}{r} - uF_r - wF_z \qquad (2\text{-}27)$$

From the tangential equation of motion, Eq. 2-2, it follows that

$$vF_\theta = u\frac{v^2}{r} + \left(u\frac{\partial}{\partial r} + w\frac{\partial}{\partial z}\right)\left(\frac{v^2}{2}\right)$$

which provides, upon substitution into Eq. 2-27,

$$p\left(u\frac{\partial}{\partial r} + w\frac{\partial}{\partial z}\right)\left(\frac{1}{\rho}\right) = \left(u\frac{\partial}{\partial r} + w\frac{\partial}{\partial z}\right)\left(\frac{p}{\rho} + \frac{q^2}{2}\right) - (uF_r + rF_\theta + wF_z) \qquad (2\text{-}28)$$

Now using Eq. 2-28 and noting that the quantity $e + (p/\rho) + (q^2/2)$ is just the stagnation enthalpy of the gas, the combined first and second laws of thermodynamics, Eq. 2-15, may be written in the form,

$$\left(u\frac{\partial}{\partial r} + w\frac{\partial}{\partial z}\right)h^0 - \omega rF_\theta = T\left(u\frac{\partial}{\partial r} + w\frac{\partial}{\partial z}\right)$$
$$+ [uF_r + (v - \omega r)F_\theta + wF_z] \qquad (2\text{-}29)$$

It will be recalled now that the force field associated with the blades has been divided into two parts: that part $F_r^{(1)}, F_\theta^{(1)}, F_z^{(1)}$ which acts normal to the relative motion and hence satisfies the condition that

$$uF_r^{(1)} + (r - \omega r)F_\theta^{(1)} + wF_z^{(1)} = 0 \qquad (2\text{-}30)$$

and that part $F_r^{(2)}, F_\theta^{(2)}, F_z^{(2)}$ which is parallel to the relative motion, according to Eq. 2-17, and is associated with the losses. In view of Eq. 2-30, therefore, Eq. 2-29 may be written

$$V_l\frac{\partial h^0}{\partial l} - \omega rF_\theta = TV_l\frac{\partial s}{\partial l} + [uF_r^{(2)} + (v - \omega r)F_\theta^{(2)} + wF_z^{(2)}] \qquad (2\text{-}31)$$

In the absence of losses the right-hand side of the equation vanishes identically, for then the shear force system $F_r^{(2)}$, $F_\theta^{(2)}$, $F_z^{(2)}$ and the entropy variation are both zero. The result of setting the left-hand side to zero gives

$$V_l \frac{\partial h^0}{\partial l} = \omega r F_\theta \tag{2-32}$$

and may be recognized as a gross application of the first law of thermodynamics. The term $V_l \partial h^0/\partial l$ represents the rate at which stagnation enthalpy of the fluid is rising and the term $\omega r F_\theta$ represents the rate at which torque applied to a particular radial element of the blade row is doing work on the system. Since the heat transfer is usually a negligible quantity, it is clear that this work must go into increasing the stagnation enthalpy of the gas. Furthermore it is evident that Eq. 2-32, being an over-all thermodynamic relation, holds regardless of whether or not losses are present. Then, of course, F_θ is the total tangential force, the sum of the normal $F_\theta^{(1)}$ and dissipative $F_\theta^{(2)}$ tangential forces. As a result of this observation it follows that the entropy variation along stream surfaces may be calculated by setting the right-hand side of Eq. 2-31 to zero. Thus

$$TV_l \frac{\partial s}{\partial l} = -[uF_r^{(2)} + (v - \omega r)F_\theta^{(2)} + wF_z^{(2)}] \tag{2-33}$$

It is clear now that only the second set of forces tends to change the entropy. Substituting from Eq. 2-17 the entropy variation may be expressed in terms of the velocity components

$$TV_l \frac{\partial s}{\partial l} = \tfrac{1}{2}C_D[u^2 + (v - \omega r)^2 + w^2]^{\frac{3}{2}} \tag{2-34}$$

so that the rate of entropy increase along the stream surface is proportional to the local values of C_D within the blade row and vanishes outside of the blade rows. In principle, at least, this relation allows computation of the entropy distribution that enters into determination of the tangential vorticity through Eq. 2-43.

Returning now to the variation of stagnation enthalpy, the moment of blade force may be eliminated from Eq. 2-32 by using Eq. 2-25, giving the change of angular momentum along a stream surface. Thus

$$V_l \frac{\partial h^0}{\partial l} = V_l \frac{\partial}{\partial l}(\omega r v) \tag{2-35}$$

which is essentially the Euler turbine equation in differential form. Clearly this may be integrated along any stream surface where the blade row in question has a fixed angular velocity ω. Across a blade row with angular velocity ω, the change Δh^0 in stagnation enthalpy is given by

$$\Delta h^0 = \omega \Delta(rv) \tag{2-36}$$

where $\Delta(rv)$ is the change in angular momentum of the fluid and all changes are measured along a fixed stream surface. When ω and $\Delta(rv)$ are of the same sign, as in a compressor, the stagnation enthalpy rises; when ω and $\Delta(rv)$ are of opposite sign, as in a turbine, the stagnation enthalpy of the gas is reduced. Across a stator, a stationary blade row, the stagnation enthalpy remains constant since $\omega = 0$. Finally in a space that is free of blades, the tangential force F_θ vanishes and consequently, according to Eq. 2-25, the angular momentum of the gas remains constant along stream surfaces. Obviously the stagnation enthalpy remains constant under this circumstance.

The tangential vorticity. A glance at the expressions for vorticity components [*8*], given by Eq. 2-5, 2-6, and 2-7 shows that, while the radial and axial vorticity components are given in terms of only the tangential velocity, the tangential vorticity components include both the radial and axial velocities. In other words the meridional velocities, the radial and axial components which make up the throughflow, may be related to the tangential vorticity. It is appropriate, then, to investigate the propagation tangential vorticity. In carrying out this analysis it will prove necessary to calculate the variation of quantities normal to the stream surfaces, and hence a length must be introduced to measure distance normal to the stream surface. This may be accomplished using the stream function ψ itself, which is defined by the properties that

$$\rho u = \rho_0 \frac{1}{r}\frac{\partial \psi}{\partial z} \tag{2-37}$$

$$\rho w = -\rho_0 \frac{1}{r}\frac{\partial \psi}{\partial r} \tag{2-38}$$

where ρ_0 is a constant reference density. It follows that, if $\delta\psi$ is the small difference in stream function between two nearby stream surfaces, then the normal distance between them is $-(1/rV_l)(\rho_0/\rho)\delta\psi$. The differential operator normal to a stream surface is just

$$\frac{w}{V_l}\frac{\partial}{\partial r} - \frac{u}{V_l}\frac{\partial}{\partial z}$$

so that, in a more compact form,

$$-\frac{\rho}{\rho_0} rV_l \frac{\partial}{\partial \psi} = \frac{w}{V_l}\frac{\partial}{\partial r} - \frac{u}{V_l}\frac{\partial}{\partial z} \tag{2-39}$$

Now it is a simple matter to calculate the variation of stagnation enthalpy normal to the stream surface, by taking $\partial h^0/\partial r$ from Eq. 2-12 and $\partial h^0/\partial z$ from Eq. 2-13. Substituting these values into the right-hand side of Eq. 2-39 gives

$$-\frac{\rho}{\rho_0} r V_l \frac{\partial h^0}{\partial \psi} = \frac{w}{V_l}\left(F_r + T\frac{\partial s}{\partial r} - w\eta + v\zeta\right) - \frac{u}{V_l}\left(F_z + T\frac{\partial s}{\partial z} - v\xi + u\eta\right)$$

$$= -V_l\eta + \frac{w}{V_l}v\zeta + \frac{u}{V_l}v\xi + \frac{w}{V_l}F_r - \frac{u}{V_l}F_z - \frac{\rho}{\rho_0} r V_l T\frac{\partial s}{\partial \psi} \quad (2\text{-}40)$$

Eq. 2-40 may be interpreted as a relation for the tangential vorticity η. The terms involving radial and axial vorticity components may, as was suggested previously, be expressed in terms of the tangential velocity. Referring again to the vorticity components, Eq. 2-5 and 2-7, it follows that

$$\begin{aligned} \frac{w}{V_l}v\zeta + \frac{u}{V_l}v\xi &= \frac{v}{r}\left[\frac{w}{V_l}\frac{\partial(rv)}{\partial r} - \frac{u}{V_l}\frac{\partial(rv)}{\partial z}\right] \\ &= -\frac{\rho}{\rho_0}V_l v\frac{\partial(rv)}{\partial \psi} \end{aligned} \quad (2\text{-}41)$$

Furthermore it appears that the force term $(w/V_l)F_r - (u/V_l)F_z$ is simply the force component normal to the stream surfaces and consequently it is convenient to define

$$F_\psi \equiv \frac{w}{V_l}F_r - \frac{u}{V_l}F_z \quad (2\text{-}42)$$

Using the results of Eq. 2-41 and 2-42, the tangential vorticity may be written, from Eq. 2-40, as

$$\eta = \frac{\rho}{\rho_0} r\left[\frac{\partial h^0}{\partial \psi} - \frac{v}{r}\frac{\partial(rv)}{\partial \psi} - T\frac{\partial s}{\partial \psi}\right] + \frac{1}{V_l}F_\psi \quad (2\text{-}43)$$

A restricted form of this was originally obtained by Bragg and Hawthorne [*19*] and the complete expression was given by Marble and Michelson [*10*]. Although this is not, in its present form, a very useful relation for determining the tangential vorticity since differentiation occurs with respect to the unknown stream function, it is, however, a very convenient guide for physical reasoning. The tangential vorticity associated with the force component F_ψ is essentially a "bound vorticity" and is of the same origin as the bound vorticity connected with the lift of a wing. If the angular momentum were invariant with ψ and the tangential vorticity depended upon h^0 and s only, then the flow outside of the blade row becomes relatively simple. Under these circumstances h^0 and s are constant along stream surfaces and hence $\eta/\rho r$ is constant along a stream surface. It is easily shown that this result follows from the fact that the circulation about a physical annular vortex tube remains constant as it moves outside of a force field.

On the other hand, if the enthalpy and entropy were uniform, and again the space outside a blade row is considered, the tangential vorticity arises only from the angular momentum. But since the angular momentum is constant along stream surfaces, then clearly the quantity $\eta r/\rho$ is constant along a stream surface. Physically this fact is related to the constancy of circulation along a helical stream tube which may have a greater or smaller component in the tangential direction depending upon how the stream tube is deformed by the flow. Within a blade row, where force components act on the fluid, the tangential vorticity responds in a manner that, while fairly complex, may be determined from Eq. 2-40 when the appropriate values of h^0, rv, and s are known.

The mathematical problem. Sufficient development of the gas dynamic details has been undertaken so that the mathematical problem of through-flow calculation can be formulated in a more or less satisfactory manner. Since the velocity components u and w are of central interest, the tangential vorticity (Eq. 2-43) will be the focus of attention. Since the stream function ψ has already been introduced it is natural to express the tangential vorticity in terms of the stream function (see Gravalos [*20*] and Marble [*10*]). From Eq. 2-6 and the definition of the stream function, it follows that

$$\eta \equiv \frac{\partial u}{\partial z} - \frac{\partial w}{\partial r} = \frac{\partial}{\partial z}\left(\frac{\rho_0}{\rho}\frac{1}{r}\frac{\partial \psi}{\partial z}\right) + \frac{\partial}{\partial r}\left(\frac{\rho_0}{\rho}\frac{1}{r}\frac{\partial \psi}{\partial r}\right) \tag{2-44}$$

To eliminate the density derivatives that naturally arise in expanding Eq. 2-44 it is necessary to introduce the equations of motion in the radial and axial direction (Eq. 2-1 and 2-3). Again utilizing the stream function Eq. 2-1 becomes

$$u\frac{\partial}{\partial r}\left(\frac{\rho_0}{\rho}\frac{1}{r}\frac{\partial \psi}{\partial r}\right) + w\frac{\partial}{\partial z}\left(\frac{\rho_0}{\rho}\frac{1}{r}\frac{\partial \psi}{\partial z}\right) - \frac{v^2}{r} = \left.\frac{\partial p}{\partial \rho}\right|_s \frac{\rho}{\rho_0}\frac{\partial}{\partial r}\left(\frac{\rho_0}{\rho}\right) - \frac{1}{\rho}\left.\frac{\partial p}{\partial s}\right|_\rho \frac{\partial s}{\partial r} + F_r \tag{2-45}$$

Here, of course, $\partial p/\partial \rho\big|_s$ is just a^2 the square of the sonic velocity and for a perfect gas

$$\frac{1}{\rho}\left.\frac{\partial p}{\partial s}\right|_\rho = (\gamma - 1)T \tag{2-46}$$

Expanding Eq. 2-45 and collecting derivatives of ρ_0/ρ gives

$$(a^2 - u^2)\frac{\partial}{\partial r}\left(\frac{\rho_0}{\rho}\right) - uw\frac{\partial}{\partial z}\left(\frac{\rho_0}{\rho}\right) = \left(\frac{\rho_0}{\rho}\right)^2\left[u\frac{\partial}{\partial r}\left(\frac{1}{r}\frac{\partial \psi}{\partial z}\right) + w\frac{\partial}{\partial z}\left(\frac{1}{r}\frac{\partial \psi}{\partial r}\right)\right] - \frac{\rho_0}{\rho}\left[F_r + \frac{v^2}{r} - (\gamma - 1)T\frac{\partial s}{\partial z}\right] \tag{2-47}$$

Treating Eq. 2-3 in an entirely similar manner yields

$$-uw\frac{\partial}{\partial r}\left(\frac{\rho_0}{\rho}\right) + (a^2 - w^2)\frac{\partial}{\partial z}\left(\frac{\rho_0}{\rho}\right)$$
$$= -\left(\frac{\rho_0}{\rho}\right)^2\left[u\frac{\partial}{\partial r}\left(\frac{1}{r}\frac{\partial\psi}{\partial r}\right) + w\frac{\partial}{\partial z}\left(\frac{1}{r}\frac{\partial\psi}{\partial r}\right)\right] - \frac{\rho_0}{\rho}\left[F_z - (\gamma - 1)T\frac{\partial s}{\partial z}\right] \tag{2-48}$$

Eq. 2-47 and 2-48 may be solved simultaneously for $(\partial/\partial r)(\rho_0/\rho)$ and $(\partial/\partial z)(\rho_0/\rho)$ and the results employed when the right-hand side of Eq. 2-44 is expanded, with the result that

$$\frac{\rho}{\rho_0}\eta = \left(1 - \frac{u^2}{a^2}\right)\frac{\partial}{\partial r}\left(\frac{1}{r}\frac{\partial\psi}{\partial r}\right) - \frac{uw}{a^2}\left[\frac{\partial}{\partial r}\left(\frac{1}{r}\frac{\partial\psi}{\partial z}\right) + \frac{\partial}{\partial z}\left(\frac{1}{r}\frac{\partial\psi}{\partial r}\right)\right]$$
$$+ \left(1 - \frac{w^2}{a^2}\right)\frac{\partial}{\partial z}\left(\frac{1}{r}\frac{\partial\psi}{\partial z}\right) - \frac{v^2}{ra^2}\frac{1}{r}\frac{\partial\psi}{\partial r} - \frac{\rho}{\rho_0}\frac{V_l^2}{a^2}\frac{F_\psi}{V_l} + \left(\frac{\rho}{\rho_0}\right)^2\frac{V_l^2}{a^2}{}_s(\gamma - 1)T\frac{\partial s}{\partial\psi} \tag{2-49}$$

Finally inserting this expression for the tangential vorticity into Eq. 2-40, the partial differential equation for the stream function is obtained

$$\left(1 - \frac{u^2}{a^2}\right)\frac{\partial}{\partial r}\left(\frac{1}{r}\frac{\partial\psi}{\partial r}\right) - \frac{uw}{a^2}\left[\frac{\partial}{\partial r}\left(\frac{1}{r}\frac{\partial\psi}{\partial z}\right) + \frac{\partial}{\partial z}\left(\frac{1}{r}\frac{\partial\psi}{\partial r}\right)\right]$$
$$+ \left(1 - \frac{w^2}{a^2}\right)\frac{\partial}{\partial z}\left(\frac{1}{r}\frac{\partial\psi}{\partial z}\right) - \frac{1}{r}\frac{v^2}{a^2}\left(\frac{1}{r}\frac{\partial\psi}{\partial r}\right)$$
$$= \left(\frac{\rho}{\rho_0}\right)^2\left\{r\frac{\partial h^0}{\partial r} - v\frac{\partial rv}{\partial\psi} - \left[1 + (\gamma - 1)\frac{V_l^2}{a^2}\right]Tr\frac{\partial s}{\partial\psi}\right\} - \frac{\rho}{\rho_0}\left(1 - \frac{V_l^2}{a^2}\right)\frac{F_\psi}{V_l} \tag{2-50}$$

The stagnation enthalpy h^0, the angular momentum rv, and the entropy s are described by Eq. 2-32, 2-25, and 2-34, respectively. The local speed of sound a, which appears in Eq. 2-50, is related to the stagnation enthalpy and the velocity components through the enthalpy integral

$$\frac{1}{\gamma - 1}a^2 = h^0 - \tfrac{1}{2}(u^2 + v^2 + w^2) \tag{2-51}$$

The density ρ, which is involved in the principle equation (Eq. 2-50) and elsewhere, follows simply from the combined first and second laws of thermodynamics

$$\frac{\gamma - 1}{\rho}\frac{\partial\rho}{\partial l} = \frac{1}{a^2}\frac{\partial a^2}{\partial l} - \frac{1}{c_v}\frac{\partial s}{\partial l} \tag{2-52}$$

The physical description of the flow field is thus completed.

So far as the stagnation enthalpy h^0, the angular momentum rv, and the entropy s are concerned, the stream surfaces themselves constitute parabolic characteristics. The differential operator for the stream function, given by Eq. 2-50, requires a little more analysis. Expanding the left-hand side of Eq. 2-50 gives the differential operator as

$$\left(1 - \frac{u^2}{a^2}\right)\frac{\partial^2\psi}{\partial r^2} - \frac{2uw}{a^2}\frac{\partial^2\psi}{\partial r\partial z} + \left(1 - \frac{w^2}{a^2}\right)\frac{\partial^2\psi}{\partial z^2} - \left(1 - \frac{u^2}{a^2}\right)\frac{1}{r}\frac{\partial u}{\partial r} - \frac{uw}{a^2}\frac{1}{r}\frac{\partial\psi}{\partial z} + \frac{v^2}{a^2}\left(\frac{1}{r}\frac{\partial\psi}{\partial r}\right) \tag{2-53}$$

The slopes of the characteristics (see Monroe [21]) may be written down directly from the coefficients of the second order terms as

$$-\frac{dz}{dr} = \frac{\frac{2uw}{a^2} \pm \sqrt{\left(\frac{2uw}{a^2}\right)^2 - 4\left(1 - \frac{u^2}{a^2}\right)\left(1 - \frac{w^2}{a^2}\right)}}{2\left(1 - \frac{u^2}{a^2}\right)}$$

$$= \frac{M^2 \sin\varphi \cos\varphi \pm \sqrt{M^2 - 1}}{1 - M^2 \sin^2\varphi} \tag{2-54}$$

where the meridional Mach number V_l/a has been denoted M and φ is the angle between the stream surface and the axis of symmetry, $\tan\varphi = u/w$. The angles between the characteristic surfaces and the stream surfaces are also easily found. For if α is the angle between characteristic surfaces and the symmetry axis, then $\tan\alpha = dr/dz$ and

$$\tan\mu = \tan(\alpha - \beta) \tag{2-55}$$

where μ is now the angle between characteristic surfaces and stream surfaces. Using $dr/dz = \tan\alpha$ from Eq. 2-54, a little calculation yields

$$\tan\mu = \pm\frac{1}{\sqrt{M^2 - 1}} \tag{2-56}$$

It is of main interest that the differential equation (Eq. 2-50) changes from elliptic to hyperbolic type accordingly as the *meridional* Mach number is less than or greater than unity. The nature of the flow then depends upon the meridional velocity component, the total velocity $\sqrt{u^2 + v^2 + w^2}$ entering only through the manner in which it affects the velocity of sound according to Eq. 2-51. Whether or not $\sqrt{u^2 + v^2 + w^2}/a$ exceeds unity has no bearing on the present problem. The characteristics, defined by Eq. 2-56, have all the properties familiar from usual axially symmetric flow problems where there is no tangential motion, so long as the Mach number M is interpreted in the appropriate manner. In the present work it will be assumed that the meridional Mach number is less than unity throughout the turbomachine.

It is now possible to make some statements, although not very conclusive ones, regarding the formulation of the inverse problem, that is where something other than the blade geometry is prescribed. They are not really conclusive because the nonlinear nature of Eq. 2-50 precludes the guarantee of a well-behaved solution under boundary conditions of any generality. The results are of interest, however, because they do delineate the maximum amount of information that must be prescribed.

If one prescribes

1. $h^0(r, z)$ or $rv(r, z)$ or $F_\theta(r, z)$ throughout the turbomachine (cf. Eq. 2-25, 2-32, and 2-35).
2. The blade shape or loading at the leading edge (cf. Eq. 2-23).
3. That the flow be tangential to inner and outer walls of given shape.
4. The values of the stream function, enthalpy, entropy, and angular momentum far upstream of any blade row.
5. That the stream function, enthalpy, entropy, and angular momentum be regular far downstream.

Then the details of the throughflow can be determined through solution of Eq. 2-50, together with those of Eq. 2-25, 2-32, 2-35, 2-43, 2-51, and 2-52 that are required for the quantity given under (1) above.

A word of further explanation may be added concerning item (2) above. Referring to Eq. 2-23 relating the three blade forces, this equation has a direct analogue when the "intrinsic" coordinates are employed and the components denoted $F_\psi^{(1)}$, $F_\theta^{(1)}$, $F_l^{(1)}$,

$$\frac{\partial}{\partial l}\left(\frac{F_\psi^{(1)}}{rF_\theta^{(1)}}\right) = \frac{\rho}{\rho_0} rV_l \frac{\partial}{\partial \psi}\left(\frac{F_l^{(1)}}{rF_\theta^{(1)}}\right) \tag{2-57}$$

But from the kinematical condition that the blade forces be normal to the relative velocity,

$$V_l F_l^{(1)} + (v - \omega r)F_\theta^{(1)} = 0 \tag{2-58}$$

and hence

$$\frac{\partial}{\partial l}\left(\frac{F_\psi^{(1)}}{rF_\theta^{(1)}}\right) = -\frac{\rho}{\rho_0} rV_l \frac{\partial}{\partial \psi}\left(\frac{v - \omega r}{rV_l}\right) \tag{2-59}$$

Now considering $F_\theta^{(1)}$ and rv as prescribed or calculable quantities according to statement (1) above, Eq. 2-59 may be integrated along a stream surface to give

$$\frac{F_\psi^{(1)}}{rF_\theta^{(1)}} - \frac{F_\psi^{(1)}}{rF_\theta^{(1)}}\left(\psi, l_0\right)\Bigg) = -\int_{l_0}^{l} \frac{\rho}{\rho_0} rV_l \frac{\partial}{\partial\psi}\left(\frac{v - \omega r}{rV_l}\right) dl \qquad (2\text{-}60)$$

where l_0 represents the leading edge of the blade on the stream surface in question. Therefore if $F_\psi^{(1)}$ is prescribed at the blade leading edge it may, in principle, be determined through Eq. 2-60 at other points of the blade. It is clear, furthermore, that the ratio of force $F_\psi^{(1)}/F_\theta^{(1)}$ represents the inclination of the leading edge with respect to a meridional plane and hence the prescription of leading edge shape will suffice also.

For the direct problem, that is the physical situation where the blade shape is prescribed, it may be considered that the components n_r, n_θ, n_z of the unit normal to the blade surface are known. In the intrinsic coordinate system this vector has components n_ψ, n_θ, n_l. Clearly this vector is normal to the velocity vector relative to the blades, so that

$$(v - \omega r)n_\theta + V_l n_l = 0 \qquad (2\text{-}61)$$

or the angular momentum must be

$$rv = r\left(\omega r - V_l \frac{n_l}{n_\theta}\right) \qquad (2\text{-}62)$$

Furthermore, since the force $F_\psi^{(1)}$, $F_\theta^{(1)}$, $F_l^{(1)}$ is parallel with the unit vector n_ψ, n_θ, n_l, consequently

$$F_\psi^{(1)} = F_\theta^{(1)} \frac{n_\psi}{n_\theta} \qquad (2\text{-}63)$$

Utilizing now the definition of $F_\theta^{(2)}$ from Eq. 2-17 and the conservation of angular momentum from Eq. 2-25, and noting that $F_\psi^{(2)} = 0$ since the velocity component normal to the stream surfaces vanishes, it follows that

$$F_\psi = \frac{n_\psi}{n_\theta}\left[\frac{V_l}{r}\frac{\partial rv}{\partial l} + \frac{1}{2} C_D(v - \omega r)\sqrt{(v - \omega r)^2 + V_l^2}\right] \qquad (2\text{-}64)$$

Consequently knowledge of n_l/n_θ and n_ψ/n_θ permits calculation of the angular momentum rv and hence the force component F_ψ normal to the stream surfaces. The direct problem is reduced, therefore, to one formally identical with the inverse problem, the solution of which has been discussed.

C,3. Linearized Treatment of Throughflow. The general throughflow problem as outlined in the previous articles is a strongly nonlinear one and, as a consequence, this exact formulation is seldom treated by other than numerical methods. The nonlinearity results from the facts that (1) the conservation relations for angular momentum, stagnation enthalpy, and entropy hold along stream surfaces and these surfaces are not known in advance of solution; (2) the coefficients in the differential equation (Eq. 2-50) depend upon the velocity components of the solution because the gas is compressible; and (3) in the direct problem the force components themselves, the angular momentum, etc., are given in terms of the final solution through their relation to the given blade surface shape. Fortunately it is usually the case, in axial turbomachine problems, that a useful linearization of the problem may be effected. The possibility of this arises because (1) the blades are relatively lightly loaded in the sense that the change in angular momentum across a blade row is small in comparison with the mean value of the angular momentum; (2) the geometrical boundaries are sufficiently simple that a reasonably accurate assumption of stream surface shape may be made in advance of detailed solution; and (3) the perturbations to mean velocity, density, and velocity of sound are sufficiently small that the compressibility terms may be calculated from conditions in the undisturbed flow.

The linearizing assumptions. The linearized problem of turbomachine throughflow may be formulated in a satisfactorily unified manner by assuming that the blades are lightly loaded. The linearization given by Marble [*8,10*] was arrived at by an iteration procedure and is inconsistent in the order of some terms neglected. A consistent linearization was first given by Rannie [*22*] for the region outside of the blade rows. In analytic form it is assumed that the force field associated with the blade row is of the form

$$\begin{aligned} F_r &= \epsilon f_r \\ F_\theta &= \epsilon f_\theta \\ F_z &= \epsilon f_z \end{aligned} \tag{3-1}$$

where $\epsilon \ll 1$ and f_r, f_θ, f_z are functions of order unity in the independent variables. It will prove convenient to employ the radial velocity component as the dependent variable in place of the stream function ψ.

The basic flow field will be assumed to be that which exists in the absence of a blade row, satisfies the boundary conditions on inner and outer surfaces, and takes on the appropriate initial values far upstream. If the inner and outer surfaces were concentric cylinders, the stream surfaces of this undisturbed flow would also be concentric cylinders. For usual axial turbomachine configurations it is adequate to assume that the inner and outer surfaces deviate from circular cylinders only in the order ϵ, so that indeed the undisturbed flow will take place along concentric cylindrical surfaces. Therefore the radial velocity component u is at most of order ϵ.

It is assumed now that all dependent variables may be expanded in powers of the small parameter ϵ which is indicative of the magnitude of the applied forces (Eq. 3-1). Denote the order of the terms by a superscript, the superscript "0" being the basic undisturbed flow, "1" the first order perturbation, etc.

$$\begin{aligned} u &= \epsilon u^{(1)} + \cdots & s &= s^{(0)} + \epsilon s^{(1)} + \cdots \\ v &= v^{(0)} + \epsilon v^{(1)} + \cdots & a^2 &= a^{2(0)} + \epsilon a^{2(1)} + \cdots \\ w &= w^{(0)} + \epsilon w^{(1)} + \cdots & \rho &= \rho^{(0)} + \epsilon \rho^{(1)} + \cdots \\ h_0 &= h^{(0)} + \epsilon h^{(1)} + \cdots & T &= T^{(0)} + \epsilon T^{(1)} + \cdots \end{aligned} \tag{3-2}$$

For consistency of the force field it is necessary to assume also that

$$C_D = \epsilon \xi_D \tag{3-3}$$

where ξ_D is a function of r and z of magnitude unity.

First order perturbation. Although it is possible to obtain the desired perturbation relations for radial velocity by differentiation of Eq. 2-50 and substitution from expansions in powers of ϵ (Eq. 3-2), it is generally simpler to accomplish this directly from the equations of motion themselves. It is convenient first, however, to derive the linearized relations for the stagnation enthalpy, the entropy, angular momentum, etc.

Recalling that $F_\theta = F_\theta^{(1)} + F_\theta^{(2)}$ it follows that to the first order in ϵ

$$F_\theta = \epsilon f_\theta^{(1)} - \tfrac{1}{2}\epsilon \xi_D (v^{(0)} - \omega r) \sqrt{(v^{(0)} - \omega r)^2 + w^{(0)2}} \tag{3-4}$$

where the expansion for $F_\theta^{(2)}$ is obtained from Eq. 2-17 using the expression for C_D given in Eq. 3-3. Then the second equation of motion, Eq. 2-2, which expresses the conservation of angular momentum of the fluid, may be written

$$\begin{aligned} (\epsilon u^{(1)} + \cdots) \frac{\partial}{\partial r} [r(v^{(0)} + \epsilon v^{(1)} + \cdots)] \\ + (w^{(0)} + \epsilon w^{(1)} + \cdots) \frac{\partial}{\partial z} [r(v^{(0)} + \epsilon v^{(1)} + \cdots)] \\ = r[\epsilon f_\theta^{(1)} - \tfrac{1}{2}\epsilon \xi_D (v^{(0)} - \omega r) \sqrt{(v^{(0)} - \omega r)^2 + w^{(0)2} + \cdots}] \end{aligned} \tag{3-5}$$

where it is assumed that $f_\theta^{(1)}(r, z)$ and $\xi_D(r, z)$ are prescribed functions. The zeroeth order relation (terms not containing the small parameter ϵ) gives simply

$$w^{(0)} \frac{\partial}{\partial z} (rv^{(0)}) = 0 \tag{3-6}$$

so that the initial undisturbed value of angular momentum $rv^{(0)}$ does not change along the axis of the turbomachine, but is transported unchanged along circular cylinders. The first order relation (terms proportional to the small parameter ϵ) may be written, taking account of Eq. 3-6, as

$$u^{(1)} \frac{\partial r v^{(0)}}{\partial r} + w^{(0)} \frac{\partial r v^{(1)}}{\partial r} = r f_\theta^{(1)} - \tfrac{1}{2} \xi_D r (v^{(0)} - \omega r) \sqrt{(v^{(0)} - \omega r)^2 + w^{(0)2}} \tag{3-7}$$

From this relation, the first order perturbation in angular momentum is seen to consist of two parts. The first part arises from the moment of tangential forces applied to the fluid, the right-hand side of Eq. 3-7, and the second part $u^{(1)} \partial r v^{(0)} / \partial r$ is due to the radial transport of the initial angular momentum $r v^{(0)}$.

Treating Eq. 2-32 for the stagnation enthalpy in exactly the same manner gives similar results for the zeroeth and first order relations:

$$w^{(0)} \frac{\partial h^{(0)}}{\partial z} = 0 \tag{3-8}$$

and

$$u^{(1)} \frac{\partial h^{(0)}}{\partial r} + w^{(0)} \frac{\partial h^{(1)}}{\partial z} = \omega r f_\theta^{(1)} - \tfrac{1}{2} \omega r \xi_D (v^{(0)} - \omega r) \sqrt{(v^{(0)} - \omega r)^2 + w^{(0)2}} \tag{3-9}$$

Similarly the entropy variations follow from Eq. 2-43 as

$$T^{(0)} w^{(0)} \frac{\partial s^{(0)}}{\partial z} = 0 \tag{3-10}$$

and

$$u^{(1)} \frac{\partial s^{(0)}}{\partial r} + w^{(0)} \frac{\partial s^{(1)}}{\partial z} = \frac{1}{2T^{(0)}} \xi_D [(v^{(0)} - \omega r)^2 + w^{(0)2}]^{3/2} \tag{3-11}$$

It is to be noticed in the preceding equations (Eq. 3-6, 3-7, 3-8, 3-9, 3-10, and 3-11) that, to calculate the first order perturbations of angular momentum, stagnation enthalpy, and entropy, only the first order perturbation in radial velocity, $u^{(1)}$, is required. Therefore it is necessary to develop a differential equation which may be solved for the radial velocity perturbation.

If the equation of motion in the radial direction be perturbed in accordance with Eq. 3-2, the resulting zeroeth order relation is

$$\frac{v^{(0)2}}{r} = \frac{1}{\rho^{(0)}} \frac{\partial p^{(0)}}{\partial r} \tag{3-12}$$

while the first order relation is just

$$w^{(0)} \frac{\partial u^{(1)}}{\partial r} = \frac{2 v^{(0)} v^{(1)}}{r} + \frac{v^{(0)2}}{r} \frac{\rho^{(1)}}{\rho^{(0)}} - \frac{1}{\rho^{(0)}} \frac{\partial p^{(1)}}{\partial r} + f_r \tag{3-13}$$

The equation of motion in the axial direction, perturbed in the same manner, yields only a first order part

$$u^{(1)}\frac{\partial w^{(0)}}{\partial r}+w^{(0)}\frac{\partial w^{(1)}}{\partial z}=-\frac{1}{\rho^{(0)}}\frac{\partial p^{(1)}}{\partial z}+f_z \tag{3-14}$$

The pressure perturbation $p^{(1)}$ may be eliminated between Eq. 3-13 and 3-14 through cross-differentiation, to give

$$w^{(0)}\left(\frac{\partial^2 u^{(1)}}{\partial z^2}-\frac{\partial^2 w^{(1)}}{\partial r\partial z}\right)-\frac{\partial w^{(0)}}{\partial r}\frac{\partial u^{(1)}}{\partial r}-\frac{\partial^2 w^{(0)}}{\partial r^2}u^{(1)}-\frac{\partial w^{(0)}}{\partial r}\frac{\partial w^{(1)}}{\partial z}$$
$$-\frac{1}{\rho^{(0)}}\frac{\partial \rho^{(0)}}{\partial r}\left(u^{(1)}\frac{\partial w^{(0)}}{\partial r}+w^{(0)}\frac{\partial w^{(1)}}{\partial z}\right)$$
$$=\frac{2v^{(0)}}{r}\frac{\partial v^{(1)}}{\partial z}+\frac{v^{(0)2}}{r}\frac{1}{\rho^{(0)}}\frac{\partial \rho^{(1)}}{\partial z}-\frac{1}{\rho^{(0)}}\frac{\partial \rho^{(0)}}{\partial r}f_z+\frac{\partial f_r}{\partial z}-\frac{\partial f_z}{\partial r} \tag{3-15}$$

To express this relation in a useable form it is necessary to eliminate (1) the perturbation axial velocity $w^{(1)}$, (2) the perturbation density $\rho^{(1)}$, (3) the perturbation angular momentum $rv^{(1)}$ and (4) the force components f_r and f_z.

The axial velocity perturbation may be eliminated through use of the continuity equation (Eq. 2-4) which gives, to the first order,

$$\frac{1}{r}\frac{\partial}{\partial z}(ru^{(1)})+\frac{\partial w^{(1)}}{\partial z}+\frac{1}{\rho^{(0)}}\left(u^{(1)}\frac{\partial \rho^{(0)}}{\partial r}+w^{(0)}\frac{\partial \rho^{(1)}}{\partial z}\right)=0 \tag{3-16}$$

The density perturbation occurs as $\partial\rho^{(1)}/\partial z$ and this may be eliminated through use of the first order equation of motion in the axial direction, Eq. 3-14. In particular the pressure p, considered as a function of ρ and s in the thermodynamic sense, may be expanded

$$\frac{1}{\rho}\frac{\partial p}{\partial z}=\frac{1}{\rho}\frac{\partial p}{\partial \rho}\bigg|_s\frac{\partial \rho}{\partial z}+\frac{1}{\rho}\frac{\partial p}{\partial s}\bigg|_\rho\frac{\partial s}{\partial z}$$
$$=a^2\frac{1}{\rho}\frac{\partial \rho}{\partial z}+\frac{\mathcal{R}T}{c_v}\frac{\partial s}{\partial z} \tag{3-17}$$

where the speed of sound is defined as $\sqrt{\partial p/\partial\rho\big|_s}=\sqrt{\gamma\mathcal{R}T}$. To the first order then, the pressure variation of Eq. 3-14 may be written

$$\frac{1}{\rho^{(0)}}\frac{\partial p^{(1)}}{\partial z}=\frac{a^{2(0)}}{\rho^{(0)}}\frac{\partial \rho^{(1)}}{\partial z}+\frac{a^{2(0)}}{c_p}\frac{\partial s^{(1)}}{\partial z} \tag{3-18}$$

Thus Eq. 3-14, written in the form

$$u^{(1)}\frac{\partial w^{(0)}}{\partial r}+w^{(0)}\frac{\partial w^{(1)}}{\partial z}=-\frac{a^{2(0)}}{\rho^{(0)}}\frac{\partial \rho^{(1)}}{\partial z}-\frac{a^{2(0)}}{c_p}\frac{\partial s^{(1)}}{\partial z}+f_z \tag{3-19}$$

may be considered an expression for $\partial\rho^{(1)}/\partial z$ and employed to eliminate the density variation that arises in Eq. 3-15. The entropy perturbation that appears in Eq. 3-19 is given by the entropy relations, Eq. 3-11, while the term $\partial v^{(1)}/\partial z$ follows from the angular momentum relation, Eq. 3-7. These substitutions into Eq. 3-15, together with some simplification, give the result:

$$
\begin{aligned}
&\frac{\partial}{\partial r}\left(\frac{1}{r}\frac{\partial ru^{(1)}}{\partial r}\right) + \frac{\partial^2 u^{(1)}}{\partial z^2} - \left(\frac{1}{w^{(0)}}\frac{\partial w^{(0)}}{\partial r}\right)\frac{\partial u^{(1)}}{\partial r} - \left(\frac{1}{w^{(0)}}\frac{\partial^2 w^{(0)}}{\partial r^2}\right)u^{(1)} \\
&\quad + \frac{\partial}{\partial r}\left[\frac{1}{1 + (w^{(0)2}/a^{2(0)})}\left(\frac{v^{(0)2}}{ra^{2(0)}}u^{(1)} - \frac{w^{(0)}}{a^{2(0)}}\frac{\partial w^{(0)}}{\partial r}u^{(1)} + \frac{w^{(0)2}}{a^{2(0)}}\frac{1}{r}\frac{\partial ru^{(1)}}{\partial r}\right)\right] \\
&\quad - \frac{1}{w^{(0)}}\left(\frac{1}{w^{(0)}}\frac{\partial w^{(0)}}{\partial r} - \frac{v^{(0)2}}{ra^{2(0)}} - \frac{1}{\rho^{(0)}}\frac{\partial \rho^{(0)}}{\partial r}\right)\left[\frac{1}{1 - a^{2(0)}/(w^{(0)2})}\right] \\
&\quad \left(-\frac{\partial w^{(0)}}{\partial r}u^{(1)} + \frac{v^{(0)2}}{rw^{(0)}}u^{(1)} + \frac{a^{2(0)}}{rw^{(0)}}\frac{\partial ru^{(1)}}{\partial r}\right) \\
&\quad + \left[\frac{2v^{(0)}}{r^2 w^{(0)}}\frac{\partial rv^{(0)}}{\partial r} + \frac{v^{(0)2}}{rw^{(0)2}}\frac{1}{\rho^{(0)}}\frac{\partial \rho^{(0)}}{\partial r} - \frac{v^{(0)}}{ra^{2(0)}}\frac{\partial rv^{(0)}}{\partial r}\frac{v^{(0)2}}{rw^{(0)2}}\right. \\
&\quad \left. - \frac{1}{\rho^{(0)}}\frac{\partial \rho^{(0)}}{\partial r}\frac{1}{w^{(0)}}\frac{\partial w^{(0)}}{\partial r} + \frac{v^{(0)2}}{ra^{2(0)}w^{(0)2}}\frac{\partial}{\partial r}\left(\frac{v^{(0)2} + w^{(0)2}}{2}\right)\right]u^{(1)} \\
&= \frac{\partial}{\partial r}\left[\frac{1}{1 - (w^{(0)2}/a^{2(0)})}\left\{\frac{\gamma - 1}{2}\frac{\xi_D}{a^{2(0)}}[(v^{(0)} - \omega r)^2 + w^{(0)2}]^{\frac{1}{2}} - \frac{w^{(0)}}{a^{2(0)}}f_z\right\}\right. \\
&\quad - \frac{1}{w^{(0)}}\left(\frac{1}{w^{(0)}}\frac{\partial w^{(0)}}{\partial r} - \frac{v^{(0)2}}{ra^{2(0)}} + \frac{1}{\rho^{(0)}}\frac{\partial \rho^{(0)}}{\partial r}\right)\left[\frac{1}{1 - (w^{(0)2}/a^{2(0)})}\right] \\
&\quad \left.\left\{\frac{\gamma - 1}{2}\frac{\xi_D}{w^{(0)}}[(v^{(0)} - \omega r)^2 + w^{(0)2}]^{\frac{1}{2}} - f_z\right\}\right] \\
&\quad - \frac{v^{(0)2}}{rw^{(0)2}}\left\{\frac{\gamma}{2}\frac{\xi_D}{a^{2(0)}}[(v^{(0)} - \omega r)^2 + w^{(0)2}]^{\frac{1}{2}} - \frac{\omega r f_\theta}{a^{2(0)}} + \frac{v^{(0)}}{a^{2(0)}}f_\theta\right\} \\
&\qquad\qquad + \frac{2v^{(0)}}{rw^{(0)2}}f_\theta - \frac{1}{\rho^{(0)}w^{(0)}}\frac{\partial \rho^{(0)}}{\partial r}f_z + \frac{1}{w^{(0)}}\left(\frac{\partial f_r}{\partial z} - \frac{\partial f_z}{\partial r}\right) \qquad (3\text{-}20)
\end{aligned}
$$

This is a second order linear partial differential equation with variable coefficients and a nonhomogeneous term. The nonhomogeneous term, the right-hand side, is given in terms of the force component, loss coefficient, and known zero order conditions. Now it is known from the investigations of Art. 2, that all force components cannot be prescribed independently but, strictly speaking, only one of them. It is necessary to express, within the present approximation f_r and f_z in terms of f_θ and ξ_D, considered prescribed, and known zeroeth order conditions.

From Eq. 2-17, giving the loss forces, it is clear that, to the present approximation,

$$f_r^{(2)} \cong 0 \tag{3-21}$$

$$f_\theta^{(2)} = -\tfrac{1}{2}\xi_D(v^{(0)} - \omega r)\sqrt{(v^{(0)} - \omega r)^2 + w^{(0)2}} \tag{3-22}$$

$$f_z^{(2)} = -\tfrac{1}{2}\xi_D w^{(0)}\sqrt{(v^{(0)} - \omega r)^2 + w^{(0)2}} \tag{3-23}$$

and hence these are known. Also to the first order, Eq. 2-16 becomes

$$(v^{(0)} - \omega r)f_\theta^{(1)} + w^{(0)}f_z^{(1)} = 0 \tag{3-24}$$

and hence, in detail, the axial force component may be written

$$f_z = f_z^{(1)} + f_z^{(2)} = -\frac{v^{(0)} - \omega r}{w^{(0)}} f_\theta^{(1)} - \left[1 + \left(\frac{v^{(0)} - \omega r}{w^{(0)}}\right)^2\right]^{\frac{1}{2}} \frac{w^{(0)2}}{2}\,\xi_D \tag{3-25}$$

Finally if Eq. 2-23, the condition of blade continuity, be written to the first order it follows, employing the ratio $f_z^{(1)}/f_\theta^{(1)}$ from Eq. 3-24, that

$$\frac{\partial}{\partial z}\left(\frac{f_r^{(1)}}{rf_\theta^{(1)}}\right) = -\frac{\partial}{\partial r}\left(\frac{v^{(0)} - \omega r}{rw^{(0)}}\right) \tag{3-26}$$

The right-hand side of Eq. 3-26 depends only upon the radius. Consequently integration from the leading edge z_0 to some arbitrary value of z gives

$$\frac{f_r^{(1)}(r, z)}{rf_\theta^{(1)}(r, z)} - \frac{f^{(1)}(r, z_0)}{rf_\theta^{(1)}(r, z_0)} = -(z - z_0)\frac{\partial}{\partial r}\left(\frac{v^{(0)} - \omega r}{rw^{(0)}}\right)$$

or

$$f_r \equiv f_r^{(1)}(r, z) = f_\theta^{(1)}\left[\frac{f_r(r, z_0)}{f_\theta^{(1)}(r, z_0)} - (z - z_0)r\frac{\partial}{\partial r}\left(\frac{v^{(0)} - \omega r}{rw^{(0)}}\right)\right] \tag{3-27}$$

where $f_r(r, z_0)/f_\theta^{(1)}(r, z_0)$ represents the inclination of the leading edge with respect to a meridional plane.

Eq. 3-25 and 3-27 are satisfactory expressions for f_z and f_r to be employed in the right-hand side of Eq. 3-20; it should be noted, in particular, that these expressions contain nothing involving the radial perturbation velocity $u^{(1)}$ and consequently do not modify the left-hand side of Eq. 3-20.

The linearized second order partial differential equation for the radial perturbation velocity may now be written in the form

$$\frac{\partial^2 u^{(1)}}{\partial r^2} + p(r)\frac{\partial u^{(1)}}{\partial r} + q(r)u^{(1)} + (1 - M_z^2)\frac{\partial^2 u^{(1)}}{\partial z^2} = \phi(r, z) \tag{3-28}$$

where $p(r)$, $q(r)$ and $\phi(r, z)$ are known and equal to

$$p(r) = \frac{1}{r}\left(1 + \gamma M_\theta^2 - \frac{r}{1 - M_z^2}\frac{1}{a^{2(0)}}\frac{\partial a^{2(0)}}{\partial r} + \frac{2rM_z^2}{1 - M_z^2}\frac{1}{w^{(0)}}\frac{\partial w^{(0)}}{\partial r}\right) \quad (3\text{-}29)$$

$$\begin{aligned} q(r) = \frac{1}{r^2}\Bigg\{ & -1 + \frac{v^{(0)2}}{w^{(0)2}}[2 - (4 - \gamma)M_z^2 + (\gamma - 1)M_\theta^2] - \frac{r^2}{w^{(0)}}\frac{\partial^2 w^{(0)}}{\partial r^2} \\ & + \frac{2rv^{(0)}}{w^{(0)2}}\frac{\partial v^{(0)}}{\partial r} - \frac{v^{(0)2} + w^{(0)2}}{w^{(0)2}(1 - M_z^2)}\frac{r}{a^{2(0)}}\frac{\partial a^{2(0)}}{\partial r} \\ & + \frac{1}{1 - M_z^2}\Bigg[1 + M_z^2 + M_\theta^2(2 - \gamma + \gamma M_z^2) + \frac{r}{a^{2(0)}}\frac{\partial a^{2(0)}}{\partial r} \\ & - 2M_z^2\frac{r}{w^{(0)}}\frac{\partial w^{(0)}}{\partial r}\Bigg]\frac{r}{w^{(0)}}\frac{\partial w^{(0)}}{\partial r}\Bigg\} \end{aligned} \quad (3\text{-}30)$$

$$\begin{aligned} \phi(r, z) = & \frac{f_z}{w^{(0)}}\left[\frac{1}{a^{2(0)}}\frac{\partial a^{2(0)}}{\partial r}\left(\frac{1}{1 - M_z^2}\right) - \frac{1}{w^{(0)}}\frac{\partial w^{(0)}}{\partial r}\left(\frac{2M_z^2}{1 - M_z^2}\right)\right. \\ & \left. - (\gamma - M_z^2)\frac{M_\theta^2}{r}\right] - \frac{1}{w^{(0)}}\left[\frac{\partial f_z}{\partial r} - \frac{\partial f_r}{\partial z}(1 - M_z^2)\right] \\ & + (1 - M_z^2)\frac{v^{(0)}}{rw^{(0)2}}f_\theta\left[2 + M_\theta^2\left(\frac{\omega r}{v^{(0)}} - 1\right)\right] \\ & + \frac{\gamma - 1}{2}M_z^2\frac{\partial}{\partial r}\left\{\frac{\xi_D}{w^{(0)2}}[(v^{(0)} - \omega r)^2 + w^{(0)2}]^{\frac{1}{2}}\right\} \\ & + \frac{\gamma - 1}{2}\frac{\xi_D}{w^{(0)2}}[(v^{(0)} - \omega r)^2 + w^{(0)2}]^{\frac{1}{2}}\left\{\frac{1}{w^{(0)}}\frac{\partial w^{(0)}}{\partial r}\left[\frac{M_z^2(3 - M_z^2)}{1 - M_z^2}\right]\right. \\ & \left. - \frac{1}{a^{2(0)}}\frac{\partial a^{2(0)}}{\partial r}\left[\frac{M_z^2(2 - M_z^2)}{1 - M_z^2}\right] + \frac{M_\theta^2}{r}\left[\frac{\gamma^2 - \gamma + 1}{\gamma - 1}M_z^2 - \frac{\gamma}{\gamma - 1}\right]\right\} \end{aligned} \quad (3\text{-}31)$$

The values of M_z^2 and M_θ^2 introduced above are defined

$$M_\theta^2 = \frac{v^{(0)2}}{a^{2(0)}} \qquad M_z^2 = \frac{w^{(0)2}}{a^{2(0)}} \quad (3\text{-}32)$$

and hence are functions of the radius in general.

To complete the problem it is necessary to prescribe boundary conditions on the differential equation (Eq. 3-28). In the first place the radial velocity disturbances should vanish far upstream and downstream of the blade row, so that

$$u^{(1)}(r, -\infty) = u^{(1)}(r, \infty) = 0 \quad (3\text{-}33)$$

Secondly, the inner and outer radii are given as functions of the axial distance down the machine and, since they can differ from circular cylindrical surfaces only by a quantity of small order (say α where $\alpha \ll 1$), they can be written

$$r_1 = r_1^{(0)} + \alpha r_1^{(1)}(z) \tag{3-34}$$

$$r_2 = r_2^{(0)} + \alpha r_2^{(1)}(z) \tag{3-35}$$

Within the linearization employed, the boundary conditions at the inner and outer radii may then be given as

$$\epsilon \frac{u^{(1)}(r_1, z)}{w^{(0)}} = \alpha \frac{dr_1^{(1)}(z)}{dz} \tag{3-36}$$

and

$$\epsilon \frac{u^{(1)}(r_2, z)}{w^{(0)}} = \alpha \frac{dr_2^{(1)}(z)}{dz} \tag{3-37}$$

Finally, the complete state of the gas is given far upstream of the blade row; the quantities $v^{(0)}(r)$, $w^{(0)}(r)$, $h^{(0)}(r)$, $s^{(0)}(r)$, etc. are assumed to be known. The formulation of the linearized problem is complete.

C,4. Incompressible Flow through Single Blade Rows with Constant Hub and Tip Radii. It is often the case that the Mach number of the flow (in the sense of Art. 2) is sufficiently low that the gas may be considered incompressible in calculating the throughflow. Furthermore, near the design point it is often a fact that the axial velocity $w^{(0)}$ far upstream of the blade row differs from a constant value only by a quantity of order ϵ. Then $w^{(0)}$ may be considered constant and any variations incorporated as perturbation quantities. These conditions, although valid only for operating points near the design, simplify the mathematical problem enormously and constitute a good starting point for investigation of more complex conditions.

When the fluid is incompressible, the losses negligible, and the axial velocity differs only slightly from a constant value $w^{(0)}$, the partial differential equation (Eq. 3-28) becomes

$$\frac{\partial^2 u^{(1)}}{\partial r^2} + \frac{1}{r}\frac{\partial u^{(1)}}{\partial r} - \frac{1}{r^2}\left[1 - \frac{1}{rw^{(0)2}}\frac{\partial}{\partial r}(rv^{(0)})^2\right]u^{(1)} + \frac{\partial^2 u^{(1)}}{\partial z^2} = \phi(r, z) \tag{4-1}$$

where

$$\phi(r, z) = -\frac{1}{w^{(0)}}\left(\frac{\partial f_z}{\partial r} - \frac{\partial f_r}{\partial z}\right) + \frac{2v^{(0)}}{rw^{(0)2}} f_\theta \tag{4-2}$$

Consider a single blade row with hub and tip shrouds consisting of concentric circular cylinders having radii r_h and r_t, respectively (see Fig. C,4a). Assume furthermore that far upstream of the blade row the axial velocity is uniform and the tangential velocity vanishes. Then the

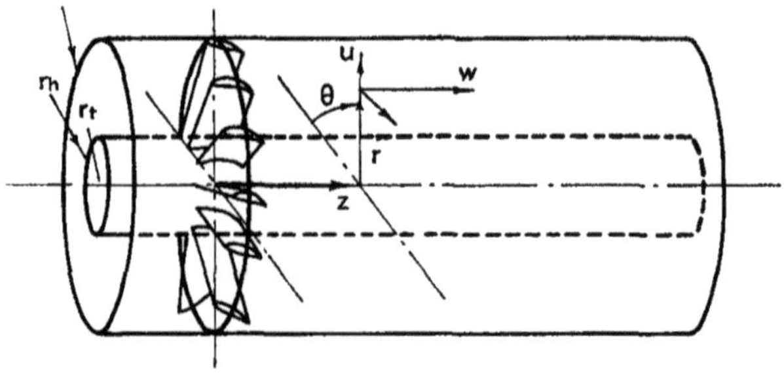

Fig. C,4a. Single blade row with hub and tip shrouds of constant radii.

tangential velocity and the axial velocity variation are only those introduced by the blade row itself so that the radial velocity perturbation is described by

$$\frac{\partial^2 u^{(1)}}{\partial r^2} + \frac{1}{r}\frac{\partial u^{(1)}}{\partial r} - \frac{1}{r^2}u^{(1)} + \frac{\partial^2 u^{(1)}}{\partial z^2} = -\frac{1}{w^{(0)}}\frac{\partial f_z}{\partial r} = -\frac{1}{w^{(0)}}\frac{\partial}{\partial z}\left(\frac{\partial h^{(1)}}{\partial r}\right) \tag{4-3}$$

where enthalpy perturbation $h^{(1)}(r, z)$ will be assumed known. The boundary conditions to be satisfied are that

$$u^{(1)}(r_h, z) = u^{(1)}(r_t, z) = 0 \tag{4-4}$$

and

$$u^{(1)}(r, -\infty) = u^{(1)}(r, \infty) = 0 \tag{4-5}$$

Theory of the actuator disk. Suppose for the moment, following [8], that the blade row be shrunk axially into a discontinuity as in Fig. C,4b so that the change in enthalpy entering into the right-hand side of Eq. 4-3 is concentrated at $z = 0$. The radial velocity perturbation is therefore a

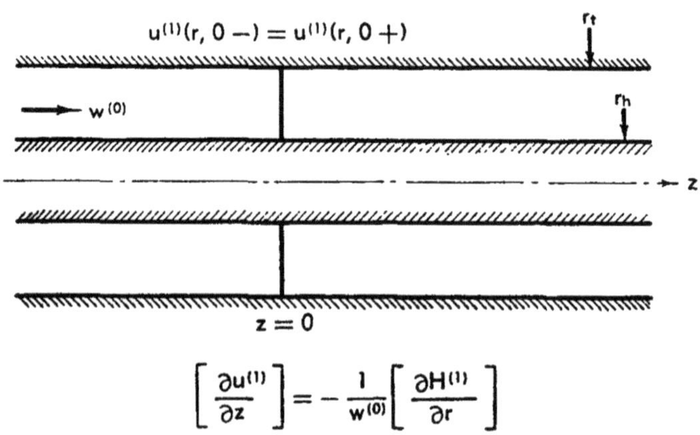

Fig. C,4b. The simple actuator disk.

solution of the homogeneous equation

$$\frac{\partial^2 u^{(1)}}{\partial r^2} + \frac{1}{r}\frac{\partial u^{(1)}}{\partial r} - \frac{1}{r^2}u^{(1)} + \frac{\partial^2 u^{(1)}}{\partial z^2} = 0 \tag{4-6}$$

both upstream and downstream of the discontinuity. Furthermore, the solution is symmetric with respect to $z = 0$ since both the differential equation and boundary conditions are symmetric. To determine the solution two matching conditions across the blade row or actuator disk must be established.

From the radial equation of motion (Eq. 2-1), it follows that the radial velocity $u^{(1)}$ can change across the actuator disk only as a result of a concentrated radial force. Within the linearized theory, therefore, the jump in radial velocity $[u^{(1)}]$ across the actuator disk may be written

$$[u^{(1)}] = \frac{1}{w^{(0)}}\int_{0-}^{0+} f_r dz \tag{4-7}$$

where we think here of the integrand as being very large, the interval very small, and the integral the total radial force of the blade row at a given point of the actuator disk. As a general rule in axial turbomachinery, the radial force component of the blades exerts a negligible influence due simply to the fact that the blade surfaces themselves are so nearly radial. This observation is so general that it is preferred to assume here that $[u^{(1)}] = 0$ and simply to note that, if a peculiar construction alters this situation, the following analysis would require modification in an obvious manner. The second matching condition follows directly from the partial differential equation (Eq. 4-3). Since the axial and radial velocity components are continuous across the actuator disk, the differential equation may be evaluated on each side of the discontinuity, integrated once with respect to z, and the results subtracted to give

$$\left[\frac{\partial u^{(1)}}{\partial z}\right]_{0-}^{0+} = -\frac{1}{w^{(0)}}\left[\frac{\partial h^{(1)}}{\partial r}\right]_{0-}^{0+} \tag{4-8}$$

a condition on the difference between values of $\partial u^{(1)}/\partial z$ evaluated on each side of the actuator disk.

Now a solution of the differential equation,

$$u^{(1)} = \sum_1^\infty C_n[J_1(\kappa_n r)Y_1(\kappa_n r_h) - Y_1(\kappa_n r)J_1(\kappa_n r_h)]e^{\kappa_n z} \qquad z < 0 \tag{4-9}$$

and

$$u^{(1)} = \sum_1^\infty C_n[J_1(\kappa_n r)Y_1(\kappa_n r_h) - J_1(\kappa_n r_h)Y_1(\kappa_n r)]e^{-\kappa_n z} \qquad z > 0 \tag{4-10}$$

satisfies the boundary conditions at $z = \pm\infty$ and at $r = r_h$, as well as gives a continuous value of $u^{(1)}$ across the actuator disk as required by the first matching condition. The boundary condition at the outer boundary, $u^{(1)}(r_t, z) = 0$, is satisfied identically by taking the characteristic values κ_n to be the roots of

$$J_1(\kappa_n r_t) Y_1(\kappa_n r_h) - J_1(\kappa_n r_h) Y_1(\kappa_n r_t) = 0 \tag{4-11}$$

and that there is a countable infinity of roots ordered as to increasing magnitude is guaranteed by the Sturm-Liouville theorem. The values of the constants follow directly from the orthogonality property of the Bessel functions and the matching condition given by Eq. 4-8. Using the solutions (Eq. 4-9 and 4-10),

$$-2 \sum_{1}^{\infty} C_n \kappa_n [J_1(\kappa_n r) Y_1(\kappa_n r_h) - J_1(\kappa_n r_h) Y_1(\kappa_n r)] = -\frac{1}{w^{(0)}} \left[\frac{\partial h^{(1)}}{\partial r}\right] \tag{4-12}$$

from which

$$C_n = \int_{r_h}^{r_t} \frac{\alpha(J_1(\kappa_n \alpha) Y_1(\kappa_n r_h) - J_1(\kappa_n r_h) Y_1(\kappa_n \alpha))}{2\kappa_n \nu_n^2 w^{(0)}} \left[\frac{\partial h^{(1)}}{\partial \alpha}\right] d\alpha \tag{4-13}$$

where ν_n is the norm of $J_1(\kappa_n r) Y_1(\kappa_n r_h) - J_1(\kappa_n r_h) Y_1(\kappa_n r)$ and is given by

$$\begin{aligned} \nu_n^2 &= \int_{r_h}^{r_t} \alpha [J_1(\kappa_n \alpha) Y_1(\kappa_n r_h) - J_1(\kappa_n r_h) Y_1(\kappa_n \alpha)]^2 d\alpha \\ &= \tfrac{1}{2}\{r_t^2 [J_0(\kappa_n r_t) Y_1(\kappa_n r_h) - J_1(\kappa_n r_h) Y_0(\kappa_n r_t)]^2 \\ &\qquad - r_h^2 [J_0(\kappa_n r_h) Y_1(\kappa_n r_h) - J_1(\kappa_n r_h) Y_0(\kappa_n r_h)]^2\} \end{aligned} \tag{4-14}$$

The radial velocity distribution is now directly calculable from the foregoing results substituted into Eq. 4-9 and 4-10 after the integrals of Eq. 4-13 have been evaluated.

The axial velocity perturbation is related to the radial velocity through the continuity condition,

$$\frac{\partial w^{(1)}}{\partial z} = -\frac{1}{r} \frac{\partial (r u^{(1)})}{\partial r} \tag{4-15}$$

which from Eq. 4-10 and 4-11 yields

$$\frac{\partial w^{(1)}}{\partial z} = -\sum_{1}^{\infty} C_n \kappa_n [J_0(\kappa_n r) Y_1(\kappa_n r_h) - J_1(\kappa_n r_h) Y_0(\kappa_n r)] e^{\pm \kappa_n z} \tag{4-16}$$

the plus or minus sign applying according to whether the point in question is upstream or downstream of the actuator disk. The integration to obtain $w^{(1)}(r, z)$ may be carried out easily to give

$$w^{(1)} = -\sum_{1}^{\infty} C_n [J_0(\kappa_n r) Y_1(\kappa_n r_h) - J_1(\kappa_n r_h) Y_0(\kappa_n r)] e^{\kappa_n z} \qquad z < 0 \tag{4-17}$$

and

$$w^{(1)} = -2\sum_{1}^{\infty} C_n[J_0(\kappa_n r)Y_1(\kappa_n r_h) - J_1(\kappa_n r_h)Y_0(\kappa_n r)]$$
$$+ \sum_{1}^{\infty} C_n[J_0(\kappa_n r)Y_1(\kappa_n r_h) - J_1(\kappa_n r_h)Y_0(\kappa_n r)]e^{-\kappa_n z} \qquad z > 0 \quad (4\text{-}18)$$

It is of interest to note here that there is a residual axial velocity perturbation far downstream of the blade row equal to $-2\sum_1^\infty C_n[J_0(\kappa_n r)Y_1(\kappa_n r_h) - J_1(\kappa_n r_h)Y_0(\kappa_n r)]$ and that the axial velocity perturbation is just twice that at the actuator disk itself. That this change should take place half upstream and half downstream of the blade row is a direct result of symmetry in the radial velocity distribution.

Since it has been assumed that the distribution of enthalpy rise is given, it follows that the first order perturbation in tangential velocity is also known. From Eq. 3-7 and 3-9 it may be inferred that

$$\omega r[v^{(1)}] = [h^{(1)}] \qquad (4\text{-}19)$$

and since this jump takes place only across the actuator disk, the tangential velocity is constant ($\equiv 0$) upstream of the blade row and constant ($\equiv v^{(1)}(r)$) downstream of the blade row. This first order solution for the velocity components is complete and agrees with that given originally in [*8*].

Since the foregoing example of the linearized treatment is one of the simplest that may be found, it is worthwhile to explore its characteristics in some detail. Take as a special case the enthalpy jump to be

$$[h^{(1)}(r)] = aw^{(0)}\omega r_t\left(\frac{r}{r_t}\right)^2 \qquad (4\text{-}20)$$

Referring to Eq. 4-19, this corresponds to imparting a tangential velocity of magnitude $aw^{(0)}r/r_t$ where a is a constant. Such a tangential velocity distribution is referred to as a solid body rotation since its magnitude is proportional to the radius.

To obtain an explicit solution, evaluation of the integrals for C_n is required:

$$C_n = \frac{aw^{(0)}\omega r_t}{\kappa_n \nu_n^2 w^{(0)} r_t}\int_{r_h}^{r_t} \alpha^2[J_1(\kappa_n\alpha)Y_1(\kappa_n r_h) - J_1(\kappa_n r_h)Y_1(\kappa_n\alpha)]d\alpha \qquad (4\text{-}21)$$

which integrates directly to give

$$C_n = \frac{-2a(\omega r_t)}{(\kappa_n r_t)^2}\left\{\frac{r_t^2[J_0(\kappa_n r_t)Y_1(\kappa_n r_h) - J_1(\kappa_n r_h)Y_0(\kappa_n r_t)] - r_h^2[J_0(\kappa_n r_h)Y_1(\kappa_n r_h) - J_1(\kappa_n r_h)Y_0(\kappa_n r_h)]}{r_t^2[J_0(\kappa_n r_t)Y_1(\kappa_n r_h) - J_1(\kappa_n r_h)Y_0(\kappa_n r_t)]^2 - r_h^2[J_0(\kappa_n r_h)Y_1(\kappa_n r_h) - J_1(\kappa_n r_h)Y_0(\kappa_n r_h)]}\right\}$$

Substitution of these coefficients into Eq. 4-9, 4-10, 4-17, and 4-18 yields the solution for radial and axial velocity perturbation upstream and downstream of a blade row represented by a discontinuity. As an illustration of the results obtained, a blade row satisfying the above enthalpy jump has been chosen having a ratio of hub radius to tip radius $r_h/r_t = 0.60$. The characteristic values may be evaluated from Eq. 4-11, and from these all necessary factors of the solution follow. The radial velocity distribution is shown in Fig. C,4c for various distances upstream (or downstream by

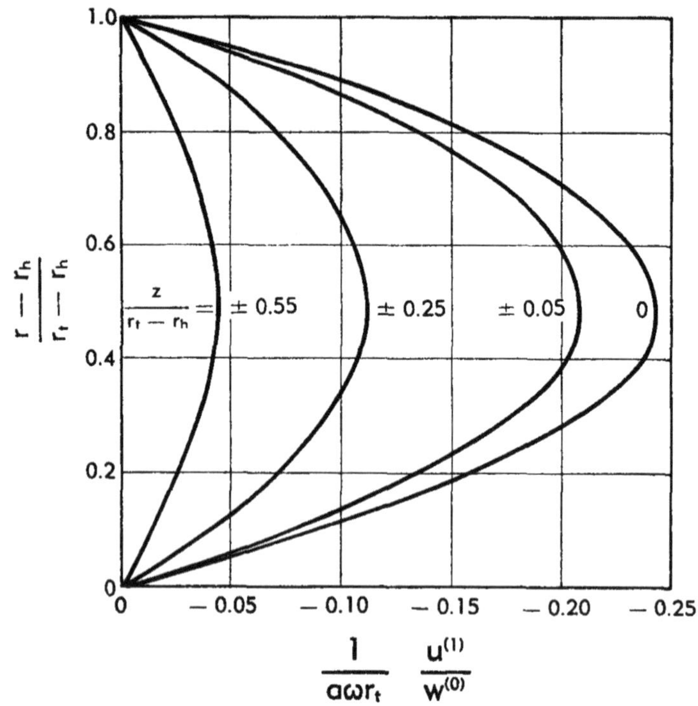

Fig. C,4c. Radial velocity patterns upstream and downstream of an actuator disk, moving with angular velocity ω and imparting tangential velocity $aw^{(0)}r/r_t$, $r_h/r_t = 0.6$.

symmetry) from the actuator disk. As is intuitively logical, the largest values of radial velocity occur near the middle of the channel. Actually, however, none of these values are large enough themselves to be of significance in turbomachine blade design. Rather it is the changes in axial velocity that are induced by this radial velocity that are of importance. The axial velocity perturbation computed from Eq. 4-17 and 4-18, using the coefficients given in Eq. 4-21, is shown in Fig. C,4d. The blade row has a quite sensible effect on the axial velocity profile at distances upstream of more than half a blade length. By virtue of the same symmetry noticed in connection with the radial velocities, distortion of the axial velocity continues to increase until it reaches a point about half a blade length downstream where the maximum and constant perturbation is achieved. These magnitudes are sufficient to distort the blade velocity diagram significantly from that which would be estimated neglecting the three-dimensional throughflow.

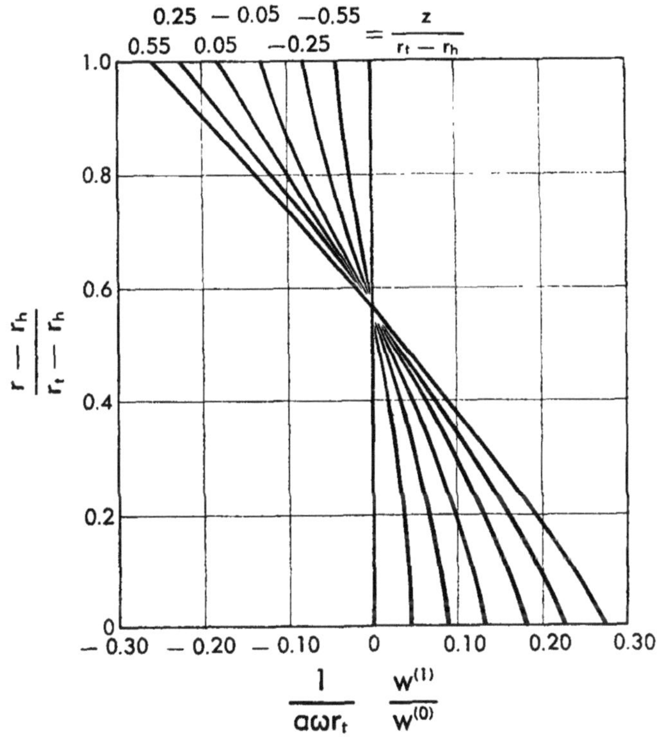

Fig. C,4d. Axial velocity perturbation patterns upstream and downstream of an actuator disk, moving with angular velocity ω and imparting tangential velocity $aw^{(0)}r/r_t$, $r_h/r_t = 0.6$.

Theory for blade row of finite chord. It is obvious that the results of actuator disk theory do not describe the flow accurately in the immediate vicinity of the blade row. The description may certainly be improved by developing the theory for a blade row of finite chord. Conceptually the blade of finite chord may be thought of as a sequence of actuator disks and, as a matter of fact, this idea has certain merit as an approximation. Mathematically the solution for the actuator disk may be employed as a unit out of which to construct the solution for a blade of finite chord by integration over the blade chord of actuator disks of the appropriate infinitesimal strength.

Consider then an actuator disk located at a point $z = \beta$ along the axis of the turbomachine. Referring to Eq. 4-9 and 4-10 it is clear that the radial velocity induced by this actuator disk may be written as

$$u^{(1)} = \sum_{1}^{\infty} C_n[J_1(\kappa_n r)Y_1(\kappa_n r_h) - J_1(\kappa_n r_h)Y_1(\kappa_n r)]e^{\kappa_n(z-\beta)}$$

upstream of the actuator disk, and as

$$u^{(1)} = \sum_{1}^{\infty} C_n[J_1(\kappa_n r)Y_1(\kappa_n r_h) - J_1(\kappa_n r_h)Y_1(\kappa_n r)]e^{-\kappa_n(z-\beta)}$$

downstream of the actuator disk. Now it will be assumed that such actuator disks are distributed continuously along the z axis at values of β between the leading and trailing edges of the blade, as shown in Fig. C,4e.

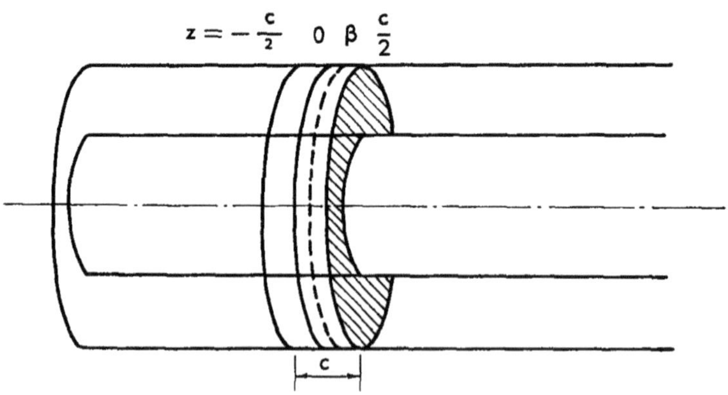

Fig. C,4e. Continuous blade row as a sequence of actuator disks.

Then again, if $h^{(1)}(r, z)$ is the perturbation of enthalpy imparted by the blade row, the function $[\partial(\partial h^{(1)}/\partial r)/\partial z]dz$ replaces the function $[\partial h^{(1)}/\partial r]$ in the determination of coefficients according to Eq. 4-13. For the elementary actuator disk situated at $z = \beta$, the flow can be written, substituting the values of C_n from Eq. 4-13,

$$u^{(1)} = \int_{r_h}^{r_t} \frac{1}{w^{(0)}} \frac{\partial^2 h^{(1)}}{\partial\alpha\partial\beta} d\beta$$

$$\sum_{1}^{\infty} \frac{\alpha[J_1(\kappa_n\alpha)Y(\kappa_n r_h) - J_1(\kappa_n r_h)Y_1(\kappa_n\alpha)][J_1(\kappa_n r)Y_1(\kappa_n r_h) - J_1(\kappa_n r_h)Y_1(\kappa_n r)]}{2\,\kappa_n\nu_n^2} e^{-\kappa_n|z-\beta|}d\alpha \quad (4\text{-}22)$$

where the order of integration and summation have been interchanged with the assurance that the series possesses the appropriate convergence properties. One of these infinitesimal discontinuities exists at each point of the interval over which the blade row exists and, since the problem is a linear one, the complete solution of the original homogeneous problem can be obtained by summing them over the blade chord. This sum takes the form of an integration from $z = -c/2$ to $z = c/2$ where, as shown in Fig. C,4e, the blade chord has the magnitude c. The complete solution is thus

$$u^{(1)} = \int_{r_h}^{r_t} \int_{-c/2}^{c/2} \frac{1}{w^{(0)}} \frac{\partial^2 h^{(1)}}{\partial\alpha\partial\beta} G(r, z; \alpha, \beta) d\alpha d\beta \tag{4-23}$$

where the function $G(r, z; \alpha, \beta)$ is just the infinite series

$$\sum_{1}^{\infty} \frac{\alpha[J_1(\kappa_n\alpha)Y_1(\kappa_n r_h) - J_1(\kappa_n r_h)Y_1(\kappa_n\alpha)][J_1(\kappa_n r)Y_1(\kappa_n r_h) - J_1(\kappa_n r_h)Y_1(\kappa_n r)]}{2\,\kappa_n\nu_n^2} e^{-\kappa_n|z-\beta|} \tag{4-24}$$

The function $G(r, z; \alpha, \beta)$ may be interpreted as proportional to the radial velocity induced at a point r, z of the turbomachine by an element of tangential vorticity bound at a point α, β of the blade row.

The complete solution for the axial velocity perturbation may be constructed in a very similar manner, for referring to Eq. 4-17 and 4-18 it is clear the axial velocity perturbation induced by an actuator disk of axial length $d\beta$ is just

$$w^{(1)} = \int_{r_h}^{r_t} \frac{1}{w^{(0)}} \frac{\partial^2 h^{(1)}}{\partial\alpha\partial\beta} d\beta$$
$$\sum_{1}^{\infty} \frac{\alpha[J_1(\kappa_n\alpha)Y_1(\kappa_n r_h) - J_1(\kappa_n r_h)Y_1(\kappa_n\alpha)][J_1(\kappa_n r)Y_1(\kappa_n r_h) - J_1(\kappa_n r_h)Y_1(\kappa_n r)]}{2\,\kappa_n\nu_n^2} e^{-\kappa_n|z-\beta|} d\alpha$$
$$z < \beta \tag{4-25}$$

and

$$w^{(1)} = 2\int_{r_h}^{r_t} \frac{1}{w^{(0)}} \frac{\partial^2 h}{\partial\alpha\partial\beta} d\beta$$
$$\sum_{1}^{\infty} \frac{\alpha[J_1(\kappa_n\alpha)Y_1(\kappa_n r_h) - J_1(\kappa_n r_h)Y_1(\kappa_n\alpha)][J_1(\kappa_n r)Y_1(\kappa_n r_h) - J_1(\kappa_n r_h)Y_1(\kappa_n r)]}{2\,\kappa_n\nu_n^2} d\alpha$$
$$- \int_{r_h}^{r_t} \frac{1}{w^{(0)}} \frac{\partial^2 h}{\partial\alpha\partial\beta} d\beta \sum_{1}^{\infty} \frac{\alpha[J_1(\kappa_n\alpha)Y_1(\kappa_n r_h) - J_1(\kappa_n r_h)Y_1(\kappa_n\alpha)] [J_1(\kappa_n r)Y_1(\kappa_n r_h) - J_1(\kappa_n r_h)Y_1(\kappa_n r)]}{2\kappa_n\nu_n^2} e^{-\kappa_n|z-\beta|} d\alpha$$
$$z > \beta \tag{4-26}$$

Again the complete solution for the blade row of finite chord is given by summing the above solution across the blade chord. For a point with any axial location z, solutions of the type given by Eq. 4-26 are summed over all values of $\beta < z$ and solutions of the type given by Eq. 4-25 for all

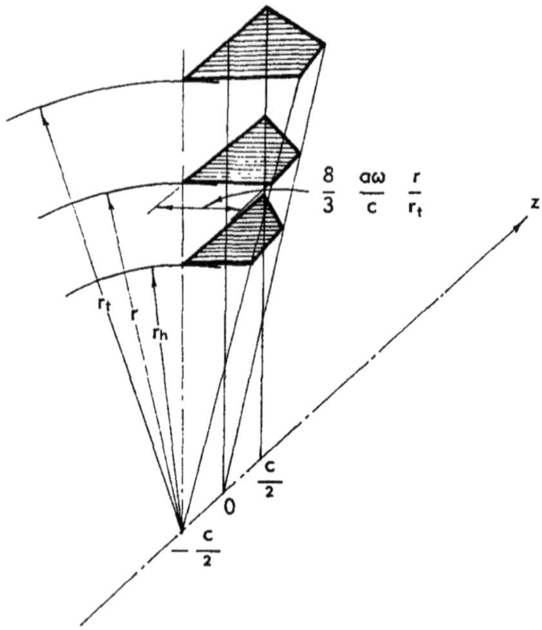

Fig. C,4f. Distribution of load on blade row of finite chord.

values of $\beta > z$. Therefore the complete solution is written as

$$w^{(1)} = \int_{r_h}^{r_t}\int_{-\infty}^{z} \frac{1}{w^{(0)}}\frac{\partial^2 h^{(1)}}{\partial\alpha\partial\beta} K(r, z; \alpha, \beta) d\alpha d\beta$$

$$- \int_{r_h}^{r_t}\int_{z}^{\infty} \frac{1}{w^{(0)}}\frac{\partial^2 h^{(1)}}{\partial\alpha\partial\beta} K(r, z; \alpha, \beta) d\alpha d\beta$$

$$+ 2\int_{r_h}^{r_t}\int_{-\infty}^{z} \frac{1}{w^{(0)}}\frac{\partial^2 h^{(1)}}{\partial\alpha\partial\beta} K(r, \beta; \alpha, \beta) d\alpha d\beta \quad (4\text{-}27)$$

where the function $K(r, z; \alpha, \beta)$ is defined as

$$\sum_{1}^{\infty} \frac{\alpha[J_1(\kappa_n\alpha)Y_1(\kappa_n r_h) - J_1(\kappa_n r_h)Y_1(\kappa_n\alpha)][J_0(\kappa_n r)Y_1(\kappa_n r_h) - J_1(\kappa_n r_h)Y_0(\kappa_n r)]\, e^{-\kappa_n|z-\beta|}}{2\,\kappa_n \nu_n^2} \quad (4\text{-}28)$$

These results complete the formal solution for radial and axial velocity perturbations induced by a blade row of finite chord. It is instructive to use the example employed for the actuator disk as an illustration of the present theory for finite blade chord so that the results may be compared. Therefore in addition to the radius ratio $r_h/r_t = 0.60$ consider a blade row of chord c with a chordwise load distribution as shown in Fig. C,4f. For

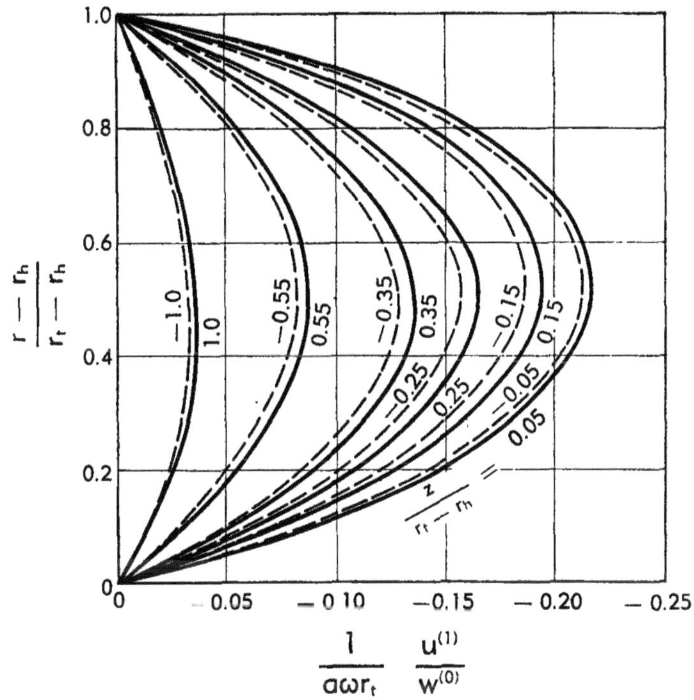

Fig. C,4g. Radial velocity patterns upstream, downstream, and within a blade row of finite chord, moving with angular velocity ω and loaded as indicated in Fig. C,4f, $r_h/r_t = 0.6$.

the first half of the blade row the value of $(1/w^{(0)})(\partial^2 h^{(1)}/\partial\alpha\partial\beta)$ is given by

$$\frac{1}{w^{(0)}} \frac{\partial^2 h^{(1)}}{\partial\alpha\partial\beta} = \frac{8}{3} a \frac{\omega}{c} \frac{\alpha}{r_t} \qquad -\frac{c}{2} \leqq \beta \leqq 0 \tag{4-29}$$

and for the second half of the blade row,

$$\frac{1}{w^{(0)}} \frac{\partial^2 h^{(1)}}{\partial\alpha\partial\beta} = \frac{8}{3} a \frac{\omega}{c} \frac{\alpha}{r_t} \left(1 - \frac{2\beta}{c}\right) \qquad 0 \leqq \beta \leqq \frac{c}{2} \tag{4-30}$$

The total work added by this distribution is identical with the total work added in the actuator disk discussed previously. The results may be obtained explicitly by integration of Eq. 4-23 and 4-27 employing the values of enthalpy given by Eq. 4-29 and 4-30. This operation involves only algebraic complexity and the formulas, because of their length, will not be quoted here. The radial velocity distributions are given in Fig. C,4g for axial stations within the blade row as well as both upstream and downstream from it. The slight asymmetry of the flow pattern caused by the unsymmetrical loading of the blade row is noticeable. Of somewhat more technical interest are the distributions of axial velocity perturbation shown in Fig. C,4h. For comparison the axial velocity perturbations for the actuator disk are shown on the same curve. The differences are certainly noticeable but not marked. Remembering then that the present results

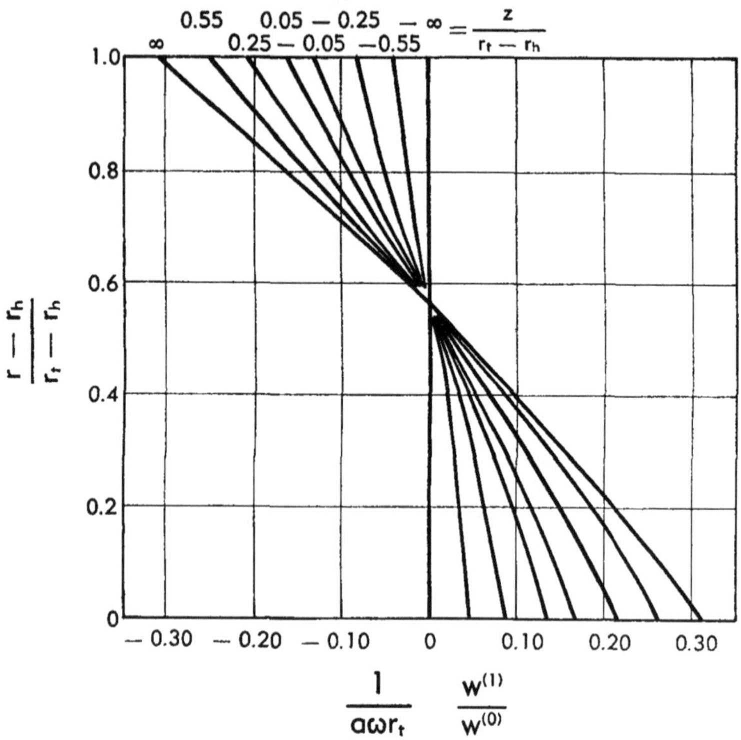

Fig. C,4h. Axial velocity perturbations upstream, downstream, and within a blade row of finite chord, moving with angular velocity ω and loaded as indicated in Fig. C,4f, $r_h/r_t = 0.6$.

hold for a blade whose ratio of length to axial extent is 2.0, it appears that for blades of high aspect ratio, such as those in the first few stages of current axial flow compressors, ordinary needs do not require the treatment of the finite blade chord. On the other hand, turbomachine blade rows with an aspect ratio of 1.0 or less do require consideration of the finite blade chord for accurate construction of velocity diagrams.

Effect of large tangential velocities. The discussion so far has been restricted to examples where the tangential velocities of the fluid were at most of the first order. This fact permitted simplification of the differential operator on the left-hand side of Eq. 4-1 and reduced the non-homogeneous term on the right-hand side to the very simple result given in Eq. 4-3. It is usually the case in multistage axial turbomachinery, however, that although the change in tangential velocity across a blade row is small, a zeroeth order tangential velocity $v^{(0)}$ exists which is of the same general magnitude as the undisturbed throughflow velocity $w^{(0)}$. The differential equation describing the radial velocity in this case is then given by

$$\frac{\partial^2 u^{(1)}}{\partial r^2} + \frac{1}{r}\frac{\partial u^{(1)}}{\partial r} + \left[\frac{1}{r^3 w^{(0)2}}\frac{\partial}{\partial r}(rv^{(0)})^2 - \frac{1}{r^2}\right] + \frac{\partial^2 u^{(1)}}{\partial z^2} = \frac{1}{w^{(0)}}\frac{\partial}{\partial z}\left[\frac{1}{r^2}\frac{\partial}{\partial r}(rv^{(0)}rv^{(1)})\right] - \frac{1}{w^{(0)}}\frac{\partial^2 h^{(1)}}{\partial r \partial z} \qquad (4\text{-}31)$$

The differential operator is complicated in an essential manner by introducing an additional variable coefficient of $u^{(1)}$; the addition of a term depending upon the tangential velocity perturbation introduces a new function on the right-hand side but does not really change the problem.

In the particular case when the zeroeth order tangential velocity is of the vortex type, that is $v^{(0)} \sim 1/r$, the term $rv^{(0)}$ is a constant and the expression $(1/r^3 w^{(0)2})\partial(rv^{(0)})^2/\partial r$ vanishes in Eq. 4-31. This leaves the differential operator, and hence the formal solution to the problem, exactly as it was for $v^{(0)} = 0$ except that now the inhomogeneous term of Eq. 4-31 replaces the simpler one of Eq. 4-3. A more interesting and significant example of large zeroeth order tangential velocity occurs when the tangential velocity is of the solid body type. Assume that the tangential velocity is given by

$$v^{(0)} = bw^{(0)}\frac{r}{r_t} \qquad (4\text{-}32)$$

where b is a numerical constant. The differential equation (Eq. 4-31) then becomes

$$\frac{\partial^2 u^{(1)}}{\partial r^2} + \frac{1}{r}\frac{\partial u^{(1)}}{\partial r} + \left[\left(\frac{2b}{r_t}\right)^2 - \frac{1}{r^2}\right]u^{(1)} + \frac{\partial^2 u^{(1)}}{\partial z^2} = b\frac{\partial}{\partial z}\left[\frac{1}{r^2 r_t}\frac{\partial}{\partial r}(r^3 v^{(1)})\right] - \frac{1}{w^{(0)}}\frac{\partial^2 h^{(1)}}{\partial r \partial z} \qquad (4\text{-}33)$$

It should be noticed in particular that the additional coefficient on the left-hand side multiplying $u^{(1)}$ is a constant $(2b/r_t)^2$. The solution may be worked out in very much the same way as it was for Eq. 4-3. It is readily determined that the radial velocity for an actuator disk located at $z = 0$ is given by

$$u^{(1)} = \sum_1^\infty C_n[J_1(\kappa_n r)Y_1(\kappa_n r_h) - J_1(\kappa_n r_h)Y_1(\kappa_n r)]e^{-\lambda_n|z|} \qquad (4\text{-}34)$$

where now the κ_n are still the roots of Eq. 4-11 and the λ_n are given by

$$\lambda_n = \sqrt{\kappa_n^2 - \left(\frac{2b}{r_t}\right)^2} \tag{4-35}$$

The constants C_n are given by

$$C_n = -\int_{r_h}^{r_t} \frac{\alpha(J_1(\kappa_n\alpha)Y_1(\kappa_n r_h) - J_1(\kappa_n r_h)Y_1(\kappa_n\alpha)}{2\lambda_n\nu_n^2 w^{(0)}} \left[\frac{b}{\alpha^2 r_t}\frac{\partial}{\partial\alpha}(\alpha v^{(1)}) - \frac{\partial h^{(1)}}{\partial\alpha}\right] d\alpha \tag{4-36}$$

The modification to the constants C_n is obvious inasmuch as it simply makes use of the new inhomogeneous term appearing in Eq. 4-33. The significant change is the modification of the exponents $e^{-\lambda_n|z|}$ appearing in the solution, Eq. 4-34. In aircraft gas turbine practice it is almost invariably true that $|2b/r_t| < \kappa_1$ and consequently all of the λ_n are real and nonvanishing. Thus the effect of a solid body rotation imposed upon the fluid far upstream of the blade row is to cause any disturbances generated by the blade row to decay more slowly upstream and downstream of the blade row than they would in the absence of a zeroeth order rotation. This effect is particularly pronounced upon the lower Bessel components corresponding to κ_1, κ_2, and κ_3 since the exponents associated with them are reduced in magnitude proportionally more than the higher Bessel components.

After having observed the nature of the modification introduced by solid body rotation it is simple to complete the solution for both the actuator disk and the blade of finite chord. For the actuator disk the axial velocity distribution is given by

$$w^{(1)} = \sum_1^\infty \frac{C_n\kappa_n}{\lambda_n}[J_0(\kappa_n r)Y_1(\kappa_n r_h) - J_1(\kappa_n r_h)Y_0(\kappa_n r)]e^{\lambda_n z} \qquad z < 0 \tag{4-37}$$

and

$$w^{(1)} = -2\sum_1^\infty \frac{C_n\kappa_n}{\lambda_n}[J_0(\kappa_n r)Y_1(\kappa_n r_h) - J_1(\kappa_n r_h)Y_0(\kappa_n r)] + \sum_1^\infty \frac{C_n\kappa_n}{\lambda_n}[J_0(\kappa_n r)Y_1(\kappa_n r_h) - J_1(\kappa_n r_h)Y_0(\kappa_n r)]e^{-\lambda_n z} \quad z > 0 \tag{4-38}$$

These demonstrate results that were intuitively clear from the previous calculation of radial velocity distribution. The over-all perturbation to the axial velocity, from far upstream of the blade row to a point far downstream, is changed because the values of $C_n\kappa_n/\lambda_n$ entering into the term

$$-2\sum_{1}^{\infty}\frac{C_n\kappa_n}{\lambda_n}\left[J_0(\kappa_n r)Y_1(\kappa_n r_h) - J_1(\kappa_n r_h)Y_0(\kappa_n r)\right]$$

are changed due to the function $(b/r^2r_t)\partial/\partial r(rv^{(1)})$ that enters into the integral for C_n given by Eq. 4-36. Referring still to Eq. 4-36, the term $[\partial h^{(1)}/\partial r]$ is a measure of the tangential vorticity generated at the actuator disk due to actual work being done by the blade row. The new term $(b/r^2r_t)\partial/\partial r(rv^{(1)})$ is of different origin. It arises from the fact that some of the axial vorticity, associated with the solid body rotation, is turned by the perturbation tangential velocity so that it becomes tangential vorticity. The second result is that this change of axial velocity profile starts farther upstream and completes farther downstream in the presence of solid body rotation than it does when no zeroeth order tangential velocity is present.

The resulting radial and axial velocities for a blade row of finite chord can be obtained immediately through simple modification of Eq. 4-23, 4-24, 4-27, and 4-28. Wherever the term

$$\frac{1}{w^{(0)}}\frac{\partial^2 h^{(1)}}{\partial r\partial z}$$

appears, it should be replaced by the term

$$\frac{1}{w^{(0)}}\frac{\partial}{\partial z}\left[-\frac{bw^{(0)}}{r^2r_t}\frac{\partial}{\partial r}(r^3v^{(1)}) + \frac{\partial h^{(1)}}{\partial r}\right]$$

wherever the characteristic value κ_n appears in the exponent $e^{-\kappa_n|z-\beta|}$, it should be replaced by λ_n to give $e^{-\lambda_n|z-\beta|}$ where λ_n is defined through Eq. 4-35. The gross results discussed above for the actuator disk apply also to the blade of finite chord. These effects are generally the same, in fact, whether the zeroeth order tangential velocity is of the solid body type or whether it is different. The particular example used is especially significant, however, since a mean solid body rotation is so frequently employed in axial compressors. Methods for treating more general distributions of zeroeth order tangential velocity will be deferred until later when the appropriate asymptotic expansions are discussed.

The entrance vane. Second order theory. The results discussed so far permit adequate treatment of every blade row in an axial turbomachine with the exception of the entrance vanes or guide vanes. Because the guide vanes are a stationary blade row the perturbation enthalpy $h^{(1)}$ vanishes and since there is no zeroeth order tangential motion upstream of the guide vanes, the term $rv^{(0)}$ vanishes also. Consequently the right-hand side of Eq. 4-31 vanishes so that the radial velocity and axial velocity perturbation vanish identically. The first order perturbation theory is inadequate therefore to deal with the guide vane problem. A second order calculation must be made, therefore, explicitly for the purposes of the guide vane. The fact that the radial velocity and axial velocity perturbation are of the second order carries no implication that these effects are small. Actually the effects are as large as or larger than those induced by ordinary rotor or stator blades; the fact that they show up as second order effects is due, as will be seen, to the quadratic nature of the perturbation term. It is also an indication that, since the change in tangential velocity across the guide vane is considerably larger than that across rotor or stator blade rows, the guide vane problem is not treated with the same accuracy of approximation by the first order analysis as are the other blade rows.

To investigate the guide vane flow, restrict conditions to a uniform zeroeth order axial velocity, $w^{(0)} = \text{const}$, along with the previous assumptions of cylindrical inner and outer boundaries, no losses, and incompressible fluid. From calculations of the previous section it is clear that the first order radial velocity and axial velocity disturbance vanish identically, that is,

$$\begin{aligned} u^{(1)} &\equiv 0 \\ w^{(1)} &\equiv 0 \end{aligned} \tag{4-39}$$

in spite of the fact that $v^{(1)} \neq 0$ and may, in fact, be rather large. The flow field is defined again through Eq. 2-1, 2-2, 2-3, and 2-4 and for purposes of the perturbation we call

$$\begin{aligned} u &= \epsilon^2 u^{(2)} \\ v &= \epsilon v^{(1)} + \epsilon^2 v^{(2)} \\ w &= w^{(0)} + \epsilon^2 w^{(2)} \\ p &= p^{(0)} + \epsilon p^{(1)} + \epsilon^2 p^{(2)} \\ F_r &= \epsilon f_r \\ F_\theta &= \epsilon f_\theta \\ F_z &= \epsilon f_z \end{aligned} \tag{4-40}$$

where account has been taken of the fact that the first order perturbation of radial and axial velocity vanish identically. Substituting into the equations of motion yields both first and second order parts. In the axial direction the first order relation is

$$\frac{1}{\rho}\frac{\partial p^{(1)}}{\partial r} = f_r \tag{4-41}$$

whereas the second order part is

$$w^{(0)}\frac{\partial u^{(2)}}{\partial z} - \frac{v^{(1)2}}{r} = -\frac{1}{\rho}\frac{\partial p^{(2)}}{\partial r} \tag{4-42}$$

From the tangential equilibrium relation the first and second order parts are respectively

$$w^{(0)} \frac{\partial}{\partial z} (rv^{(1)}) = rf_\theta \tag{4-43}$$

and

$$w^{(0)} \frac{\partial}{\partial z} (rv^{(2)}) = 0 \tag{4-44}$$

In the axial direction the first and second order parts are respectively

$$\frac{1}{\rho} \frac{\partial p^{(1)}}{\partial z} = f_z \tag{4-45}$$

and

$$w^{(0)} \frac{\partial w^{(2)}}{\partial z} = -\frac{1}{\rho} \frac{\partial p^{(2)}}{\partial z} \tag{4-46}$$

It is interesting to note that the radial and axial force field produces a first order pressure field within the blade row itself but no radial or axial velocity field. Moreover the tangential velocity distribution is entirely accounted for by the first order relation, Eq. 4-43, and the second order part $v^{(2)}$ vanishes identically according to Eq. 4-44 and the initial condition that the tangential velocity vanishes identically ahead of the guide vane. The second order velocity field is then defined by Eq. 4-42 and 4-46 in addition to the continuity equation

$$\frac{\partial u^{(2)}}{\partial r} + \frac{u^{(2)}}{r} + \frac{\partial w^{(2)}}{\partial z} = 0 \tag{4-47}$$

It is a simple matter to eliminate the second order pressure $p^{(2)}$ and axial velocity perturbation to give

$$\frac{\partial^2 u^{(2)}}{\partial r^2} + \frac{1}{r} \frac{\partial u^{(2)}}{\partial r} - \frac{u^{(2)}}{r^2} + \frac{\partial^2 u^{(2)}}{\partial z^2} = \frac{1}{w^{(0)}} \frac{\partial}{\partial z} \left(\frac{v^{(1)2}}{r} \right) \tag{4-48}$$

where the tangential velocity distribution $v^{(1)}$ is either prescribed or is known from prescribed tangential blade force or blade shape. The mathematical problem is therefore exactly that treated in the solution of Eq. 4-3 where now the known function

$$\frac{1}{w^{(0)}} \frac{\partial}{\partial z} \left(\frac{v^{(1)2}}{r} \right)$$

replaces the term

$$-\frac{1}{w^{(0)}} \frac{\partial}{\partial z} \left(\frac{\partial h^{(1)}}{\partial r} \right)$$

The corresponding solutions for the actuator disk and for the blade row of finite chord may be carried over directly with the above substitution.

For example consider a guide vane, approximated by an actuator disk, which imparts a solid body rotation to the fluid. The jump in tangential velocity across the blade row may then be written in the form,

$$[v^{(1)}] = bw^{(0)}\frac{r}{r_t} \tag{4-49}$$

The term of interest here, corresponding to the right-hand side of Eq. 4-8, is

$$\frac{1}{w^{(0)}}\left[\frac{v^{(1)2}}{r}\right] = \frac{b^2w^{(0)}r}{r_t^2} \tag{4-50}$$

so that the equation for the coefficient C_n to be employed in expansions (Eq. 4-9, 4-10, 4-17, and 4-18) may be written as

$$C_n = \frac{-b^2w^{(0)}}{2\kappa_n\nu_n^2r_t^2}\int_{r_h}^{r_t}\alpha^2[J_1(\kappa_n\alpha)Y_1(\kappa_nr_h) - J_1(\kappa_nr_h)Y_1(\kappa_n\alpha)]d\alpha \tag{4-51}$$

These coefficients differ only in the multiplicative constant from the example of a rotating blade row treated earlier, the results of which are shown in Fig. C,4c and C,4d. Comparison with the appropriate coefficients of that example, given by Eq. 4-21, shows that by replacing $a\omega r_t$ in Fig. C,4c and C,4d by the product $-b^2w^{(0)}$ from the present problem, the curves in these figures apply equally well to the present example of a guide vane. The radial and axial velocity distortions are therefore similar in the two situations but of opposite sign. Exactly the same parallel exists between the solutions of Eq. 4-48 for a continuous blade row imparting solid body rotation and detailed solutions for a rotating blade row given in Fig. C,4g and C,4h.

Asymptotic expansion of the Bessel functions. In a great number of instances the hub ratio r_h/r_t is sufficiently large that the characteristic values κ_n, the roots of Eq. 4-11, may be approximated by their asymptotic representations with good accuracy. This technique offers considerable simplicity in the calculation of some complicated summations involved.

The asymptotic representations of the Bessel function for large values of the argument κ_nr may be written as

$$J_1(\kappa_nr) = \frac{\sin(\kappa_nr) - \cos(\kappa_nr)}{\sqrt{\pi\kappa_nr}}\left[1 + \frac{15}{128(\kappa_nr)^2} + \cdots\right] - \frac{\sin(\kappa_nr) + \cos(\kappa_nr)}{\sqrt{\pi\kappa_nr}}\left[\frac{3}{8(\kappa_nr)} + \cdots\right] \tag{4-52}$$

and

$$Y_1(\kappa_nr) = -\frac{\sin(\kappa_nr) + \cos(\kappa_nr)}{\sqrt{\pi\kappa_nr}}\left[1 - \frac{15}{128(\kappa_nr)^2} + \cdots\right] + \frac{\sin(\kappa_nr) - \cos(\kappa_nr)}{\sqrt{\pi\kappa_nr}}\left[\frac{3}{8(\kappa_nr)} + \cdots\right] \tag{4-53}$$

Furthermore the Bessel functions of zero order may be written in asymptotic form

$$J_0(\kappa_n r) = \frac{\cos(\kappa_n r) + \sin(\kappa_n r)}{\sqrt{\pi\kappa_n r}}\left[1 + \frac{1}{128(\kappa_n r)^2} + \cdots\right] + \frac{\sin(\kappa_n r) - \cos(\kappa_n r)}{\sqrt{\pi\kappa_n r}}\left[\frac{1}{8\kappa_n r} + \cdots\right] \tag{4-54}$$

$$Y_0(\kappa_n r) = \frac{\sin(\kappa_n r) - \cos(\kappa_n r)}{\sqrt{\pi\kappa_n r}}\left[1 + \frac{1}{128(\kappa_n r)^2} + \cdots\right] - \frac{\cos(\kappa_n r) + \sin(\kappa_n r)}{\sqrt{\pi\kappa_n r}}\left[\frac{1}{8\kappa_n r} + \cdots\right] \tag{4-55}$$

In general it will be sufficient to retain only the terms in the brackets that are independent of the argument. On this basis the function that occurs in the solution for the radial velocity may be written

$$J_1(\kappa_n r)Y_1(\kappa_n r_h) - J_1(\kappa_n r_h)Y_1(\kappa_n r) \approx \frac{-2}{\pi\kappa_n\sqrt{rr_h}}\sin\kappa_n(r - r_h) \tag{4-56}$$

The characteristic values for the asymptotic calculations are then, according to Eq. 4-11, given by

$$\frac{-2}{\pi\kappa_n\sqrt{r_t r_h}}\sin\kappa_n(r_t - r_h) = 0 \tag{4-57}$$

which then establishes them as

$$\kappa_n \approx \frac{n\pi}{r_t - r_h} \tag{4-58}$$

The accuracy with which these characteristic values check the true roots of Eq. 4-11 is a reasonable measure of the accuracy that can be expected from the asymptotic solution. Consequently the values of the characteristic numbers $\kappa_n r_h$ are tabulated below for the first 10 roots calculated numerically and estimated from the asymptotic result for the radius ratio $r_h/r_t = 0.6$. For all practical purposes the asymptotic values are identical

n	$\kappa_n r_h$	$\frac{n\pi}{r_t/r_h - 1}$
1	4.758051	4.712389
2	9.448369	9.424778
3	14.182998	14.137167
4	18.861456	18.849556
5	23.571475	23.561945
6	28.282281	28.274334
7	32.993535	32.986723
8	37.705076	37.699113
9	42.416800	42.411502
10	47.128604	47.123891

with the exact characteristic numbers for $n > 4$. Furthermore the error involved in employing the asymptotic values entirely is not very large and usually well within the error made in applying the throughflow analysis in the first place. In this case the functions for the radial velocity distribution may be written

$$J_1(\kappa_n r)Y_1(\kappa_n r_h) - J_1(\kappa_n r_h)Y_1(\kappa_n r) \approx \frac{-2\left(\frac{r_t}{r_h} - 1\right)}{\sqrt{\frac{r}{r_t}}} \sin n\pi\left(\frac{r - r_h}{r_t - r_h}\right) \tag{4-59}$$

Using an argument that is exactly similar, the functions appropriate to the axial velocity perturbation may be written asymptotically as

$$J_0(\kappa_n r)Y_1(\kappa_n r_h) - J_1(\kappa_n r_h)Y_0(\kappa_n r) = \frac{-2\left(\frac{r_t}{r_h} - 1\right)}{\sqrt{\frac{r}{r_h}}} \cos n\pi\left(\frac{r - r_h}{r_t - r_h}\right) \tag{4-60}$$

The use of these asymptotic functions simplifies the numerical calculation of a given problem in two ways. First, the actual determination of the characteristic values is made simpler and the evaluation of the functions is trivial. Second, the infinite series involved may often be summed directly to give a closed result. For example, the function $G(r, z; \alpha, \beta)$ appearing within the integral giving the radial velocity for a blade row of finite chord, Eq. 4-23 and 4-24, may be written asymptotically as

$$G(r, z; \alpha, \beta) \approx \frac{1}{8\pi r}\sqrt{\frac{\alpha r}{r_h}} \ln\left[\frac{\cosh \pi\left(\frac{z - \beta}{r_t - r_h}\right) + \cos \pi\left(\frac{\alpha - r}{r_t - r_h}\right)}{\cosh \pi\left(\frac{z - \beta}{r_t - r_h}\right) + \cos \pi\left(\frac{\alpha - r_h}{r_t - r_h} + \frac{r - r_h}{r_t - r_h}\right)}\right] \tag{4-61}$$

This demonstrates the fact, which was intuitively clear before, that the function $G(r, z; \alpha, \beta)$ has a logarithmic singularity at $\alpha = r$, $z = \beta$. This is equivalent to the statement that an annular vortex ring at the point α, β induces infinite radial velocities in the immediate neighborhood of the ring. In a similar manner the function $K(r, z; \alpha, \beta)$ involved in calculation of the axial velocity perturbation, Eq. 4-29 and 4-30, yields the asymptotic expression

$$K(r, z; \alpha, \beta) \approx \frac{1}{8\pi r}\sqrt{\frac{\alpha r}{r_h}} \ln\left\{\left[\cosh\left(\frac{z - \beta}{r_t - r_h}\right) + \cos \pi\left(\frac{\alpha - r}{r_t - r_h}\right)\right]\left[\cosh \pi\left(\frac{z - \beta}{r_t - r_h}\right) + \cos \pi\left(\frac{\alpha - r_h}{r_t - r_h} + \frac{r - r_h}{r_t - r_h}\right)\right]\right\} \tag{4-62}$$

In exactly the same fashion as the radial velocity, the axial velocity perturbation induced by an annular vortex element is logarithmically infinite in the neighborhood of the element. Eq. 4-61 and 4-62 are extremely useful in problems that require consideration of a blade row of finite chord. Using these results the integrands of Eq. 4-23 and 4-27 may be evaluated algebraically and the integration carried out by conventional numerical means, insuring the appropriate treatment of the logarithmic singularity.

For many examples the series involved in a calculation cannot be summed directly but they can be placed in a convenient closed form for calculation. Consider, for instance, the actuator disk problem which was solved previously; the radial velocity is given by Eq. 4-9 and 4-10, and the coefficients are given explicitly by Eq. 4-21 et seq. Formal substitution of the coefficient into Eq. 4-9 and 4-10 yields a series which converges as $1/n^2$ but which cannot be summed directly into a closed form. The first derivative of this series can be summed easily, however, so that the radial velocity component appears as a definite integral. After carrying out these processes and some simplification, the radial velocity becomes

$$u^{(1)} = -\frac{a\omega}{\pi}\sqrt{\frac{r_h}{r}}\left(\frac{r_h}{r_t}+\sqrt{\frac{r_t}{r_h}}\right)\int_{r_h}^{r_t}\ln\left[1+e^{\frac{-2\pi|z|}{r_t-r_h}} + 2e^{\frac{-\pi|z|}{r_t-r_h}}\cos\pi\left(\frac{\xi-r_h}{r_t-r_h}\right)\right]d\xi \quad (4\text{-}63)$$

The integrand is regular within the domain of interest and elementary numerical integration may be used to determine the radial velocity at any point of the field. The asymptotic representations for the Bessel functions permit convenient computational procedures of this sort in many instances.

C,5. Solutions for Variable Hub and Tip Radii. It is nearly always the case in actual turbomachines that hub radius, tip radius or both vary along the direction of flow. In many instances, for example in the early stages of a multistage compressor, the hub and tip radii vary so much that the change of radius through a given blade row must be taken into account in determining the throughflow. It is true that a significant portion of this change in radius may be to compensate for density changes in the fluid which, in the present section, are being neglected. However, the general flow pattern is not so greatly changed by this compressibility effect but that the incompressible flow patterns give most of the necessary information.

In order that the conditions of linearization be satisfied it is necessary that the distortion of the hub and tip contours induce only first order radial velocities. Since the axial velocity is of zeroeth order this restriction requires that the slopes of the surfaces be of order ϵ at the most. Then it is clear that the problem of throughflow with variable hub and tip radii may be stated

$$\frac{\partial^2 u^{(1)}}{\partial r^2} + \frac{1}{r}\frac{\partial u^{(1)}}{\partial r} - \frac{1}{r^2}\left[1 - \frac{1}{rw^{(0)2}}\frac{\partial(rv^{(0)})^2}{\partial r}\right] + \frac{\partial^2 u^{(1)}}{\partial z^2} = \phi(r, z) \quad (5\text{-}1)$$

with the boundary conditions

$$u^{(1)}(r, -\infty) = u^{(1)}(r, \infty) = 0 \quad (5\text{-}2)$$

and

$$\frac{u^{(1)}}{w^{(0)}} = \frac{dr_t}{dz} \quad \text{on} \quad r = r_t \quad (5\text{-}3)$$

$$\frac{u^{(1)}}{w^{(0)}} = \frac{dr_h}{dz} \quad \text{on} \quad r = r_h \quad (5\text{-}4)$$

where the quantities $r_t(z)$ and $r_h(z)$ refer to the variable hub and tip radii given by Eq. 3-34 and 3-35. Since the problem is linear it can be solved in two parts. The solution of the inhomogeneous differential equation with homogeneous boundary conditions, $u^{(1)}(r_h, z) = u^{(1)}(r_t, z) = 0$, is the first part and this was discussed in the section covering constant hub and tip radii. The second part is the solution of the homogeneous differential equation:

$$\frac{\partial^2 u^{(1)}}{\partial^2 r^2} + \frac{1}{r}\frac{\partial u^{(1)}}{\partial r} - \frac{1}{r^2}\left[1 - \frac{1}{rw^{(0)2}}\frac{\partial(rv^{(0)})^2}{\partial r}\right] u^{(1)} + \frac{\partial^2 u^{(1)}}{\partial z^2} = 0 \quad (5\text{-}5)$$

with inhomogeneous boundary conditions given in Eq. 5-2, 5-3, and 5-4. The sum of these two solutions satisfies both the inhomogeneous right-hand side of Eq. 5-1 and the inhomogeneous boundary conditions.

Throughflow with variable hub radius. Consider the particular instance where $dr_t/dz = 0$ and $dr_h/dz = f_h(z)$ in Eq. 5-3, 5-4 above. Assume furthermore that the zeroeth order tangential velocity $v^{(0)}$ is either zero or of the vortex type. The most convenient technique of solution is the Fourier transform with respect to z, the axial direction. Denote

$$U(r, k) = \frac{1}{\sqrt{2\pi}}\int_{-\infty}^{\infty} u^{(1)}(r, z)e^{-ikz}dz \quad (5\text{-}6)$$

the Fourier transform of the radial velocity component. Then the homogeneous differential equation (Eq. 5-5) becomes

$$\frac{d^2U}{dr^2} + \frac{1}{r}\frac{dU}{dr} - \left(k^2 + \frac{1}{r^2}\right)U = 0 \quad (5\text{-}7)$$

while the boundary conditions are

$$U(r_t, k) = 0 \quad (5\text{-}8)$$

$$U(r_h, k) = \frac{w^{(0)}}{\sqrt{2\pi}}\int_{-\infty}^{\infty} f_h(z)e^{-ikz}dz \equiv w^{(0)}F_h(k) \quad (5\text{-}9)$$

Now the differential equation (Eq. 5-7) has solutions $J_1(ikr)$ and $Y_1(ikr)$. By employing the Fourier inversion theorem it is not difficult to show that the radial velocity induced by the hub radius variation is

$$\frac{u^{(1)}(r, z)}{w^{(0)}} = \frac{1}{\sqrt{2\pi}} \int_{-\infty}^{\infty} F_{\rm h}(k) \frac{J_1(ikr)Y_1(ikr_{\rm t}) - J_1(ikr_{\rm t})Y_1(ikr)}{J_1(ikr_{\rm h})Y(ikr_{\rm t}) - J_1(ikr_{\rm t})Y_1(ikr_{\rm h})} e^{ikz} dk \tag{5-10}$$

Explicit solutions for particular forms of the hub shape are obtained by evaluation of the integral in Eq. 5-10, most generally this involves contour integration.

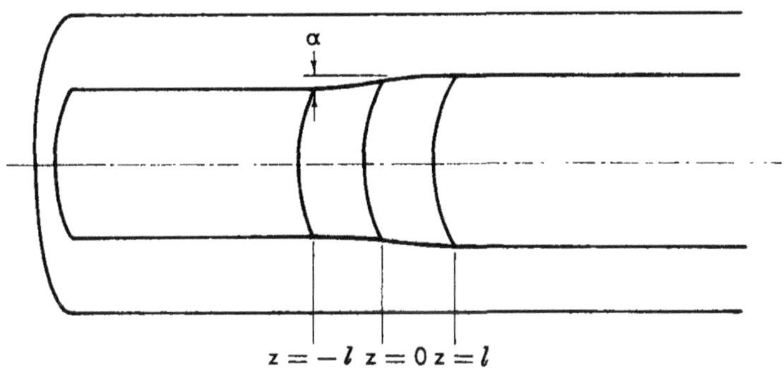

Fig. C,5a. Axial turbomachine with variable hub radius.

Consider a particular example where the hub slope is given by

$$\begin{aligned} f_{\rm h}(z) &= \frac{\pi r_{\rm h}}{2l} \cos \frac{\pi z}{2l}; \qquad |z| \leqq l \\ f_{\rm h}(z) &= 0; \qquad |z| \geqq l \end{aligned} \tag{5-11}$$

as shown in Fig. C,5a. Then the Fourier transform of the boundary shape yields

$$F_{\rm h}(k) = \left(\frac{r_{\rm h}}{l}\right) \frac{1}{l} \left(\frac{\pi}{2}\right)^{\frac{1}{2}} \frac{\cos kl}{(\pi/2l)^2 - k^2} \tag{5-12}$$

In anticipation of employing the technique of contour integration to evaluate the radial velocity component, it is worthwhile to note that the integral in Eq. 5-10 may be rewritten in terms of the Hankel functions $H_1^{(1)}(ikr)$ and $H_1^{(2)}(ikr)$ of the first and second kind to give

$$\frac{u^{(1)}(r, z)}{w^{(0)}} = \frac{r_{\rm h}}{l} \frac{\pi}{4l} \int_{-\infty}^{\infty} \frac{\cos kl}{(\pi/2l)^2 - k^2} \left[\frac{H_1^{(1)}(ikr)H_1^{(2)}(ikr_{\rm t}) - H_1^{(1)}(ikr_{\rm t})H_1^{(2)}(ikr)}{H_1^{(1)}(ikr_{\rm h})H_1^{(2)}(ikr_{\rm t}) - H_1^{(1)}(ikr_{\rm t})H_1^{(2)}(ikr_{\rm h})} \right] e^{ikz} dk \tag{5-13}$$

But if the complex variable σ is denoted $\sigma = \zeta + ik$, then the complex integral to consider is

$$\frac{-r_h}{l}\frac{\pi i}{4l}\int \frac{\cosh \sigma l}{(\pi/2l)^2+\sigma^2}\left[\frac{H_1^{(1)}(\sigma r)H_1^{(2)}(\sigma r_t) - H_1^{(1)}(\sigma r_t)H_1^{(2)}(\sigma r)}{H_1^{(1)}(\sigma r_h)H_1^{(2)}(\sigma r_t) - H_1^{(1)}(\sigma r_t)H_1^{(2)}(\sigma r_h)}\right] e^{\sigma z}d\sigma \tag{5-14}$$

and it is necessary to evaluate this over the imaginary axis. Now the integral has poles at $\sigma = \pm i\pi/2l$ and at the roots of the denominator

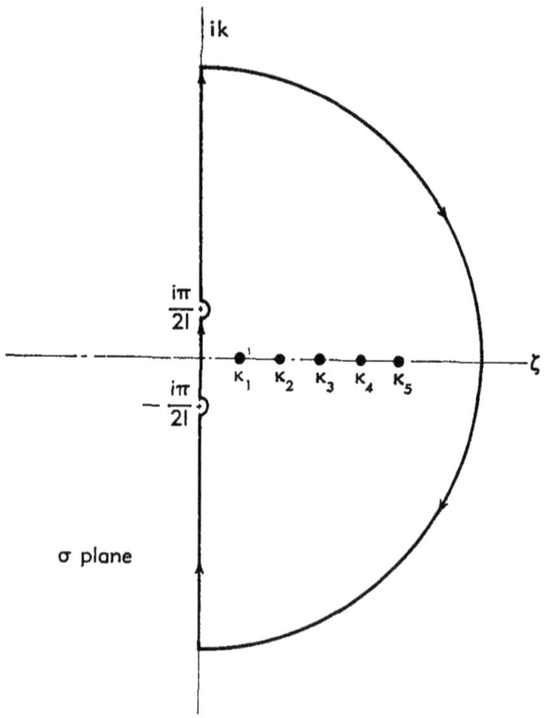

Fig. C,5b. Contour for evaluation of radial velocity integral, $z \leqq -l$.

$H_1^{(1)}(\sigma r_h)H_1^{(2)}(\sigma r_t) - H_1^{(1)}(\sigma r_t)H_1^{(2)}(\sigma r_h) = 0$. There are an infinite number of these roots on the positive real axis; they are in fact just the roots κ_n of

$$J_1(\kappa_n r_h)Y_1(\kappa_n r_t) - J_1(\kappa_n r_t)Y_1(\kappa_n r_h) = 0$$

which occurred previously in the problem for constant hub and tip radii. Using the contour of Fig. C,5b indented about the two singularities on the imaginary axis, the integral along the imaginary axis may be evaluated in terms of the residues on the real axis provided the integral along the arc vanishes. This condition is assured when $z < -l$, that is, upstream of the wall curvature. Evaluation of the residues gives the value of the definite integral in Eq. 5-13 to be

$$\frac{u^{(1)}}{w^{(0)}} = -\frac{r_h}{l}\frac{\pi^2}{2l}$$
$$\sum_1^\infty \frac{\cosh \kappa_n l}{(\pi/2l)^2 + \kappa_n^2} e^{\kappa_n z} \left\{\frac{J_1(\kappa_n r)Y_1(\kappa_n r_t) - J_1(\kappa_n r_t)Y_1(\kappa_n r)}{\begin{array}{l} r_h[J_0(\kappa_n r_h)Y_1(\kappa_n r_t) - J_1(\kappa_n r_t)Y_0(\kappa_n r_h)] \\ + r_t[J_1(\kappa_n r_h)Y_0(\kappa_n r_t) - J_0(\kappa_n r_t)Y_1(\kappa_n r_h)]\end{array}}\right\} \quad (5\text{-}15)$$

In exactly the same manner, with the exception that the arc of the contour is drawn in the left half plane, the solution for $z > l$ may be written

$$\frac{u^{(1)}}{w^{(0)}} = -\frac{r_h}{l}\frac{\pi^2}{2l}$$
$$\sum_1^\infty \frac{\cosh \kappa_n l}{(\pi/2l)^2 + \kappa_n^2} e^{-\kappa_n z} \left\{\frac{J_1(\kappa_n r)Y_1(\kappa_n r_t) - J_1(\kappa_n r_t)Y_1(\kappa_n r)}{\begin{array}{l} r_h[J_0(\kappa_n r_h)Y_1(\kappa_n r_t) - J_1(\kappa_n r_t)Y_0(\kappa_n r_h)] \\ + r_t[J_1(\kappa_n r_h)Y_0(\kappa_n r_t) - J_0(\kappa_n r_t)Y_1(\kappa_n r_h)]\end{array}}\right\} \quad (5\text{-}16)$$

To find an appropriate solution in the region $-l \leqq z \leqq l$ requires somewhat more consideration. The complex integral

$$I_1 = -\frac{r_h}{l}\frac{\pi i}{4l}$$
$$\int \frac{(\cosh \sigma l + \sinh \sigma l)}{(\pi/2l)^2 + \sigma^2}\left[\frac{H_1^{(1)}(\sigma r)H_1^{(2)}(\sigma r_t) - H_1^{(1)}(\sigma r_t)H_1^{(2)}(\sigma r)}{H_1^{(1)}(\sigma r_h)H_1^{(2)}(\sigma r_t) - H_1^{(1)}(\sigma r_t)H_1^{(2)}(\sigma r_h)}\right] e^{\sigma z} \quad (5\text{-}17)$$

converges when evaluated over a contour consisting of the entire imaginary axis and large semicircle in the right half plane, provided that $z > -l$. Similarly the integral

$$I_2 = -\frac{r_h}{l}\frac{\pi i}{4l}$$
$$\int \frac{(\cosh \sigma l - \sinh \sigma l)}{(\pi/2l)^2 + \sigma^2}\left[\frac{H_1^{(1)}(\sigma r)H_1^{(2)}(\sigma r_t) - H_1^{(1)}(\sigma r_t)H_1^{(2)}(\sigma r)}{H_1^{(1)}(\sigma r_h)H_1^{(2)}(\sigma r_t) - H_1^{(1)}(\sigma r_t)H_1^{(2)}(\sigma r_h)}\right] e^{\sigma z} \quad (5\text{-}18)$$

converges when evaluated over a contour with its large semicircle in the left half plane provided $z < l$. These two integrals have the common region of convergence $-l \leqq z \leqq l$ and half of their sum is equal to the integral required, that is $u^{(1)}/w^{(0)} = \frac{1}{2}(I_1 + I_2)$. Carrying out the evaluation in detail gives the radial velocity distribution

$$\frac{u^{(1)}(r, z)}{w^{(0)}} = \frac{\pi r_h}{2l}\left[\frac{I_1\left(\frac{\pi r}{2l}\right)K_1\left(\frac{\pi r_t}{2l}\right) - I_1\left(\frac{\pi r_t}{2l}\right)K_1\left(\frac{\pi r}{2l}\right)}{I_1\left(\frac{\pi r_h}{2l}\right)K_1\left(\frac{\pi r_t}{2l}\right) - I_1\left(\frac{\pi r_t}{2l}\right)K_1\left(\frac{\pi r_h}{2l}\right)}\right]\cos\frac{\pi z}{2l}$$
$$-\frac{r_h}{l}\frac{\pi^2}{2l}\sum_1^\infty \frac{\cosh \kappa_n z}{(\pi/2l)^2 + \kappa_n^2} e^{-\kappa_n l}\left\{\frac{J_1(\kappa_n r)Y_1(\kappa_n r_t) - J_1(\kappa_n r_t)Y_1(\kappa_n r)}{\begin{array}{l} r_h[J_0(\kappa_n r_h)Y_1(\kappa_n r_t) - J_1(\kappa_n r_t)Y_0(\kappa_n r_h)] \\ + r_t[J_1(\kappa_n r_h)Y_0(\kappa_n r_t) - J_0(\kappa_n r_t)Y_1(\kappa_n r_h)]\end{array}}\right\}$$
(5-19)

It is an elementary matter to develop the corresponding relationship for the axial velocity perturbations. From the continuity equation it follows that

$$\frac{w^{(1)}}{w^{(0)}} = -\int_{-\infty}^{z} \frac{1}{r}\frac{\partial}{\partial r}\left(\frac{ru^{(1)}}{w^{(0)}}\right) dz \tag{5-20}$$

where the appropriate representation of $u^{(1)}/w^{(0)}$ must be employed in each of the three regions. Assuming the axial velocity undistorted far upstream of the hub curvature, the axial velocity perturbation in the region $-\infty \leqq z \leqq -l$ may be written, using Eq. 5-15, as

$$\frac{w^{(1)}}{w^{(0)}} = \frac{r_h}{l}\frac{\pi^2}{2l} \sum_{1}^{\infty} \frac{\cosh \kappa_n l}{(\pi/2l)^2 + \kappa_n^2} e^{\kappa_n z} \left\{ \frac{J_0(\kappa_n r)Y_1(\kappa_n r_t) - J_1(\kappa_n r_t)Y_0(\kappa_n r)}{r_h[J_0(\kappa_n r_h)Y_1(\kappa_n r_t) - J_1(\kappa_n r_t)Y_0(\kappa_n r_h)] + r_t[J_1(\kappa_n r_h)Y_0(\kappa_n r_t) - J_0(\kappa_n r_t)Y_1(\kappa_n r_h)]} \right\}$$

$$-\infty \leqq z \leqq -l \tag{5-21}$$

Similarly

$$\frac{w^{(1)}}{w^{(0)}} = \frac{r_h}{l}\frac{\pi}{2}\left[\frac{I_0\left(\frac{\pi r}{2l}\right)K_1\left(\frac{\pi r_t}{2l}\right) - I_1\left(\frac{\pi r_t}{2l}\right)K_0\left(\frac{\pi r}{2l}\right)}{I_1\left(\frac{\pi r_h}{2l}\right)K_1\left(\frac{\pi r_t}{2l}\right) - I_1\left(\frac{\pi r_t}{2l}\right)K_1\left(\frac{\pi r_h}{2l}\right)}\right]\left(1 + \sin\frac{\pi z}{2l}\right)$$

$$+ \frac{r_h}{l}\frac{\pi^2}{2l} \sum_{1}^{\infty} \frac{1 + e^{-\kappa_n l}\sinh(\kappa_n z)}{(\pi/2l)^2 + \kappa_n^2} \left\{ \frac{J_0(\kappa_n r)Y_1(\kappa_n r_t) - J_1(\kappa_n r_t)Y_0(\kappa_n r)}{r_h[J_0(\kappa_n r_h)Y_1(\kappa_n r_t) - J_1(\kappa_n r_t)Y_0(\kappa_n r_h)] + r_t[J_1(\kappa_n r_h)Y_0(\kappa_n r_t) - J_0(\kappa_n r_t)Y_1(\kappa_n r_h)]} \right\}$$

$$-l \leqq z \leqq l \tag{5-22}$$

holds in the region where the actual wall curvature exists. Finally, downstream of the wall distortion,

$$\frac{w^{(1)}}{w^{(0)}} = \frac{r_h}{l}\pi\left[\frac{I_0\left(\frac{\pi r}{2l}\right)K_1\left(\frac{\pi r_t}{2l}\right) - I_1\left(\frac{\pi r_t}{2l}\right)K_0\left(\frac{\pi r}{2l}\right)}{I_1\left(\frac{\pi r_h}{2l}\right)K_1\left(\frac{\pi r_t}{2l}\right) - I_1\left(\frac{\pi r_t}{2l}\right)K_1\left(\frac{\pi r}{2l}\right)}\right]$$

$$+ \frac{r_h}{l}\frac{\pi^2}{2l} \sum_{1}^{\infty} \frac{2 - e^{-\kappa_n z}\cosh \kappa_n l}{(\pi/2l)^2 + \kappa_n^2} \left\{ \frac{J_0(\kappa_n r)Y_1(\kappa_n r_t) - J_1(\kappa_n r_t)Y_0(\kappa_n r)}{r_h[J_0(\kappa_n r_h)Y_1(\kappa_n r_t) - J_1(\kappa_n r_t)Y_0(\kappa_n r_h)] + r_t[J_1(\kappa_n r_h)Y_0(\kappa_n r_t) - J_0(\kappa_n r_t)Y_1(\kappa_n r_h)]} \right\}$$

$$l \leqq z \tag{5-23}$$

Up to the present time, exact calculations utilizing these results have not been carried out, partly for the reason that an adequate approximation to this solution may be made. This will be indicated in a later article where an example will be presented.

Throughflow with variable tip radius. When it is the tip radius rather than the root radius that varies along the direction of flow, the procedure for calculating the flow is changed only slightly. If now $dr_t/dz = f_t(z)$ and $dr_h/dz = 0$, Eq. 5-9 is replaced by

$$U(r, k) = \frac{w^{(0)}}{\sqrt{2\pi}} \int_{-\infty}^{\infty} f_t(z)e^{-ikz}dz = w^{(0)}F_t(k) \tag{5-24}$$

Similarly the Fourier inversion gives, corresponding to Eq. 5-10,

$$\frac{u^{(1)}(r, z)}{w^{(0)}} = \frac{1}{\sqrt{2\pi}} \int_{-\infty}^{\infty} F_t(k) \left[\frac{J_1(ikr_h)Y_1(ikr) - J_1(ikr)Y_1(ikr_h)}{J_1(ikr_h)Y_1(ikr_t) - J_1(ikr_t)Y_1(ikr_h)}\right] e^{ikz}dk \tag{5-25}$$

Note here that only the numerator of the bracketed term is changed from its value in Eq. 5-10 where the hub radius is varying. Hence, in the ensuing contour integration, the only differences that appear from the previous case are modifications of the numerator. If, for example, the slope of the tip contour is given as

$$\begin{aligned} f_t(z) &= \frac{\pi r_t}{2l} \cos \frac{\pi z}{2l}; \qquad |z| \leq l \\ f_t(z) &= 0; \qquad |z| \geq l \end{aligned} \tag{5-26}$$

while the hub diameter remains constant, the appropriate solutions for the radial and axial velocities can be obtained by substituting the expressions,

$$\begin{gathered} J_1(\kappa_n r_h)Y_1(\kappa_n r) - J_1(\kappa_n r)Y_1(\kappa_n r_h) \\ J_1(\kappa_n r_h)Y_0(\kappa_n r) - J_0(\kappa_n r)Y_1(\kappa_n r_h) \\ I_1\left(\frac{\pi r_h}{2l}\right) K_0\left(\frac{\pi r}{2l}\right) - I_0\left(\frac{\pi r}{2l}\right) K_1\left(\frac{\pi r_h}{2l}\right) \end{gathered} \tag{5-27}$$

respectively, in place of the expressions,

$$\begin{gathered} J_1(\kappa_n r)Y_1(\kappa_n r_t) - J_1(\kappa_n r_t)Y_1(\kappa_n r) \\ J_0(\kappa_n r)Y_1(\kappa_n r_t) - J_1(\kappa_n r_t)Y_0(\kappa_n r) \\ I_0\left(\frac{\pi r}{2l}\right) K_1\left(\frac{\pi r_t}{2l}\right) - I_1\left(\frac{\pi r_t}{2l}\right) K_0\left(\frac{\pi r}{2l}\right) \end{gathered} \tag{5-28}$$

at the appropriate positions in the numerators of Eq. 5-15, 5-16, 5-19, 5-21, 5-22, and 5-23.

It is clear also that, since the problem is a linear one, the situation where both hub and tip radii vary can be treated by superposition of the perturbations caused by hub variation only and tip variation only.

C,6. Effects of Upstream Conditions and Compressibility. The throughflow with strong tangential velocities imposed far upstream was treated earlier only when the tangential velocity distribution was that corresponding to either a vortex ($v^{(0)} \sim 1/r$) or a solid body rotation ($v^{(0)} \sim r$). In practical cases the tangential velocity is often of a more complex nature and cannot be treated easily within the framework of the theory developed so far, since the variable coefficients that occur in the partial differential equation for radial velocity give rise to special functions whose type depends upon the velocity distribution. This is quite unsatisfactory for the development of a useful theory. Furthermore it is not unusual that a strong variation of the axial velocity $w^{(0)}(r)$ should be imposed far upstream of the blade row. This introduces a similar situation to the general tangential velocity distribution. Both of these upstream velocity distributions have in common the property that, in the absence of compressibility effects, they influence only the coefficient $q(r)$ in Eq. 3-28 and leave the coefficient $p(r) = 1/r$.

On the other hand, when the tangential and axial Mach number values are of significant size to merit consideration, the coefficient $p(r)$ and the coefficient of $\partial^2 u^{(1)}/\partial z^2$ are involved so that the problem becomes a good bit more complicated. Since it is almost never possible to consider the Mach number values M_θ and M_z as independent of the radius, it is necessary to deal with the situation where all of the coefficients in Eq. 3-28 are, to some extent, functions of the radius.

It is clear that anything to be done must, moreover, be of an approximate nature since an exact treatment would prove intractable. The method to be developed will utilize an asymptotic solution of the ordinary differential equation

$$\frac{d^2U}{dr^2} + p(r)\frac{dU}{dr} + [\kappa_n^2(1 - M_z^2) + q(r)]U = 0$$

that arises from Eq. 3-28. The method, first applied to this problem by Rannie [22], rests on the fact that the first characteristic value κ_1 is large in some sense. Physically this requires that the hub ratio r_h/r_t should not be too small, since $\kappa_1 \cong \pi/(r_t - r_h)$. Under these conditions the problem may be treated by the well-developed procedures described by Erdelyi [23].

Variation of tangential and axial velocities. Suppose that flow may be considered incompressible but that large and arbitrary variations with radius occur in the tangential velocity $v^{(0)}(r)$ and axial velocity $w^{(0)}(r)$ prescribed far upstream of the blade row. Then, writing the coefficients in detail, Eq. 3-28 becomes

$$\frac{\partial^2 u^{(1)}}{\partial r^2} + \frac{1}{r}\frac{\partial u^{(1)}}{\partial r} + \left[-\frac{1}{r^2} + \frac{2v^{(0)}}{r^2 w^{(0)2}}\frac{d(rv^{(0)})}{dr} - \frac{r}{w^{(0)}}\frac{d}{dr}\left(\frac{1}{r}\frac{dw^{(0)}}{dr}\right)\right]u^{(1)} + \frac{\partial^2 u^{(1)}}{\partial z^2} = \varphi(r, z) \quad (6\text{-}1)$$

If solutions are of the form $e^{\pm\kappa_n z}U_n(r)$, as was appropriate for the blade row in a channel of constant hub and tip radius, the function $U_n(r)$ satisfies the equation

$$\frac{d^2U_n}{dr^2} + \frac{1}{r}\frac{dU_n}{dr} + \left[\kappa_n^2 - \frac{1}{r^2} - \frac{r}{w^{(0)}}\frac{d}{dr}\left(\frac{1}{r}\frac{dw^{(0)}}{dr}\right) + \frac{2v^{(0)}}{r^2w^{(0)2}}\frac{d(rv^{(0)})}{dr}\right]U_n = 0 \tag{6-2}$$

where the characteristic value κ_n^2 is to be considered large in comparison with other quantities in the problem. For convenience, introduce the new dependent variable $V_n = \sqrt{r}\,U_n$; the differential equation transforms into

$$\frac{d^2V_n}{dr^2} + \left[\kappa_n^2 - \frac{3}{4r^2} - \frac{r}{w^{(0)}}\frac{d}{dr}\left(\frac{1}{r}\frac{dw^{(0)}}{dr}\right) + \frac{2v^{(0)}}{r^2w^{(0)2}}\frac{d(rv^{(0)})}{dr}\right]V_n = 0 \tag{6-3}$$

so that the first order differential is suppressed. For any distributions $v^{(0)}(r)$ and $w^{(0)}(r)$ the coefficient of $V_n(r)$ is a certain known function $F(r)$ which may be written explicitly as

$$F(r) = -\frac{3}{4r^2} - \frac{r}{w^{(0)}}\frac{d}{dr}\left(\frac{1}{r}\frac{dw^{(0)}}{dr}\right) + \frac{2v^{(0)}}{r^2w^{(0)2}}\frac{d(rv^{(0)})}{dr} \tag{6-4}$$

The differential equation to be solved is then

$$\frac{d^2V_n}{dr^2} + [\kappa_n^2 + F(r)]V_n = 0 \tag{6-5}$$

which in the case of constant hub and tip radii must satisfy the conditions

$$V_n(r_t) = V_n(r_h) = 0 \tag{6-6}$$

Now this may be considered as a differential equation with large parameter κ_n and solved by an asymptotic method. Choose the solution to be of the form,

$$V_n(r) = e^{\alpha\kappa_n r + \sum_0^\infty \varphi_j(r)\kappa_n^{-j}} \tag{6-7}$$

that is, the argument of the exponential function is expanded in inverse powers of the large parameter κ_n. Substitution into the differential equation (Eq. 6-5) gives the formal result

$$\sum_0^\infty \kappa_n^{-j}\varphi_j'' + (\alpha\kappa_n)^2 + 2\alpha\kappa_n\sum_0^\infty \kappa_n^{-j}\varphi_j' + \left(\sum_0^\infty \kappa_n^{-j}\varphi_j'\right)^2 + \kappa_n^2 + F(r) = 0 \tag{6-8}$$

and since κ_n sets the order of magnitude, the coefficients of each power of κ_n must vanish identically. The coefficient of κ_n^2, the greatest power of κ_n, gives $\alpha^2 + 1 = 0$, or

$$\alpha = \pm i \tag{6-9}$$

Similarly the coefficient of κ_n may be written as $2\alpha\varphi_0' = 0$ so that the coefficient φ_0 is a constant. Proceeding to the power κ_n^0, the coefficient reads

$$\varphi_0'' + 2\alpha\varphi_1' + (\varphi_0')^2 + F(r) = 0 \tag{6-10}$$

Taking account of the constancy of φ_0, this gives the function φ_1 as

$$\varphi_1 = -\frac{1}{2\alpha}\int_a^r F(r)dr \tag{6-11}$$

where the lower limit of the integral is left arbitrary for the moment. Further coefficients may readily be obtained in a similar manner, but those obtained so far are adequate to demonstrate the technique. To the order of $1/\kappa_n$ then, the functions V_n become

$$V_n(r) = e^{\pm i\left[\kappa_n + \varphi_0 + \frac{1}{2\kappa_n}\int_a^r F(r)dr\right]}$$

Since it is required that the V_n vanish at hub and tip radii, choose the linear combination

$$V_n(r) = \sin\left[\kappa_n(r - r_h) + \frac{1}{2\kappa_n}\int_{r_h}^r F(r)dr\right] \tag{6-12}$$

where values of φ_0 and a have been selected to make the V_n vanish at $r = r_h$. The large parameter κ_n must be chosen so that

$$\kappa_n(r_t - r_h) + \frac{1}{2\kappa_n}\int_{r_h}^{r_t} F(r)dr = n\pi$$

To the same order of accuracy, then, the value of κ_n may be calculated as

$$\kappa_n = \frac{n\pi}{r_t - r_h}\left[1 - \frac{r_t - r_h}{2n^2\pi^2}\int_{r_h}^{r_t} F(r)dr\right] \tag{6-13}$$

Transforming back now to the original dependent variable and denoting $\frac{1}{2}\int_{r_h}^{r_t} F(r)dr \equiv g(r)$ the appropriate characteristic functions of Eq. 6-1 may be written

$$\frac{1}{\sqrt{r}}\sin\left\{n\pi\frac{r - r_h}{r_t - r_h} + \frac{r_t - r_h}{n\pi}\left[g(r) - \frac{r - r_h}{r_t - r_h}g(r_t)\right]\right\} e^{\pm\frac{n\pi z}{r_t - r_h}\left[1 - \frac{r_t - r_h}{n^2\pi^2}g(r_t)\right]} \tag{6-14}$$

where terms of higher order than $[(r_t - r_h)/n\pi]^2$ have been deleted. It is now a simple matter to build solutions from these in precisely the manner that the Bessel function combination was used in the development following Eq. 4-9. The functions possess the appropriate orthogonality property since they satisfy a problem of the Sturm-Liouville type.

Before discussing particular solutions, two features of this asymptotic solution should be observed. In the first place, suppose either that the function $F(r) = 0$ or that the expansion is cut off at terms of the order $1/\kappa_n$. Then the functions U_n, asymptotic solutions to Eq. 6-2, are of the form

$$U_n \sim \frac{1}{\sqrt{r}} \sin n\pi \left(\frac{r - r_h}{r_t - r_h}\right) \tag{6-15}$$

and hence are identical, with the exception of an irrelevant constant factor, to the asymptotic form of the Bessel function combination introduced in Eq. 4-59. In fact exactly the same approximation is being made in each case.

In the second place consider the upstream flow where $w^{(0)} = \text{const}$ and $v^{(0)} = bw^{(0)}r/r_t$. Then

$$F(r) = -\frac{3}{4r^2} + \frac{4b^2}{r_t^2}$$

and the integral

$$g(r) \equiv \frac{1}{2}\int_{r_h}^{r}\left(-\frac{3}{4r^2} + \frac{4b^2}{r_t^2}\right)dr = \frac{2b^2(r - r_h)}{r_t^2} - \frac{3}{8}\left(\frac{1}{r_h} - \frac{1}{r}\right) \tag{6-16}$$

The characteristic values are then

$$\kappa_n = \frac{n\pi}{r_t - r_h}\left[1 - \frac{2b^2}{n^2\pi^2}\left(1 - \frac{r_h}{r_t}\right)^2 + \frac{3}{8n^2\pi^2}\frac{r_t}{r_h}\left(1 - \frac{r_h}{r_t}\right)^2\right] \tag{6-17}$$

and the characteristic functions may be written down explicitly as

$$\frac{1}{\sqrt{r}} \sin n\pi \left\{\frac{r - r_h}{r_t - r_h}\left[1 - \frac{3}{8n^2\pi^2}\left(\frac{r_t}{r_h} - 1\right)^2\left(\frac{r_h}{r} - \frac{r_h}{r_t}\right)\right]\right\}$$
$$e^{\pm\frac{n\pi z}{r_t - r_h}\left[1 - \frac{2b^2}{n^2\pi^2}\left(1 - \frac{r_t}{r_t}\right)^2 + \frac{3}{8n^2\pi^2}\frac{r_t}{r_h}\left(1 - \frac{r_t}{r_t}\right)^2\right]} \tag{6-18}$$

Note that the radial dependence of the characteristic functions does not contain any influence of rotation parameter b; the term having coefficient $3/8n^2\pi^2$ arises only in the approximation to the Bessel functions. The upstream solid body rotation only affects the axial dependence (exponential decay) of the solutions. Speaking generally, the influence of an imposed upstream solid body rotation reduces the rate of decay upstream and downstream of a disturbance from

$$\exp\left(\pm\frac{n\pi z}{r_t - r_h}\right) \text{ to } \exp\left\{\pm\frac{n\pi z}{r_t - r_h}\left[1 - \frac{2b^2}{n^2\pi^2}\left(1 - \frac{r_h}{r_t}\right)^2\right]\right\}$$

This is precisely the effect observed when this problem was treated exactly and may be compared with the results following Eq. 4-34. The solutions obtained here are, in fact, the correct asymptotic representations of those obtained previously when b is not large.

Now, having been satisfied that the asymptotic solutions to the differential equation (Eq. 6-1) agree with exact solutions under circumstances where they can be compared, the asymptotic method may be applied to problems where exact solutions can be obtained only numerically. As an example consider the case of a blade row operating downstream of a guide vane that imparts a solid body rotation given again by

$$v^{(0)} = w_0 b \frac{r}{r_t} \tag{6-19}$$

where w_0 is the mean axial velocity. In general, when a homogeneous fluid of constant stagnation enthalpy flows through a guide vane of this type, the axial velocity distribution is disturbed although its mean value remains constant. It will be shown later that far downstream of the guide vane the axial velocity profile is given with reasonable accuracy as

$$w^{(0)} = w_0 \left\{ 1 + k^2 \left[1 + \left(\frac{r_h}{r_t}\right)^2 - 2\left(\frac{r}{r_t}\right) \right]^2 \right\} \tag{6-20}$$

where actually $k^2 = b^2/2$. It is now straightforward to calculate the functions $F(r)$ and $g(r)$. Substitution into Eq. 6-4 gives, after a little manipulation,

$$F(r) = -\frac{3}{4r^2} + \frac{4b^2}{r_t^2} \frac{1}{[a^2 - k^2(r/r_h)^2]^2} \tag{6-21}$$

where the quantity a^2 has been used to denote

$$a^2 = 1 + \frac{1}{2}k^2 \left[1 + \left(\frac{r_t}{r_h}\right)^2 \right] \tag{6-22}$$

The function $g(r) \equiv \frac{1}{2}\int_{r_h}^{r} F(r)dr$ may also be evaluated explicitly as

$$g(r) = \frac{3}{8}\left(\frac{1}{r} - \frac{1}{r_h}\right) + \frac{b^2}{a^2}\frac{r_h}{r_t^2} \left[\frac{\dfrac{r}{r_h}}{a^2 - \left(k\dfrac{r}{r_h}\right)^2} - \frac{1}{a^2 - k^2} \right]$$

$$+ \frac{b^2}{2a^3} \frac{r_h/r_t^2}{k} \ln \left[\left(\frac{a + k\dfrac{r}{r_h}}{a + k} \right) \left(\frac{a - k}{a - k\dfrac{r}{r_h}} \right) \right] \tag{6-23}$$

Substitution of this result into Eq. 6-14 gives the characteristic function in terms of which may be expanded the asymptotic solution for any blade row far downstream of an entrance guide vane.

The effect on the radial velocity distribution is probably of less interest than is the effect on rate of decay of the disturbance upstream and downstream of the blade row. Referring again to Eq. 6-14, the modification to the decay rate depends only upon the value of $g(r_t)$. From Eq. 6-23 this becomes directly

$$g(r_t) = -\frac{3}{8}\left(\frac{1}{r_h} - \frac{1}{r_t}\right) + \frac{b^2}{a^2}\frac{r_h}{r_t^2}\left[\frac{\frac{r_t}{r_h}}{a^2 - \left(k\frac{r_t}{r_h}\right)^2} - \frac{1}{a^2 - k^2}\right]$$

$$+ \frac{b^2}{2a^3k}\frac{r_h}{r_t^2}\ln\left[\left(\frac{a + k\frac{r_t}{r_h}}{a + k}\right)\left(\frac{a - k}{a - k\frac{r_t}{r_h}}\right)\right] \quad \text{(6-24)}$$

As was observed before, the term $-\frac{3}{8}[(1/r_h) - (1/r_t)]$ arises from asymptotic approximation to the original Bessel functions and is generally unimportant. All of the other terms in Eq. 6-24 are positive since $r_t/r_h > 1$. As a result the argument of the exponential function in Eq. 6-14 is decreased; therefore the range of the disturbance upstream and downstream of the blade row is extended over that without initial rotation or axial velocity variation. When the axial velocity is constrained to remain uniform, that is $k = 0$ and $a^2 = 1$, the result reduces to that given in the exponent of Eq. 6-18. On the other hand, when the axial velocity alone is present (i.e. $b = 0$), the resulting function reduces to the first two terms of Eq. 6-24.

This particular result given by Eq. 6-23 and 6-24 for the upstream conditions distorted by a zeroeth order rotational velocity and axial velocity variation will prove very useful in the construction of approximate solutions applicable in compressor design procedures.

Effect of compressibility. Compressibility influences that affect the throughflow directly are usually not of major concern in compressor design. The reason for this is that, although the Mach number relative to the rotating blades may be in the transonic regime, the meridional Mach number, upon which the change from elliptic to hyperbolic equation depends, almost invariably will be in the subsonic region. Thus although compressibility influences in the throughflow do exist and exert their influence upon the flow field, it must be kept in mind that some of the most significant compressibility effects are those associated with the detailed blade geometry.

To illustrate the compressibility influences without becoming unnecessarily enmeshed in detailed calculations, consider a rather over-simplified example. Suppose that both the zeroeth order axial velocity $w^{(0)}$ and sonic velocity $a^{(0)}$ are constant and that the tangential velocity is a solid body rotation, $v^{(0)} = bw^{(0)}r/r_t$. This may be realized physically with no difficulty but, it must be admitted, the example differs somewhat from the circumstances usually encountered downstream of a guide vane. Solutions must be obtained to the homogeneous partial differential equation,

$$\frac{\partial^2 u^{(1)}}{\partial r^2} + p(r)\frac{\partial u^{(1)}}{\partial r} + q(r)u^{(1)} + (1 - M_z^2)\frac{\partial^2 u^{(1)}}{\partial z^2} = 0 \tag{6-25}$$

where the functions $p(r)$ and $q(r)$ are given by Eq. 3-29 and 3-30. These functions, as well as the coefficient of $\partial^2 u^{(1)}/\partial z^2$, involve the Mach numbers $M_z = w^{(0)}/a^{(0)}$ and $M_\theta = v^{(0)}/a^{(0)}$. Clearly the axial Mach number M_z is a constant and the tangential Mach number may be expressed

$$M_\theta^2 = \frac{b^2 w^{(0)2}}{a^{(0)2}}\frac{r^2}{r_t^2} = M_z^2 b^2 \left(\frac{r}{r_t}\right)^2 \tag{6-26}$$

in terms of the axial Mach number and the radius. With these observations, the coefficients in Eq. 6-25 may be written explicitly as

$$p(r) = \frac{1}{r}\left[1 + \gamma M_z^2 b^2 \left(\frac{r}{r_t}\right)^2\right] \tag{6-27}$$

$$q(r) = -\frac{1}{r^2} + \frac{b^2}{r_t^2}[4 - (4 - \gamma)M_z^2] + (\gamma - 1)\frac{b^4}{r_t^4}M_z^2 r^2 \tag{6-28}$$

If solutions of the differential equation (Eq. 6-25) are sought in the form $U_n(r)e^{\pm\kappa_n z}$, the ordinary differential equation for $U_n(r)$ is

$$\frac{d^2U_n}{dr^2} + \frac{1}{r}\left[1 + \gamma M_z^2 b^2\left(\frac{r}{r_t}\right)^2\right]\frac{dU_n}{dr} + \left\{(1 - M_z^2)\kappa_n^2 + 4\frac{b^2}{r_t^2}\left[1 - \left(1 - \frac{\gamma}{4}\right)M_z^2\right] - \frac{1}{r^2} + (\gamma - 1)M_z^2\frac{b^4}{r_t^4}r^2\right\}U_n = 0 \tag{6-29}$$

The term involving the first derivatives may be eliminated in the conventional manner; the transformation is simply

$$U_n(r) = \frac{1}{\sqrt{r}}V_n(r)e^{-\frac{\gamma}{4}M_z^2 b^2\left(\frac{r}{r_t}\right)^2} \tag{6-30}$$

which reduces, of course, to the transformation used previously when the axial Mach number M_z, and hence the tangential Mach number M_θ, vanish. After substitution of this relation into the differential equation (Eq. 6-29) and carrying out some simplification, the differential equation for $V_n(r)$ becomes

$$\frac{d^2V_n}{dr^2} + \left\{(1 - M_z^2)\kappa_n^2 - \frac{3}{4r^2} + \frac{b^2}{r_t^2}\left[4 - \left(4 + \frac{3\gamma}{2}\right)M_z^2\right] + \frac{b^4}{r_t^4}M_z^2\left[(\gamma - 1) - \frac{\gamma^2 M_z^2}{2}\right]r^2\right\}V_n = 0 \tag{6-31}$$

Now calling

$$\bar{\kappa}_n^2 = (1 - M_z^2)\kappa_n^2 \tag{6-32}$$

and

$$F(r) = -\frac{3}{4r^2} + \frac{b^2}{r_t^2}\left[4 - \left(1 + \frac{\gamma}{2}\right)M_z^2\right] + \frac{b^4}{r_t^4}M_z^2\left[(\gamma - 1) - \frac{\gamma^2}{2}M_z^2\right]r^2 \tag{6-33}$$

The above differential equation assumes the form

$$\frac{d^2V_n}{dr^2} + [\bar{\kappa}_n^2 + F(r)]V_n = 0 \tag{6-34}$$

which is exactly the equation treated previously in the paragraphs following Eq. 6-5. The results of that analysis may be carried over and applied directly to the present problem. It is a simple calculation to show then that

$$g(r) = \frac{3}{8}\left(\frac{1}{r} - \frac{1}{r_h}\right) + \frac{2b^2}{r_t^2}\left[1 - \left(1 + \frac{3\gamma}{8}\right)M_z^2\right](r - r_h) + \frac{b^4}{6r_t^4}M_z^2\left[(\gamma - 1) - \frac{\gamma^2 M_z^2}{2}\right](r^3 - r_h^3) \tag{6-35}$$

Treating the problem as before, it is clear that the appropriate solutions of Eq. 6-34 are

$$V_n = \sin\left[\bar{\kappa}_n(r - r_h) + \frac{1}{\bar{\kappa}_n}g(r)\right]$$

so that, to the appropriate order of magnitude,

$$\bar{\kappa}_n = \frac{n\pi}{r_t - r_h}\left[1 - \frac{r_t - r_h}{n^2\pi^2}g(r_t)\right] \tag{6-36}$$

which gives

$$V_n = \sin\left\{n\pi\left(\frac{r - r_h}{r_t - r_h}\right) + \frac{r_t - r_h}{n\pi}\left[g(r) - \frac{r - r_h}{r_t - r_h}g(r_t)\right]\right\}$$

very much as before. The characteristic functions of the problem may then be written, taking account of the transformation given by Eq. 6-30,

$$\frac{1}{\sqrt{r}}e^{-\frac{\gamma}{4}M_z^2b^2\left(\frac{r}{r_t}\right)^2}\sin\left\{n\pi\left(\frac{r - r_h}{r_t - r_h}\right) + \frac{r_t - r_h}{n\pi}\left[g(r) - \frac{r - r_h}{r_t - r_h}g(r_t)\right]\right\} e^{\pm\frac{n\pi}{\sqrt{1-M_z^2}}\frac{z}{r_t - r_h}\left[1 - \frac{r_t - r_h}{n^2\pi^2}g(r_t)\right]} \tag{6-37}$$

The compressibility correction thus enters into the problem in a fairly complicated manner, affecting both the radial variation and the axial decay of the disturbance. The axial decay term is worth examining in some detail. The decay factor is explicitly

$$\exp\left(\pm\frac{n\pi}{\sqrt{1-M_z^2}}\frac{z}{r_t-r_h}\left\{1-\frac{2b^2}{n^2\pi^2}\left(1-\frac{r_h}{r_t}\right)^2\left[1-\left(1+\frac{3\gamma}{8}\right)M_z^2\right]\right.\right.$$
$$-\frac{b^4}{6n^2\pi^2}M_z^2\left(\gamma-1-\frac{\gamma^2M_z^2}{2}\right)\left[1-\left(\frac{r_h}{r_t}\right)^3\right]\left(1-\frac{r_h}{r_t}\right)$$
$$\left.\left.+\frac{3}{8n^2\pi^2}\frac{r_t}{r_h}\left(1-\frac{r_h}{r_t}\right)^2\right\}\right) \quad (6\text{-}38)$$

In the absence of any swirl velocity component, that is where $b = 0$, the only correction is the obvious scale contraction in the z direction given by $z/\sqrt{1-M_z^2}$. This effect tends to crowd all disturbances caused by the blade into a region close to the blade row. It is simply the Prandtl-Glauert contraction and is associated with any compressible flow predominantly in the direction of z and has no peculiar association with the turbomachine problem. This scale contraction was first applied to the axial turbomachine problem by Horlock [*13*]. When a tangential velocity of significant magnitude is present, which is invariably the case in any practical example, other Mach number corrections enter. For example the extension of the disturbed region, as controlled by the term

$$\frac{-2b^2}{n^2\pi^2}\left(1-\frac{r_h}{r_t}\right)^2\left[1-\left(1+\frac{3\gamma}{8}\right)M_z^2\right]$$

may be modified by 15 per cent or more by the compressibility influence $[1+(3\gamma/8)]M_z^2$. Here again the compressibility effect tends to decrease the extent of the disturbance. Except for rather large values of b (ratio of tangential to axial velocities) corresponding to severe off-design conditions, the effect of axial scale contraction $z/\sqrt{1-M_z^2}$ dominates the other compressibility influences.

It is perhaps appropriate to reiterate the fact that the compressibility effect discussed here is often not the one of primary interest to compressor or turbine design. Local blade channel choking and the accompanying losses depend strongly upon the local blade geometry and consequently are properly treated under cascade theory. However, it is of extreme importance to have a good calculation of the axially symmetric throughflow in order to know gas velocities and flow angles in the neighborhood of the blades. Thus there exists one compressibility effect in calculating the throughflow and another in determining the flow about a blade cascade placed in that throughflow.

There appears to exist one class of problems in which the effects of compressibility may not be segregated into the two familiar categories of throughflow and blade characteristics. This example is the transonic compressor where the blade tips operate at a relative Mach number larger than unity while the blade roots operate at a subsonic Mach number.

A recent investigation by McCune [*17*] for a finite number of blades with thickness, but without lift, has shown that the detailed flow about blades at the rotor tip influence in a major way detailed flow about the blades at the root. This result, one which is not an uncommon one in transonic flow fields, may be interpreted as meaning that the roots and tips of the blades may not be considered as independent cascades operating in their particular flow fields. Since this flow may not be divided, even approximately, in the manner of which the flow has been treated in the foregoing work, the transonic compressor represents a strictly three-dimensional problem that must be treated in complete detail. It should be mentioned that although these results were obtained for nonturning blades it is probable that similar, though less drastic, results would be obtained for the turning blade row.

C,7. Approximations to the Throughflow. It is amply clear that while the results of the preceding sections are ideal for investigating the throughflow in detail for one given blade row, the general complexity of the calculations makes it desirable to search out simple and approximate techniques for routine use in turbomachine work, particularly for consideration of multiple blade rows. One approximation [*2,3*] which has been used to considerable extent is the so-called radial equilibrium theory. Here one considers only the flow far upstream and far downstream of a blade row. At these sections all curvature of the meridional flow has ceased and the flow is determined by equilibrium between the radial pressure gradient and the centripetal acceleration caused by the motion of gas elements about the axis of symmetry. The shortcoming of this analysis is that no description is given of how near to the blade row the modification of axial velocity takes place. Information of this sort is particularly necessary when blade rows are spaced closely enough so that the flow fields of adjacent blade rows overlap. In such cases of mutual interference an approximation to the flow field is required also. It will prove possible to approximate the detailed development of the axial velocity profile with considerable accuracy and with sufficient simplicity for use in the many complex situations. Finally it will prove possible to approximate the effects of variation in hub and tip radii as well as the modification due to compressibility.

Radial equilibrium theory. Of the many possible ways to approach radial equilibrium theory, one will be chosen here [*10*] that parallels the linearized development given in the previous section. Excluding the blade force normal to the stream surfaces, which need only be considered in very special cases, the tangential vorticity is given by Eq. 2-43 as

$$\frac{\partial u}{\partial z} - \frac{\partial w}{\partial r} = \frac{\rho}{\rho_0} r \left\{ \frac{\partial h}{\partial \psi} - \frac{v}{r} \frac{\partial (rv)}{\partial \psi} - T \frac{\partial s}{\partial \psi} \right\}$$

For the present considerations, neglect the entropy variation. Far downstream of a single blade row, say at $z = \infty$, the radial velocities vanish and the variation in stream function $\delta\psi = -r \frac{\rho}{\rho_0} w\delta r$. Then, since all other quantities become functions of the radius only, it follows that

$$w \frac{dw}{dr} = \frac{dh^0}{dr} - \frac{v}{r} \frac{d}{dr} (rv) \tag{7-1}$$

Now if the total enthalpy h^0 and the angular momentum rv were known functions of the radius at $z = \infty$, it would be a simple matter to integrate this relation. Both the enthalpy and the angular momentum are transported along the stream surfaces and change only upon passage through the blade row. Each of these quantities consists then of two parts, that transported from far upstream ($z = -\infty$) and that imparted by the blade row. Consider for the moment the transport of angular momentum, given by Eq. 2-25. To the order of ϵ, the order of magnitude of the blade forces, Eq. 2-25 possesses a zeroeth order part,

$$w^{(0)} \frac{\partial(rv^{(0)})}{\partial z} = 0 \tag{7-2}$$

and a first order part,

$$u^{(1)} \frac{\partial(rv^{(0)})}{\partial r} + w^{(0)} \frac{\partial(rv^{(1)})}{\partial z} = rf_\theta \tag{7-3}$$

From Eq. 7-2 the initial angular momentum is unchanged along the axial direction so that the quantity $rv^{(0)}$ is known far downstream. The first order contribution $rv^{(1)}$ follows from Eq. 7-3. Integration with respect to z gives simply

$$rv^{(1)} = \frac{1}{w^{(0)}} \int_{-\infty}^{z} rf_\theta dz - \frac{1}{w^{(0)}} \frac{d(rv^{(0)})}{dr} \int_{-\infty}^{z} u^{(1)} dz \tag{7-4}$$

The first integral is just the angular momentum imparted by the blade row. Since the tangential force f_θ vanishes outside of the blade row, it is sufficient to extend the integration only across the blade chord and

$$\frac{1}{w^{(0)}} \int_{-\infty}^{z} rf_\theta dz = \frac{1}{w^{(0)}} \int_{-c/2}^{c/2} rf_\theta dz \equiv r\Delta v^{(1)} \tag{7-5}$$

for any point downstream of the blade row. This change in angular momentum across the blade row, $r\Delta v^{(1)}$, will be considered prescribed. The second integral in Eq. 7-4 represents the perturbation of angular momentum due to the radial transport of the angular momentum from far upstream. It is clear, in fact, that $\int_{-\infty}^{z} (u^{(1)}/w^{(0)})dz$ is equal to the radial distance that a stream surface moves in passing from far upstream to a point z. The angular momentum is transported along these stream surfaces so that the angular momentum which exists at a distance z downstream and radius r is not $rv^{(0)}$ but rather

$$v^{(0)}\left(r - \epsilon \int_{-\infty}^{z} \frac{u^{(1)}}{w^{(0)}}\, dz\right)$$

To first order, then, an angular momentum perturbation,

$$-\frac{d(rv^{(1)})}{dr}\,\epsilon \int_{-\infty}^{z} \frac{u^{(1)}}{w^{(0)}}\, dz$$

exists due to the radial transport of initial angular momentum. This is the origin of the last term of Eq. 7-4.

It is not really convenient to express the radial transport in terms of the radial perturbation velocity $u^{(1)}$ inasmuch as the axial perturbation velocity is the quantity of main interest. The radial and axial velocity components are related through the continuity equation, but the integral $\int_{-\infty}^{z} (u^{(1)}/w^{(0)})dz$ which is required can be given in terms of $w^{(1)}$ by some direct physical reasoning. Since the stream surface bounds a constant mass flow of fluid between the hub and local radius of the stream surface, the surface must be displaced to accommodate a variation in mass flow $\epsilon \int_{r_h}^{r} \rho w^{(1)} 2\pi r dr$ where $w^{(1)}$ is the axial velocity perturbation from the flow when the stream surface was cylindrical. This mass flow variation is compensated by decreasing the radius of the stream surface by an amount $-\Delta r$ through which the mass flow is, to the first order, $-\rho w^{(0)} 2\pi r \Delta r$. The mass flow integral and this last expression must be equal for the stream surface to bound a constant mass flow. Consequently

$$\Delta r \equiv \int_{-\infty}^{z} \frac{u^{(1)}}{w^{(0)}}\, dz = -\frac{1}{rw^{(0)}} \int_{r_h}^{r} w^{(1)} r dr \tag{7-6}$$

With the results of Eq. 7-5 and 7-6 it is possible to express the perturbation angular momentum, far downstream of the blade row, as

$$rv^{(1)} = r\Delta v^{(1)} + \frac{1}{rw^{(0)}} \frac{d(rv^{(0)})}{dr} \int_{r_h}^{r} w^{(1)} r dr \tag{7-7}$$

The same consideration as given in Eq. 7-2 et seq. gives a similar expression for the stagnation enthalpy perturbation far downstream:

$$h^{(1)} = \omega r \Delta v^{(1)} + \frac{1}{rw^{(0)}} \frac{dh^{(0)}}{dr} \int_{r_h}^{r} w^{(1)} r dr \tag{7-8}$$

where the fact has been used that the enthalpy perturbation $\Delta h^{(1)}$ across the blade row is equal to $\omega r \Delta v^{(1)}$, as may be deduced from Eq. 2-36.

Now Eq. 7-7 and 7-8 give the information necessary to express Eq. 7-1, for the axial velocity, in terms of known quantities. It is appropriate to split this equation into its zeroeth and first order parts. The zeroeth order part is a trivial statement of the conservation of zeroeth order enthalpy, angular momentum, and axial velocity profile along cylindrical surfaces. The first order part is the one of interest in computing the perturbation in the axial velocity profile. It may be written

$$w^{(0)}\frac{dw^{(1)}}{dr}+\left[\frac{1}{r^4w^{(0)}}\frac{d}{dr}(rv^{(0)})^2-\frac{d}{dr}\left(\frac{1}{r}\frac{dw^{(0)}}{dr}\right)\right]\int_{r_h}^{r}w^{(1)}rdr \tag{7-9}$$

where now the expressions for $rv^{(1)}$ and $h^{(1)}$ may be entered from Eq. 7-7 and 7-8. After some simplification this gives

$$w^{(0)}\frac{dw^{(1)}}{dr}+\left[\frac{1}{r^4w^{(0)}}\frac{d}{dr}(rv^{(0)})^2-\frac{d}{dr}\left(\frac{1}{r}\frac{dw^{(0)}}{dr}\right)\right]\int_{r_h}^{r}w^{(1)}rdr$$
$$=-\left(\frac{v^{(0)}}{r}-\omega\right)\frac{d}{dr}(r\Delta v^{(1)})-\frac{1}{r^2}\frac{d}{dr}(rv^{(0)})(r\Delta v^{(1)}) \tag{7-10}$$

where the zeroeth order part of Eq. 7-1 has been employed in the reduction.

In the particular case that the upstream axial velocity is uniform and the upstream tangential velocity vanishes, then

$$\frac{dw^{(2)}}{dr}=\frac{\omega}{w^{(0)}}\frac{d}{dr}(r\Delta v^{(1)}) \tag{7-11}$$

which may be integrated directly. As before, it appears that a stationary blade row produces no distortion, to the first order, when there is no initial distortion far upstream. Thus the guide vane requires special treatment, and a second order analysis of this particular problem gives

$$\frac{dw^{(1)}}{dr}=-\frac{r\Delta v^{(1)}}{r^2w^{(0)}}\frac{d}{dr}(r\Delta v^{(1)}) \tag{7-12}$$

It is justified to utilize this second order analysis in the case of the entrance vane because the angular momentum increment $r\Delta v^{(1)}$ across it is generally much larger than that across other blade rows. Hence the axial velocity disturbance $w^{(2)}$ calculated in this particular case may be sufficiently large that it becomes the upstream distortion $w^{(0)}$ for succeeding blade rows.

Returning to the general case of Eq. 7-10, this integro-differential equation may be converted to a second order differential equation by differentiation and subsequent elimination of the integral. To do this it is convenient to denote

$$\frac{1}{w^{(0)}}\left[\frac{1}{r^4w^{(0)}}\frac{d}{dr}(rv^{(0)})^2-\frac{d}{dr}\left(\frac{1}{r}\frac{dw^{(0)}}{dr}\right)\right]=h(r) \tag{7-13}$$

$$\frac{1}{w^{(0)}}\left[\left(\frac{v^{(0)}}{r}-\omega\right)\frac{d}{dr}(r\Delta v^{(1)})+\frac{1}{r^2}\frac{d}{dr}(rv^{(0)})r\Delta v^{(1)}\right]=k(r) \tag{7-14}$$

so that Eq. 7-10 becomes

$$\frac{dw^{(1)}}{dr} + h(r) \int_{r_h}^{r} w^{(1)} r dr = -k(r) \tag{7-15}$$

The corresponding second order differential equation is

$$\frac{d^2w^{(1)}}{dr^2} - \frac{d}{dr}[\ln h(r)] \frac{dw^{(1)}}{dr} + rh(r)w^{(1)} = k(r) \frac{d}{dr} \ln \left[\frac{h(r)}{k(r)}\right] \tag{7-16}$$

The solution to this must satisfy two conditions. One is simply that the perturbation axial velocity does not modify the mass throughflow; thus

$$\int_{r_h}^{r_t} w^{(1)} r dr = 0 \tag{7-17}$$

The second condition is that the solution satisfy Eq. 7-15, which it did before differentiation. Since the solution satisfies Eq. 7-16 by definition, it need in addition satisfy Eq. 7-15 at only one point. It is convenient to choose this point as $r = r_t$ inasmuch as here the integral vanishes. Therefore it is sufficient that

$$\frac{dw^{(1)}}{dr}(r_t) = -k(r_t) \tag{7-18}$$

Eq. 7-16 and the conditions given by Eq. 7-17 and 7-18 complete the mathematical problem for the throughflow far downstream of an arbitrary blade row. It should be noted here that there is no difficulty in accounting for compressibility in this calculation. The density then enters through the continuity relation into Eq. 7-6 et seq. and in the condition of Eq. 7-7.

In general Eq. 7-16 is difficult to integrate and numerical analysis is required. In fact, for numerical analysis, it is often more convenient to work with Eq. 7-15. An iterative solution is applicable, for example, where the first approximation is obtained by integrating $-k(r)$ and neglecting the integral $\int_{r_h}^{r} w^{(1)} r dr$. This first order solution is then used in the integral to obtain a more accurate second approximation. Some examples will now help to illustrate the calculation of the throughflow far downstream from a blade row.

Flow downstream of an entrance vane. Consider an entrance vane system that imparts a solid body rotation

$$\Delta v^{(1)} = bw_0 \frac{r}{r_t} \tag{7-19}$$

Upstream of the guide vane the flow is axial with a uniform velocity $w^{(0)} \equiv w_0$. Let the station far downstream of the guide vane be denoted by subscript 1. Then according to Eq. 7-12

$$\frac{dw^{(2)}}{dr} = -b^2 w_0 \left(\frac{2r}{r_t^2}\right) \tag{7-20}$$

Integrating this and satisfying the integral condition given by Eq. 7-17 yields, after a very elementary calculation,

$$w_1^{(2)} = \frac{b^2 w_0}{2}\left[1 + \left(\frac{r_h}{r_t}\right)^2 - 2\left(\frac{r}{r_t}\right)^2\right] \tag{7-21}$$

A typical profile corresponding to this calculation is shown in Fig. C,7a.

It should be noted here that the small quantity, denoted ϵ in the formal perturbation analysis, is the constant b which gives the ratio of

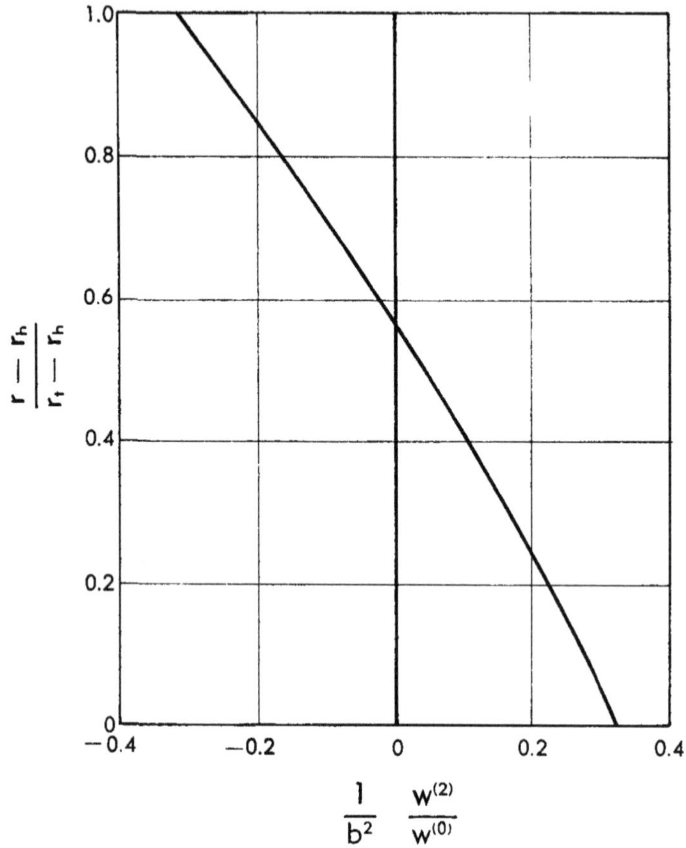

Fig. C,7a. Axial velocity profile far downstream of entrance guide vane, $v^{(1)} = bw^{(0)}r/r_t$, $r_h/r_t = 0.6$.

the tangential velocity at the blade tip to the undisturbed axial velocity. In guide vanes this need not be small and may, in fact, be about unity. In spite of this, the procedure gives a very satisfactory approximation to the flow far downstream.

Rotor far downstream from entrance vane. With the results that have been developed it is possible to continue discussion of throughflow in an axial compressor, considering the blade rows to be spaced sufficiently far apart that the development of each axial velocity profile is complete before the next blade row is encountered. Consider the rotor, then, which imparts a vortex rotation

$$\Delta v_2^{(1)} = aw_0 \frac{r_t}{r} \tag{7-22}$$

and hence imparts the same enthalpy rise at all radii. So far as the rotor is concerned the upstream conditions are not uniform but have a zeroeth order axial and tangential velocity distortion,

$$v_1^{(0)} = bw_0 \frac{r}{r_t} \tag{7-23}$$

$$w_1^{(0)} = w_0 \left\{1 + \frac{b^2}{2}\left[1 + \left(\frac{r_h}{r_t}\right)^2 - 2\left(\frac{r}{r_t}\right)^2\right]\right\}$$

A word of explanation should be added here inasmuch as these were considered perturbations in the guide vane calculation and now appear as zeroeth order quantities. The fact is that the tangential velocity disturbance is not strictly a perturbation, although it was treated as such to the second order, and may not be omitted from calculations of the ensuing stages. The axial velocity disturbance is generally of much smaller magnitude, depending upon the hub ratio, and it may be neglected at appropriate steps of the work. Then according to Eq. 7-13 and 7-14 the functions $h(r)$ and $k(r)$ become

$$h(r) = \frac{4b^2}{r}\left\{1 + \frac{b^2}{2}\left[1 + \left(\frac{r_h}{r_t}\right)^2 - 2\left(\frac{r}{r_t}\right)^2\right]\right\}^{-2} \tag{7-24}$$

$$k(r) = \frac{2abw_0}{r}\left\{1 + \frac{b^2}{2}\left[1 + \left(\frac{r_h}{r_t}\right)^2 - 2\left(\frac{r}{r_t}\right)^2\right]\right\}^{-1} \tag{7-25}$$

In these expressions the axial velocity disturbance far upstream of the rotor is characterized by the presence of the term

$$\left[1 + \left(\frac{r_h}{r_t}\right)^2 - 2\left(\frac{r}{r_t}\right)^2\right]$$

Its influence is usually a small one, becoming less significant as the hub ratio r_h/r_t approaches unity. Neglecting these terms, Eq. 7-16 for the axial velocity perturbation far downstream of the rotor blade row may be written

$$\frac{d^2w_2^{(1)}}{dr^2} + \frac{1}{r}\frac{dw_2^{(1)}}{dr} + \left(\frac{2b}{r_t}\right)^2 w_2^{(1)} = 0 \tag{7-26}$$

where the subscript 2 has been employed to denote conditions downstream of the rotor. The solution consists of Bessel functions of order zero, and may be written as the linear combination

$$w_2^{(1)} = AJ_0\left(2b\frac{r}{r_t}\right) + BY_0\left(2b\frac{r}{r_t}\right) \tag{7-27}$$

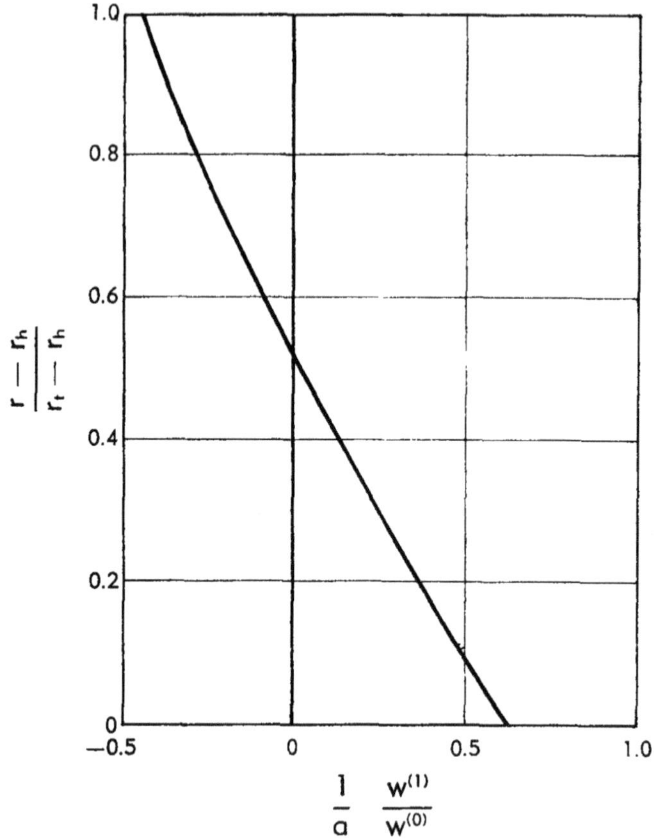

Fig. C,7b. Axial velocity far downstream of rotor blade row, $v^{(1)} = aw^{(0)}r_t/r$, $r_h/r_t = 0.6$.

where the constants are to be determined from the conditions given by Eq. 7-17 and 7-18. Direct substitution gives, for Eq. 7-18,

$$AJ_1(2b) + BY_1(2b) = w_0 a \tag{7-28}$$

On the other hand, the integral condition (Eq. 7-17) gives

$$A\left[J_1(2b) - \frac{r_h}{r_t}J_1\left(2b\frac{r_h}{r_t}\right)\right] + B\left[Y_1(2b) - \frac{r_h}{r_t}Y_1\left(2b\frac{r_h}{r_t}\right)\right] \tag{7-29}$$

Some calculation then gives the solution for the axial velocity perturbation downstream of the rotor, where the quantity $2b/r_t$ is denoted κ

$$w_2^{(1)} = aw_0\left\{\frac{\left[Y_1(\kappa r_h) - \frac{r_t}{r_h}Y_1(\kappa r_t)\right]J_0(\kappa r) + \left[J_1(\kappa r_h) - \frac{r_t}{r_h}J_1(\kappa r_t)\right]Y_0(\kappa r)}{J_1(\kappa r_t)Y_1(\kappa r_h) - J_1(\kappa r_h)Y_1(\kappa r_t)}\right\} \tag{7-30}$$

This particular distortion is shown in Fig. C,7b. The complete axial velocity profile downstream of the rotor is made up of the uniform axial velocity w_0, the disturbance $w_1^{(2)}$ (Eq. 7-21) caused by the inlet guide vane, and the additional distortion (Eq. 7-30) caused by the rotor.

Additional rotors and stators. In the normal multistage turbomachine the succeeding rotor and stator blade rows, respectively, add and subtract rotational velocity components of about the same magnitude. It will be assumed in the present example that tangential velocity changes across rotor and stator blade rows are equal and opposite and that the tangential velocity change induced always has the vortex distribution. Thus across

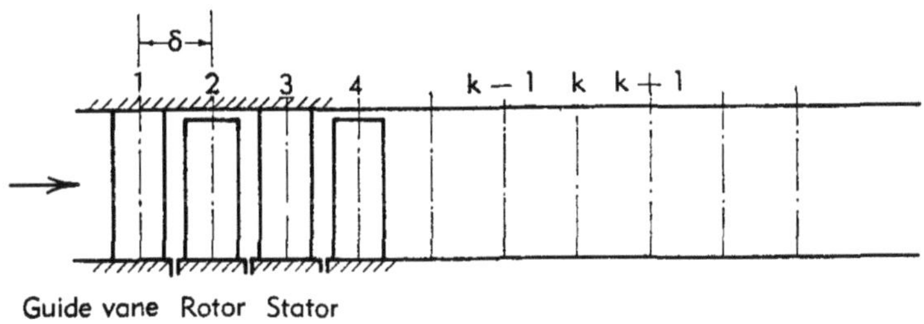

Fig. C,7c. Schematic diagram of multistage axial compressor showing blade row numbering.

the stator following the rotor discussed above the change in tangential velocity is

$$\Delta v_3^{(1)} = -aw_0 \frac{r_t}{r} \tag{7-31}$$

and since the changes in axial and tangential velocity across the first rotor are definitely of first order, the axial and tangential velocities "far upstream" of the first stator are identical with those "far upstream" of the first rotor. Consequently the axial velocity distortion across the first stator is exactly the negative of that across the first rotor, that is

$$w_3^{(1)} = -w_2^{(1)} \tag{7-32}$$

It is simple to see, then, what the complete axial velocity profile becomes downstream of each blade row of the axial compressor, a schematic diagram of which is shown in Fig. C,7c. Upstream of the guide vane the axial velocity is uniform and equal to w_0. Between the guide vane and rotor, which are supposed here to be far apart as compared with the distance required for the change in axial velocity to take place, the axial velocity is just

$$w_1(r) = w_0 \left\{1 + \frac{b^2}{2}\left[1 + \left(\frac{r_h}{r_t}\right)^2 - 2\left(\frac{r}{r_t}\right)^2\right]\right\} \tag{7-33}$$

Between the following rotor and guide vane, the total axial velocity is

$$\frac{w_2(r)}{w_0} = 1 + \frac{b^2}{2}\left[1 + \left(\frac{r_h}{r_t}\right)^2 - 2\left(\frac{r}{r_t}\right)^2\right]$$
$$+ a\left\{\frac{\left[Y_1(\kappa r_h) - \frac{r_t}{r_h}Y_1(\kappa r_t)\right]J_0(\kappa r) + \left[J_1(\kappa r_h) - \frac{r_t}{r_h}J_1(\kappa r_t)\right]Y_0(\kappa r)}{J_1(\kappa r_t)Y_1(\kappa r_h) - J_1(\kappa r_h)Y_1(\kappa r_t)}\right\} \tag{7-34}$$

where the rotor has introduced the additional vortex motion indicated by Eq. 7-22. Downstream of the ensuing stator row the vortex motion is removed and the axial velocity profile reverts to that downstream of the guide vane,

$$w_3(r) = w_1(r) \tag{7-35}$$

Similarly, downstream of the second rotor, the axial velocity profile becomes exactly what it was downstream of the first rotor, that is

$$w_4(r) = w_2(r) \tag{7-36}$$

The velocity profile continues to oscillate in this manner, attaining one definite axial velocity profile downstream of each rotor, a different one downstream of each stator.

Development of axial velocity profiles. The basic deficiency of the radial equilibrium analysis given in the preceding subarticle is that, although it gives the equilibrium axial velocity profiles far upstream and downstream of the blade row, it fails to describe the manner in which this change takes place. There is no clue in radial equilibrium theory as to whether the transition takes place very close to the blade row in question or whether it is spread out over many blade chord lengths. This question becomes of extreme importance in a multistage compressor where the change in axial velocity caused by one blade row may not be complete before the next blade row is reached. Hence the radial equilibrium velocity profiles are never actually attained and the entrance and exit angles to the blade row must be designed to axial velocity profiles that depend upon the flow fields of the neighboring blade rows. This phenomenon is known as the mutual interference between blade rows.

The methods developed for detailed calculation of the flow field about a blade row of definite chord or an actuator disk may be employed to study the development of the axial velocity profile. The general complexity of the method prohibits its application to multistage compressor problems and actually makes an undesirable amount of labor out of the routine design of a single blade row. Therefore some method for approximating the development of the axial velocity profile is needed, particularly for use in multistage turbomachines [*9,10,11,12*].

Toward this end it should be remembered that, so far as the actuator disk is concerned, the axial velocity profile is known at three stations. Far upstream of the blade row the perturbations associated with the blade row in question vanish. Far downstream of the blade row, the axial velocity perturbation is calculated from the radial equilibrium theory according to Eq. 7-11, 7-12, or 7-16. From detailed calculations carried out previously, in fact from the observation that the radial velocity is symmetrical about the actuator disk, it follows that the change in axial velocity is half completed at the plane of the actuator disk. If $w^{(0)}(r)$ is the axial velocity far upstream of the blade row and $w^{(1)}(r, z)$ is the perturbation axial velocity, it is appropriate to denote $w^{(1)}(r, \infty)$ the perturbation axial velocity far downstream of the blade row calculated from radial equilibrium theory. Hence, making the origin $z = 0$ in the plane of the actuator disk and taking $w^{(1)}(r, \infty)$ as being known from radial equilibrium theory, the axial velocity at $z = -\infty$ is

$$w^{(0)}(r) \tag{7-37}$$

at $z = 0$

$$w^{(0)}(r) + \tfrac{1}{2}w^{(1)}(r, \infty) \tag{7-38}$$

and at $z = \infty$

$$w^{(0)}(r) + w^{(1)}(r, \infty) \tag{7-39}$$

Now it is clear upon inspection of the complete solution for axial velocity profile, such as that given in Eq. 4-27 and 4-28, that each Fourier-Bessel component of the axial velocity disturbance decays upstream of the actuator disk by an exponential factor $e^{-\kappa_n|z|}$ where κ_n is the appropriate characteristic value for the Fourier-Bessel component considered. For an approximation it is sufficiently accurate to assume that all components decay at the same rate or, in other words, that the entire axial velocity perturbation decays as some exponential function $e^{\lambda\frac{z}{r_t - r_h}}$ when $z < 0$. Referring again to Eq. 7-37 and 7-38 it is appropriate to write

$$w = w^{(0)}(r) + \tfrac{1}{2}w^{(1)}(r, \infty)e^{\lambda\frac{z}{r_t - r_h}} \tag{7-40}$$

which satisfies the requirement at both the actuator disk and far upstream. For the remainder of the axial velocity profile to develop in a manner symmetrical with the development upstream, it is necessary that

$$w = w^{(0)} + w^{(1)}(r, \infty)(1 - \tfrac{1}{2}e^{-\lambda\frac{z}{r_t - r_h}}) \tag{7-41}$$

This approximation clearly satisfies both conditions given by Eq. 7-38 and 7-39.

The problem yet remains to choose the constant λ in such a manner as to make this approximation a good one. Clearly, if the perturbation axial velocity $w^{(1)}(r, \infty)$ consisted of only the first Fourier-Bessel term in the expansion (cf. Eq. 4-27 and 4-28) the value of λ would be very nearly π since the asymptotic value of $\kappa_1 = \pi/(r_t - r_h)$. Although axial velocity disturbances encountered in practice invariably include many more than the first harmonic component, the first harmonic does constitute a large portion of the disturbance. Consequently, for all but the region very near the actuator disk, the first harmonic component dominates the flow pattern and $\lambda = \pi$ is a very reasonable value to use. Calculations were carried out by Marble [10] to determine what value of λ provided the least mean square error from an exact solution. For a blade row imparting a solid body rotation it was found that this optimum λ had the value 3.25 for a hub ratio $r_h/r_t = 0.5$ and rapidly approached the value π as r_h/r_t increased toward unity. Hence in consideration of all other approximations connected with the throughflow theory, it seems completely adequate to choose $\lambda = \pi$, or in general the first characteristic value.

Using this exponential approximation to the detailed development of the throughflow it is a simple matter to write down the solution for flow through an entrance vane imparting a solid body rotation. Referring to the example following Eq. 7-22 for which radial equilibrium flow was computed,

$$w^{(0)}(r) = w^{(0)} \qquad \text{and} \qquad w^{(1)}(r, \infty) = \frac{w_0 b^2}{2}\left[1 + \left(\frac{r_h}{r_t}\right)^2 - 2\left(\frac{r}{r_t}\right)^2\right]$$

If again this entrance vane is represented by an actuator disk located at $z = 0$, then the exponential approximation to the axial velocity field becomes

$$w(r, z) = w_0 \left\{1 + \frac{b^2}{4}\left[1 + \left(\frac{r_h}{r_t}\right)^2 - 2\left(\frac{r_h}{r_t}\right)^2\right] e^{\pi\frac{z}{r_t - r_h}}\right\} \qquad z \to 0 \qquad (7\text{-}42)$$

and

$$w(r, z) = w_0 \left\{1 + \frac{b^2}{2}\left[1 + \left(\frac{r_h}{r_t}\right)^2 - 2\left(\frac{r}{r_t}\right)^2\right]\left(1 - \frac{1}{2}e^{-\pi\frac{z}{r_t - r_h}}\right)\right\} \qquad (7\text{-}43)$$

The development of this velocity profile is shown in Fig. C,7d for various distances upstream and downstream of the actuator disk. Comparison with the exact (linearized) solution indicates negligible error of this approximation except in the immediate vicinity of the actuator disk, that is for $|z/(r_t - r_h)| < 0.1$. The whole actuator disk concept breaks down in this region anyway because this distance is usually well within the blade chord. Consequently the exponential approximation, Eq. 7-42 and 7-43, may be used wherever the actuator disk itself is valid.

The general utility of the exponential approximation may be extended by superposition of actuator disk solutions, using exponential approximation, to make up a blade of finite chord. Actually this is an unnecessarily complex procedure because a quite adequate approximation to a blade row of finite chord may be obtained with two, or at most three, actuator disks situated at appropriate axial positions along the blade chord.

Using the exponential approximation it is possible to discuss the problems of mutual interference between blade rows in a multistage axial

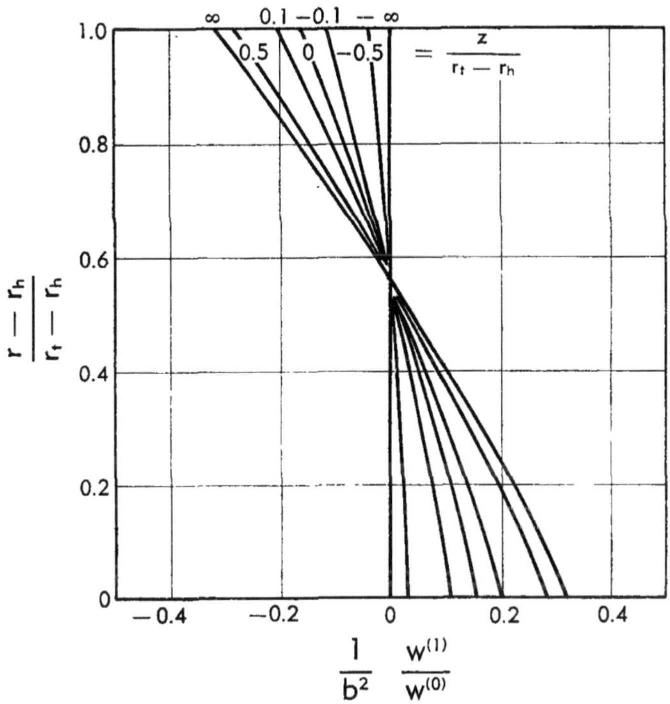

Fig. C,7d. Development of axial velocity profile upstream and downstream of entrance guide vane, calculated using exponential approximation, $v^{(1)} = bw^{(0)}r/r_t$, $r_h/r_t = 0.60$.

compressor. Consider, for example, the multistage compressor for which the radial equilibrium axial velocity profiles were determined in the work following Eq. 7-19. Let the individual blade rows be separated by a distance δ so that the entrance guide vane is represented by an actuator disk at $z = 0$, the first rotor by an actuator disk at $z = \delta$, the first stator at $z = 2\delta$, and so on, as indicated in Fig. C,7c. The change in axial velocity imposed by each of these blade rows is known and, hence, to determine the complete flow field it is merely necessary to superpose the effects of individual blade rows in the appropriate manner.

It is convenient to denote the radial equilibrium axial velocity perturbation across the guide vane by

$$\Delta w_z^{(2)} \equiv w_0 \frac{b^2}{2}\left[1 + \left(\frac{r_h}{r_t}\right)^2 - 2\left(\frac{r}{r_t}\right)^2\right] \tag{7-44}$$

and the magnitude of radial equilibrium axial velocity perturbation across each of the other rows by

$$\Delta w^{(1)} \equiv$$

$$w_0 a \left\{\frac{\left[Y_1(\kappa r_h) - \frac{r_h}{r_t} Y_1(\kappa r_t)\right] J_0(\kappa r) + \left[J_1(\kappa r_h) - \frac{r_t}{r_h} J_1(\kappa r_t)\right] Y_0(\kappa r)}{J_1(\kappa r_t) Y_1(\kappa r_h) - J_1(\kappa r_h) Y_1(\kappa r_t)}\right\} \tag{7-45}$$

The perturbation given by Eq. 7-45 will appear with a positive sign for a rotor blade row and with a negative sign for a stator. The perturbation axial velocity flow field for the guide vane is then

$$\Delta_1 w(r, z) = \begin{Bmatrix} \Delta w_z^{(2)} \cdot \frac{1}{2} e^{\pi \frac{z}{r_t - r_h}} & z \leq 0 \\ \Delta w_z^{(2)}\left(1 - \frac{1}{2} e^{-\pi \frac{z}{r_t - r_h}}\right) & z \geq 0 \end{Bmatrix} \tag{7-46}$$

while that for the first rotor is just

$$\Delta_2 w(r, z) = \begin{Bmatrix} \Delta w^{(1)} \cdot \frac{1}{2} e^{\pi\left(\frac{z-\delta}{r_t - r_h}\right)} & z \leq \delta \\ \Delta w^{(1)}\left(1 - \frac{1}{2} e^{-\pi\left(\frac{z-\delta}{r_t - r_h}\right)}\right) & z \geq \delta \end{Bmatrix} \tag{7-47}$$

The first stator imparts a disturbance which is the negative of that due to the first rotor and is displaced a distance δ farther downstream. A similar procedure will give the flow field for any of the subsequent rotors and stators. In general the perturbation associated with any of the blade rows downstream of the stator may be written

$$\Delta_n w(r, z) = \begin{Bmatrix} (-1)^n \Delta w^{(1)} \cdot \frac{1}{2} e^{\pi\left[\frac{z-(n-1)\delta}{r_t - r_h}\right]} & z \leq (n-1)\delta \\ (-1)^n \Delta w^{(1)}\left(1 - \frac{1}{2} e^{-\pi\left[\frac{z-(n-1)\delta}{r_t - r_h}\right]}\right) & z \geq (n-1)\delta \end{Bmatrix} \tag{7-48}$$

It is now simply a matter of summation over the perturbations given by Eq. 7-46 and 7-48 to find the axial velocity field at any point of the multistage axial turbomachine. Suppose that there are a total of N stages and hence N rotors and N stators following an entrance guide vane. The various blade rows are then numbered from 1 through $2N + 1$. If the axial velocity distribution is desired in the section between the kth and $(k + 1)$th blade rows, it may be written down using the appropriate solutions upstream and downstream of this section. When k is unity or larger, the complete axial velocity may be written

$$w(r, z) = w_0 + \Delta w_g^{(2)}\left(1 - \tfrac{1}{2}e^{-\pi\frac{z}{r_t - r_h}}\right)$$
$$+ \sum_{n=2}^{k} (-1)^n \Delta w^{(1)}\left(1 - \tfrac{1}{2}e^{-\pi\left[\frac{z-(n-1)\delta}{r_t - r_h}\right]}\right) + \sum_{k+1}^{2N+1} (-1)^n \Delta w^{(1)} \cdot \tfrac{1}{2}e^{\pi\left[\frac{z-(n-1)\delta}{r_t - r_h}\right]}$$
$$k\delta \leqq z \leqq (k+1)\delta \quad (7\text{-}49)$$

For the velocity distribution upstream of the guide vane, i.e. the particular circumstance where $k = 0$, the solution may be written

$$w(r, z) = w_0 + \Delta w_g^{(2)} \cdot \tfrac{1}{2}e^{\pi\frac{z}{r_t - r_h}}$$
$$+ \sum_{n=2}^{2N+1} (-1)^n \Delta w^{(1)} \cdot \tfrac{1}{2}e^{\pi\left(\frac{z-(n-1)\delta}{r_t - r_h}\right)} \qquad z \leqq 0 \quad (7\text{-}50)$$

Using Eq. 7-49 and 7-50 the flow field through the entire turbomachine may be found by computing the individual sections corresponding to particular values of k. It will be of particular interest to determine the behavior of the flow field in the vicinity of the first few blade rows to observe just how many stages are involved in the transition region until a steady repeating flow pattern is achieved. Then it will be of interest to calculate what this steady repeating pattern is for a stage deeply embedded in the compressor. From this latter it will be possible to observe the effects of mutual interference of blade rows.

The transition from undisturbed flow far upstream to the periodic pattern developed through the stages far aft of the guide vane is observed most easily by studying the shape of a stream surface which lies at the middle of the annulus far upstream of the guide vane. The radial velocity distribution is required to compute streamline shape, for the local slope is just $u^{(1)}(r, z)/w_0$ where $u^{(1)}(r, z)$ is the perturbation radial velocity. Since the continuity equation relates the radial and axial velocities as

$$\frac{1}{r}\frac{\partial ru}{\partial r} = -\frac{\partial w}{\partial z}$$

it is an elementary integration to obtain the radial velocity distribution from Eq. 7-49 and 7-50. In this manner it follows that, to the accuracy of the exponential approximation,

$$u(r, z) = \sum_{1}^{2N+1} \Delta_n u_{max} e^{-\pi\left|\frac{z-(n-1)\delta}{r_t - r_h}\right|} \quad (7\text{-}51)$$

where $\Delta_n u_{max}$ is calculated from the $\Delta_n w^{(1)}$ as

$$\Delta_n u_{max} = \frac{-\pi}{2\frac{r}{r_t}\left(1 - \frac{r_h}{r_t}\right)} \int_{r_h}^{r} \Delta_n w^{(1)}(r) r dr \quad (7\text{-}52)$$

Now since $u(r, z)/w_0$ is the flow angle, the shape of the desired stream surface is easily constructed from Eq. 7-51 either graphically or by direct integration with respect to z. The values of $\Delta_n w^{(1)}(r)$ are known and the result is shown in Fig. C,7e. The scale of the vertical motion of the streamline is grossly magnified to show the effects. It is seen that the periodic flow is established surprisingly quickly, essentially by the time the second rotor is reached. The transient state caused by the inlet vanes and the first rotor is of very short duration, partly because the distortion due to the rotor is of the same sense as that due to the guide vanes and assists in completing rapidly the distortion due to the guide vanes.

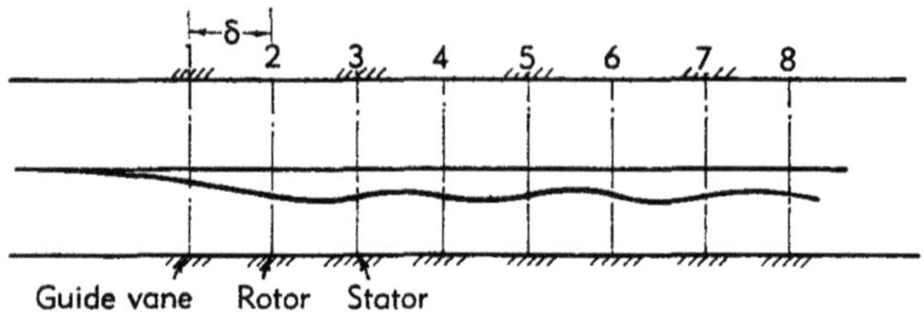

Fig. C,7e. Deflection of middle stream surface near entrance of multistage axial compressor, vertical scale magnified.

It may be deduced from these results that most of the blade rows in a multistage turbomachine are deeply imbedded, in the sense that they do not feel the influence of the compressor ends. Consequently the second point mentioned above, the steady repeating pattern for a deeply embedded stage, takes on considerable significance. To investigate this flow pattern it is convenient to transform the axis to the plane of the kth blade row, that is to introduce a variable ζ such that $z = \zeta + (k - 1)\delta$. Eq. 7-49 then becomes

$$\begin{aligned} w(r, \zeta) = w_0 &+ \Delta w_g^{(2)}\left(1 - e^{-\pi\left[\frac{\zeta+(k-1)\delta}{r_t - r_h}\right]}\right) \\ &+ \sum_{n=2}^{k} (-1)^n w^{(1)}\left(1 - \tfrac{1}{2} e^{-\pi\left[\frac{\zeta+(k-n)\delta}{r_t - r_h}\right]}\right) \\ &+ \sum_{k+1}^{2N+1} (-1)^n \Delta w^{(1)} \tfrac{1}{2} e^{\pi\left[\frac{\zeta+(k-n)\delta}{r_t - r_h}\right]} \end{aligned} \tag{7-53}$$

In anticipation of making an infinite number of blade rows both upstream and downstream of the case in question, it is appropriate to introduce new indices for each of the two sums in Eq. 7-53. In the sum $\sum_{n=2}^{k}$ call $j = k - n$ and in the sum $\sum_{k+1}^{2N+1}$ call $j = n - k - 1$, then with a little rearrangement,

$$w(r, \zeta) = w_0 + \Delta w_g^{(2)}\left(1 - \tfrac{1}{2}e^{-\pi\left[\frac{\zeta + (k+1)\delta}{r_t - r_h}\right]}\right) + \tfrac{1}{2}[1 + (-1)^k]\Delta w^{(1)}$$

$$+ (-1)^{k-1}\frac{\Delta w^{(1)}}{2}\left(e^{-\pi\zeta}\sum_0^{k-2}(-1)^j e^{-j\pi\delta} + e^{\pi\zeta}e^{-\pi\delta}\sum_0^{2N-k}(-1)^j e^{-j\pi\delta}\right) \quad (7\text{-}54)$$

where the factor $\frac{1}{2}[1 + (-1)^k]$ is unity when k is even (the section under consideration is just downstream of a rotor) or zero when k is odd (the section is just downstream of a stator). Now if the stage is deeply embedded, one may consider that there are an infinite number of similar stages upstream and downstream of the section being investigated. Thus both k and $2N - k$ approach ∞ so that the coefficient of $\Delta w_g^{(2)}$ becomes unity and both series may be summed simply. After a little manipulation the result may be written

$$w(r, \zeta) = w_0 + \Delta w_g^{(2)} + \frac{\Delta w^{(1)}}{2} + (-1)^k\frac{\Delta w^{(1)}}{2}\left[1 - \frac{\cosh \pi\left(\frac{\zeta - \frac{1}{2}\delta}{r_t - r_h}\right)}{\cosh \pi\left(\frac{\frac{1}{2}\delta}{r_t - r_h}\right)}\right] \quad (7\text{-}55)$$

The flow thus consists of a mean velocity given by $w^{(0)} + \Delta w_g^{(2)} + (\Delta w^{(1)}/2)$ and a component,

$$(-1)^k\frac{\Delta w^{(1)}}{2}\left[1 - \frac{\cosh \pi\left(\frac{\zeta - \frac{1}{2}\delta}{r_t - r_h}\right)}{\cosh \pi\left(\frac{\frac{1}{2}\delta}{r_t - r_h}\right)}\right] \quad (7\text{-}56)$$

which fluctuates along the z axis. The mean velocity profile is given by the sum of the average velocity w_0, the distortion $\Delta w_g^{(2)}$ caused by the guide vanes and half of the distortion produced by the next rotor. The fluctuating part reaches its maximum value halfway between the two blade rows, that is where $\zeta = \frac{1}{2}\delta$. Here it is equal to

$$(-1)^k\frac{\Delta w^{(1)}}{2}\left[1 - \frac{1}{\cosh \pi\left(\frac{\frac{1}{2}\delta}{r_t - r_h}\right)}\right] \quad (7\text{-}57)$$

Clearly then, if the blade spacing becomes very large,

$$\cosh\frac{\pi}{2}\frac{\delta}{r_t - r_h} \to \infty$$

and the fluctuation in axial velocity profile is just

$$(-1)^k\frac{\Delta w^{(1)}}{2} \quad (7\text{-}58)$$

This is, of course, just the result of simple radial equilibrium theory which, as already pointed out, does hold exactly if the blade rows are spaced infinitely far apart along the z axis.

When the blade rows are not spaced a large distance apart, the flow fields of adjacent blade rows tend to cancel each other to a certain extent, that is they interfere. The amplitude of the fluctuation in axial velocity profile from one section to the next is reduced below that given by radial equilibrium theory, Eq. 7-58, by just the factor

$$1 - \frac{1}{\cosh \frac{\pi}{2}\left(\frac{\delta}{r_t - r_h}\right)} \tag{7-59}$$

appearing in Eq. 7-57. The second term of Eq. 7-59 is designated the

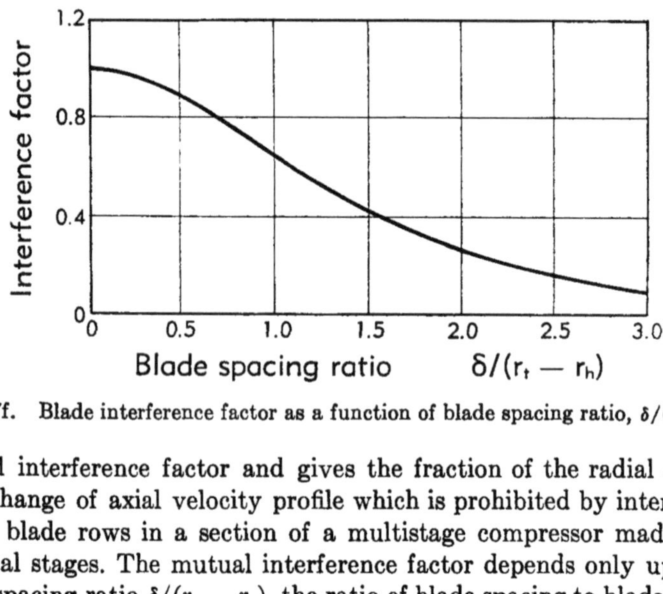

Fig. C,7f. Blade interference factor as a function of blade spacing ratio, $\delta/(r_t - r_h)$.

mutual interference factor and gives the fraction of the radial equilibrium change of axial velocity profile which is prohibited by interference of the blade rows in a section of a multistage compressor made up of identical stages. The mutual interference factor depends only upon the blade spacing ratio $\delta/(r_t - r_h)$, the ratio of blade spacing to blade length. A plot of this factor is shown in Fig. C,7f. It appears then that when the blade spacing is large, of the order of twice as large as the blade length, practically the whole radial equilibrium shift takes place after one blade row, before the next one is encountered. If the blade spacing is small, say about half of the blade length or less as in the early stages of axial compressors, then only about one third of the radial equilibrium may take place before the field of the next blade row takes effect. In view of the rapidity with which this oscillating pattern develops, as discussed in connection with Fig. C,7e, this mutual interference factor may be used with considerable assurance beyond the fourth blade row.

Effects of upstream conditions and compressibility. A few words may well be added here concerning some factors which were found to affect the exponential coefficients in the asymptotic expansions used to investigate large upstream tangential and axial velocity distortions and also compressible throughflow. In the first place it may be recalled that investigation following Eq. 4-31 showed that when the flow upstream of a blade row possessed a strong solid body rotation given by $v^{(0)} = bw_0r/r_t$, the exponential decay of this perturbation was given by $e^{-\lambda_n|z|}$, cf. Eq. 4-34 and 4-35, where

$$\lambda_n \cong \sqrt{\left(\frac{n\pi}{r_t - r_h}\right)^2 - \left(\frac{2b}{r_t}\right)^2} \tag{7-60}$$

For use in the exponential approximation under these circumstances, the ppropriate factor in the exponent should not be simply π, but rather

$$\pi\sqrt{1 - \left(\frac{2b}{\pi}\right)^2\left(1 - \frac{r_h}{r_t}\right)^2} \tag{7-61}$$

In cases of practical importance, such as mutual interference in the axial compressor discussed in the previous section, the value of the radical may be as low as 0.80. Reference to Eq. 7-59 shows that the influence of upstream solid body rotation tends to reduce the effective blade spacing δ, in the present example to 0.80 of its geometrical value. The physical effect is to increase the interference factor, as may be seen from Fig. C,7f, and consequently to reduce the axial velocity distortion from one blade row to another.

In general the effects of upstream tangential velocity and axial velocity distortion were treated by the method of asymptotic solution of the differential equation. The appropriate exponential factors were determined explicitly in Eq. 6-13. Thus when the upstream axial velocity and tangential velocity are given as $w^{(0)}(r)$ and $v^{(0)}(r)$ respectively, the appropriate factor in the exponent is

$$\pi\left\{1 - \frac{r_t - r_h}{2\pi^2}\int_{r_h}^{r_t}\left[\frac{2v^{(0)}}{r^2w^{(0)2}}\frac{d(rv^{(0)})}{dr} - \frac{r}{w^{(0)}}\frac{d}{dr}\left(\frac{dw^{(0)}}{rdr}\right) - \frac{3}{4r^2}\right]dr\right\} \tag{7-62}$$

instead of simply π. Since this expression is only a constant to be employed in the exponential approximation, the integral may be evaluated numerically without difficulty.

The effect of compressibility has been treated also by means of asymptotic methods in a previous section but not in a completely general manner. For the case of a solid body rotation $v^{(0)} = bw_0r/r_t$ and a uniform axial velocity $w^{(0)} = w_0 = \text{const}$, the asymptotic solution for the characteristic functions was carried out in the development following Eq. 6-25. In particular it was found that the decay factor, the exponential part of the characteristic functions, was given in Eq. 6-38. From this then it is clear that the appropriate constant to employ in the exponential approximation to the throughflow is not exactly π, but

$$\frac{\pi}{\sqrt{1 - M_z^2}} \left\{ 1 - \frac{2b^2}{\pi^2} \left(1 - \frac{r_h}{r_t}\right)^2 \left(1 - \frac{\gamma + 2}{8} M_z^2\right) - \frac{b^4}{6\pi^2} M_z^2 \left[(\gamma - 1) - \frac{\gamma M_z^2}{2} \right] \left(1 - \frac{r_h}{r_t}\right)^3 \left(1 - \frac{r_h}{r_t}\right) - \frac{3}{8\pi^2} \frac{r_t}{r_h} \left(1 - \frac{r_h}{r_t}\right)^2 \right\} \tag{7-63}$$

This is simply a number for any set of operating conditions and reduces to the former value for incompressible flow with upstream solid body rotation when the Mach numbers approach zero. So far as compressibility is concerned, the strongest effect is usually the factor $\sqrt{1 - M_z^2}$ in the denominator. As pointed out previously this is simply the Prandtl-Glauert correction associated with the axial velocity. Effects of the rotational velocities appear only in the other terms of the numerator. The predominant effect of compressibility is to increase the constant in the exponential approximation, opposite the influence of the solid body rotation far upstream. With reference to the multistage compressor discussed previously, the compressibility correction tends to increase the effective blade row spacing along the axis. For ordinarily encountered axial velocities the compressibility correction may increase the value of the appropriate exponential constant from π to 1.15π. It is also worth noting that in a conventional compressor the two main corrections to the exponential constant, the solid body rotation imparted by the guide vane and the compressibility correction, tend very nearly to compensate each other and make the appropriate constant π.

Variable hub and tip radii. It has recently been pointed out to the author by G. Oates that it is possible to develop an exponential approximation for the effect of variable hub and tip radii, the exact solution for which was discussed previously in detail. It has not as yet been developed to the extent of exponential approximation for perturbations due to blade rows and has not at all found its way into practice. Consequently the discussion here will be limited to the example of hub radius variations which was worked out in detail following Eq. 5-11. Take the amplitude of the sine wave to be αr_h so that the hub radius is r_h upstream of the contraction and $r_h(1 + 2\alpha)$ downstream. If now the axial velocity profile is uniform and the tangential velocity vanishes (or has the distribution of a vortex) far upstream of the contraction, then the axial velocity profile is certainly uniform far downstream of the profile. By continuity, the uniform perturbation on the axial velocity far downstream is

$$w^{(1)}(r, \infty) = \frac{2r_h^2\alpha}{r_t^2 - r_h^2} \tag{7-64}$$

and according to Eq. 5-23, it is clear then that

$$\frac{2r_h^2}{r_t^2 - r_h^2} = \frac{\pi r_h}{l}\left[\frac{I_0\left(\frac{\pi r}{2l}\right)K_1\left(\frac{\pi r_t}{2l}\right) - I_1\left(\frac{\pi r_t}{2l}\right)K_0\left(\frac{\pi r}{2l}\right)}{I_1\left(\frac{\pi r_h}{2l}\right)K_1\left(\frac{\pi r_t}{2l}\right) - I_1\left(\frac{\pi r_t}{2l}\right)K_1\left(\frac{\pi r_h}{2l}\right)}\right]$$

$$+ \frac{r_h}{l}\frac{\pi^2}{2l}\sum_1^\infty \frac{2 - e^{-\kappa_n z}\cosh \kappa_n \zeta}{(\pi/2l)^2 + \kappa_n^2}\left\{\frac{J_0(\kappa_n r)Y_1(\kappa_n r_t) - J_1(\kappa_n r_t)Y_0(\kappa_n r)}{r_h[J_0(\kappa_n r_h)Y_1(\kappa_n r_t) - J_1(\kappa_n r_t)Y_0(\kappa_n r_h)] - r_t[J_1(\kappa_n r_h)Y_0(\kappa_n r_t) - J_0(\kappa_n r_t)Y_1(\kappa_n r_h)]}\right\} \tag{7-65}$$

In particular the radial dependence of the two functions on the right-hand side of Eq. 7-65 cancels to give a constant result. Employing this relation, it appears upon writing Eq. 5-22 at the point $z = 0$ in the center of the converging section, that

$$\frac{w^{(1)}(r, 0)}{w^{(0)}} = \frac{r_h^2 \alpha}{r_t^2 - r_h^2} \tag{7-66}$$

and consequently the axial velocity is uniform here also. This result could also have been obtained directly by a symmetry argument with respect to $z = 0$.

The exponential approximation in this case consists in noting that, in addition to the above observations, the terms of principal importance in the summations are those corresponding to the first characteristic value, $\kappa_1 \approx \pi/(r_t - r_h)$. Now denoting

$$T(r) = -\frac{\pi r_h}{l}\left[\frac{I_0\left(\frac{\pi r}{2l}\right)K_1\left(\frac{\pi r_t}{2l}\right) - I_1\left(\frac{\pi r_t}{2l}\right)K_0\left(\frac{\pi r}{2l}\right)}{I_1\left(\frac{\pi r_h}{2l}\right)K_1\left(\frac{\pi r_t}{2l}\right) - I_1\left(\frac{\pi r_t}{2l}\right)K_1\left(\frac{\pi r_h}{2l}\right)}\right] \tag{7-67}$$

The axial velocity distribution may be denoted in the following approximate form:

$$\frac{w^{(1)}(r, z)}{w^{(0)}} = \alpha\left[\frac{1}{(r_t/r_h)^2 - 1} + \frac{1}{2}T(r)\right]\frac{1}{2}\cosh \pi\frac{l}{r_t - r_h}\, e^{\pi\frac{z}{r_t - r_h}} \qquad z \leqq l \tag{7-68}$$

$$\frac{w^{(1)}(r, z)}{w^{(0)}} = \alpha\left[\frac{1}{(r_t/r_h)^2 - 1} + \frac{1}{2}T(r)\right]\left(e^{-\pi\frac{l}{r_t - r_h}}\sinh \pi\frac{z}{r_t - r_h} - \sin \pi\frac{z}{2l}\right) + \frac{\alpha}{(r_t/r_h)^2 - 1}\left(1 + \sin \pi\frac{z}{2l}\right) \qquad -l \leqq z \leqq l \tag{7-69}$$

$$\frac{w^{(1)}(r, z)}{w^{(0)}} = \alpha\left[\frac{2}{(r_t/r_h)^2 - 1} + T(r)\right]\left(1 - \frac{1}{2}\cosh \pi\frac{l}{r_t - r_h}\, e^{-\pi\frac{z}{r_t - r_h}}\right) + \frac{2\alpha}{(r_t/r_h)^2 - 1} \qquad l \leqq z \tag{7-70}$$

Since $T(r)$ can be evaluated directly, this approximation for the axial velocity perturbation is very simple to use. Its accuracy is also adequate for any but extremely unusual needs.

The axial velocity profiles have been calculated at several axial positions using the exponential approximation given above, and the results

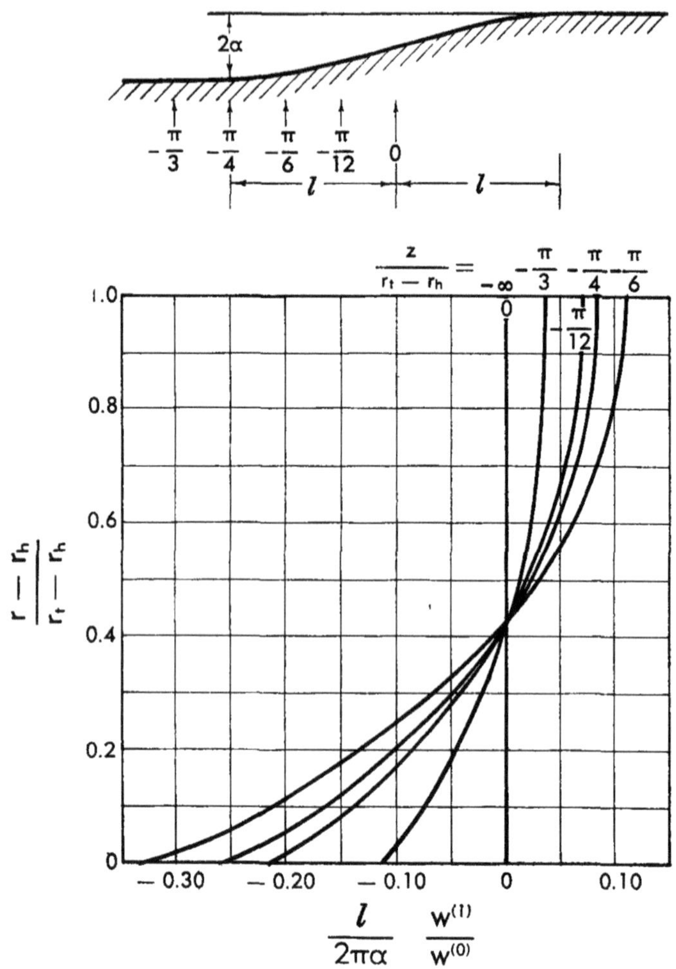

Fig. C,7g. Approximate axial velocity profiles caused by variation in hub radius, $r_h/r_t = 0.6$.

are shown in Fig. C,7g. Comparison with the exact solution suggests that they are more reliable than a significant number of terms in the Fourier-Bessel expansion and, of course, incomparably easier to calculate. Combined with the axial velocity distortion associated with an actuator disk, the results give a more adequate description of the throughflow for a blade row in channel of varying cross section. It is obvious that a similar approximation may be developed for a variable tip radius.

In summary it may be said that the exponential approximation combined with a radial equilibrium theory provides all of the throughflow information and the required accuracy for turbomachine technology. While some specific problems may require the detailed exact treatment, it is probable that, when these problems have been investigated thoroughly, they too will permit approximate treatment similar to the exponential approximation.

C,8. Cited References

1. Ruden, P. Untersuchungen über einstufige Axialgeblase. *Luftfahrtforschung 14*, 325–346, 458–473 (1937). (Transl. *NACA Tech. Memo. 1062*, 1944.)
2. Traupel, W. *Neue Allegemeine Theorie der Mehrstufigen Axialen Turbomaschine.* Leeman, Zurich, 1942.
3. Bowen, J. T., Sabersky, R. H., and Rannie, W. D. Theoretical and experimental investigations of axial flow compressors. *Summary Rept., Contract N6-ori-102, Task Order 4, Office of Nav. Research, Calif. Inst. Technol. Mech. Eng. Lab.*, Jan. 1949.
4. Meyer, R. Beitrag zur Theorie feststehender Schaufelgitter. *Mitt. Inst. Aerodynamik (E. T. H., Zurich) 11*, 1946.
5. Wu, C. H. General through-flow theory of fluid flow with subsonic or supersonic velocity in turbomachines of arbitrary hub and casing shapes. *NACA Tech. Note 2302*, 1951.
6. Wu, C. H., and Brown, C. A. Method of analysis for compressible flow past arbitrary turbomachine blades on general surface of revolution. *NACA Tech. Note 2407*, 1951.
7. Wu, C. H. A general theory of three-dimensional flow in subsonic and supersonic turbomachines of axial-, radial-, and mixed-flow types. *NACA Tech. Note 2604*, 1952.
8. Marble, F. E. The flow of a perfect fluid through an axial turbomachine with prescribed blade loading. *J. Aeronaut. Sci. 15*, 473–485 (1948).
9. Marble, F. E. Some problems concerning the three-dimensional flow in axial turbomachines. Presented at Annual Meeting, *Inst. Aeronaut. Sci.*, Jan. 1949. *Preprint 182.*
10. Marble, F. E., and Michelson, I. Analytical investigation of some three-dimensional flow problems in turbomachines. *NACA Tech. Note 2614*, 1952.
11. Railly, J. W. The flow of an incompressible fluid through an axial turbo-machine with any number of rows. *Aeronaut. Quart. 3*, Sept. 1951.
12. Horlock, J. H. Some actuator disc theories for the flow of air through axial turbo-machines. *Brit. Aeronaut. Research Council Repts. and Mem. 3030*, 1952.
13. Horlock, J. H. The compressible flow through actuator disks. *Aeronaut. Quart. 9*, 110–130 (1958).
14. Wu, C. H., and Wolfenstein, L. Application of radial-equilibrium conditions to axial-flow compressor and turbine design. *NACA Rept. 955*, 1950. (Formerly issued as *NACA Tech. Note 1795.*)
15. Weske, J. R. Fluid dynamic aspects of axial-flow compressors and turbines. *J. Aeronaut. Sci. 14*, 651–656 (1947).
16. Hawthorne, W. R. Rotational flow through cascades. *Quart. J. Mech. and Appl. Math. 8, Part 3*, Sept. 1955.
17. McCune, J. E. The transonic flow field of an axial compressor blade row. *J. Aeronaut. Sci. 25*, 616–626 (1958).
18. Bauersfeld, W. Zuschrift an die Redaktion. *Z. Ver. deut. Ing. 49*, 2007, 2008 (1905).

19. Bragg, S. L., and Hawthorne, W. R. Some exact solutions of the flow through annular cascade actuator discs. *J. Aeronaut. Sci. 17*, 243–249 (1950).
20. Gravalos, F. G. A laminar theory of the flow through a turbomachine. *Rensselaer Polytech. Inst. Bull., Eng. and Sci. Series 62*, Jan. 1950.
21. Monroe, G. M. *A Study of Compressible Perfect Fluid Motion in Turbomachines with Infinitely Many Blades*. Ph.D. Thesis. Grad. School of Aeronautics, California Inst. Technol., 1951.
22. Rannie, W. D. Three dimensional flow in axial turbomachines with large free stream vorticity. Proceedings of "The Traveling Seminar," June 1954. Published by *Advisory Group for Aeronaut. Research and Development, NATO*, Paris.
23. Erdelyi, A. *Asymptotic Expansions*. Dover, 1956.

SECTION D

EXPERIMENTAL TECHNIQUES

JOHN R. ERWIN

D,1. Introduction. The design of an aircraft gas turbine component, whether air inlet, compressor, diffuser, combustor, turbine, afterburner, exhaust nozzle, or silencer, is concerned with the compressible three-dimensional flow of viscous, turbulent gases through passages or rows of closely spaced blades having high aerodynamic and structural loading. The theoretical analysis of gas turbine element detail has not yet progressed to a position permitting complete design from values obtained solely by calculation. Compressor and turbine design will probably be based largely on experimental results for many years because of the differences in behavior between real fluids and the ideal, nonviscous, incompressible fluids more readily dealt with theoretically.

From the standpoint of developing high performance power plants, the gas turbine has an advantage over the piston engine in that the intake, compression, combustion, expansion, and exhaust phases of the thermal cycle are primarily steady flow processes that occur in component parts: the inlet, compressor, combustor, turbine, afterburner, and exhaust nozzle, which can be separated for individual study. The compressor and turbine components can be separated into single-stage units for more thorough instrumentation and tests over a wider range of the many variables than is possible with complete components. These single stages can be combined into components with a degree of success that will improve as a more complete understanding of the flow processes is achieved. Axial flow compressor and turbine blade sections can be treated two-dimensionally if the annular blade rows are "unwrapped" into linear cascades.

For the purpose of illustration, the methods of attacking the design problem experimentally can be separated into two approaches. The first approach is that used perforce in the design of early engines: to build the best compressor or turbine or complete engine that one can from existing knowledge, and then attempt to determine from test results where faults occurred and what remedy to use for improvement. From a large number of tests, dependent both on fortune and the quality of engineering judgment brought to bear on the problem, a satisfactory design procedure can be obtained in this manner. This can be said: Because this approach was used, engines were developed in a much shorter period of time than had a more cautious procedure been followed. Whether the further evolution of the gas turbine would advance most rapidly through this attack alone is not clearly answerable, but it seems unlikely.

The second experimental approach is that of isolating effects to permit a close study of separate phenomena in the hope of understanding each individually, and then to combine the separate simpler flows to help determine proper design for the complex flows which occur in turbomachines. In this approach one proceeds from the study of diffusers and nozzles to duct bends, two-dimensional cascades, three-dimensional cascades, single-stage components, several-stage and multistage compressors and turbines—first at low and then at high speeds. This would seem to be a process that would never reach the desired conclusion because the earlier work on which later phases were based would become obsolete before the completion of the process. However, several or most of these phases can, and should, proceed simultaneously, for the information obtained from one source can be a direct aid in understanding results obtained from the other sources. In addition, a number of successful multistage axial flow compressors have been designed solely on information obtained in low speed two-dimensional cascades. In practice, of course, both engine and component approaches are employed, in conjunction with theoretical analysis, with differing emphasis by the various groups engaged in gas turbine design.

The performance of gas turbines can be determined by tests of models of the components or by testing full scale components separately. The advantages of model testing are many: individual effects can be isolated; less power is required to drive compressors or is produced by turbines; smaller compressed air and exhausting facilities can be used; high combustor and turbine temperatures can be avoided so that simpler apparatus and instrumentation are permissible; more readily workable materials may be used; and rotational speeds, and hence stresses, can be reduced. It would appear that model testing would be much quicker and a great deal less expensive than full scale component or engine testing. While this is generally true for low speed tests of simple models, experience has shown that it is not necessarily cheaper and usually takes longer to build and test a model axial flow compressor than a full scale compressor for an engine. This result has been due more to the emphasis and priority placed by management on directly salable engines rather than to technical differences of construction.

In order that model tests be valid, significant similarity coefficients must have the same value in the model as in the full scale component. The model must be geometrically similar to the actual turbine or compressor. The flow coefficient, Mach number, Reynolds number, and the ratio of specific heats should be the same for model and full scale working fluids. If the model is geometrically similar and if the similarity coefficients have the same value as in the full scale installation, the quantities and qualities defining gas turbine performance can be accurately determined. By disregarding one or more of the necessary conditions, the test equipment required and the difficulty of making tests can be greatly reduced; for example, if Mach number similarity is disregarded, inexpensive cascades or stage rigs can be used which greatly simplify model testing. Good judgment must be exercised in extrapolating such results to the conditions of the application, however.

The conditions under which tests are conducted can influence the performance of the engine or component to be studied, and the type of instrumentation employed can determine the accuracy of measurements. A discussion of the methods of obtaining such results is therefore pertinent to this volume.

D,2. Elements of Experimental Installations. Installations for studying gas turbines or their components are usually designed to permit tests under steady state conditions. This is the usual operating condition for which the components are designed, even though engines must accelerate and though the flow through the compressor and turbine blades, which experience rapidly moving wakes of upstream blades, might more properly be considered unsteady. The flow through air intakes during supersonic flight or during flight maneuvers, through diffusers and combustion chambers following compressors or turbines, or through exhaust nozzles and silencers, is not steady. The unsteady flow case is difficult to treat either theoretically or experimentally, although considerable progress has been made in the study of rotating stall. More emphasis will certainly be placed on the study of unsteady phenomena in the future. At present, satisfactory results are being obtained by designing to minimize intermittent effects in gas turbines and assuming that the flow obeys an averaging process.

Settling chamber and intake installations intended to permit the testing of aircraft gas turbines or their components are usually designed to produce uniform flows entering the element under study. To accomplish this, settling chambers of cross-sectional area 10 to 20 times as large as the flow area of the unit under test are used. The plenum chamber usually contains several screens to reduce velocity variations followed by a honeycomb straightener to remove rotational component of the flow. Because the velocity is low, total temperature and pressure measurements can be made accurately so that surveys of the flow are unnecessary. The transition from the plenum to the test section is accomplished with bellmouths or fairings having large radii of curvature to prevent or minimize local high velocity regions and to insure uniform flow into the unit to be studied In [*1*], recommended settling chamber and intake configurations are presented. These configurations can be used directly in annular cascade and single- and multistage turbine and engine test apparatus. The principles apply to linear cascade, duct, and diffuser tests.

For axial flow compressors, the following general setup is recommended in [*1*] to be applied to that type, insofar as is practical.

Inlet. An inlet settling chamber designed to insure smooth entry of the air into the compressor should be placed immediately ahead of the compressor. In order to secure the desired uniformity of flow, one or more screens and a honeycomb straightener are placed within the depression tank. A bell-mouthed inlet should be provided within the plenum chamber for the compressor. A typical settling chamber is illustrated in Fig. D,2a.

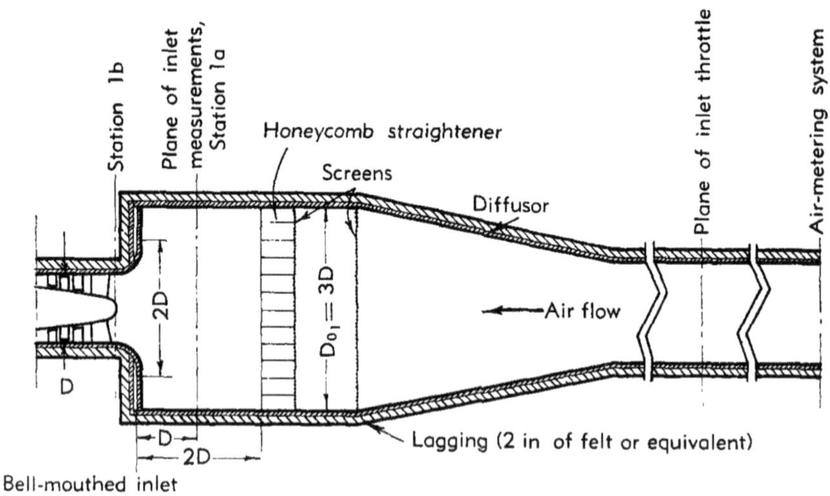

Fig. D,2a. Setup of inlet depression tank for axial flow compressor [*1*].

Outlet. The compressor may be tested either with or without a diffuser. The discharge collector should be designed so that the static pressure variations around the circumference at the outlet measuring station are less than 5 per cent of the mean dynamic pressure and so that the collector and ducting losses are low enough to permit the desired operating range of pressure ratios and flows. In [*2*], a series of tests were made on the collector illustrated in Fig. D,2b in which six different outlet configurations were used. By using outlet pipes with different diameters and by using either one or two pipes, the ratio of outlet-pipe cross-sectional area to the collector-inlet cross-sectional area was varied from 0.52 to 3.12. From these tests it was determined that the number of outlet pipes apparently had little effect on the static pressure distribution at the collector inlet. The total pressure loss was smaller with one outlet pipe than with two, for the same total flow area, except for the smallest outlet configuration investigated; the principal total pressure losses occurred at the collector inlet because of the sudden expansion in flow area.

Air facilities. Whenever possible, altitude-exhaust and dry refrigerated or heated inlet air should be provided to obtain a wide range of inlet pressures and temperatures, in order to simulate the conditions encountered in high altitude flight. At altitude, the Mach number or corrected speed and Reynolds number of an engine are different than in sea level flight, and significant changes in engine performance can occur. At extreme altitudes, satisfactory combustion is not easily obtained, particularly if altitude starting is required. The necessary information is difficult

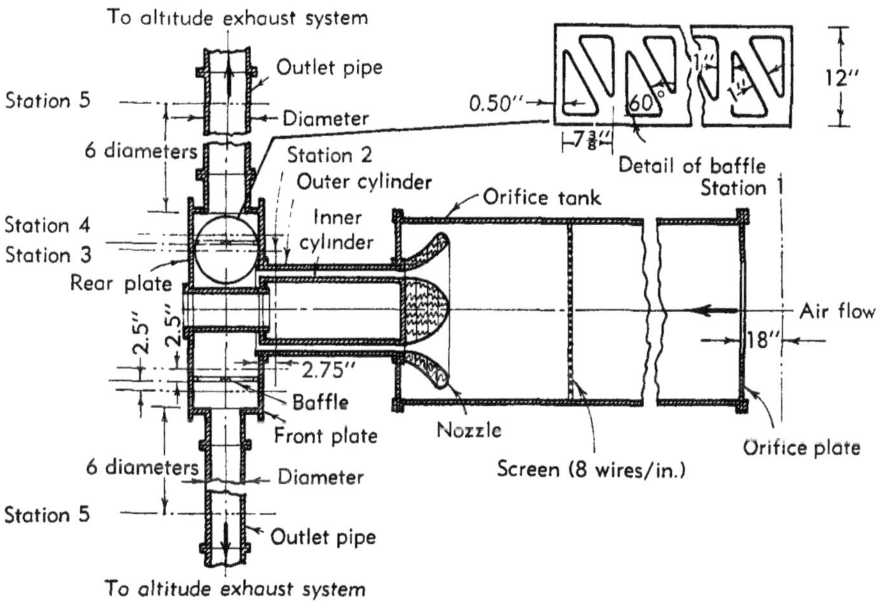

Fig. D,2b. Schematic diagram of research unit for axial flow compressor collector showing location of measuring stations [*2*].

to obtain from flight tests, yet it must be secured for the design of military aircraft. Extensive ground test facilities for this purpose are therefore justified.

The weight flow of the test media is measured by devices such as variable thin plate orifices, standard nozzles, or venturis, or by calibrated entrance bells. Accurate results are most readily obtained when a settling chamber containing a honeycomb to remove rotational components from the flow and several screens to equalize the velocity of the flow into the meter is used. If the operating conditions are not greatly different from those under which the calibration was made, satisfactory results can be obtained with standard meters if the installation is made as specified [*3*].

Flow measurements. In making flow measurements downstream of single-stage compressors or turbine rotors, a survey is usually made along a single radial line. Wakes from struts or blades upstream of the rotor are usually not fully eliminated in passing through a single blade row. Erroneous readings can result if proper precautions are not taken. Whenever practical, duplicate measurements should be taken along more than one radial line in the plane of the measuring station to ascertain that true sample values are obtained. The accuracy of the results is best checked through independent determination by separate methods. For example, in high speed single-stage compressor tests, the power input can be obtained by measuring the temperature rise, momentum and pressure increase, and the torque applied to the rotor. As the total pressure rise and the mass flow produced by a rotor can ordinarily be measured accurately, the efficiency can be obtained by three separate and essentially independent methods. Although not positive evidence of test validity, weight flow checks at several stations often provide a clue that instrument calibrations have changed or other improper conditions exist.

Independent variation of Reynolds number and Mach number. In order to study the operation of a gas turbine component or a model thereof as the Reynolds number is varied, either the density, the velocity, the scale of the model, or the viscosity of the test medium may be changed. Usually it is not convenient to vary the Reynolds number independently of Mach number over a significant range in test installations designed to permit rapid measurements and quick changes of the model engine component. An attempt to study similar compressor designs over a range of Reynolds number was made in Germany by Eckert [*4*]. In this investigation a compressor rotor and stator design was built in several sizes, all of the same blade camber and setting, chord-spacing ratio, and hub-tip radius ratio. As the same inner and outer diameter was maintained for all tests, the blade aspect ratio decreased as the blade chord increased. Thus, in this study, geometrical similarity was not obtained, and hence comparison of the results suffered.

The most widely used method of varying compressor Reynolds number and Mach number separately is to study the engine component in a closed circuit, permitting selection of the reference density. Auxiliary pumping equipment of high capacity and power would be required to accomplish the same result with open cycle operation. Closed circuit operation extends the usefulness of a given driving unit because compressor power requirements can be matched by varying the system density. A further advantage of the closed cycle is that fluids other than air may be used as the test medium. Gases in which sound travels more slowly than in air permit compressors to operate at desired Mach numbers with reduced rotational speed and stress. Freon 12 (CCl_2F_2) has been used for this purpose because it is nontoxic and stable under usual testing conditions. Due to the low value of γ (the ratio of specific heats = 1.125) compared to air (γ = 1.4), the use of Freon 12 should be limited to single stages of low pressure ratio unless the component is designed for operation in this gas.

D,3. Instrumentation. Of equal importance to that of establishing valid test conditions is the necessity of providing proper instrumentation, if significant results are to be obtained. Results obtained from tests are only as good as the instrumentation which provided the data. In any comparatively new field of endeavor as that of the gas turbine, the development of special testing equipment and instrumentation becomes necessary. The high speed, unsteady flows that exist in the narrow passages between the compressor and the turbine blade rows have introduced new problems of measurement. Since erroneous conclusions can result from measurements made through the use of inadequate instrumentation, each new configuration should be analyzed to determine the type of instrument, its location and the quantities which must be measured to obtain reliable information. An excellent study of these problems has been made by Dean [*5*].

Many instruments used in the gas turbine field are neither unique nor complex. Indicating devices and methods of using them for measurements of time, weights, and forces; humidity, pressure, and temperature of still media; surface smoothness; and shaft speed to a high degree of accuracy have been in existence for some time and need not be mentioned herein. Sensing elements and indicators peculiar to, or of special interest in, this field will be discussed briefly in this section.

The technique of instrumentation for the measurement of the overall component performance characteristics has a history dating back to the early development period of the steam turbine. Today, however, there is very much interest in detail measurements to determine the performance of individual stages or particular radial sections throughout the axial flow compressor or turbine. Through such procedures the optimum values of the aerodynamic design parameters can be determined to provide a basis for modifying existing machines and for the development of improved future engines. Thus those measurements of the gas flow which will completely define all its intrinsic values are necessary. In addition, the machine speeds, temperatures, and in some cases, stresses, should be observed to obtain optimum mechanical design to insure safe operation with minimum weight.

Design of Probes. Considering first those measurements required to completely define the aerothermodynamic gas quantities at a point, it is usually sufficient that the static and total pressure, the total temperature, the direction of flow, and the gas composition be known. For the calculation of the flow relative to any rotating element, the rotational speed must also be known to relate the measurements made in the stationary or absolute coordinate system to that in the relative or rotating coordinate system. In addition, the turbulence level and vorticity are of interest for correlation with cascade test results. Because the fluid passages in jet propulsion engines are small, the adaptation of sensing probes designed for general flight or wind tunnel experimentation is not always possible. The knowledge gained in older fields of aerodynamic research has been used as a guide in the specialized field of gas turbine instrumentation. Instruments which protrude into the gas stream must be compact and rugged to prevent erroneous readings from aerodynamic interference effects and instrument deflections. Factors which affect the size and design of the instruments are the accuracy, the reaction speed, and the ease of manufacture that is desired or necessary.

Rate of response. To permit an estimation of the rate of response of instruments to pressure changes, a series of tests using small diameter tubes was made [*6*]. The tubes were connected to a manometer and a pressure difference of 50 inches of alcohol was imposed. The manometer deflection was recorded at even time intervals. Tubes having inside diameters of 0.040, 0.020, and 0.010 in. and 12, 6, and 3 in. long were tested (Fig. D,3a). Short lengths of tubes as small as 0.020-inch inside diameter can be used in instruments for surveying flows which remain steady for 30 to 40 sec. The rate of response of Langley prism and pyramid probes, described later in this section, is also indicated in Fig. D,3a. These probes have 3-inch lengths of 0.020-inch inside diameter tubing attached inside the probe stem to 12-inch lengths of 0.040-inch tubing.

A theoretical relationship for determining the lag in tubing of radius r, length l, into an instrument cavity of volume V has been derived in [*7*] and [*8*]:

$$\frac{\Delta p}{dp_2/dt} = \frac{8V\mu l}{\pi r^4 \kappa p_{\mathrm{m}} g}$$

The volume term includes one half of the tubing volume. The term p_2 is the pressure at the instrument and p_{m} is the mean pressure in the tube. The constant κ is unity for air undergoing isothermal change. The equation was derived for conditions which do not include the slip and molecular flow regions. For relationships in these regions, see [*9*] and [*10*].

Total pressure. A wide variety of total pressure tubes have been reported by Gracey [*11*] to determine the effect of angle of attack on the pressure measured by the probes over a range of subsonic, transonic, and supersonic speeds. The probe designs investigated included a wide range of external and internal details (Fig. D,3b). The first group, designated series A, had cylindrical exteriors. The second group, series B, C, and D, had 15°, 30°, and 45° conical exteriors. The E series had ogival noses. A group of shielded probes, designated series A_s, was also investigated as

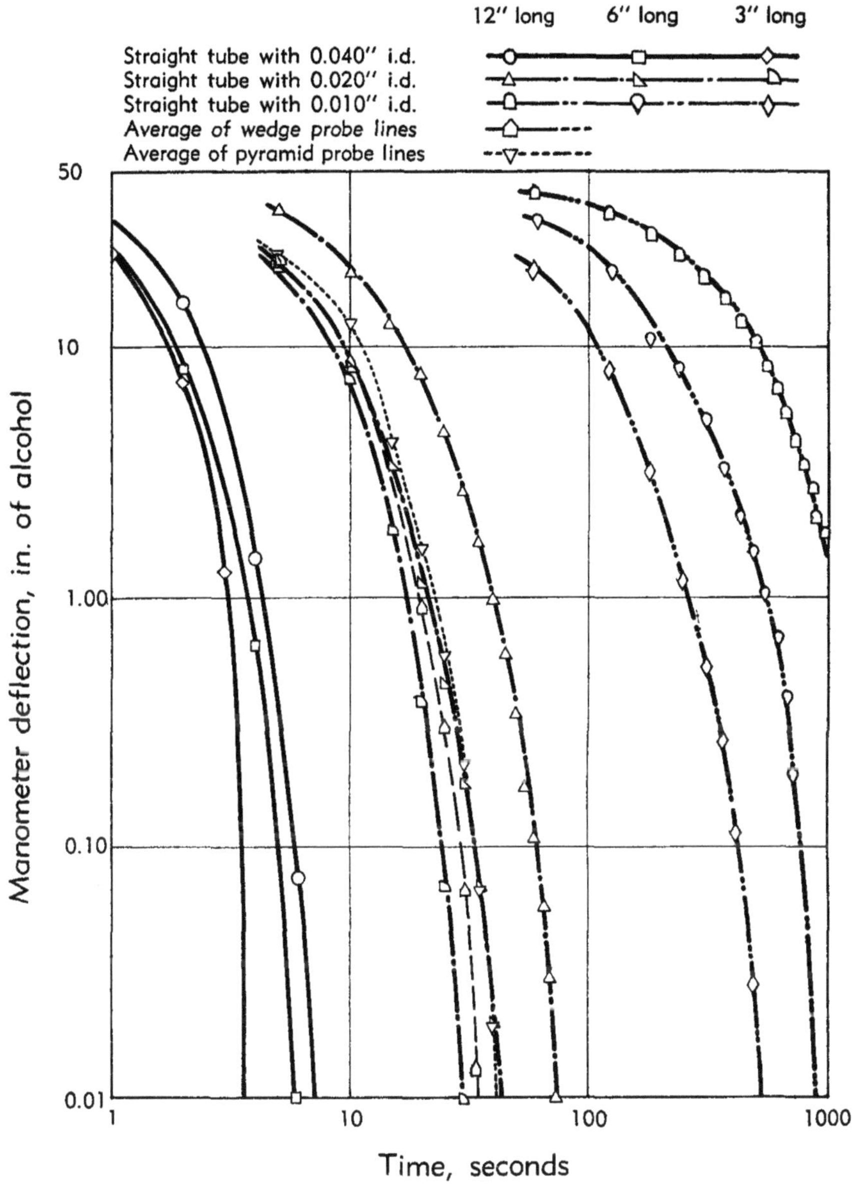

Fig. D,3a. Manometer tube deflection versus time for various size tube flow restrictions as compared to the prism and pyramid probe reaction characteristics.

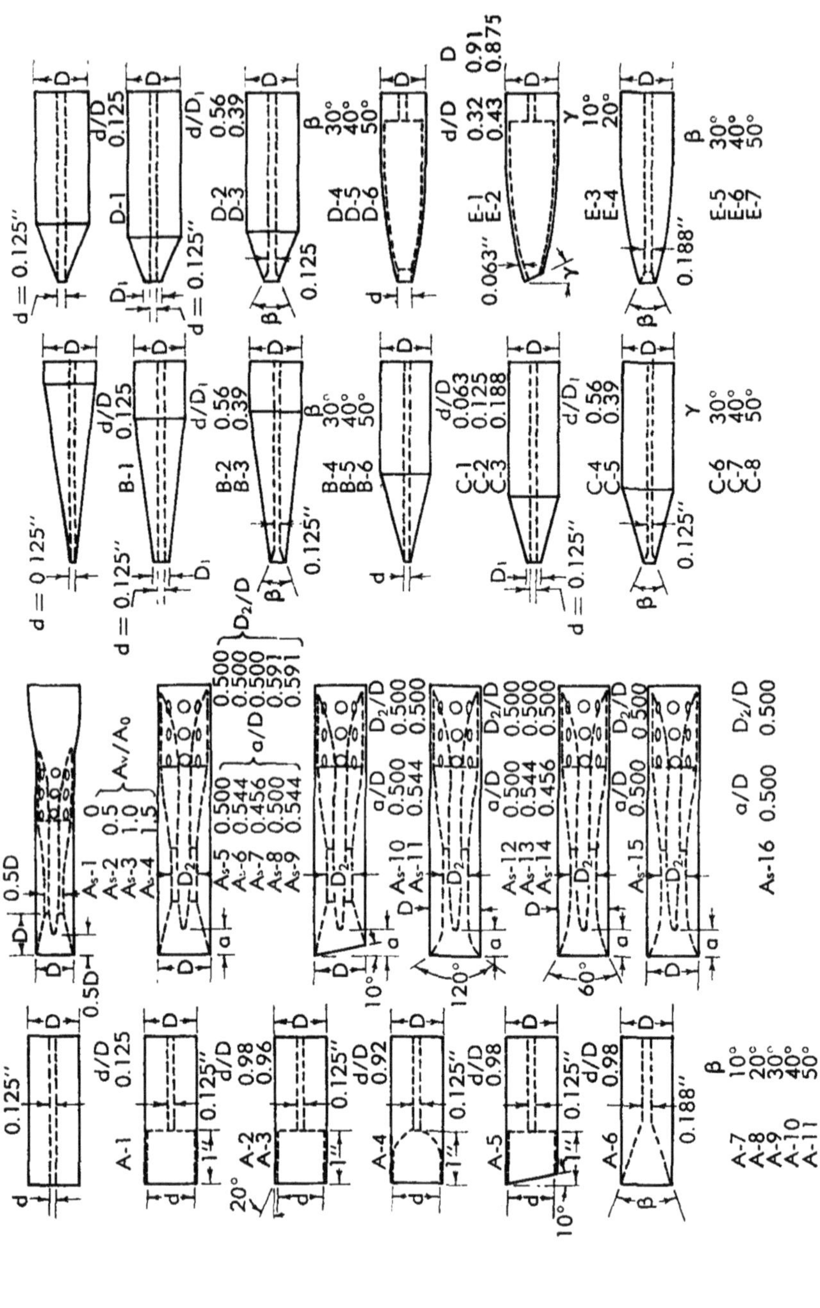

Fig. D,3b. Diagrams of total-pressure tubes. (a) Series A (cylindrical) and A_s (shielded). (Vent areas A_v–A_0 of tubes A_s-4 to A_s-16 is 1.5.) (b) Series B (15° conical); series C (30° conical); series D (45° conical); and series E (ogival).

part of this program. Most of the probes were tested over a wide range of angle of attack at a Mach number of 0.26. Several of the probes were tested at Mach numbers of about 1.6 and several were tested in the Mach number range around 0.90.

The results of this investigation are presented in terms of the sensitivity angle. The sensitivity angle is defined as the angle of attack at

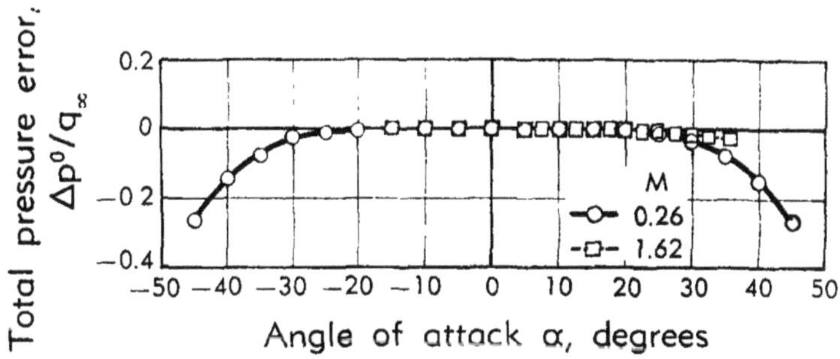

Fig. D,3c. Variation of total-pressure error with angle of attack of tube A-2 at subsonic and supersonic speeds.

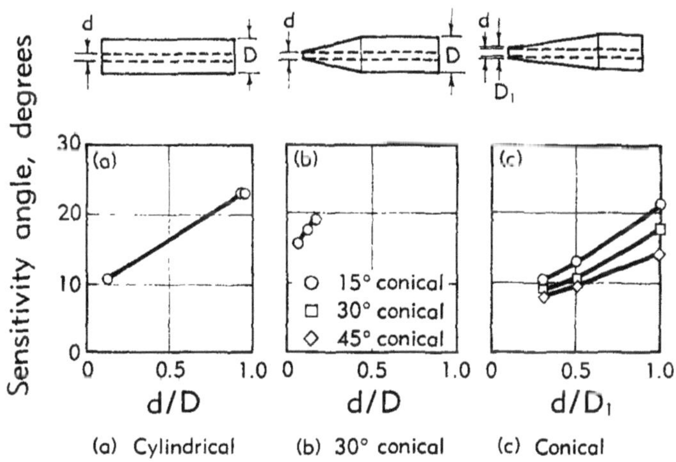

Fig. D,3d. Variation of sensitivity angle with size of impact opening of cylindrical and conical tubes. $M = 0.26$.

which the total pressure error reaches a value of 1 per cent of the indicated impact pressure. An example of the data from which the sensitivity angle of the probes was determined is given in Fig. D,3c for tube A-2. As determined from these curves the sensitivity angle is 23° at $M = 0.26$ and 29° at $M = 1.62$.

The effect of varying the ratio of the impact opening diameter to the probe diameter is shown in Fig. D,3d for typical probes at $M = 0.26$.

For subsonic speeds, enlarging the impact opening diameter with respect to the probe diameter increases the angle range over which an accurate value of total pressure can be measured. The effect of the internal chamber shape on the performance of the probes can be obtained from a comparison of the results of tubes A-2 through A-11 as given in Table D,3a. All of these probes had a large ratio of impact opening diameter to probe diameter and all had a chamber depth equal to the probe diameter. The performance of all of these probes is very good. The results indicate that the conical internal chamber has about a 1.5° greater range than the

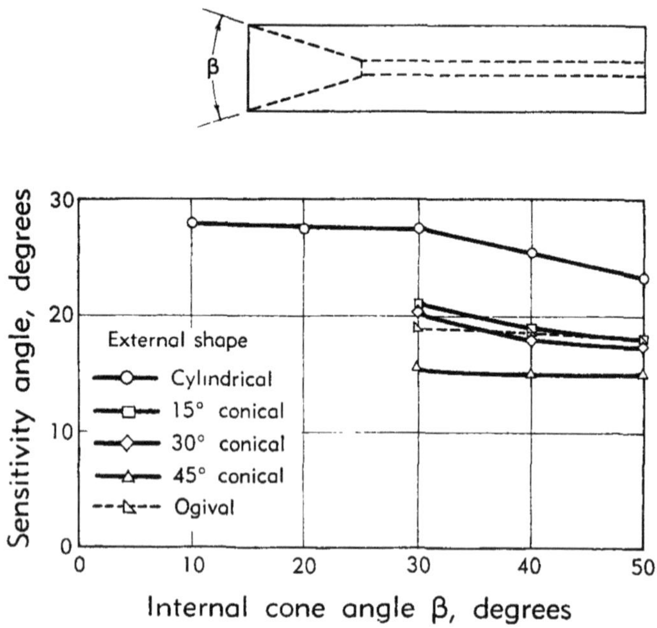

Fig. D,3e. Variation of sensitivity angle with internal cone angle for cylindrical, conical, and ogival tubes. $M = 0.26$.

cylindrical chamber and the hemispherical chamber has a range about 2.5° greater than the cylindrical chamber. The effect of varying the internal cone angle on the performance of cylindrical, conical, and ogival probes is shown in Fig. D,3e. It is apparent from this figure that internal cone angles of 30° or less are beneficial for providing maximum angle range. Fig. D,3f also provides a comparison of the effect of external contour on the performance of the probes at $M = 0.26$. For subsonic speeds the cylindrical exterior is superior to the conical and ogival tubes.

From the results presented, it can be seen that the best combination of external shape, relative impact opening diameter, and internal configuration would be a cylindrical tube with a large impact opening and a small internal conical angle. Such a tube produced a total pressure reading correct to within 1 per cent of the indicated impact pressure over a range of $\pm 27.5°$ at $M = 0.26$.

Tests of tubes E-6, E-3, and E-4 having slanted impact openings indicated that the total angle of attack range over which the tube is not sensibly affected by angle of attack remains essentially constant but that the insensitive range is shifted in the expected direction by an amount very nearly equal to the angle of inclination of the impact opening.

Effect of geometry on performance of shielded tubes. Significant increases in the sensitivity angle can be obtained by venting Kiel-type

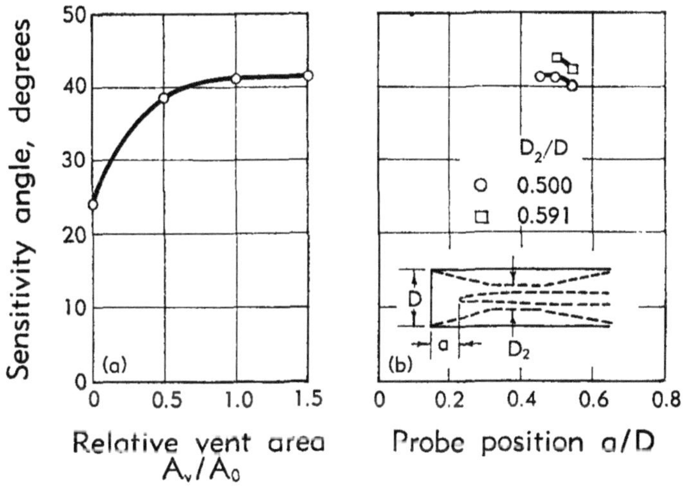

Fig. D,3f. Variation of sensitivity angle with vent area, probe position, and throat diameter of shielded tubes. $M = 0.26$. (a) Vent area. (b) Probe position and throat diameter.

shielded tubes as shown in Fig. D,3f for $M = 0.26$. As seen in part (b) of this figure, the effect of the probe position within the shield and the ratio of the shield internal to external diameter have little effect on the range of a vented Kiel tube. As in the case of the unshielded tubes, the effect of a slant on the front of the shielded tubes had the result of shifting the center of the unaffected range in the expected direction by an amount almost equal to the angle of slant. The effect of internal entry angle into shielded tubes is shown on Fig. D,3g. Tube A_s-5 had a conical internal configuration with an internal cone angle of 28°. Tubes A_s-12 and A_s-15 were constructed with curved internal passages with internal entry angles of 12° and 60° respectively. From Fig. D,3g, it can be seen that a significant increase in working range can be achieved by increasing the internal entry angle of shielded probes. A vented, shielded tube having a highly curved entrance exhibited acceptable performance over an angle of attack range of $\pm 63°$ at $M = 0.26$.

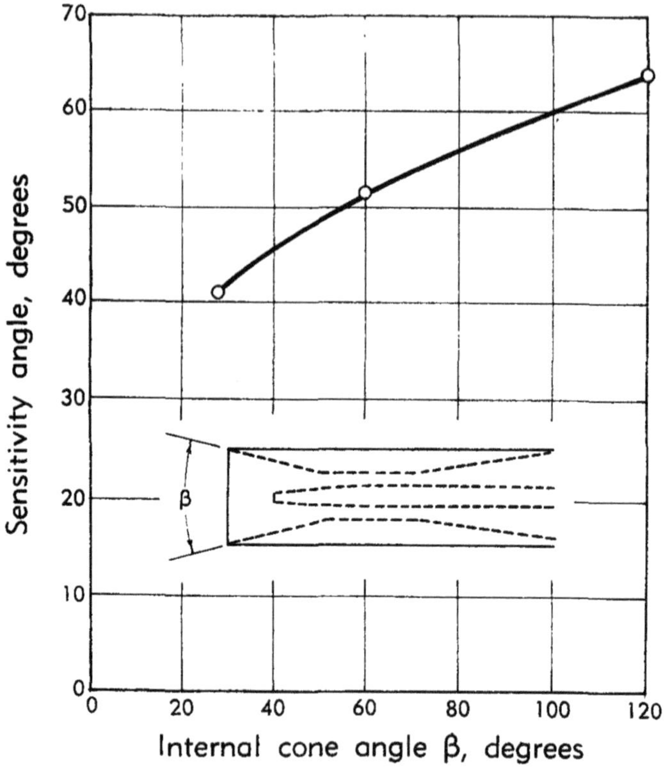

Fig. D,3g. Variation of sensitivity angle with angles of entry of shield of three shielded tubes at $M = 0.26$. Tubes A_s-5, A_s-12, and A_s-15.

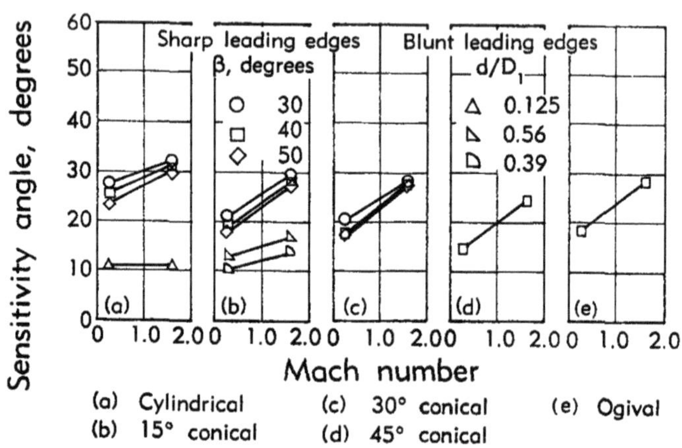

Fig. D,3h. Variation of sensitivity angle with Mach number of tubes having cylindrical, conical, and ogival external shapes with sharp and blunt leading edges. (a) Cylindrical. (b) 15° conical. (c) 30° conical. (d) 45° conical. (e) Ogival.

Effect of Mach number. In Fig. D,3h, the sensitivity of cylindrical, conical, and ogival tubes of varying impact openings and leading edge angles at $M = 1.62$ are compared with values at 0.26. It is seen that in general the effect of Mach number is to increase the useful range for probes of this type. That some justification exists for drawing a straight line between Mach numbers of 0.26 and 1.62 is illustrated on Fig. D,3i, which presents results for tube E-2 for Mach numbers from 0.6 to 1.1. Also shown on Fig. D,3i are values estimated for the sensitivity of tube E-2, estimated from test data taken from similar tubes at Mach numbers of 0.26 and 1.62. From this figure there is evidence that the sensitivity angle of nonshielded tubes varies linearly with Mach number in the range from $M = 0.26$ to 1.62.

The effect of Mach number on performance for shielded tubes in the subsonic range is shown on Fig. D,3j. A decrease of about 10 per cent in

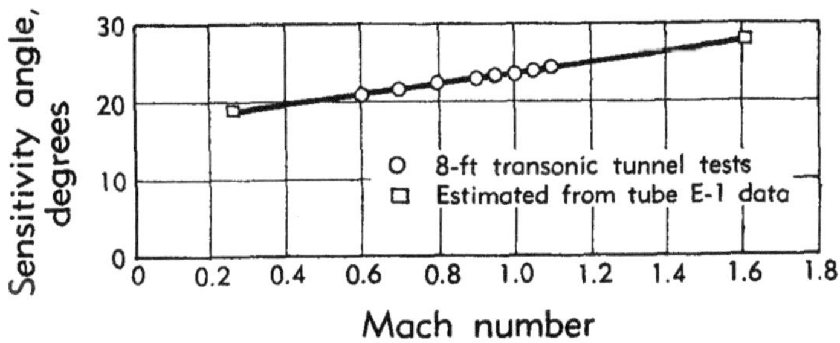

Fig. D,3i. Variation of sensitivity angle with Mach number of ogival tube E-2.

sensitivity angle is observed as the Mach number is increased from 0.26 to 0.9. The variation of sensitivity angle with Mach number for a particular shielded tube, A_s-12, is shown in Fig. D,3k, over the Mach number range from 0.26 to 1.62. The behavior of this shielded probe at $M = 1.62$ was different from the performance of the several unshielded probes tested in this Mach number range. The difference in performance exhibited by this probe was to indicate too high a value of total pressure at large angles of attack whereas the unshielded probes deviated in the direction of indicating a low total pressure at high angles of attack. Whether this behavior is typical of shielded probes or pertained only to the particular probe tested is not clear. Since the performance of probe A_s-12 reversed at about 50° angle of attack and produced a zero error at 55°, and negative values beyond 55°, it is possible that a different design of shielded probe would produce a larger sensitivity angle at $M = 1.62$. However, investigations which will require total pressure indication at angles of attack greater than ± 40 at a Mach number of 1.62 are likely to be rather few in number. Further data on Mach and Reynolds number effects on typical probes are presented in [*12*]

Total pressure tubes in struts. The measurement of gas properties in small passages such as occur in turbomachinery is particularly difficult due to the blockage which the instrument itself provides. One method of attacking this problem is to bury the instrumentation as much as possible

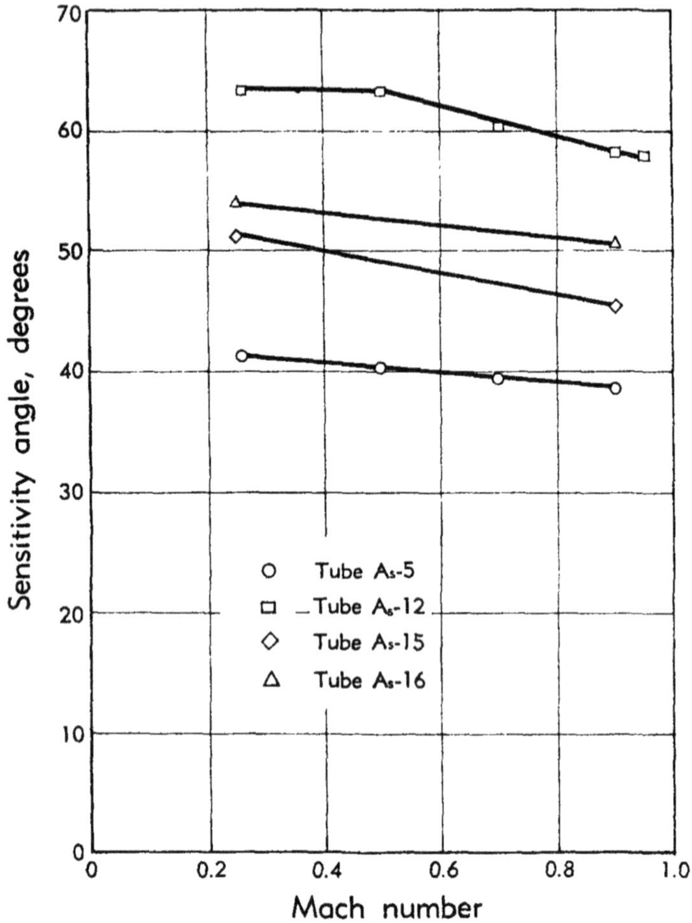

Fig. D,3j. Variation of sensitivity angle with Mach number of four tubes with shields having conical and curved entries.

within stationary objects which naturally occur in the flow path. Typical examples of this would be stationary blades in compressors or turbines or the struts which are used for support. Total pressure tubes installed in the leading edge of such objects would provide a minimum blockage to the air flow. The use of a flush tube in the leading edge of a compressor blade is appealing from the practical standpoint. However, if the stagnation point location cannot be predetermined accurately or shifts during operation, flush orifices do not provide an accurate measure of the total pressure since occurrence of the stagnation point at any location other than directly over the orifice will cause the orifice to act to some degree as a static pressure measuring device. Tests of 0.062″ outside diameter tubes located in the leading edge of an isolated airfoil having a 0.125″ leading edge diameter have been conducted to determine how far forward such a tube must extend to be insensitive to angle of attack. Protrusion of the tube one tube diameter forward of the airfoil leading edge yielded total pressure readings within 1% of the impact pressure at a Mach number of 0.50 over an angle of attack range from +4 to −9°. Protrusion of the total pressure tube 2.7 diameters increased the insensitive range to

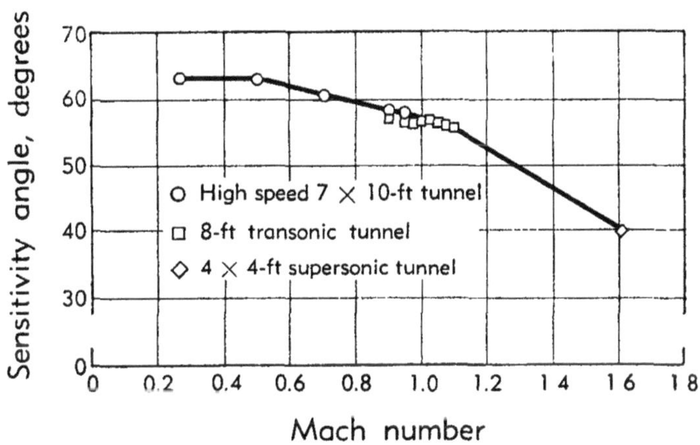

Fig. D,3k. Variation of sensitivity angle with Mach number of shielded tube A_s-16.

+9 and −13° at a Mach number of 0.5. Chamfering the internal contour of the total pressure tube with a total angle of 60° increased the insensitive range in both directions by 2° for a protrusion of 2.5 diameters. Increasing the protrusion of the chamfered tube to 6 diameters increased the range of insensitivity to ±18°, and a further extension to 8 diameters did not improve the insensitive range.

Time lag and averaging effects. Probes do not necessarily indicate a true, time-averaged value of total pressure when subjected to unsteady flows. In [*13*] an analysis is presented of the effect of square-wave pressure fluctuations on the reading of typified total pressure probes. Analytical results are compared with experimental results for probes of varying diameters and lengths inserted and withdrawn from a high speed air stream at a rate of from 3 to 50 cycles per second. The ratio of the time at maximum pressure to the time of one cycle was varied from zero to one, and the difference between maximum and minimum pressure was varied from 0.05 to 0.8 of the minimum pressure. The experimental results were in good agreement with the theory for Reynolds numbers less than 8000, when *Re* is based on the inside diameter of the probe. The averaging error was minimized when the inside diameter of the probe was made small, and the length was large. A typical comparison is shown in Fig.

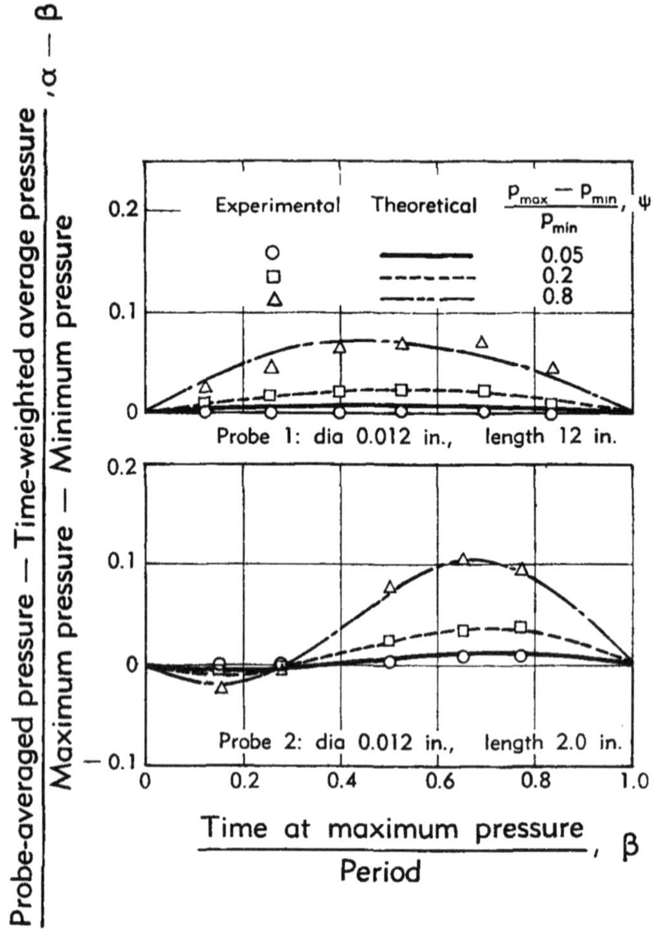

Fig. D,31. Comparison of calculated and experimental probe error. Top, probe 1, diameter = 0.012 in.; length = 12 in.; bottom, probe 2, diameter = 0.012 in.; length = 2.0 in.

D,31. When the Reynolds number of an impact probe (based on the probe radius) decreases to a value below 100, due to viscosity, a true reading of total pressure is not obtained even under steady flow conditions. At a Reynolds number of 50, the measured values are increased by about 12 per cent of the dynamic pressure for incompressible flow, or by about 2 per cent of the total pressure for compressible flow.

An investigation of the frequency response of total pressure probes having several lengths of $\frac{1}{8}$-in. inside diameter tubes leading to a $\frac{1}{2}$-in. diameter pressure transducer has been conducted by Westley [*14*]. The frequency response was measured by impressing sound waves of 130 db sound pressure level into a cavity 1.95 in. in diameter and 0.9 in. in depth. The frequency range imposed was varied from 50 to 3000 cps. The response using a tube length of 1.151 in. rose rapidly at frequencies about 1000 cps, peaking around 1700 cps, and dropped sharply to low values at 2000 cps. With a tube length of 2.558 in., the peak occurred at about 950 cps. With a 5.806-in. tube, the peak occurred around 450 cps, with a lower amplitude peak observed also at about 1300 cps. A significant damping effect was obtained by filling the tube with cotton in tests using the 2.558-in. extension, although some loss in sensitivity occurred at higher frequencies. These results, including a comparison with a hypodermic boundary layer probe tested at a sound pressure level of 140 db, are summarized in Table D,3b.

Table D,3b. Resonant frequencies, sensitivities and frequency range.

Probe	Length of tube, in.	Resonant frequency cycles/sec	Sensitivity at resonance, mv/v per lb/in.²	Maximum frequency for sensitivity within 0.157 ± 0.05 mv/v per lb/in.², cycles/sec
D	0	—	0.157	2600
A	1.151	1700	1.135	1050
B	2.558	960	1.320	460
C	5.806	450	1.510	205
Boundary layer probe	11.05	75	0.225	60

An investigation of the effect on the frequency response of the volume of the chamber between a 9-inch long total pressure tube having a 0.031 inch inside diameter and the face of a microphone is reported in [*15*]. By the use of a suitable chamber so that the probe tube terminated in its characteristic impedance, the frequency response was smoothed significantly over the range of calibration from about 100 cps to about 8000 cps.

Static pressure. Wall taps are the most reliable means of measuring static pressure within turbomachines. Sharp-edged, 0.040-in. diameter orifices drilled at 90 ± 10° to the surface are simple and satisfactory, although careful inspection to avoid burrs is in order. Fig. D,3m provides an indication of the errors encountered with several configurations of wall static pressure orifices [*16*]. Several commonly used designs of static pressure probes and their performance are discussed in [*17*]. This work indicates that the static pressure error is less than 0.1% of the

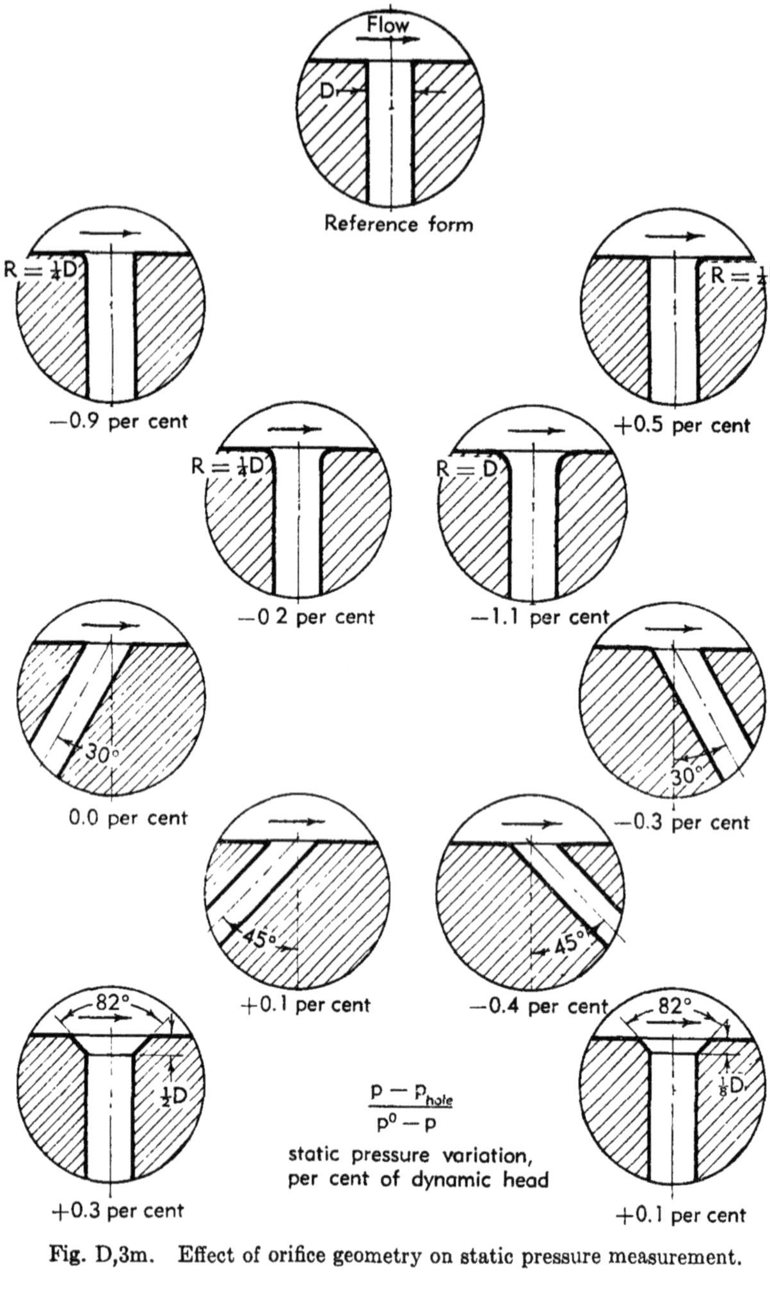

Fig. D,3m. Effect of orifice geometry on static pressure measurement.

dynamic pressure if the orifices are 5 or more probe diameters downstream of the nose and 10 or more stem diameters upstream of the stem. A hemispherical nose is desirable for subsonic probes; ogival or conical tips for supersonic velocities. From this and other investigations, the orifice in a probe or other surface has been found to yield the most accurate readings if made very small (0.1 probe diameters or less) and twice as deep as the orifice diameter. For the small diameter probes necessary for gas turbine research, orifices of this size are difficult to make and indicate slowly. Orifices about twice the diameter indicated by the above rule may be employed with negligible error, although 0.015 in. is considered optimum.

Static pressures must sometimes be measured in flows of extreme fluctuation. Under these circumstances a high order of accuracy cannot be expected, but comparative results can be obtained by shielding the orifice from the dynamic pressure by a high resistance screen. A porous sphere, perhaps of sintered metal, appeals as an obvious shield for the orifice. Fine mesh metal filter cloth rolled into a $\frac{1}{4}$-in. diameter cylinder $\frac{1}{2}$ in. long and sealed on one end and soldered to a $\frac{1}{4}$-in. stem works reasonably well in incompressible flow.

A method of preventing standing waves in pressure leads between an orifice and a transducer is described in [*18*]. The effect of an infinite extension of the pressure lead tubing was obtained by providing a spiral passage about 300 tube diameters long containing an absorbent material after the transducer. A resonance-free response up to 8000 cps was observed.

Flow direction. The direction of flow relative to a principal axis of the machine may be obtained with relative ease with the calibrated spherical-head type of probe discussed in [*19*]. The more direct method of measurement, the so-called null type, requires rotation of the probe so that the pressures at two symmetrical openings in the instrument head are equal. The null-type yaw head is especially useful in combination instruments designed to measure the flow angle and the static and total pressure simultaneously since the probe is aligned with the flow when the readings are taken; however, such probes usually require calibration to determine their mutual interference effects.

A cylindrical tube having pressure taps at three circumferential positions at the same axial location is a practical null-type instrument for direction and total pressure measurement. A calibration curve for a typical cylindrical probe having the yaw orifices oriented 55° from the total pressure opening is presented in Fig. D,3n from [*20*]. The directional sensitivity of such probes is satisfactory, as the pressure difference in the two yaw orifices is about 6 per cent of the dynamic pressure per degree of yaw, and the total pressure varies only 1 per cent of the dynamic pressure for either 5° of yaw or 5° of pitch. The advantages of this type of probe are simplicity, ease of installation, and ease of maintaining accuracy due to the lack of extended arms. The prime disadvantage is that, in practice, accurate static pressure measurement cannot be made.

In the investigation reported in [6], a study was made of the basic elements of combination probes intended to measure the total and static

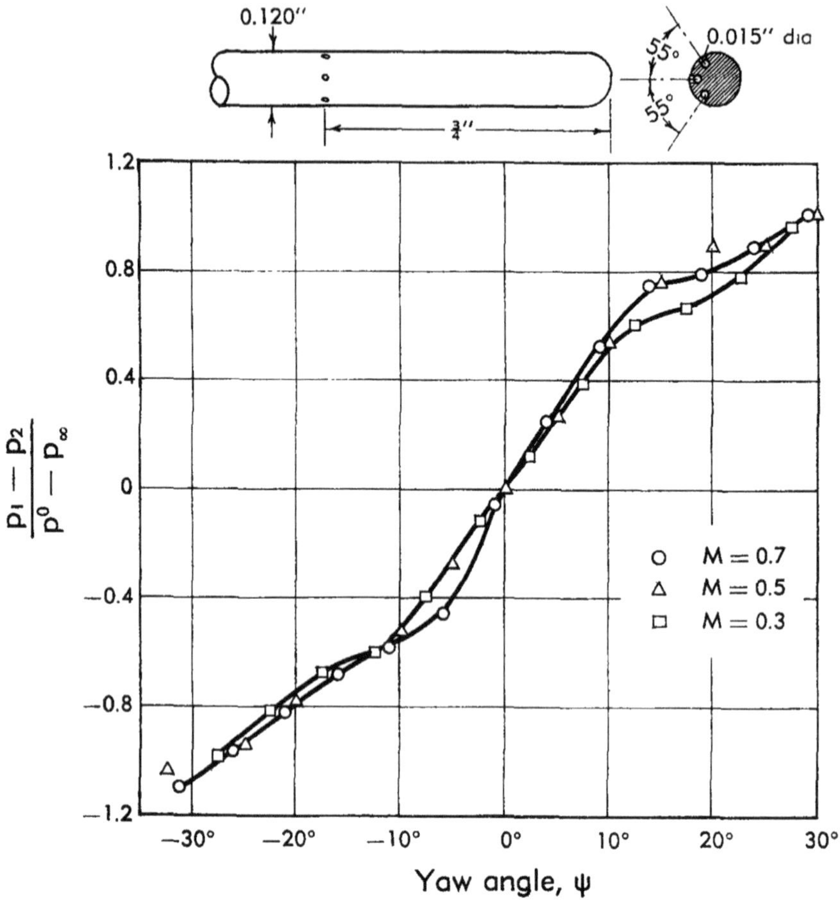

Fig. D,3n. Change of probe pressure difference with yaw angle.

pressure and the direction of the flow (see Fig. D,3o). To determine the most sensitive design for null-type probes, the pressures indicated by tubes having ends slanted 30, 45, 60, and 90° from the tube axis were recorded as the yaw angle was varied from +180° to −180°. The results of these tests are presented in Fig. D,3p. The sensitivity of all four tubes was very good. The yaw tube pressure difference ranged from 2.3 to 3.0 per cent dynamic pressure per degree as the angle of cutoff was increased from 30 to 90°. The sensitivity of claw-type yaw probes (Fig. D,3q) was obtained by testing instruments having the configurations indicated in Fig. D,3r. Probes having a 15 to 45° included angle between the slanted probe ends and a 120° included angle between the arms were found to be very sensitive, indicating a pressure difference per degree of yaw angle of about 9 per cent dynamic pressure.

Thermal flag. The usual type of flow direction measuring instrument employed in aerodynamic measurements relies on sensing a fraction of

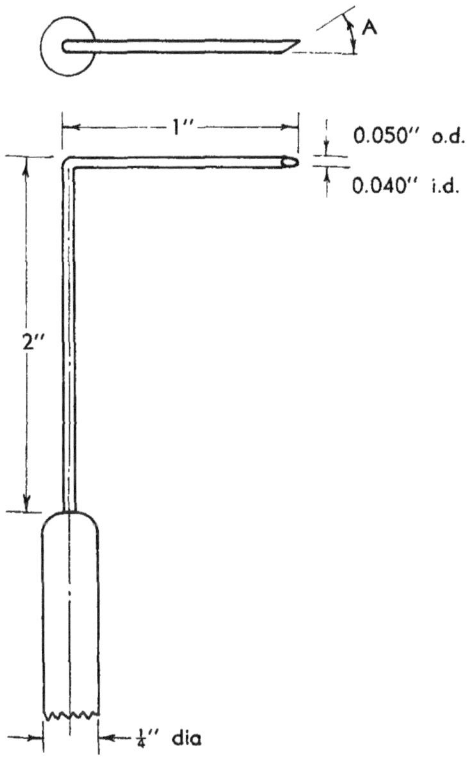

Fig. D,3o. Details of yaw element probes.

the dynamic pressure of the flow. In the usual case, two orifices are located symmetrically a finite distance apart in space or on the surfaces of a sphere or wedge. In the presence of a velocity or total pressure gradient, such instruments are inherently unable to read the true flow direction because of the finite spacing between the orifices. Hot-wire anemometers have been used in attempting to overcome this problem but are too delicate for most practical applications. The hot film probe is more rugged but suffers from the effect of the finite distance between measuring surfaces. Flow visualization techniques have also been used, particularly tufts of materials having little mass but high flow resistance.

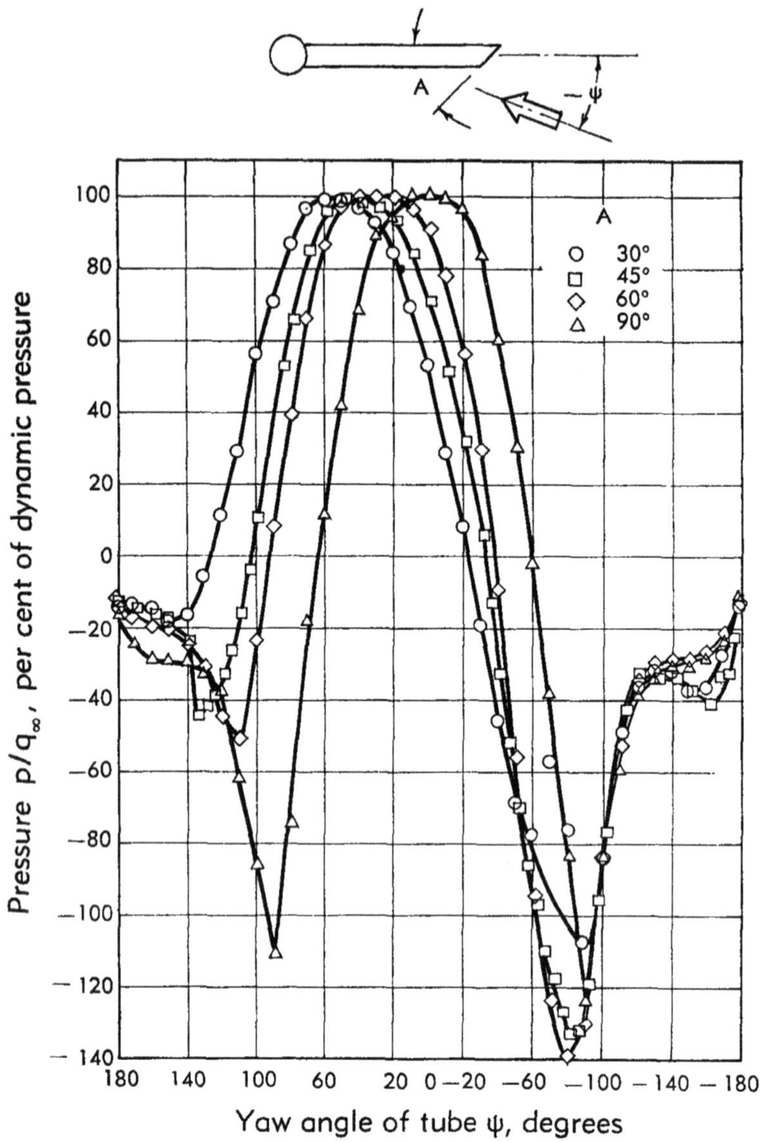

Fig. D,3p. Indicated pressure versus yaw angle of four yaw element probes. 295 ft/sec.

A device which has recently received attention in an attempt to solve this problem is the thermal flag. This probe uses a thermister of small diameter (typically 0.014 in.) to detect the wake of a warm wire having a diameter of 0.002 in. Although these wires are small, they are larger by an order of magnitude than the wires employed with hot-wire anemometers. Thermal flags have been used to survey the exit flow from a

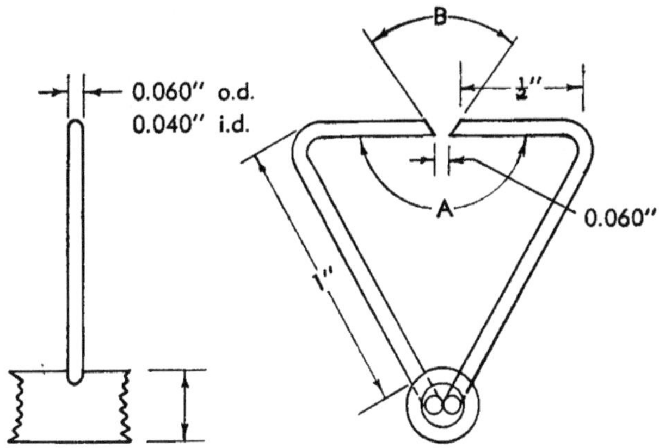

Fig. D,3q. Details of claw-type yawmeters.

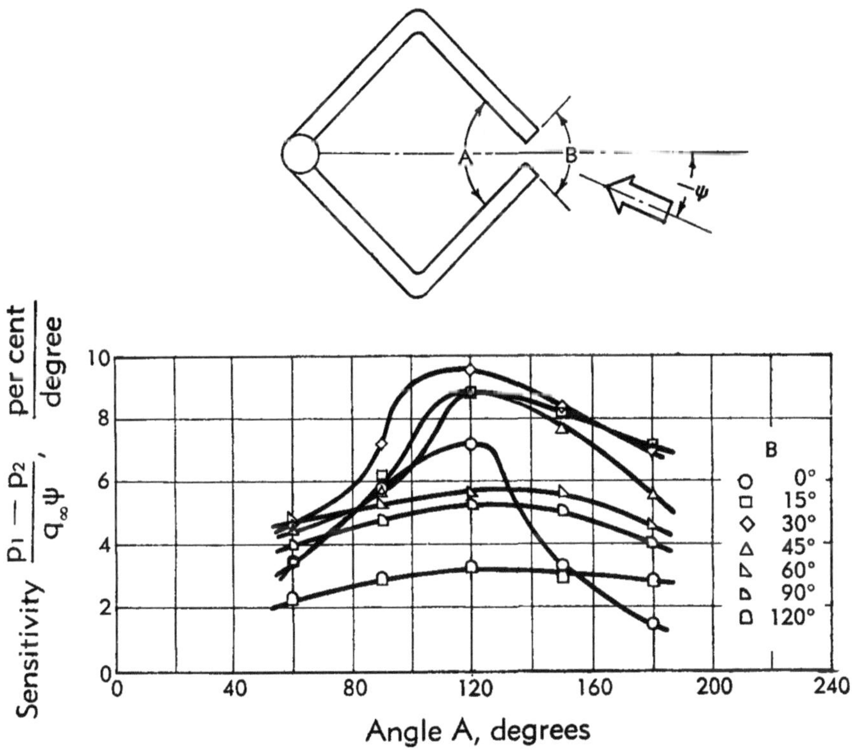

Fig. D,3r. Yaw sensitivity of claw-type instruments. 295 ft/sec.

long pipe having a turbulent boundary layer. Some advantage of thermal flag yaw probes over conventional cobra and wedge probes was observed. However, a true comparison of insensitivity to gradients cannot be made from the results of this preliminary investigation due to differing stem details of the probes tested, and stem design was shown to be significant.

Combination probes. A widely used combination probe consists of a central claw-type yaw head with total and static pressure probes located

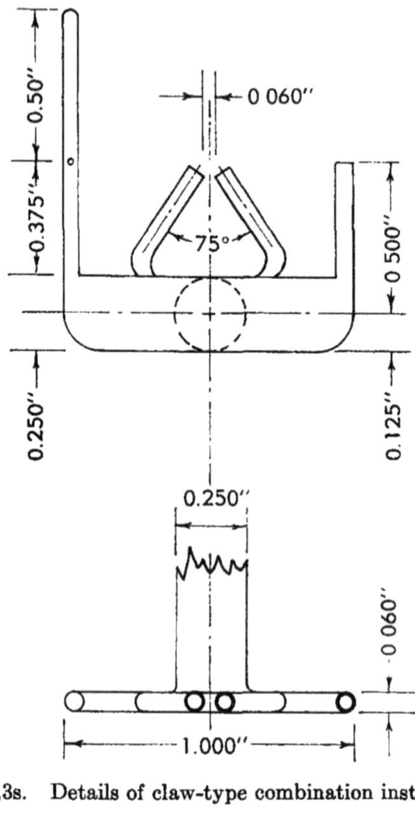

Fig. D,3s. Details of claw-type combination instrument.

on the sides. While such instruments are relatively easy to make, the extended tubes are easily bent, and the static pressure calibration factor is usually not constant as the speed is changed (Fig. D,3s). Further, the readings are not taken at a single point in the flow; therefore, in streams of varying pressure or direction, a true indication of existing conditions is not obtained (Fig. D,3t).

A combination instrument intended to measure flow direction and pressure at a small point is illustrated in Fig. D,3u. In this instrument, the prism probe, the flow direction, and total and static pressure are obtained by orifices in the front and sides of a 0.090-in. outside diameter tube. Prism probes of this size require somewhat closer work to install the tubes than do similar claw-type probes. The prism probe is, however, much less prone to lose its calibration. The sensitivity of a typical prism-type probe to yaw and to pitch is shown in Fig. D,3v and D,3w. The

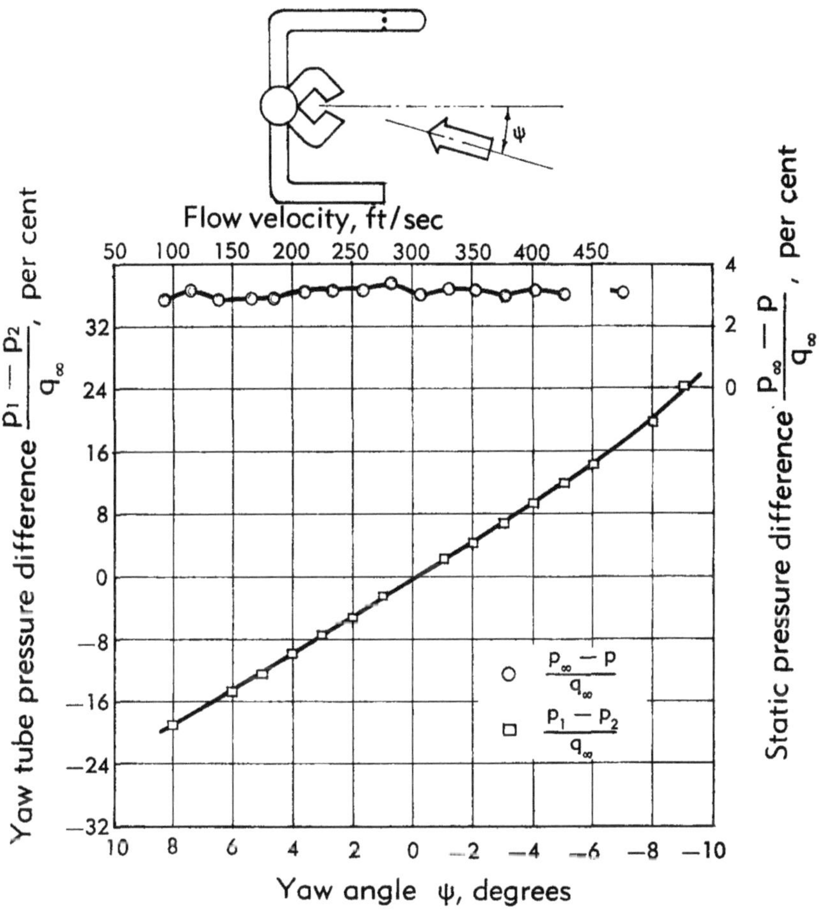

Fig. D,3t. Variation of yaw tubes pressure difference with yaw angle. Difference in static pressure reading between instrument (claw type) and standard static probe with flow velocity.

static pressure calibration varied less than ½ per cent dynamic pressure, and the yaw null point by less than ¼° as the velocity was varied over the test range from 100 to 500 ft/sec.

A comparison has been made between the turning angle calculated from measurements made by a stationary prism probe and that measured directly by a prism probe rotating with the rotor of a 28-in. tip diameter axial flow compressor. Readings were taken at three diameters over the stable range of the compressor. This comparison is presented in Fig. D,3x. Although the instrument rotating with the rotor was placed midway between blade wakes and was not traversed in the circumferential direction, excellent agreement was obtained.

Recently, emphasis has been placed on the study of secondary flows and end losses in turbomachines. The gas motions in such flows rarely lie in a single plane. To measure three-dimensional flows with the more usual instruments would be a laborious task because consecutive surveys using

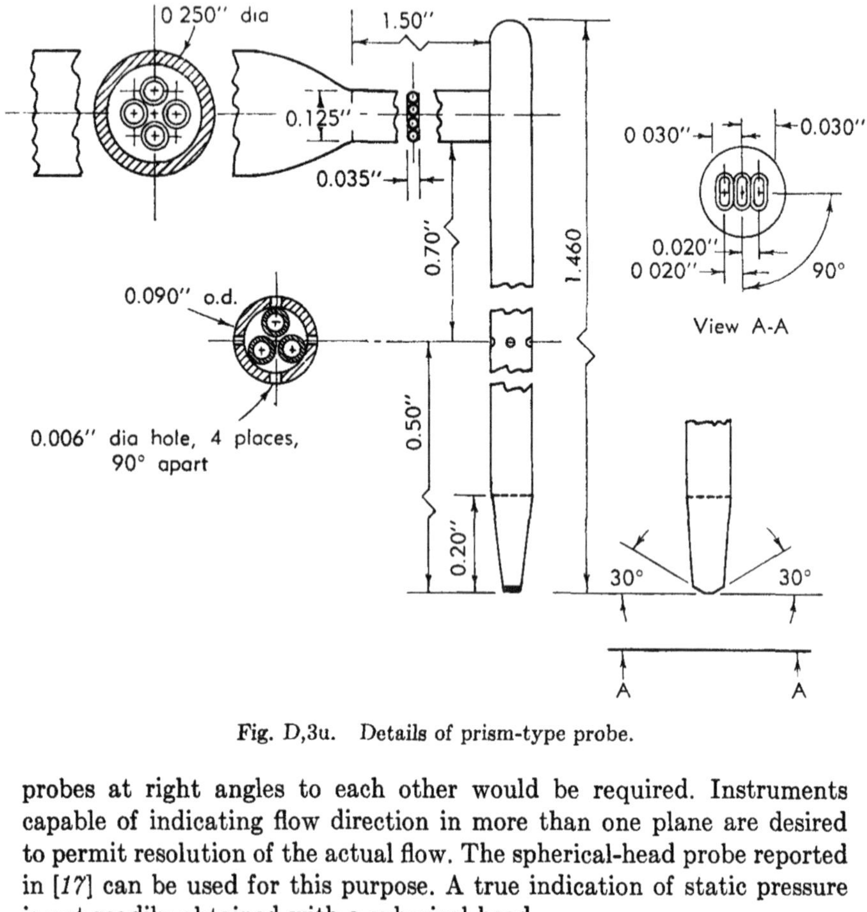

Fig. D,3u. Details of prism-type probe.

probes at right angles to each other would be required. Instruments capable of indicating flow direction in more than one plane are desired to permit resolution of the actual flow. The spherical-head probe reported in [*17*] can be used for this purpose. A true indication of static pressure is not readily obtained with a spherical head.

An extension of the prism-type probe designed to indicate flow direction in two planes and total and static pressure has been developed. With this instrument, the pyramid probe, readings are obtained by orifices in the front and sides of a 0.090-in. outside diameter tube (Fig. D,3y). Considerable skill is required to construct pyramid probes of this size, but only 30 man-hours were required to make the first of this type. Calibration curves for a pyramid instrument in yaw are presented in Fig. D,3z and in pitch in Fig. D,3aa. When this probe was rotated $\pm 10°$ in roll at zero yaw and zero pitch angle, the static pressure calibration varied less than $\frac{1}{2}$ per cent of the dynamic pressure and the pitch and yaw null

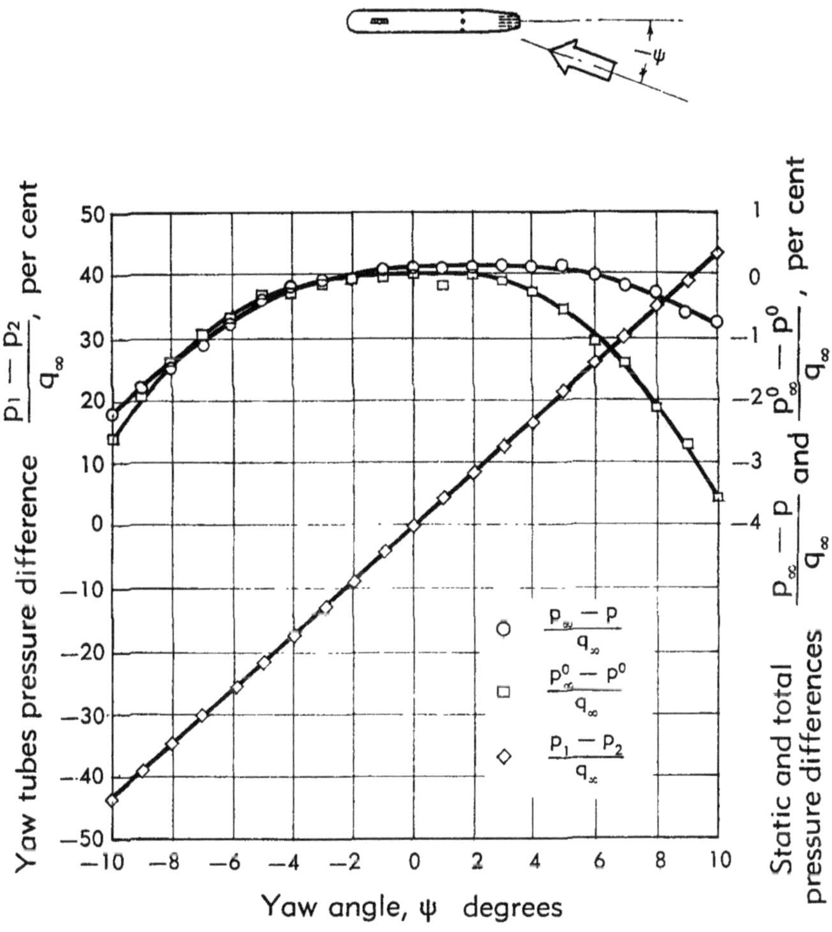

Fig. D,3v. Variation of yaw tube pressure difference. Differences in static and total pressure readings between prism-type probe and standard static and total pressure probes with yaw angle. 390 ft/sec.

points changed less than $\frac{1}{2}°$. Over the test velocity range of from 100 to 500 feet per second, the calibration held equally constant at zero yaw, pitch, and roll angle. A study of probes for three-dimensional flow measurement is presented in [*21*].

A probe designed to permit null-type pitch angle measurements over a 60° range without rotating the entire probe in that plane is shown in Fig. D,3bb. The probe is rotated about the stem axis in the usual manner for yaw angle measurement. For pitch angle reading, the lead screw is actuated until equal pressures are observed from the "pitch" orifices. The motion of the lead screw is then read and converted to angular values.

Extensive calibration of a hemispherical-headed probe containing a central total pressure orifice and two pairs of orifices located in perpendicular planes for flow direction measurement has been conducted at subsonic and transonic speeds [22]. The results indicate a smooth variation

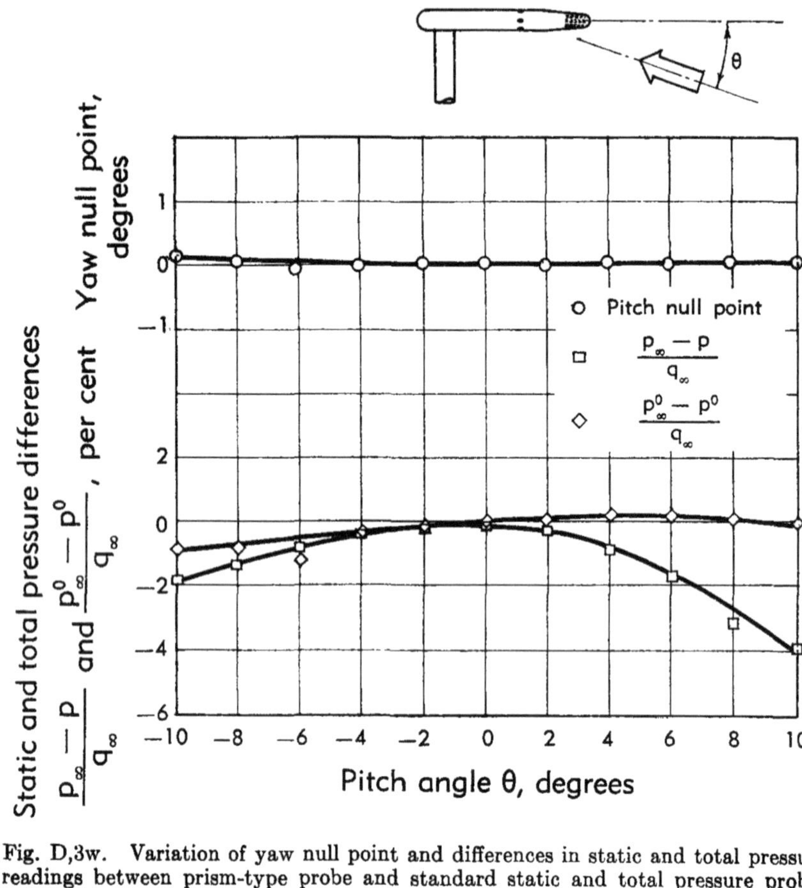

Fig. D,3w. Variation of yaw null point and differences in static and total pressure readings between prism-type probe and standard static and total pressure probes with pitch angle. 308 ft/sec.

in the sensitivity to flow angularity of the probe in the speed and angle range investigated.

In small scale model tests, insertion of a multitube probe may significantly disturb the flow. If the flow is reasonably steady, total pressure and direction can be measured and static pressure approximated by recording of pressure and angle as a single tube containing a surface orifice is rotated about the axis of the tube. For surveying a passage, continuous rotation and recording with rapid response equipment can be used as the flow passage is traversed by the orifice.

Temperature. Temperature measurements in turbomachinery are ordinarily made with thermocouples, although mercury-in-glass and resistance thermometers are used with high accuracy when the size of the sensing element can be relatively large. For high sensitivity over

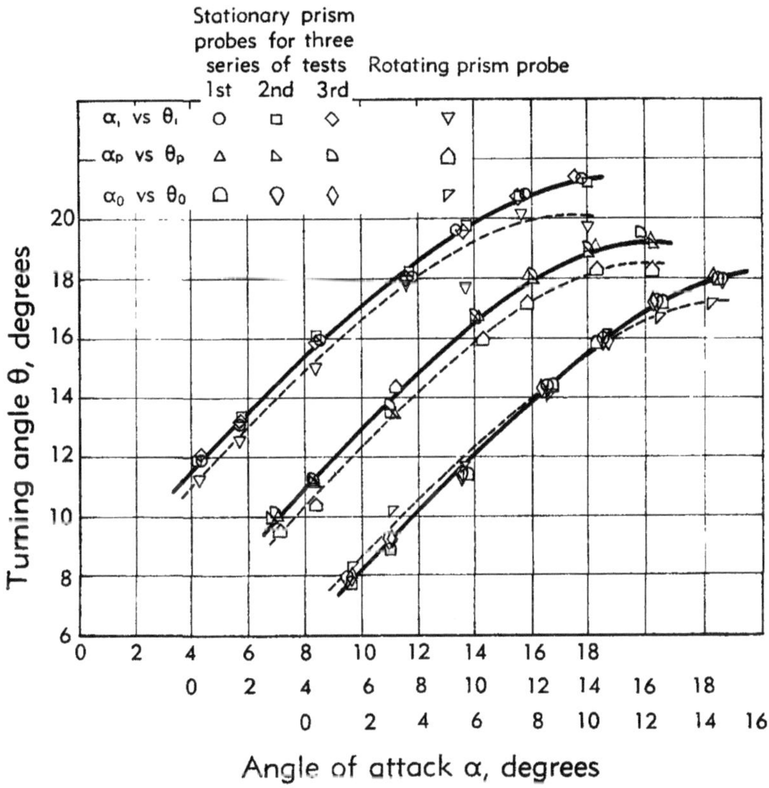

Fig. D,3x. Comparison of flow angles at three radii of a compressor rotor as measured by both stationary and rotating prism probes during three similar series of tests.

narrow temperature ranges the thermistor is superior up to its maximum value, currently about 600°F.

The working ranges of various temperature measuring systems are presented in Fig. D,3cc, taken from [*23*]. This figure also provides an indication of the uncertainty resulting when temperatures are measured by these various devices. In addition to the thermocouple materials presented in Fig. D,3cc, iron-constantan is widely used in the lower temperature range and chromel-alumel in the middle range. For higher temperatures, iridium and silicon carbide can be used to 4000°F, as in the design shown in Fig. D,3dd. Platinum and its alloys and tungsten have also been used for high temperature thermocouples. Water-cooled probes are often necessary when high temperature gas streams are being measured. The sensing element may be shielded from the cooled walls for minimum calibration factors, or may receive the gas after it has been cooled, (Art. 9).

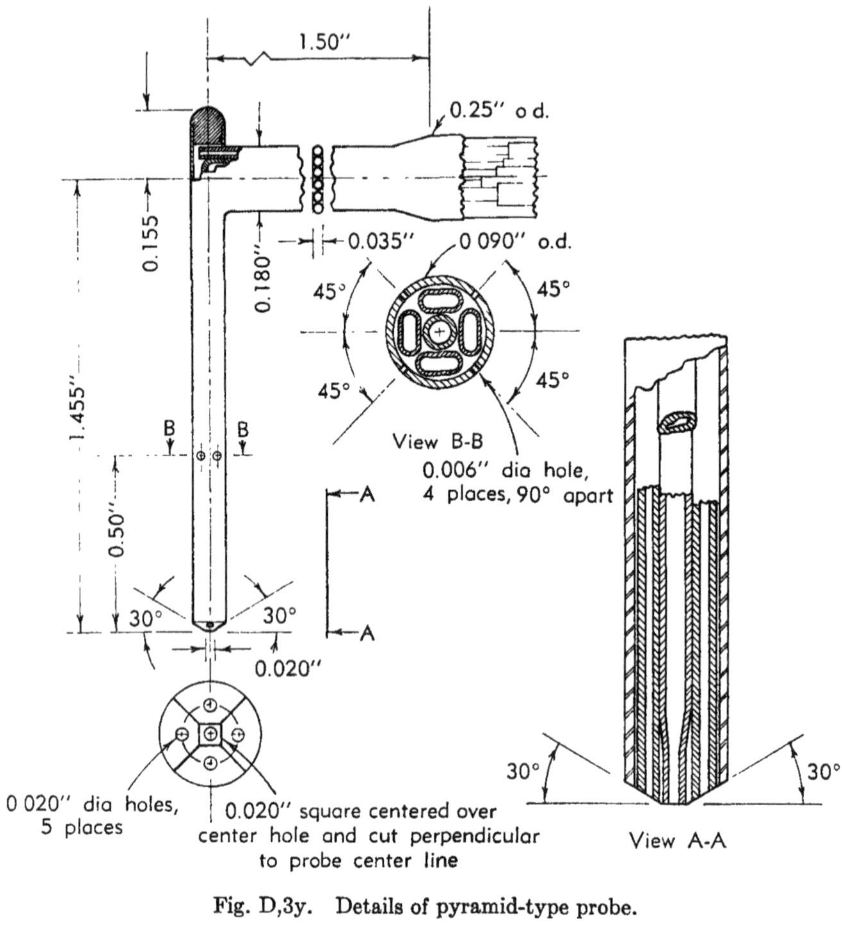

Fig. D,3y. Details of pyramid-type probe.

By vibration-filling the insulation into pencil-type probes, Fig. D,3dd, it has become practical to fabricate these coaxial thermocouples with outside diameters of 0.030 in. Such probes having wall thicknesses of from 0.003 to 0.005 in. can be subjected to a bend radius of 0.05 in. without affecting their performance. Lengths of ten feet can be fabricated reliably.

The National Advisory Committee for Aeronautics has reported a series of tests [*24*] of the temperature indication and response rate of 3 bare-wire and 3 shielded thermocouples (Fig. D,3ee). The probes were tested at subsonic and supersonic speeds with a total pressure of one atmosphere and a total temperature in the range from 70 to 100°F.

Twelve samples of simple butt-welded, exposed wire thermocouples (probe 1) exhibited a roughly linear variation of recovery ratio in the subsonic range, varying from $R_0 = 0.997$ at $M = 0.2$ to $R_0 = 0.971$ at

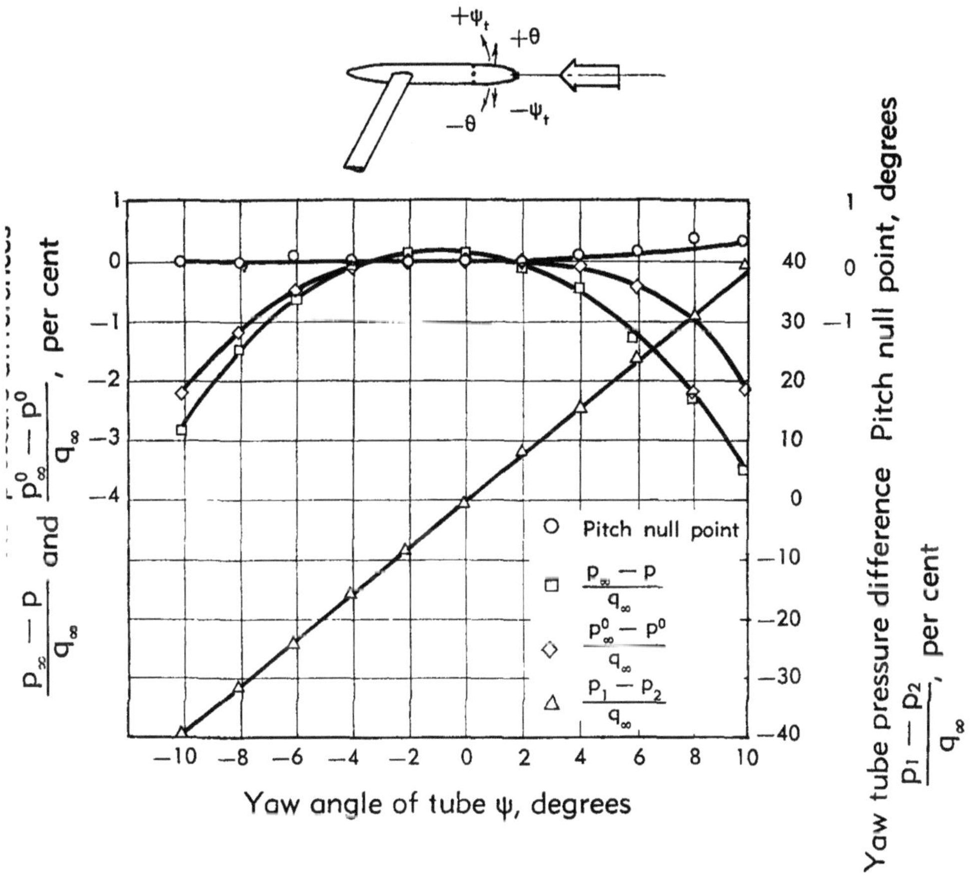

Fig. D,3z. Variation of pitch null point, yaw tube pressure difference, and differences in static and total pressure readings between pyramid-type probe and standard static and total pressure probes with yaw angle. 402 ft/sec.

$M = 0.8$. In the supersonic range, five samples of probe 1 indicated a linear change of R_0 from 0.9593 at $M = 1.28$ to 0.9564 at $M = 2.21$.

Probe 2 indicated only a small and consistent decrease in recovery ratio with increasing Mach number. Ninety probes exhibited R_0 values from 0.9985 at $M = 0.2$ to 0.9905 at $M = 0.8$ with little scatter. At supersonic speeds, the trend of R_0 with M continued for the two probes tested, but at a reduced rate until the probe tip intersected the bow shock wave of the probe support. When this intersection occurred, at Mach numbers between 1.4 and 1.7, a rapid and nonlinear decrease in R_0 with M occurred.

The recovery of twenty probes of type 3 exhibited the same value of R_0 at $M = 0.2$ as probe 2. At higher speeds, R_0 decreased more rapidly with $R_0 = 0.9835$ at $M = 0.8$, probably because the flow velocity in the

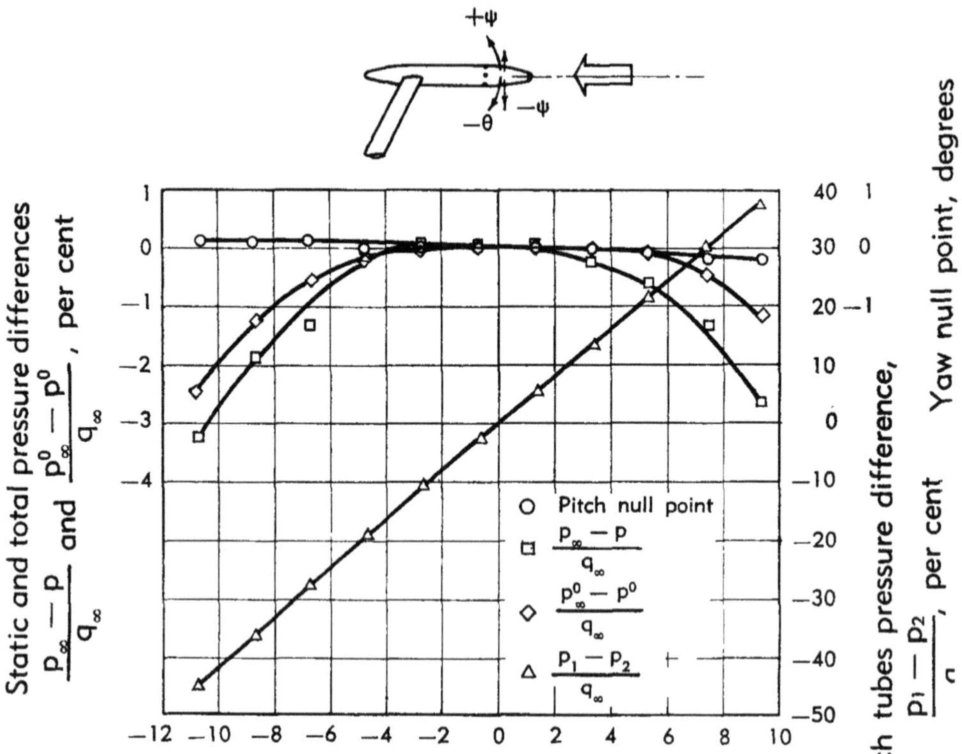

Fig. D,3aa. Variation of pitch tube pressure difference, yaw null point, and differences in static and total pressure readings between pyramid-type probe and standard and total pressure probes with pitch angle. 402 ft/sec.

region of the thermocouple junction of probe 3 is higher than for probe 2, due to the greater obstruction of the flow afforded by the supporting shaft of probe 2. Less discontinuity in the variation of R_0 with M was observed with probe 3 in the supersonic range, R_0 decreasing from about 0.965 at $M = 1.2$ to about 0.945 at $M = 2.2$.

The total temperature indication of the shielded probes 4, 5, and 6 (Fig. D,3ff) was very satisfactory and, particularly in the case of one hundred type-5 probes, consistent.

Changes in the difference between the true total temperature and the reading of the probes were observed under conditions other than those of the original tests which were run at a total pressure of one atmosphere and with yaw and pitch angles equal to zero. The change observed is expressed as a ratio to the difference measured under the original test conditions. The effects of varying total pressure, yaw angle, and pitch angle

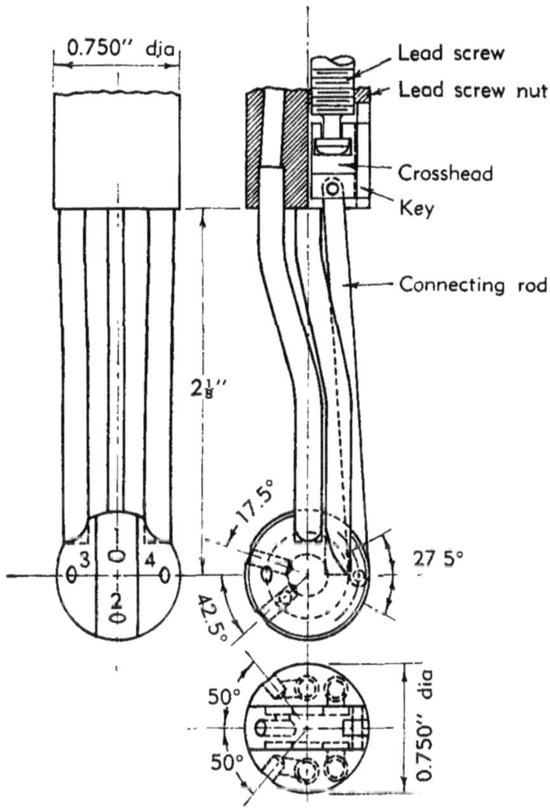

Fig. D,3bb. Details of 4-hole yaw and pitch probe.

are presented in Fig. D,3gg, D,3hh, and D,3ii, respectively. The shielded probes indicate larger changes in the parameter $(R_0 - R)/(1 - R_0)$, in general, than the unshielded probes. Since $1 - R_0$ is smaller for the shielded probes, the change in temperature reading is not necessarily larger for the shielded probes, however.

Survey Mechanisms. For surveys of flow through an annulus, the measuring instruments must be supported in suitable holders to permit accurate radial, angular, and, in some cases, circumferential positioning. For most low and medium speed studies, manual operation is the most accurate and the least complicated. A typical instrument carriage and combination probe used at the NACA Langley Laboratory for low speed annular cascade and compressor research is presented in Plate D,3a. The traversing drive screw has 20 threads per inch. The indicating head used to turn the drive screw is graduated in hundredths of an inch. The yaw

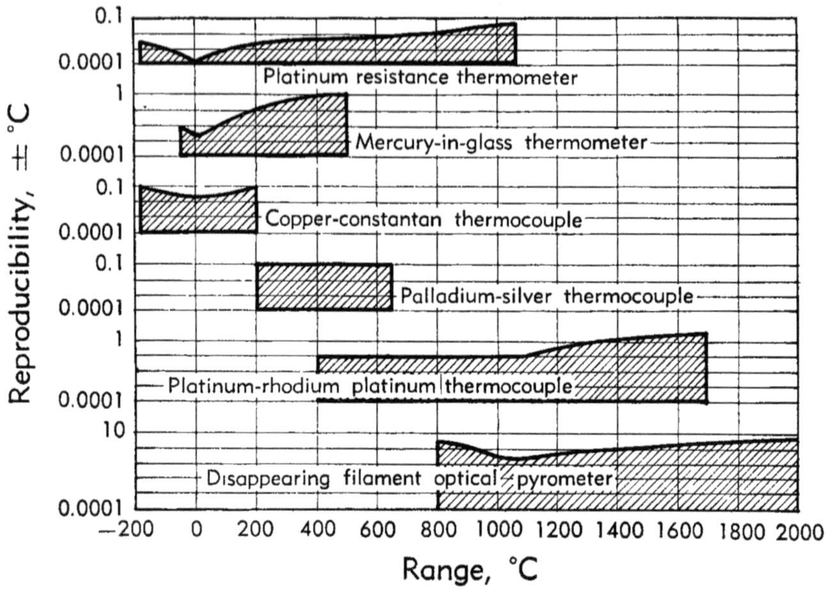

Fig. D,3cc. Range and reproducibility of various temperature-measuring instruments under best conditions.

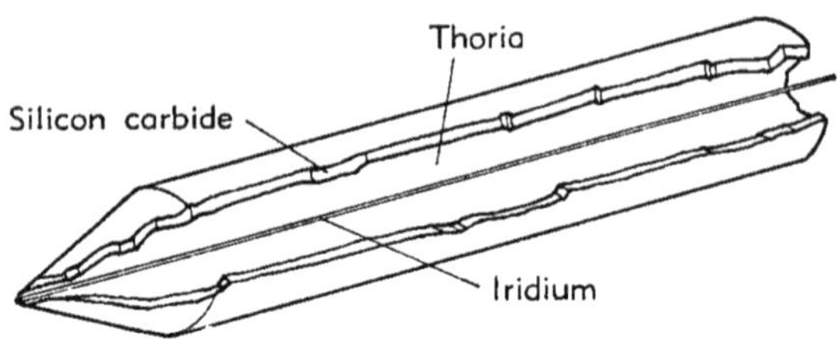

Fig. D,3dd. Pencil-type thermocouple probe in which silicon carbide serves both as a supporting tube and as an active element.

arm attached to the probe carries a vernier scale permitting angular measurements in tenths of a degree over a range of 180 degrees.

During operation of experimental engines or high speed components, danger to personnel close to the installation is ever present. To reduce the hazard, remote-controlled and indicating survey mechanisms have been developed. In the compact carriage shown in Plate D,3b, one electric motor drives the lead screw for traversing motion and another rotates the barrel to which the probe is attached. Indication of linear and annular position is obtained by self-balancing potentiometers which measure the resistances of a helical potentiometer geared to the carriage barrel. Instrument mounts that indicate remotely by means of selsyns have been used successfully.

A remotely controlled, remote-indicating carriage for surveying the flow in cascade tunnels is illustrated in Fig. D,3jj [*25*]. The yawmeter assembly slides along two rails parallel to the cascade. The carriage, or saddle, is moved along the rails by a lead screw which is driven, through reduction gearing, by a repeater motor. Manual control can be used for detailed work or for checking purposes. The yawmeter extends through a wormwheel with a worm that is spring-loaded to avoid backlash. The worm is driven by a repeater motor. The yawmeter is accurately keyed to the wormwheel so that correct alignment is always obtained to permit removal of the instrument for recalibration. The yawmeter can be moved manually to permit traverse in the spanwise direction of the test blades. The motions of the main lead screw and the wormwheel can be transmitted to a drum and scriber to produce a permanent record, automatically, of air direction against distance along the cascade. Motor-driven units can be made to align null-type yaw probes automatically. This is accomplished by connecting the pressure leads from the yaw tubes to a sensitive pressure switch. A pressure switch developed at Langley Field consists of a small metal disk attached to a thin (0.008-in.) rubber diaphragm about 1 in. in diameter. A fine wire soldered to the disk forms part of the motor circuit. Wires which complete the left rotation and right rotation circuits of the motor are attached to pointers located a few hundredths of an inch away from the disk, one on either side of the diaphragm. The pressure lead from one leg of the yaw head is connected to a small chamber on one side of the diaphragm and the lead from the other yaw leg is connected to the chamber on the opposite side. If unequal pressures exist, the diaphragm moves to one side, the disk and pointer contact, and the proper circuit is made to cause the probe to be rotated toward the null position. When operating correctly, the instrument hunts rapidly about the null direction with an angular motion of $\pm\frac{1}{2}°$. Difficulty has been experienced in obtaining pressure switches that are sufficiently sensitive for operation with low dynamic pressures and strong enough for use with high dynamic pressures. A pressure switch developed at the National Gas Turbine Establishment, Farnborough, England, uses a 0.0025-in., phosphor-bronze diaphragm 5 in. in diameter. The sealing disks which form the pressure chambers each carry two insulated contacts connected to low speed and high speed circuits of the drive motor. The low speed contacts are set to within 0.001 in. of platinum contacts on the diaphragm. A pressure difference of about $\frac{1}{20}$ in. of water is sufficient to actuate the switch. The gap in the high speed circuit is about 0.003 in. To prevent hunting, the contact spacing should not be set too small, and the air volume in the pressure leads and chambers should be kept to a minimum.

MEASURING AND RECORDING METHODS. The testing of a gas turbine or turbine component usually requires the recording of many pressures. Liquid manometers, either U-tube or multitube, are commonly employed for indication in lengths of from 3 to 6 feet. Alcohol, water, tetrabromo-

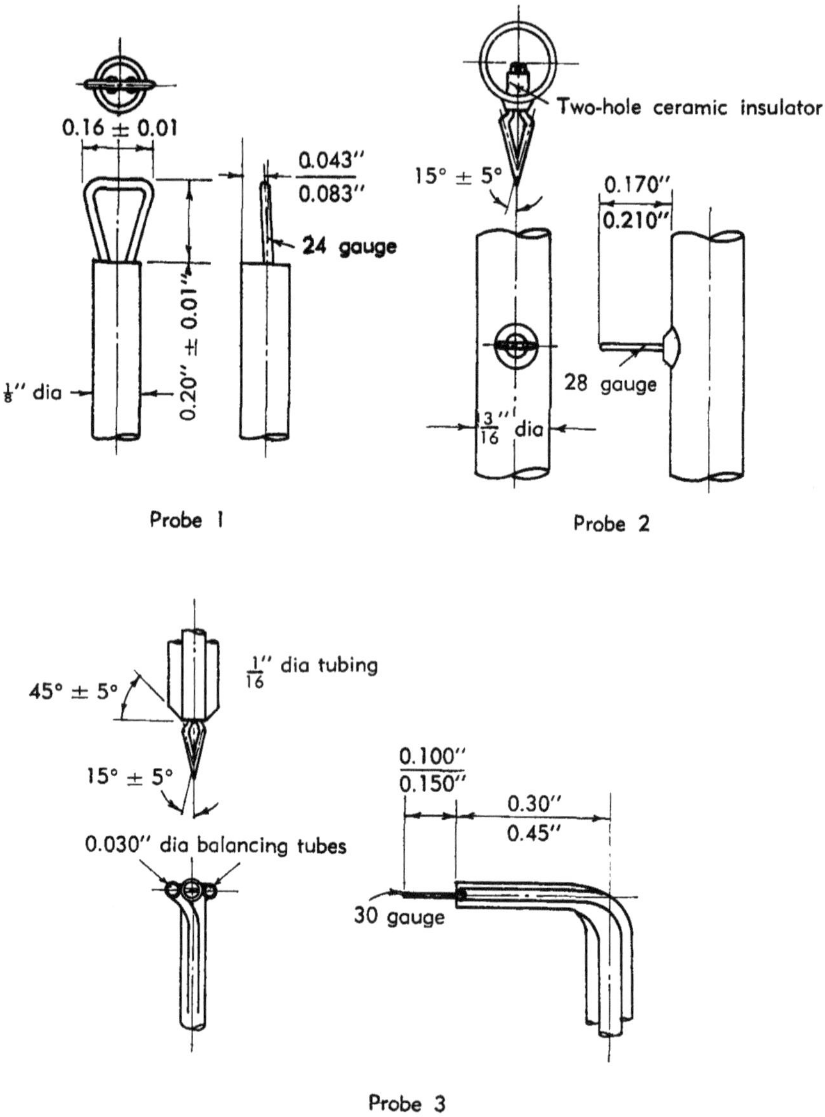

Fig. D,3ee. Thermocouple probe details. Tubing diameters given in nominal size; wire sizes in American wire gauge. (All dimensions in inches.)

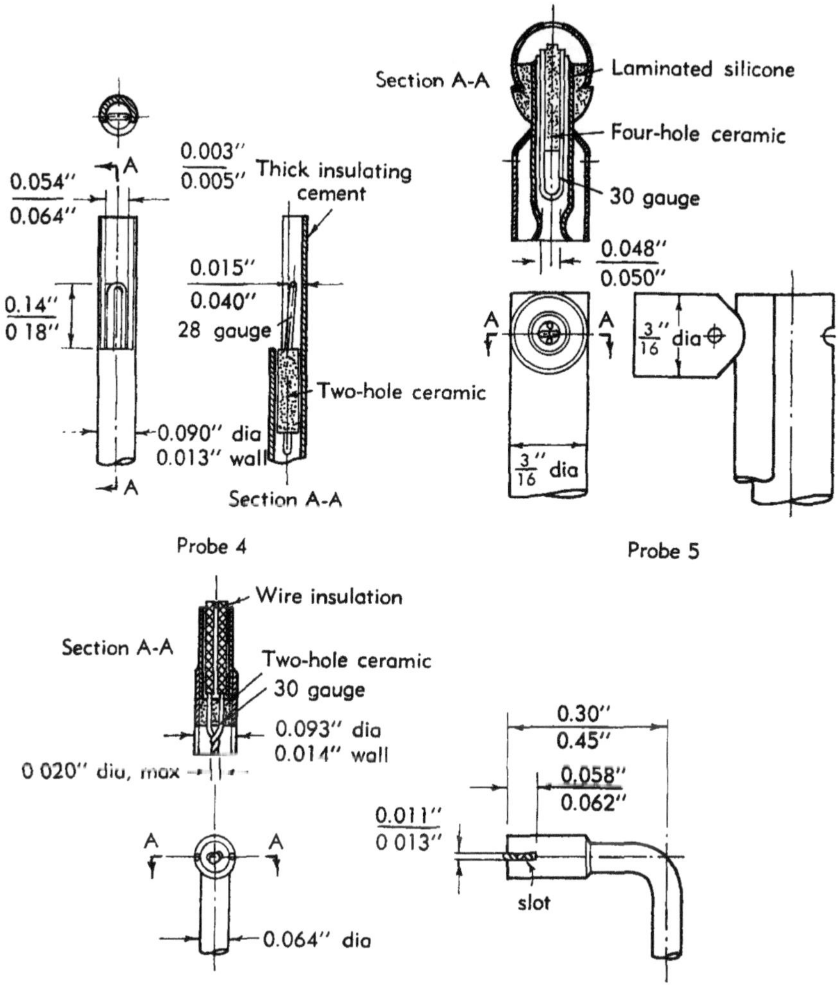

FIG. D3,ee. (*Continued*)

ethylene, and mercury are the most frequently used fluids. Discussions of various manometer types and their fluids may be found in [*26,27*]. Information on conventional U-tubes (multitube, projection, and tilting) and micromanometers is given. Mechanical, optical, and electrical types of pressure capsules are also described for uses where special conditions make them desirable. When many pressures must be recorded simultaneously, a recorder described in [*28*] will provide a trace on paper proportional to the magnitude of the pressure for sixty or more readings. Pressure-sensitive switches have been successfully employed by the NACA [*29,30*] for rapid recording of many channels of data. This NACA technique and others are described in [*31*].

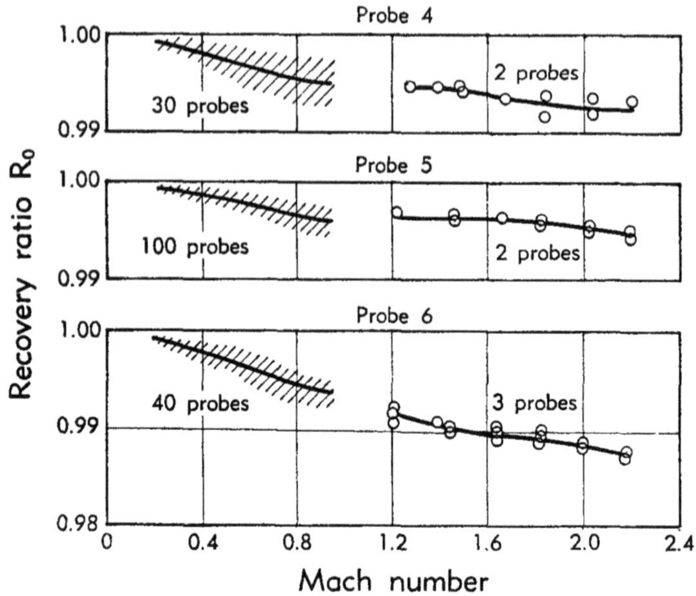

Fig. D,3ff. Recovery ratio as function of free stream Mach number at reference conditions. Top, probe 4; middle, probe 5; bottom, probe 6.

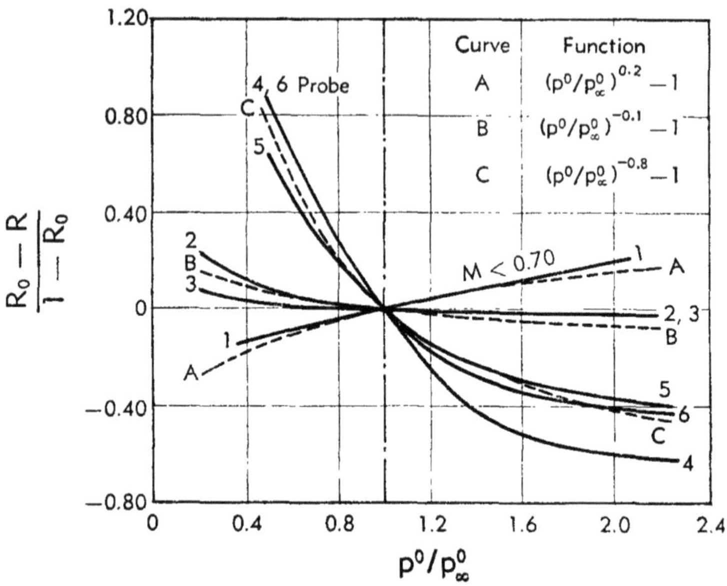

Fig. D,3gg. Variation of recovery ratio with total pressure at zero angle of attack.

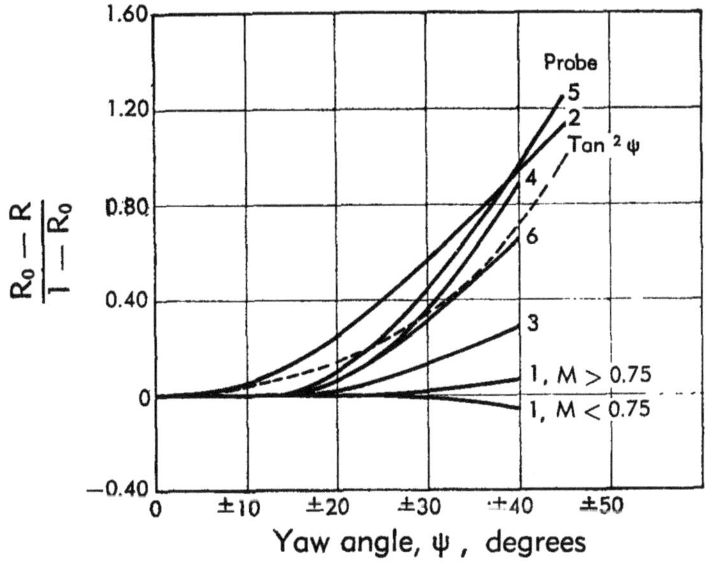

Fig. D,3hh. Variation of recovery ratio with yaw angle at 1 atm total pressure and zero pitch angle.

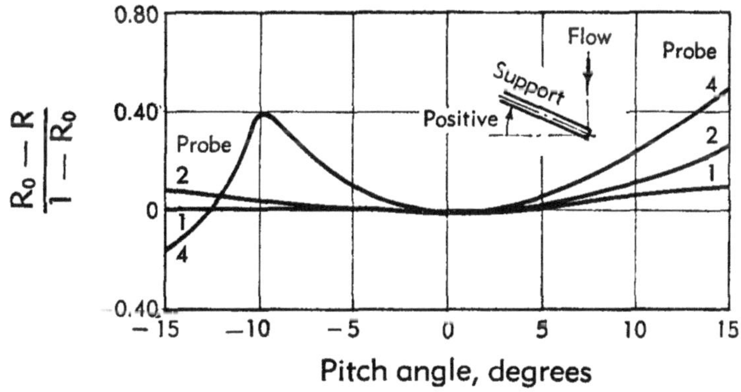

Fig. D,3ii. Variation of recovery ratio with pitch angle at 1 atm total pressure and zero yaw angle.

Composition of Test Medium. The composition of the test medium must be known for the accurate computation or correlation of performance characteristics. With air as the test fluid, this is a simple matter since the measurement of the only usual important impurity, water vapor, can be made by sensitive hygrometers. If gases other than air are used as test media, the velocity of sound in the fluid must be known. In [*32*], the use of a Helmholtz resonator to measure the velocity of sound in a

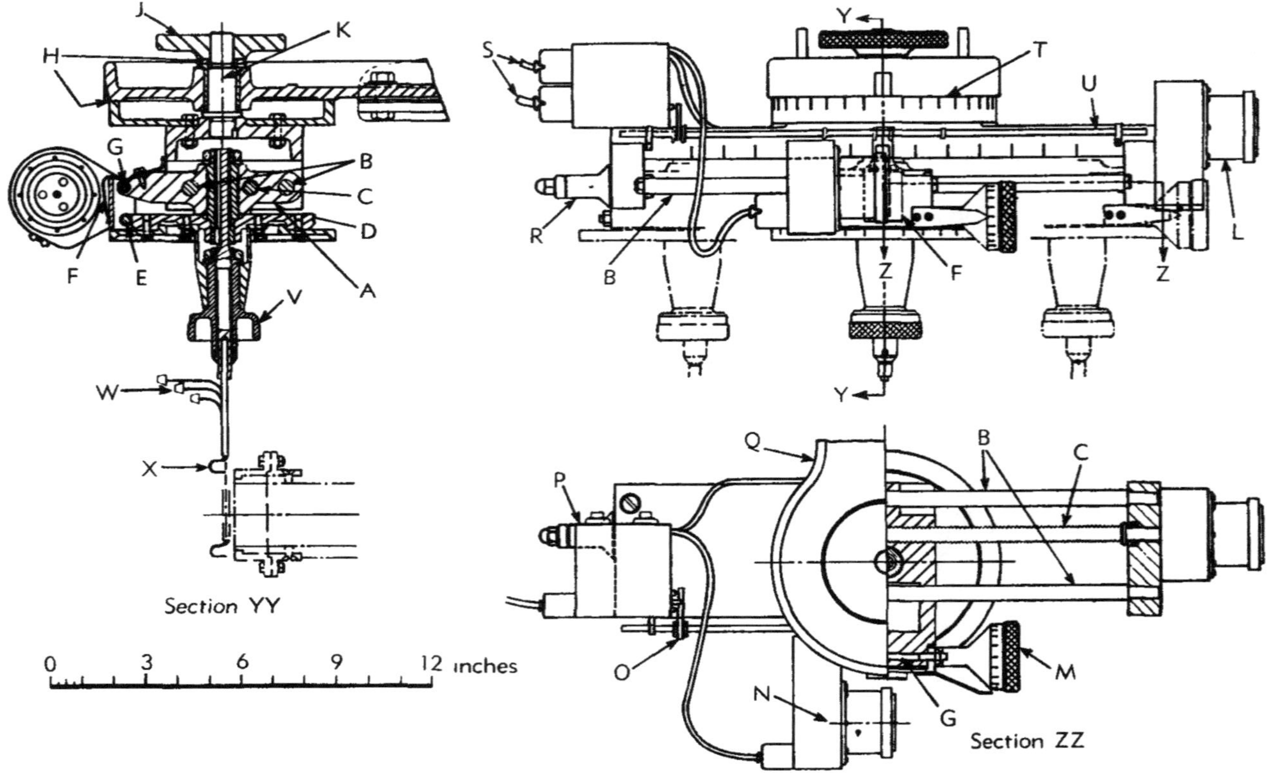

Fig. D,3jj. Traverse gear for National Gas Turbine Establishment No. 3 High Speed Cascade Tunnel. A, Saddle; B, Guide rods; C, Main traverse leadscrew; D, Wormwheel carrying yawmeter; E, Worm; F, Spring holding worm in close mesh with wormwheel so avoiding backlash; G, Pivot for yawmeter motor, gearbox, and worm; H, Traverse girdle assembly locked in position by friction between these surfaces; J, Lock nut for locating guide assembly; K, Guide assembly rotates about this center line; L, Main traverse repeater motor and gear box; M, Micrometer for manual control of yawmeter; N, Yawmeter repeater motor and gearbox; O, Traverse limit switches; P, Box containing limit switches and local control buttons; Q, Main bracket for mounting traverse gear on tunnel; R, Micrometer for manual control of traverse; S, Connections to director unit and control panel; T, Scale for setting traverse direction parallel to cascade; U, Rod carrying limit [illegible] parallel to blade (manual control only); W, Yawmeter and Pitot connections; X, Yawmeter-

gas is described. In essence, the device consists of two chambers connected by a small orifice spanned by a hot wire. A sample of the gas is introduced to both chambers. A diaphragm forming one end of one cell is driven by an audio oscillator over a range of frequencies. At a frequency of resonance, an oscillating flow occurs across the orifice; the hot wire circuit is affected and so indicates. Theoretically this meter would not require calibration, but in practice calibration is necessary due to nonideal diaphragm action and to the sensitivity of the device to slight construction inaccuracies, particularly with regard to the shape and dimensions of the orifice.

Mechanical Measurements. To determine mechanical performance, the measurements of interest in gas turbine development are rotational speed, temperature, stress, vibration, contour accuracy, transmitted force or torque, and running clearances. The temperature measurement is usually made using a thermal sensitive element at the point of the desired temperature reading with an indicating device some distance away. Three general types, the vapor pressure, liquid expansion, and electrical are described in [*33*]. Units of these types have been in use for some time, and good accuracies can be obtained with standard instruments and procedures. For measurements made on rotating elements, the electrical type with slip rings is almost universally used. Stress is nearly always found using strain measurements and the known stress-strain relationship of the material. The wire straingauge unit is superior to other methods, and is in general use. Description and test results of 6 commercial kinds are given in [*34*]. Although care is required in their use, the effects of prestraining, temperature change, and weathering have been found to produce small and easily explainable errors [*35*]. If vibration measurements are required, three basic instruments that may be applied are the vibrograph which records flexural vibrations in any direction, the torsiograph which normally indicates torsional vibration of nonrotating elements, and the vibration indicator which employs a feeler unit that can be attached to any vibrating element to record effective amplitudes [*36*]. All three have been designed to provide the vibration record on magnetic tape or photographic film for observation.

Torque measurement. Methods of determining the torque transmitted by a rotating shaft fall into two classes. The first class includes those systems in which the reaction of the rotating members on stationary casings is measured. In most cases, this type is mechanically simpler. The most widely used example is the electrical dynamometer, used either as a power source or as an absorber. A torque arm is attached to the stator casing which is supported in bearings. If the torque transmitted is low and steady, the magnitude can be determined by resting the end of the arm on a platform scale. In those applications where the torque is apt to fluctuate somewhat, the torque reading is obtained by measuring the oil or air pressure on a diaphragm or piston or the resistance of straingauges mounted on a link resisting rotation of the casing.

In the second class are those systems in which the torque in the rotating shaft is determined by measuring the shaft twist. These methods are used in engine testing, or in component testing when gears or the design of the power source or absorber make reaction methods impractical or inaccurate. The twist in a shaft can be deduced by attaching straingauges to the shaft with electrical connections to stationary meters made through slip rings [*37*], or by detecting the relative angular motion between two sections of the shaft some distance apart. There are several methods of accomplishing the latter. If two toothed wheels are rigidly attached to the shaft, a fluctuating emf will be induced in magnetic pickups close to the wheels as the shaft rotates. As the shaft twists under load, a phase shift in the two induced voltages proportional to the torque transmitted will occur [*38,39*]. By the use of suitable instruments the phase shift can be measured, and the twist and torque obtained. If means are provided to rotate the pickups, a new position can be found such that the phase difference is nulled. The angular difference between the pickups at zero load and at the operating load should be proportional to the twist of the shaft and to the torque.

The twist in a section of a shaft can also be determined by measuring the change in inductance or capacitance in suitable circuits, caused by a change in the spacing between surfaces attached to the shaft. In practice, this is accomplished by attaching a sleeve containing one surface to each of the twisting sections. The surfaces that are part of the electrical or electromagnetic circuits lie in axial planes with a small air gap between them. As the shaft twists, the gap, and hence the capacitance or the inductance, changes. The change in capacitance, and therefore the shaft twist and torque, can be measured by a stationary circuit through slip ring connections. When the air gap is part of an electromagnetic circuit, the torque which is related to the change in inductance can be determined by a stationary circuit without the use of slip rings [*40*].

An optical method of torque measurement investigated by Rebeske was found to be inferior to straingauge torquemeters [*41*]. Other systems rely on the change of magnetic permeability or electrical resistance of the shaft material or special coatings under load to provide an indication of transmitted torque.

Clearance measurement. With the development of light weight engines the measurement of clearance between compressor or turbine blade tips and casing has become more important and more difficult. Because lighter casings are more flexible, they are more susceptible to distortion from vibration, pressure variations, and flight accelerations. Also, in general, the permissible size of the sensor becomes increasingly restricted as the engine hardware is packed closer together and as the casing thickness

Plate D,3a. Combination instrument and instrument carriage for low speed testing.

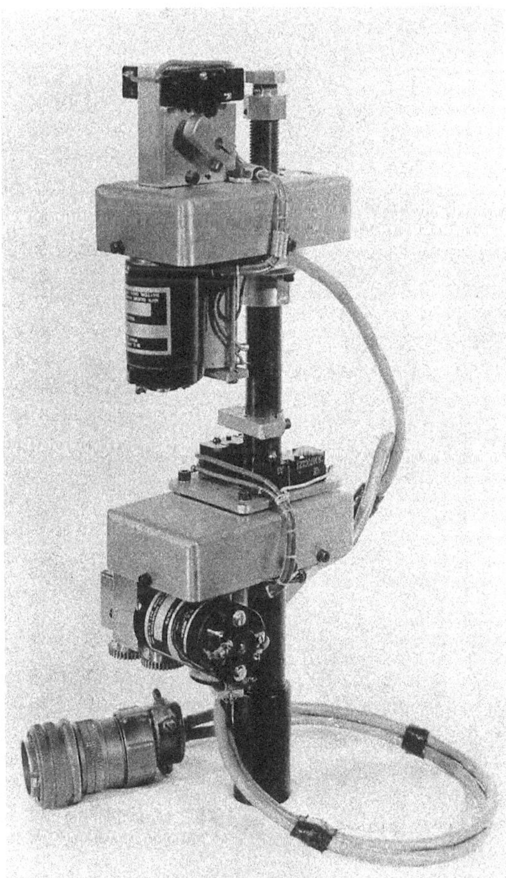

Plate D,3b. Remote control instrument carriage.

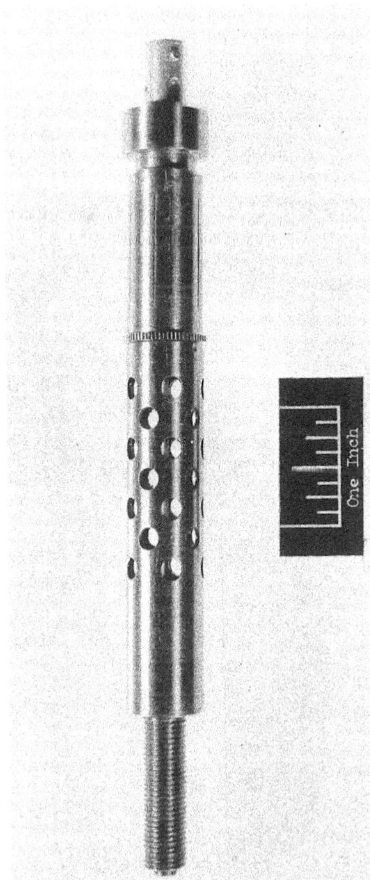

Plate D,3c. Clearanceometer probe.

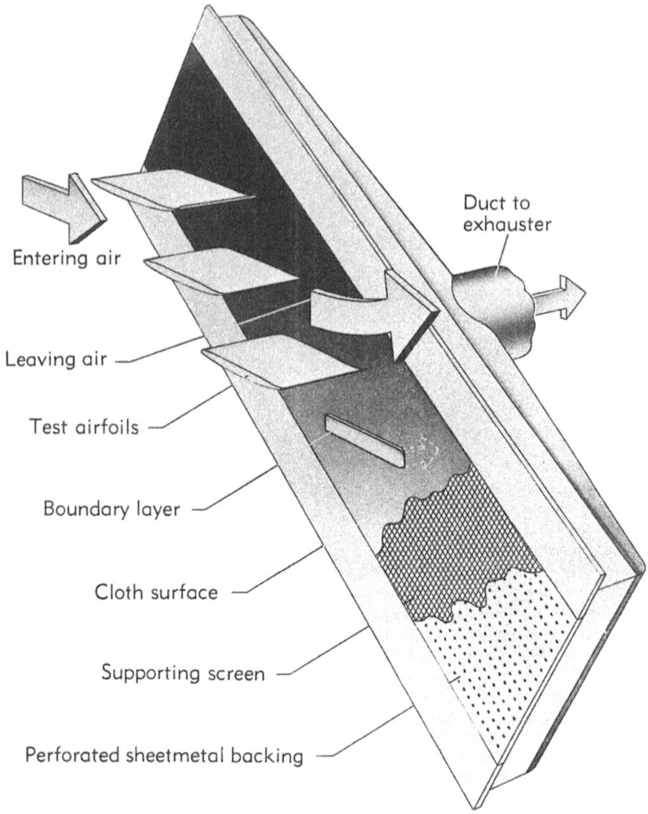

Plate D,4a. Construction of porous wall test section.

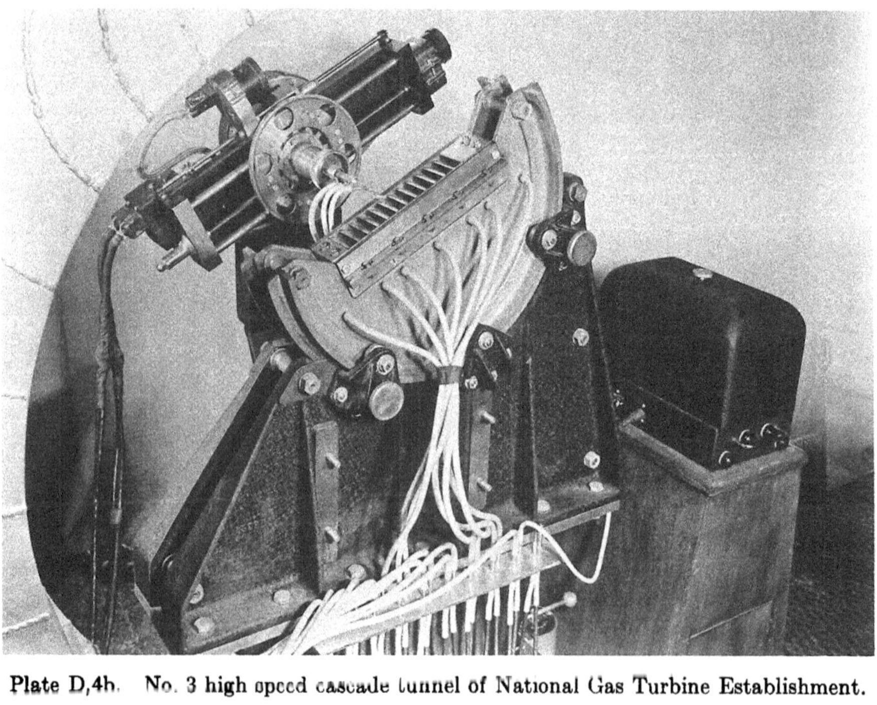

Plate D,4b. No. 3 high speed cascade tunnel of National Gas Turbine Establishment.

Plate D,4c. Langley 7-inch high speed cascade tunnel.

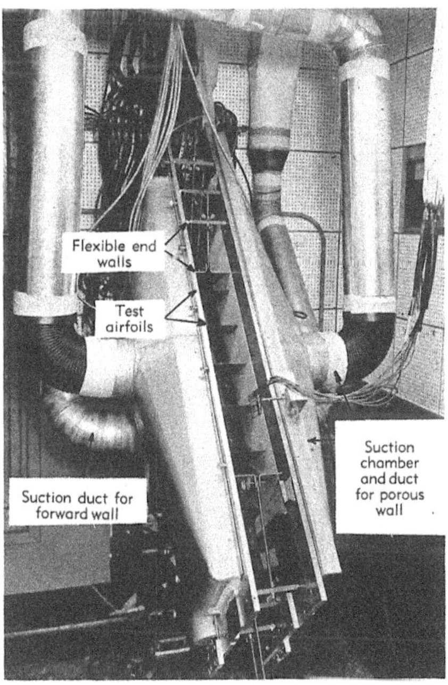

Plate D,4d. 5-inch cascade tunnel equipped with porous walls.

Plate D,4e. Cascade balance.

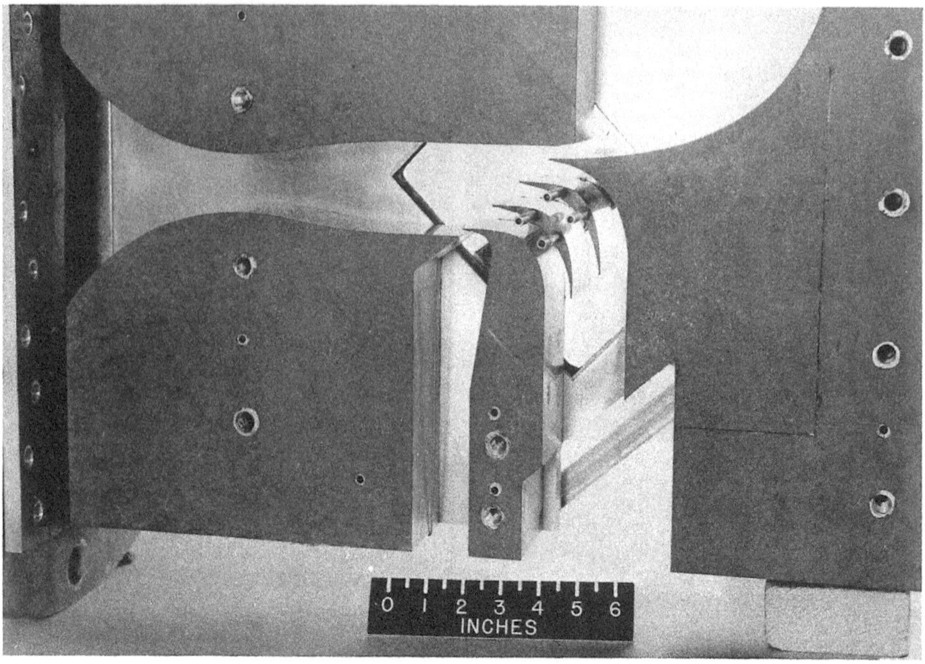

Plate D,4f. Langley supersonic cascade.

Plate D,4g. Supersonic cascade at University of Toledo.

Plate D,4h. Toledo supersonic cascade tunnel with side plate removed.

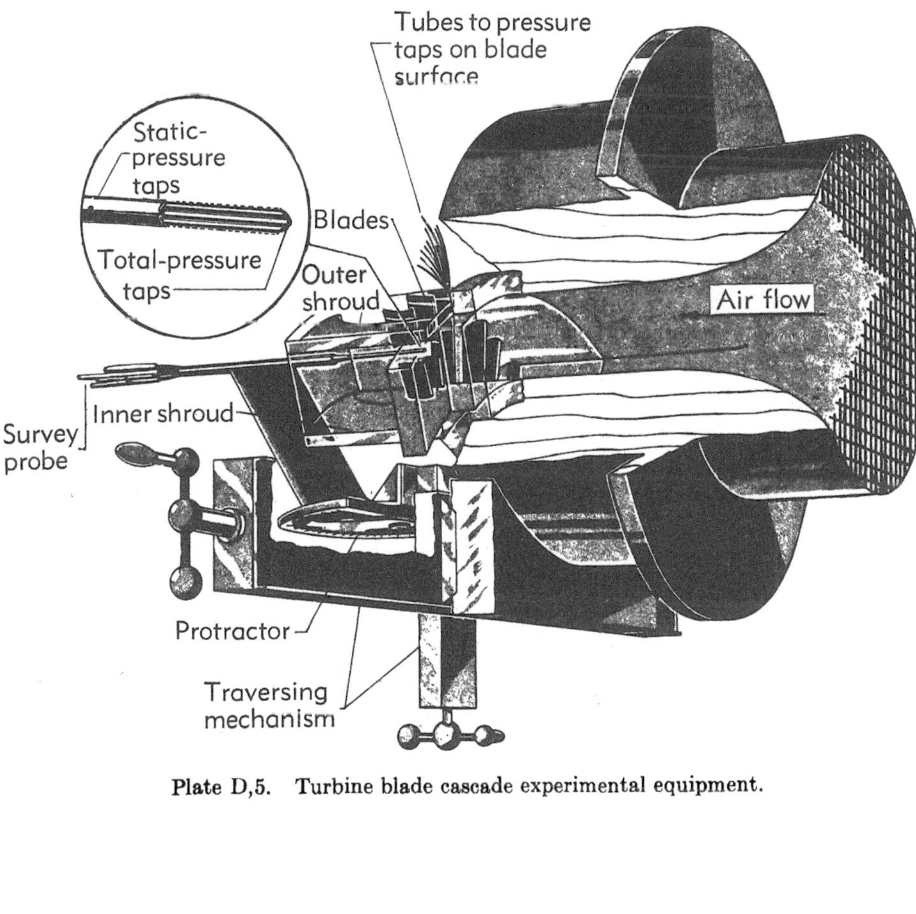

Plate D,5. Turbine blade cascade experimental equipment.

Plate D,6a. Pressure transfer device with top half of casing removed.

Plate D,6b. Installation of large-capacity centrifugal compressor.

Plate D,6c. Radial-inlet impeller used in experimental investigation of flow within rotating passages.

Plate D,7a. Stator assembly showing installation of total and static pressure orifices.

Plate D,7b. 60-inch diameter low speed research compressor of General Electric Company.

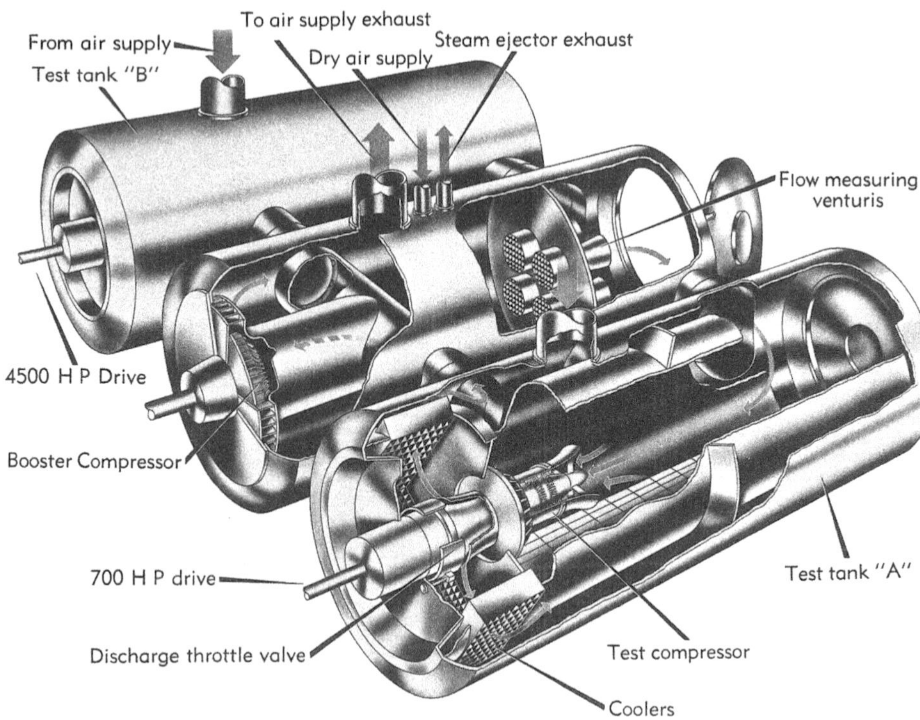

Plate D,7c. Scale model compressor testing facility of General Electric Company.

Plate D,8a. Ainley's test turbine.

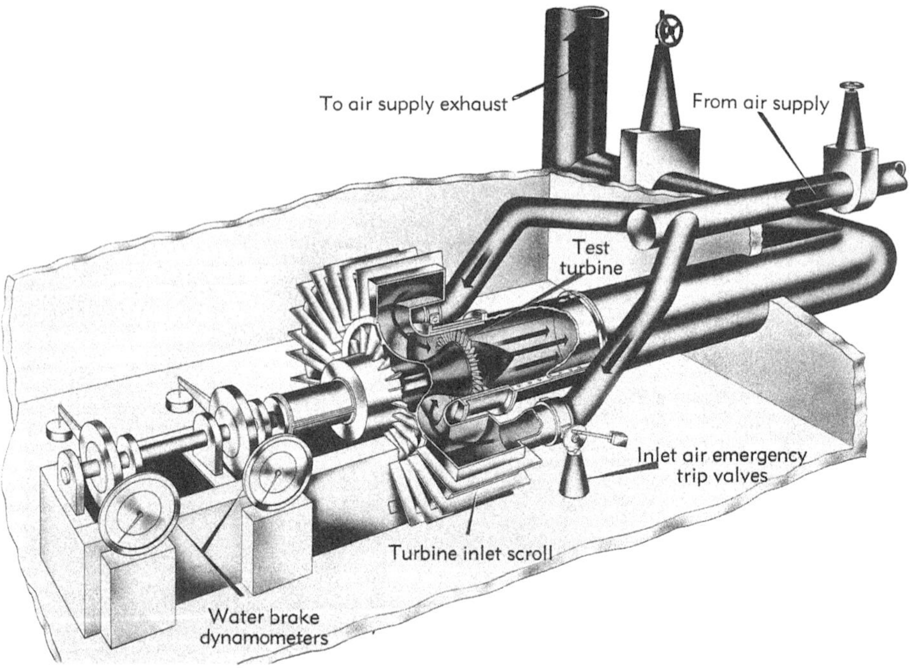

Plate D,8b. Cold air turbine testing facility of General Electric Company.

Plate D,8c. High temperature turbine test facility of General Electric Company.

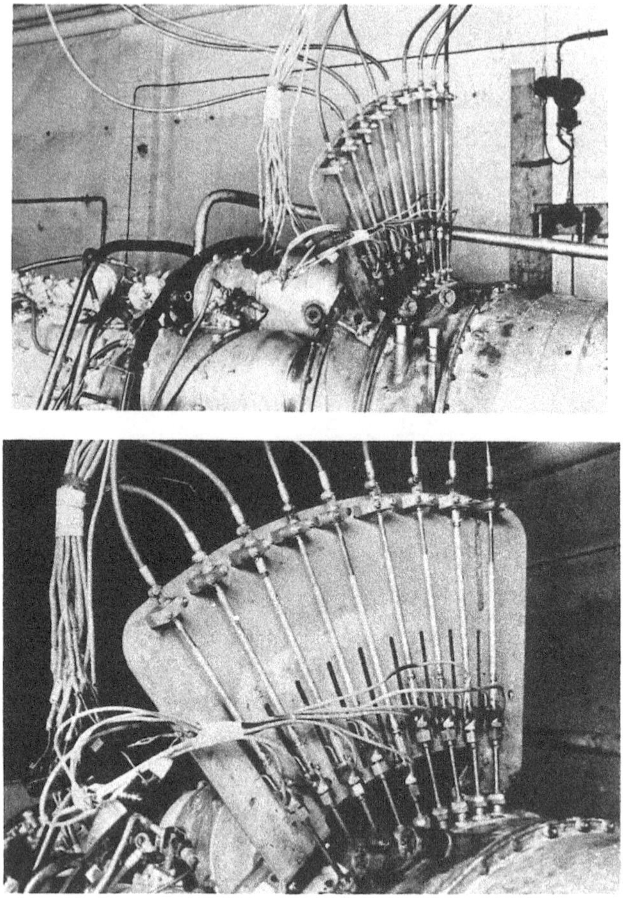

Plate D,9a. Engine tests for combustion chamber exit temperature traverse.

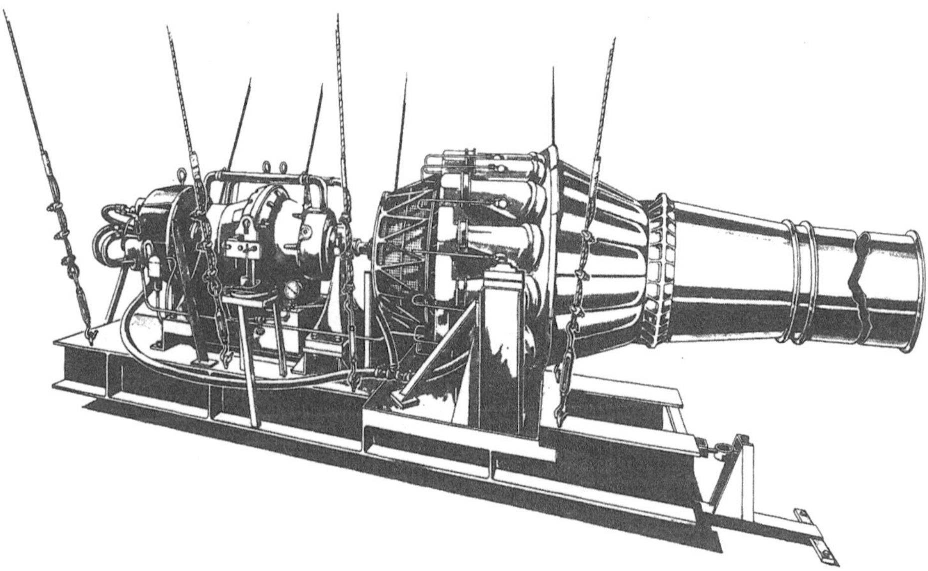

Plate D,9b. Turbine-propeller engine and suspension arrangement.

is decreased. A continuously indicating clearance measurement system has been developed which utilizes a small, lightweight sensing element (Plate D,3c). This device measures changes in electrical capacitance between the probe and the running blades. The system is capable of measuring clearances in the range of from 0 to about 0.180 inches for a normal compressor or turbine blade. The accuracy of the measurement is estimated as being within ±5 per cent. A standard probe can be used with engine temperatures up to 900°F and a special high temperature clearanceometer probe which is equipped with cooling air has been used for turbine measurements at temperature up to 1500°F. The probe is calibrated by static measurements, and the indication is most satisfactorily obtained through the use of a calibrated oscilloscope. However, the indication of tip clearance may also be recorded on magnetic tape.

The principle of operation of this clearanceometer is that the probe operates as a resonant electrical circuit. A coil in one end of the probe forms part of the resonant circuit and the capacitance between the end of the probe and compressor or turbine blade forms another part. As the clearance between the probe and the bucket changes, the tuning of the resonant circuit is affected and the resonant frequency changes. The indication system consists of a crystal-controlled oscillator, a radio frequency amplifier, and a phase-sensitive discriminator. The change in capacitance as caused by the change in clearance is quite small so the circuit must be very sensitive. The probe must be mechanically stable to prevent significant changes in the physical distance between the probe and the blade end. A serious consideration in the design of this type of clearance measuring device is the frequency response required to maintain the wave shape of the blade passing pulses with correct indication of the peak amplitude over the engine speed range. In a typical case the blade passing frequency is often 12,000 cycles per second or greater. A frequency response which is linear to at least 20 times the fundamental is required to assure accurate measurement of this wave form.

A radio frequency carrier system was selected as the method of detecting a clearance-sensing capacitance for several reasons. This circuit is sensitive to static changes in capacitance, which permits steady state clearance measurements and calibration and permits the instrument to be calibrated statically without access to the engine. The radio frequency system used is relatively insensitive to extraneous signals. A frequency of 10 megacycles was used in this system to assure adequate frequency response and sensitivity to small capacitance changes.

The metal parts of the clearanceometer probe were constructed of Invar, which has a very low temperature coefficient of expansion. Of several materials tested, fused quartz was selected as the most satisfactory for the coil form and dense boron nitride for the spacers. These materials were found to be adequately stable both mechanically and electrically for dynamic measurements over the temperature range to which they were exposed. The major difference between the standard and the high temperature probe is the provision for passing cooling air through the probe and around the sensing plate in the high temperature probe. Ionized gas is a sufficiently good conductor to serve as a ground for the probe and the output indication is then a function of the clearance between the probe and the ionized air surface rather than the true bucket clearance. The stream of cool air flowing around the sensing plate deflects the ionized gas from the immediate vicinity of the capacitance plate, thus permitting metal to metal clearance measurement.

Straingauges. The mechanical design of the various components of gas turbine engines has probably made more progress and advanced closer to perfection than the aerodynamic design of compressors and turbines or than the aero-thermodynamic design of combustors. A significant part of this advance has been due to the application of advanced theory, but this is equally true in the fields of aerodynamics and combustion. A factor which has enabled the mechanical designer to progress rapidly is the recent increase in sophistication of the straingauges, thermocouples, clearanceometers, transfer devices, recording methods, and analysis systems which have been developed. In the early phases of a new engine development several dozen straingauges may be applied to the blades and disks of the compressor and turbine rotors and stators. Turbine disk and blade temperatures can be recorded at a dozen or more strategic locations. The rotor blade tip clearances of one or more stages of the compressor and turbine can be indicated and recorded continuously during tests. By real time monitoring and electronic integration of the stress history, the mechanical analyst has a very good indication of how close to failure the critical elements are operating and how many more high stress test points, such as compressor stall, can be tolerated.

Certainly one of the most useful measuring devices employed in gas turbine research and development is the straingauge. Small, light, and accurate straingauges have been developed which can be used for short periods to 1800°F and for sustained periods up to 1500°F. With the application of slip-ring assemblies, straingauges can be employed on rotating elements of either large or small jet engines. Bonded wire straingauges are now considered superior to other types. This type of gauge consists of a fine filament of resistance wire attached to, but electrically insulated from, the base material by a suitable adhesive. As the base material strains, either in compression or tension, the strain is transmitted through the binding material to the gauge. Due to the property of materials to change their electrical resistance as they are physically strained, the resistance of the wire gauge changes as it is strained. The resistance wire is not used as a straight line element, but rather in the form of an accurately spaced flat coil or grid. Currently there are in general use three methods of fabricating the gauge grid. The first method produces an etched foil gauge by the use of photographic techniques. The second method consists in forming the grid by hand winding the sensitive wire (typically 0.001 inches in diameter) around pins mounted on a jog. The

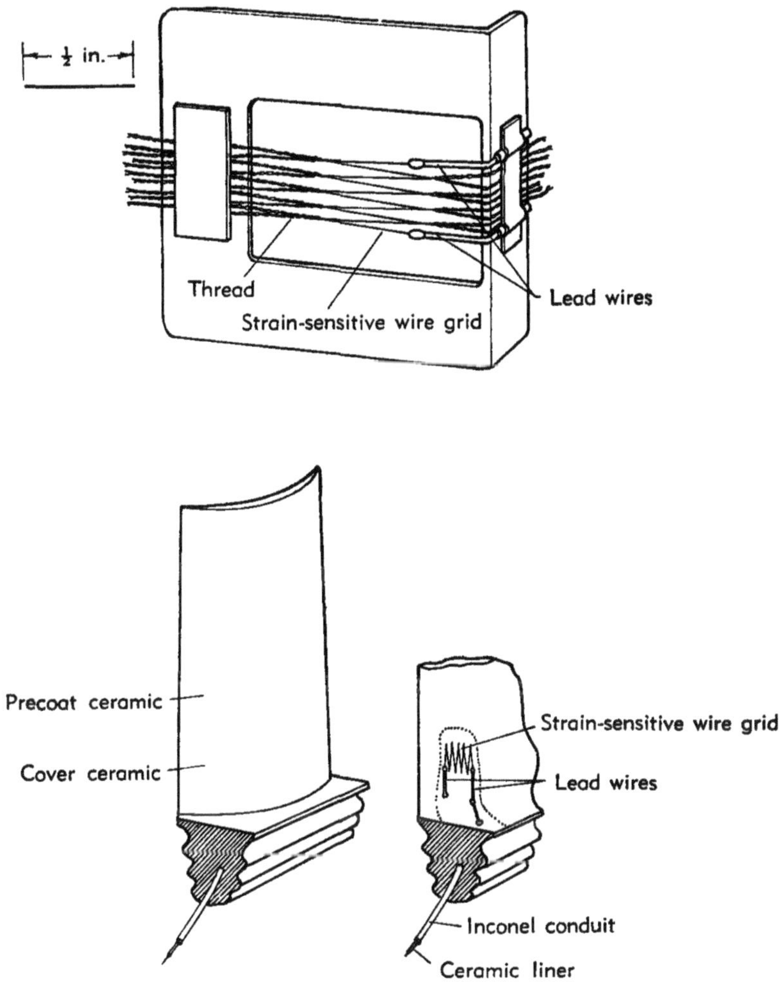

Fig. D,3kk. High temperature straingauge installation on turbine blade. Top, wire grid held in jig; bottom, gauge mounted on turbine blade.

gauge is flattened and set by squeezing between hard surfaces. The third method [*42*] employs a frame (Fig. D,3kk) to hold two small rods in the proper position. The wire is wound around the rods and the lead wires held in place and attached. Fine silk thread is then used to tie each wire loop to the frame so that the rods may be removed. After tightening the loops by pulling the silk strands, the gauge is ready for cementing in place.

The thread is removed after the first cover coat of adhesive has been applied and allowed to set.

The junction between the gauge resistance wire and the remainder of the electrical circuit is frequently a source of difficulty. In order to improve the relatively low fatigue life, gauges are now constructed with tabs before being applied to the test part. The tab consists of a $0.001 \times \frac{1}{64}$-in. ribbon folded and welded to the strain wire. Material used to make the tab is the same as that of the resistance wire in order to avoid thermocouple action at the junction.

In preparing to attach a straingauge to a surface which is to be measured, it is important that the surface be sand blasted and degreased by either a solvent or washing with a strong detergent. The surface preparation often makes the difference between failure and success, particularly in applications on rotating parts. Before the straingauges are mounted to aluminum alloy materials it is desirable to clean the surface with a detergent and by vapor degreasing after the sand blast operation. Ceramic cements are in general use for bonding straingauges to metallic surfaces. The curing cycle for these cements consists of a room temperature drying cycle of from one to four hours depending upon the atmospheric conditions and the thickness of the coating, which should be less than 0.003 of an inch. Following the air drying the period of heating at from 150 to 175°F for four hours is recommended. The final step is a cure at 600 to 1000°F for a minimum of 30 minutes. In preparing a test specimen a precoat of ceramic cement is applied to the surface and cured as previously indicated. The straingauge is then laid over the test specimen at the desired position and held in place with cellulose tape. The cement is brushed over the grid leaving the ends of the tab exposed for attachment of the lead wires. Usually two coatings are necessary to completely cover the grid. The thickness of the ceramic matrix after final curing is between 0.003 and 0.005 in. Lead wires in the form of a 0.001×0.016-in. ribbon can be applied to the airfoil surface in much the same manner. The junction between lead wire and straingauge tab is made by welding.

Blade contour measurement. It is often desirable to know accurately the contour shape of compressor or turbine blades. Since the airfoils are usually quite small, special precautions must be taken to find errors in construction. A point-by-point check along the airfoil surface can be made through the use of a small feeler moved along the surface of the airfoil by means of two micrometer drives mounted at right angles to each other. When the feeler touches the metal surface, a low voltage circuit is completed and a signal light flashes. Systems which indicate changes in electrical capacitance between the model and the measuring device are used to avoid actual contact. Other methods utilizing optical leverage systems provide a continuous trace of the surface at a scale several times that of the model. If the blades can be cut and sections removed for examination, a number of optical-enlarging systems are commercially available. One method of sectioning blades and yet retaining the angular relation between sections is to cast a rectangular volume of plastic material around the blade before cutting. After cutting and finishing to the inspection station, the rectangular surfaces of the cast material can be used to determine the angular relationship between sections.

Special Techniques. In the usual aircraft gas turbine, it is very difficult to obtain true readings or true average readings of static pressure, static temperature, or direction of the flow between blade rows. For these reasons, it would appear likely that new approaches to solving the problems of determining the flow velocity and direction will eventually evolve. If it were possible, for example, to measure the average velocity of sound across a circumferential segment equal to one or multiple blade spacings, this information would be very useful since the average static temperature could be determined directly. If it were further possible to measure the average tangential component of flow velocity across this same segment, the interstage information desired would be readily obtainable through the use of continuity and equilibrium relations, since the weight flow, total temperature, and total pressure can ordinarily be obtained with acceptable accuracy and reliability. The average velocity of sound and the tangential component of the flow velocity might be obtained by measuring the time of travel of a sharp sound pulse between a transmitter and receivers mounted on either side of the transmitter along the same circumference. The average axial velocity across a rotor blade row might conceivably be obtained in a similar manner. If it were possible to measure individual particle velocities, the desired interstage information would be directly obtainable.

Occasionally, when more usual techniques of investigation fail to provide the information desired from an experimental setup, special methods are employed. In this category are means of making air flows visible, tuft studies, lampblack patterns of boundary layer directions, methods of determining the point of transition from laminar to turbulent boundary layers, hot wire anemometers, uncambered and untwisted propellers for detecting shed vortices, and optical methods to observe density differences or the path of gases having indices of refraction different from the primary test medium.

Flow visualization. The method of using lampblack for flow visualization in cascades, blowers, compressors, or turbines consists of painting a mixture of lampblack and oil or kerosene over the surface being investigated and conducting the test while the vehicle dries. This method has been used successfully to make visible the occurrence of boundary layer transition and separation from the blade surface in cascade tests [*43*]. It is useful in compressor testing to give some insight into the effects of struts and supports on the over-all flow pattern [44]. In conjunction with tuft studies, lampblack patterns have been useful in investigating stalling of airfoils [45] and in examining backflow phenomena in centrifugal compressors [46].

In making tuft studies on stationary blades or walls by photographing them at an exposure greater than the period of tuft oscillation, an indication of the degree of turbulence present and the direction of the mean flow can be obtained. Tufts on rotors may be observed by illuminating them with stroboscopic lights which are triggered by a contactor on the rotor shaft. Another means of flow visualization is the use of smoke resulting from hot kerosene vapor [47], smoke candles, or titanium tetrachloride.

The use of smoke as a quantitative method as well as a means of observing flow behavior has been markedly advanced by Brown of the University of Notre Dame [48]. A very low turbulence and vibration-free wind tunnel having a contraction ratio of 48 to 1 for high speeds (up to 220 feet per second) and seven 14 × 18 mesh bronze screens followed by five 20 × 20 mesh nylon screens is used. The smoke is produced by coking (burning with insufficient air) grain straw in a closed low pressure vessel. The smoke is cooled in water-jacketed pipes and filtered before being introduced upstream of the anti-turbulence screens. Photographs are taken with 30 microsecond-duration flash lamps. In another facility, Brown has photographed smoke streamlines through shock waves at low supersonic speeds. These techniques make possible visual studies at Reynolds numbers larger by a factor of 10 than older methods and open a new avenue of approach to many aerodynamic problems.

Schlieren and interferometer methods. The apparatus and techniques for schlieren photography and interferometer studies in general aerodynamics are described in Vol. IX. In the gas turbine field these methods have been applied to two-dimensional cascades of compressor and turbine blades. The schlieren method is of definite value and has been applied extensively. The question arises as to when the interferometer becomes a practical and worthwhile instrument in turbomachine research and development. Ordinarily, the boundary layers and the secondary flows are much smaller in turbine cascades than in compressor cascades of similar physical proportions. The use of interferometer techniques might therefore be practical for studying the flow through turbine cascades because two-dimensional flow is more closely approached, so that less distortion of the fringes due to three-dimensional flows results [49]. Where compressor cascades of high aspect ratio can be tested this method can be of value [50]. The advantage of obtaining experimental data in this manner is that the physical condition of the flow in the entire field is obtained. If the idealized flow over the entire field is known by potential flow mapping or other methods, a quantitative measure of the difference between the real and the ideal flows can be obtained for the complete flow field.

Unsteady flows. The hot wire anemometer has been of value in studies of the effects of turbulence and Reynolds number on boundary layer behavior (see IX, F,1 and F,2). The instrumentation developed for these investigations is not well suited to the study of the large velocity fluctuations that occur in axial and centrifugal compressors and turbines. Under certain conditions of air flow—including those of present interest—the instantaneous air flow velocity is related to the derivative of the hot wire potential; consideration of the wire heat capacity might lead one to expect this result, but a rigorous proof requires a considerable amount of analysis. The differentiated signal may be amplified (a direct-coupled amplifier is necessary to avoid phase distortion) and presented on an oscilloscope. The hot wire probe used in turbulence work may be modified so as to measure the average and instantaneous directions and magnitudes of air flow under the conditions of large velocity changes that occur in turbomachines [*51*].

The hot wire responds (approximately) only to that component of air velocity perpendicular to the wire axis; the exact behavior with respect to wire inclination must be determined by calibration. If the velocity pattern repeats (as happens under steady operating conditions in a compressor) then velocity observations along the coordinate axes of the system may be taken in sequence; for transient phenomena, simultaneous measurements are necessary (obtained by the use of two oscilloscopes with locked synchronization). In the instrument of [*52*], the hot wire element is etched tungsten wire, 1 mm × 0.0088 mm, copper plated at its ends and soldered to a pair of steel needles which are pressed together during soldering so as to maintain the wire under 20 gm of tension. At a distance of 5 mm from the wire, the needles are embedded in dental cement at the end of a long supporting tube; they are here connected to two pairs of wires, one pair serving as a current supply for the hot wire and the other pair being used as potential leads. Also projecting from the tube is a fine wire thermocouple used to measure air temperature in the neighborhood of the hot wire. If a fixed current (from 100 to 200 ma) is passed through the wire, the potential across it is a measure of the air velocity. Because of a change in wire characteristics, this potential at a fixed air velocity will vary in time; it is found, however, that recalibration is not necessary since a simple cold-resistance measurement allows an automatic compensation process to be carried out. The analysis of this compensation requires consideration of such factors as the temperature distribution along the wire, possible oxidation and plastic flow of the wire, and the mean-value effect of velocity fluctuations.

The theory used has been checked by a large number of critical tests—for example, velocity fluctuations at one set of wire orientations may be used to calculate the fluctuation at another orientation and this calculated fluctuation may be compared with experiment. In general, experimental agreement is excellent.

D,4. Two-Dimensional Cascades.

INTRODUCTION. Although there are certain problems that can only be studied with single-stage or multistage components, such as interstage matching or the stalling of a complete compressor, there is much information that can best be obtained with stationary models in two-dimensional cascade tunnels. The effects of a particular variable can be studied independently by isolating the model from the maelstrom occurring within a gas turbine. As instrumentation problems are reduced, flow conditions can be held constant, and since the cost of operating most cascade tunnels is low, more detailed measurements can be made. Cascade models and test equipment are simple and usually not highly stressed, so that the construction is rapid and inexpensive.

The information desired from two-dimensional cascades of airfoils falls into several distinct categories. It is generally assumed that the work done and the loss incurred by conventional compressor and turbine blade sections will vary in a consistent manner as the geometric and aerodynamic parameters of the cascade are varied. The first category of information desired is the over-all pattern of performance of a particular family of blade sections.

Once the pattern of performance is established, it becomes of interest to extend the limits of operation by testing other types of blading. The particular limits frequently encountered are the relative Mach number, the static pressure change or axial velocity ratio, and flow deflection.

There is a need for basic data on boundary layer growth under varying conditions of pressure gradient, surface curvature, Reynolds number, and Mach number. There is also a need for information of the effects on blade performance of practical construction: surface roughness, leading and trailing edge radii, and deviation from design coordinates.

The primary variables that affect the flow turning angle and losses through airfoils in cascade are the profile, camber, chord-to-spacing ratio, inlet angle, angle of attack, static pressure ratio, Mach number, and Reynolds number. The desired information can be obtained by measuring the pressures or forces exerted on one or more of the test blades, or by direction and pressure surveys of the flow upstream and downstream of the blades. To provide data applicable to the design of axial flow compressors and turbines, a series of blade sections having constant thickness but of varying camber is customarily tested over the range of conditions to be encountered in application. The effects of section thickness, section thickness distribution, and mean-line type are investigated separately.

Such tests provide useful data for the design of gas turbines, but are ordinarily insufficient to indicate the direction toward improved designs.

The design of a two-dimensional cascade tunnel capable of testing guide vane, compressor blade, turbine nozzle, or turbine blade sections is complicated by the number of variables that must be incorporated. A wide range of entering and leaving air directions and velocities, chord angle, blade spacing, and axial velocity ratio must be accommodated. Provision for schlieren observation of the flow and for special instrumentation is desirable in high speed cascades. The many cascades that have been built differ greatly in general layout, size, systems of changing variables, and methods of taking measurements. These differences, the use of different blade types, and methods of presenting data have hindered the correlation of results from the various sources.

Design of Test Apparatus.

General objective. The general objective in the design of a two-dimensional cascade tunnel is to simulate the flow through a row of an infinite number of infinitely long blades. This objective could be approached by using cascades of many very long blades, but the air supply necessary to drive high speed tunnels of large size would require excessive power. For this reason, cascades of from 5 to 15 blades of aspect ratios from 2 to 5 are used. The side and end wall boundaries of the test section are treated with special care to minimize the differences between an infinite cascade and a cascade of practical size and power. The problems of treating the cascade boundaries can be separated into three phases: the entrance to the blade row, the test section, and the exit from the blade row.

The desired condition at the entrance to the test section is that the flow be uniform in static and total pressures and in wall boundary layer thickness, and correspond in direction to the flow through an infinite cascade. Boundary layer removal slots are usually provided on the side walls a chord length or so upstream of the blades to produce an even boundary layer thickness. Slots in the end walls (Fig. D,4a [*25*]) are used to control the upstream flow direction as well as to reduce or eliminate the low energy flow entering the cascade. In some tunnels, flexible end walls are used upstream of the blades to permit shaping the end boundaries to contours simulating streamlines of an infinite cascade.

Two-dimensional flow through the test section of a turbine cascade can be closely approached with blades of aspect ratio 2 without difficulty. With compressor cascades, the rapid growth and interaction of the wall and blade boundary layers in a strong adverse pressure gradient causes considerable difficulty with blades of aspect ratios as high as 5. Fig. D,4b [*25*] presents a simplified picture of the flow contraction caused by side wall boundary layer growth. The static pressure rise through a compressor blade cascade is reduced because of the effective contraction of the pas-

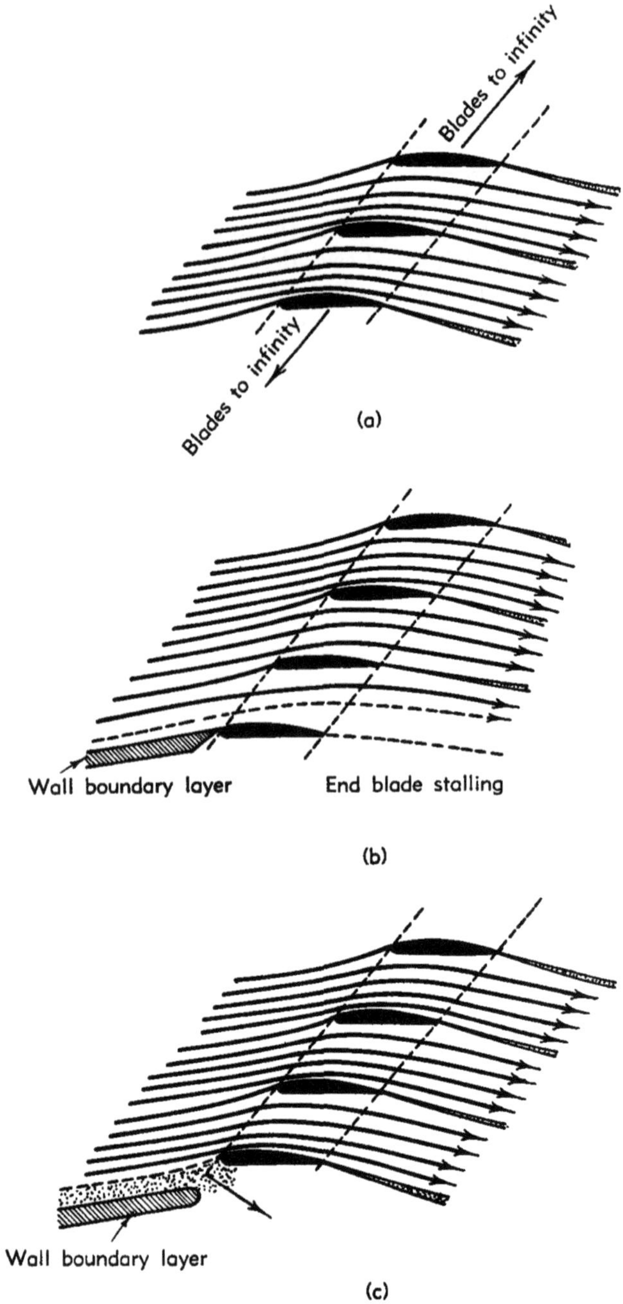

Fig. D,4a. Streamline flow through cascades. (a) Infinite cascade (diagrammatic). (b) Finite cascade with no suction. (c) Finite cascade with suction.

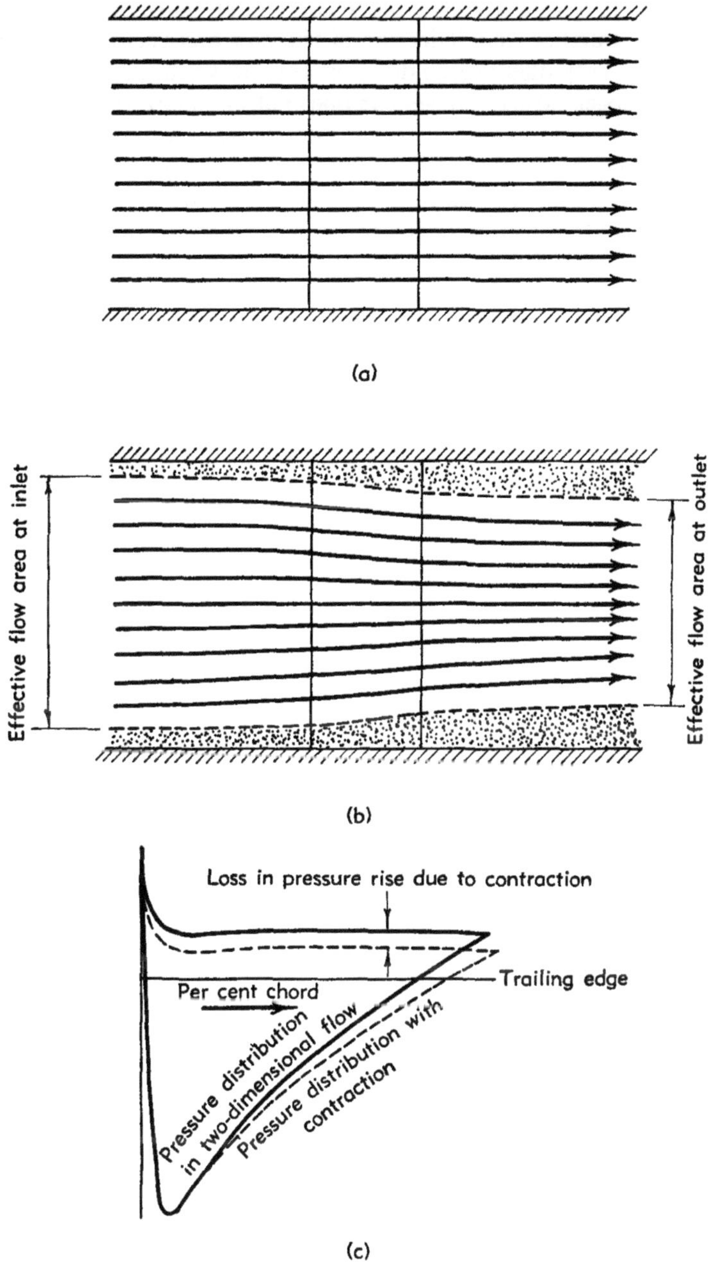

Fig. D,4b. Effect of boundary layer on streamline flow. (a) Uniform flow in absence of boundary layer. (b) Contraction of streamlines due to growth of boundary layer. (c) Effect of contraction on center-line pressure distribution.

sage, and the pressures over the blade surfaces are affected as illustrated in Fig. D,4b. Further, the deflection of the flow and the drag and lift of the blades are very likely to be different than in an infinite cascade. The use of high aspect ratios reduces but does not eliminate these discrepancies. Boundary layer removal through porous test-section side walls has been employed in attempts to eliminate these difficulties [*53*]; a close approach to two-dimensional flow was obtained with compressor blades of aspect ratio 1.0. A sketch of the construction used in this investigation is shown in Plate D,4a.

The discharge from the cascade should be uniform in static pressure and direction. The total pressure should be constant spanwise. With porous side walls, these requirements pose no particular problems. With solid test-section walls, a uniform discharge flow is difficult to obtain, and the case of the infinite cascade can only be approximated. If the tunnel side walls extend downstream of the blades, the end walls should extend a similar distance in the direction of the primary flow.

Entrance design. A settling chamber having a cross-sectional area at least 9 times as large as the test-section area and 1 to 3 diameters in length is provided. At the upstream end, a honeycomb having a depth-spacing ratio of from 2 to 6 is employed to remove residual rotation from the flow source. Several screens of from 30 to 60 mesh, spaced several inches from the honeycomb and from each other, are installed to equalize the flow velocity into the entrance. At Langley, contracting the flow area in two successive steps has been found to be satisfactory for either rectangular or circular settling chambers. The first fairings reduce the passage width from that of the chamber to the test-section width. Where the first fairings become parallel, the second set begins, converging to the test-section height. This construction facilitates moving the second set of fairings with the tunnel end walls to accommodate the various tunnel heights required and avoids having a skewed entrance.

Two methods are used for varying the angle between the incoming air and the axial direction of the cascade. In the simpler method, the inlet air angle is changed by replacing a section of the tunnel between the settling chamber and the test blades. This system is used in Langley and National Gas Turbine Establishment (N.G.T.E.) low speed cascades and in the United Aircraft Corporation high speed cascade. With this system, a series of runs at constant inlet angle is normally made. The N.G.T.E. No. 3 high speed cascade (Fig. D,4c [*25*] and Plate D,4b), the Langley 7-inch high speed tunnel (Plate D,4c) and the Deutsche Forschungsanstalt fur Luftfahrt (D.F.L.) high speed cascade at Braunschweig (Fig. D,4d [*54*]) are fitted with a turntable supporting the profiles. With this configuration, the blades can remain fixed in relation to the axial direction of the cascade as the inlet air direction is changed. As the flow into compressor and turbine rows ordinarily varies in this

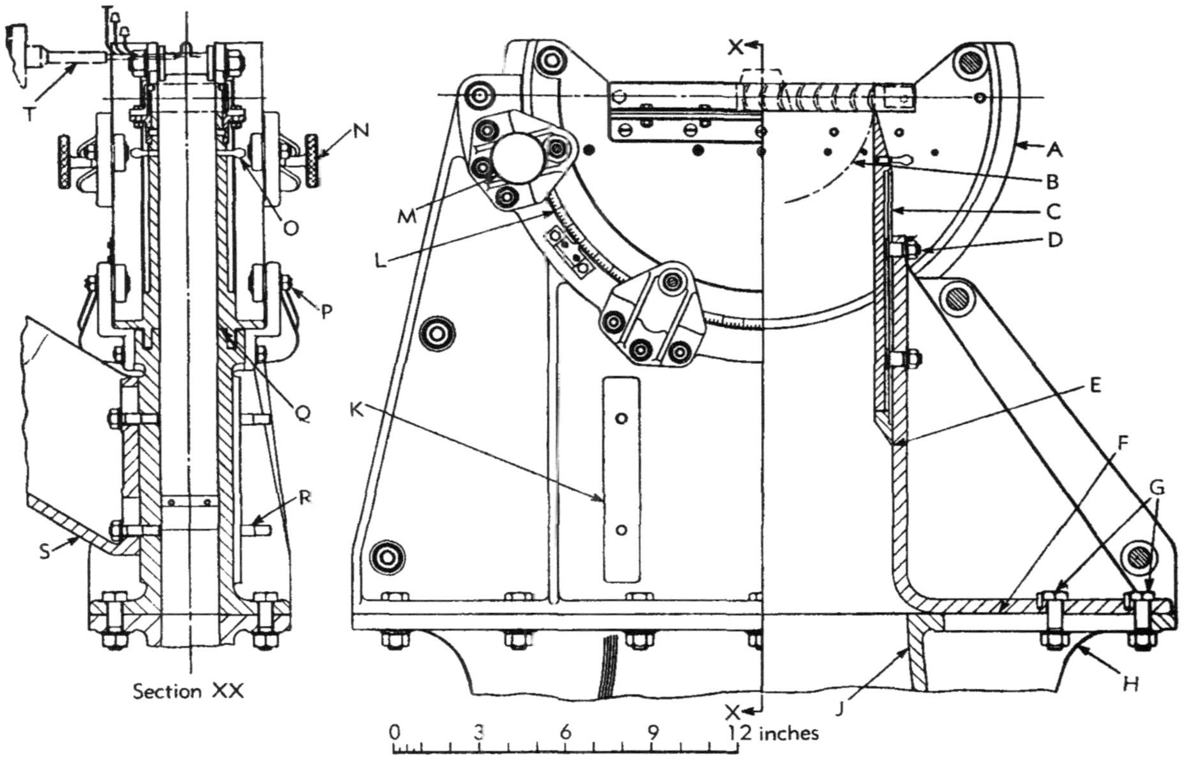

Fig. D,4c. Universal tunnel end for National Gas Turbine Establishment No. 3 High Speed Cascade Tunnel. A, Turntable carrying cascade; B, Locus of leading edge of end blade in cascade. Wall tip must follow this path; C, Vertical sliding portion of adjustable wall; D, Nuts for locking item C in position; E, Acceleration provided at sliding joint; F, Complete wall assembly slides horizontally on this surface; G, Bolts for locking wall assembly in position; H, Fixed inlet section; J, Acceleration section; K, Pad for mounting traverse-gear bracket (removed); L, Scale for setting cascade angle relative to air inlet direction; M, Clamp; N, Cascade mounting flange; O, Static tappings; P, Turntable guides; Q, Sliding butt joint accurately machined; R, Studs for mounting traverse gear. Note these are provided on both sides of the tunnel for calibration purposes; S, Traverse-gear bracket; T, Traverse gear.

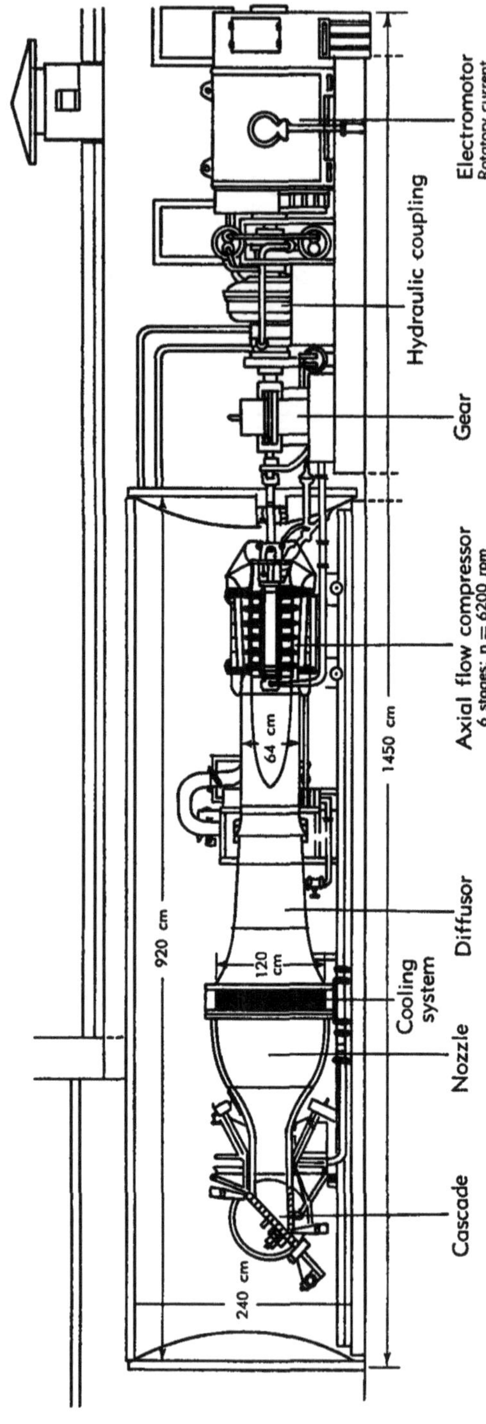

Fig. D,4d. Sketch of the Deutsche Forschungsanstatt für Luftfahrt tunnel. (For particulars, see Table D,4.)

manner, predictions as to compressor and turbine performance over a range of flow quantity coefficients can be made with less interpolation from data so obtained. A disadvantage of this system of changing variables is that either the number of test airfoils or the tunnel height must change when the inlet angle is changed, thus either interconnecting mechanisms between the tunnel turntables and end walls must be provided, or provision made to add or remove blades.

Test blade aspect ratio, size, and number. Factors to be considered in choosing a suitable size for a cascade tunnel are instrumentation difficulties, the Reynolds number range, and the power required to drive the tunnel. There are a number of important advantages in the use of test blades of the same chord as the application is likely to be. Using compressed air as the test fluid, and discharging to atmospheric pressure, the Reynolds and Mach number relationship which will exist in the cascade will be very similar to that encountered in typical aircraft compressors at high altitudes and speeds. Further advantages gained in using small blades lie in the relatively low capacity air-supply compressors and the lower power required to obtain tests at the usual operating Mach numbers with a large number of blades. The blade aspect ratio, usually 2 to 5, is selected as high as is practical considering power requirements and blade strength.

The difficulties encountered in studying the flow about small blades lie in the obtainable accuracy of manufacture and alignment in cascade, the greater care of flow measurement required to obtain the desired accuracy, and the difficulty of obtaining detailed information concerning the flow about the blades. It is impractical to measure the pressure distribution about small (less than 2-in. chord) blades by means of static taps in the surfaces of the blades. This is unfortunate, because a knowledge of the surface pressures is helpful in developing improved blade shapes and in selecting optimum operating conditions. Blade surface pressure distributions provide a direct comparison between calculated and actual flows and, by integration to obtain the lift force, a check on the accuracy of turning angle and pressure or force measurements.

The modern British tendency has been to test cascades of from 7 to 16 blades having chord lengths of from 0.75 to 1.50 in., and aspect ratios from 2 to 3. A Reynolds number of about 200,000 is obtained with a Mach number of 0.5. A turbulence factor, usually of the order of 2.0, is employed to estimate the effective Reynolds number. The high speed cascade tunnel of the NACA Langley Laboratory accommodates seven blades of 3.5-in. chord with aspect ratio of 2.0. This cascade is equipped with a porous side-walled test section and boundary layer removal slots about one chord length upstream on all four walls. Use of a diffuser downstream of the test section for turbine blade testing reduces the power required. Plate D,4c shows this tunnel partly assembled. Blade surface pressures are measured on the central blade by means of static orifices in the surfaces. Schlieren photographs of the flow can be made by replacing the porous walls with glass.

The high speed cascade tunnel at D.F.L. Braunschweig (Fig. D,4d) has a test section 300 mm wide and variable in height from 250 to 500 mm.

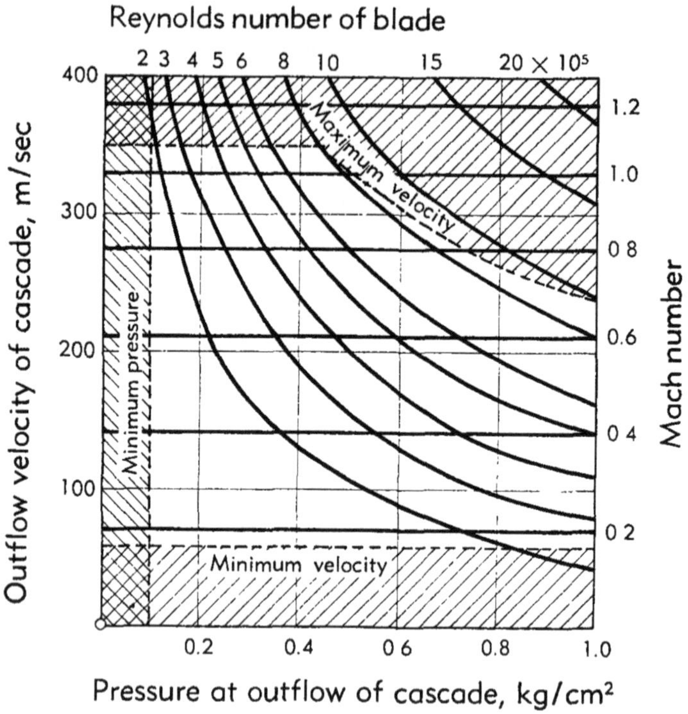

Fig. D,4e. Reynolds number and Mach number range of the tunnel. Reynolds number is based on a blade chord c of 60 mm ($\cong$ 2.5 in.).

The tunnel is enclosed in a large tank which can be evacuated. The attainable Mach number and Reynolds number are illustrated in Fig. D,4e, and further information is presented in Table D,4.

Table D,4. Particulars of the variable density high speed cascade wind tunnel of the Deutsche Forschungsanstalt für Luftfahrt, Braunschweig.

Power installed	1300 kw
Air intake of compressor	30 m^3/sec
Head of compressor	7800 m
Pressure ratio	2.14
Maximum Mach number	1.1
Pressure of outflow of cascade (= pressure in the evacuated tank)	0.1 to 1.0 kg/cm^2
Maximum dynamic pressure at outflow of cascade	0.5 kg/cm^2

The possibilities of applying transonic wind tunnel techniques (slotted, perforated, or porous walls) to cascade tunnels have been investigated only superficially. A cascade of compressor blade sections produces much stronger disturbances in the air stream than does the usual aircraft model in a large transonic wind tunnel. The preliminary results indicate that, while valuable information can be obtained, the difficulties of obtaining reliable data in the transonic range are serious. The step between high speed and transonic cascade testing is about equal in difficulty to the step between low speed and high speed cascade testing.

A transonic cascade tunnel having a 5 × 20 in. test section has been designed and constructed by the General Electric Company (Fig. D,4f). Perforated walls and floors are used to generate transonic flow and to cancel shock and expansion waves originating from the test airfoils. A relatively large number (from 12 to 36) of test blades are employed to

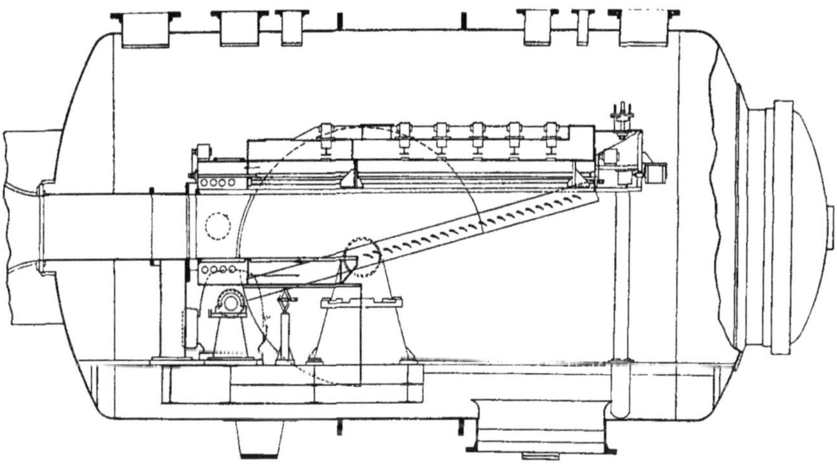

Fig. D,4f. Transonic cascade tunnel of General Electric Company.

simulate actual entrance conditions into a transonic compressor. Because the test-section height is fixed, airfoils are removed or added to the test cascade as the inlet flow angle is varied. Schlieren photographs of the flow can be taken through glass walls of the tank in which the tunnel is enclosed to permit independent variation of Mach number and Reynolds number.

Porous-wall, test-section design. The application of the porous-wall technique permits a close approach to two-dimensional flow in cascades of low aspect ratio. In this method, the surface of the side walls from about one half chord upstream of the test blades to one half chord downstream of the blades consists of a permeable material (Plate D,4a). For high pressure rise cascades, the flexible end walls are also constructed to permit area suction of the boundary layer. Suction chambers of relatively large cross section are provided to insure that the pressure is uniform on the discharge side of the permeable surfaces. The porosity of the surface must be such that the desired amount of flow can be drawn through the walls at a suction chamber pressure lower than the lowest pressure on the blades. If the exhauster capacity or intake pressure is limited, the material used as the porous surface must be properly selected. The permeability coefficients and mechanical properties of several materials are presented in a following discussion.

With the porous test-section side walls in the Langley Aeronautical Laboratory low speed cascades, the static pressure rise through the cascade can be controlled, within limits, by the amount of flow drawn through the permeable walls. When the pressure rise was adjusted to the two-dimensional value, the pressures on the blade surfaces near the tunnel wall were similar to those along the tunnel center line, the force obtained by integrating the pressure distribution nearly equaled the value associated with the measured momentum and pressure change across the cascade, and the wakes downstream of the test blades were constant in the spanwise direction, except for a small region near the wall. The porous-wall technique has not yet been applied extensively in published high speed cascade research, but preliminary results indicate that this method can be of value in establishing two-dimensional flows and in varying axial velocity through high speed linear cascades.

Exit section. In order to provide a means to simulate the curved streamlines that would exist in an infinite cascade at the position of the end walls, flexible extensions were attached to the rigid end walls approximately one chord length upstream of the Langley cascades. The flexible end walls (see Plate D,4d) continued through the test section and about one chord length downstream of the test cascade. The curvature of these extensions could be varied during operation of the facility. Usually a gap equal to about half the blade spacing is allowed between the end airfoils and the flexible walls. An experienced operator can adjust the end conditions to achieve uniform entering and exit flow in a few minutes. The direction of the flexible extensions downstream of the cascade is set parallel to the direction of the exit flow.

To eliminate the time required to survey the exit flow in order to measure the exit direction, in United Aircraft Corporation cascades the discharge direction of the flow is measured by the angle of two parallel exit plates, or tailboards, ten chords long, hinged to the trailing edges of the outer blades. Static pressure taps are located parallel to the cascade one and one-half chords downstream of the blade trailing edges. During operation, the exit plates are revolved until an angle is found at which, ideally, the static pressures are the same at all orifice locations. Good comparisons between the values measured by point-to-point surveys and those recorded by the tailboards have been obtained for all but high inlet angles. This technique has been used with good results on turbine blade cascades in the Langley 7-in. high speed cascade [*55*].

Force measurement. In the early researches of Christiani, Harris and Fairthorne, and Keller, the forces exerted by the central airfoil of the test cascade were measured by standard wind tunnel balance methods. In this system the central airfoil is supported separately from the walls of the test section. As the velocity into the cascade is increased to the test value, the floating airfoil moves in response to the forces being exerted. The force necessary to restore the test blade to its original position can be measured. By measuring the required force in two directions, the resultant force can be obtained and resolved into desired components. In the very thorough investigation of a guide vane cascade carried out by Sawyer [*43*] under the guidance of Ackeret, an application of the Amsler principle was employed to construct a balance free from the effects of static friction. According to Amsler, there is no static friction parallel to the axis of a cylindrical bearing when the shaft rotates. In Sawyer's balance the central airfoil was attached to a platform supported by bearings riding on two rotating shafts (Fig. D,4g and Plate D,4e). Thus, a force exerted on the platform parallel to the shaft axes can be measured with little error due to friction. Arrangement is made to permit measuring the force in any direction in the plane of the shaft axes by rotating the platform assembly, so that the shafts are parallel to the desired direction.

Porous materials. Porous materials have found several important uses in connection with experimental techniques of flow through turbines and compressors. One significant application of porous materials is in two-dimensional cascade testing. Porous materials have been used as surfaces of blades, for area suction and for sweat cooling. These surfaces are more easily applied to stationary blades, such as stators and turbine nozzles, than to rotating blades since the mechanical problems are less complex. Area suction has another application in diffusers attempting a high pressure rise per unit length, where boundary layer separation limits the rate and amount of diffusion that can be accomplished efficiently.

A permeability coefficient, derived from Darcy's law, can be used as an approximate measure of the porosities of the different types of porous materials:

$$a = \frac{q p_0 \mu L}{(p_1^2 - p_0^2)\rho_0}$$

where a = permeability coefficient (in.^{2})
q = weight rate of flow per unit area (lb/in.^{2}/sec)
p_1 = pressure entering sample (lb/in.^{2})
p_0 = pressure leaving sample (lb/in.^{2})
L = thickness of the sample (in.)
μ = dynamic viscosity of the fluid (lb-sec/in.^{2})
ρ_0 = weight of the fluid per unit volume (lb/in.^{3})

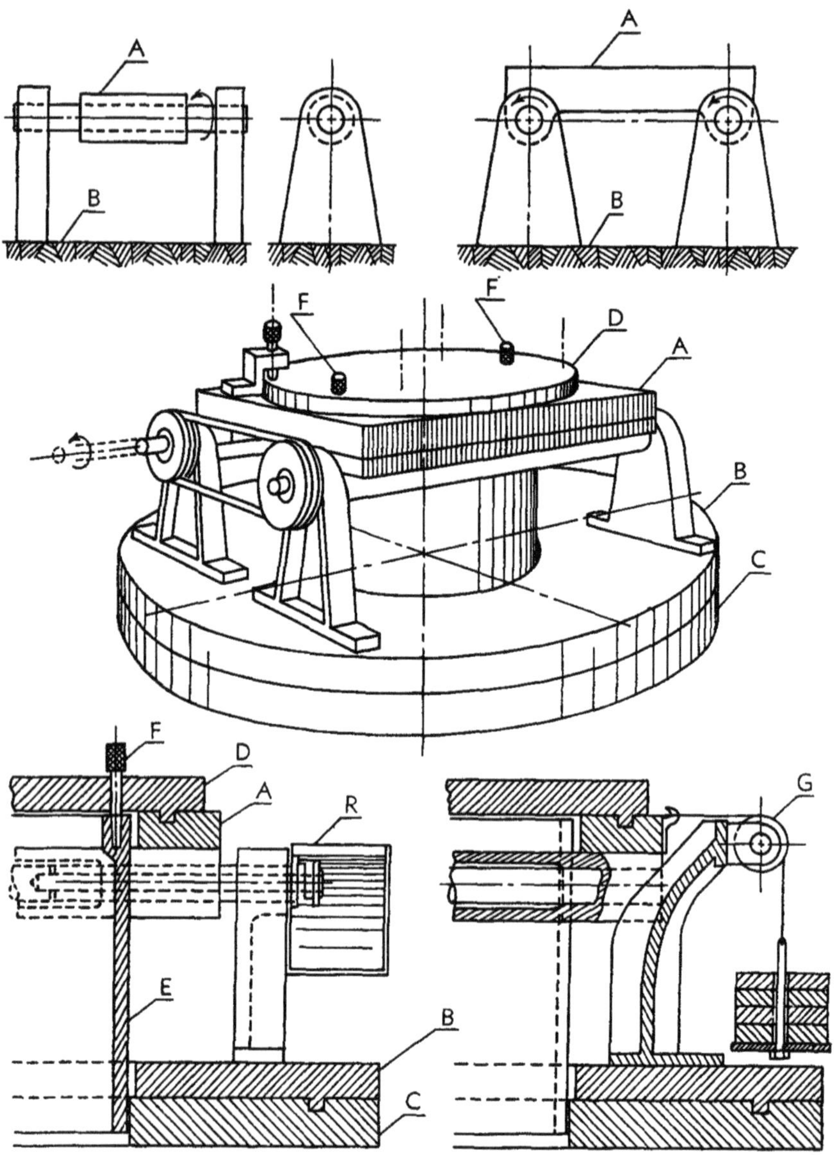

Fig. D,4g. Schematic drawing of cascade balance (simplified), showing steps in development. A, axially "floating" platform of basic unit; B, shaft carrier plate; C, foundation plate; D, auxiliary plate upon which blade is mounted; E, cylinder rigidly fixed to foundation plate C; F, taper pins locking D to E; G, pulley; R, oil reservoirs for main bearing lubrication.

Since this relation is valid only for viscous laminar flows of a gas through a porous medium, all permeability coefficients to be mentioned herein apply in that range.

The types of porous materials which have been used or investigated in attempting to find suitable porous walls for cascade testing include the following:

1. woven meshes, metallic or fabric
2. mechanical perforated metal screens
3. sintered metal particles
4. electro-deposited metal screens
5. porous plastic or glass

Woven meshes, metallic or fabric, are easy to produce and inexpensive: the metallic meshes and screens have proved to be the most suitable because of their workability and smoothness. A commercially available material, twill-dutch double-weave filter cloth, usually woven of Monel wire, having 250 fill wires of 0.008-in. diameter per inch and 30 warp wires of 0.010-in. diameter per inch was found to be very satisfactory for porous walls when commercially calendered from the as-woven thickness of 0.027 in. to 0.018 in. Porous materials of this type are relatively homogeneous in character and have permeability coefficients which range from 0.2×10^{-10} in.^{2} to 1.1×10^{-10} in.^{2} at a thickness of 0.020 in.

The sintered materials require a more involved process of manufacture including the production of fine particles of desired shape, grading, mixing with a fusing material, and subjecting the mixture to temperatures and pressure. In small lots, this process makes for an expensive product. Some sintered materials have a rather smooth surface, but the pores or openings are not as homogeneous in size and shape as those in the woven, perforated, and electro-deposited materials. The porosity of sintered materials tend toward lower values with permeability coefficients ranging from 1.5×10^{-10} in.^{2} to 15×10^{-10} in.^{2} at a thickness of $\frac{1}{8}$ in.

The mechanically perforated metals may be produced with a very wide range of sizes and shapes of perforations; they are fairly homogeneous and inexpensive, having permeability coefficients ranging roughly from 0.2×10^{-10} in.^{2} to 4.0×10^{-10} at a thickness of 0.020 in. Hammering and rolling processes have been used successfully on perforated materials to reduce the porosity and increase the surface smoothness.

Electro-deposited porous materials are the most uniform of the aforementioned porous materials. Due to the method of manufacture, the thickness of fine meshes of this porous material is limited to rather small values—usually less than 0.025 in.; this means that, for most practical applications, electro-deposited porous materials need a supporting perforated material. The porosities of this type of material tend toward higher values with permeability coefficients ranging from 1.0×10^{-10} in.^{2} to 20×10^{-10} in.^{2} at a thickness of 0.010 in.

Porous plastic materials have been little studied for aerodynamic applications; the openings tend to vary in size much the same as sintered materials, although smooth surfaces and good machineability can be obtained. Photosensitive glass has been developed permitting etching of 0.030-in. diameter holes in 0.20-in. thick glass plates. This material holds promise of allowing schlieren photographs to be made through porous surfaces.

SUPERSONIC CASCADES.

General discussion. Sections for supersonic compressor blades must be tested at the Mach numbers of operation, and preferably at similar Reynolds numbers. Fewer blades are required than for high speed cascade tests, but higher pressure ratio air supply equipment is required. Because supersonic compressors have not demonstrated high efficiency, no systematic cascade tests of supersonic compressor blade sections have yet been made using suitable test apparatus. There is reason to believe that supersonic compressors will benefit as much from such research as subsonic compressors have.

Supersonic cascade tunnels can be considered to be of two general types, one having the test-section entrance open to the atmosphere and the other having a completely enclosed test section. The open jet type is easier to adjust because the test section can be constructed as a separate unit and mounted independently. This type of cascade (Plate D,4f) is ordinarily used when the air supply is a high pressure storage tank and there is no necessity of recovering the dynamic pressure of the stream. Open jet cascades must be relatively small to permit sufficient running time, but the Reynolds numbers are high because of the high static pressure.

Closed test-section cascades (Plate D,4g and D,4h) are used when continuous operation is intended. Because the circuit is closed, low pressure, and hence lower power operation, are possible. The pressure level and therefore the Reynolds number can be varied independently of the Mach number. Greater mechanical complication is required to maintain a continuous flow path while permitting some flexibility of test-section geometry. Either type of test section can be used to test shock-in-rotor or impulse rotor blade sections or supersonic stator blade sections.

SUPERSONIC CASCADE TUNNEL DESIGN. The component parts of a supersonic cascade tunnel are similar in function to those of a subsonic cascade; each requires the use of a settling chamber, an accelerating nozzle, and a test section wherein the model blades are mounted. The supersonic cascade differs in two essential respects from its subsonic counterpart: (1) the need of a throttle downstream of the shock-compression rotor and stator sections to provide the necessary back pressure to position the normal shock within the blade passages; and (2) the immobility of the nozzle floors with respect to each other when, as is usual, the nozzle is designed for only one Mach number. However, in the supersonic case, it is no longer necessary to set the nozzle depth equal to an integral number of blade pitch spacings multiplied by the cosine of the air inlet angle, since there may be no mutual interference effects between leading portions of supersonic profiles when at their design incidence. Therefore it is possible to construct a cascade of only two blades to determine the behavior of the flow within the passage and measure the basic performance parameters, total pressure loss, and exit Mach number and direction. Usually four or more blade sections are used to insure that representative conditions have been established. The test models are usually mounted between glass side walls to permit schlieren or interferometer photographs of the flow to be made.

To insure the proper functioning of a supersonic cascade, the test section must be designed so that the starting area contraction ratio is not exceeded in any of the individual flow passages formed by the cascade blades. This is especially true for shock-in-rotor blade sections in which the maximum permissible area contraction is desired to obtain the lowest shock losses. Extreme care must be exercised in mounting the component blades of the cascade, since one slightly misaligned or incorrectly spaced blade can disrupt the entire flow or completely choke the nozzle. The phenomenon of choking (the inability to force the normal shock into the cascade) controls the lowest supersonic Mach number for which the cascade is operable. Supersonic compressor blade sections are generally tested at their design Mach number or slightly higher values representative of overspeed compressor operation. Testing at these higher Mach numbers is facilitated (for sections whose velocity normal to the cascade leading edge is subsonic) by merely rotating the entire cascade so that the flow is expanded about the leading edge of the foremost blade. The upper Mach number limit is reached when the inclination of the most forward blade is such that a detached shock is formed or the reflection of the expansion waves from the opposite nozzle wall extends upstream of the cascade. Similarly the test Mach number may be reduced through the use of oblique compression shocks from the leading blade. Thus it is not necessary to have a great number of separate nozzles or contour changes with flexible walls in order to test over a range of Mach numbers. The method used to raise the back pressure behind a rotatable supersonic cascade requires attention. With a fixed wall fairing the area contraction between the end blades and the fairing must be considerably less than that between the test blades, because of the boundary layers. Increasing the pressure ratio across the test section by reducing the speed of the air supply compressor or by contracting the passage well downstream of the cascade will force a normal shock through the passages formed by the outermost blades and the nozzle wall fairing, before full throttling can be accomplished within the cascade. For this reason a throttle that affects the flow in only the central passages is necessary to obtain maximum pressure recovery in the test model.

D,5. Three-Dimensional or Annular Cascades. An intermediate facility between the two-dimensional cascade tunnel and the single-stage test compressor is the three-dimensional cascade. In this apparatus a complete blade row, either specially constructed or a rotor or stator from a multistage machine, is subjected to a flow similar to that encountered in operation. Thus, effects not simulated in the linear cascade can be studied, but under more closely controllable and widely variable conditions than in the complete engine. A primary advantage is the ability to provide more complete instrumentation than in the actual machine. Boundary layer flows and end effects are not the same as in running rotors or unshrouded stators. Nevertheless, again applying the principle of isolating separate effects, this device can provide information concerning the behavior of blade profiles of known two-dimensional performance in three-dimensional arrangement. The annular cascade thus can be a useful tool to the gas turbine researcher and its results of value to the designer.

There are several reasons why three-dimensional and annular cascades are a natural extension of two-dimensional cascades. These facilities retain part of the ease of obtaining detailed measurements and the simplicity of model construction of the two-dimensional cascade tunnel. The study of a number of additional variables that occur in turbomachines is permitted: (1) the effects of a spanwise (radial) variation of circulation on the turning angle and loss, (2) the boundary layer flow on guide vanes and stators may be observed, and (3) the differences in blade section performance between two and three dimensions. Considerable differences can exist between the performance of given blade sections in a two-dimensional cascade and in a three-dimensional blade row. For example, it would seem quite reasonable to assume that the flow in a guide vane having an appreciable pressure drop would be little affected by Reynolds number, Mach number, or three-dimensional effects. Quite frequently, however, it is discovered that flow conditions leaving three-dimensional guide vanes are quite different than predicted on the basis of two-dimensional cascade tests of the same sections.

Because of the necessity for spanwise surveys of flow conditions, the three-dimensional cascade does not lend itself to investigations wherein the effects of a number of geometric variables are determined over a range of aerodynamic parameters. Three-dimensional or annular cascades appear to be best suited for detailed investigations of flows which cannot be adequately simulated in two-dimensional cascades and as a source of basic information to extend existing theories.

In addition to being a logical extension of the two-dimensional cascade, three-dimensional and annular cascades exhibit unique features which permit conditions to be established that cannot be accomplished or can only be accomplished with difficulty in two-dimensional cascades. An excellent example is the use of an annular cascade to study rotating stall. Propagating stalls have been produced in two-dimensional cascades. Study of the phenomena was necessarily restricted by the finite length of the cascade. Three-dimensional cascades can more readily be used to simulate transonic flow into a blade row. In two-dimensional cascades, the wave pattern is not the same upstream of all the blades due to the finite number of blades. This can be readily seen when it is considered that the blade nearest to the settling chamber will have only its own bow wave upstream of it, while the second blade will have two waves, and so on. If a sufficiently large number of blades can be provided in a two-dimensional cascade, this problem can be minimized, but in a three-dimensional cascade it can be eliminated entirely.

Compressor and turbine blade cascades. A number of annular tunnels designed to permit tests of compressor and turbine blade cascades have been built in the United States. In those with which the author is familiar, the general plan of the several cascades is quite similar. The intent was to establish a rotating flow of constant angular momentum entering the test section. In order to avoid wakes from swirl vanes near the test blades, the vanes were placed well upstream and in a region of low velocity to avoid an extended region of high velocity flow with attendant wall boundary layer growth.

The tangential velocity that must be imparted to the flow by the swirl vanes can be reduced by placing the vanes at a radius larger than that of the blades to be studied. The design of an annular cascade built at the NACA Langley Laboratory is illustrated in Fig. D,5. This tunnel is an induction type, the motive power being compressed air. The main stream is drawn from the atmosphere through bellmouthed entrance rings, past a closely spaced straightening grid, caused to rotate by the swirl vanes, then accelerated tangentially and axially to the test section. The flow direction and pressures were measured across the annulus at several circumferential positions to insure uniformity of entering conditions. The test airfoils can be equipped with static orifices on their surfaces. A ring carrying direction- and pressure-measuring instruments is provided downstream from the blades. The ring can be rotated to permit readings at any circumferential position in addition to the usual radial traverse.

The swirl vanes are adjustable in unison. Several sets of different cambers are provided to cover the desired range, estimated to be inlet angles, β, from 0 to 75°. With small values of β, from 0 to 30°, the flow at the test blades was quite uniform. With larger angles, thick boundary layers, particularly on the inner casing, were present. This difficulty was encountered in the similar annular cascades previously mentioned. At the NACA Lewis Laboratory, somewhat improved flows were obtained by suction slots located in the entrance cones. The United Aircraft Research Division improved their three-dimensional rig by providing an annular slot about one-fourth of the tunnel span wide, just upstream of the test section. The flow on the inner casing was permitted to pass into this opening. As yet no adequate means of obtaining flow with a high inlet angle and thin boundary layers at the test section of annular cascades is at hand.

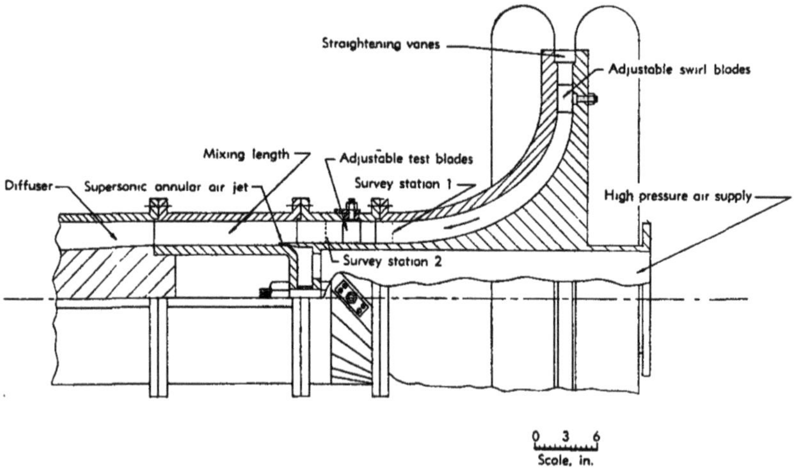

Fig. D,5. Langley Laboratory annular cascade.

Guide and nozzle vane cascades. Three-dimensional cascades intended primarily for the study of guide vanes and turbine nozzles have been operated successfully in Germany, England, and the United States. The flow into guide vanes and nozzles ordinarily has zero or very small inclination to the engine axis; therefore provision to swirl the flow upstream of the blades is unnecessary. The problem of producing uniform flow into the test section is considerably simplified. In the British vortex wind tunnel and the Langley guide-vane tunnel, both low speed tunnels, axial entry is used. Provision in the former is made to permit radial surveys at several stations upstream and downstream of the blades. In the Langley tunnel, a rotatable ring downstream of the test section makes surveying a sector of the flow possible. Many readings are required when one or more complete passages are studied. Yaw meters of the usual type tend to be in error when subjected to flows having total pressure gradients. With combination probes—and they are widely used—an error in direction results in an error in the static pressure indication because of misalignment. It is therefore desirable to survey a representative sector of the flow so that valid readings can be selected.

An annular cascade was constructed in Germany to test compressor guide vanes. In order to vary the inlet air direction to some extent, a radial flow section was incorporated in the inlet. The design was quite similar to the annular cascades previously described. One difference was that the entry to the tunnel was axial rather than radial. This feature would seem detrimental, for greater wetted surface results and the flow must make an additional 90° turn. Insufficient data is provided to evaluate the performance of this inlet configuration. The test blades were mounted on a hub attached to a central shaft supported by bearings. The test assembly was prevented from rotating by a torque arm fastened to the shaft at one end and resting on a weighing scale at the other. The tip diameter of the blades was 8.66 in. and the hub-to-tip ratio was 0.73. A rather low aspect ratio, about 0.6, was used. A two-stage axial flow compressor was used to draw air through the cascade. Test velocities up to 820 ft/sec could be obtained with this apparatus [*56*].

At the NACA Lewis Laboratory, turbine nozzle blades have been tested in a sector of an annular cascade (Plate D,5). The outer shroud diameter is 23.4 in. and the inner shroud diameter is 18.0 in. Five blades of about 2-in. chord were mounted in the annular sector. Measurements of the exit flow direction and total and static pressure were made at 11 circumferential stations and at 5 to 9 radial positions for total-to-discharge static pressure ratios from 1.124 to 1.682. The static pressure distributions about the blade surfaces were measured at three diameters.

Good agreement between the surface velocities calculated by an approximation of the stream filament method and the measured values was observed at speeds near the critical velocity ratio on the convex surface [*57*].

D,6. Single-Stage Compressors.

General discussion. Axial flow compressor rotors are tested alone or with stators to obtain information not easily or accurately obtainable from multistage machines. Single rotors or stages are used to provide design data for complete machines, blade-section performance data for comparison with linear and annular cascade results, and to study three-dimensional, tip clearance, aspect ratio, Reynolds number, and Mach number effects under known steady flow conditions.

Single-stage axial flow compressors in laboratories of gas turbine manufacturers are usually tested at full scale and high speed. The information obtained can be applied to gas turbine design with a minimum of extrapolation or correction. Closed circuits, intakes through turbines to refrigerate the air, or auxiliary facilities to supply cold air are used.

Exhausters to lower the pressure are employed so that flight Reynolds numbers may be simulated. Testing at reduced pressures directly lowers the power required to drive the compressor. The blade stresses due to the air loads are also lowered.

The information usually desired from tests of single rotors or from single-stage compressors consists of the efficiency and rate of energy transfer at various combinations of flow quantity and rotational speed under standard entering conditions. These general performance values are obtained by calculation from measured values of fluid direction, total temperature, total and static pressure upstream and downstream of the various elements under test, torque, rotational speed, and flowmeter pressure drop.

Because of centrifugal force effects, flows through blade tip clearances, secondary flows, and the difference in the boundary layer entering a rotor or stator compared to that entering a cascade of airfoils, the flow through compressor rotor and stator blades is not likely to be identical to that through a cascade, even though the same blade geometry and inlet air directions are considered.

In order to examine the differences, and to obtain more detailed information concerning flow through rotating blades, special devices have been developed. The hot wire anemometer has been applied to this problem [*51*] although few results have appeared in the literature. One major difficulty is the extremely rapid response required of the equipment if useful results from high speed tests are to be obtained. For example, consider a rotor having 50 blades running at 6000 rpm. A stationary instrument will then have 5000 blade wakes impinging upon it per second. The width of a typical wake may be $\frac{1}{5}$ of the spacing between wakes, so that the time of passing is 1/25,000 second. If the instrument is to produce a signal proportional to the velocities in the wakes with minimum fidelity, perhaps 5 to 10 points per wake, it would have to respond linearly at frequencies in the range from 125,000 to 250,000 cps. This is very far beyond the capabilities of any hot wire-amplifier combination known to the author. Qualitative information can be obtained under conditions similar to those of the example, however, and quantitative results can be obtained at lower wake-impingement rates.

Another method of studying the flow through running blades is to permit probes and orifices to rotate with the wheel. In this manner, the blade surface pressures and the regions of total pressure loss can be measured. These readings can be taken most readily by means of sealed cells to permit the transfer of pressures from the rotating system to stationary manometers. Several transfer devices have been developed. These can be separated into two groups, single-cell units which transfer pressure axially and multicell devices which transfer radially from the axis of rotation. Mercury labyrinths and rubber seals have been applied to both groups.

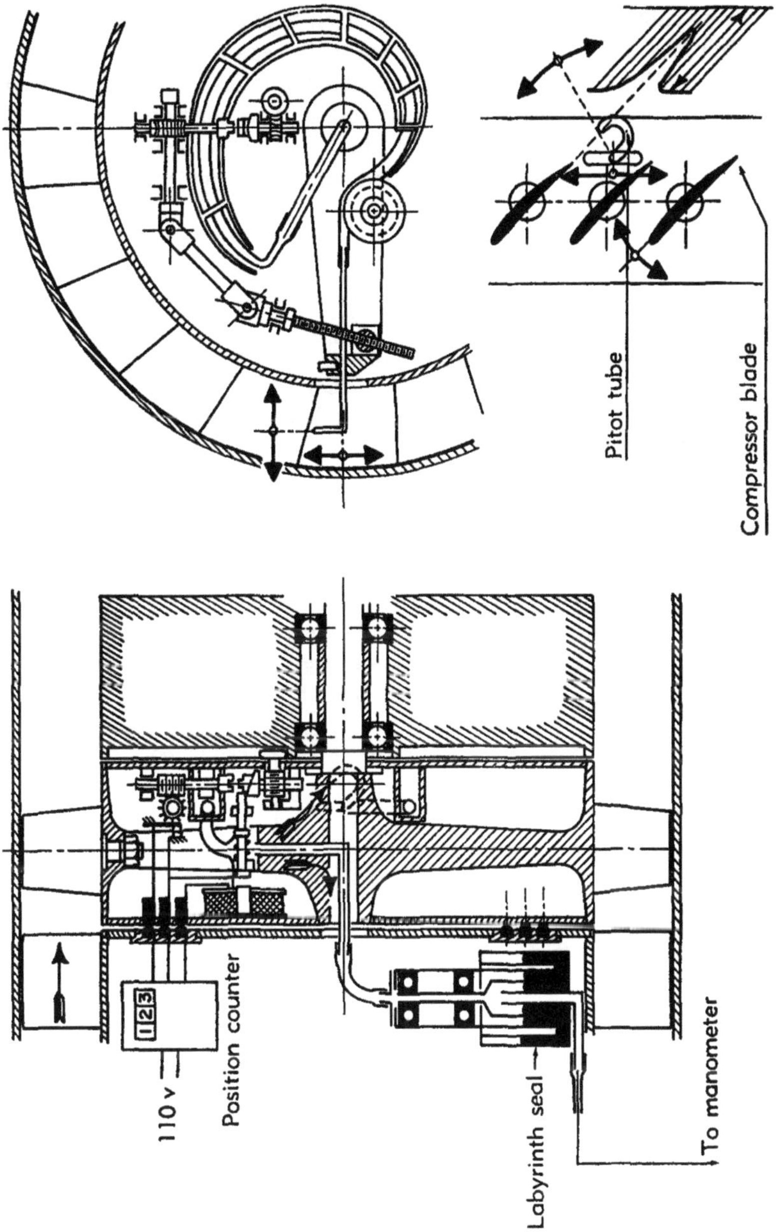

Fig. D,6a. Schematic presentation of the mechanism for a rotating Pitot tube.

A mercury-sealed, single-cell device used by Muhlemann [*58*] for measuring wake distributions is shown in Fig. D,6a. The device was used successfully in the speed range from 0 to 1000 rpm. Before a test, the single Pitot tube used was set at the desired radius in the proper direction. While the test was in progress, the tube was traversed across the wake by a friction drive, through an electromagnetically operated clutch and a system of worm gears to a lead screw. The position of the wake was determined by an electric counter which indirectly recorded the revolutions of the lead screw. A rubber-sealed, single-cell device developed by

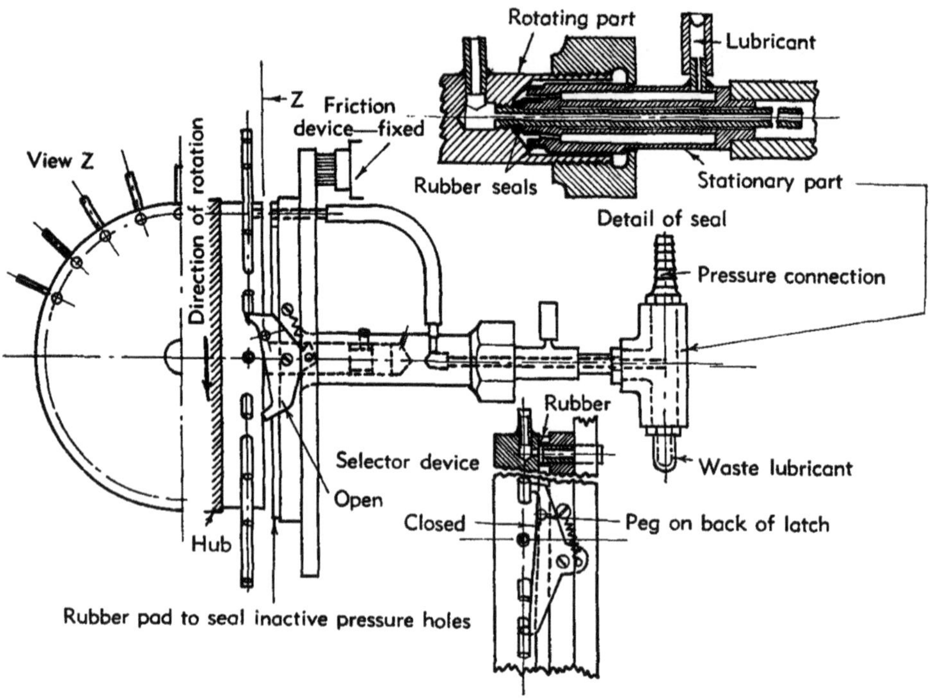

Fig. D,6b. Selector mechanism and seal.

Weske [*59*] for measuring blade pressure distributions and wakes is shown in Fig. D,6b. For measuring pressure distributions, this device was used in conjunction with a selector mechanism consisting of an escapement and friction drive to connect the cell to one of the blade pressure orifices. For wake measurements, the survey instrument was moved across the wake by a spring-loaded lever and pulley system. Radial adjustment was made before each test.

In order to reduce the running time where the power requirements are high or to increase the reliability of the results by taking all the measurements at the same instant, a multicell transfer device is desirable.

A mercury-sealed multicell device was designed at the Langley Laboratory of the NACA [*60*]. This device, presented schematically in Fig. D,6c is partially filled with mercury while rotating. The grooves on the rotating disks drive the mercury out into spinning annular seals capable of sustaining large pressure differences. Since the transfer is in a radial direction, as many cells as desired can be placed along the axis. The stationary casing is enclosed in a water jacket to prevent overheating due to fluid and bearing friction. The 24-cell unit shown in Plate D,6a was used successfully up to 2500 rpm.

A rubber-sealed multicell device developed by Davey [*61*] is believed to be superior to the other devices in simplicity and ease of maintenance.

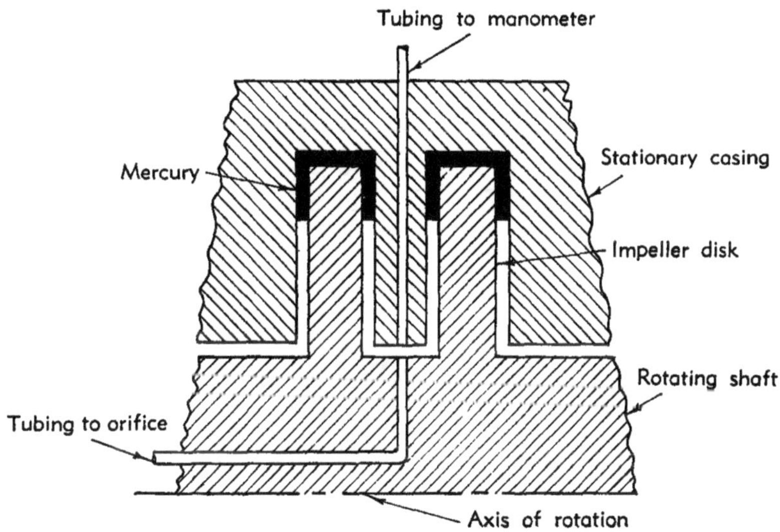

Fig. D,6c. Schematic diagram of pressure transfer element.

It consists of a row of commercially available ball bearings having built-in flexible dust covers (Fig. D,6d). The device has been run 10 hours at 6000 rpm without any difficulties. At 4000 rpm, a seal life of from 30 to 90 hours has been realized. A life of 500 hours is predicted at speeds below 2200 rpm. A somewhat similar device is employed by Leist [*62*].

Another method of measuring many pressures on a rotor is to use a commercially available tube-switching device, the scanivalve. Locating the pressure sampling unit on the axis of rotation permits measuring the pressures of 48 orifices. By proper arrangement, only one seal is required between the rotating parts and the indicating system. Since switching from one tube to the next is accomplished by electrical signal through slip rings to the scanivalve motor, readings can be taken as slowly as the operator desires or rapidly, up to about 24 readings per minute. The pres-

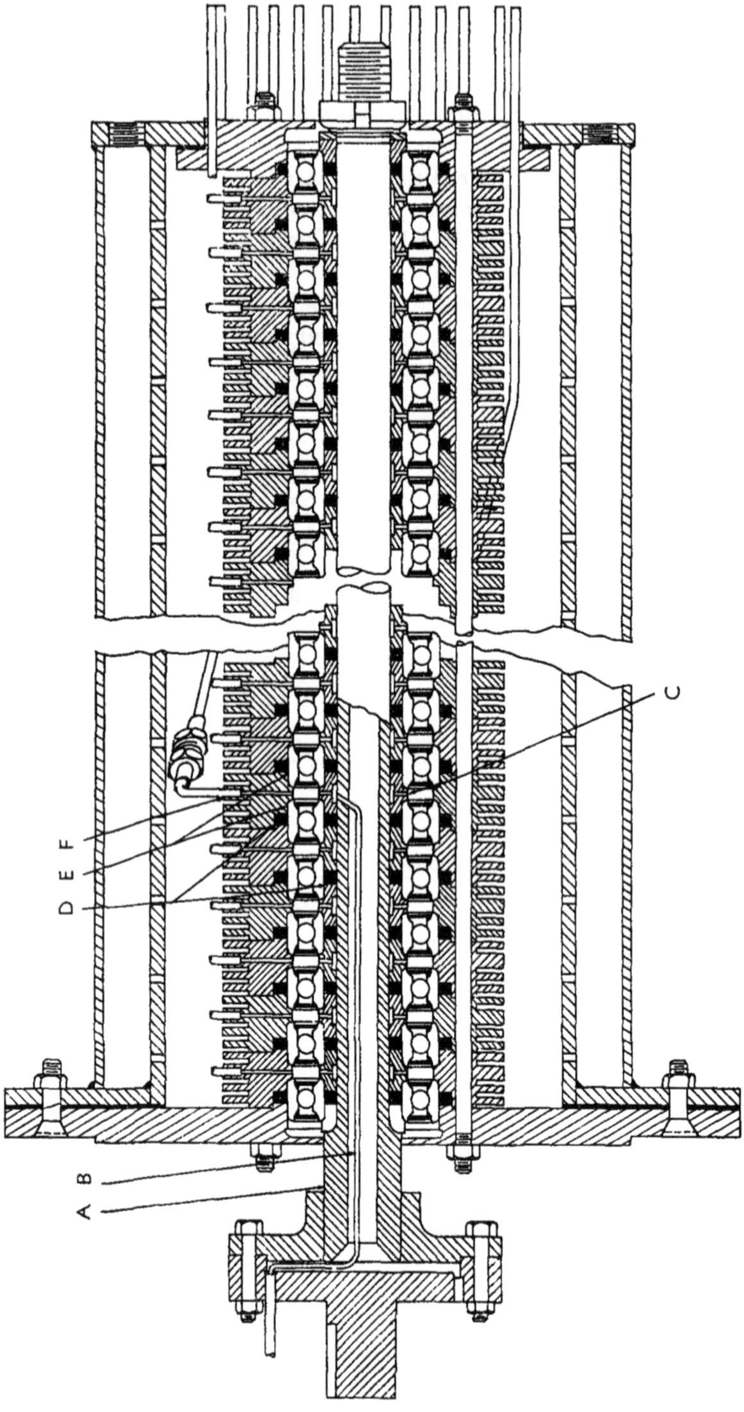

Fig. D,6d. Sealed ball bearing pressure transfer device. A, hollow shaft; B, rotating pressure tubes; C, rotating spacer; D, "O" ring gaskets; E, two Synthe-Seal ball bearings; F, stationary spacer.

sure reading is indicated from stationary equipment, so a wide variety of meters, including standard manometers can be used. In bench tests, satisfactory operation of this system has been obtained at speeds up to 10,000 rpm.

At NACA Langley, a 42-in. tip diameter test compressor equipped with a 24-cell mercury pressure transfer device is used in the study of secondary flows and boundary layer behavior in rotating blade rows, and to compare cascade and rotor blade performance. Guide vanes turning the flow in a direction opposed to that of the rotor are used. The dynamic pressure distribution over NACA 65-series blade surfaces in a two-dimensional cascade tunnel and in the rotor are presented in Fig. D,6e [*63*]. The comparison indicates that the blades sections operate similarly in cascade and in a vortex-type rotor at design conditions. The relationship between the angle of attack and the fluid turning angle of the blades in cascade was similar to that in the compressor at the mean diameter and at the outboard section one inch from the outer case. At the inboard station of the rotor, one inch from the inner casing, the turning angle increased 1.5° for a 1° increase of angle of attack. The rate interpolated from cascade tests would be one third of this. At flow coefficients other than design, secondary flow and boundary layer centrifuging effects caused the blades to perform differently in the compressor than in the cascade.

Centrifugal compressors. The NACA Subcommittee on Supercharger Compressors has recommended standard procedures for rating and testing centrifugal compressors [*1*]. The test setups described are adequate for superchargers which get their flow through small ducts of considerable length. For valid testing of centrifugal compressors for aircraft gas turbines, settling chambers similar to those recommended for axial flow compressors should be used. Sometimes the attachment of a plenum chamber to the compressor intake is impractical, as with double-entry units. In this case the room in which the test bed is located can be sealed except for an air intake and an exhaust discharge. The air flow can be metered at the intake to the room which acts as a large settling chamber. A single venturi has been found to be more suitable than several. Walker has noted in [*64*] that the air must not impinge on the engine predominantly from either the fore or aft direction; otherwise an error in gross thrust reading will result. Untreated brick or concrete walls are slightly porous, and enough leakage can occur even through sealed walls to cause a significant weight flow error.

In testing impellers alone, a vaneless radial diffuser is frequently used. The inlet conditions are determined by total temperature and total and static pressure measurements in the inlet pipe two diameters upstream of the impeller. Similar measurements are made at the impeller discharge. If the over-all pressure is desired, the instruments in the diffuser discharge

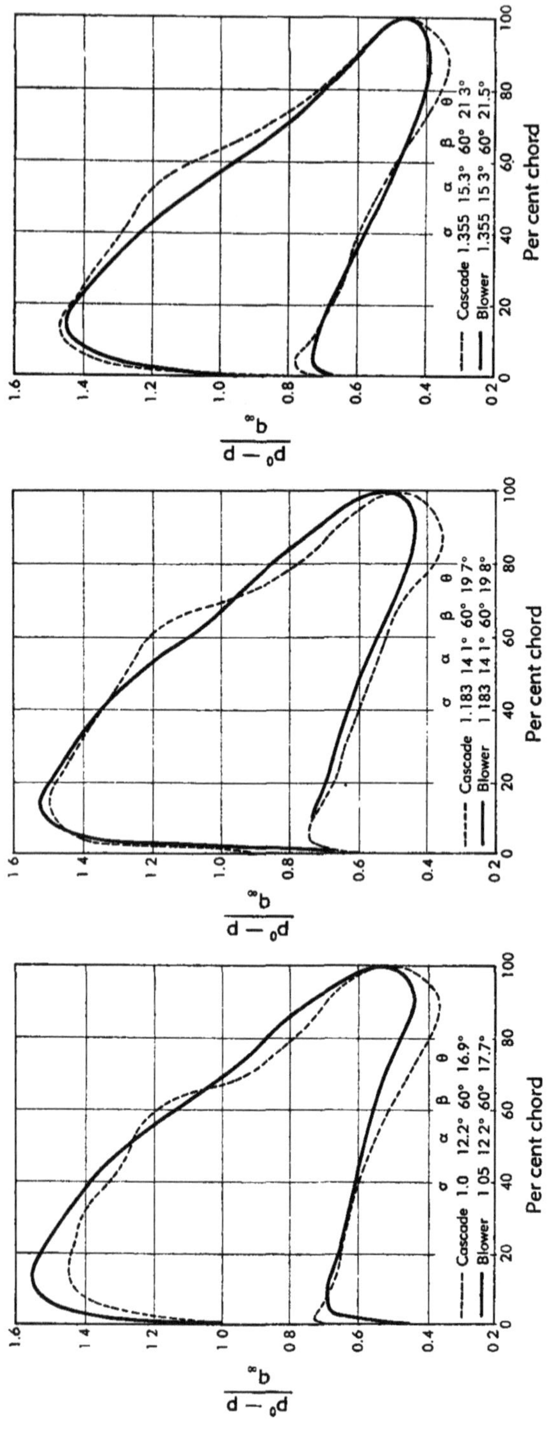

Fig. D,6e. Comparison of blower and cascade pressure distributions at design angle of attack. Left, 65—(10.96) 10 (outboard section); center, 65—(12)10 (mean section); right, 65—(13.04)10 (inboard section).

must be withdrawn, as the performance of vaneless diffusers has been shown to be markedly affected by such apparently minor obstructions. The static pressure distribution along the surface of the impeller casing and the diffuser provide an indication of flow conditions.

The over-all pressure ratio and the efficiency of a complete compressor are determined from inlet pipe readings and from measurements in the discharge pipes 12 diameters downstream of the collector. The installation of a large capacity centrifugal compressor is shown in Plate D,6b. An air-tight steel tank 6 feet in diameter and $13\frac{1}{2}$ feet in length is used as a stagnation chamber.

At the NACA Lewis Laboratory, a 48-in. tip diameter compressor has been constructed (Plate D,6c) to provide detailed information as to the actual flows within centrifugal compressor impellers. By means of a multi-cell pressure transfer device, the total and static pressures of the flow at many points throughout the passages can be obtained. The impeller vanes are removable to facilitate instrumentation and to permit varying the vane shape. The maximum tip speed attainable with this rig is about 1000 ft/sec but, with an intensively instrumented impeller, the maximum running speed is kept below 700 ft/sec.

D,7. Component Testing of Multistage Axial Flow Compressors. As mentioned in Art. 1, there are a number of approaches leading to the design of axial flow compressors. However, no matter what design method is used, in the present state of the art the operating characteristics of a compressor, which is significantly different from its predecessors, are difficult to predict. This is particularly true in the starting range for high pressure ratio compressors. The compressor performance at low speeds must be known in order to determine the power and speed of the cranking motor and whether special provisions must be made to permit the engine to start and idle.

The necessary information is best obtained by testing full scale compressors under simulated conditions of actual operation. To accomplish full scale, full speed testing, drive motors or steam turbines of up to 50,000 horsepower are required. Refrigerated, low pressure air sources are necessary to simulate flow conditions encountered in high altitude flight.

Performance measurements. The values necessary to define the over-all compressor performance are the pressure ratio, the weight flow, and the efficiency over the attainable range of weight flows at rotational speeds from 10 per cent to 115 per cent design. The simulation of temperatures and pressures encountered in high altitude flight is very desirable. The over-all pressure ratio can be obtained from total pressure surveys in the entrance and exit ducts. The air weight flow measurements are usually made in the intake piping by standard metering methods. The compressor efficiency can be calculated from measurements of temperature, pressure, and humidity in the intake settling chamber, and temperature and pressure surveys in the discharge annulus. Shaft power measurements are useful in estimating turbine torque requirements. When corrected for bearing, seal, and windage losses, power input measurements provide a check on efficiencies obtained by temperature surveys. The NACA Subcommittee on Compressors has recommended standard procedures for rating and testing multistage axial flow compressors [*1*]. Suitable instruments similar to those described in Art. 3 are also presented.

Research measurements. In order to compare a rotor or stator tested as a single stage with the performance of a similar unit in a multistage compressor, the velocity and direction of the flow in a representative sector of the particular annulus should be ascertained. To accomplish this, the total and static pressure and the temperature and direction of the flow must be measured or deduced. Total temperature and pressure and yaw probes can be made small and rigid enough to be inserted into the stator blades. The static pressure along the outer casing is easily obtained. When stators having inner shrouds are used, the static pressures at the hub can be measured without insurmountable difficulty by running leads through the blades. A photograph of a compressor stator illustrating the installation of total and static pressure tubes in a stator assembly is presented in Plate D,7a. The static pressure at radial positions between the inner and outer casings is more difficult to measure due to the curvature of the flow and the resulting static pressure gradients in the circumferential direction.

Little information is available to indicate the proper locations of static pressure orifices in compressor casings. Sawyer [*43*] has presented the pressure field measured at the side wall of a low speed guide-vane cascade. The best measure of the upstream static pressure in this case (staying within 5 per cent chord in the axial direction) would be obtained by an orifice 20 per cent of the blade spacing from the concave surface and 5 per cent chord upstream of the leading edge. The optimum downstream position would be 40 per cent of the spacing from the concave surface and in line with the trailing edges. Westphal and Dunavant [*65*] obtained the streamline and equipotential line distributions through typical turbine blades using a wire-mesh potential-flow-plotting device. In the distributions presented and from other turbine flow plots obtained by this method, the optimum position for an orifice to measure upstream static pressure would be about 35 per cent of the spacing from the concave surface, in line with the leading edges. Near the trailing edges, the static pressure appears to be quite uniform across the passage and a region in the middle is suggested as the logical location to measure the downstream static pressure.

The local dynamic pressure ratios q/q_1 in the field of flow about a two-dimensional cascade of NACA 65-series compressor blade sections are presented in Fig. D,7a. For this cascade the best location for a static pressure orifice to measure the upstream static pressure would be about 40 per cent of the blade spacing from the concave surface. A similar location would be optimum in the plane of the trailing edges to observe the exit static pressure. Although the position should change as the air angles, blade camber, and geometry are varied, it is estimated from the foregoing that the proper location for static pressure orifices near compressor blades is probably nearer to the concave surface than to the convex. As smaller pressure gradients usually occur in the trailing edge region, more reliable readings can be obtained of the downstream rather than the upstream static pressure. A close approach to the static pressure entering the tip section of a rotor blade row can be obtained from an orifice placed in the

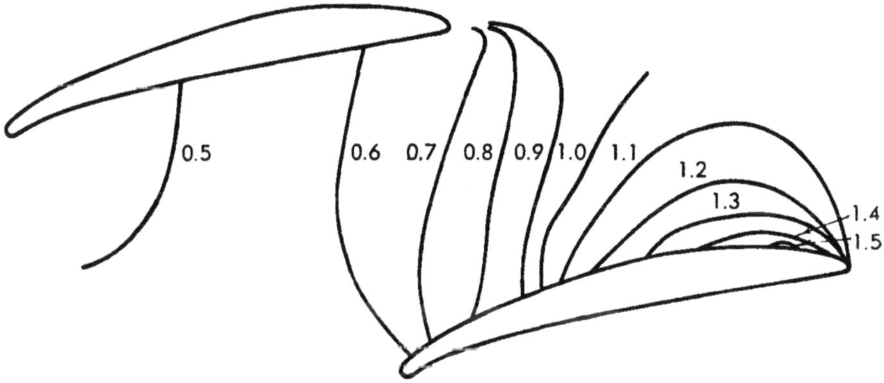

Fig. D,7a. Ratio of local dynamic pressure to entering dynamic pressure.

outer casing about 30 per cent chord downstream from the rotor tip leading edge.

Investigation of interstage flows. A method of estimating the flow distribution downstream of axial flow compressor rotors or stators when a minimum of experimental data is available has been presented in [*66*]. If the weight flow, total temperature, and total pressure can be measured, and the flow direction can be deduced or measured across the radial height of the compressor annulus, the static pressure and axial velocity distribution can be determined. Iterative or trial-and-error solutions are required, however, and the effect of low energy regions in reducing the effective flow area can only be approximated, unless detailed survey data are available. A method of measuring or deducing the radial static pressure distribution directly would be advantageous.

A method of obtaining the static pressure at desired stations suggested by observations of compressor cascade and rotor blade pressure distributions follows: A given cascade, that is, similar blades whose geometrical relation to each other is fixed, exhibit the rather surprising property that the ratio between the velocity over the forward portion of the convex surface at one point and the entering velocity remain the same as the inlet air direction changes. This phenomenon is borne out strikingly in Fig. D,7b [*63*], which presents the pressure distribution observed over

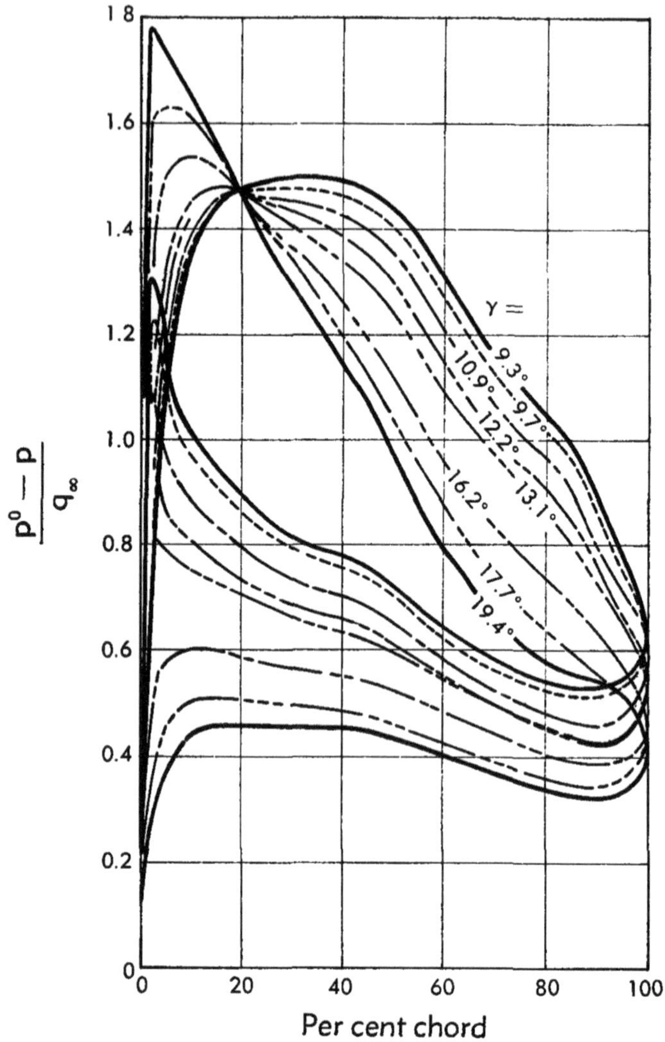

Fig. D,7b. Pressure distributions at various angles of attack. (Mean section.)

the surfaces of a compressor rotor tested at low speeds. Examination of pressure distribution obtained from compressor airfoils in low speed cascades substantiates this happening; the position of constant velocity ratio varies from one cascade arrangement to another, but, for one cascade, a point on the convex surface somewhere between 15 and 25 per cent chord from the leading edge maintains the same velocity ratio as the inlet angle is varied. This phenomenon has not yet been confirmed nor denied at high speeds; however, if this behavior continues to exist under compressible flow conditions, it would be possible to determine dynamic and static pressures at desired stations by orifices located at the proper chordwise position of the stator blades.

Research compressors. The details of flow through multistage compressors are not fully known. The best theoretical treatments presently available are incapable of dealing with the compressible, three-dimensional, viscous flows through the blade rows. There is a great need for thorough research on the radial flows, boundary layer growth on the blade and casing surfaces, interference between adjacent rows and stages, and stalling and surging mechanisms within turbomachines. Many of these phenomena can be readily investigated if incompressible flows are employed. If the effect of Mach number is neglected, multistage compressors can be constructed and studied with modest shop and experimental equipment. Because of the continuing need for information and the many advantages that accrue from disregarding the Mach number similarity requirement, a number of low speed multistage compressors have been built in the United States and many other countries.

In order to reduce the power and rotational speed required, several methods can be employed to test multistage compressors under conditions satisfying several or all of the significant similarity coefficients. The system of using gases having low sound velocities in place of air has been discussed in Art. 6. Relatively large compressors operating with atmospheric inlet attain flight Reynolds numbers at low rotational speed. Small compressors running in water experience Reynolds numbers of large aircraft components at very low rotational speeds and small powers.

An ideal test compressor would be sufficiently flexible to be used in the study of many of the problems now associated with multistage axial flow compressors:

1. The relative advantage of various types of tangential velocity distribution, i.e. vortex, solid-body, etc.
2. Tangential, axial, and radial flow direction and velocity measurements over a representative sector downstream of each blade row in order to compare the actual with the single stage or theoretical performance of the blading.
3. Investigations of the flow in the various stages as the Mach number, Reynolds number, or flow coefficient is varied.
4. The effect of blade loading, tip clearance, and axial spacing between successive blade rows on over-all performance.
5. Stalling and surging and the development of methods of delaying or preventing these occurrences.

Hot wire measurements and visual observations of smoke or tufts to verify and extend the measurements of standard instruments would be desirable.

The desirable features of research compressors are: sufficient size and axial spacing between blade rows to permit interstage measurements, the ability to reset or replace rotor and stator blades easily, a speed range permitting tests over the Reynolds number range from 50,000 to 400,000, rotor shaft positioning controllable radially so that tip clearances in the annulus sector where instruments are located can be varied without machine work, rugged construction so that extended stall-operation study can be made without damage, and provision to survey representative sectors of the annulus with probes capable of measuring flow directions in more than one plane.

A test compressor in use at the National Gas Turbine Establishment has several of these features [*67*]. This compressor was built primarily for investigating the variation of axial flow compressor performance at low speeds and low Reynolds numbers, but the extreme flexibility which has been achieved in the design renders it suitable for a large variety of research investigations into the nature of the air flow through compressor blading. Both the number of stages built into the compressor and the axial spacing between blade rows may be varied, the maximum number of stages which may be accommodated being eight, with one-sixth mean chord axial spacing. The tip diameter is 20 inches with a hub-to-tip radius ratio of 0.75. Provision is made for traversing Pitot static and yaw meter tubes across the annulus at any desired circumferential or axial position.

The California Institute of Technology test compressor, illustrated in Fig. D,7c from [*68*], was also designed for flexibility in low speed testing. The compressor was designed so that each blade and each blade row may be removed or adjusted individually. The blade angle settings are adjustable by $\frac{1}{2}°$ increments. The compressor size was chosen sufficiently large so that the use of instruments of a reasonable size would not cause undue flow interference. A tip diameter of 36 inches was chosen with a constant hub-to-tip radius ratio of 0.60. The compressor speed range selected is up to 2000 rpm (315 ft/sec tip speed) to produce sufficient pressure differences to be measured with good accuracy, and yet low enough to avoid structural difficulties and to keep within the available power limitation of 100 hp. The maximum number of stages was chosen as three plus entrance guide vanes to permit duplication of the conditions typical of the middle and last stages of a multistage compressor. In addition, two rows of rear recovery vanes may be included. Since it was felt that the effects of axial spacing between blade rows had been established by other investigations, fixed axial positions were included for all blade rows. Blade aspect ratios and solidities typical of modern high performance compressors are used.

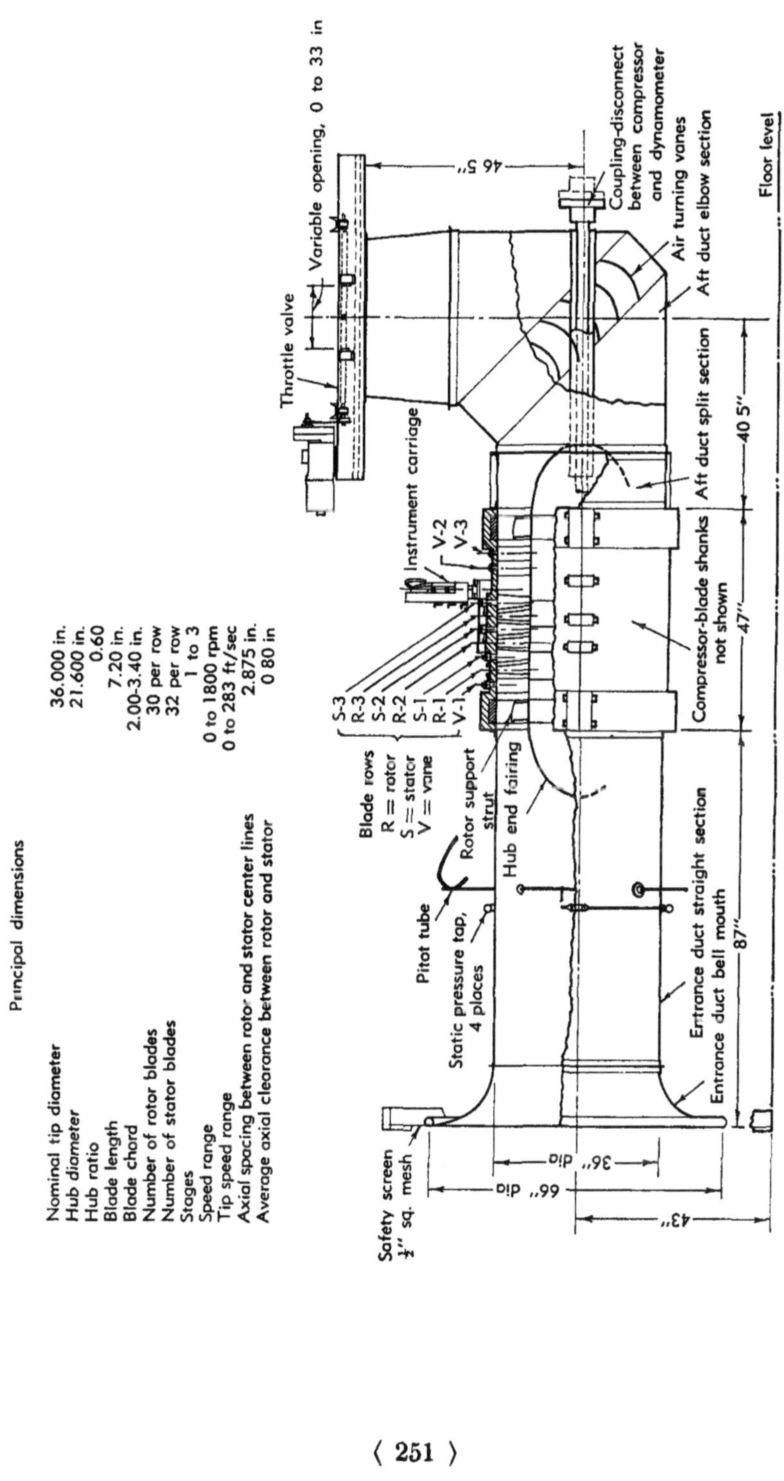

Fig. D,7c. Assembly drawing of the California Institute of Technology test installation.

Instrumentation was provided to allow detailed measurements near the blade rows as well as over-all performance measurements. Over-all performance measurements are made by a tachometer, torquemeter (dynamometer torque), and a Prandtl-type Pitot tube. The detailed measurements are made directly downstream of each blade row by several types of yaw, Pitot-static, and total head probes, each of which may be mounted in a special instrument carriage. This special carriage consists of a base and movable head. The base fits into any one of a system of access ports in such a way that bottom surface is flush with the inside of the compressor casing. The probe is accurately located and held in the carriage head and may be positioned by means of three wheels which move the probe radially, rotate the probe about its own axis, and move the entire carriage head in the circumferential direction. Corresponding counters give the position of the probe. A flexible tape covers the slit in the carriage base when the head is moved circumferentially. A circumferential angle of 15° can be covered by the traversing probe. This angle is about 30 per cent greater than the $11\frac{1}{4}°$ stator blade spacing.

A 60-in. diameter low speed compressor has been constructed by the General Electric Company at Evendale, Ohio (Plate D,7b). The compressor is mounted vertically to reduce inlet distortion and permit access to the transparent ports in the casing. Provision is made for 4 stages, although the number of stages can be varied. A 400-hp steam turbine drives the rotor over a speed range from 100 to 800 rpm through reduction gears.

For closed circuit testing of small multistage compressors, the General Electric Company utilizes the facility illustrated in Plate D,7c. Through the use of pressure-sensitive switches similar to those described in [*30*], 195 pressure and 60 temperature readings can be recorded in about one minute. Rotational speeds up to 20,000 rpm can be provided in either test cell. Dry, refrigerated air flows of 100 pounds per second can be supplied at inlet pressures from 0.1 to 1.7 atmospheres. Continuous monitoring and recording of straingauges mounted on stationary and rotating parts is provided.

Running a compressor in water reduces the size, rotational speed, stresses, and power required to very low values. If transparent casings are used the flow can be studied visually. Because the viscosity of water increases by a factor of six from boiling to freezing, the Reynolds number can be varied over a wide range. The National Gas Turbine Establishment has constructed a 3-stage axial compressor to operate in water [*69*]. The tip diameter is 12 in. with a hub diameter of 9 in. At the normal running speed, about 150 rpm, $\frac{1}{4}$ hp is required to drive the rotor assembly. Based on 1.5-in. chord blades, the estimated Reynolds number is in the range of 100,000 at a water temperature of 120°F.

D,8. Component Testing of Turbines. Because turbine blade rows operate with strong pressure drops in general, the design problem might seem much less severe than for compressor blade rows. This situation would probably exist were it not so important from a weight, cost, and strategic materials usage standpoint to obtain maximum power output from the turbine. For this reason, the same need for thorough research noted in the previous article as being required for compressors is required for turbines.

The efficiency of the turbine is of importance in effecting engine fuel economy. Further, the temperature at which the turbine can operate for hundreds of hours without exceeding a very small growth strongly influences the thrust or shaft power produced. The turbine must not only drive the compressor and accessories over a wide range of rotational speeds but also must produce the necessary power for the cruise condition at peak efficiency and maximum continuous allowable temperature. This must be accomplished with the weight flow and pressure provided by the compressor in its small range of high efficiency at cruise rpm. For good efficiency, close radial and axial clearances must be maintained between the turbine rotor and casings at cruise power output. At other powers, and when changing speed, the relative dimension changes must be such as to prevent rubs. The expansion and contraction of the turbine wheel and casings under the various operating conditions are difficult to predict.

As noted in the introduction, major alterations of engines on the test stand are impractical. To provide reasonable assurance of satisfactory engine operation, the efficiency, weight flow, and torque characteristics of the turbine over the range of inlet temperatures, pressures, pressure ratios, and rotational speeds to be encountered should be determined by tests at full speed and power before assembly on the engine.

The facilities required to permit full scale turbine tests under operating temperatures, pressures, and speeds are of considerable cost, size, power, and complexity. The laboratories of the major gas turbine manufacturers in the United States and in England are equipped with electrical, steam, or water absorption dynamometers of from 10,000- to 30,000-hp capacity. The air is supplied to the turbine stands by compressors requiring a maximum of from 9400 to 25,000 hp. The water pumped during full capacity operation varies from 8500 to 120,000 gal/min. Large quantities of hot exhaust gases and extreme noise levels must be dissipated. Much attention must be devoted to maintaining safe procedures and conditions. Viewing the multitude of secondary considerations necessary, the possibility of obtaining many hours of running at desired conditions would seem small. However, in his concise description of the Westinghouse Gas Turbine Laboratory [*70*], W. R. New reported that a very high use factor was obtained in the first eight months of operation.

Research turbines. In research or early development studies, the emphasis is on obtaining valid data from a number of turbine designs rather than to insure exact matching of engine complements and proper mechanical operation under actual conditions.

In such investigations the difficulty and cost of testing single or multi-stage turbines can be reduced by taking advantage of the difference between the velocity of sound in cold air and that in the hot gases which are the working fluid of the aircraft turbine. The criteria of flow similarity can be closely satisfied by tests of reduced scale models using cold air at the appropriate pressure as the test media [*71*]. Assuming that the geometrical similarity is observed, the flow relationships that must be the same in the model and full scale turbines to obtain representative results are the Mach number, Reynolds number, the flow coefficient or ratio of axial to peripheral velocity, and the ratio of specific heats of the working fluids.

All of these similarity coefficients can be satisfied by proper selection of model size, rotational speed, and inlet air pressure and temperature with the exception of the ratio of specific heats γ. The steady state values of γ are lower for hot combustion gases than for cold air, so that the rates at which the various physical properties of the different fluids change with respect to each other are different. For conventional subsonic turbines the calculated differences are small, however, and are usually neglected. Further, the finite time required for some of the component gases of combustion products to change their heat capacity suggests that in the rapid flow processes of gas turbines, the effective heat capacity may be nearer to that of cold air [*32,72*]. Thus the slight differences aforementioned may in the actual process be of even less significance. Lovesey [*73*] shows excellent agreement between test results obtained from model turbines run on cold air and full scale engines.

The advantages of cold air testing are many and the disadvantages, for noncooled blades, are few. The primary advantages are:

1. The difficulties of instrumentation and test setup required to make hot gas tests are avoided.
2. The air supply can be smaller by a factor of from 2 to 20.
3. The power produced is less by a factor of from 5 to 100.
4. The test apparatus, including the model turbine, can be significantly smaller.
5. Easily machined, forged, or cast materials for blades and disks can be used.

The apparatus and instrumentation necessary to obtain the data required to compute turbine performance is similar to compressor stage instrumentation. A typical rig described by Ainley in [*38*] is presented in Fig. D,8. A photograph of the test rotor is reproduced in Plate D,8a.

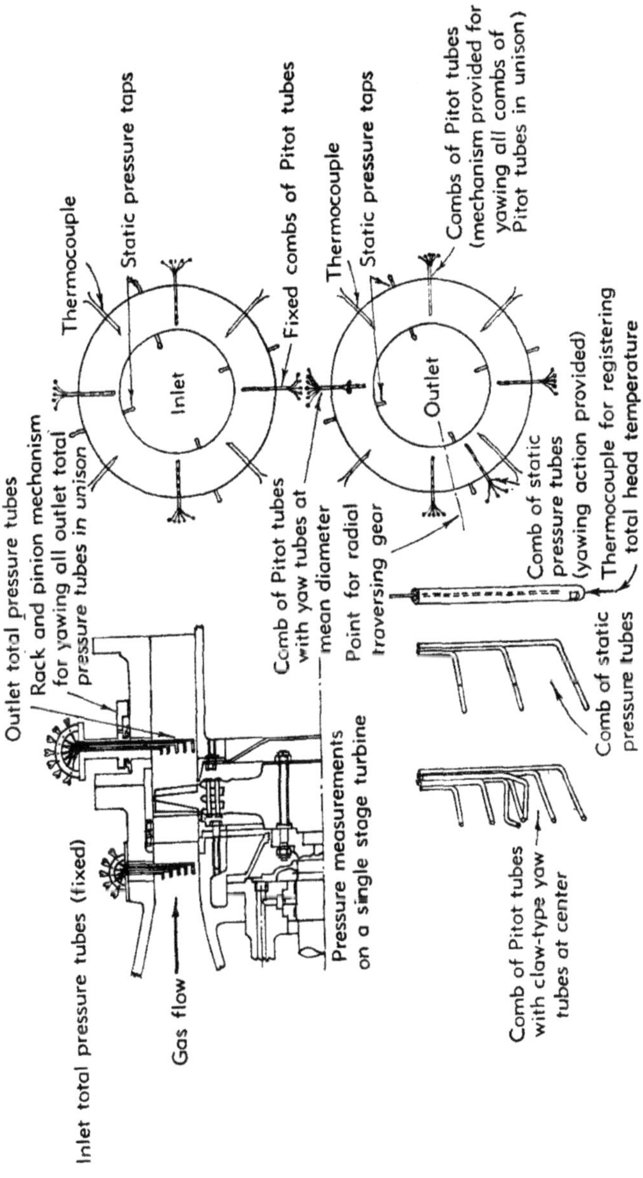

Fig. D,8. **Instrumentation at inlet and outlet of a turbine equipped for testing.**

Cold air and hot gas turbine testing facilities of the General Electric Company are illustrated in Plate D,8b and D,8c. The air turbine test facility is used for aerodynamic testing of full size and scale model single and multistage turbines under simulated flight conditions at low temperatures, 100–375°F. Production or developmental engine turbine parts can be adapted to the facility as well as specially designed research vehicles. This facility is equipped with independent, remotely operated inlet and discharge control valves plus air flow measuring equipment for both main and cooling air. The main air may be dried as well as heated to eliminate any condensation shock or discharge icing problems. The facility is interconnected electrically with an instrumentation center for automatic recording and plotting of interstage traverse data, and observation of turbulence phenomena. The control room is equipped with photomanometer panels for instantaneous photographic logging of a large number of pressures.

The hot gas turbine testing facility is used in evaluating new designs of turbine components for small gas turbine engines. A 2800 horsepower load absorbing dynamometer and a high temperature air supply (adequate to meet the requirements of small aircraft engines) is provided along with the necessary fuel, lubricating oil, control, instrumentation, and atmospheric pressure exhaust systems. A direct fired combustion system provides cooling air over the ranges of 100°F to 1000°F with flows from 0 to 1.5 lb/sec. The control room is equipped with instrumentation necessary for the measurement of torque, speed, flows, temperatures, pressures, vibration, and strain. The control center also contains all of the manual and automatic controls and safety devices necessary to assure satisfactory operation, mechanical reliability, and safety of the facility and test vehicle.

Running turbines have been studied in detail with standard temperature and pressure probes, but, to the author's knowledge, no attempt has been made to examine turbine rotor flows by pressure transfer devices. Apparently, satisfactory turbine design methods have been derived, partly from the vast steam turbine experience that exists, but largely from cascade and gas turbine experience of recent years. Perhaps as more highly loaded turbines of increased flow capacity become necessary and efficiencies must be maintained or increased, efforts to study turbine rotor flows in greater detail will be in order.

D,9. Gas Turbine Engine Testing. Ideally, a new gas turbine engine is an assembly of tested components and need only make its appearance on the complete engine test stand long enough to prove its ability to start and operate satisfactorily over the range of conditions and the period specified for official-type tests. This ideal has not often been achieved, however; usually the engine is tested before the components and consequently much engine-testing time is devoted to determining where the aerodynamic, mechanical, and control troubles are occurring and the nature of the difficulties.

One important purpose of engine testing is to determine how well the compressor and turbine are matched; that is, whether their weight flow, pressure ratio, and speed and efficiency characteristics are compatible. It would appear desirable to have both the compressor and turbine working at their optimum conditions at the design point, but a compromise must usually be made to provide surge margin and to permit acceleration and operation over a wide range of speeds.

The over-all performance values that define the performance of a jet propulsion engine are the net thrust and fuel consumption. The quantities usually measured on a static test bed are the air flow, fuel rate, gross thrust, and bearing and tail pipe temperatures. The air flow can be measured by standard orifices, nozzles, or venturis in the intake piping, but in less elaborate installations a calibrated bellmouth intake is used. Fuel flowmeters of sufficient accuracy are commercially available for most applications; in some instances fuel weighing or volume tanks are employed as a positive check when runs of considerable duration are made. The gross thrust is usually obtained by suspending the engine by means of flex-plates or parallel-arm bearing systems and opposing the thrust by air or fluid pressure on balancing diaphragms equipped with damping devices. Gross thrust can also be obtained by tailpipe total and static pressure and total temperature measurements. This is a more involved and usually less accurate method, but in some cases, as in flight tests, it is the most practical system. Iron-constantan and chromel-alumel thermocouples are widely used to observe bearing and tailpipe temperatures.

When gas turbines are run during the development period, more detailed information is desired. Total pressure surveys in the intake duct or just upstream of the compressor guide vanes insure that the inlet system is satisfactory. Static pressure measurements of each compressor stage permit comparison with design values to check the matching of the individual stages with each other and whether the desired pressure ratios are being attained. Compressor exit surveys of total and static pressure and total temperature not only aid in evaluating compressor performance but are also useful in determining whether the flow distribution is likely to be a source of combustor or turbine troubles.

The accurate measurement of high turbine inlet temperatures is difficult because of radiation and conduction and the severe temperature gradients often encountered. This problem is treated in Vol. IX; therefore, only the sonic orifice and tailpipe temperature plus compressor work methods of obtaining turbine inlet temperatures are outlined here. In the sonic orifice method, a sample stream of combustion products is drawn through an orifice of known size. The gas is then cooled to a known temperature and drawn through another orifice to measure the rate of flow. From the rate of flow at known temperature and pressure, the mass flow per unit area at the sonic throat of the hot orifice can be calculated. The total pressure at the turbine inlet is readily obtained, so that the total temperature can be deduced. When care is taken, temperatures accurate to within 2 per cent in the temperature range from 3400 to 4000°R can be obtained using an instrument developed by Blackshear [*74,75*].

In [*76*], equations are derived which evaluate the effects of thermal vibrational relaxation and specific heat changes on pneumatic-probe pyrometer measurements in the region to 3000°K and Mach number <2.0. By using these derived equations, a simpler working set of equations is given which is shown to be adequately accurate. Methods are given by which the results of these equations may be used to determine the error in temperature measurement if different assumptions are made as to the state of thermal equilibrium of the gas in the probe. By using these methods, examples are given for a typical probe. Different assumptions must be made, depending upon the velocity, probe size, and gas constituents. The results are in partial agreement with the experimental evidence obtained in other reports. The relaxation error entailed by using any of the methods of [*76*] will not be large for the case of hydrocarbon combustion in air in the region of free stream static temperature <2500°K and free stream Mach number <1.2.

For most purposes, a sufficiently accurate indication of the inlet temperature can be obtained by adding the temperature drop through the turbine to the tailpipe temperature, which can be measured accurately in the cooler tailpipe because of reduced radiation and conduction. In jet propulsion engines the enthalpy drop through the turbine can be calculated from the enthalpy increase and air flow in the compressor plus the power delivered to accessories and the estimated losses due to bearings, seals, and windage.

The turbine inlet total temperature distribution over a sector covering the discharge from one combustor can is useful data when this type of burner is used, even if it is only of comparative value. Plate D,9a illustrates an installation to permit such measurements. The radial position of the probes is remotely controlled; total pressure heads can be installed if desired [*73*].

The measurements just described are useful in determining the overall performance of the engine and in deciding whether or not the various components are performing satisfactorily. However, more detailed information is necessary to study the actual flow conditions within the components. Providing instruments to detect the pressures, temperatures, and directions of the flow in the compressor and turbine without changing the flow is a difficult task because of the high speed of the flow, the small size of the blades, and the close axial clearances normally used. The testing of complete gas turbines involves the same instrument problems as those of multistage compressor and turbine component testing plus those arising from the assembly of the separate elements. In addition, such instrumentation requires a longer setup and running time, thus adding to the cost of an already expensive process. Due to these difficulties, very little detailed information was obtained during tests of early engines. Because data so obtained is of value in evaluating the interaction between and within the compressor and turbine under operating conditions, interstage measurements have become routine on development engines. If high power component testing stands are not available, engine testing is the only way to verify or to provide a basis for turbomachine design and theory.

Although the turbine inlet total temperature can be determined by one means or another, several other measurements and assumptions must be made to estimate the prime factor that limits engine speed and thrust—the turbine blade temperature. Even though an accurate calculation can be made of the total temperature and gas velocity relative to the turbine rotor blade, other factors, i.e. heat transfer and conduction, which affect the blade temperature and hence its strength and elongation, are either difficult to determine or vary with operating conditions. The positive method of running rotors to destruction is hardly practical for current gas turbines; therefore, direct methods of measuring or deducing blade temperatures and stresses find considerable favor. Lovesey [*73*] indicates a 100°F difference at the blade tip and root between the temperature obtained by the Brinell recovery method and the calculated total temperature relative to the blades. In this method a few specially tempered blades are mounted in a wheel and run under steady conditions for about one-half hour. These blades are then removed and their hardness, hence, temperature of annealing, measured over the entire surface.

Methods have been devised to make electrical connections to one or more rotating thermocouples with a minimum of error. To avoid the generation of extraneous voltages that exist if different metals are used due to the thermoelectric effect, copper slip rings and brushes are used under controlled scoring conditions. Junctions between the thermocouple metals and copper leads to the slip rings of the rotating circuits and between compensating thermocouple metals and copper leads to the brushes of the stationary circuit are placed in a region of controlled temperature (Fig. D,9a and D,9b). For high speed rotors, the number of thermocouples that can be accommodated if the standard multiple slip-ring system is used is limited. For mechanical reasons, only about six sets of leads can be inserted in a protection tube having the proper immersion depth of 10 diameters. A remote-control rotating switch having contacts of thermocouple metals has been designed and built. As only one set of compen-

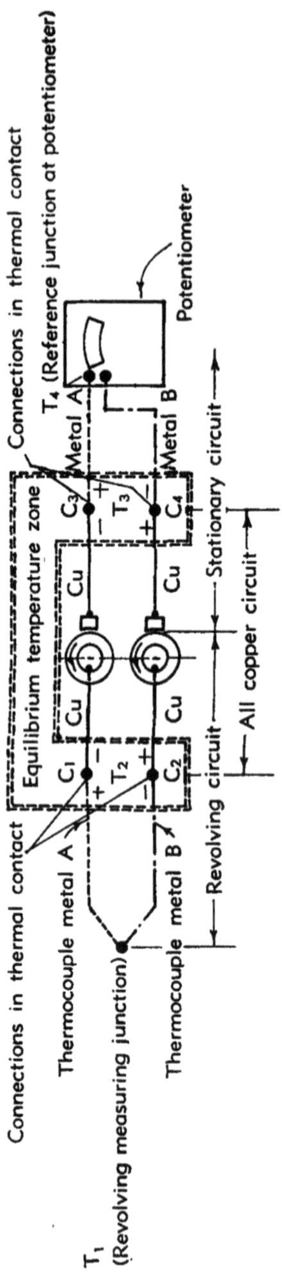

Fig. D,9a. Schematic diagram of compensated thermocouple pickup system.

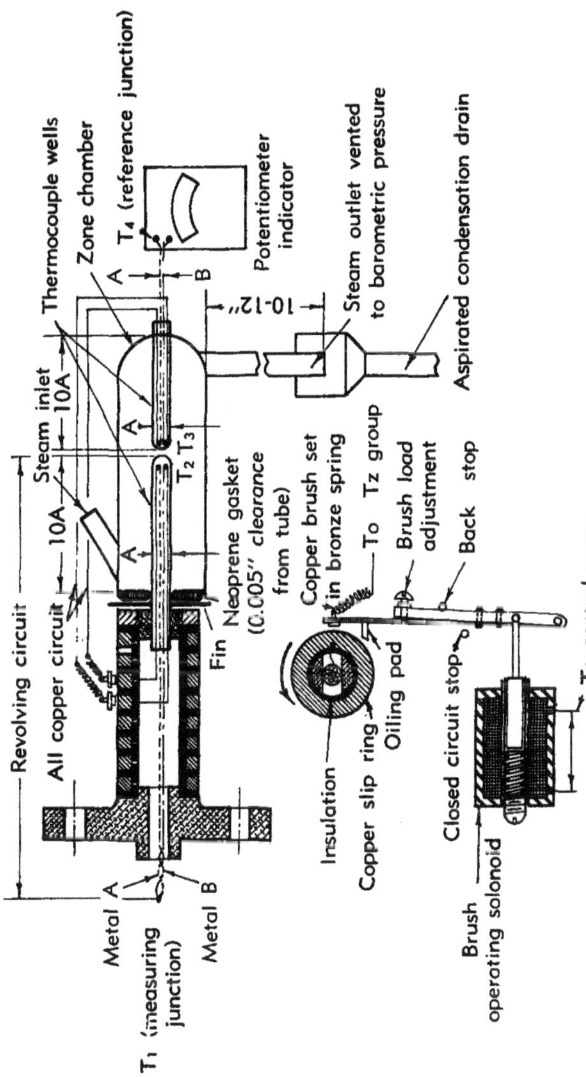

Fig. D,9b. Arrangement of parts in practical shaft-end pickup of compensated type.

sating junctions is required, the limitation is removed so that many thermocouples can be installed and read [*77*].

ENGINE TESTING UNDER FLIGHT CONDITIONS. In addition to determining the performance of a gas turbine under sea level conditions, the study of high altitude operation, starting, windmilling speeds and drag, idling thrust, and icing is mandatory for production engines. In general the temperature and pressure range encountered in high altitude flight cannot be duplicated on static test stands. A number of different methods of testing gas turbines under flight conditions are in use [*78*]. The most direct method is to install the engine in large aircraft. Test stands have been located on the summit of a mountain. Simulated conditions are obtained by methods varying from the addition of a few simple elements to static stands to the construction of large wind tunnels especially designed to achieve this purpose.

Actual flight. The "flying test bed" is widely used in the United States and abroad. The test engine is mounted in one of the nacelles, fore or aft in the fuselage, to one side of the fuselage, or suspended in the bomb bay of a multiengined aircraft. Flight altitude, attitude, and forward speed effects on aerodynamic thermopropulsive and mechanical performance can be observed with satisfactory accuracy. If the engine is mounted separately from the airframe, direct measurement of the net thrust can be accurately obtained. If the engine is installed in a wing nacelle or within the fuselage, direct thrust readings are difficult to obtain. With such installations, the thrust is usually calculated from pressure and temperature measurements at the jet nozzle exhaust. The mature aircraft used in the past for these studies have not been able to reach the extreme altitudes and speeds desired; however, flight testing does provide a means of "proving out" an engine with a minimum of special equipment.

Simulated flight. The simplest method of simulating some flight conditions is to diffuse the exhaust from the engine to lower the discharge static pressure. An altitude corresponding to the exhaust pressure is thus simulated. Because the temperature-pressure variation of the atmosphere with altitude does not follow the isentropic relationship, true flight conditions would not exist unless the inlet flow were throttled or heated. For example, assuming isentropic flow relationships, if a static nozzle exit pressure of 973 psf corresponding to 20,000 feet could be obtained, a 519°R stagnation temperature into the engine would result from a forward speed of 920 ft/sec at altitude. The stagnation pressure at this flight speed would be 1630 psf, or considerably less than sea level atmospheric pressure. By throttling the inlet, or by raising the temperature to simulate higher speeds, conditions encountered in flight could be represented. If throttling alone is used, one flight speed can be simulated at each altitude, varying from zero flight speed at sea level to sonic at 24,000 feet, etc.

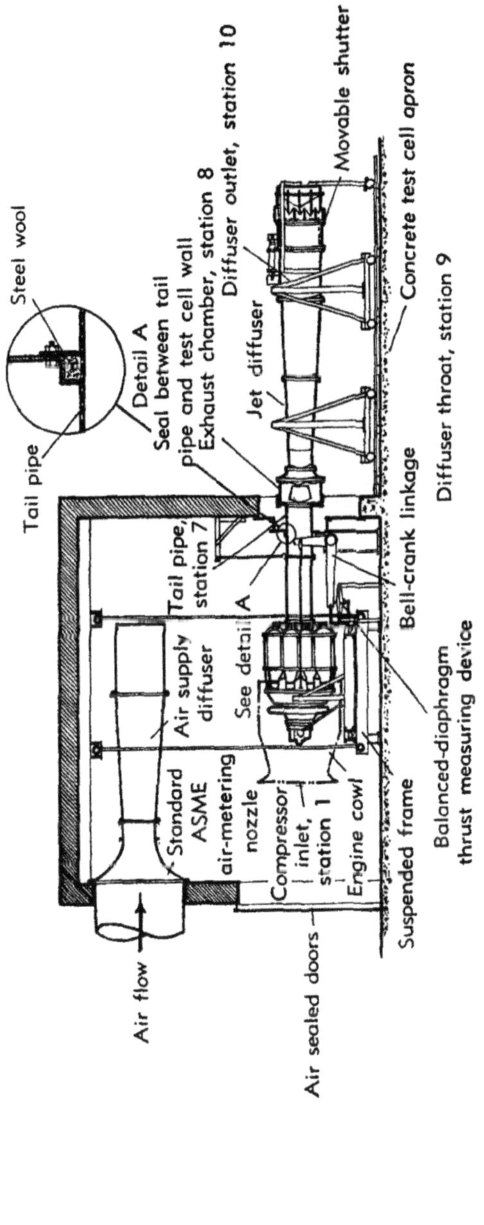

Fig. D,9c. Schematic diagram of apparatus for jet diffuser.

If a suitable diffuser is attached to the jet nozzle, Fig. D,9c from [*79*], the exit static pressure can be reduced to less than one half of atmospheric pressure corresponding to an altitude of over 20,000 feet. With atmospheric intake, the ram pressure ratio would thus be over 2.0. By throttling the intake or diffuser discharge, intermediate flight altitudes and ram pressure ratios can be simulated.

If means are provided to reduce the temperature of the air entering the engine, other conditions than the unique altitude-flight speed relation of intake throttling can be studied. One system to accomplish this is to draw the intake air through a turbine. The turbine can be wastefully loaded, or can be used to drive a compressor aiding the jet diffuser to

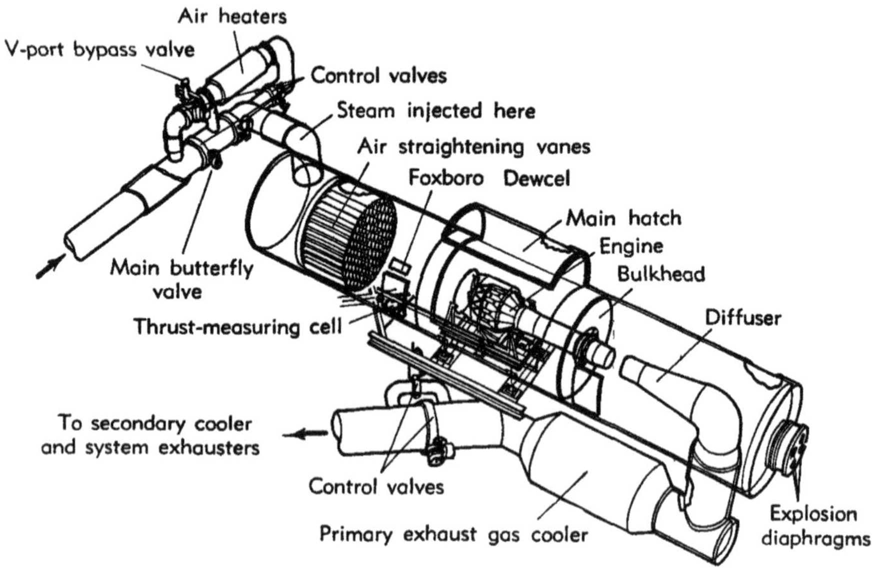

Fig. D,9d. Altitude chamber with engine installed in test section.

maintain a low pressure at the jet nozzle exhaust. To reduce icing troubles in the refrigerating turbine, dry air should be supplied. Some drying can be accomplished by passing the charge air from the atmosphere through a heat exchanger cooled by the cold charge air leaving the turbine. If sufficient cooling capacity is available, icing conditions can be studied. Windmilling speeds and the ease of starting under altitude flight conditions can be determined if steam ejectors or other means are available to reduce the engine tailpipe pressure to low values (Fig. D,9d).

The net thrust of a turbojet can be measured directly by mounting the engine in a suitable nacelle in a wind tunnel whose diameter is several times as large as that of the nacelle. The exhaust gases are usually removed from the tunnel through a duct, cooled by water sprays, pumped up to atmospheric pressure if necessary, and discharged.

The reduction of the engine noise is a serious problem in most gas turbine test installations. The sounds heard upstream of an engine consist mostly of high frequency whine, very annoying when several engine lengths away but not very distracting 20 to 30 lengths in front of the engine. Toward the back, and particularly in a 60° cone at the rear of the exhaust nozzle, the predominant noise exhibits very rapid uneven bursts, which overpower the whine observed forward, and are excruciating to the unprotected ear within 10 to 20 lengths and quite disturbing at a hundred yards. Complaints have been received from persons several miles distant from tests of large engines. As a result, the engine test stands in the various industrial and government laboratories are equipped with extensive mufflers. At the Wilgoos laboratory of Pratt and Whitney Aircraft, the exhaust gases are cooled by sprayed water, then pass through an underground labyrinth into large concrete expansion chambers before being discharged into the atmosphere.

Range of engine tests. In testing turbojet engines, or the gas generating units of turboprop engines, a major difficulty exists in determining whether the components are operating at optimum conditions because compressor-combustor-turbine units tend to operate at only one compressor pressure ratio and turbine inlet temperature at each speed. There are several means by which the operating range of an engine can be varied from this single point on the compressor and turbine characteristic curves at one speed or from single line operation over the range of speeds. A variable-area exhaust nozzle permits changing the tailpipe pressure, thereby increasing or decreasing the turbine pressure ratio and the rotational speed. To maintain the speed, a different turbine inlet temperature and pressure ratio is required. Thus, within the temperature and surge limits of the engine, a variable tail cone permits a change in the operating conditions at a given speed. The operating conditions can be changed in a slightly different manner by blowing off or bleeding some of the air from the compressor. Neither of these methods nor a combination of the two permits a very wide range of conditions to be studied, however, as is illustrated in Fig. D,9e, taken from [*80*]. In most cases where the characteristics of the compressor and turbine are not known from component tests, the small change of conditions thus attainable will not be sufficient to provide the desired information. This is particularly true near the design point, usually selected at the highest permissible temperature and pressure, where the information is most needed.

In conventional gas turbines, the turbine nozzle operates under choked flow conditions for normal operating speeds. The effective flow area of the nozzle and the total pressure and total temperature approaching it determine the weight flow of the engine. A variable turbine nozzle can therefore permit the operation of an engine over a range of conditions. Variable nozzle turbines have not been applied widely, however.

A method making possible the operation of a turbojet engine over a wider range of conditions is to add or subtract shaft power from the engine. In some gas turbines this can be done readily by attaching a dynamometer to existing auxiliaries (Plate D,9b). By freeing the turbine from the usual steady state requirement of providing the power to drive the compressor and the accessories, the turbine temperature and pressure ratio can be varied at constant speed. An analysis of the effects of ex-

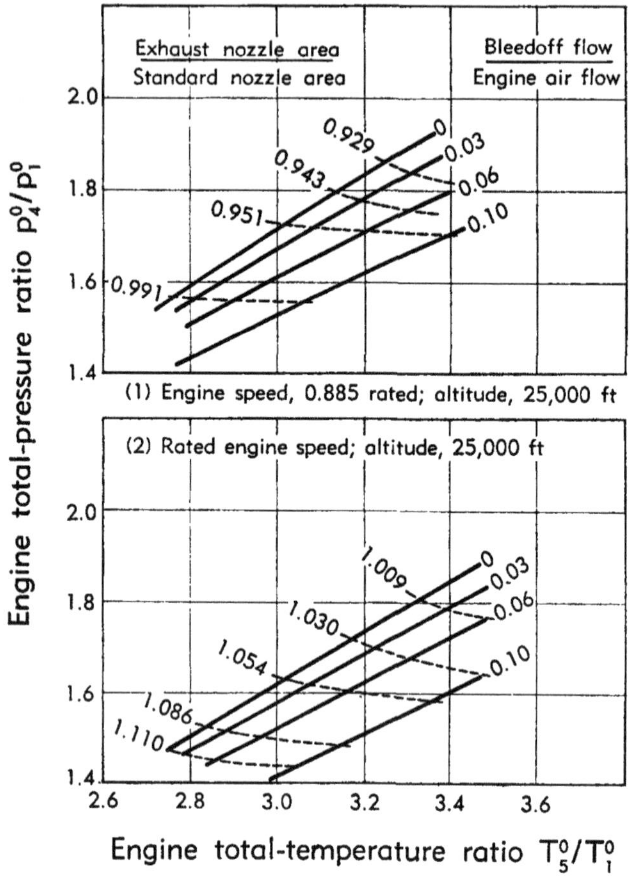

Fig. D,9e. Effect of compressor outlet bleed-off and nozzle area on engine pumping characteristics. Flight Mach number, 0.53.

tracting power from the shaft on turbojet engine performances is presented in [*81*].

D,10. Cited References.

1. NACA Subcommittee on Compressors. Standard procedures for rating and testing multistage axial-flow compressors. *NACA Tech. Note 1138*, 1946.
2. Moses, J. J., and Kazberovich, T. I. Effect of size and number of outlet pipes on design of collectors for rating and testing axial-flow compressors. *NACA Tech. Note 1607*, 1948.
3. Flow measurement by means of thin plate orifices, flow nozzles and venturi tubes. *Chapter 4, Am. Soc. Mech. Engrs., Power Test Codes 19.5; 4*, 1959.

4. Eckert, B. Summary of the results of research on axial flow compressors at the Stuttgart Research Institute for Automobiles and Engines. *Navy Dept. Buships 338, Vol. 3, Part A.*, 1946.
5. Dean, R. C., Jr. Aerodynamic measurements, *Gas Turbine Laboratory, Mass. Inst. of Technol.* The Mass. Inst. of Technol. Press, Cambridge, 1953.
6. Schulze, W. M., Ashby, G. C., Jr., and Erwin, J. R. Experimental development of flow-surveying instruments applicable to gas turbine research. *NACA Tech. Note 2830*, 1952.
7. Charnley, W. J. Note on a method of correcting for lag in aircraft Pitot-static systems. *Brit. Aeronaut. Research Council Repts. and Mem. 2352*, 1950.
8. Wildhack, W. A. Pressure drop in tubing in aircraft instrument installations. *NACA Tech. Note 593*, 1937.
9. Davis, W. T. Lag in pressure systems at extremely low pressures. *NACA Tech. Note 4334*, 1958.
10. Brown, G. P., DiNardo, A., Cheng, G. K., and Sherwood, T. K. The flow of gases in pipes at low pressures. *J. Appl. Phys. 17*, 802–813 (1946).
11. Gracey, W. Wind tunnel investigation of a number of total pressure tubes at high angles of attack, subsonic, transonic, and supersonic speeds. *NACA Rept. 1303*, 1957.
12. Thompson, J. S., and Holder, D. W. Notes on wind tunnel pressure measurements from operator's point of view. *AGARD Rept. 164*, 1958.
13. Johnson, R. C. Averaging of periodic pressure pulsations by a total pressure probe. *NACA Tech. Note 3568*, 1955.
14. Westley, R. An investigation of the frequency response of Pitot probes and preliminary measurements of the Pitot pressure fluctuations in the NAE 5-inch pilot supersonic wind tunnel. *AGARD Rept. 163*, 108 (1958).
15. Franklin, R. E., and Archbold, R. B. A small-bore pick-up for the measurement of fluctuating pressures, *The Aeronautical Quarterly 11*, 348–354 (1960).
16. Rayle, R. E. *S. M. Thesis*, Dept. of Mech. Eng., Mass. Inst. Technol., 1949.
17. Merriam, K. G., and Spaulding, E. R. Comparative tests of Pitot-static tubes. *NACA Tech. Note 546*, 1935.
18. Goff, K. W., Mercer, D. M. A., and the staff of Bolt, Beranek and Newman, Inc. Probe tube microphone for use in high temperatures and high velocity air streams, *WADC Tech. Note 56–59, AD 110432*, October, 1956.
19. Eckert, B. Experiences with flow-direction instruments. *NACA Tech. Mem. 969*, 1941.
20. Baker, V. D. Characteristics of a 3 hole yaw probe. *Pratt and Whitney Aircraft Corp. Rept. PWA-618*, 1946.
21. Bryer, D. W., Walshe, D. E., and Garner, H. C. Pressure probes selected for three-dimensional flow measurement. *Brit. Aeronaut. Research Council Repts. and Mem. 3037*, 1958.
22. Hutton, P. G. Static response of a hemispherical-headed yawmeter at high subsonic and transonic speeds. *Brit. Aeronaut. Research Council Tech. Rept. CP 401*, 1958.
23. Hall, J. A. The accurate measurement of temperature. *Research 11*, 147–151 (1958).
24. Stickney, T. M. Recovery and time-response characteristics of six thermocouple probes in subsonic and supersonic flow. *NACA Tech. Note 3455*, 1955.
25. Carter, A. D. S., Andrews, S. J., and Shaw, H. Some fluid dynamic research techniques. *Proc. Inst. Mech. Engrs. 163* (*W.E.P. 60*), 1950.
26. Klein, M. Micromanometers. II: Testing equipment and instruments. *Gt. Brit. Ministry of Air Production Volkenrode* (*VG-252*) *Repts. and Transls. 952*, 1947.
27. Ower, E. *The Measurement of Air Flow.* Chapman and Hall, 1949.
28. Campbell, P. J. A multiple recording manometer. *J. Aeronaut. Sci. 10*, 313–317 (1943).
29. NACA Lewis Staff. Central automatic data-processing system. *NACA Tech. Note*, 1958.

30. Tozier, R. E. Centralized recording and computation of propulsion system research data. *AGARD Avionics Panel Symposium,* Aachen, 1959.
31. Dimeff, J. A survey of new developments in pressure measuring techniques in the NACA. *AGARD Rept. 166,* 1958.
32. Huber, P. W., and Kantrowitz, A. A device for measuring sonic velocity and compressor Mach number. *NACA Tech. Note 1664,* 1948.
33. Sontag, H., and Brombacher, W. G. Aircraft power-plant instruments. *NACA Rept. 466,* 1933.
34. Campbell, W. R. Tests of six types of bakelite-bonded wire strain gages. *NACA Tech. Note 1656,* 1948.
35. Campbell, W. R. Errors in indicated strain for a typical wire strain gage caused by prestraining, temperature changes and weather. *NACA Tech. Note 1011,* 1946.
36. Theodorsen, T., and Gelalles, A. G. Vibration response of airplane structures. *NACA Rept. 401,* 1934.
37. Rebeske, J. J., Jr. Investigation of a NACA high-speed strain-gage torquemeter. *NACA Tech. Note 2003,* 1950.
38. Ainley, D. G. The performance of axial-flow turbines. Lectures on internal combustion turbines, *Proc. Inst. Mech. Engrs. 159,* 1948.
39. Encke, W. Experimental technique. *Gt. Brit. Ministry of Air Production Volkenrode (VG-236) Repts. and Transls. 936,* 1947.
40. Godsey, F. W., Jr., and Langer, B. F. Aircraft engine torque meters. *Amer. Inst. Elec. Engrs. 63,* 1944.
41. Rebeske, J. J., Jr. Investigation of an NACA high-speed optical torque-meter. *NACA Tech. Note 2118,* 1950.
42. Abbott, I. H. Some applications of strain gages in aeronautical research, *AGARD Memo. AG 2/M1,* 1952.
43. Sawyer, W. T. Experimental investigation of a stationary cascade of aerodynamic profiles. *Verlag Leemann, Zurich,* 1949.
44. Roepcke, F. A., Burtt, J. R., and Medeiros, A. A. Performance of Westinghouse 19B six-stage axial-flow compressor. *NACA Memo. Rept. E5K07,* 1945.
45. Abbott, I. H., and Sherman, A. Flow observations with tufts and lampblack of the stalling of four typical airfoil sections in the NACA variable-density tunnel. *NACA Tech. Note 672,* 1938.
46. Benser, W. A., and Moses, J. J. An investigation of backflow phenomenon in centrifugal compressors. *NACA Rept. 806,* 1945.
47. Preston, J., and Sweetling, N. E. An improved smoke generator for use in the visualisation of air-flow, particularly boundary layer flow at high Reynolds numbers. *Natl. Gas Turbine Establishment, Repts. and Mem. 2023,* 1943.
48. Brown, F. N. M. The organized boundary layer. *Proceedings of the Sixth Annual Conference on Fluid Mechanics,* 1959.
49. Cullen, R. E., and Poe, H. R. Preliminary studies of the nonsteady flow of air through a radial, single-stage turbine using the Mach-Zehnder interferometer. *Univ. of Michigan Research Institute Report No. 2388-13-P,* 1958.
50. Kriebel, A. R., Seidel, B. S., and Schwind, R. G. Stall propagation in a cascade of airfoils. *NACA Tech. Note 4134,* 1958.
51. Fessler, T. E., and Hartmann, M. J. Preliminary survey of compressor rotor-blade wakes and other flow phenomena with a hot-wire anemometer. *NACA RM E-56A13,* 36 (1956).
52. Pearson, C. E. Measurements of instantaneous vector air velocity by hot-wire methods. *J. Aeronaut. Sci. 19,* 73–82 (1952).
53. Erwin, J. R., and Emery, J. C. Effect of tunnel configuration and testing technique on cascade performance. *NACA Tech. Note 2028,* 1950.
54. Schlicting, H. The variable density high speed cascade wind tunnel of the Deutsche Forschungsanstalt für Luftfahrt, Braunschweig. *AGARD Rept. 91,* 1956.
55. Dunavant, J. C., and Erwin, J. R. Investigation of a related series of turbine-blade profiles in cascade. *NACA Tech. Note 3802,* 1956.
56. Weinig, F., and Eckert, B. Measurements on compressor-blade lattices. *NACA Tech. Mem. 1193,* 1948.

57. Huppert, M. C., and MacGregor, C. Comparison between predicted and observed performance of gas-turbine stator blade designed for free-vortex flow. *NACA Tech. Mem. 1810*, 1949.
58. Muhlemann, E. Experimentelle Untersuchungen an einer axialen Geblasestufe. (Experimental studies of an axial blower stage.) *Mitt. Inst. Aerodynamik*, 1946.
59. Weske, J. R. An investigation of the aerodynamic characteristics of a rotating axial-flow blade grid. *NACA Tech. Note 1127*, 1947.
60. Runckel, J. F., and Davey, R. S. Pressure-distribution measurements on the rotating blades of a single-stage axial-flow compressor. *NACA Tech. Note 1189*, 1947.
61. Davey, R. S. Multiple pressure transfer device. *Instruments 23*, 350 (1950).
62. Leist, I. K. An experimental arrangement for the measurement of the pressure distribution on high speed rotating blade rows. *ASME Paper No. 56-GTP-13*, 1956.
63. Westphal, W. R., and Godwin, W. R. Comparison of NACA 65-series compressor-blade pressure distributions and performance in a rotor and in cascade. *NACA Tech. Note 3806*, 1957.
64. Walker, D. N. Technique of testing gas turbine engines. Development of the British Gas Turbine Jet Unit. *Inst. Mech. Engrs., London*, 472–483 (1947).
65. Westphal, W. R., and Dunavant, J. C. Application of the wire-mesh plotting device to incompressible cascade flows. *NACA Tech. Note 2095*, 1950.
66. Finger, H. B., Schum, H. J., and Buckner, H. A., Jr. Experimental and theoretical distribution of flow produced by inlet guide vanes of an axial-flow compressor. *NACA Tech. Note 1954*, 1949.
67. Howell, A. R. Fluid dynamics of axial compressors. Lectures on the Development of the British Gas Turbine Jet Unit. *Inst. Mech. Engrs., London*, 441–452 (1947).
68. Bowen, J. T., Sabersky, R. H., and Rannie, W. D. Theoretical and experimental investigations of axial-flow compressors. *Calif. Inst. Technol. Mech. Eng. Lab.*, 1949.
69. Howell, A. R. Overall and stage characteristics of axial flow compressors. *Proc. Inst. Mech. Engrs., London*, 163–249 (1950).
70. New, W. R. A laboratory for gas-turbine development. *Mech. Eng. 70*, 195–204 (1948).
71. Eckert, E. Cold air model tests with gas turbines. *Natl. Research Council of Canada, Div. of Mech. Eng. Tech. Transl. TT-42*, 1947.
72. Spooner, R. B. Effect of heat-capacity lag on a variety of turbine-nozzle flow processes. *NACA Tech. Note 2193*, 1950.
73. Lovesey, A. C. Modern methods of testing aero-engines and power plants. *J. Roy. Aeronaut. Soc. 54*, 327–358 (1950).
74. Blackshear, P. L., Jr. Sonic-flow orifice temperature probe for high-gas-temperature measurements. *NACA Tech. Note 2176*, 1950.
75. Simmons, F. S., and Glawe, G. E. Theory and design of a pneumatic temperature probe and experimental results obtained in a high temperature gas stream. *NACA Tech. Note 3893*, 1957.
76. Kuhns, P. W. Effects of thermal relaxation and specific-heat changes on measurements with a pneumatic-probe pyrometer. *NACA Tech. Note 4026*, 1957.
77. Tarr, P. R. Methods for connection to revolving thermocouples. *NACA Research Memo. E50j23a*, 1951.
78. Coleman, W. J., and Desanto, D. F. Measuring in-flight thrust of a turbojet-powered aircraft. *AGARD Report 197*, 1958.
79. Essig, R. H., Bohanon, H. R., and Gabriel, D. S. Jet diffuser for simulating ram-pressure and altitude conditions on a turbo-jet-engine static test stand. *NACA Tech. Note 1687*, 1948.
80. Fleming, W. A., Wallner, L. E., and Wintler, J. T. Effect of compressor-outlet bleedoff on turbo-jet-engine performance. *NACA Research Mem. E50E17*, 1950.
81. Koutz, S. L., Hensley, R. V., and Rom, F. E. Effect of heat and power extraction on turbojet-engine performance. III: Analytical determination of effects of shaft-power extraction. *NACA Tech. Note 2202*, 1950.

SECTION E

FLOW IN CASCADES

A. R. HOWELL

E,1. Ideal Flow in Cascades.

Introduction. A detailed knowledge of the flow through cascades is necessary in the design of blading for axial compressors and turbines, and cascade performance can be obtained either directly from cascade wind tunnel tests and theoretical considerations or indirectly from overall compressor and turbine performance. It is with data from theoretical considerations and from subsonic cascade tunnel tests that this section is mainly concerned, though the articles include brief accounts of secondary effects and of heat transfer to blades in cascades. Much of the subject matter is also covered in other sections of the volume, but in this section the emphasis has been placed on general considerations of the experimental and theoretical cascade information available, and on comparisons between the various methods used to obtain this information, with particular reference to its useful application to the design of compressors and turbines.

Art. 1 deals with ideal flow in cascades, giving a general review of the position and then indicating what can be deduced from such flow calculations. For a more detailed discussion of the general theories of potential flow, reference should be made to Sec. B on two-dimensional flow of an ideal fluid in turbomachines.

Nomenclature. The letter symbols used conform with those used throughout the volume. Most symbols are also defined in the text when they first occur. As there are at the moment several systems of notation and sign convention available for defining cascade properties, the symbols used are illustrated for typical compressor and turbine cascades in Fig. E,1a. Here α_1 and α_2 denote the inlet and exit flow angles, respectively, relative to what would be the axis of rotation in an actual machine, and γ is the stagger angle. The stagger angle is the angle of the line joining the leading and trailing edges as defined by the mean center or camber line and it is not the tangential chord angle. The blade angles corresponding to α_1 and α_2 are given by α_1' and α_2'. The chord is c and the spacing or pitch between adjacent blades is s. The symbol o which is the opening between the adjacent nozzles or blades at the passage exit is also used as the throat width at the passage inlet to the compressor cascades.

When making comparisons between cascade airfoils from different sources, the airfoils are taken to be defined by their main geometrical characteristics as in [*1*]. These are the blade camber angle θ equal to $\alpha_1' - \alpha_2'$, the maximum thickness-to-chord ratio t/c, the position of maximum camber defined by distance from the leading edge divided by the chord a/c, the station of maximum thickness as per cent of chord, and the leading and trailing edge radii as per cent of maximum thickness. For given values of the above main characteristics, airfoils from different sources are very similar, and the small detail differences in shape are usually within the tolerances (± 0.002 in. to ± 0.005 in. for 1-inch chord blades) necessary for reasonable costs for the hundreds of blades necessary in a gas turbine power plant.

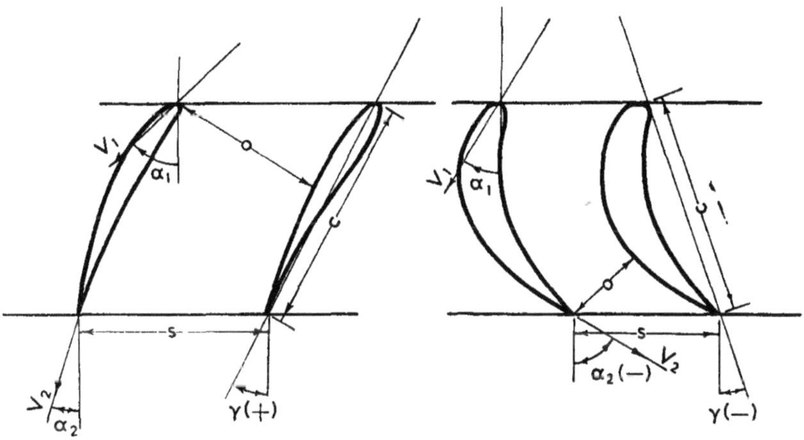

Fig. E,1a. Cascade notation.

Most compressor blades follow the usual airfoil practice [*2,3,4*] of wrapping as a fairing a basic thickness form symmetrically around the mean camber line which serves as a skeleton of the airfoil. Turbine blades [*5*] are often, if not usually, thought of in terms of passages or channels rather than of airfoil shapes, and in addition may be simply constructed of circular arcs and straight lines instead of uniformly changing curvatures as with airfoils. However, these can also be reduced to the above main geometrical characteristics for comparison purposes [*6*].

Ideal flow methods and uses. Many methods are now available for obtaining with varying degrees of accuracy and complication the ideal flow conditions past airfoils in cascade, with the aim of obtaining the pressure and velocity distributions, and/or the turning angles or lift coefficients, these often being called the "direct" problems. More recently, an increasing number of attempts have been made at satisfactory solutions for the "inverse" problem of the airfoil shape in cascades, corresponding to specified pressure or velocity distributions. Tyler [*7*] reviews many of these methods, and [*8–23*] are direct solutions while [*24,25,26, 27,28*] and also [*15,16,23*] are inverse solutions. More recent German investigations are considered by Schlichting [*29*]. All of the previous theories and methods are for incompressible flows. Approximate and limited methods for compressible flows are given in [*27,30*]. In a mathematical sense, ideal flow problems may be said to be solved at least for incompressible flows, but from a practical standpoint they are still of only limited application to compressor and turbine design.

For direct solutions, normally involving at least advanced if not complicated numerical computation, the alternative of the electrolytic tank, using an electrical analogy, is worthy of consideration. Malavard [*31*], de Haller [*32*], Hargest [*33*], and others have successfully used such tanks for cascade work. De Haller [*32*] gives examples of pressure distributions, streamlines, and exit flow angles. He finds that the tank exit flow angles α_2 are about 1° lower than cascade tunnel test values, or, in other words, the test deviation $\delta = \alpha_2 - \alpha_2'$ is 1° higher than the theoretical or tank solution. It is also shown that it is not essential to do tests at every incidence to deduce the general flow for many incidences.

The most useful deductions from ideal flows are the exit flow angles α_2 and the pressure distributions for various incidences $\iota = \alpha_1 - \alpha_1'$. The α_2 for given α_1 gives the turning angles $\tau = \alpha_1 - \alpha_2$, the lift coefficients C_L, and the deviations $\delta = \alpha_2 - \alpha_2'$. The pressure or velocity distributions can be used to estimate the critical Mach number [*1,19,34*], or to indicate roughly when stalling is likely to occur [*1,19,35*], or again to estimate heat transfer to blades [*36*]. When there is likely to be a considerable amount of laminar flow as in a nozzle cascade in a wind tunnel, estimates of drag coefficients or total pressure losses can be made from the theoretical velocity distributions, but it is not known whether such theoretical or experimental results give the same losses as those occurring in the more turbulent flow in an actual machine. Unfortunately, from a calculation point of view, the Reynolds numbers *Re* in which one is interested are 5×10^4 to 5×10^5 and so include the critical Reynolds number range of 1 to 3×10^5, the *Re* being based on the larger inlet and exit velocities. With inverse solutions there is the possibility of applying to cascades the large amount of data available from isolated airfoil practice on the more desirable forms of pressure and velocity distribution around the airfoils.

Exit flow and deviation angles. Agreements, when unstalled, to within about 1° between experimental and theoretical results for the exit flow angle α and the deviation δ are obtained by Howell [*1,13*], de Haller [*32*], Carter and Hughes [*19*], and Sawyer [*28*], so that the agreement might be said to be good enough for most purposes. The theoretical deviation increases slightly with incidence ($\iota = \alpha_1 - \alpha_1'$) of the order of $\frac{1}{2}$° to 1° for a 10° increase in ι or the turning angle τ. The closer the blades the smaller this increase will be.

A simple rule for deviation based largely on theoretical results, but generally agreeing with British experimental information, is that given by Carter and Hughes [*19*] and Carter [*37*] where

$$\delta = m\theta\sqrt{\frac{s}{c}} \quad \text{for compressors} \tag{1-1}$$

$$\delta = m\theta\left(\frac{s}{c}\right) \quad \text{for turbines} \tag{1-2}$$

and m is a function of stagger and the position of maximum camber a/c, and is given in [*19,37*]. Here θ is the camber angle equal to $\alpha_1' - \alpha_2'$ and s/c is the spacing-to-chord ratio. Typical values of m are given in Table E,1. This rule is a fairly recent adaptation of the one given by Constant and referred to in [*1*].

Table E,1

Stagger γ	30°	0°	−30°
$m\left(\frac{a}{c} = 0.4\right)$	0.20	0.13	0.09
$m\left(\frac{a}{c} = 0.5\right)$	0.26	0.21	0.19

An interesting confirmation of the above deviation rule comes from the turbine nozzle example calculated and tested by Sawyer [*28*]. The values of s/c, θ, a/c, γ, and α_2' for this airfoil are 0.69, 40°, 0.5, −15°, and −35° respectively. From [*37*] m is then 0.2, and since the blade is a turbine one the δ from Eq. 1-2 is 5.5°, giving α_2 equal to −29.5° compared with the test value of −29.5° and Sawyer's calculated value of −30°. Another confirmation is from the tests done by Finger, Schum, and Buckner [*38*] on compressor inlet guide vanes where in the middle of the blade s/c, θ, a/c, and γ are approximately 0.8, 30°, 0.4, and −16° respectively. Then from [*37*] m is 0.1 and from Eq. 1-2, since inlet guide vanes are effectively turbine-type blades, the deviation is 2.4°, which again compares favorably with test values of 2° to 3°.

Pressure distributions and critical Mach numbers. As with the exit flow and deviation angles, reasonably good agreement is found between theoretical and experimental pressure or velocity distributions in [*13,19, 28*]. In [*39*] Katzoff, Bogdonoff, and Boyet find large differences between theory and experiment, the latter showing lift coefficients only 65 to 80 per cent of the theoretical values, although the general form of the distributions was in reasonably good agreement. This kind of discrepancy usually occurs due to stream contraction [*1,40*] and to secondary flows, and the low aspect ratios used might have made matters worse in [*39*].

A reasonable approximation in such cases to the experimental velocity distribution on the airfoil can be made if the experimental lift coefficient is known, either by arbitrarily adjusting the circulation on the airfoil or by using the Pinkerton distortion [*41*].

Fig. E,1b gives an example of theoretical and experimental velocity distributions deduced from the results of [*19*]. The values of θ, γ, s/c, a/c,

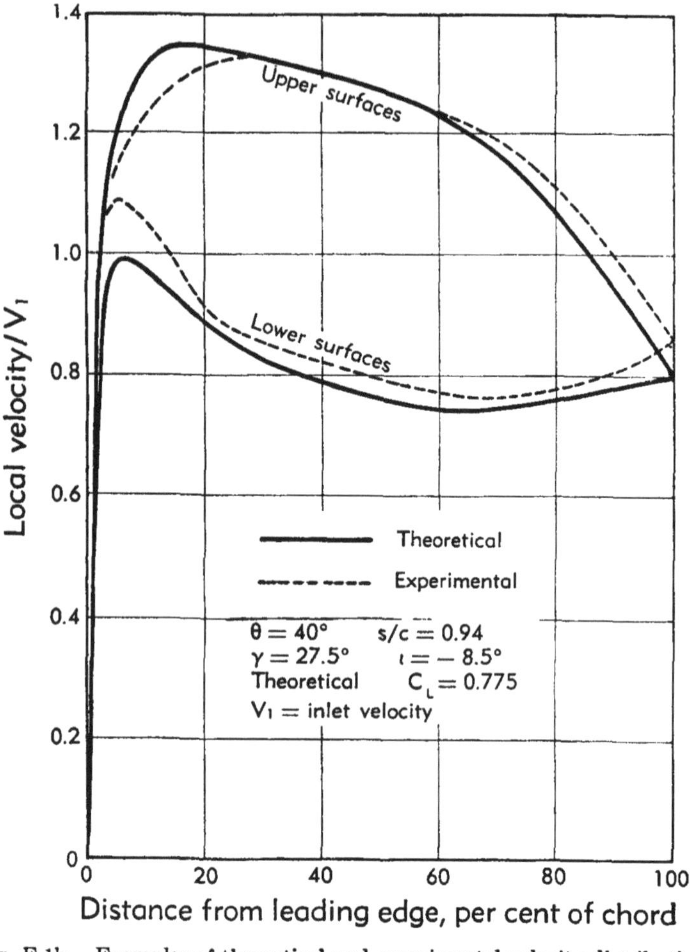

Fig. E,1b. Examples of theoretical and experimental velocity distributions.

and ι are 40°, 27.5°, 0.94, 0.5, and −8.5° respectively, and the theoretical lift coefficient C_L is 0.775. The theoretical pressure distribution of this same compressor cascade is given in Fig. E,1c, compared with the pressure distribution for a turbine cascade from [*28*] with a somewhat similar airfoil. The values of θ, γ, s/c, a/c, and ι for the turbine nozzle example are 40°, −15°, 0.69, 0.5, and −5° respectively and the theoretical C_L is

0.76. Both airfoils have 10 per cent chord maximum thicknesses, but the position of maximum thickness is at 37.5 per cent chord for the turbine blade compared with 33 per cent chord for the compressor blade. In Fig. E,1c, the maximum suction point is at about 0.5*c* for the turbine cascade compared with 0.15*c* for the compressor cascade. This difference for what are in effect very similar airfoils is, according to Carter [*37*, Fig. 7], what one would expect to find from the change of stagger from compressor to turbine cascade. For general comparisons of compressor, turbine, and

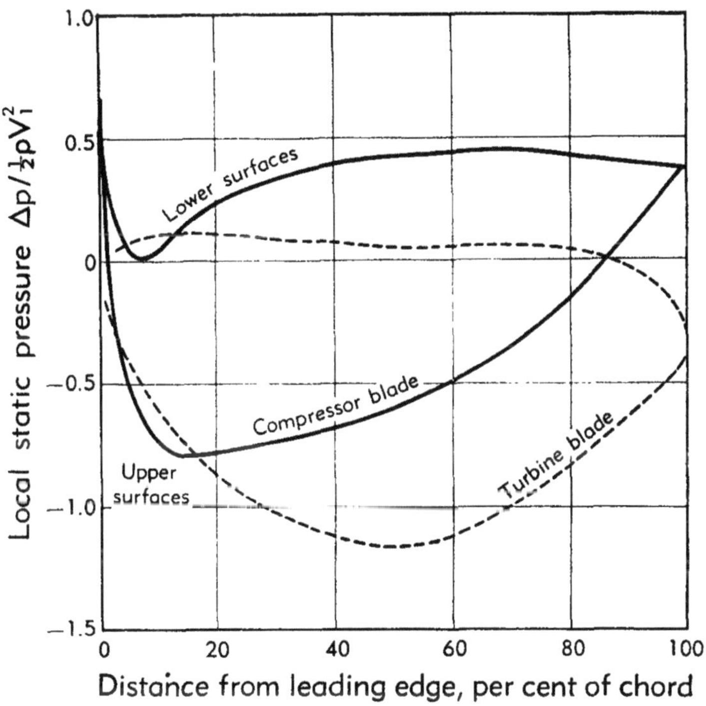

Fig. E,1c. Typical compressor and turbine pressure distributions.

isolated airfoils, Zweifel [*42*], Howell and Carter [*43*], and Carter [*37*] suggest that it is better to base the pressure or velocity distributions on the outlet rather than on the inlet dynamic head or velocity.

Theory and experiment suggest that the critical Mach number M_{cr} is reached for most practical applications when the maximum local velocity on the surface of the airfoil equals the local speed of sound. Approximate conditions in compressible flow can be obtained from the well-known, Glauert-Ackeret formula.

$$\frac{\Delta p}{\frac{1}{2}\rho V^2} = \frac{(\Delta p/\frac{1}{2}\rho V^2)_{M=0}}{\sqrt{1 - M^2}} \tag{1-3}$$

For the maximum velocity to be equal to the local velocity of sound there is a definite relationship [*1,34,44*] between $(\Delta p/\frac{1}{2}\rho V^2)$ and M, and hence between M and the incompressible $(\Delta p/\frac{1}{2}\rho V^2)_{M=0}$. Details of these relationships, as applied to cascades, and of the somewhat more recent correction by von Kármán [*44*] are given in [*1,34*]. Using these methods the M_{cr} for two compressor airfoils, having approximately the same inlet blade angles and exit flow angles from [*19*], have been calculated and are plotted in Fig. E,1d.

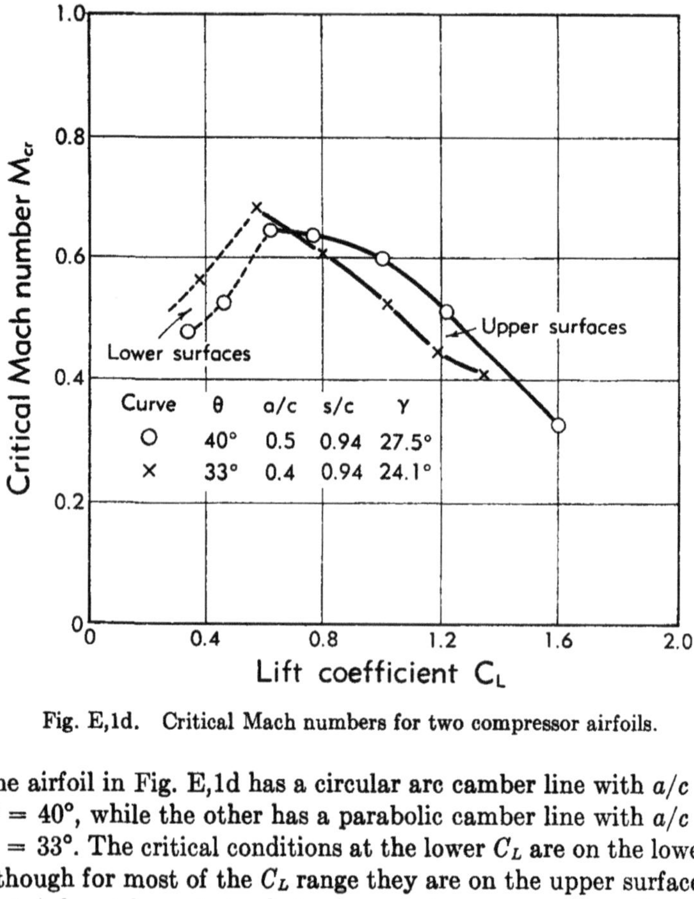

Fig. E,1d. Critical Mach numbers for two compressor airfoils.

One airfoil in Fig. E,1d has a circular arc camber line with $a/c = 0.5$ and $\theta = 40°$, while the other has a parabolic camber line with $a/c = 0.4$ and $\theta = 33°$. The critical conditions at the lower C_L are on the lower surface, though for most of the C_L range they are on the upper surface. The Glauert-Ackeret formula tends to give a value comparable with the drag M_{cr} corresponding to say a 50 per cent increase in drag or losses, while the von Kármán correction, which gives M_{cr} values about 0.02 lower, tends to correspond more to sonic conditions on the surface. Bogdonoff [*45*] also defines a force-break Mach number M_{fb} which is higher than the ordinary critical, but one cannot tell from [*45*] how M_{fb} would compare with those deduced from incompressible flow calculations.

E,2 Two-Dimensional Performance.

General discussion. The actual two-dimensional flow through a cascade differs from ideal incompressible flow in two main respects: the boundary layers on the blades and the compressibility effects when the Mach numbers (M) are large, say M over 0.6 near design conditions and M over 0.4 away from design conditions. Tests carried out at Mach numbers up to 0.2 to 0.3 can usually be regarded as low speed incompressible flow results. Interest in the boundary layers centers around the resulting losses or drag coefficients, and on indications of both the positive and negative stalling incidences. On the basis of local velocity gradients, Carter and Hughes [*19*] and Howell [*1*] suggest reasonable agreement between values deduced from diffuser tests and those obtained from compressor cascades. More recently Goldstein and Mager [*35*] have investigated theoretically the attainable circulation of airfoils in cascade and have obtained results for compressor and turbine cascades which are also somewhat comparable with experimental experience, as has Schlichting [*29*]. Generally, however, resource is still made to experiment to determine both the loss or drag coefficients and the stalling characteristics of cascades.

Cascade tunnel experimental techniques are covered in Sec. D; here we consider the test results that have been obtained. Experimental data are reported in [*1,5,6,19,28,37,38,39,40,42,43,* and *45–56*], and although these comprise a fair number of references, test data have not on the whole been reported as fully as the theoretical data discussed in Art. 1. Much of the data has been analyzed in a generalized way to produce design curves, so that only a few of the actual test results used are given by way of example. Generalized cascade data will be considered later (Art. 5). Before proceeding further with the two-dimensional performance it is necessary to define the various cascade relationships that are used for plotting and analyzing the results of cascade test work.

Cascade relationships. The lift and drag coefficients (C_L and C_D) for airfoils in cascade can be obtained by actual force measurements as done by Harris and Fairthorne [*46*], Keller [*40*], and Sawyer [*28*] or by deduction from turning angles ($\tau = \alpha_1 - \alpha_2$) and loss measurements, most of the more recent cascade information being obtained by the latter method. Assuming two-dimensional flow conditions with the same axial velocity at inlet and exit, there are definite mathematical relationships [*1,40,57*] between the two methods. In the relationships to be deduced below, the flow is regarded as incompressible and the notation will be the same as that previously used except for the introduction of $\bar{\omega}$ for the mean loss of total head pressure across the cascade. This loss ($\bar{\omega}$) is usually divided by $\frac{1}{2}\rho V_1^2$, the inlet velocity head for compressor cascades, and it is referred to $\frac{1}{2}\rho V_2^2$, the exit velocity head for turbine cascades.

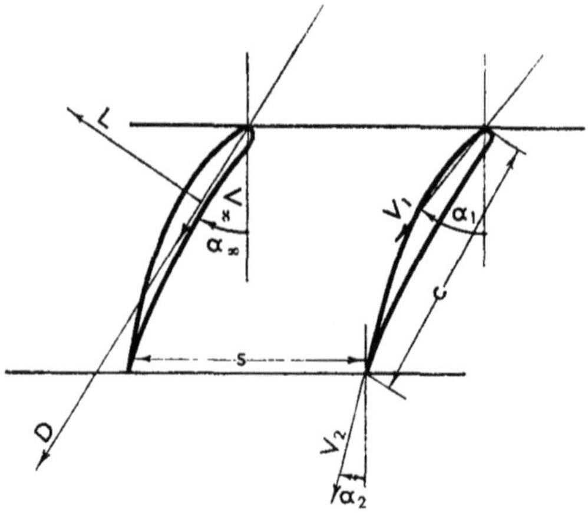

Fig. E,2a. Cascade relationships.

Consider first the cascade with no losses. The force on one blade of unit height along or tangential to the compressor cascade of Fig. E,2a is given by the rate of change of momentum in this direction, that is, the mass flow past the blade multiplied by the change in whirl velocity, and so

$$\begin{aligned}\text{Tangential force} &= (s\rho V_z)(V_z \tan \alpha_1 - V_z \tan \alpha_2)\\ &= \left[\frac{s\rho V_z^2(\tan \alpha_1 - \tan \alpha_2)}{\cos \alpha_\infty}\right] \cos \alpha_\infty\end{aligned}$$

The axial force per unit height is $s\Delta p$ where Δp is the static pressure rise, giving

$$\begin{aligned}\text{Axial force} &= s\Delta p\\ &= \tfrac{1}{2}s\rho(V_1^2 - V_2^2)\\ &= \tfrac{1}{2}s\rho V_z^2(\tan^2 \alpha_1 - \tan^2 \alpha_2)\\ &= \left[\frac{s\rho V_z^2(\tan \alpha_1 - \tan \alpha_2)}{\cos \alpha_\infty}\right] \sin \alpha_\infty\end{aligned}$$

where α_∞ is the direction of the vector mean velocity V_∞ to V_1 and V_2 and

$$\tan \alpha_\infty = \tfrac{1}{2}(\tan \alpha_1 + \tan \alpha_2) \tag{2-1}$$

It will be seen that the above two forces have a resultant L perpendicular to the vector mean direction α_∞ and so the lift

$$L = C_{L_{\text{th}}} \tfrac{1}{2} c\rho V_\infty^2 = \frac{s\rho V_z^2(\tan \alpha_1 - \tan \alpha_2)}{\cos \alpha_\infty}$$

and then

$$C_{L_{\text{th}}} = 2\left(\frac{s}{c}\right)(\tan \alpha_1 - \tan \alpha_2) \cos \alpha_\infty \tag{2-2}$$

since $V_z = V_\infty \cos \alpha_\infty$.

The above equations (Eq. 2-1 and 2-2) can also be deduced from circulation considerations and they apply equally to compressor and turbine applications. They give the required relationship between the flow angles α_1 and α_2 and the theoretical lift coefficient $C_{L_{th}}$.

The mean total head pressure loss $\bar{\omega}$ for the compressor cascade can be regarded as an opposing axial force $\bar{\omega}s$ which can be resolved along the vector mean direction to give the drag, and perpendicular to this direction to reduce the lift. The resulting expressions obtained are

$$C_D = \left(\frac{s}{c}\right)\left(\frac{\bar{\omega}}{\frac{1}{2}\rho V_\infty^2}\right)\cos\alpha_\infty \tag{2-3}$$

and

$$C_L = C_{L_{th}} - C_D \tan\alpha_\infty \tag{2-4}$$

Alternative formulas for drag coefficients are

$$C_D = \left(\frac{s}{c}\right)\left(\frac{\bar{\omega}}{\frac{1}{2}\rho V_1^2}\right)\left(\frac{\cos^3\alpha_\infty}{\cos^2\alpha_1}\right) \tag{2-5}$$

and

$$C_D = \left(\frac{s}{c}\right)\left(\frac{\bar{\omega}}{\frac{1}{2}\rho V_2^2}\right)\left(\frac{\cos^3\alpha_\infty}{\cos^2\alpha_2}\right) \tag{2-6}$$

where Eq. 2-5 and 2-6 are used for compressor and turbine cascade analyses respectively.

Eq. 2-4 gives the true value of C_L but it is generally more convenient to use its theoretical value $C_{L_{th}}$ which is a function only of the turning angle τ and α_1 or α_2, and does not include a loss component. In the unstalled region the difference between the two lift coefficients is rarely greater than 2 per cent. The blade efficiencies η_b are given [*52*] by

$$\eta_b = 1 - \frac{2}{\sin 2\alpha_\infty}\frac{C_D}{C_{L_{th}}} \quad \text{(compressors)} \tag{2-7}$$

and

$$\eta_b = \frac{1}{\left(1 - \dfrac{2}{\sin 2\alpha_\infty}\dfrac{C_D}{C_{L_{th}}}\right)} \quad \text{(turbines)} \tag{2-8}$$

and they are simple formulas because $C_{L_{th}}$ and not C_L has been used. In the remainder of this article the theoretical lift coefficient, Eq. 2-2, will always be used; hence the subscript $_{th}$ will be dropped and it will be denoted simply by C_L.

Test results at low speed. Most of the experimental work published on cascades has been done at low speeds. The fact that there has been very little published information at high speeds is partly due to the ease with which low speed results can be obtained. If used with care, low speed test results can be applied to most compressor and turbine designs even when their Mach numbers are high, as the incidences for maximum lift/drag are little affected by the Mach number up to the critical values, the main effect of the Mach number being to reduce the working incidence range for low loss or high efficiency.

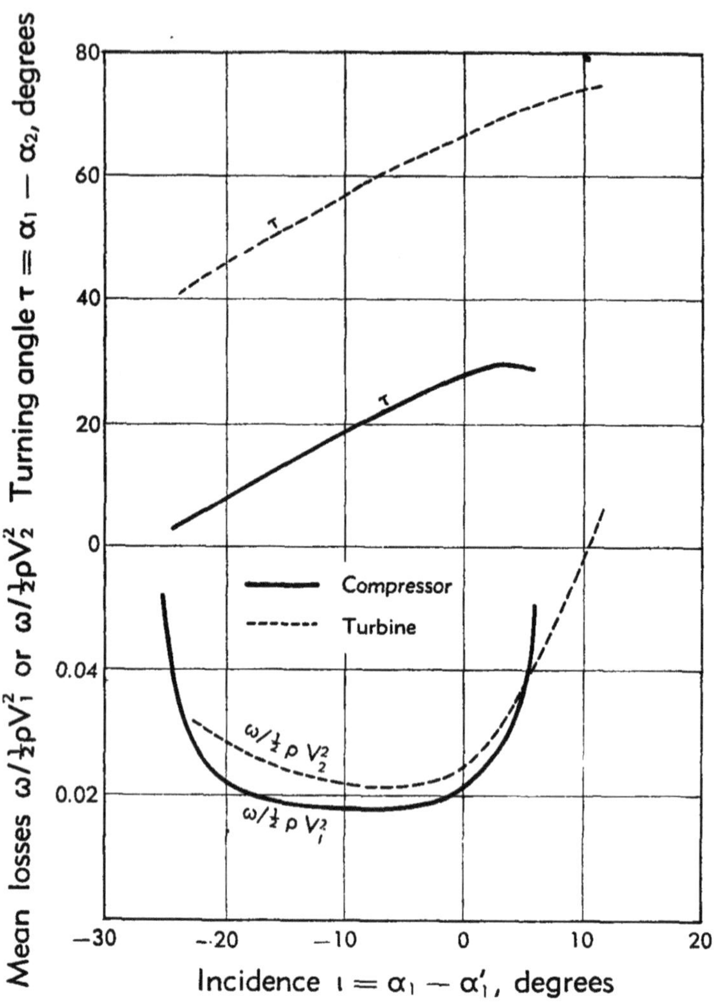

Fig. E,2b. Turning angles and losses for typical low speed compressor and turbine cascades.

Typical low speed compressor and turbine results from [43] are compared in Fig. E,2b and E,2c. The θ, s/c, a/c, and α_2' for the compressor cascade are 33°, 0.94, 0.4, and 14° respectively, while the corresponding values for the turbine cascade are 85°, 0.94, 0.5, and −66.5°. In Fig. E,2b, the turning angle $\tau = \alpha_1 - \alpha_2$ and the losses expressed as $\bar{\omega}/\frac{1}{2}\rho V_1^2$ and $\bar{\omega}/\frac{1}{2}\rho V_2^2$ are plotted against incidence $\iota = \alpha_1 - \alpha_1'$. In Fig. E,2c, the corresponding values of the theoretical lift coefficient C_L and the drag coefficient C_D are plotted against the incidence. In particular it will be noticed that the attainable turbine turning angles and lift coefficients are over twice those for the compressor cascade.

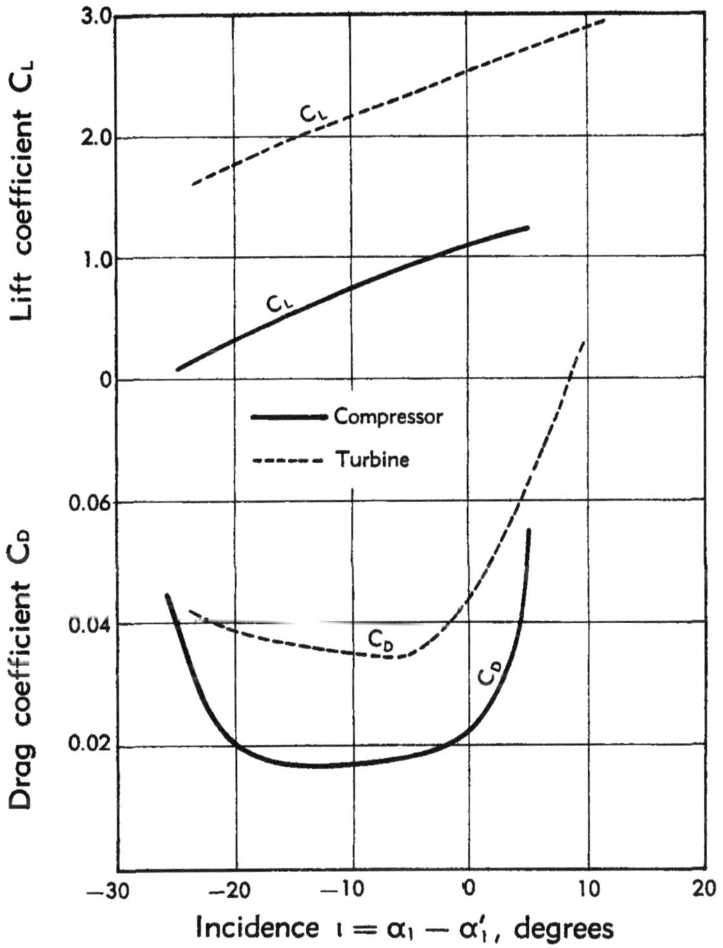

Fig. E,2c. Lift and drag coefficients for typical low speed compressor and turbine cascades.

For design purposes it is generally desirable to operate at as high a turning angle as possible provided the losses or drag coefficients do not increase too much and the blades do not become stalled. Stalling of a cascade can be taken to occur when the blade loss is twice its minimum value, but for design it is obviously inadvisable to operate right on the point of stalling, so that for practical use a nominal turning angle τ^*, as in [*1,56*], is taken which is equal to 0.8 of the stalling turning angle.

The superscript * is used to denote nominal conditions. Another suitable criterion for design can be obtained without reference to the actual stalling turning angles, that is, by taking conditions corresponding to the maximum lift-drag ratio for the cascades. Values for the two criteria for the two previous cascades are given in Table E,2. The maximum L/D

Table E,2

Cascade	Nominal conditions τ^*	C_L^*	Maximum (L/D) conditions τ	C_L	L/D
Compressor	24°	0.95	26°	1.03	53
Turbine	57°	2.17	61°	2.31	68

quoted are at least comparable if not better than those for isolated airfoils [*43*,*58*], at the Reynolds numbers of 3 to 5 $\times$ 10^5 of these test results. Reynolds number is referred to inlet velocities for compressor airfoils and to exit velocities for turbine cascades.

As with pressure distributions in Art. 1 it is sometimes more convenient to express the lift and drag coefficients in terms of the exit instead of the vector mean velocities, this being done in [*37*,*42*,*43*]. Again in conjunction with γ it is possible to define another form of incidence ι_∞ which corresponds more to isolated airfoil practice, and is taken to be equal to the difference between the stagger angle and the vector mean angle, giving $\iota_\infty = \alpha_\infty - \gamma$.

The best values to take for nominal or maximum (L/D) turning angles and lift coefficients are discussed in Art. 5 under generalized cascade data. Away from nominal values the losses or drag coefficients and the turning angles or deviations can be found by the approximate curves given by Carter [*37*] for compressors and turbines, and by Howell [*1*,*56*] for compressors. In Fig. E,2d, the values from [*56*] are given, but plotted in a somewhat different manner. The deviation δ is plotted as $(\delta - \delta^*)/\tau^*$ and it and the drag coefficient are plotted against $(\iota - \iota^*)/\tau^*$ where ι^*, τ^*, and δ^* are the nominal values of incidence, turning angle, and deviation respectively.

It is often found (Keller [*40*]; Howell [*1*]), when no boundary layer suction is used on the tunnel walls [*1*,*28*,*59*], that with compressor cascades the air stream in passing through the cascade contracts and gives a considerably reduced pressure rise to that expected. This effect does not usually occur with turbine cascades due to their more favorable pressure gradients. Such contractions appear [*1*] to have only a small effect on measurements of loss and turning angle, even though the effect on pressure rise may be appreciable. Therefore it is suggested that with reasonable blade aspect ratios the direct measurements of loss and turning angles only should normally be used in analysis, and that lift and drag coefficients should be deduced from them as if the flow were properly two-dimensional. With the smaller blade aspect ratios, corrections to turning angles may be necessary as discussed in Art. 3 on secondary effects.

Test results at high speed. The high speed performances of compressor and turbine cascades can be markedly different. For example,

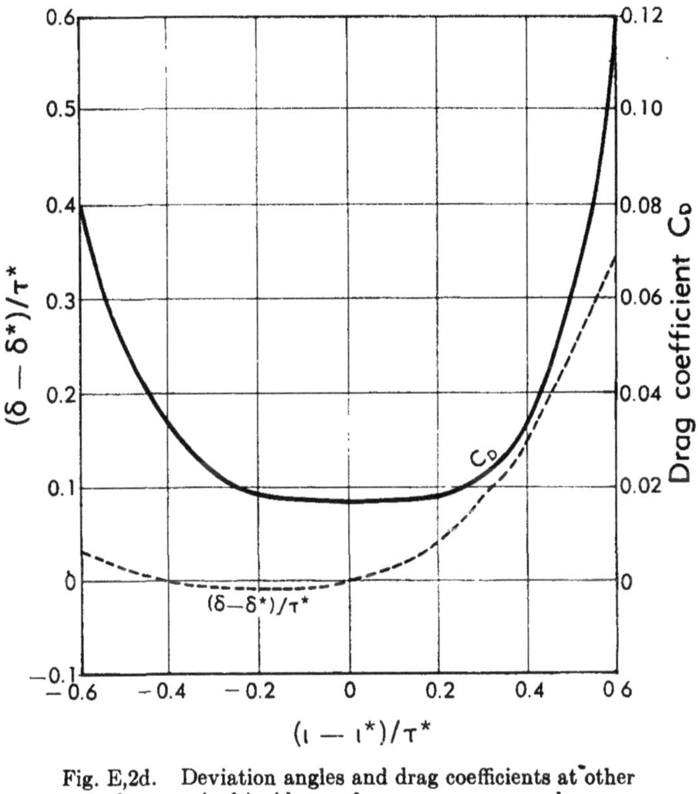

Fig. E,2d. Deviation angles and drag coefficients at other than nominal incidences for compressor cascades.

efficient operation is unlikely above inlet Mach numbers of about 0.7 for normal compressor blades, while the efficiency of a turbine operating at 0.9 exit Mach number might actually be better than that at 0.5 exit Mach number. Fig. E,2e illustrates some of the variations to be expected in blade loss with M for compressor and turbine cascades. The compressor cascade losses are deduced from an example given by Howell [*52*] and show a marked increase of loss between 0° and 10° incidence for a change of M from 0.3 to 0.7. The turbine results of Ainley [*6*] are for Mach numbers 0.5 and 0.9 and show little compressibility effects. Actually the losses for the impulse blade at $M = 0.9$ are mostly lower than those at $M = 0.5$, a result which was first noticed by Todd [*51*]. Todd also illustrates his result by photographs of the shock wave formations involved at these Mach numbers. It should be pointed out that the higher losses in Fig. E,2e for the turbine blades do not necessarily mean lower

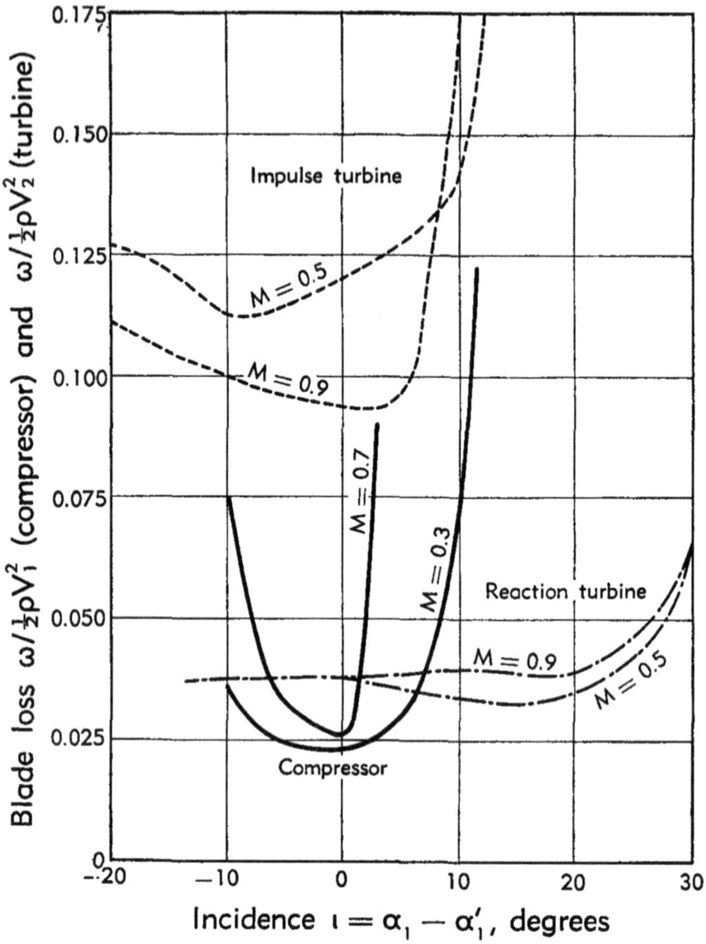

Fig. E,2e. Variation of blade loss with Mach number.

efficiencies than the compressor blades would give since the turbine blade work done is much greater for given velocities.

On compressor blades up to the critical Mach number, Bogdonoff [*45*] shows that there is less than 1° change in deviation or turning angle for given incidences with increasing M, so that Mach number effects on δ or τ can normally be neglected. However, on turbines, Ainley [*6*] suggests that the deviations or exit flow angles may vary 2° to 6° between $M = 0.4$ and $M = 1.0$, and he gives curves for these variations. Todd

[*51*], over a similar range, shows 3° to 4° change in angles. All the turbine variations give smaller deviations or higher turning angles at the higher Mach numbers. For conventional turbine blades the curves given by Ainley [*6*] are such that at $M = 1.0$, the outlet angle α_2 is given by the common "steam turbine" rule of cos α_2 equal to (opening/pitch) or (o/s). Though with compressible flow there are many possible ways of defining the lift coefficient, Eq. 2-2 giving the low speed C_L is convenient to use even at high speed, but physically it then becomes purely another but useful way of expressing the amount of turn given to the fluid.

It has already been stated that compressibility effects appear to be more critical for compressor than for turbine cascades. The critical Mach number for compressor cascades can be defined as that Mach number at which local sonic conditions are first attained on the blade surface, and it is obtained either from theory (Art. 1) or by schlieren photographs [*45*]. With actual high speed cascade experimental measurements it is convenient to define a drag critical Mach number corresponding to, say, a 50 per cent increase in loss, or a force-break Mach number where the turning angle falls off as in Bogdonoff [*45*], or again by the point where the pressure rise drops off (Bowden and Jefferson [*54*] and Howell [*1*]). There is as yet no general correlation of Mach number effects, and reference must be made to the actual test results as given in, for example, [*1,45,52,54,56*]. With further increases in M the losses may increase to such an extent that zero pressure rise results; in [*1,56*] this limit is defined as the "choking" Mach number M_{ch}. There is a theoretical expression for M_{ch} dependent on the value of ($o/s \cos \alpha_1$) where o is the throat width or opening at inlet to the cascade (Fig. E,1a), s the blade spacing, and α_1 the inlet flow angle. The expression [*1*] is as follows for a ratio of specific heats of the fluid equal to 1.4:

$$\frac{o}{s \cos \alpha_1} = M_{ch}\left(\frac{1.2}{1 + 0.2M_{ch}^2}\right)^3 \tag{2-9}$$

Fig. E,2f shows a comparison between the test results from [*1,56*] and the theoretical values given by Eq. 2-9. The more recent values in [*56*] are the ones recommended. The $M = 1.0$ limit is usually a cascade tunnel one and higher values than 1.0 could be obtained in an actual compressor under certain conditions.

Reynolds number effects. Most cascade tests have been carried out in the ranges of direct interest. However, for operation under altitude conditions or for scaling components up and down in size, it is important to know the Reynolds number effects on the results that have already been discussed. Also, from a fundamental research viewpoint it is desirable to separate Reynolds and Mach number effects. Unfortunately there is little published cascade data on this subject, so that few comparisons can be made. Results from actual compressor and turbine tests are reported in [*6,60,61,62,63,64*], and from turbine cascades in [*6,28,51*].

In Fig. E,2g some of the available results are plotted. The losses or $(1 - \eta)$ are given as relative to their values at a Reynolds number *Re* of 10^5, the *Re* for compressor and turbine blades being based on the inlet and exit flow velocities respectively. The turbine cascade losses deduced from Ainley [*6*], Sawyer [*28*], and Todd [*51*] are such that they vary between $Re^{-0.5}$ to $Re^{-1.0}$ over the *Re* range of 0.3×10^5 to 2×10^5 with

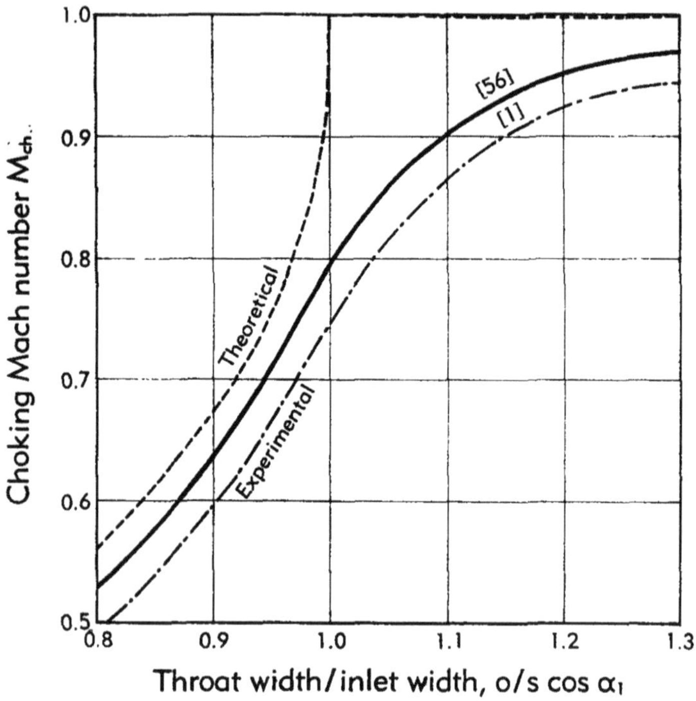

Fig. E,2f. Choking Mach number and throat width for compressor cascades.

low turbulence in the cascade tunnels. Sawyer [*28*] also shows that increasing the degree of turbulence reduces the Reynolds number at which the losses become large. It was not possible from these results to deduce definite rules or to discriminate effectively between the reaction- and impulse-type turbine blades. Compressor cascades could be expected to give somewhat comparable variations of losses with Reynolds number.

Actual compressors and turbines usually have much smaller variations in efficiencies or losses with *Re* than have cascades, due to the greater turbulence of the flows and to the secondary losses probably not varying much with *Re*. A curve for turbines deduced from Moyes and Pennington [*62*], and a mean curve for compressors from d'Epinay [*61*] and Moyes

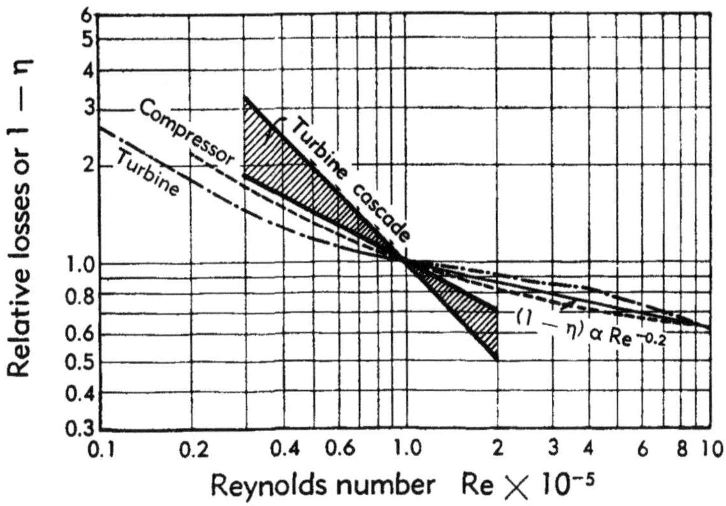

Fig. E,2g. Variation of losses with Reynolds number.

and Pennington [62] are also given in Fig. E,2g. In the *Re* range of 0.6×10^5 to 10^6 the $1 - \eta$ is very roughly proportional to $Re^{-0.2}$.

E,3. Secondary Effects.

Description. In use, a cascade of blades has a finite length or height and is bounded by end walls on which there is normally an appreciable boundary layer. In addition there may also be a tip clearance at one end. The presence of this boundary layer and tip clearance gives rise to secondary effects which are reasonably well known from a qualitative point of view. Theoretically, certain aspects are understood, but considerably more experimental data are required before quantitative values can be accurately established for these effects. The two-dimensional results already discussed are applied to the design of the individual airfoil sections which make up the blades in compressors and turbines, but the over-all performances of the machines have to be obtained, at the present time, by the use of empirical correction factors which take into account these secondary effects. In this article, only the secondary effects in simple cascades are considered, and reference should be made to Sec. C, F, and H for the more complex three-dimensional flows in actual compressors and turbines.

During cascade tests with low aspect ratio blades and/or thick boundary layers on the end walls, it has been observed that the exit flow angles obtained were different than those for high aspect ratio blades with thin boundary layers. Normally the exit flow angle α_2 at mid-blade height was larger, so that the deviation $\delta = \alpha_2 - \alpha_2'$ was higher than expected for two-dimensional flow. On the basis of wing theory, Carter and Cohen [65],

Hausmann [66], and Carter [67] investigated the induced velocities due to the trailing vortices and obtained values for the induced angles necessary to correct the observed test readings to values corresponding to true infinite or two-dimensional flow conditions. The methods are somewhat similar to those employed in the case of a single airfoil projecting into a wind tunnel wall boundary layer (see Preston [68]). In addition, in [65,67] an expression is given for the induced drag. So far, this method of attack, from wing theory, has given no quantitative indication of the actual secondary flow in the blade passage or duct.

Another approach which will be referred to as the "duct theory" is given by Squire and Winter [69] and as a more generalized result by Hawthorne [70]. In [69] the secondary flow is regarded as a perturbation on an essentially two-dimensional motion. In the simplest case of the impulse cascade the secondary vorticity is deduced to be quantitatively equal to twice the turning angle τ multiplied by the incoming vorticity. Excellent correlation is obtained between the theoretical and experimental secondary or induced velocities in the plane of the trailing edge of the cascade. The method does not directly give the mean exit flow angles some distance downstream of the cascade, as usually observed during cascade testing, although a form of local flow angle near the trailing edge of the cascade can be obtained. A drag coefficient associated with the secondary flows can be deduced, but it is not known how much of this drag occurs as loss when applied to an actual compressor or turbine; it is not loss that would be measured by a Pitot tube in normal experimental work. More detailed theoretical aspects of vorticity and circulation are given in [71,72,73,74].

The nature of secondary flow is illustrated in Fig. E,3a. Because of the turning of the stream, there is a pressure gradient across the blade passage to balance the centrifugal forces and there is a relatively high pressure on the lower surface and a low pressure on the upper surface of the blade. Near the walls the velocities are smaller, so that the required pressure gradient is reduced, which gives on the lower blade surface a smaller pressure near the wall than in the middle of the blade with a resulting secondary flow toward the wall on this surface. On the upper surface the corresponding flow is away from the wall. This motion is superimposed on the main flow and results in a helical motion in the passage. Test and theoretical values of the secondary or induced velocities in the passage have been given for an impulse cascade by Squire and Winter [69].

Immediately downstream of the trailing edge of any blade, there is a surface of discontinuity of velocity, equivalent to a vortex sheet. This vortex sheet is unstable and rolls up into two trailing vortices. Traversing with a Pitot tube a short distance downstream of the cascade reveals the two trailing vortices as cores of high total head loss. Such traverses showing the cores of high total head loss and the other losses associated with secondary flows have been given by Howell [*56*], Carter and Cohen [*65*], Carter [*67*], Sawyer [*28*], and Todd [*75*]. A comparison between the reaction turbine cascade result of Sawyer [*28*] and the compressor cascade results of [*56,65,67,75*] shows how much smaller are the secondary losses for reaction turbine blades. The displacements of the boundary layers have been discussed theoretically by Weske [*76*], while Scholz [*77*] reports some measured secondary flow losses at the ends of a reaction turbine cascade. Visualization studies of secondary flows made by the National Advisory Committee for Aeronautics are given in [*78,79*].

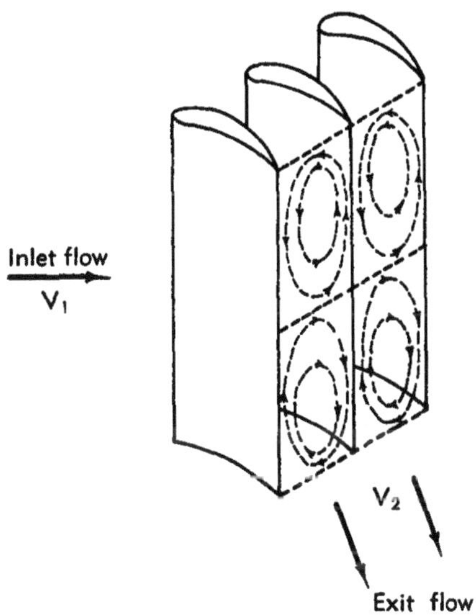

Fig. E,3a. Nature of secondary flow in a cascade.

Wing theory—Induced angles. In the simplest wing theory the vortex sheet is assumed to roll up into two trailing vortices, distance b' apart, where b' is of course somewhat less than the blade height b. By considering the appropriate image system, Carter and Cohen [*65*] and Carter [*67*] have deduced the following simple expression for the induced incidence or downwash angle ϵ to be applied to the vector mean angle α_∞ at mid-blade height.

$$\epsilon = -\Delta\alpha_\infty = \frac{1}{4}\frac{C_L}{s/c}\left(1 - \frac{b'}{b}\right) \tag{3-1}$$

In Eq. 3-1, $\Delta\alpha_\infty$ is the change in α_∞, C_L is the lift coefficient (Art. 2), s/c the spacing-to-chord ratio, and in addition s/b is assumed to be small. The value of b'/b varies with the boundary layer thickness, with the aspect ratio, and with compressor or turbine cascades, the normal variation being between 0.85 and 0.95. For example, with a value of 0.88 for b'/b, ϵ is equal to $0.03C_L/(s/c)$ radians or $1.7C_L/(s/c)$ degrees. If α_∞ is the actual test vector mean angle, then the effective or two-dimensional angle is $\alpha_\infty - \epsilon$.

Test results are corrected to two-dimensional flow conditions by taking note of the change ϵ or $\Delta\alpha_\infty$ in α_∞ for the given C_L and then working out the corresponding changes in α_1 the inlet flow angle and in α_2 the exit flow angle, by the use of the cascade relationships of Art. 2. When

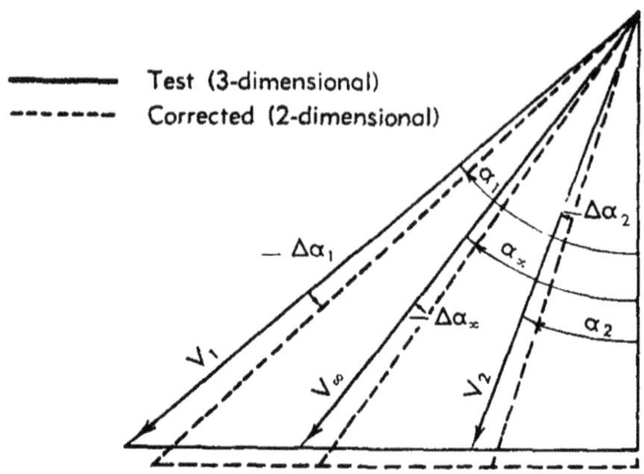

Fig. E,3b. Test and corrected angles for a compressor cascade.

ϵ or $\Delta\alpha_\infty$ is small, as they usually are, then the changes in α_1 and α_2 are given by

$$\frac{\Delta\alpha_1}{\Delta\alpha_\infty} = \left(\frac{1 + \tan\alpha_1 \tan\alpha_\infty}{1 + \tan^2\alpha_1}\right) \tag{3-2}$$

$$\frac{\Delta\alpha_2}{\Delta\alpha_\infty} = \left(\frac{1 + \tan\alpha_2 \tan\alpha_\infty}{1 + \tan^2\alpha_2}\right) \tag{3-3}$$

These changes are illustrated in Fig. E,3b, for an arbitrary compressor cascade.

To show the magnitude of the induced angles, consider the two compressor cascades used in [*65*]. The camber angle θ, the inlet blade angle α_1' and the outlet blade angle α_2' were 30°, 37°, and 7° respectively. The aspect ratio and the value of b'/b were 2.0 and 0.88 respectively. Table E,3a shows the results obtained for two spacing-to-chord ratios, and gives finally the corrected or two-dimensional flow angles. The test or three-dimensional deviation angles δ at mid-blade height are 1.8° and 1.5° greater than the corrected or two-dimensional deviations. Smaller boundary layers on the cascade tunnel walls can reduce these differences by making b'/b nearer to the desirable value of unity.

Table E,3a.

Spacing/chord (s/c)	0.5	1.0
Test α_1	38.0°	37.5°
Test α_∞	26.8°	28.2°
Test α_2	13.0°	17.0°
Test deviation δ	6.0°	10.0°
C_L	0.49	0.81
$\Delta\alpha_1$	1.5°	1.3°
$\Delta\alpha_\infty$	1.7°	1.4°
$\Delta\alpha_2$	1.8°	1.5°
Corrected α_1	36.5°	36.2°
Corrected α_∞	25.1°	26.8°
Corrected α_2	11.2°	15.5°
Corrected deviation δ	4.2°	8.5°

The wing theory method of Hausmann [*66*] is more complex and it is difficult to make a direct comparison with the above simple theory. Compared with the previous theory, Hausmann's analysis shows more serious induced effects with increasing boundary layer thickness and with reduced aspect ratios. There is no great change in induced angles for between 4 and 16 blades in the cascade tunnel, and aspect ratios of 2 to 4 give very small changes for the example given in [*66*].

Wing theory—Induced drag. For secondary flows, the kinetic energy of the secondary or induced velocities can be expressed in terms of drag coefficients, but it is not known how much of this "drag" really occurs as loss in a compressor or turbine or even in a cascade tunnel. With wing theory, on a similar basis to the induced drag of finite isolated wings, an induced drag coefficient C_{D_i} can be defined equal to ϵC_L. In the calculations for ϵ the effective blade length is taken as b', but conventional drag coefficients are based on the true blade length b, so that C_{D_i} becomes $\epsilon C_L(b'/b)$, which when combined with Eq. 3-1 becomes

$$C_{D_i} = \frac{1}{4}\frac{C_L^2}{s/c}\left(1 - \frac{b'}{b}\right)\left(\frac{b'}{b}\right) \tag{3-4}$$

With the value of b'/b of 0.88 for the example of Table E,3a, $C_{D_i} = 0.0264C_L^2/(s/c)$. For the example quoted, the values of C_{D_i} are 0.013 and 0.017 for the spacing-to-chord ratios s/c of 0.5 and 1.0 respectively.

Duct theory. The aim in duct theory is to ascertain the vorticity pattern in a plane perpendicular to the main stream at the exit of a duct or a cascade for a given, but nonuniform, upstream main velocity distribution, and then to obtain the secondary velocities from the vorticity pattern. Squire and Winter [*69*] deduced the following expression for the downstream vorticity ξ:

$$\xi = -2\tau \frac{dV}{dz} \tag{3-5}$$

where τ is the turning angle, V is the upstream main velocity at a distance z along the span or blade height, and dV/dz is the incoming vorticity. Hawthorne [70], Preston [71], and others have also deduced Eq. 3-5. A more generalized result is given by Hawthorne [70].

The actual works [69,70,71] should be consulted for rigorous deductions of Eq. 3-5 and for the more generalized result of [70]. Below is given a very simple, but less rigorous, solution which gives the same results. In a cascade or duct, consider an element of a curved stream tube of circular cross section, the radius a of which is very small compared with the local radius of curvature r of the tube. Let the main stream velocity at the center o of the tube be V_o and its variation with z, which is perpendicular to the turning plane, be given by $V = V_o + \lambda z$ as shown in Fig. E,3c.

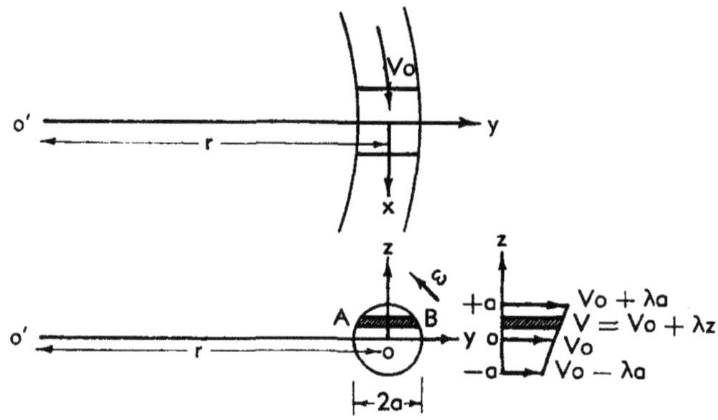

Fig. E,3c. Simple duct theory for secondary flows.

In this simple solution the forces tending to rotate the stream tube element are considered, and it is assumed that they are the same as when the rotation is of the "solid body" type. If ω is the resulting angular velocity, the secondary vorticity ξ, which we need to know, is equal to 2ω. The density and depth (along the x axis) of the element are both taken as unity. The strip AB of fluid of width dz and length l will experience an outward force dF given by

$$dF = ldz\frac{V^2}{r} = ldz\frac{(V_o + \lambda z)^2}{r}$$

$$= \frac{ldz}{r}V_o(V_o + 2\lambda z)$$

if λz is small compared with V_o.

The couple about the center of gravity o of the forces dF is expressed as

$$\int_{-a}^{+a} zdF = \frac{2V_o\lambda}{r}\int_{-a}^{+a} z^2 ldz = \frac{2V_o\lambda}{r} I_{yy}$$

where I_{yy} is the moment of inertia of the section about the y axis and is equal to one half of I_{xx}, the moment of inertia about the x axis. The previous couple is in the opposite direction to the angular velocity ω in the figure and so

$$I_{xx}\frac{d\omega}{dt} = -\frac{2V_o\lambda}{r}I_{yy}$$

giving

$$\frac{d\omega}{dt} = -\frac{V_o\lambda}{r}$$

where dt is a time increment. As $\xi = 2\omega$ and $\lambda = dV/dz$, then

$$\frac{d\xi}{dt} = -\frac{2V_o\lambda}{r} = -2\frac{V_o}{r}\frac{dV}{dz}$$

If τ is the turning angle of the stream tube, then $V_o/r = d\tau/dt$ giving

$$\frac{d\xi}{dt} = -2\frac{dV}{dz}\frac{d\tau}{dt}$$

or

$$d\xi = -2\frac{dV}{dz}d\tau \tag{3-6}$$

so that when dV/dz is a constant in the stream tube considered,

$$\xi = -2\tau\frac{dV}{dz} \tag{3-7}$$

This is the result given in Eq. 3-5 and obtained in [*69,70,71*]. The more general solution of Hawthorne [*70*] can be obtained in a somewhat similar manner.

Eq. 3-5 and 3-7 strictly apply only to the turning of a constant area passage, such as with a typical impulse cascade; Eq. 3-6 should be integrated for other applications. However, after allowances have been made for blade thickness effects, the final integrated results for compressor and turbine cascades in general are not very different from those given by Eq. 3-5 and 3-7. Another effect which has to be taken account of in more accurate estimates is the fact that the main velocity profile deteriorates in passing through a compressor cascade while it improves with a reaction turbine cascade. Again, stalling at the blade ends affects the secondary flows.

Having obtained the downstream vorticity ξ, let v and w be the secondary velocities in the y and z direction respectively of Fig. E,3c. Then

$$\xi = \frac{\partial w}{\partial y} - \frac{\partial v}{\partial z} = -2\tau\frac{dV}{dz}$$

and continuity gives

$$\frac{\partial v}{\partial y} + \frac{\partial w}{\partial z} = 0$$

If ψ is the stream function of the secondary motion, $v = \partial\psi/\partial z$ and

$w = -\partial\psi/\partial y$, and

$$\frac{\partial^2\psi}{\partial y^2} + \frac{\partial^2\psi}{\partial z^2} = -\xi = 2\tau \frac{dV}{dz} \tag{3-8}$$

which is Poisson's equation for the secondary vorticity. The solution of Eq. 3-8 is difficult and involved, although it can be solved by a series

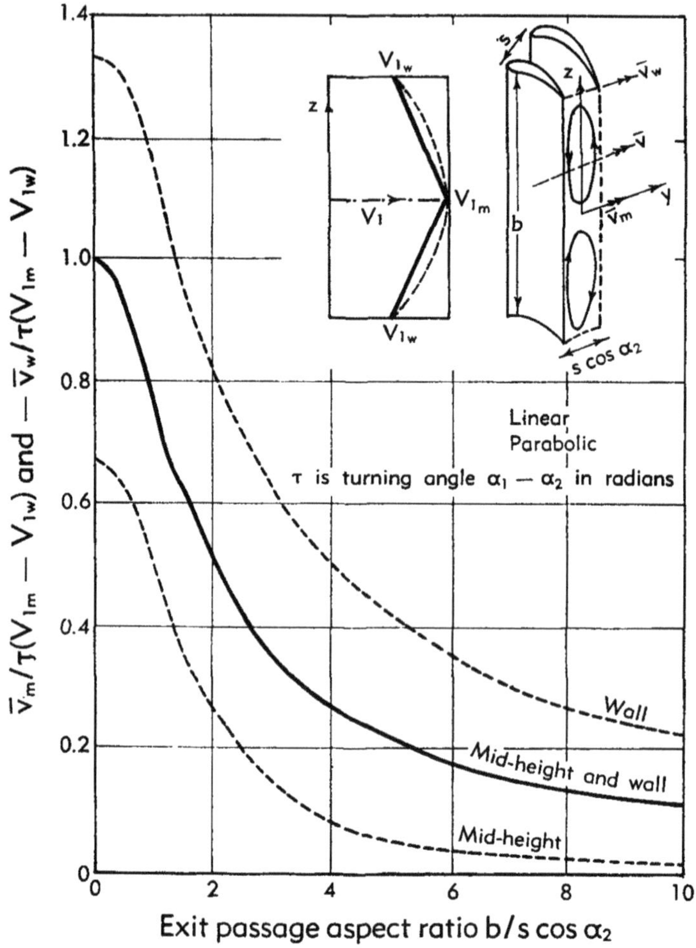

Fig. E,3d. Secondary velocities at mid-height and wall.

method [*69*], by relaxation methods [*73*], and by electrical analogues [*80*]. In the following subarticle, the application of an approximate and limited solution by the author is given.

Application of duct theory. From any secondary flow theory, a blade designer wishes finally to estimate the three-dimensional exit flow angles α_2. A form of local flow angle near the trailing edge of a cascade can be found by taking the secondary flow angle component $\Delta\alpha_2 = \tan^{-1}(\bar{v}/V_2)$, where $\bar{v}$ is the mean local secondary velocity perpendicular to the span or blade height and V_2 is the two-dimensional exit velocity. The three-dimensional exit flow angle is equal to the two-dimensional exit flow angle plus $\Delta\alpha_2$. $\Delta\alpha_2$ is also equal to $-\Delta\tau$, the change in the turning angle τ. To indicate the likely effects, the author has deduced some approximate expressions for the mean secondary velocities $\bar{v}_m$ and $\bar{v}_w$ at the mid-height and wall positions respectively. In the expressions, V_{1_m} and V_{1_w} are the values of the inlet or upstream velocity V_1 at mid-height and wall respectively, τ is the turning angle in radians and is equal to $\alpha_1 - \alpha_2$, b is the blade height, s is the blade spacing, and α_2 is the exit flow angle.

For a linear velocity distribution upstream, Eq. 3-9 applies approximately

$$\frac{\bar{v}_m}{\tau(V_{1_m} - V_{1_w})} = \frac{-\bar{v}_w}{\tau(V_{1_m} - V_{1_w})}$$

$$= \left(\frac{1.045s\cos\alpha_2}{b}\right)\tanh\left(\frac{b}{1.045s\cos\alpha_2}\right) \quad (3\text{-}9)$$

For a parabolic velocity distribution upstream, Eq. 3-10 and 3-11 apply approximately:

$$\frac{\bar{v}_m}{\tau(V_{1_m} - V_{1_w})} = \frac{4s^2\cos^2\alpha_2}{b^2}\left[1 - \frac{(\sqrt{3}b/s\cos\alpha_2)}{\sinh(\sqrt{3}b/s\cos\alpha_2)}\right] \quad (3\text{-}10)$$

$$\frac{\bar{v}_w}{\tau(V_{1_m} - V_{1_w})} = \frac{4s^2\cos^2\alpha_2}{b^2}\left[1 - \frac{(\sqrt{3}b/s\cos\alpha_2)}{\tanh(\sqrt{3}b/s\cos\alpha_2)}\right] \quad (3\text{-}11)$$

These expressions for the secondary velocities $\bar{v}_m$ and $\bar{v}_w$ are plotted in Fig. E,3d, for various values of the exit passage aspect ratio $b/s \sin\alpha$.

Eq. 3-9, 3-10, and 3-11, although only approximate in general, are accurate for very small aspect ratios and are reasonably accurate for very large aspect ratios. In addition, Eq. 3-9 for the linear velocity distribution gives the correct answer for a passage aspect ratio $b/s \sin\alpha$ of 2.0.

Tables E,3b and E,3c give some results from the above expressions for an impulse cascade with $\alpha_1 = 45°$, $\alpha_2 = -45°$, τ (2-dim.) $= 90° = \pi/2$ radians, $s/c = 0.5$, $V_{1_w}/V_{1_m} = 0.5$, and for aspect ratios (b/c) of 2, 3, and 4.

Table E,3b. Linear velocity distribution upstream.

Aspect ratio (b/c)	2	3	4
$b/s\cos\alpha_2$	5.66	8.48	11.31
$\bar{v}_m/\tau(V_{1_m} - V_{1_w})$	+0.185	+0.123	+0.092
$\bar{v}_w/\tau(V_{1_m} - V_{1_w})$	−0.185	−0.123	−0.092
$\Delta\tau = -\Delta\alpha_2$ (mid-height)	−8.2°	−5.5°	−4.1°
$\Delta\tau = -\Delta\alpha_2$ (wall)	+16.2°	+10.9°	+8.2°

Table E,3c. **Parabolic velocity distribution upstream.**

Aspect ratio (b/c)	2	3	4
$b/s \cos \alpha_2$	5.66	8.48	11.31
$\bar{v}_m/\tau(V_{1_m} - V_{1_w})$	+0.0416	+0.0185	+0.0104
$\bar{v}_w/\tau(V_{1_m} - V_{1_w})$	−0.366	−0.254	−0.193
$\Delta\tau = -\Delta\alpha_2$ (mid-height)	−1.9°	−0.8°	−0.5°
$\Delta\tau = -\Delta\alpha_2$ (wall)	+29.9°	+21.8°	+16.9°

For the aspect ratio $b/c = 3.0$ with the previous impulse cascade, approximate values of $-\Delta\alpha_2$ and $\Delta\tau$ along the span or blade height have

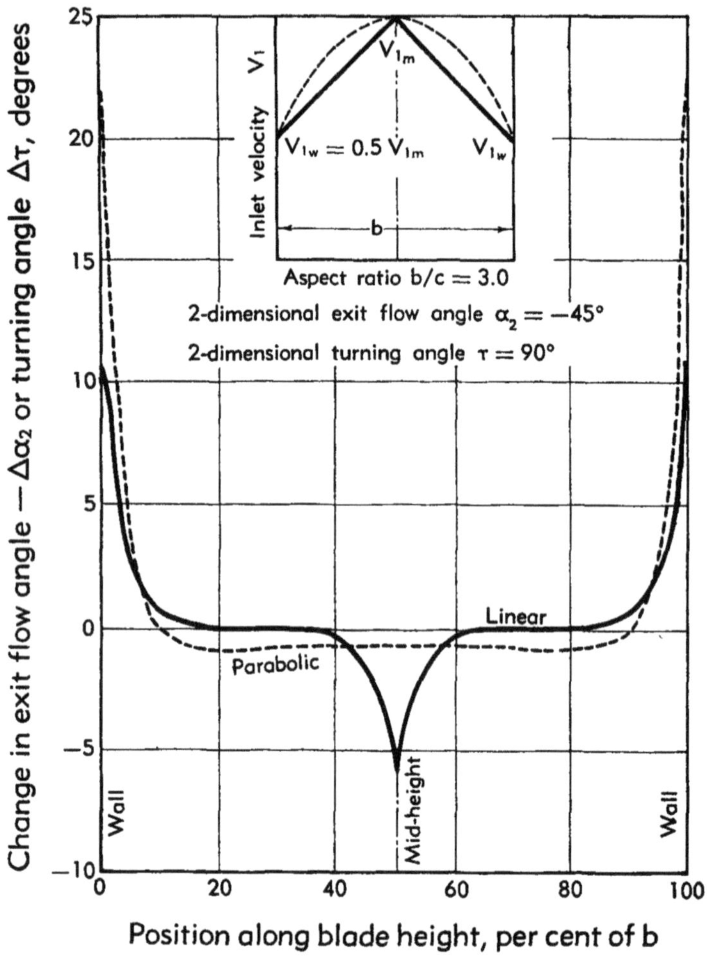

Fig. E,3e. **Secondary exit flow angles along blade height.**

been deduced and are plotted in Fig. E,3e. It will be noticed that the parabolic distribution involves much smaller angle changes near the mid-height position, but that at the wall the parabolic distribution gives about twice the angle change for the linear velocity distribution.

E,4. Heat Transfer to Blades in Cascade.

General considerations. Information on heat transfer to blades in cascade is necessary for estimating the amount of cooling required for the blade rows of cooled turbines in general and also the quantity of heat needed to de-ice the inlet guide vanes and the first few blade rows of compressors in aircraft engines. Much of the earlier German turbine-cooling investigation was aimed at the use of normal gas temperatures, but with inferior blade materials. More recently the general trend has been toward the use of higher temperature turbines with the material of the blades at more normal temperatures. A considerable amount of data now exist on the subject, as shown by [*36,81–105*]. In particular [*36,90, 105*] give good general surveys of the problems of interest to readers of this section and include British, American, and German information. [*81,82,84*] are useful standard references to heat transfer in general, while [*36,91,92*] give results on compressor blades. The application of turbine cooling is considered in more detail in Vol. XI.

All cooling or heating of blades requires a knowledge of the rates of heat transfer between a gas stream and the blades, though the conditions are more involved with effusion or transpiration cooling. These rates of external heat transfer are inescapable and tend to dominate the resulting characteristics of any cooling system. Heat transfer coefficients h are defined as the heat quantity transferred per second per unit surface area per degree temperature difference between the gas and the surface. It is convenient to express them nondimensionally in the form of a Nusselt number $Nu = hc/k$, where c is the blade chord and k is the thermal conductivity of the gas stream. It will be seen later that Nu is primarily a function of Reynolds number (Re) and cascade geometry. The internal cooling and the taking of heat away from a turbine blade can be done by air or liquid or a combination of both. In this article, consideration is given only to simple internal air-cooling with the passages arranged along the spanwise direction. Generally, the methods of Smith [*36*] and Ainley [*105*] are followed.

External heat transfer coefficients. The external heat transfer coefficients h_g from gas to blade and their Nusselt numbers Nu vary around the blade profile. High local rates of heat transfer occur at the leading edge where the boundary layer is thin and laminar. The values decrease as the boundary layer builds up around the surface and then increase sharply when the boundary layer becomes turbulent [*36,105*]. However, average values are usually taken which are measured experimentally or are deduced from theory. The mean Nusselt number Nu varies with the Reynolds number Re, approximately in proportion to Re^{χ} where χ varies between 0.5 (for a wholly laminar flow) and 0.8 (for a wholly turbulent flow). Ainley [*105*] has deduced the following simple expression for

$Nu = h_g c/k$:

$$Nu = Nu^* \left(\frac{Re}{2 \times 10^5}\right)^{\chi} \left(\frac{T_g}{T_b}\right)^{0.14} \tag{4-1}$$

and has given curves for Nu^* and χ for turbine blades.

In Eq. 4-1, Nu^* is the mean Nusselt number at $Re = 2 \times 10^5$ and T_g and T_b are the gas and mean blade absolute temperatures. Actually T_g is an "effective" gas temperature, the temperature that an uncooled body would attain in the same gas stream, there being no conduction or

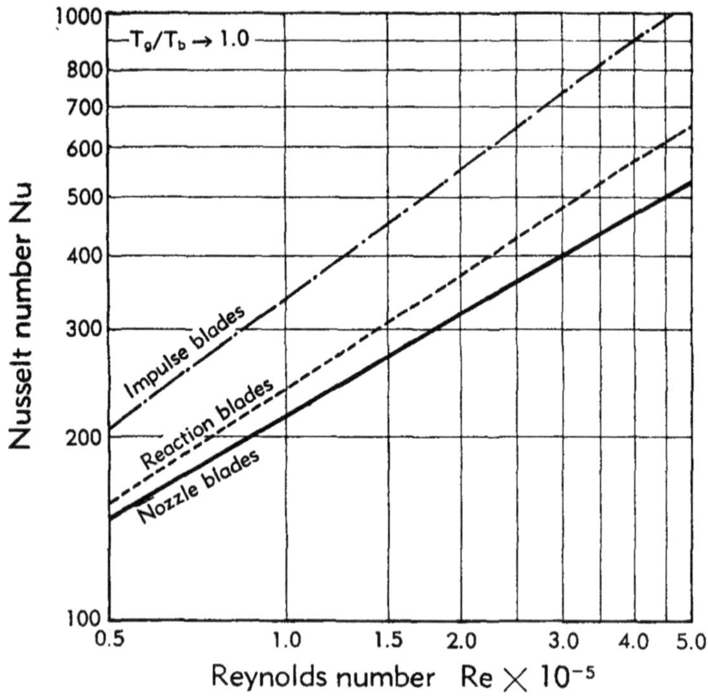

Fig. E,4a. Variation of Nusselt number with Reynolds number for turbine cascades.

radiation of heat to or from the body. T_g is approximately equal to the total gas temperature minus 0.15 of the mean kinetic gas temperature around the blade. The viscosity μ and the gas thermal conductivity k are conveniently taken at effective gas temperature conditions. In this article, Re is always based on the exit flow velocity V_2. Some values of Nu, for turbine blades, taken from the mean curves of [*105*] are given in Table E,4a, and the Nusselt numbers are plotted against Re in Fig. E,4a, all for $T_g/T_b \to 1.0$.

Table E,4a.

Blade	α_1	Nu^*	χ	Nu
Nozzle	0°	320	0.55	$0.38Re^{0.55}$
Reaction	$-\alpha_2/2$	370	0.625	$0.18Re^{0.625}$
Impulse	$-\alpha_2$	550	0.7	$0.106Re^{0.7}$

Values for T_g/T_b other than 1.0 can be obtained by multiplying the previous cascade values of Nu by the factor $(T_g/T_b)^{0.14}$ in Eq. 4-1. This correction is not very large, being a factor of 1.058 for $T_g/T_b = 1.5$, the latter being rarely exceeded. Actual results from turbine tests, as distinct from cascade tests, suggest somewhat higher values of Nu^* and χ due to greater turbulence and unsteady flows in the turbine [*105*].

In applying Nusselt numbers it is often easier to consider $\Delta H/H$ where ΔH is the heat flow to the blade and H is the maximum possible heat transfer equal to $w_g c_p(T_g - T_b)$, where w_g is the gas flow through the passage between two blades. With Re based on the exit velocity V_2, [*36*] gives $\Delta H/H$ as

$$\frac{\Delta H}{H} = \left(\frac{Nu}{Re\,Pr}\right)\left(\frac{S_g}{s\cos\alpha_2}\right) \tag{4-2}$$

$$\Delta H = h_g S_g b(T_g - T_b) \tag{4-3}$$

$$H = w_g c_p(T_g - T_b) \tag{4-4}$$

where Pr is the Prandtl number $\mu c_p/k$, S_g is the blade perimeter, b is the blade height, and c_p is the gas specific heat at constant pressure. Pr is nearly constant at 0.71 for air and S_g is approximately equal to 2.3c for turbine blades, and so

$$\frac{\Delta H}{H} = 3.24\left(\frac{Nu}{Re}\right)\left(\frac{c}{s\cos\alpha_2}\right) \tag{4-5}$$

Values of $\Delta H/H$ for three typical turbine blades in cascade are shown in Table E,4b and Fig. E,4b.

Table E,4b.

Blade	s/c	α_1	α_2	$\Delta H/H$	$\Delta H/H$ at $Re = 2 \times 10^5$
Nozzle	0.75	0°	−60°	$3.3Re^{-0.45}$	0.014
Reaction	0.625	25°	−50°	$1.44Re^{-0.375}$	0.015
Impulse	0.5	45°	−45°	$0.98Re^{-0.3}$	0.025

An added interest in $\Delta H/H$ is that by Reynolds' analogy [*36*] it is roughly one half the blade loss expressed in terms of $\bar{\omega}/\frac{1}{2}\rho V_2^2$ as in Art. 2. Such estimates are considered in [*36*].

No general correlation seems to have been made for compressor blades, and it is suggested that Eq. 4-1 should be used without the $(T_g/T_b)^{0.14}$ term and with χ equal to 0.8. Very rough values of Nu^* would be about 400 for inlet guide vanes and about 600 for rotor and stator blades, the limited information available indicating large variations in Nu^*. The Re is based here on exit velocity V_2.

Heat transfer in cooling passages. In the simple internal air-cooling of turbine blades by passages arranged along the spanwise direction, it is desirable for generally efficient operation to limit the quantity of cooling air per row to values of 1 per cent to 3 per cent of the hot gas flow. In these spanwise passages with the cooling air flowing from blade root to tip, it is necessary to devise cooling passage shapes such that the internal cooled surface area is large. At the same time the cooling air must scrub the internal surfaces at a high velocity, which implies that the cross-sectional flow area of the passages must be small. Ainley [*105*] has produced a useful figure of merit for comparing the relative efficiency of

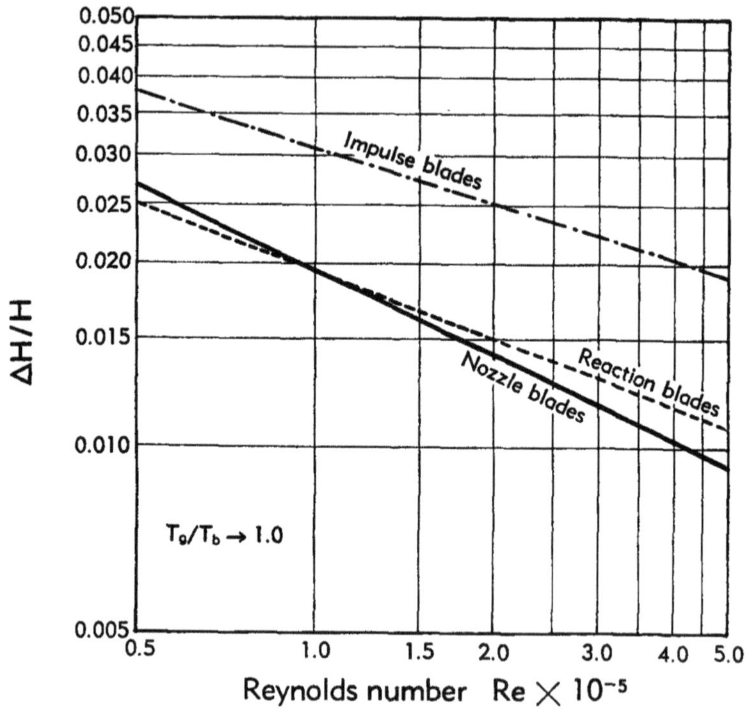

Fig. E,4b. Variation of $\Delta H/H$ with Reynolds number for three turbine cascades

various cooling passage configurations with a blade of given external shape, which he termed the "Z factor," defined as

$$Z \text{ factor} = \frac{(S_o/c)^{1.2}}{(A_o/c^2)} \tag{4-6}$$

where S_o is the total wetted periphery of cooling passages in one blade at any spanwise position, A_o is the total cross-sectional area of the cooling passages in one blade, and c is the blade chord. For given external conditions the Z factor determines the relative blade temperatures, and if, in addition, the hydraulic diameters of the cooling passages being compared are equal, then the pressure losses in the passages also remain the same.

In comparing the cooling characteristics of different turbine cascades,

Smith and Pearson [*90*], Ainley [*100,105*], and others use the concept of a relative blade temperature equal to $(T_b - T_c)/(T_g - T_c)$, where T_g, T_b, and T_c are the absolute temperatures of the gas, the blade, and the cooling air respectively. T_b is usually the mean chordwise value of the blade temperature at the given spanwise position. As with the external heat transfer coefficients, it is convenient to think first in terms of the isothermal conditions where $T_g/T_c \rightarrow 1.0$, the smallish effect of T_g/T_c being regarded as a correction, though it is not proposed here to deduce the necessary corrections. With $T_g/T_c \rightarrow 1.0$, it is possible to find easily a simple relationship for the relative blade temperature by equating the external heat transfer to a section of the blade to that for the corresponding section of the cooling passages, neglecting metal conduction in the spanwise direction and radiation effects. The other assumptions made are that the gas and cooling air viscosities are the same, which is true for $T_g/T_c \rightarrow 1.0$; that the Prandtl number is 0.71; that the cooling passage Nusselt number (Nu_c) [*36,82*] is given by Eq. 4-7 for turbulent flow, which is the more normal condition of the cooling air; and that S_g is approximately equal to 2.3*c*.

$$Nu_c = 0.020(Re_c)^{0.8} \tag{4-7}$$

The Re_c is based on the hydraulic diameter which is equal to four times the hydraulic mean depth, that is, equal to $(4A_c/S_c)$. The expressions obtained for the local relative blade temperatures at some given spanwise position are as follows:

$$\frac{T_b - T_c}{T_g - T_c} = \frac{1}{1 + X} \tag{4-8}$$

and

$$X = \frac{T_g - T_b}{T_b - T_c}$$

$$= 0.0066\left(\frac{Z}{Nu_g}\right)\left(Re_g \frac{w_c}{w_g}\frac{b}{c}\frac{s}{c}\cos\alpha_2\right)^{0.8} \tag{4-9}$$

It will be noticed that the local relative blade temperatures, for given external conditions Nu_g, Re_g, b/c, s/c, and $\cos\alpha_2$ are only dependent on the *Z* factor of Eq. 4-6 and on the ratio of the cooling air quantity w_c to the gas flow quantity w_g. Eq. 4-9 applies at a given spanwise position and actually the cooling air temperature T_c increases from its value T_{c_r} at the root as the air moves up the cooling passage, due to the heat transfer involved. Taking T_{c_r} to be the original cooling air temperature it is again possible to deduce simple expressions for the relative blade temperature $(T_b - T_{c_r})/(T_g - T_{c_r})$, as given by Eq. 4-10 and 4-11.

$$\frac{T_b - T_{c_r}}{T_g - T_{c_r}} = 1 - \frac{X}{1 + X}e^{-Y} \tag{4-10}$$

and

$$Y = \frac{Nu_g}{Re_g} \frac{w_g}{w_c} \frac{c}{s} \frac{3.24}{\cos \alpha_2} \frac{X}{1 + X} \frac{y}{b} \tag{4-11}$$

where y is the spanwise position measured from the root, and b is the blade height. An example of this calculation is shown in Fig. E,4c, for $Z = 200$, $Re_g = 2 \times 10^5$, $Nu_g = 320$, $s/c = 0.75$, $b/c = 2.0$, $\alpha_1 = 0°$, $\alpha_2 = -60°$, and $T_g/T_{c_r} \rightarrow 1.0$. The results are plotted for three cooling

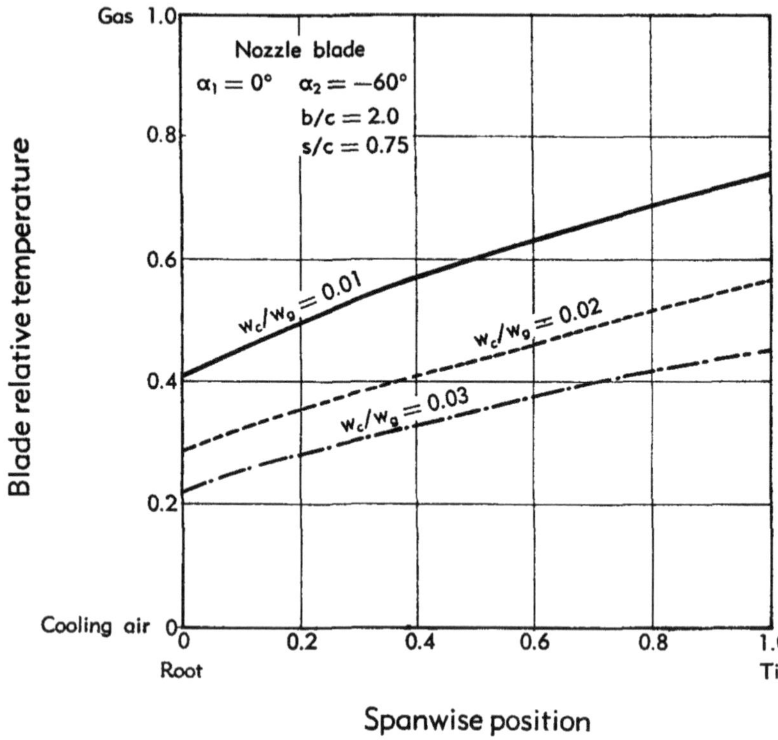

Fig. E,4c. Spanwise variations of blade temperatures.

air flow-to-gas flow ratios (w_c/w_g) of 0.01, 0.02, and 0.03. The results for T_g/T_c greater than 1.0 would give somewhat higher relative blade temperatures.

To illustrate how the Z factor controls the number of cooling passages for different hole shapes, Fig. E,4d has been prepared. For a Z factor of 200 and a hydraulic diameter of $0.0266c$, one can have 50 circular holes of diameter $0.0266c$ or 13 rectangular holes of length $0.146c$ and width $0.0146c$. Both cooling passage configurations would be expected to give similar cooling and pressure loss characteristics.

More generally, measurements of average heat transfer and friction coefficients for subsonic flow or air in smooth tubes at high surface fluid temperatures have been explored comprehensively by the NACA [*86*]. Internal cooling passages require thicker trailing edges than normal and some results on an experimental investigation of a thick airfoil nozzle cascade are given in [*89*].

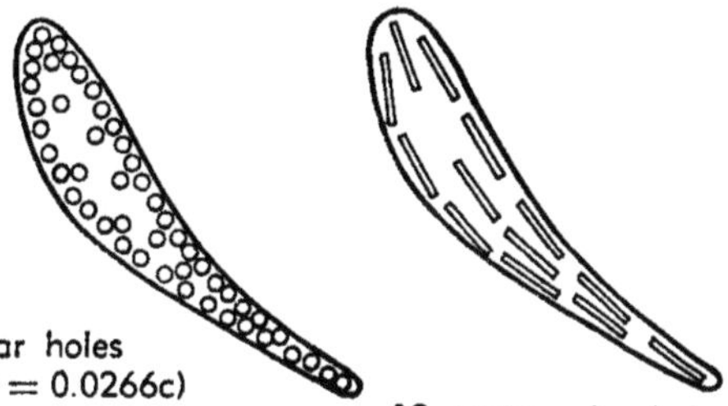

Fig. E,4d. Two cooling passage configurations giving similar cooling and pressure loss characteristics.

E,5. Generalized Cascade Data.

Methods of correlation and prediction. Cascades have more variation in them than do isolated airfoils; for example, their blade cambers θ and blade shapes vary much more and they have the additional variables of stagger angle γ and spacing-to-chord ratio s/c. It is not surprising then that attempts are made to correlate and predict cascade performance in a relatively simple way. A blade designer wishes to know what turning angles or lift coefficients he is likely to obtain from given cascades over a design incidence ι range of about $-5°$ to $+5°$, with efficient operation. Efficient operation is defined by the use of "nominal" or maximum lift/drag conditions already mentioned in Art. 2. One of the earliest attempts at correlation over a wide range of conditions for compressor blades was made by Howell [*1*] in 1942; more modern versions of the curves are given in [*56,106*]. Excluding the earlier steam turbine investigations, Zweifel [*42*] in 1945 deduced rules for the spacing of turbine blades based on the fact that the optimum spacing was specified by a loading coefficient or factor equivalent to the lift coefficient based on the exit velocity V_2. Independently, Howell and Carter [*43*] in 1946 used a theoretical approach similar to Zweifel's for compressor and turbine cascades, and used it again in [*52,107*].

All the previously mentioned, generalized cascade data are in the nature of recommended design deflections and lift coefficients and are not of universal application. In [*37*] Carter gives probably the most universal attempt to generalize low speed cascade data and covers a very wide range of compressor and turbine conditions. Again, Ainley and Mathieson [*108*] consider turbine cascade correlation in their examination of the flow and pressure losses in blade rows of axial flow turbines. More advanced methods of prediction of lift coefficient and losses are those dependent on actual theoretically determined pressure distributions around the airfoils such as in [*19,29,35,39*]. Davis [*53*] suggests a method of correlating axial flow compressor cascade data for airfoils of particular families.

Some comparisons between available data. Although, as in Art. 2, there are small differences between the turning angles τ and the lift coefficients C_L for nominal and maximum L/D conditions, it is assumed here that they are near enough to be considered the same. The nominal turning angle τ^* is taken equal to 0.8 of the low speed stalling turning angle. The values to be discussed can be thought of as suitable general design conditions with blade cambers chosen to give reasonable incidences. In Fig. E,5a are shown compressor cascade turning angles τ plotted for exit flow angles α_2 between 50° and 0°, the latter angle being in the axial direction. The curves given are with a spacing-to-chord ratio $s/c = 1.0$ for the results from [*1,43,56*]; the results are reasonably comparable. Also given are the turning angles for zero incidence $\iota = 0$ deduced from the more universal generalizations of [*37*]. These turning angles are considerably lower than those for the former results at the lower α_2, and therefore, according to [*37*], the former design curves should require higher blade cambers θ than those given by a simple zero incidence criterion. The various methods give somewhat different variations of turning angles with spacing-to-chord ratio, though the values of τ in Fig. E,5a are typical for the middle part of normal compressor blades.

Design rules for turbine blades are much more varied, as such blades work satisfactorily over a much wider range of incidence. Because of this, personal choice plays a more important part. Ainley and Mathieson [*108*] give some comparisons with their own rules. For example, for an impulse cascade with $\alpha_1 = 50°$, $\alpha_2 = -50°$, and $\tau = 100°$, the rules suggested by Zweifel [*42*] and Howell and Carter [*43*] both happen to give a recommended spacing-to-chord ratio s/c of about 0.4 while the data of Ainley and Mathieson [*108*] suggest an s/c of about 0.7. It is stated, however, in [*108*] that a considerable variation of s/c is possible with a turbine cascade without appreciably affecting the performance. The lift coefficient based on the exit velocity in [*42*] is 0.8, while in [*43*] the corresponding values are 0.9, 1.05, and 1.125 for s/c of 0.5, 0.75, and 1.0 respectively.

Turning angles and lift coefficients. To illustrate the turning angles and lift coefficients to be expected from various compressor and turbine cascades, the method of [*43,52,107*] is taken. This method has a simple theoretical conception, based on pressure distributions, which is discussed in [*43*], and it applies with reasonable accuracy for both compressors and turbines over a wide range of conditions. Briefly, it involves a loading coefficient or factor equivalent to the lift coefficient, based on the exit

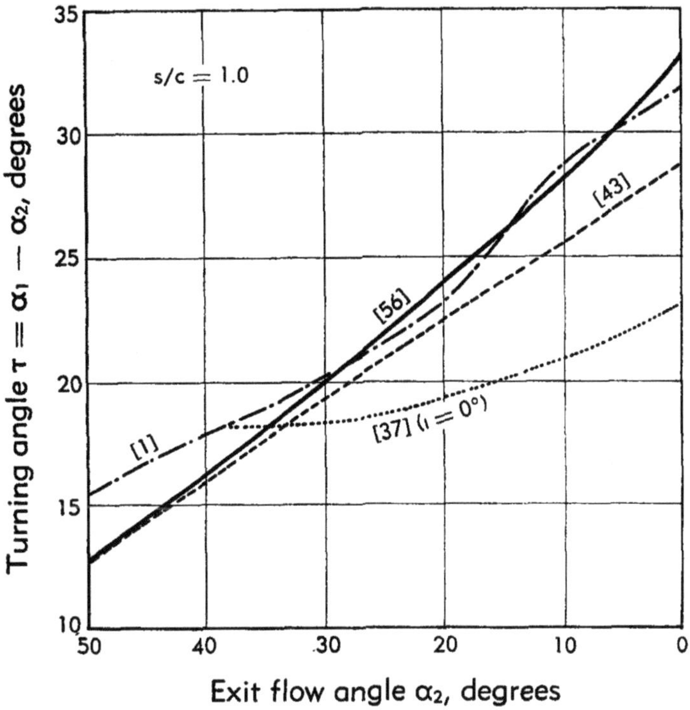

Fig. E,5a. Comparisons of compressor cascade turning angles for different exit flow angles.

velocity V_2 with a correction for spacing-to-chord ratio s/c, due to the "blockage" effect of finite thickness blades. The final relationship produced is

$$C_L \left(\frac{V_\infty}{V_2}\right)^2 = 1.125 \left[\frac{6\left(\frac{s}{c}\right) - 1}{5\left(\frac{s}{c}\right)}\right] \tag{5-1}$$

or with Eq. 2-2,

$$C_L = 2\left(\frac{s}{c}\right)(\tan \alpha_1 - \tan \alpha_2)(\cos \alpha_\infty)$$

$$= 1.125 \left(\frac{V_2}{V_\infty}\right)^2 \left[\frac{6\left(\frac{s}{c}\right) - 1}{5\left(\frac{s}{c}\right)}\right]$$

$$= 1.125 \left(\frac{\cos \alpha_\infty}{\cos \alpha_2}\right)^2 \left[\frac{6\left(\frac{s}{c}\right) - 1}{5\left(\frac{s}{c}\right)}\right] \tag{5-2}$$

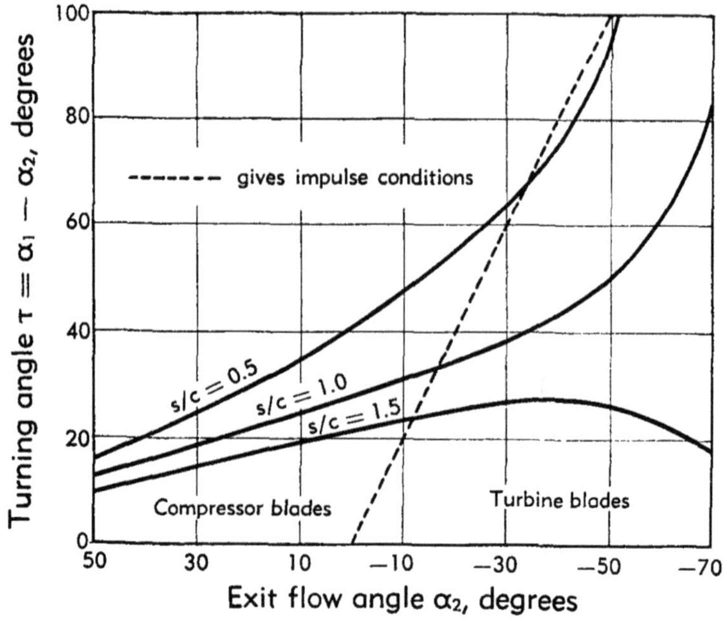

Fig. E,5b. Generalized data on turning angles.

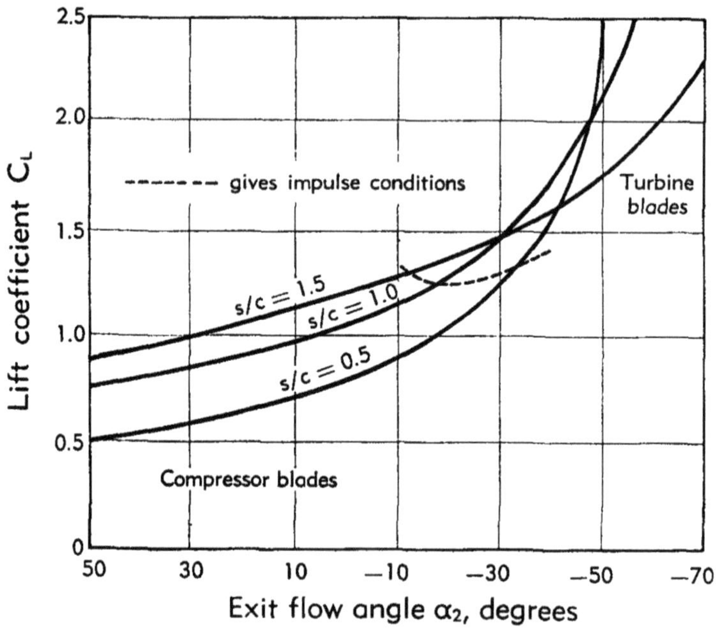

Fig. E,5c. Generalized data on lift coefficients.

From Eq. 5-2 both the turning angle $\tau = \alpha_1 - \alpha_2$ and the normal lift coefficient C_L can be obtained, and the respective results of such a calculation are given in Fig. E,5b and E,5c for the exit flow angle α_2 range of 50° to −70° and for three spacing-to-chord ratios s/c of 0.5, 1.0, and 1.5. The much higher turning angles and lift coefficients for the turbine blades compared with the compressor blades will be noted. The C_L's are of the order of 1.5 to 2.5 for turbines and 0.6 to 1.0 for compressors, depending somewhat on the exit flow angles and the spacing-to-chord ratios for the cascades. These results apply to Reynolds numbers *Re* of about 2 to 5×10^5 when based on inlet velocities for compressors and exit velocities for turbines.

In applying cascade data to compressors, the work done per unit flow or the stage temperature rise is proportional [*56*] to the product of the tangential velocity of the rotor, the axial velocity, and $\tan \alpha_1 - \tan \alpha_2$. In [*56*] the approximate rule given below is suggested:

$$\tan \alpha_1 - \tan \alpha_2 = \frac{1.55}{1 + 1.5(s/c)} \qquad (5\text{-}3)$$

Eq. 5-3 is said to apply, for α_2 between 40° and 0°, with fair accuracy to the nominal turning angle τ^* curves of [*56*]. This expression suggests that the tangent difference of the inlet angle α_1 and the exit angle α_2 is largely independent of α_2 for normally used compressor cascades.

E,6. Cited References and Bibliography.

Cited References

1. Howell, A. R. The present basis of axial flow compressor design. Part I: Cascade theory and performance. *Brit. Aeronaut. Research Council Repts. and Mem. 2095*, 1942.
2. Millikan, C. B. *Aerodynamics of the Airplane.* Wiley, 1941.
3. Abbott, I. H., von Doenhoff, A. E., and Stivers, L. S. Summary of airfoil data. *NACA Rept. 824*, 1945.
4. Abbott, I. H., and von Doenhoff, A. E. *Theory of Wing Sections.* McGraw-Hill, 1949.
5. Talbert, J. E., and Smith, J. C. Aerothermodynamic design of turbines for aircraft power plants. *J. Aeronaut. Sci. 15*, 556 (1948).
6. Ainley, D. G. The performance of axial-flow turbines. *Proc. Inst. Mech. Eng. 159*, 230–244 (1948).
7. Tyler, R. A. The available theoretical analyses of two dimensional cascade flow. *Natl. Research Council of Canada Rept. AN-4*, 1948.
8. Kawada, S. A contribution to the theory of latticed wings. *Proc. Third Intern. Congress Appl. Mech.*, Stockholm, 1930.
9. Betz, A. Diagramme zur Berechnung von Flügel-reihen. *Ing. Arch. 2*, 359 (1931). Transl. in *NACA Tech. Mem. 1022*, 1942.
10. Weinig, F. *Die strömung um die Schaufeln von Turbomaschinen.* Barth, Leipzig, 1935.
11. Pistolesi, E. Sul calcolo di shiere infinite di ali sottili. *Aerotechnica 17*, 484–506 (1937). (Transl. in *NACA Tech. Mem. 968*, 1941.)
12. Merchant, W., and Collar, A. R. Flow at an ideal fluid past a cascade of blades, Part 2. *Brit. Aeronaut. Research Council Repts. and Mem. 1893*, 1941.

13. Howell, A. R. A theory of arbitrary aerofoils in cascade. *Phil. Mag. 39*, (7), 913 (1948).
14. Garrick, I. E. On the plane potential flow past a lattice of arbitrary airfoils. *NACA Rept. 788*, 1944.
15. Mutterperl, W. The conformal transformation of an airfoil into a straight line and its application to the inverse problem of airfoil theory. *NACA Wartime Rept. L113*, Dec. 1944.
16. Mutterperl, W. A solution of the direct and inverse potential problems for arbitrary cascades of airfoils. *NACA Wartime Rept. L81*, Dec. 1944.
17. Traupel, W. Calcul de l'écoulement potential d'un fluide dans une grille d'aubes. *Sulzer Tech. Rev. 1*, 1945. (Also available from *Sixth Intern. Congress Appl. Mech.*, Paris, 1946.)
18. Lighthill, M. J. A mathematical method of cascade design. *Brit. Aeronaut. Research Council Repts. and Mem. 2104*, 1945.
19. Carter, A. D. S., and Hughes, H. P. A theoretical investigation into the effect of profile shape on the performance of aerofoils in cascade. *Brit. Aeronaut. Research Council Repts. and Mem. 2384*, 1946.
20. Katzoff, S., Finn, R. S., and Laurence, J. C. Interference method for obtaining the potential flow past an arbitrary cascade of airfoils. *NACA Tech. Note 1252*, 1947. Also *NACA Rept. 879*, 1947.
21. Legendre, R. *Tracé des Aillettes pour Fluides à Densité Légèrement Variables.* Assoc. Tech. Maritime et Aéronaut., Paris, 1947.
22. Vazsonyi, A. On the aerodynamic design of axial-flow compressors and turbines. *J. Appl. Mech. 15*, 53 (1948).
23. Woolard, H. W. The incompressible flow about a cascade of airfoils. *Cornell Aeronaut. Lab. Rept. AF-734-A-1*, Nov. 1950.
24. Ackeret, J. The design of closely spaced blade grids. *Schweiz. Bauzit. 120*, 103–108 (1942). (Available as *Brit. RTP Trans. 2007.*)
25. Diesendruck, L. Iterative interference methods in the design of thin cascade blades. *NACA Tech. Note 1254*, 1947.
26. Goldstein, A. W., and Jerison, M. Isolated and cascade aerofoils with prescribed velocity distribution. *NACA Tech. Note 1308*, 1947. Also *NACA Rept. 869*, 1947.
27. Costello, G. R. Method of designing cascade blades with prescribed velocity distribution in compressible potential flows. *NACA Tech. Note 1970*, 1949. Also *NACA Rept. 978*, 1949.
28. Sawyer, W. T. Experimental investigations of a stationary cascade of aerodynamic profiles. *ETH Inst. Aerodynam., Zurich, 17*, 1949.
29. Schlichting, H. Problems and results of investigations on cascade flow. *J. Aeronaut. Sci. 21*, 163 (1954).
30. Alpert, S. Design method for two-dimensional channels for compressible flow with application to high-solidity cascades. *NACA Tech. Note 1931*, 1949.
31. Malavard, L. The use of rheo-electrical analogies in certain aerodynamical problems. *J. Roy Aeronaut. Soc. 51*, 441 (1947).
32. de Haller, P. Application of electrical analogy to the investigation of cascades. *Sulzer Tech. Rev. 3*, 4 (1947).
33. Hargest, T. J. An electric tank for the determination of theoretical velocity distributions. *Brit. Aeronaut. Research Council Repts. and Mem. 2699*, 1949.
34. Davis, H. A new method for the aerodynamic design of multistage axial-flow compressors. *J. Aeronaut. Sci 15*, 41 (1948).
35. Goldstein, A. W., and Mager, A. Attainable circulation of airfoils in cascade. *NACA Tech. Note 1941*, 1949. Also *NACA Rept. 953*, 1949.
36. Smith, A. G. Heat flow in the gas turbine. *Proc. Inst. Mech. Eng. 159*, 245–254, 1948.
37. Carter, A. D. S. The low speed performance of related aerofoils in cascade. *Brit. Aeronaut. Research Council Current Paper 29*, 1949.
38. Finger, H. B., Schum, H. J., and Buckner, H. A. Experimental and theoretical distribution of flow produced by inlet guide vanes of an axial-flow compressor. *NACA Tech. Note 1954*, 1948.

39. Katzoff, S., Bogdonoff, H. E., and Boyet, H. Comparisons of theoretical and experimental lift and pressure distributions on airfoils in cascade. *NACA Tech. Note 1376*, 1947.
40. Keller, C. *Axialgebläse vom standpunkt der Traflugeltheori.* Leeman, Zurich, 1934.
41. Pinkerton, R. M. Calculated and measured pressure distributions over the midspan section of the NACA 4412 airfoil. *NACA Rept. 563*, 1936.
42. Zweifel, O. The spacing of turbo-machine blading, especially with large angular deflection. *Brown Boveri Rev. 32*, 12, 1945.
43. Howell, A. R., and Carter, A. D. S. Fluid flow through cascades of aerofoils. *Sixth Intern. Congress Appl. Mech.*, Paris, 1946.
44. von Kármán, Th. Compressibility effects in aerodynamics. *J. Aeronaut. Sci. 8*, 337–356 (1941).
45. Bogdonoff, S. M. NACA cascade data for the blade design of high-performance axial-flow compressors. *J. Aeronaut. Sci. 15*, 89 (1948).
46. Harris, R. G., and Fairthorne, R. A. Windtunnel experiments with infinite cascades of aerofoils. *Brit. Aeronaut. Research Council Repts. and Mem. 1206*, 1928.
47. Christiani, K. Experimentalle Untersuchung eines Tragflügelprofils bei Gitteranordung. *Luftfahrtforschung 2*, 1928.
48. Collar, A. R. Some experiments with cascades of aerofoils. *Brit. Aeronaut. Research Council Repts. and Mem. 1768*, 1937.
49. Shimoyana, Y. Experiments on rows of aerofoils for retarded flow. *Mem. Fac. Eng., Kyushu Univ., Japan, 8, (4)*, 1938.
50. Reeman, J. The turbine for the simple jet propulsion engine. *Proc. Inst. Mech. Eng. 153*, 495–504 (1945).
51. Todd, K. W. Practical aspects of cascade wind tunnel research. *Proc. Inst. Mech. Eng. 157*, 482, 1947.
52. Howell, A. R. Aerodynamics of the gas turbine. *J. Roy. Aeronaut. Soc. 52*, 329 (1948).
53. Davis, H. A method of correlating axial flow compressor cascade data. *Trans. Am. Soc. Mech. Engrs. 70*, 951 (1948).
54. Bowden, A. T., and Jefferson, J. L. The design and operation of the Parsons experimental gas turbine. *Proc. Inst. Mech. Eng. 160*, 454–469 (1949).
55. Weinig, F., and Eckert, B. Measurements on compressor-blade lattices. *NACA Tech. Mem. 1193*, 1948.
56. Howell, A. R. Fluid dynamics of axial-flow compressors. *Proc. Inst. Mech. Eng. 153*, 441–452 (1945).
57. Wislicenus, G. F. *Fluid Mechanics of Turbomachinery.* McGraw-Hill, 1947.
58. Jacobs, E. N., and Sherman, A. Aerofoils section characteristics as affected by variations of the Reynolds Number. *NACA Rept. 586*, 1937.
59. Carter, A. D. S., Andrews, S. J., and Shaw, H. Some fluid dynamic research techniques. *Proc. Inst. Mech. Eng. 163*, 249–263 (1950).
60. Eckert, B. Überblick über Forshungargebnisse des Forschungsinstitutes für Kraftfahrwesen and Fahrzeugmotoren, Stuttgart, an axial durchströmten—Verdichtern. *Lilienthal—Gesellschaft, Bericht 171*, 1943.
61. d'Epinay, J. L. Aerodynamic methods applied to turbo-machine research. *Brown Boveri Rev. 37, (10)*, 1950.
62. Moyes, S. J., and Pennington, W. A. The influence of size on the performance of turbo-jet engines. *Third Anglo-American Aeronaut. Conference*, Brighton, England, Sept. 1951.
63. Ainley, D. G., Petersen, S. E., and Jeffs, R. A. Overall performance characteristics of a four-stage reaction turbine. *Brit. Aeronaut. Research Council Repts. and Mem. 2416*, 1946.
64. Samaras, D. G., and Tyler, R. A. The relative merits of rotary compressors. *J. Aeronaut. Sci. 15*, 625 (1948).
65. Carter, A. D. S., and Cohen, E. M. Preliminary investigation into the three dimensional flow through a cascade of aerofoils. *Brit. Aeronaut. Research Council Repts. and Mem. 2339*, 1946.

66. Hausmann, G. F. The theoretical induced deflection angle in cascades having wall boundary layers. *J. Aeronaut. Sci. 15*, 686 (1948).
67. Carter, A. D. S. Three-dimensional flow theories for axial compressors and turbines. *Proc. Inst. Mech. Eng. 159*, 1948.
68. Preston, J. H. The interference on a wing spanning a closed tunnel, arising from boundary layers on the side walls, with special reference to the design of two-dimensional tunnels. *Brit. Aeronaut. Research Council Repts. and Mem. 1924*, 1944.
69. Squire, H. B., and Winter, K. G. The secondary flow in a cascade of airfoils in a non-uniform stream. *J. Aeronaut. Sci. 18*, 271 (1951).
70. Hawthorne, W. R. Secondary circulation in fluid flow. *Proc. Roy. Soc. London 206*, 374 (1951).
71. Preston, J. H. A simple approach to the theory of secondary flows. *Aeronaut. Quart. 5*, 218 (1954).
72. Hawthorne, W. R. Rotational flow through cascades. Part I: The components of vorticity. *Quart. J. Mech. and Appl. Math. 8*, 266 (1955).
73. Hawthorne, W. R., and Armstrong, W. D. Rotational flow through cascades. Part II: The circulation about the cascade. *Quart. J. Mech. and Appl. Math. 8*, 280 (1955).
74. Kronauer, R. E. Secondary flow in fluid dynamics. *Proc. First U.S. Natl. Congress of Appl. Mech.*, 747–756 (1952).
75. Todd, K. W. Some developments in instrumentation for air flow analysis. *Proc. Inst. Mech. Eng. 161*, 213–221 (1949).
76. Weske, J. R. Fluid dynamic aspects of axial-flow compressors and turbines. *J. Aeronaut. Sci. 14*, 651 (1947).
77. Scholz, N. Secondary flow losses in turbine cascades. *J. Aeronaut. Sci. 21*, 707 (1954).
78. Herzig, H. Z., Hansen, A. G., and Costello, G. R. A visualization study of secondary flows in cascade. *NACA Rept. 1163*, 1953.
79. Rohlik, H. E., Kofskey, M. G., Allen, H. W., and Herzig, H. Z. Secondary flows and boundary layer accumulations in turbine nozzles. *NACA Rept. 1168*, 1953.
80. Ehrich, F. F. Secondary flows in cascades of twisted blades. *J. Aeronaut. Sci. 22*, 51 (1955).
81. Squire, H. B. Heat transfer calculations for aerofoils. *Brit. Aeronaut. Research Council Repts. and Mem. 1986*, 1942.
82. McAdams, W. H. *Heat Transmission*. McGraw-Hill, 1942.
83. Ellerbrock, H. H. NACA investigations of gas turbine blade cooling. *J. Aeronaut. Sci. 15*, 721 (1948).
84. Eckert, E. R. G. *Introduction to the Transfer of Heat and Mass*. McGraw-Hill, 1950.
85. Livingood, J. N. B., and Brown, W. B. Analysis of spanwise temperature distribution in three types of air-cooled turbine blade. *NACA Rept. 994*, 1950.
86. Humble, L. V., Lowdermilk, W. H., and Desmon, L. C. Measurements of average heat-transfer and friction coefficients for subsonic flow of air in smooth tubes at high surface and fluid temperatures. *NACA Rept. 1020*, 1950.
87. Eckert, E. R. G., and Low, G. M. Temperature distribution in internally heated walls of heat exchangers composed of non-circular flow passages. *NACA Rept. 1022*, 1950.
88. Livingood, J. N. B., and Brown, W. B. Analysis of temperature distribution in liquid-cooled turbine blades. *NACA Rept. 1066*, 1950.
89. Andrews, S. J., and Schofield, N. W. An experimental investigation of a thick aerofoil nozzle cascade. *Brit. Aeronaut. Research Council Repts. and Mem. 2883*, 1950.
90. Smith, A. G., and Pearson, R. D. The cooled gas turbine. *Proc. Inst. Mech. Eng. 163*, 221–234 (1950).
91. Gray, V. H., and Bowden, D. T. Icing characteristics and anti-icing heat requirements for hollow and internally modified gas-heated inlet guide vanes. *NACA Research Mem. E50I08*, 1950.

92. von Glahn, U., and Blatz, R. E. Investigation of power requirements for ice prevention and cyclical de-icing of inlet guide vanes with internal electric heaters. *NACA Research Mem. E50H29*, 1950.
93. Staniforth, R. Contribution to the theory of effusion cooling of gas turbine blades. *Proc. Inst. Mech. Eng.—General Discussion on Heat Transfer*, 1951.
94. Ellerbrock, H. Some NACA investigations of heat transfer of cooled gas turbine blades. *Proc. Inst. Mech. Eng.—General Discussion on Heat Transfer*, 1951.
95. Brown, T. W. F. Effect of the radiation correction on cooling loss in high temperature cooled gas turbines. *Proc. Inst. Mech. Eng.—General Discussion on Heat Transfer*, 1951.
96. Hahnemann, H. W. Stationary rig experiments on the heat extracting power of closed thermosyphon cooling holes. *Brit. Aeronaut. Research Council Current Paper 152*, 1951.
97. Eckert, E. R. G., and Livingood, J. N. B. Method for calculation of laminar heat transfer in air flow around cylinders of arbitrary cross section (including large temperature differences and transpiration cooling). *NACA Rept. 1118*, 1952.
98. Rogers, G. F. C. Factors influencing the optimum aerodynamic design of cooled turbines. *Brit. Aeronaut, Research Council Current Paper 155*, 1952.
99. Staniforth, R. A theoretical note on effusion cooled gas turbine blades. *Brit. Aeronaut. Research Council Current Paper 165*, 1952.
100. Reeman, J., Buswell, R. W., and Ainley, D. G. An experimental single-stage air-cooled turbine. *Proc. Inst. Mech. Eng. 167*, 1953.
101. Brown, T. W. F. High temperature turbine machinery for marine propulsion. *Proc. Inst. Mech. Eng. 26th Thomas Lowe Gray Lecture*, 1954.
102. Wilson, D. G., and Pope, J. A. Convective heat transfer to turbine blade surfaces. *Proc. Inst. Mech. Eng. 168*, 861 (1954).
103. Hubbart, J. E., Stone, H. O., and Arne, V. L. Method for rapid determnationi of pressure change for one-dimensional flow with heat transfer, friction, rotation and area change. *NACA Tech. Note 3150*, 1954.
104. Cohen, H., and Bayley, F. J. Heat transfer problems of liquid cooled gas turbine blades. *Proc. Inst. Mech. Eng. 169*, 53 (1955.)
105. Ainley, D. G. The high temperature turbo-jet engine. *J. Roy. Aeronaut. Soc. 60*, 563 1956.
106. Howell, A. R., and Bonham, R. P. Overall and stage characteristics of axial flow compressors. *Proc. Inst. Mech. Eng. 163*, 235–248 (1950).
107. Carter, A. D. S. *The Axial Compressor. Sec. 5: Gas Turbine Principles and Practice.* George Newnes, London, 1955.
108. Ainley, D. G., and Mathieson, G. C. R. An examination of the flow and pressure losses in blade rows of axial flow turbines. *Brit. Aeronaut. Research Council Repts. and Mem. 2891*, 1951.

Bibliography

Andrews, S. J. Tests related to the effect of profile shape and camber-line on compressor cascade performance. *Brit. Aeronaut. Research Council Repts. and Mem. 2743*, 1949.

Bridle, E. A. Some high speed tests on turbine cascades. *Brit. Aeronaut. Research Council Repts. and Mem. 2697*, 1949.

Carter, A. D. S. Some tests on compressor cascades of related aerofoils having different positions of maximum camber. *Brit. Aeronaut. Research Council Repts. and Mem. 2694*, 1948.

Cornell, W. G. The stall performance of cascades. *Proc. Second U.S. Natl. Congress Appl. Mech.*, June 1954.

Costello, G. R., Cummings, R. L., and Sinnette, J. T. Detailed computational procedure for design of cascade blades with prescribed velocity distributions in compressible potential flows. *NACA Rept. 1060*, 1950.

Detra, R. W. The secondary flow in curved pipes. *Mitt. ETH Inst. Aerodynam., Zurich, 20*, 1953.

Dunavant, J. C., and Erwin, J. R. Investigation of a related series of turbine-blade profiles in cascade. *NACA Tech. Note 3802*, 1953.

Dunsby, J. A. Schlieren tests on some conventional turbine cascades. *Brit. Aeronaut. Research Council Repts. and Mem. 2728*, 1949.

Ehrich, F. F., and Detra, R. W. Transport of the boundary layer in secondary flow. *J. Aeronaut. Sci. 21*, 136 (1954).

Eichenberger, H. Secondary flow within a bend. *J. Math. and Phys. 32*, 34 (1953).

Erwin, J. R., and Emery, J. C. Effect of tunnel configuration and testing technique on cascade performance. *NACA Rept. 1016*, 1949.

Erwin, J. R., Savage, M., and Emery, J. C. Two-dimensional low-speed cascade investigation of NACA compressor blade sections having a systematic variation in mean-line loading. *NACA Research Mem. L53I30b*, 1953.

Hansen, A. G., and Yohner, P. L. A numerical procedure for designing cascade blades with prescribed velocity distributions in incompressible potential flow. *NACA Tech. Note 2101*, 1950.

Hargest, T. J. The theoretical pressure distributions around some conventional turbine blades in cascade. *Brit. Aeronaut. Research Council Repts. and Mem. 2765*, 1950.

Herrig, L. J., Emery, J. C., and Erwin, J. R. Effect of section thickness and trailing edge radius on the performance of NACA 65-series compressor blades in cascade at low speeds. *NACA Research Mem. L51J16*, 1951.

Herrig, L. J., Emery, J. C., and Erwin, J. R. Systematic two dimensional cascade tests of NACA 65-series compressor blades at low speed. *NACA Research Mem. L51G31*, 1951.

Keast, F. H. High speed cascade testing techniques. *Am. Soc. Mech. Engrs., Gas Turbine Power Division Meeting, Paper 51-SA-31*, June 1951.

Libby, P. A., and Visich, M. Laminar heat transfer in two-dimensional subsonic effusors. *J. Aeronaut. Sci. 22*, 425 (1955).

Lin, C. C. On the subsonic flow through circular and straight lattices of airfoils. *J. Math. and Phys. 28*, 117–130 (1949).

Martin, B. W., and Cohen, H. Heat transfer by free convection in an open thermosyphon tube. *Brit. J. Appl. Phys. 5*, 91 (1954).

Schirakura, M. A theory on cascades built up of arbitrary blade sections. *Hydraulics and Hydraulic Machinery Div., Japan Soc. Mech. Engrs.*, Jan. 1951.

Scholz, N. On the calculation of the potential flow around airfoils in cascade. *J. Aeronaut. Sci. 18*, 68 (1951).

Scholz, N. Strömungsuntersuchungen an Schaufelgittern. *Ver. deut. Ing. Berlin Forschungsheft 442, (B20)*, 1954.

Spurr, R. A., and Allen, H. J. A theory of unstaggered airfoil cascades in compressible flow. *NACA Rept. 888*, 1947.

Stanitz, J. D. Design of two-dimensional channels with prescribed velocity distributions along the channel walls. *NACA Rept. 1115*, 1951.

Stanitz, J. D., and Sheldrake, L. J. Application of a channel design method to high solidity cascades and tests of an impulse cascade with 90° of turning. *NACA Rept. 1116*, 1951.

Woolard, H. W. A note on the subsonic compressible flow about airfoils in a cascade. *J. Aeronaut. Sci. 17*, 379 (1950).

Yeh, H. The development of cascade profiles for high subsonic potential flows. *J. Aeronaut. Sci. 19*, 630 (1952).

SECTION F

THE AXIAL COMPRESSOR STAGE

W. D. RANNIE

F,1. Introduction. The aerodynamic design of axial compressors is accomplished by an appropriate synthesis of the results of two-dimensional cascade investigations and of three-dimensional flow theory for the turbomachine. In principle, the procedure is similar to the combination of two-dimensional airfoil theory and experiment with three-dimensional wing theory for the design of airplane wings.

The analogy between the turbomachine theory and the wing theory has already been pointed out in the introduction to Sec. C. In both theories, the primary problem is that of finding the velocity induced at an arbitrary point on an airfoil surface by the vorticity field associated with all other airfoil surfaces. In both theories, linearizing approximations are introduced to give tractable mathematical problems. Although the techniques employed in the two theories are necessarily quite different, the solutions of the problems have similar character. In both, the first order approximations are quite adequate for general application, and there is little need for more exact analysis.

If one compares the two-dimensional cascade data required in the design of compressors with the two-dimensional airfoil data required in the design of airplane wings, similarities are again evident.

Experiments are essential for both, although potential theory and boundary layer theory are invaluable as guides and for interpretation of the experiments. Theory is somewhat less useful for cascades than for airfoils because the range of Reynolds numbers for compressor blades is much lower than for airplane wings; hence potential flow gives a poorer representation of actual flow for a compressor blade than for a wing.

In spite of the close similarities in the information required and the procedures used in the respective designs, further examination shows that the problems of compressor design are very different from those of wing design. Put in its crudest and simplest terms, this results from the fact that the airplane wing is an airfoil of finite span, whereas a compressor blade is an airfoil with infinite end plates. Because of induced drag, the airplane wing has its maximum lift-to-drag ratio at a moderate lift coefficient. There is no induced drag on a compressor blade, so its maximum lift-to-drag ratio occurs at essentially maximum lift. One can readily imagine the problems that would beset an airplane designer if the optimum condition for cruise were just on the verge of stall; for the compressor designer, this is the normal state of affairs.

The conditions for stalling of the compressor blades are of such dominating importance in compressor design that it is essential to understand how they arise. The aspect ratios of compressor blades are usually in the range of 2 to 4, although lower or higher values are required occasionally. It has been found that the combination of three-dimensional flow theory and two-dimensional cascade characteristics gives an adequate description of the flow over the central 80 per cent or more of the blade height. At each end of the blade is a region of flow that will be called the wall layer, where the flow phenomena are so complex that they cannot be described even approximately in terms of the three-dimensional theory or the two-dimensional cascade characteristics. The thickness of the wall layer is somewhat indefinite and it varies with circumstances, but under normal conditions it will not amount to more than a rather small fraction of the blade chord.

The circulation around the blade vanishes at the wall, whether a tip clearance is present or not. However, the velocity induced by the shed vorticity is small outside of a highly localized region very close to the wall. In order to satisfy the wall boundary condition, the flow behaves as if equal, and opposite vorticity appears on the other side of the wall; the combined effects die out rapidly with distance and have no appreciable influence outside the wall region. In addition to this vorticity field within the wall layer, there are strong distortions in velocity produced by the wall boundary layer passing through the blade row. The direction of flow may be quite different from that outside the wall region and may vary.

It is fortunate indeed that the very complex flow within the wall layer is not the dominating influence on compressor performance, otherwise analysis would be almost impossible. As was mentioned above, the wall layer thickness is normally a fraction of the blade chord length; hence the static pressure within the wall layer cannot be very different from the pressure at the edge of the central flow region. The flow within the wall layer simply conforms to the pressure distribution at its outer boundary, as long as the outer flow is sufficiently stable to small disturbances. Local separation may occur within the wall layer, and losses are certainly severe in that region. However, the wall layer thickness does not increase beyond some limiting thickness from one stage to the next if the central part of the blade is not too heavily loaded. Since there is a strong adverse pressure gradient through each blade row, this behavior is quite different from that of a boundary layer in a simple diffuser. Evidently the action of the rotor blade rows keeps the thickness of the wall layer constant by direct addition of energy or a very effective mixing process.

The flow over the central 80 per cent or so of the blade height controls the wall layer flow, at least to the extent of preventing unlimited growth in the layer thickness, only if the incidence angles for all blade sections in the central region are well below the stalling angles. However, a compressor operates at best efficiency if it is designed so that the blades are almost stalled. The incentive to push the blade loading to the limit is so great that the prediction of stall becomes the most important factor in compressor design.

Two-dimensional cascade experiments are quite reliable for predicting compressor blade section performance below stalling incidence, but not so reliable for determination of the stall point, even under the most favorable conditions. As in single airfoils, the stalling angle is sensitive to variations of Reynolds number and to turbulence and other unsteadiness in the approaching stream; and it is almost impossible to reproduce the free stream conditions of a compressor in a two-dimensional cascade test. The influence of the wall layer becomes greater as the flow outside the layer approaches stalling incidence. Local regions of separated flow always occur within the wall layer, and these become potent triggering sources as the flow at the edge of the layer nears an unstable condition. When stall does occur, starting at one end of the compressor blade, it is abrupt and severe, with a large patch of periodically separating flow extending over an appreciable fraction of the blade height, or even over the entire blade.

The prospect of understanding all details of the flow in axial compressors is certainly hopeless if the description in the preceding paragraphs is even approximately correct. One can say that all details of the flow are not of equal importance and we should be concerned only with those that are significant. Howell [*1*] wrote a 10-page paper on the design of axial compressors in 1945, published by the Institution of Mechanical Engineers (London). It was widely read and accepted as a useful design manual. In 1956, the NACA [*2*] put out a design manual of over 1000 pages. A direct comparison is not quite fair, but it does give an indication that in the intervening years our knowledge increased tremendously. When one examines the new knowledge, however, it becomes apparent that there has been a much more rapid accumulation of knowledge of things that might be important than elimination of things found to be unimportant. A discussion of compressor design problems must be selective and the selection is still very much an individual choice.

F,2. Cascade Characteristics. The axial flow turbomachine consists of one or more rows of rotating and stationary blades in an annular duct. A typical blade row is shown in Fig. F,2a. Since the flow is confined by the cylindrical inner and outer walls, all streamlines lie on approximately cylindrical surfaces. Advantage is taken of this situation to introduce a simplifying approximation into the theory of the turbomachine. One imagines a cylindrical surface through the blade row, at radius r say, to be cut and straightened so that the contours of the blade sections originally on the cylindrical surface become those for a two-dimensional cascade. The derived cascade is assumed to have an infinite number of blades, since there is no beginning or end for the annular cascade. The two-dimensional flow characteristics of the derived cascade are identified with the flow characteristics of the blade sections at radius r in the annular cascade of the turbomachine.

There are obvious limitations on the validity of identifying the blade characteristics at each radius with those of a corresponding two-dimensional cascade (see Sec. B). The interference effects of the blades are not the same in the annular cascade as in the straight cascade because of the divergence of the blades in the former. However, if the aspect ratio (blade height/blade chord) is large, one would expect the error to be small. For

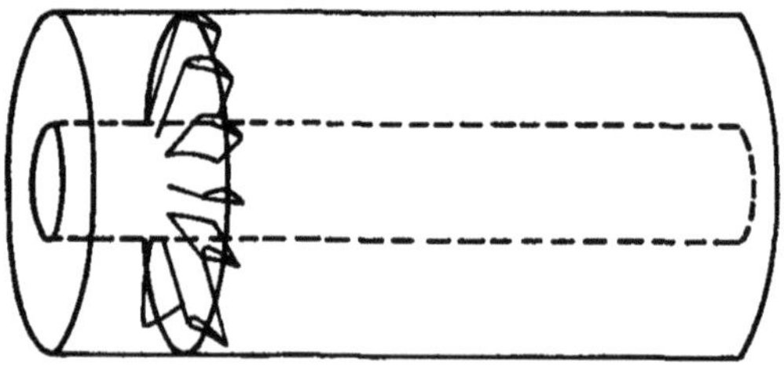

Fig. F,2a. Typical annular blade row.

low solidity (blade surface area/annulus area) the divergence angle of the blades is large, but the interference effects are small. For high solidity, only nearby blades cause interference on any particular blade and for these the divergence angle is low. Decreasing hub ratio will also tend to make the identification of blade element characteristics in the annular cascade and the corresponding two-dimensional cascade less valid.

The comparisons above were made with the assumption that the streamlines through the annular cascade lay on circular cylinders. Such a condition is not satisfied in general. If the flow approaching the blade row has vorticity or if the blades shed vorticity into the downstream flow, radial components of velocity occur in and near the blade row, giving rise to a three-dimensional character of the flow through the blades. Again, if the aspect ratio of the blades is large so that there is very little *variation* in radial velocity through the axial length of the blade, the assumption of two-dimensional character of the local flow appears a reasonable one.

The flow in the neighborhood of the blade ends does not have two-dimensional character under any circumstances, because of wall friction and the presence of wall boundary layers. However, if the aspect ratio is high, the fraction of the blade height that is strongly influenced by wall effects is quite small. The fluid in the boundary layers on the blades is subject to different centrifugal action than the fluid outside the boundary layers. The resulting radial forces in the boundary layers are usually stronger in rotor blade rows than in stators. Since the radial displacement in the boundary layer flow is proportional to the chord, the displacement as a fraction of blade height decreases as the aspect ratio increases.

The possibility of relating the blade characteristics at each blade radius to the local flow direction and velocity immediately in front of the blade row, independently of the blade geometry at other radii and independently of the flow conditions elsewhere, introduces a great simplification in the theory of turbomachines. From the discussion of the preceding paragraphs, it is seen that the approximation is better with higher blade aspect ratio and higher hub ratio. Many axial flow compressors satisfy the conditions sufficiently well. Axial flow pumps, because of strength requirements, typically have few blades, low hub ratio, and high solidity, so that the blade aspect ratio is low. The validity of the procedure of using two-dimensional cascade theory for the blade characteristics of axial flow pumps is dubious. The procedure is, of course, quite invalid for mixed flow and centrifugal machines.

The characteristics of two-dimensional cascades of blades for axial turbomachines are discussed in Sec. E in considerable detail. In the treatment to follow, the behavior of cascades will be given in simple parametric form in terms of constants that must be determined from a detailed knowledge of blade geometry and flow conditions. Two relations equivalent to the two momentum equations are required to specify cascade characteristics. For turbomachine applications, the most convenient relations give leaving flow direction and total pressure loss as functions of inlet angle.

The notation that will be used is shown in Fig. F,2b. The axial and whirl components V_z and V_θ are averaged values in front of, or behind, the cascade. The averaging process is carried out so that the flow rate perpendicular to the cascade and the flux of momentum parallel to the cascade are correct. For incompressible, two-dimensional flow, the axial velocity components, V_{z_1} upstream and V_{z_2} downstream, are equal. The cascade characteristics can be specified by the two equations:

$$\tan \alpha_2 = A + F(\tan \alpha_1) \tag{2-1}$$

$$\frac{2\Delta p^0}{\rho V_1^2} = K(\tan \alpha_1) \tag{2-2}$$

where Δp^0 is the loss in total pressure through the cascade, A is a constant, $F(\tan \alpha_1)$ and $K(\tan \alpha_1)$ are functions of the inlet angle α_1. In general, the coefficients on the right-hand side of Eq. 2-1 and 2-2 are dependent on inlet Mach number and Reynolds number; however, this dependence is slight except in the transonic and supersonic regimes, where the validity of the two-dimensional approximation to blade characteristics of a turbo-machine is doubtful anyway.

The flow leaving angle. The theory for irrotational, two-dimensional flow of a perfect fluid through the cascade is not always realistic, but it does give a systematic method for estimating the leaving angle. According to two-dimensional potential flow theory, the leaving angle is given by a relation of the form

$$\tan \alpha_2 = A + B \tan \alpha_1 \tag{2-3}$$

where A and B, depending on the blade shape, solidity, and stagger angle, are constants for a given cascade. Unfortunately, the computational effort

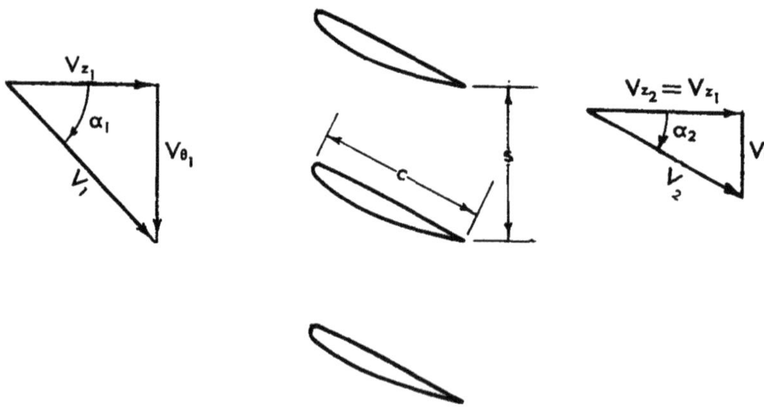

Fig. F,2b. Velocity diagrams for stationary cascade.

required to obtain these two constants is in general so great that it is not worthwhile. However, if the blade sections can be approximated by thin, slightly cambered arcs of approximately parabolic shape, the constants A and B in Eq. 2-3 above can be evaluated in terms of camber, solidity, and stagger angle without much difficulty. The theory is given in Sec. B. The potential theory for thin, slightly cambered airfoils is frequently surprisingly accurate, although this apparent accuracy is probably fortuitous. By neglecting real fluid effects, the lift, and hence the magnitude of turning, is over-estimated. On the other hand, neglect of the effect of thickness under-estimates the turning. In single airfoils, or cascades of low solidity, the reduction of lift from real fluid effects is larger than the increase of lift from thickness effects, so the lift calculated from potential flow around an airfoil of zero thickness is too high. For solidities of the order of unity, the two effects appear to be of comparable magnitude and the leaving angle from the potential theory is often very close to the measured value. Unfortunately, one cannot count on such cancellation as a general rule.

For cascades of very high solidity, the leaving angle is almost independent of the entering flow direction, so $F(\tan \alpha_1) \cong 0$. Potential flow theory for cascades with solidity of unity, or even lower, gives leaving angles that are nearly independent of inlet angles; however, in practice, the relative insensitivity of the leaving angle is confined to the rather narrow range of the inlet angle corresponding to very slight separation on the blades. For very high solidity, the leaving angle is essentially independent of the entering angle even when separation is severe.

Total pressure loss. The dependence of the total pressure loss coefficient K on $\tan \alpha_1$ resembles the dependence of the drag coefficient of an isolated airfoil on the angle of attack; $K(\tan \alpha_1)$ is quite small over a range

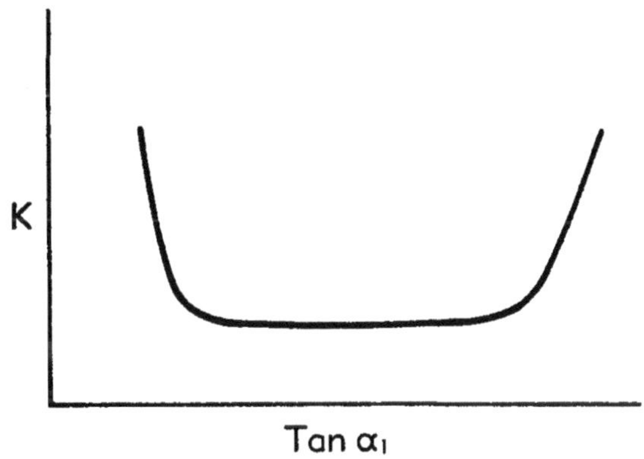

Fig. F,2c. Typical pressure loss coefficient.

of 10° or so in α_1 and rises sharply outside of this range as the flow separation becomes severe. The general behavior is indicated in Fig. F,2c (see also Fig. E,2b). The location of the low loss region along the axis of $\tan \alpha_1$, and to a lesser extent the width of the region, depend upon the stagger angle and camber of the blade section. The magnitude of the loss within this region depends on the blade thickness and cascade solidity as well as the stagger and camber. There appear to be no simple universal expressions for losses in terms of cascade parameters, so that appeal must be made to test results for precise information. The minimum value of K varies over a wide range. For low solidities, K can be estimated from single airfoil data (see below) and, for very high solidities, it can be estimated by calculating friction in the blade channels.

Maximum pressure rise. The important part of the loss curve for turbomachine practice is the region around the sharp increase at the higher value of $\tan \alpha_1$. This corresponds to positive stall for a single airfoil, and in the cascade represents the condition for maximum turning and maximum pressure rise at low total pressure loss. Since the purpose of a compressor cascade is to develop pressure increase by conversion of kinetic energy, it is pertinent to establish the maximum pressure rise that can be attained for a given magnitude of inlet velocity, assuming one is free to choose the most favorable cascade configuration. At very low solidity, the flow separates from the blades when the turning angle is low and hence the pressure increase small. At very high solidity, friction in the blade passages absorbs most of the pressure increase from diffusion, so the pressure increase is again small. At solidity of the order of unity (say, 0.7 to 1.3), cascade experiments have shown that the pressure rise attainable is a maximum; hence this is the range of solidity that is always used where the number of blade rows is kept to a minimum. Strictly, it is not the maximum pressure rise that is so important in practice, but rather the maximum pressure that can be obtained with moderate losses. However, there is usually little difference between these two conditions.

In the turbomachine analysis that follows, it will be convenient to have a simple criterion for maximum attainable pressure rise. Experience with compressors indicates that, unless $V_2/V_1 > \nu$ where $\nu \cong 0.65$, the losses become excessive. Actually, ν depends on stagger angle and blade shape, and criteria that are more satisfying from a physical point of view have been developed. The rule above is certainly over-simplified, but it is probably sufficiently accurate for a general analysis of turbomachine performance.

Lift and drag. An alternative method of specifying cascade characteristics is in terms of lift and drag coefficients analogous to those for isolated airfoils. It can be shown that the vector average of the velocities upstream and downstream of the cascade plays the role of the velocity at infinity for an isolated airfoil, since the blade force is normal to this velocity for an inviscid fluid. In the velocity diagram shown in Fig. F,2d, the mean velocity and the directions of lift and drag are indicated. From momentum balance across the cascade and the geometry of the diagram, it is readily shown that

$$C_L = \frac{s}{c}\left[2\,\frac{V_{\theta_1} - V_{\theta_2}}{V_\infty} - \frac{V_{\theta\infty}V_1}{V_\infty^2}\,K\right]$$

$$C_D = \frac{s}{c}\,\frac{V_z V_1}{V_\infty^2}\,K$$

Hence the drag coefficient C_D is directly proportional to the pressure loss coefficient, while both the lift and drag contribute to turning.

For applications to the calculation of flow in axial turbomachines, it is more convenient to specify cascade characteristics in terms of leaving angle and pressure loss coefficient than in terms of lift and drag coefficients. The advantages are particularly marked in calculation of the performance of specified blading. In addition, the leaving angle is so nearly constant for solidities of the order of unity or higher that it is the natural parameter for cascade characteristics. At low solidities, the lift coefficient is a useful parameter because the isolated airfoil characteristics can be transferred almost directly to the cascade. However, the leaving angle is still more convenient for determination of turbomachine performance.

Moving cascades. The discussion of characteristics above was restricted to stationary blades. As long as the flow is two-dimensional, the results can be applied directly to cascades moving with constant speed.

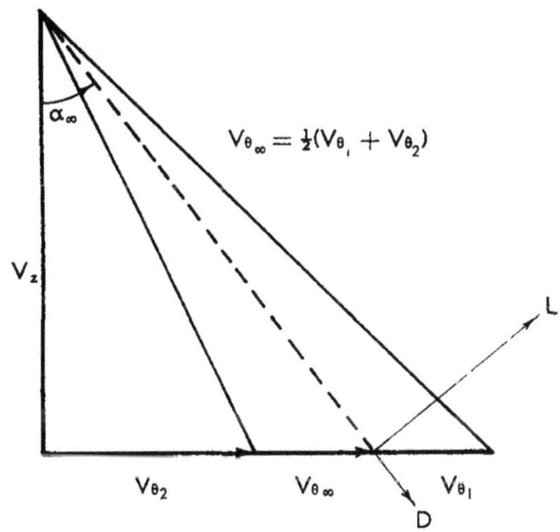

Fig. F,2d. Lift and drag forces.

If the cascade moves along its axis with a constant velocity ωr, representing the speed of a rotor at radius r, the velocities relative to the cascade are shown in Fig. F,2e. The leaving angle β_2 and the total pressure loss Δp^0 relative to the cascade are given by relations similar to those for a stator

$$\tan \beta_2 = A + F(\tan \beta_1) \tag{2-4}$$

$$\frac{2\Delta p^0}{\rho W_1^2} = K(\tan \beta_1) \tag{2-5}$$

With the configuration as shown, both pressure and kinetic energy increase through the cascade, whereas in a stator, of course, the pressure increases at the expense of kinetic energy.

The rate of increase of tangential momentum per unit area through the cascade is

$$\rho V_z(W_{\theta_1} - W_{\theta_2}) = \rho V_z(V_{\theta_2} - V_{\theta_1})$$

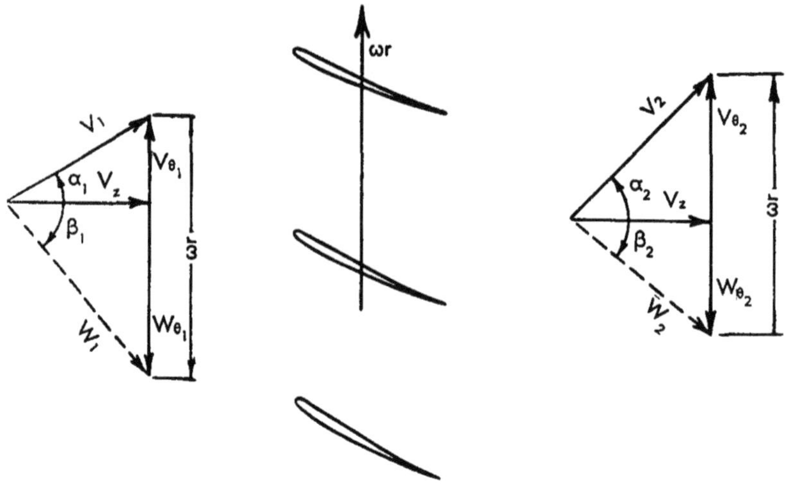

Fig. F,2e. Velocity diagrams for a moving cascade.

and since the motion of the cascade is in the tangential direction alone, the work input per unit area is

$$\rho V_z(V_{\theta_2} - V_{\theta_1})\omega r$$

the familiar Euler result. The static pressure rise through the cascade is, from the Bernoulli equation with the addition of a loss term,

$$\begin{aligned} p_2 - p_1 &= \tfrac{1}{2}\rho(W_{\theta_1}^2 - W_{\theta_2}^2 - W_1^2K\,(\tan\beta_1)) \\ &= \rho\omega r(V_{\theta_2} - V_{\theta_1}) - \tfrac{1}{2}\rho(V_{\theta_2}^2 - V_{\theta_1}^2) - \tfrac{1}{2}\rho W_1^2K\,(\tan\beta_1) \end{aligned}$$

Transferring the second term on the right to the other side, the increase in total pressure is seen to be

$$p_2^0 - p_1^0 = \rho\omega r(V_{\theta_2} - V_{\theta_1}) - \tfrac{1}{2}\rho W_1^2K\,(\tan\beta_1) \tag{2-6}$$

and the efficiency η can be defined in terms of total pressure rise as

$$\eta = \frac{p_2^0 - p_1^0}{\rho\omega r(V_{\theta_2} - V_{\theta_1})} = 1 - \frac{1}{2}\frac{W_1^2}{\omega r(V_{\theta_2} - V_{\theta_1})}K\,(\tan\beta_1) \tag{2-7}$$

F,3. Axial Compressors with Large Hub Ratio. The simplest introduction to the analysis of performance characteristics of axial turbomachines can be gained from a study of machines with large hub ratio. In such machines, the rotor blade speed is nearly constant over the blade height so that there is no advantage in twisting the blades. Thus performance can be deduced directly from two-dimensional cascade characteristics and the complexities of the three-dimensional flow theory can be avoided. The over-all performance of compressors with moderate hub ratio, say 0.7 or more, is predicted remarkably well by an analysis based on the two-dimensional cascade characteristics at *mid-blade* height; hence, the methods described in this article have somewhat wider applicability than the title implies.

An examination of cascade characteristics (e.g. Sec. E) shows that there is a range of incidence angles of 10° or so in which the losses are small, with sharply increasing losses at both ends of the range. As long as all blade sections in a compressor operate within the low loss incidence range, the performance of the machine can be predicted from the known cascade characteristics quite satisfactorily. When the compressor operating point requires flow incidence angles that are outside the low loss range, the over-all performance prediction becomes less reliable; this is a result of poor representation of compressor blade characteristics by two-dimensional cascades as flow separation progresses. The compressor operating point at which the efficiency is highest corresponds almost invariably to the sharp knee of the loss curve at the higher incidence angle. Normally, this is the "design point" for the compressor.

From the remarks above, it is evident that at least the two-dimensional losses must be included in any discussion of compressor performance; otherwise the discussion has little connection with reality. In particular, there would be no definite criterion for choice of a design point if losses were ignored. Other refinements to the analysis such as the effects of three-dimensional flow, secondary flows, and tip clearances can be deferred without drastic influence of the results, but the two-dimensional losses are essential. The performance of various blade arrangements for compressors is discussed in general terms in what follows. The flow is assumed to be two-dimensional, and particular attention is paid to the problems of choosing the design point of operation.

Performance parameters. It is convenient to introduce dimensionless coefficients to describe the turbomachine performance characteristics. The flow coefficient ϕ is defined as

$$\phi = \frac{Q}{A\omega r} \tag{3-1}$$

where Q is the volume flow rate through the machine, A is the cross-sectional area of the annulus, ω is the angular velocity of the rotor, and r is the radius of the mean blade height. The cross-sectional area of the annulus $A = 2\pi rb$, where b is the blade height. If the flow rate per unit area is constant over the annulus, then $Q = AV_z$ where V_z is the axial component of flow velocity and

$$\phi = \frac{V_z}{\omega r} \tag{3-2}$$

The work coefficient Ψ is defined as

$$\Psi = \frac{2P}{\rho Q(\omega r)^2} \tag{3-3}$$

where P is the power input to the rotor and ρ is the fluid density. The work coefficient is simply the work done per unit mass of fluid reduced to dimensionless form by the factor $\frac{1}{2}\omega^2r^2$. When the flow is two-dimensional, that is, when V_z is uniform over the annulus,

$$P = \rho Q\omega r(V_{\theta_2} - V_{\theta_1}) \tag{3-4}$$

where $V_{\theta_2} - V_{\theta_1}$ is the change in the whirl component of velocity through the rotor. Under these circumstances, the work coefficient becomes

$$\Psi = 2\,\frac{V_{\theta_2} - V_{\theta_1}}{\omega r} \tag{3-5}$$

where $V_{\theta_2} - V_{\theta_1}$ is the change in the whirl component of velocity through the rotor.

It is important to realize that the expressions for ϕ and Ψ given by Eq. 3-2 and 3-3 are valid only for two-dimensional flow, when V_z is uniform over the annulus. At zero flow rate, i.e. $Q = 0$, the power required to drive the compressor is zero, according to Eq. 3-4. In reality, however, the power required to drive a compressor at zero flow rate (shut-off) is never zero, and frequently is even larger than the power required at design flow rate for the same rotational speed. The reason for this is that the flow is not two-dimensional at zero flow rate, but is positive over some parts of the annulus and negative over other parts. Since the power input P and the volume flow rate Q can be measured readily, the coefficients ϕ and Ψ as defined by Eq. 3-1 and 3-3 can always be determined without ambiguity. The interpretation of these coefficients as given by Eq. 3-2 and 3-5 is valid only when the flow is known to be two-dimensional. For instance, Ψ approaches infinity as Q approaches zero in direct measurement, whereas Eq. 3-5 predicts (incorrectly) a finite value for Ψ at shut-off.

A third coefficient, closely connected with the work coefficient, and actually more widely used, is the pressure coefficient ψ defined as

$$\psi = \frac{2(p_2^0 - p_1^0)}{\rho\omega^2r^2} \tag{3-6}$$

where $p_2^0 - p_1^0$ is the increase in stagnation pressure produced by the rotor. If there were no losses, and if the flow were two-dimensional, the pressure coefficient ψ would be equal to the work coefficient Ψ; hence, in general,

$$\psi = \eta\Psi \tag{3-7}$$

where η is the efficiency of the rotor.

For two-dimensional flow, the work coefficient depends only on the leaving angle from the rotor blade row, and, as was pointed out in Art. 2, the leaving angle generally can be estimated from blade geometry with considerable confidence, more so than the losses. The losses must be estimated in order to find the pressure coefficient, so that the latter is derived from the work coefficient rather than determined separately. In fact, experienced designers frequently estimate over-all efficiency directly from the blade shape, Reynolds number, and Mach number without detailed breakdown of the various sources of loss.

Rotor with axial inlet flow. The simplest type of compressor consists of a single rotor blade row in an annular duct. If the flow approaching

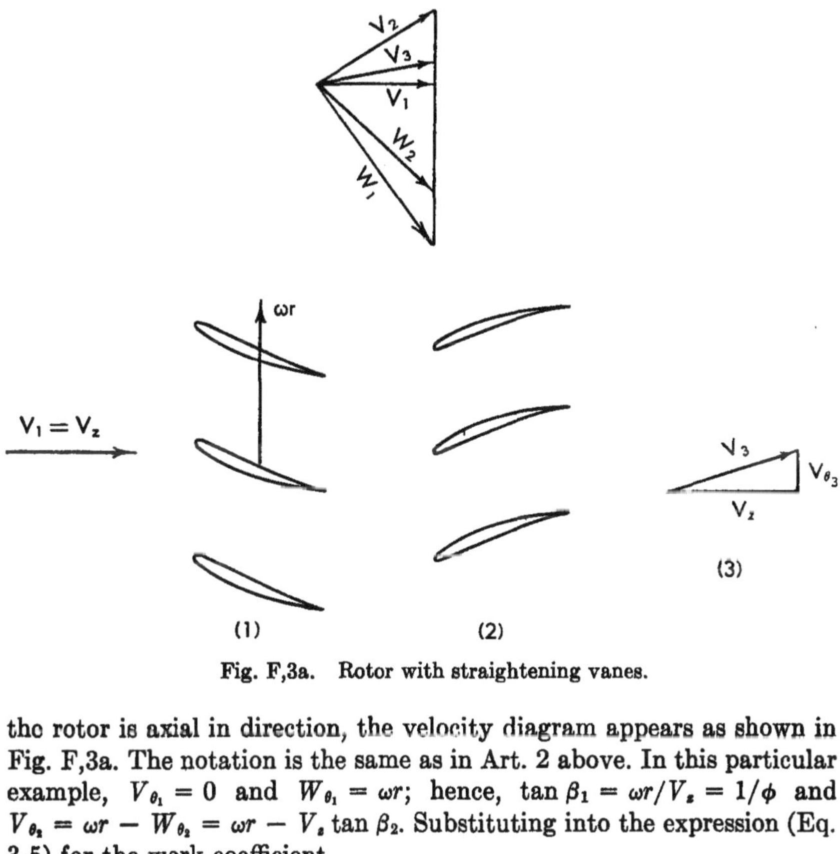

Fig. F,3a. Rotor with straightening vanes.

the rotor is axial in direction, the velocity diagram appears as shown in Fig. F,3a. The notation is the same as in Art. 2 above. In this particular example, $V_{\theta_1} = 0$ and $W_{\theta_1} = \omega r$; hence, $\tan \beta_1 = \omega r / V_z = 1/\phi$ and $V_{\theta_2} = \omega r - W_{\theta_2} = \omega r - V_z \tan \beta_2$. Substituting into the expression (Eq. 3-5) for the work coefficient,

$$\Psi = 2(1 - \phi \tan \beta_2) \tag{3-8}$$

where $\tan \beta_2$ is a function of ϕ alone, of the form

$$\tan \beta_2 = A + F\left(\frac{1}{\phi}\right)$$

In order to simplify the discussion of performance, it will be assumed that $F(1/\phi) = 0$, corresponding to high solidity. The minor modifications to the performance characteristics resulting from a more general relation between $\tan \beta_2$ and ϕ will be mentioned later.

With the simplification that $\tan \beta_2 = \text{const}$, the relation between Ψ and ϕ is linear and $\Psi = 2 - 2\phi \tan \beta_2$. The work coefficient Ψ and a quantity proportional to the power, $\phi\Psi = 2P/(\rho A \omega^3 r^3)$ are sketched in Fig. F,3b. The curves in the figure are good representations of actual rotor performance when the blade solidity is high, except for low values of ϕ whereas the shut-off condition is closely approached the two-dimensional flow breaks down and the theory is no longer valid. When the blade solidity is not large, the leaving angle can be represented approximately by the relation $\tan \beta_2 = A + B \tan \beta_1$ (see Art. 2), and then $\Psi = 2(1 - B) - 2A\phi$. The performance curves of Fig. F,3b are the same shape, but the intercept on the axis $\phi = 0$ is smaller. More generally,

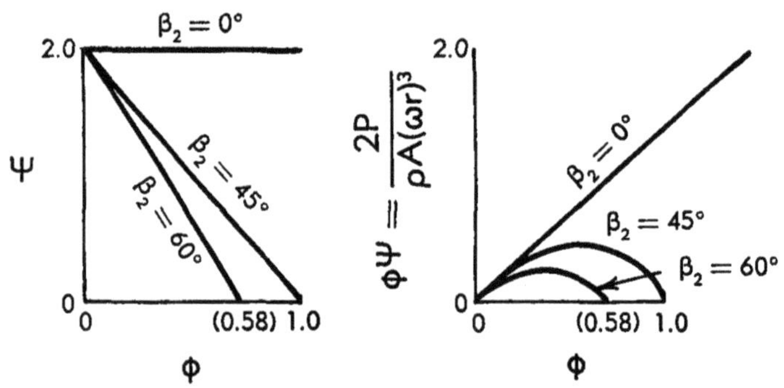

Fig. F,3b. Work and power characteristics for constant leaving angle.

with $\tan \beta_2 = A + F(\tan \beta_1)$, the curve of Ψ vs. ϕ will have a slight curvature, with $d^2\Psi/d\phi^2$ always less than zero.

A single rotor in a duct with axial approach flow imparts a circumferential component of velocity to the flow. Although a fraction of the velocity head of the whirl component may be converted to pressure downstream, the fraction is generally small unless a stator blade row behind the rotor turns the flow back to the axial direction, at the same time converting as much of the whirl velocity head to pressure as the efficiency of the stator allows. The stator, of course, does not change the work coefficient of the rotor.

The losses in the compressor can be separated into three groups, associated with the rotor blades, the stator blades, and the duct. Although the losses in any one of the groups are not strictly independent of the configuration of components of the other groups, the interaction is generally slight. With the notation of Art. 2, the total pressure loss through the rotor can be expressed in the form

$$\Delta p_r^0 = \tfrac{1}{2}\rho W_1^2 K_r$$

where the subscript r refers to the rotor. The loss coefficient K_r is a function of $\tan \beta_1$ alone and, hence, in this example, is a function of $\omega r/V_a = 1/\phi$. In the most convenient dimensionless form, the loss through the rotor is then

$$\frac{2\Delta p_r^0}{\rho\omega^2 r^2} = (1 + \phi^2)K_r(\phi) \tag{3-9}$$

Now K_r is small over a range of a few degrees in the relative inlet angle β_1, and rises sharply at the ends of this range. Hence the dimensionless loss will have a similar behavior as a function of ϕ; Fig. F,3c illustrates a typical example. The detailed shape of the curve can be found from two-dimensional cascade data for the particular blade shape.

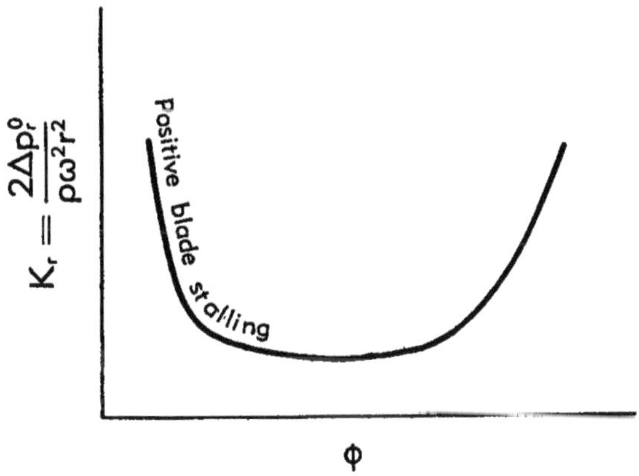

Fig. F,3c. Rotor loss coefficient.

The loss in stagnation pressure through the stator can be expressed as

$$\Delta p_s^0 = \tfrac{1}{2}\rho V_2^2 K_s$$

where the loss coefficient K_s is a function of $\tan \alpha_2 = (1/\phi) - \tan \beta_2$. In dimensionless form, this loss becomes

$$\frac{2\Delta p_s^0}{\rho\omega^2 r^2} = [(1 - \phi \tan \beta_2)^2 + \phi^2]K_s(\phi) \tag{3-10}$$

Since K_s is small over a range of ϕ corresponding to a few degrees in α_1, and further the factor multiplying K_s in Eq. 3-10 does not vary rapidly with ϕ, the dimensionless loss through the stator can be represented by a graph similar to that for the rotor in Fig. F,3c. If the rotor and stator are well matched, the regions of low loss, particularly the lower ends, will coincide.

The magnitude of the duct loss has a definite meaning only if the application of the turbomachine is specified. When a diffusor, elbow, or other component is an integral part of the turbomachine system, it is appropriate to include the component loss in a procedure for choosing blades. The stagnation pressure loss for the system, exclusive of rotor and stator blades, will be proportional to the square of the through-flow velocity, i.e.,

$$\frac{2\Delta p_d^0}{\rho\omega^2 r^2} = K_d\phi^2 \tag{3-11}$$

where K_d is generally constant.

Introducing the notation of Eq. 3-6 and 3-7, and collecting the individual terms: Eq. 3-9, 3-10, and 3-11,

$$\Psi - \psi = (1 - \eta)\Psi = K_r(1 + \phi^2) + K_s[(1 - \phi \tan \beta_2)^2 + \phi^2] + K_d\phi^2 \tag{3-12}$$

Dividing through by $\Psi = 2 - 2\phi \tan \beta_2$,

$$1 - \eta = \frac{K_r(1 + \phi^2) + K_s[(1 - \phi \tan \beta_2)^2 + \phi^2] + K_d\phi^2}{2 - 2\phi \tan \beta_2} \tag{3-13}$$

The numerator in the latter expression is small in the low-loss range of ϕ, provided that K_d is not a very large number, and generally increases with ϕ although the first term multiplying K_s decreases with ϕ. The denominator increases with ϕ decreasing; hence $1 - \eta$ decreases with ϕ down to the value of ϕ corresponding to the sharp rise in K_r and K_s, reaches a minimum, and begins to increase. Since $\psi = \eta\Psi$, the maximum value of ψ occurs at a lower value of ϕ than that corresponding to maximum η.

The preceding analysis is appropriate for estimating the performance of a compressor with given blading. The choice of blading to produce a given performance is limited by the restrictions on maximum pressure rise discussed in Art. 2. If one accepts the simple rule for velocity ratio as applicable to both rotor and stator, then the restrictions on the velocity diagram are, for a stator designed for axial exit flow,

$$\frac{W_2}{W_1} > \nu \qquad \frac{V_3}{V_2} > \nu \tag{3-14}$$

where $\nu \cong 0.65$. From the velocity diagram, Fig. F,3a,

$$W_1^2 = \omega^2 r^2(1 + \phi^2) \qquad W_2^2 = \omega^2 r^2 \phi^2 + \left(1 - \frac{\Psi}{2}\right)^2$$

$$V_2^2 = \omega^2 r^2\left(\phi^2 + \frac{\Psi^2}{4}\right) \qquad V_3^2 = \omega^2 r^2 \phi^2$$

and substituting into Eq. 3-14, conditions on ϕ and Ψ become, with some rearrangement,

$$\frac{(\Psi - 2)^2}{4\nu^2} + \frac{\phi^2}{\nu^2/(1 - \nu^2)} > 1 \quad \text{(rotor)} \tag{3-15}$$

$$\Psi < 2\frac{\sqrt{1 - \nu^2}}{\nu}\phi \quad \text{(stator)} \tag{3-16}$$

where Eq. 3-15 is a restriction on the flow through the rotor and Eq. 3-16 is applicable to the stator. Then the design point Ψ, ϕ must lie outside the ellipse in Fig. F,3d (Eq. 3-15) and to the right of the straight line (Eq. 3-16) to satisfy the condition of moderate loss. The figure is drawn

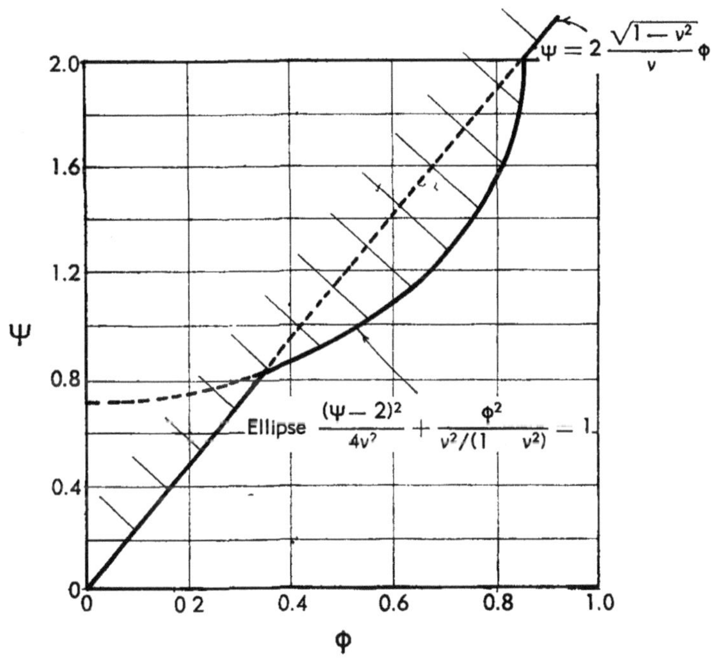

Fig. F,3d. Design point limits for rotor with straightening vanes.

with $\nu = 0.65$, but the shape of the limiting line is similar for other values of ν.

The design point for a compressor of this particular type may be placed anywhere within the unshaded portion of Fig. F,3d, although the efficiency will vary from point to point. The solidity for highest efficiency will be of the order of unity along the limiting line and will decrease as the design point moves away from the shaded region. The equation for efficiency (Eq. 3-13) can be written in terms of ϕ and Ψ as

$$\eta = 1 - \frac{K_r(1 + \phi^2) + K_s(\frac{1}{4}\Psi^2 + \phi^2) + K_d\phi^2}{\Psi} \tag{3-17}$$

Along the limiting line of Fig. F,3d, Ψ is related to ϕ so that η becomes a function of ϕ alone. The coefficients K_r and K_s are functions of ϕ as well, since the optimum rotor and stator cascades vary with the design point. However, the variation of K_r or K_s with ϕ is not very great, and an examination of the expression (Eq. 3-17) shows that η is small when ϕ is very small and small again when ϕ is large. Hence η is a maximum at some intermediate value of ϕ, although the precise value cannot be determined unless K_r, K_s, and K_d are specified. Since these coefficients are not

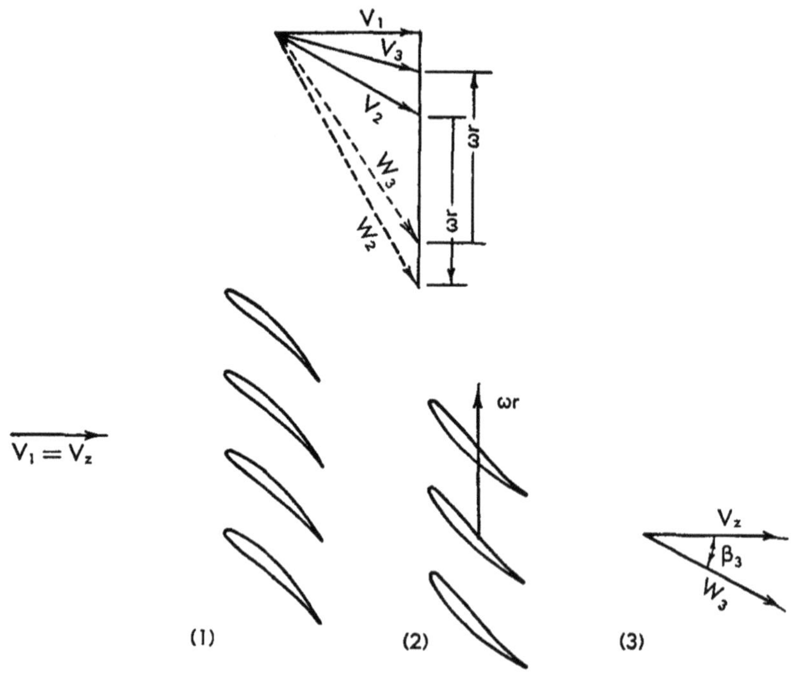

Fig. F,3e. Rotor with prerotation vanes.

readily expressed in analytical form as functions of ϕ, the calculation of maximum η is best done graphically. Usually the maximum efficiency occurs in the range $0.3 < \phi < 0.5$.

Rotor with prerotation vanes. A second, very common type of single-stage blower has a row of prerotation vanes in front of the rotor. The advantage of this arrangement is that it is possible to obtain a higher relative velocity at the rotor and hence reduce the rotor rotational speed required for a given pressure rise. The cascade arrangement and velocity diagram are shown in Fig. F,3e. The convention used above, measuring V_θ positive upwards and W_θ positive downwards, is retained. Normally, at design operation, the absolute exit velocity from the rotor, V_3, would be in the axial direction.

The work done by the rotor is given by the work coefficient Ψ where

$$\Psi = 2\frac{V_{\theta_3} - V_{\theta_2}}{\omega r}$$

for two-dimensional flow. Since $V_{\theta_2} = V_z \tan \alpha_2$ (where α_2 is negative in this example) and $V_{\theta_3} = \omega r - V_z \tan \beta_3$ where β_3 is the leaving angle relative to the rotor,

$$\Psi = 2 - 2(\tan \beta_3 + \tan \alpha_2)\phi \tag{3-18}$$

This is the general result for prescribed leaving angles from the prerotation vanes and the rotor. Since the direction of flow entering the

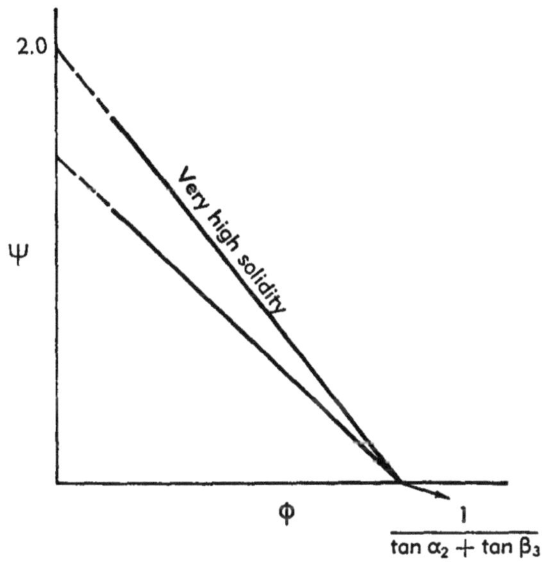

Fig. F,3f. Work coefficient for rotor with prerotation vanes.

vanes is constant (axial), $\tan \alpha_2$ is independent of the flow rate. The leaving angle from the rotor is given by a relation of the form $\tan \beta_3 = A + F(\tan \beta_2)$ where $\tan \beta_2 = (\omega r + V_{\theta_2})/V_z = (1/\phi) + \tan \alpha_2$.

If the rotor cascade is such that the leaving angle is approximated by the potential flow theory, so $\tan \beta_3 = A + B \tan \beta_2 = A + B \tan \alpha_2 + (B/\phi)$, then

$$\Psi = 2(1 - B) - 2[A + (B + 1) \tan \alpha_2]\phi \tag{3-19}$$

and the performance curve is again a straight line in the Ψ, ϕ diagram. For a rotor of very high solidity, $B = 0$ as shown in Fig. F,3f, the efficiency can be expressed in terms of ϕ and Ψ as for the previous example, although the form is modified slightly and the contribution of residual swirl downstream of the rotor is neglected.

$$1 - \eta = \frac{(K_s + K_r + K_d)\phi^2 + \frac{1}{4}K_r(\Psi + 2)^2}{\Psi} \tag{3-20}$$

The limitation on the maximum amount of diffusion through the rotor cascade can be applied as above. There is no restriction on the prerotation vanes, since they accelerate the flow. Assuming that the rotor operates at design condition so $\tan \beta_3 = 1/\phi$, the limiting criterion is $W_3/W_2 > \nu$, or in terms of ϕ and Ψ:

$$\frac{(\Psi + 2)^2}{4/\nu^2} - \frac{\phi^2}{1/(1 - \nu^2)} < 1 \tag{3-21}$$

Hence the limiting line is a branch of a hyperbola with center at $\phi = 0$, $\Psi = 2$; semiaxes $2/\nu$ and $1/\sqrt{1 - \nu^2}$ and asymptote $\Psi + 2 = 2\sqrt{1 - \nu^2}/\nu\ \phi$. Any point below and to the right of the hyperbola represents a possible design point (see Fig. F,3g for $\nu = 0.65$). Comparison

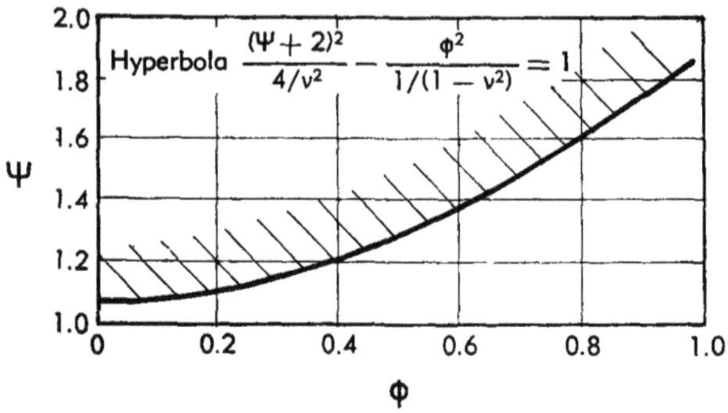

Fig. F,3g. Design point limits for rotor with prerotation vanes.

with Fig. F,3d shows that the compressor with prerotation vanes followed by a rotor has considerably more flexibility in choice of the design point than the compressor with rotor followed by straightening vanes if $\phi < 0.8$. At higher values of ϕ, the latter type allows larger Ψ; however, this advantage is rather academic because the losses are generally so severe at flow coefficients greater than 0.8 that such design points are undesirable.

Compressor with repeating flow. The two types of compressor discussed above are suitable for single-stage design with axial entering and leaving flow. For multistage design, it is not necessary to have axial flow between any pair of blade rows; so an extra degree of freedom is introduced. The diagram, Fig. F,3h, shows a typical arrangement, with the flow repeating its pattern after each two rows. The two types above are special cases of this more general blade arrangement, the first corresponding to $\alpha_1 = 0$ in Fig. F,3h, the second to $\alpha_2 = 0$.

The work coefficient for the repeating stage is given by the relations

$$\Psi = 2\frac{W_{\theta_1} - W_{\theta_2}}{\omega r} = 2\frac{V_{\theta_2} - V_{\theta_1}}{\omega r} = 2\frac{\omega r - V_{\theta_1} - W_{\theta_2}}{\omega r}$$

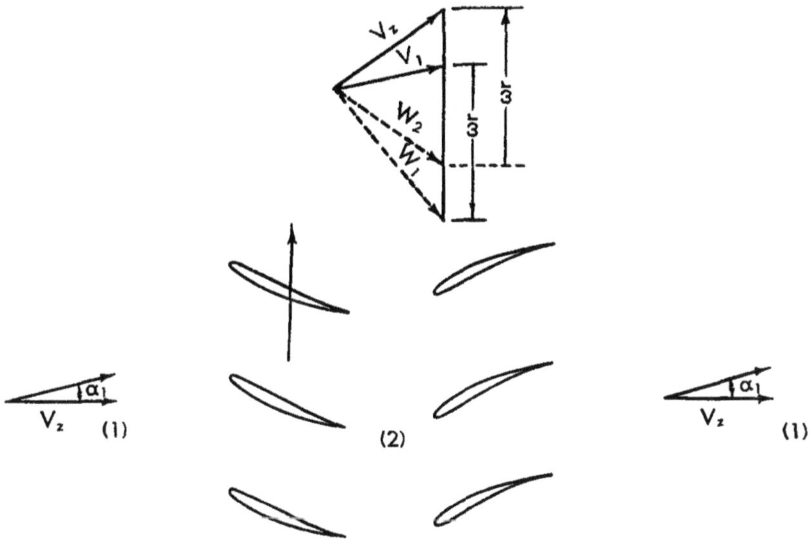

Fig. F,3h. Rotor and stator for repeating flow.

or in terms of the leaving angles β_2 and α_1 for rotor and stator, respectively,

$$\Psi = 2 - 2(\tan \beta_2 + \tan \alpha_1)\phi \tag{3-22}$$

If $\tan \beta_2 = A_r + B_r \tan \beta_1$ and $\tan \alpha_1 = A_s + B_s \tan \alpha_2$ with A's and B's constant, as is suggested by potential theory, Ψ is again linear in ϕ

$$\Psi = 2\frac{1 - B_r - B_s + B_rB_s}{1 - B_rB_s} - 2\frac{A_r + A_s - A_sB_r - A_rB_s}{1 - B_rB_s}\phi \tag{3-23}$$

where $B_r = B_s = 0$ for high solidity. The stage efficiency can be expressed in terms of ϕ and Ψ as

$$\eta = 1 - \frac{K_r[\phi^2 + (1 - \phi \tan \alpha_1)^2] + K_s[\phi^2 + (\frac{1}{2}\Psi + \phi \tan \alpha_1)^2] + K_d\phi^2}{\Psi} \tag{3-24}$$

The duct loss coefficient K_d is small if there are many stages, since the major duct losses occur in front of, and behind, the blade rows.

The limitation on maximum diffusion through the rotor is $W_2/W_1 > \nu$ and through the stator $V_1/V_2 > \nu$. In terms of ϕ and Ψ, these conditions become

$$\frac{\phi^2 + (1 - \phi \tan \alpha_1 - \frac{1}{2}\Psi)^2}{\phi^2 + (1 - \phi \tan \alpha_1)^2} > \nu^2 \quad \text{(rotor)} \tag{3-25}$$

$$\frac{\phi^2(1 + \tan^2 \alpha_1)}{\phi^2 + (\phi \tan \alpha_1 + \frac{1}{2}\Psi)^2} > \nu^2 \quad \text{(stator)} \tag{3-26}$$

The inequalities above contain a parameter $\tan \alpha_1$, so that they give a family of limiting lines as $\tan \alpha_1$ is varied. The stator limit is a straight line, whereas the rotor limit is an ellipse or hyperbola with axes at an angle to the axes of ϕ and Ψ. A particularly simple situation arises if rotor and stator limits are met simultaneously. Then $\tan \alpha_1 = (1 - \frac{1}{2}\Psi)/2\phi$

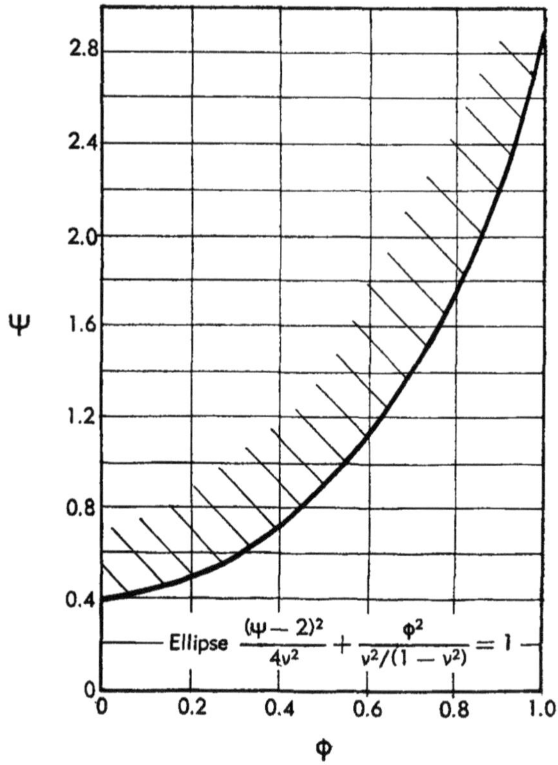

Fig. F,3i. Design point limits for symmetrical blading.

and the velocity diagram becomes symmetric, with conditions (Eq. 3-25 and 3-26) reducing to

$$\frac{\left(\Psi - 2\frac{1+\nu^2}{1-\nu^2}\right)^2}{\left(\frac{4\nu}{1-\nu^2}\right)^2} + \frac{\phi^2}{\left(\frac{\nu}{1-\nu^2}\right)^2} > 1 \tag{3-27}$$

The limiting line with $\nu = 0.65$ is shown in Fig. F,3i for comparison with Fig. F,3d and F,3g. The efficiency along the limiting line is obtained from Eq. 3-24 on substituting for $\tan \alpha_1$ and applying Eq. 3-27 with an equality sign:

$$\eta = 1 - \frac{1}{1-\nu^2}K_r - K_d\frac{\phi^2}{\Psi} \tag{3-28}$$

For the stage efficiency in a multistage compressor, the last term on the right should be dropped. Then the efficiency is not a direct function of the flow coefficient, but depends on ϕ only through the loss coefficient K_r, which has in turn a weak dependence on ϕ.

F,4. Axial Compressors with Small Hub Ratios. The hub ratios of compressors have decreased over the years as the volume flow rate for a given frontal area was raised. This trend led to difficulties, because it was desirable to keep the total pressure rise uniform from hub to casing, and the large difference of rotor blade velocity over the blade height was incompatible with constant work input. Three-dimensional flow theories played a major role in the development of compressors with low hub ratio. Actually, this is perhaps the sole instance in the history of axial compressors where an important design concept came from purely theoretical investigations.

The three-dimensional flow theories are developed in detail in Sec. C. The applications of all ramifications of these theories to compressor design would require a very lengthy treatment; hence we will discuss only the major applications, with relatively simple examples illustrating their importance.

Vortex flow pattern. The simplest flow pattern that is of any interest for compressors is a simple vortex whirl distribution with a constant axial velocity. The total pressure in such a flow is independent of radius, and the flow is two-dimensional on cylindrical surfaces. As the flow passes through a blade row the strength of the circulation of the basic vortex is changed. This was the type of flow pattern that was used in all low speed fans and pumps and is still the most practical for most single-stage machines. The first axial compressor designs were based on vortex flow patterns, and there are some advantages to this. In particular, if the blade height is chosen to give constant axial velocity throughout the machine, the same blades can be used for all stages if similar blade sections are kept at the same radius.

The major disadvantage with the vortex flow pattern appears only if the hub ratio is small. For instance, Fig. F,4a shows the velocity diagrams at three radii for a multistage compressor with a hub ratio of 0.5. It is convenient to introduce the dimensionless radius $\xi = r/R$ and dimensionless velocity components $\phi = V_z/\omega R$, $\lambda_1 = V_{\theta_1}/\omega R$, $\lambda_2 = V_{\theta_2}/\omega R$ where R is the rotor tip radius. Then the work coefficient is $\Psi = 2\xi(\lambda_1 - \lambda_2)$ and the radius ξ_s at which the velocity diagram is symmetrical is given by the condition $\xi_s = \lambda_1 + \lambda_2$. Hence for the vortex flow pattern

$$\xi\lambda_1 = \tfrac{1}{2}(\xi_s^2 + \tfrac{1}{2}\Psi)$$

$$\xi\lambda_2 = \tfrac{1}{2}(\xi_s^2 - \tfrac{1}{2}\Psi)$$

The velocity diagrams shown correspond to $\phi = 0.5$, $\Psi = 0.4$, and $\xi_s = 0.75$. It is clear that the rotor hub would be in serious difficulties. Decreasing ξ_s to pull the rotor hub triangle to the left gives some improvement, but the conditions for the rotor are still unacceptable. The velocity entering the rotor at the hub is lower than is needed and the velocity entering the rotor at the tip is higher than needed; what would be most desirable is a shift of axial flow velocity from the casing to the hub.

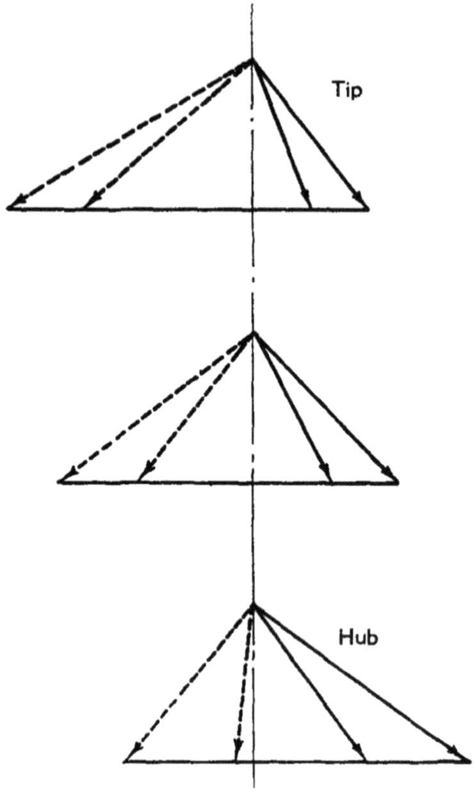

Fig. F,4a. Velocity diagrams for vortex flow.

$\phi = 0.5 \qquad \psi = 0.4 \qquad \xi_s = 0.75$

Rotational flow patterns. The application of rotational flow to the control of velocity profiles can be illustrated by physical considerations only. Suppose that a uniform axial flow approaches a row of stationary vanes that impart a whirl to the flow. The distributions of axial and whirl components of velocity for a vortex flow downstream of the vanes are shown in Fig. F,4b, left. The difference in pressure between hub and casing requires that the velocity at the hub be higher than at the casing because the total pressure is uniform behind the vanes, which are here assumed frictionless. Fig. F,4b, middle, shows the distributions when the whirl velocity is the same as in Fig. F,4b, left, at the mean radius but now increases with the radius. The pressure difference between hub and casing is approximately the same as before; hence the resultant velocities at hub and casing cannot be much different. With the whirl velocities now small at the hub and large at the casing, the axial component must increase at the hub and decrease outwards to give the same static and total pressure as for the vortex flow pattern. The reverse occurs for a whirl distribution that decreases outward faster than for a vortex, Fig. F,4b, right.

As can be seen from this illustration, fixing the whirl distribution determines the direction of flow behind the vanes as well as the axial velocity distribution. Subject only to the continuity restriction on the

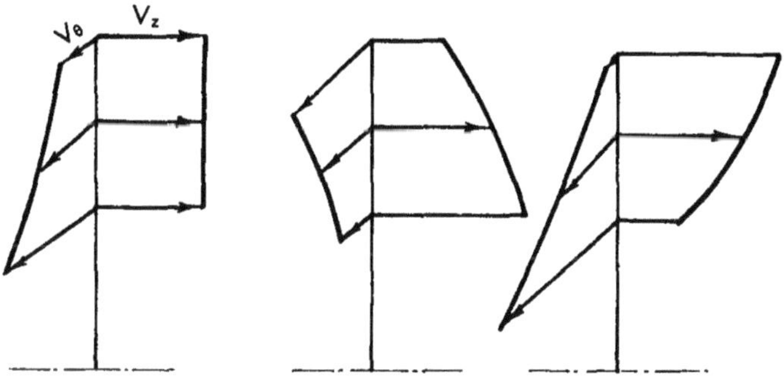

Fig. F,4b. Components of velocity leaving stator.

axial component of velocity, any one of the three can be chosen arbitrarily, thereby determining the other two. Suppose now that the flow from the vanes passes through a rotor that adds the same total pressure increment at all radii, again without losses. This is equivalent to adding a whirl increment that is inversely proportional to the radius. The pressure difference from hub to casing is larger than in front of the rotor. The vortex flow leaves the rotor as another vortex flow, as shown in Fig. F,4c. The flow with a whirl increasing with the radius before entering the rotor has the same whirl increment upon leaving, but as it is the square of the whirl component that enters in the Bernoulli equation, the contribution of the whirl component is less at the root and greater at the tip than for the original vortex flow. Hence the axial component increases at the hub and decreases at the casing to compensate, as shown in Fig. F,4c, middle. Again, the reverse occurs in the third example.

In a rotational flow with constant total pressure everywhere, the vortex lines are parallel to the streamlines. This may be shown by the following argument. The vorticity components here are

$$\omega_z = \frac{1}{r}\frac{d}{dr}(rV_\theta)$$

in the axial direction and

$$\omega_\theta = -\frac{dV_z}{dr}$$

in the whirl direction; the condition that the streamlines and vortex lines are parallel is

$$\frac{V_\theta}{r}\frac{d}{dr}(rV_\theta) + V_z\frac{dV_z}{dr} = 0$$

and since $dp/dr = \rho V_\theta^2/r$, we have $p/\rho + \frac{1}{2}(V_\theta^2 + V_z^2) = \text{const.}$ Radial components of velocity have been neglected in the analysis although it is clear they must appear if the axial velocity distribution changes. For the time being it will be assumed that blade rows are separated sufficiently so the radial components of velocity are zero at intermediate stations

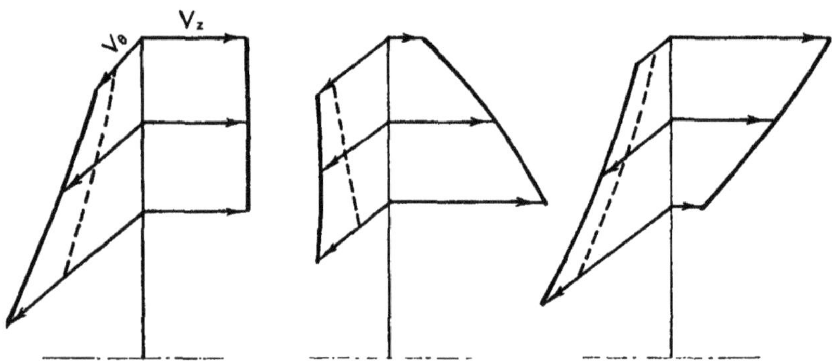

Fig. F,4c. Components of velocity leaving rotor.

between the rows. Behind the rotor the streamlines, and hence the vortex lines, are at a larger angle to the axis than in front of the rotor. Hence the whirl component of vorticity is larger behind and therefore the axial velocity gradient increases after the rotor.

Clearly, a whirl velocity that increases with radius is the type needed to improve the velocity diagram in Fig. F,4a. Choosing $V_{\theta_1}/R\omega = \lambda_1 = \nu_1\xi$ where ν_1 is a constant, then

$$\phi_1^2 = K_1 - \nu_1^2\xi^2 \tag{4-1}$$

with a constant K_1 that must satisfy the continuity equation

$$\int_{\xi_i}^{\xi} \sqrt{K_1 - \nu_1^2\xi^2}\,\xi d\xi = \tfrac{1}{2}(1 - \xi_i^2)\bar{\phi} \tag{4-2}$$

where $\bar{\phi}$ is the average flow coefficient in the annulus. Then for a constant work coefficient Ψ, $\lambda_2 = \lambda_1 + (\Psi/2\xi) = \nu_1\xi + (\Psi/2\xi)$ and

$$\phi_2^2 = K_2 - \int \frac{\lambda_2}{\xi}\frac{d}{d\xi}(\xi\lambda_2) = K_2 - \frac{\nu_1^2}{2}\xi^2 - \Psi\nu_1 \ln \xi \tag{4-3}$$

where K_2 is determined from the continuity condition by numerical integration. The velocity diagrams with $\bar{\Phi} = 0.50$, $\Psi = 0.4$, and $\lambda_1 + \lambda_2 = \xi$ at $\xi = \xi_s$ for comparison with Fig. F,4a are shown in Fig. F,4d. The improvement over the previous example, particularly in the rotor, is very obvious.

Another choice of whirl velocity distribution would lead to different velocity diagrams, but once a satisfactory type of distribution is found, there will be little to gain from minor modifications. An extra degree of

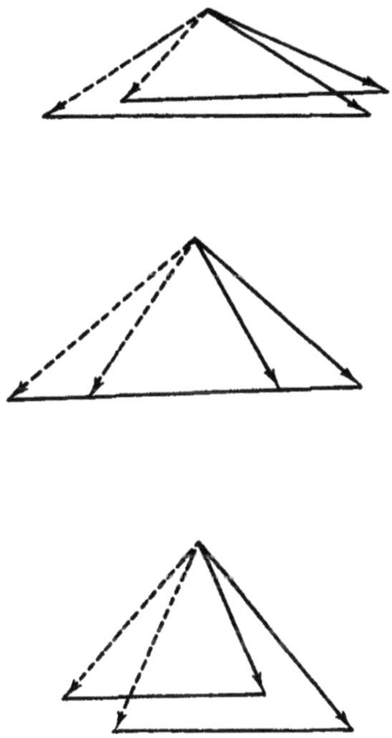

Fig. F,4d. Velocity diagrams for flow $\lambda_1 \sim \xi$.
$\bar{\phi} = 0.5$, $\bar{\psi} = 0.4$, $\lambda_1 + \lambda_2 = \xi$ at $\xi = \xi_s$

flexibility is added if the total pressure can be varied radially. However, this can be accomplished efficiently only through a rotor, so that there is no advantage for the flow approaching the first rotor where conditions are usually most critical.

Axial velocities near blade rows in rotational flow. The axial velocity changes in going through the rotor or stator blade rows if the flow is rotational; if one chooses distributions far upstream and far downstream of the row, one must know the distribution at the blade row in order to design blading to produce the desired flow. From linearized three-dimensional flow theory, it can be shown that the axial velocity at an actuating disk representing the blade row is the average of upstream and downstream axial components. If the blade extends over a finite axial distance, one needs to know the axial velocity at leading and trailing edges, at least, for design of the blading. The three-dimensional flow theory gives the velocities everywhere, but in its most general form it involves elaborate calculation. A very simplified form of the theory, however, can give results that are sufficiently accurate for all practical purposes.

Assume that we have a two-dimensional cascade of finite span, but one in which the blades are twisted. Represent the blade row by an actuating plane for the moment, and assume that the flow is rotational. Then the axial and "radial" components of velocity can be represented by

$$\left.\begin{aligned} V_x &= -\sum_{n=1}^{\infty} A_n e^{\pi n x} \cos \pi n y + V_{x_1} && x < 0 \\ V_x &= \sum_{n=1}^{\infty} A_n e^{-\pi n x} \cos \pi n y + V_{x_2} && x > 0 \\ V_r &= \sum_{n=1}^{\infty} A_n e^{\pi n x} \sin \pi n y && x < 0 \\ V_r &= \sum_{n=1}^{\infty} A_n e^{-\pi n x} \sin \pi n y && x > 0 \end{aligned}\right\} \tag{4-4}$$

where $V_{x_1}(y)$ and $V_{x_2}(y)$ are the axial components far upstream and far downstream, respectively. The channel width is then taken equal to unity in these expressions. The coefficients A_n can be determined from

$$V_{x_2} - V_{x_1} = -2 \sum_{r=1}^{\infty} A_n \cos \pi r y \tag{4-5}$$

equivalent to the continuity condition across the simulated blade row. Eq. 4-4 satisfy the conditions

$$\frac{\partial V_x}{\partial x} + \frac{\partial V_r}{\partial y} = 0 \tag{4-6}$$

$$\left.\begin{aligned} \frac{\partial V_x}{\partial y} - \frac{\partial V_r}{\partial x} &= \frac{dV_{x_1}}{dx} && x < 0 \\ \frac{\partial V_x}{\partial y} - \frac{\partial V_r}{\partial x} &= \frac{dV_{x_2}}{dx} && x > 0 \end{aligned}\right. \tag{4-7}$$

Hence they satisfy the linearized equations of motion upstream and downstream. The last pair of equations states that the vorticity is constant along the mean streamlines, as for linearized rotational flow.

If $V_{z_2} - V_{z_1}$ is linear in y and of the form $V_{z_2} - V_{z_1} = K(y - \frac{1}{2})$, then the Fourier coefficients from Eq. 4-5 are

$$A_n = \frac{2K}{\pi^2}\frac{1}{n^2} \quad \text{for } n \text{ odd}$$

and

$$A_n = 0 \qquad \text{for } n \text{ even}$$

Hence the terms in the series decrease very rapidly, and retaining the first term alone will give an error of only 10 per cent. Then we have

$$V_z = V_{z_2} + \frac{2K}{\pi^2} e^{-\pi x} \cos \pi y \qquad x > 0$$

but this term gives a poor approximation to V_z at $x = 0$, since then $V_z = \frac{1}{2}(V_{z_1} + V_{z_2})$. However, if we put

$$V_z = V_{z_2} - \frac{V_{z_2} - V_{z_1}}{2} e^{-\pi x} \tag{4-8}$$

the value of V_z is correct at $x = 0$, it is correct at $x = \infty$, and it contains the exponential factor that was shown to be most important in the Fourier expansion; there is a similar formula for x less than zero, i.e.

$$V_z = V_{z_1} + \frac{V_{z_2} - V_{z_1}}{2} e^{+\pi x} \qquad x < 0$$

Suppose that the blade row extends axially from $x = 0$ to $x = a$ and is loaded in the x direction with a distribution $(1/a)f(x)$ where $(1/a)\int_0^a f(x)dx = 1$. Then superimposing the actuator disk solutions, an elementary calculation gives

$$V_z = V_{z_1} + \tfrac{1}{2}(V_{z_2} - V_{z_1})e^{\pi x/b} F\left(\frac{b}{a}\right) \qquad x < 0$$

$$V_z = V_{z_2} - \tfrac{1}{2}(V_{z_2} - V_{z_1})e^{-\pi x/b} G\left(\frac{b}{a}\right) \qquad x > a$$

where

$$F\left(\frac{b}{a}\right) = \frac{1}{a}\int_0^a f(\xi)e^{-\pi\xi/b}d\xi$$

$$G\left(\frac{b}{a}\right) = \frac{1}{a}\int_0^a f(\xi)e^{-\pi(a-\xi)/b}d\xi$$

The functions F and G are plotted in Fig. F,4e, for simple loadings shown in the shaded diagrams

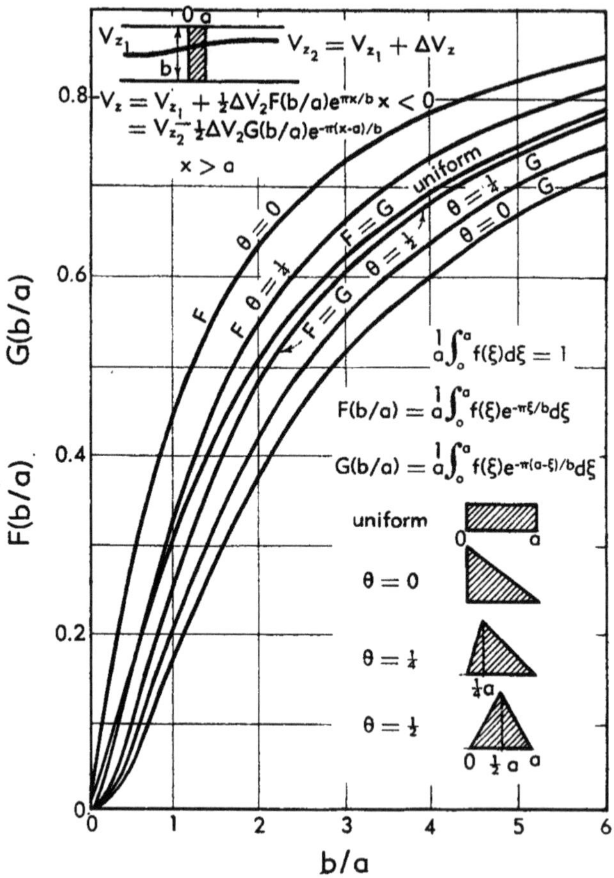

Fig. F,4e. Blade row interference factors.

F,5. Secondary Flows and Stalling in Axial Compressors. The previous articles have given an account of the flow in axial compressors based on an axially symmetric flow distribution and an infinite number of blades. The major flow patterns for unstalled operation are described remarkably well in such terms but, of course, it is necessary to take into account the finite number of blades, since the interaction of the wall boundary layers with the individual blade channels, the tip clearance leakage, and the behavior of the blade boundary layers cannot be represented adequately by an axially symmetric flow. Some of the asymmetric flows are discussed in this article, with particular attention given to those that are best understood, or if not as yet well understood, of major importance. A number of the asymmetric phenomena have been studied separately, although in practice they are not simply additive and may have strong interactions. The literature in this field is very extensive;

unfortunately, very little of it can be considered definitive at the present time.

Secondary flow. The assumption of an infinite number of blades in turbomachine theory is quite satisfactory if the radial variation of velocity is small over a radial distance comparable with the blade gap, since two-dimensional cascade theory can be applied locally at each radius. However, the thickness of the wall boundary layer is generally of the same order as the blade gap, and with a large variation of velocity through the boundary layer the influence of the finite blade gap becomes important.

The nature of the flow distribution through the blade row can be understood qualitatively from a simple model. Imagine a boundary layer flow approaching a two-dimensional cascade, as shown in Fig. F,5a. We

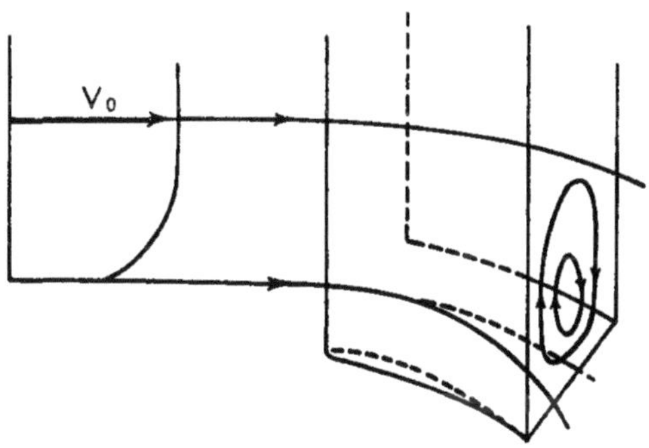

Fig. F,5a. Boundary layer flow through a blade channel.

will neglect all viscous effects (including separation), which will be important only very close to the wall and the blade surfaces. If the boundary layer thickness is of the same order of magnitude or less than the blade gap, the static pressure distribution at the wall will tend to be controlled by the flow outside the boundary layer. In particular, the pressure gradient within the blade row normal to the streamlines will be $\partial p/\partial n \cong \rho V^2/R$, where V is the velocity outside the boundary layer and R is the radius of curvature of a streamline. The fluid in a stream filament within the boundary layer has a lower approach velocity and lower velocity within the channel, and hence must have a smaller radius of curvature of the streamline to balance the prescribed pressure gradient. As a result, a velocity component perpendicular to the outside streamlines appears within the blade row, directed from the high pressure side of the channel to the low pressure side. Since the normal velocity on the blade surfaces must remain zero, a circulatory flow of the type shown in Fig. F,5a develops in each blade channel.

A general analysis of the flow described above is intractable because the flow is rotational. However, the problem can be linearized if restricted to small deviations from the basic two-dimensional flow outside the boundary layer as was done by Squire and Winter [3] and Hawthorne [4]. The method we will use is somewhat different from theirs, although the results are essentially the same.

We choose an orthogonal coordinate system based on the two-dimensional flow outside the boundary layer with dimensionless coordinates s along the streamlines, n normal to the streamlines, and ζ normal to the wall. The characteristic length used for defining the dimensionless coordinates is the blade gap (the ratio gap/chord $\sim 0(1)$ in what follows). A triad of unit vectors $\mathbf{i}, \mathbf{j}, \mathbf{k}$ is chosen so that $\mathbf{i}$ and $\mathbf{j}$ are in a plane parallel to the wall, with $\mathbf{i}$ in the direction of s and $\mathbf{j}$ in the direction of n. Then the two-dimensional flow through the blade row is given by the velocity vector $V\mathbf{i}$ where V and $\mathbf{i}$ are functions of s and n but independent of ζ. Far upstream, the velocity vector is $V_0\mathbf{i}_0$ and far downstream $V_1\mathbf{i}_1$ where both vectors are constants. The continuity and irrotationality restrictions are satisfied when the unit vectors $\mathbf{i}$ and $\mathbf{j}$ satisfy the conditions

$$\begin{aligned} \frac{\partial \mathbf{i}}{\partial s} &= \frac{1}{V}\frac{\partial V}{\partial n}\mathbf{j} & \qquad \frac{\partial \mathbf{i}}{\partial n} &= -\frac{1}{V}\frac{\partial V}{\partial s}\mathbf{j} \\ \frac{\partial \mathbf{j}}{\partial s} &= -\frac{1}{V}\frac{\partial V}{\partial n}\mathbf{i} & \qquad \frac{\partial \mathbf{j}}{\partial n} &= \frac{1}{V}\frac{\partial V}{\partial n}\mathbf{i} \end{aligned} \tag{5-1}$$

We now add a perturbation to the approaching flow so the velocity vector far upstream is given by

$$V_0\mathbf{i}_0 + \epsilon u_0(\zeta)\mathbf{i}_0 + \epsilon v_0(\zeta)\mathbf{j}_0$$

where u_0 and $v_0 \sim O(V_0)$ and are functions of ζ alone, and ϵ is a small number. The component $\epsilon v_0\mathbf{j}_0$ is added to allow for the skewness of the boundary layer one would expect to find when there is relative motion of the blade rows. The vorticity in the approaching flow is then

$$\omega_0 = \nabla \times (V_0\mathbf{i}_0 + \epsilon u_0\mathbf{i}_0 + \epsilon v_0\mathbf{j}_0) = -\epsilon\frac{dv_0}{d\zeta}\mathbf{i}_0 + \epsilon\frac{du_0}{d\zeta}\mathbf{j}_0 \tag{5-2}$$

It is essential to be specific about the magnitudes of the gradients $du_0/d\zeta$ and $dv_0/d\zeta$. We will take $du_0/d\zeta$ and $dv_0/d\zeta \sim O(V_0)$, corresponding to a boundary layer thickness of the same order as the blade gap. If $du_0/d\zeta$ and $dv_0/d\zeta \sim O(\epsilon V_0)$, the "boundary layer" is much thicker than the blade gap and the problem is similar to that studied in the previous article, where the flow at each radius is treated as two-dimensional. If $du_0/d\zeta$ and $dv_0/d\zeta \sim O(1/\epsilon V_0)$, the boundary layer is thin compared with the blade gap, and a treatment different from either of the two cases above is required; however, the problem for the thin boundary layer is not of much interest in the present connection.

Within the blade row and downstream, assume that the velocity vector can be represented by

$$V\mathbf{i} + \epsilon\mathbf{u} \tag{5-3}$$

where $|\mathbf{u}| \sim O(V_0)$; then the vorticity vector is

$$\boldsymbol{\omega} = \epsilon\nabla \times \mathbf{u} \tag{5-4}$$

The equations of motion for a steady incompressible inviscid flow can be written in the form

$$(V\mathbf{i} + \epsilon\mathbf{u}) \cdot \nabla\boldsymbol{\omega} - \boldsymbol{\omega} \cdot \nabla(V\mathbf{i} + \epsilon\mathbf{u}) = 0 \tag{5-5}$$

This is the exact form upon eliminating the pressure. If $\boldsymbol{\omega}$ and its derivatives are of order ϵV_0, as suggested by the conditions upstream, the second order terms in Eq. 5-5 can be neglected and one obtains

$$V\frac{\partial}{\partial s}(\boldsymbol{\omega}) - \boldsymbol{\omega} \cdot \nabla(V\mathbf{i}) = 0 \tag{5-6}$$

Letting $\boldsymbol{\omega} = \omega_1\mathbf{i} + \omega_2\mathbf{j} + \omega_3\mathbf{k}$, substituting into Eq. 5-6, carrying out the operations with the help of the relations (Eq. 5-1) and separating into components, one readily obtains

$$V\frac{\partial\omega_1}{\partial s} - \frac{\partial V}{\partial s}\omega_1 - 2\frac{\partial V}{\partial n}\omega_2 = 0 \tag{5-7}$$

$$V\frac{\partial\omega_2}{\partial s} + \frac{\partial V}{\partial s}\omega_2 = 0 \tag{5-8}$$

$$V\frac{\partial\omega_3}{\partial s} = 0 \tag{5-9}$$

These equations are in a convenient form for integration along the streamlines.

Since $\omega_3 = 0$ far upstream, Eq. 5-9 shows that

$$\omega_3 = 0 \tag{5-10}$$

everywhere. Eq. 5-8 can be integrated immediately to give

$$\omega_2 = \epsilon\frac{V_0}{V}\frac{du_0}{d\zeta} \tag{5-11}$$

satisfying the condition upstream. Eq. 5-7 can then be written

$$V^2\frac{\partial}{\partial s}\left(\frac{\omega_1}{V}\right) = 2\epsilon\frac{\partial V}{\partial n}\frac{V_0}{V}\frac{du_0}{d\zeta}$$

and integrating along the streamlines,

$$\omega_1 = 2\epsilon V_0\frac{du_0}{d\zeta}V\int_{-\infty}^{s}\frac{1}{V^3}\frac{\partial V}{\partial n}ds - \epsilon\frac{V}{V_0}\frac{dv_0}{d\zeta} \tag{5-12}$$

Hence the vorticity distribution can be evaluated everywhere if the two-dimensional unperturbed flow is known. Since $\partial V/\partial n = -V/R$ where R is the radius of curvature of the streamline (with n increasing away from the center of curvature), the integral in Eq. 5-12 can be replaced by $\int_0^\theta (1/V^2)d\theta$ where θ is the angle the streamline makes with its initial direction. Then the vorticity distribution downstream of the blade row can be approximated by

$$\omega_1 \cong -2\epsilon \frac{du_0}{d\zeta} \frac{V_0 V_1}{\bar{V}^2} \Delta\theta - \epsilon \frac{dv_0}{d\zeta} \frac{V_1}{V_0}$$

$$\omega_2 = \epsilon \frac{V_0}{V_1} \frac{du_0}{d\zeta}$$

$$\omega_3 = 0 \tag{5-13}$$

where $\bar{V}$ is the average velocity through the blade row and $\Delta\theta$ is the total turning angle. To this approximation ω is a function of ζ alone downstream of the blade row.

Although the expressions (Eq. 5-13) above represent a good approximation to the vorticity distribution downstream of a blade passage, these alone do not solve the problem completely. The complete solution obtained by integrating the vorticity equations through the blade row must include an arbitrary irrotational flow as well. The irrotational part of the solution is required to satisfy the boundary condition of zero normal velocity on the blade surfaces and results from bound vorticity in the blades and trailing vortex sheets that are induced by the nonuniform approach flow. Apparently the problem has not yet been solved in an entirely satisfactory manner.

The irrotational perturbation induced within the blade row must disappear a short distance downstream, so one can expect that the perturbation velocities behind the trailing edges of the blades are approximated reasonably well by the vorticity distribution given by Eq. 5-13. The assumption made in [3] and implicitly assumed in most analyses of the problem is that the boundaries of the flow that emerges from a particular blade are nearly plane cylindrical surfaces normal to the wall. Then the flow in a normal cross section of the extension of the blade channel downstream is approximately that resulting from a vorticity distribution prescribed by the first of Eq. 5-13 with the boundary conditions that the normal component of the velocity is zero on the boundary of the semi-infinite channel as shown in Fig. F,5b, where the coordinate denoted previously by n is replaced by η and b is the channel width. The resulting flow field is the one that would be produced if long, straight parallel walls extended downstream from the blade trailing edges. However, the vortex sheets between channels may not be represented very well by the flow field, because the boundary condition on the vortex sheets is that the pressure is continuous across them, rather than the normal velocity vanishing.

The two-dimensional flow resulting from the vorticity distribution ω_1 in the semi-infinite channel can be found readily by applying the appropriate Green's function. In the approximate form for ω_1 in Eq. 5-13, the

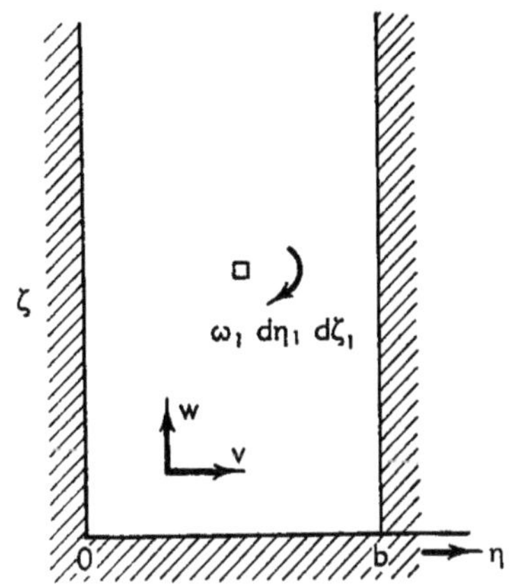

Fig. F,5b. Channel cross flow downstream of blade passage.

vorticity is independent of η and depends on ζ only through the functions $du_0/d\zeta$ and $dv_0/d\zeta$. Then the components of velocity are given by

$$v = \frac{2}{\pi}\epsilon \sum_{n=1}^{\infty} \frac{\sin (2n+1)\pi/b\eta}{2n+1} \left\{ 2 \sinh \frac{(2n+1)\pi}{b} \zeta \int_{\zeta}^{\infty} \omega_1(\zeta') \exp\left[-\frac{(2n+1)\pi}{b} \zeta' \right] d\zeta' + 2 \exp\left[-\frac{(2n+1)\pi}{b} \zeta \right] \int_0^{\zeta} \omega_1(\zeta') \sinh\left[\frac{(2n+1)\pi}{b} \zeta' \right] d\zeta' \right\} \quad (5\text{-}14)$$

$$w = \frac{2}{\pi}\epsilon \sum_{n=1}^{\infty} \frac{\cos (2n+1)\pi\eta/b}{2n+1} \left\{ 2 \sinh \frac{(2n+1)\pi}{b} \zeta \int_{\zeta}^{\infty} \omega_1(\zeta') \exp\left[-\frac{(2n+1)}{b} \pi\zeta' \right] d\zeta' - 2 \exp\left[-\frac{2n+1}{b} \pi\zeta \right] \int_0^{\zeta} \omega_1(\zeta') \sinh\left[\frac{(2n+1)\pi}{b} \zeta' \right] d\zeta' \right\} \quad (5\text{-}15)$$

The series in these expressions converge rapidly, and ordinarily only the first one or two terms are required.

As an example, the velocity profiles for $\epsilon = 1$, $v_0 = 0$, and $u_0 =$

$-V_0(1 - \zeta)$ in $0 < \zeta < 1$ and $u_0 = 0$ for $\zeta > 1$ are shown in Fig. F,5c. There is an appreciable overturning in the middle part of the channel close to the wall.

The analysis above breaks down very close to the wall where viscous effects predominate, but agrees remarkably well with experiment over the outer part of the wall boundary layer as shown by Squire and Winter [*3*], Hawthorne [*4*], Armstrong [*5*], and others. Briceland [*6*] found almost perfect agreement between theory and experiment when he took into account the shift of the streamlines by an approximate calculation. The shift of the streamlines is a second order effect in the theory, but Hawthorne and Armstrong [*7*] have shown that it is more important to allow

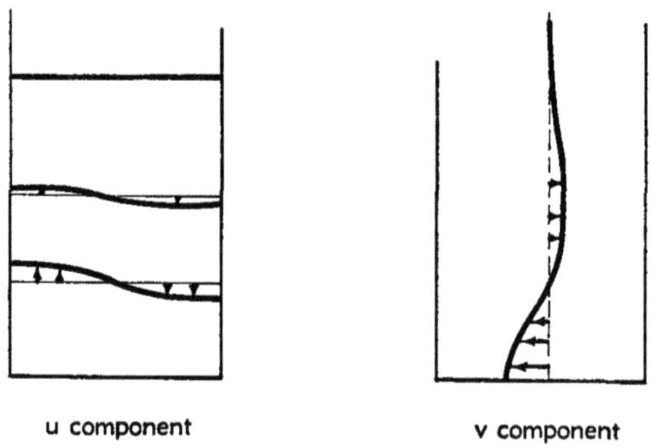

Fig. F,5c. Secondary flow velocities downstream of a blade channel. $u_0 = -(V_0/2)(1 - \zeta)$, $v_0 = 0$, $\epsilon = 1$, $\Delta\theta = 25°$, and $V_1/V_0 = 0.75$.

for the shift than to refine the calculations of the nominally first order effects.

Three-dimensional boundary layer theory has been applied to the flow conditions that approximate those in compressors, for instance by Mager [*8*] and Sears [*9*]. Many configurations of boundary layer flow were examined visually at the NACA by Hansen, Herzig, and Costello [*10*], and others. These experiments showed clearly the character of the flow very close to the wall. In most applications to compressors at high Reynolds numbers, so little of the total flow through the compressor is contained in the regions very close to the boundaries that the direct viscous effects seem unimportant compared with other phenomena.

Flow through blade tip clearances. The losses that arise from leakage through the clearances at the tips of unshrouded blades in multistage axial compressors are frequently quite large in comparison with other losses resulting from viscous effects. Mechanically, it is difficult to avoid large tip clearances in high speed compressors because of differences in thermal expansion in the rotor and stator.

The discussion of tip clearance flows below will be confined to clearance gaps that are a small fraction of the blade chord. The lifting line approximation to the blade is certainly not applicable; rather the clearance flow is analogous to the flow through a narrow chordwise gap in an infinite wing. If viscous effects are neglected, the plane of symmetry through the center of the gap in the wing represents the wall boundary for the blade clearance. Considering the infinite wing with a narrow gap, it is clear that the pressure distribution near the gap is close to that of the two-dimensional flow far from the gap. As a result, a jet issues from the suction surface of the wing with a velocity component normal to the chord, which is the result of the pressure difference between pressure and suction surfaces. Since the fluid in the jet comes from the lower side of the wing, it will have a component of velocity in the chordwise direction equal to that on the lower side of the wing. Hence, the resultant velocity within the jet is equal to the velocity outside the jet, but has a different direction.

The jet is bounded by vortex sheets, where, as can be verified readily, the vortex strength of any strip of the sheet is exactly equal to the vortex strength of the corresponding element of the wing chord. Hence the entire circulation of the wing is shed in each of the vortex sheets; however, since the vortex lines of the sheets are of equal and opposite strengths, the influence of the shed vorticity is confined to a small region in its immediate vicinity.

The vortex sheets that form the sides of the jet will not be in equilibrium in general, and hence will tend to roll up. This rolling-up starts at the edge of the sheet that emerges from the leading edge of the wing and progresses rapidly as one goes away from the chord. A similar situation arises in three-dimensional wing theory. The trailing vortex sheet behind a wing with elliptic loading is in equilibrium when the sheet is flat, but is unstable to small disturbances and eventually rolls up. If the loading is not elliptic, the sheet is not in equilibrium and will start to roll up from the edges immediately after leaving the wing. The vortex sheets forming the edges of the jet in the wing with a slot could be in equilibrium for a particular chordwise loading distribution, although this too would be unstable, most likely, and would only delay the rolling-up process. The phenomena described here can be observed in a striking manner in an axial flow pump operating under appropriate cavitation conditions. Concentrated vortices cavitate and are made visible as a streak of bubbles. The photograph, Plate F,5, is a typical example. The small streaks indicate local concentrations of vorticity resulting from irregularities in the distribution in the sheet. The large cavitating core increases in diameter away from the blade showing the increasing strength of the rolled-up sheet.

Although the behavior of the vortex sheet as it moves away from the blade is interesting, it does not affect appreciably the flow close to the end of the blade. It is difficult to observe or measure the flow within the tip clearance, but it seems reasonable to assume that the flow perpendicular to the chord is similar to free streamline flow through a sharp-edged, two-dimensional slot. The solution of this problem is well known, and if δ is the clearance height, the final width of the jet is $\pi\delta/(\pi + 2)$, as shown in Fig. F,5d where p_l is the pressure on the lower (pressure) side of the chord and p_u is the pressure on the upper (suction) side. Then the mass flow leaking over the end of the blade at any chordwise position is $[\pi/(\pi + 2)]\delta\sqrt{2\rho(p_l - p_u)}$ by direct application of Bernoulli's equation.

The force normal to the chord exerted by the blade is modified by the clearance. With no clearance, the force is $p_l - p_u$ per unit area of

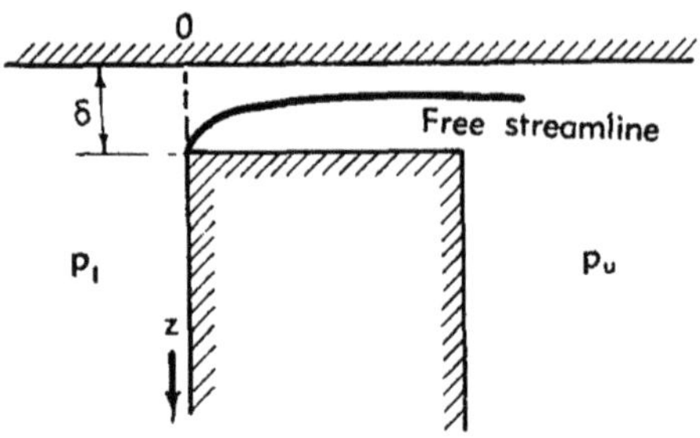

Fig. F,5d. Flow across blade thickness in tip clearance.

blade surface. With the clearance, there is of course no blade surface in the clearance, but in addition the pressure on the lower side of the blade drops from p_l to p_u at $z = \delta$. From the free streamline model it is a simple matter to show that

$$\int_0^\delta (p_l - p)dz = (p_l - p_u)\frac{\pi - 2}{\pi + 2}\delta \tag{5-16}$$

Hence the effective blade shortening, as far as blade force is concerned, is

$$\delta_e = \delta + \frac{\pi - 2}{\pi + 2}\delta = \frac{2\pi}{\pi + 2}\delta$$

i.e. 22 per cent more than the geometrical clearance. If the tip clearance is constant along the chord, the same relation holds for the entire blade.

As a result of the tip clearance, the work done by the rotor blades is less than it would be for no clearance. With the model above, the power input to the rotor is

$$P = \pi\rho\omega^3R^5\int_{\xi_i}^{1-\delta_0/R}\Psi\phi\xi d\xi \tag{5-17}$$

where Ψ is the work coefficient and ϕ the flow coefficient at the dimensionless radius $\xi = r/R$, with R the outer radius of the annulus. Defining

$$\bar{\phi} = \frac{2}{1-\xi_i^2}\int_{\xi_i}^{1} \phi\xi d\xi \qquad \bar{\Psi} = \frac{2}{1-\xi_i^2}\frac{1}{\bar{\phi}}\int_0^1 \psi\phi\xi d\xi \tag{5-18}$$

the tip clearance reduces the average work coefficient by $\Delta\bar{\Psi}$ where

$$\frac{\Delta\bar{\Psi}}{\bar{\Psi}} = \frac{2\pi}{\pi+2}\frac{2}{1-\xi_i^2}\frac{\delta}{R}\frac{\Psi_1}{\bar{\Psi}}\frac{\phi_{1_w}}{\bar{\phi}} \tag{5-19}$$

with Ψ_1 and ϕ_1 the work coefficient and flow coefficient near the blade tip. Now if we assume that the pressure field near the blade tip is determined principally by the flow through the blades outside the wall boundary layer, we should evaluate Ψ_1 by extrapolation from the main stream flow rather than by the measured local value of $2\xi\Delta V_\theta/\omega R$. The flow coefficient ϕ, on the other hand, should be evaluated close to the wall, say at a distance $\delta/2$ from the wall, since ϕ_1 is a measure of the amount of fluid that would be subjected to work corresponding to Ψ_1 if there were no clearance. For this reason, the notation ϕ_{1_w} is introduced into Eq. 5-19.

The losses resulting from the tip clearance flow are difficult to identify in a perfect fluid model. Certainly the rolled-up vortex sheet is the principal mechanism for dissipation of kinetic energy of the jet through the clearance. To determine the dissipation, it would be necessary to follow the motion of the rolled-up vortex in great detail as it went downstream, and this is impractical in the complex flow field in which it is imbedded.

We will make the assumption that the kinetic energy of the velocity component normal to the chord is "lost" through dissipation, recognizing that this may not be quite correct quantitatively, but that it should give at least the correct form of dependence on compressor design parameters. The kinetic energy of the component of velocity normal to the chord line is $\frac{1}{2}v^2 = \Delta p/\rho$ per unit mass flow rate where $\Delta p = p_l - p_u$ is the static pressure differential over the blade tip at any particular chordwise station. Since the mass flow rate per unit chord length is $[\pi\delta/(\pi+2)]\rho v$, the total kinetic energy for N blades is

$$\Delta E = \frac{1}{\sqrt{2}}\frac{\pi}{\pi+2}N\delta\rho\int_0^c \left(\frac{\Delta p}{\rho}\right)^{\frac{3}{2}} dx \tag{5-20}$$

and hence the decrease in efficiency resulting from this dissipation is

$$\Delta\eta = \frac{\Delta E}{P} = \sqrt{2}\frac{\pi}{\pi+2}\frac{2}{1-\xi_i^2}\frac{c}{s}\frac{1}{\bar{\Psi}\bar{\phi}}\int_0^1 \left(\frac{\Delta p}{\rho\omega^2R^2}\right) d\left(\frac{x}{c}\right)$$

where s is the blade gap. Define the number θ as

$$\theta = \int_0^1 (\Delta p)^{\frac{3}{2}} d\left(\frac{x}{c}\right) \Big/ (\overline{\Delta p})^{\frac{3}{2}} \tag{5-21}$$

where $\overline{\Delta p}$ is the average value of Δp along the chord. Then θ depends on the chordwise load distribution; for uniform loading, $\theta = 1$, and for a linear loading, maximum at the leading edge and dropping to zero at the trailing edge, for instance, $\theta = 1.13$. In this more convenient form, the decrease in efficiency becomes

$$\Delta\eta = \sqrt{2}\,\frac{\pi}{\pi + 2}\,\frac{2}{1 - \xi_i^2}\,\frac{c}{s}\,\frac{\theta}{\bar{\Psi}\bar{\phi}}\left(\frac{\overline{\Delta p}}{\rho\omega^2 R^2}\right)^{\frac{3}{2}} \tag{5-22}$$

Since $\overline{\Delta p}$ is the average pressure difference over the blade, it should be evaluated in terms of the velocity triangles ignoring the wall boundary layer, again assuming that the pressure field near the blade tip is determined by the blade action outside the wall layer.

The normal force $\overline{\Delta p}c$ on the blade is the same as the lift force L in Fig. F,2d. Using the procedure discussed in Art. 2, one finds

$$\frac{\overline{\Delta p}}{\rho\omega^2 R^2} = \frac{s}{c}(\lambda_1 - \lambda_2)\left[\phi^2 + \left(\frac{\lambda_1 + \lambda_2}{2}\right)^2\right]^{\frac{1}{2}} \left[1 - \frac{K}{2}\,\frac{1}{\lambda_1 - \lambda_2}\,\frac{\frac{1}{2}(\lambda_1 + \lambda_2)(\phi^2 + \lambda_1^2)}{\phi^2 + \left(\frac{\lambda_1 + \lambda_2}{2}\right)^2}\right] \tag{5-23}$$

where λ_1 and λ_2 are the dimensionless whirl components of velocity and K is the loss coefficient for the cascade. For application to a rotor, it is necessary to replace λ_1 and λ_2 by $\xi - \lambda_2$ and $\xi - \lambda_1$, respectively, and with some rearrangement Eq. 5-23 becomes

$$\frac{\overline{\Delta p}}{\rho\omega^2 R^2} = \frac{s}{c}\,\frac{\Psi}{2\xi}\left[\phi^2 + \left(1 - \frac{\lambda_1 + \lambda_2}{2}\right)^2\right]^{\frac{1}{2}} \left[1 - (1 - \eta)\,\frac{1 - \frac{1}{2}(\lambda_1 + \lambda_2)}{\phi^2 + \left(1 - \frac{\lambda_1 + \lambda_2}{2}\right)^2}\right] \tag{5-24}$$

For a complete stage, one must take the sum of two terms of the form given in Eq. 5-22, one evaluated from Eq. 5-23, and the other from Eq. 5-24.

The analysis above estimates the decrease of power input and efficiency for variations of tip clearance alone, ignoring all other effects. Reliable measurements of the effect of tip clearance in the range of interest for high performance machines are scarce since it is quite difficult to achieve the accuracy required. For a set of rotor experiments by Williams [*11*] at four different clearances, each at five flow rates, the experimental results for the work coefficient averaged 20 per cent lower than predicted value from Eq. 5-19 with a scatter of ± 20 per cent and for decrease of efficiency averaged 12 per cent higher than predicted by Eq. 5-22 with a scatter of ± 20 per cent. Earlier and less extensive measurements by Rains [12] gave poorer agreement with the analysis.

The relative motion between the blade tips and the casing or hub leads to a scraping action of the blades on the fluid at the wall and the fluid carried with the wall. In compressors this action tends to increase the pressure on the lower surface of the blade very close to the wall, and hence tends to increase the losses; however, no very convincing quantitative analysis has appeared.

Shroud rings with labyrinth seals fitting into circumferential troughs are used frequently on stator blades, less frequently on rotor blades. With the same running clearances, the unshrouded blades generally give the better efficiency. The shroud rings often require a large axial clearance to allow for differential axial motion between rotor and stator. The circumferential troughs have the effect of a severe roughness on the flow at the wall, and in addition, the varying pressure field induced by the blades causes radial flow in and out of the slot. This can enlarge the separated zone that is invariably found near the corner on the suction side of a blade meeting a stationary wall. The external fluid flow over shrouds is more complicated, less well understood, and less predictable than the flow near the tip of an unshrouded blade.

Stalling of axial compressors. The most significant phenomenon affecting the performance of axial flow compressors is the stall. The difficulties in predicting the flow rate at which stall first occurs have been mentioned previously. The stalled flow itself is very complex and has been studied intensively only in the last ten years. In the following, a brief description of the general characteristics of stalled flow will be given first, then a discussion of the difficulties in theories of the stall, and finally, a more detailed description of the flow phenomena.

The axial compressor almost invariably stalls in an asymmetric manner, with well-defined regions of the annulus severely stalled and the remainder of the annulus unstalled. The stalled patches rotate with uniform angular velocity in the direction of the rotor rotation at a speed that is usually 30 to 60 per cent of the rotor speed. An example of typical behavior is shown in Fig. F,5e.

The flow within the shaded regions is very strongly retarded, if not blocked completely. The axial velocity outside the stalled patches is the same as if the compressor were operating unstalled at a flow coefficient $\bar{\phi} = \phi_s$ where ϕ_s corresponds to an axial velocity slightly greater than that for maximum pressure. The transitions between the steady state flow patterns, depicted in the figure, occur by a process of splitting or combining of the stalled patches, but the transitions are rapid and unsteady. As the compressor is throttled, the flow rate through the unstalled area of the annulus is constant and the decreasing average flow rate through the machine is accomplished by increasing the area of the stalled regions. At shut-off, when the net flow through the machine is zero, periodic traveling disturbances still occur although the flow in the "unstalled" portion is not always smooth.

The description above is typical of the behavior of a compressor with solidity of the order of unity, but each compressor has its own minor variations. The phenomenon is called propagating or rotating stall; the smaller stall patches covering only a part of the radius are termed partial or partial span stall, while a stalled region extending from hub to tip is termed full or full span stall. The partial stall can occur at the hub or casing, depending on the relative blade loading at root or tip in a particular compressor. Partial stall always seems to occur first as the flow rate is decreased in compressors with a blade aspect ratio larger than 3 or 4; for aspect ratios less than 2, only full span stall is observed, although

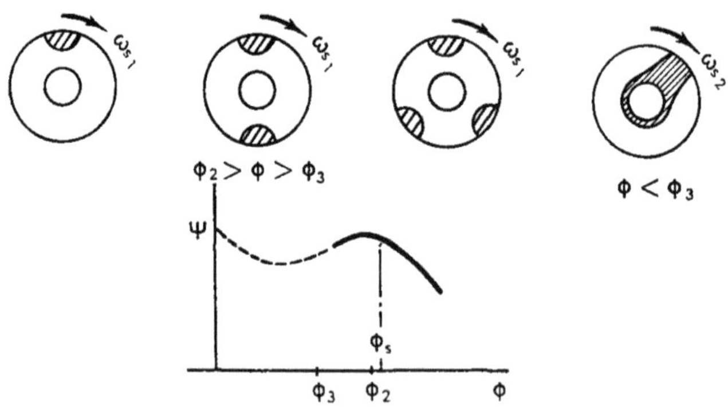

Fig. F,5e. Typical rotating stall cell patterns.

there may be two stalled regions instead of the one that is almost invariably found with a higher aspect ratio. As the solidity is decreased, a propagating disturbance is still observed as the blades stall, although the sharp demarcation between stalled and unstalled flow disappears.

One of the first reports of propagating stall was made in 1945 by Cheshire [*13*], who described the phenomenon in a centrifugal compressor. Propagating stall in axial compressors was not described in the open literature until 1953 and 1954 [*14,15,16*], although the phenomenon had been recognized earlier. It was obvious that the stalling behavior resulted from a nonlinear mechanism, and several theories [*17,18,19*] were advanced to explain it. In these theories, the equations of motion are linearized, limiting the analysis strictly to small disturbances, and the nonlinear character is introduced in various ways, for instance, by a time lag or an abrupt change in lift coefficient. These theories predict a propagating speed with a value that depends on the numerical value of parameters in the theory, but give little more information. Since the observed disturbances are always of large amplitude, the applicability of the theories is somewhat in question.

Propagating disturbances of small amplitude were found by Benenson [*20,21*] in annular stator cascades at suitable angles of attack. The values of the propagating speeds, as well as the wavelengths and amplitudes of the disturbances, fitted the assumptions postulated by the present author [*21*] for a quasi-steady linearized stall theory. This theory was more general than previous linearized theories, but was not essentially different in character. An irrotational disturbance moving at a constant speed was postulated in front of a two-dimensional cascade. Downstream, two moving disturbances were assumed, one irrotational, the other rotational. Satisfying the momentum equations and the continuity equation across the cascade gave the three conditions required to determine the magnitudes and phase relationship of two of the three disturbances in terms of the other and to determine the propagating speed. The influence of the nonlinear effects was introduced into the theory by assuming that the pressure loss through the cascade increased sharply, although continuously, at a particular angle of attack. The quasi-steady assumption would certainly be valid for a disturbance wavelength equivalent to many blade gaps; in one of the experiments, the wavelength was equal to 60 blade gaps.

The most striking difference between small amplitude propagating disturbances and those found in compressors was that the flow through a part, at least, of all blade channels was irrotational at all times. The "stalling" was confined to moderate periodic separation of the boundary layers on the suction sides of the blades. Hence the blade channels ran almost full instead of being blocked, as typical of compressors. To the present author, these findings seem important, although indirectly; because they show that small self-propagating disturbances can occur and can be described in all important details by a consistent linearized theory, they therefore indicate that the large amplitude stall, similar only in its self-propagating property, cannot be represented by the linearized equations of motion. A few theories have been proposed with allowance for large disturbance velocities, but so far none has progressed beyond a very speculative stage.

Detailed measurement of the flow within stall cells is very difficult because of the extremely turbulent nature of the flow. The hot wire anemometer can be used to determine instantaneous values of velocity, but these values alone, without the corresponding directions, are of little use. However, one serious effort to explore the flow patterns within the stall cells in a multistage compressor has been made with anemometer arrangements to measure direction as well as velocity [*22*] and some significant facts have emerged. In the central part of a full-span stall cell, there was a forward component of velocity of small magnitude. The direction of flow between the blade rows was consistent with a small axial component from back to front. Bordering the central portion of the stall cell and separating it from the unstalled flow outside were two transition regions where the velocity fluctuations were more violent than in the central portion. Because of the very irregular velocity fluctuations it was not possible to find a net flow in any direction, if such a flow existed. The circumferential extent of the transition regions corresponded to one or two blade gaps. Similar measurements in the partial stall cells, which extended over two or three blade gaps circumferentially, gave results over the entire cell similar to those in the transition regions of the full stall cells. Again, no interpretation in terms of net flow in any particular direction could be made.

The stall cells, both partial and full span, extended axially through the three stages of the compressor with no significant variation in cross-sectional shape or size (the compressor had identical blading in all stages and was run at low speed). There was no apparent spiraling of the stall cells through the machine, i.e. the stalled regions behaved like rigid axial wakes rotating at a constant angular velocity. This behavior is surprising if one examines the general flow pattern expected at different radii. Let ω be the angular velocity of the rotor, ω_s the angular velocity of the stall cell, and $\bar{\omega}$ the average angular velocity of a fluid element in the unstalled flow at radius r. Taking a coordinate system moving with velocity $\omega_s r$, the stall pattern is reduced to rest, and unrolling a cylindrical section at radius r the general pattern can be represented schematically as in Fig. F,5f. If the flow is two-dimensional, and the stalled region is of constant width, the edges of the cell must be parallel to the direction of the velocity in the unstalled regions; otherwise one must accept the unlikely result that the fluid within the stalled zone, with very large turbulent fluctuations and low average velocity, can make a rapid transition to join the unstalled smooth flow outside.

The average angular velocity of the fluid elements in the unstalled flow varies with the radius in general, while ω_s is constant. For instance, in the experiments described above, the compressor blades were designed for vortex flow and hence $\bar{\omega} \sim 1/r^2$. The observations show, however, that the stall cell extends axially at all radii; that is, it behaves as if $\bar{\omega} = \omega_s$. Evidently the propagating speed is close to the rotating speed of the fluid elements outside the stall at approximately the average radius of the stall cell. Radial flow on both sides of the cell produce the necessary adjustments to satisfy the conditions that $\bar{\omega} \cong \omega_s$ at larger and smaller radii.

Although one can estimate the rotating speed of a stall cell from the mean whirl speed of fluid elements at essentially design flow rate and the average radius of a stall cell, there are no reliable ways of predicting the size, shape, and number of stall cells. The partial stalls that are the first to appear upon reducing flow rate of the compressor below its design value usually extend circumferentially over a length of two or three blade gaps and radially about the same distance. With 30 blades, one, two, three, and four cells may appear in succession as the flow rate is reduced. With a much larger number of blades, the increments in number may not be a single integer and the maximum number may be 10 or more. The full span stall replaces the partial stalls very abruptly and increases in

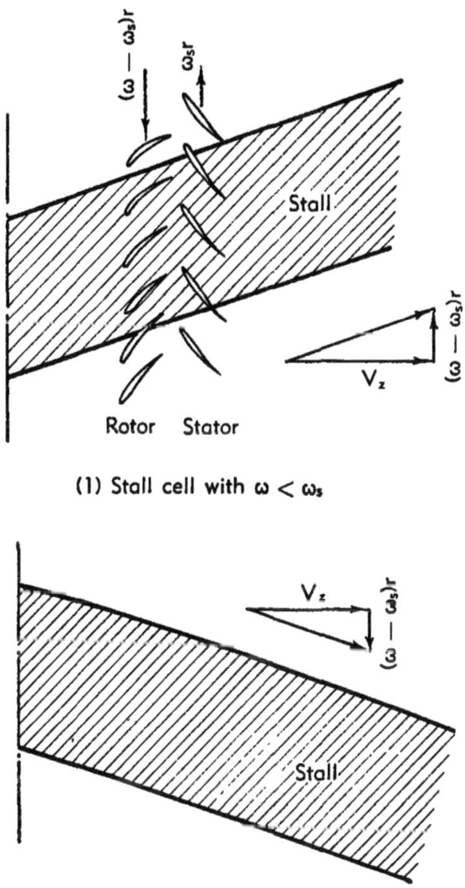

(2) Stall cell with $\omega > \omega_s$

Fig. F,5f. Stall cell in coordinate system moving with the stall.

cross-sectional area with further reduction of flow rate. The total area of the stalled portion of the annulus, in both partial and full stall regimes, can be estimated by assuming that the flow is at design in the unstalled portion and that there is no flow through the stalled portion; however, this rule is seldom exact. Upon increasing the flow rate through the compressor from the stalled regime, the cell patterns repeat in reverse, but the transitions occur at a higher flow rate than for decreasing flow. The difference is small for the partial stall transitions but is usually very marked between the regimes of full and partial stall.

The compressor performance is of course strongly influenced by the stall, the efficiency more than torque, but no methods for estimating the effects of stall have been developed so far. To calculate the torque, it is necessary to determine forces on the blades as they cross the stall cell (see Fig. F,5f), and clearly this will be difficult. The pressure rise and efficiency in the stalled regime depend on the mixing losses and recovery of the rotating sectors of unstalled flow behind the compressor and hence will be strongly influenced by the geometry of the receiver; indeed, it is scarcely possible to make a definitive measurement of over-all static or total pressure rise in stalled operation.

The performance of axial compressors designed for high pressure ratio is particularly sensitive to stall, since the first stages are stalled during starting and part speed operation. At part speed the volume flow rate is less than design in the first stages and greater than design in the last stages, because the density ratio does not match the annulus areas. Although the volume flow rate in the middle stages is close to the design value, and these stages might be expected to have good performance, the rotating stall cells developed in the first stages extend far back into the compressor, frequently to the last stages, and degrade the performance much more than would be expected from a combination of stages at their appropriate volume flow rates. The deterioration of part speed performance becomes so severe as the design pressure ratio is increased that it becomes impractical to attempt construction of a single shaft compressor for a pressure ratio of more than 8:1. For higher pressure ratios, it is necessary to use two shafts that can operate over a range of speed ratio or to incorporate stator blades that can be adjusted in angle during operation.

Perhaps more important than the influence of stall on performance is the effect on blade stresses. The rotating stall cells produce periodic forces on both rotor and stator blades that are at least as large in magnitude as the dynamic head of the flow velocities at the design flow angles. The large magnitude of the periodic forces combined with the numbers of possible stall cells makes it difficult to avoid blade resonance and fatigue failure, even when the stalled condition occurs only for brief intervals of time while the compressor is brought up to full speed. The possible disturbing frequencies can be estimated at least approximately by the methods described above. It is particularly important to keep away from blade resonance in the stall at the higher rotative speeds, where the aerodynamic forces are large. As an added precaution, it is apparently standard practice to measure stall frequencies on prototype machines and make adjustments of blade frequency if necessary.

Surging is an oscillating disturbance characterized by average flow rate varying with time through the compressor and is quite distinct from the rotating stall where the flow rate averaged over the annulus is constant. The frequency of the surging oscillation is the natural frequency of the gas contained in the compressor and the associated ducting or reservoirs. The surging oscillations may occur with very large amplitudes; hence the natural frequency is not necessarily identical with the frequency for small amplitude sound waves. Since the surging oscillations are highly dissipative, they can be maintained only if the compressor can supply energy to the oscillation. The hysteresis effects mentioned previously, in particular the transition between partial and full stall, are the most likely mechanisms for introducing energy into the surging oscillation.

A typical pressure-flow characteristic curve is sketched in Fig. F,5g; tracing the hysteresis loop in the direction shown by the arrows clearly

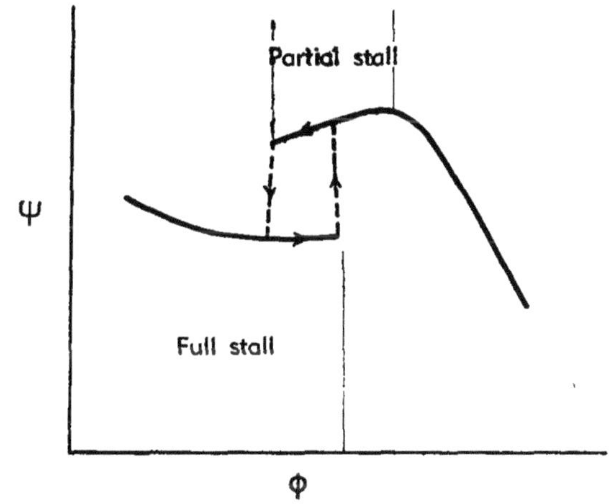

Fig. F,5g. Compressor operating line with hysteresis loop.

gives energy input into the system. The transitions from partial to full stall, and vice versa, take place through unsteady flow processes which require a time that is comparable with, although usually shorter than, the period of the rotating stall cells. If the natural frequency of the system external to the compressor is low compared with the transition time, the hysteresis loop is traced out in a quasi-steady fashion and energy is fed into the oscillation, until the amplitude reaches a value corresponding to a balance of input and dissipative energies. The area of the hysteresis loop is constant for quasi-steady oscillations corresponding to constant energy input and the dissipation is proportional to the square of the amplitude of the oscillation, so that the oscillation reaches a very definite amplitude.

In some configurations, the limiting amplitude is so great that the net flow through the compressor is reversed in a portion of the cycle. If the natural frequency of the system external to the compressor is high compared with the stall pattern transition time, the hysteresis loop collapses to some intermediate operating line and there is not sufficient energy fed into the oscillation to maintain it; under such circumstances surging does not occur. Hence, surging can be prevented if the natural frequency of the reservoirs connected with the compressor can be made sufficiently high by keeping the volumes small. The critical transition time is inversely proportional to the rotative speed of the compressor; whereas the natural frequency of the connected reservoir capacity will be relatively insensitive to rotative speed. Hence a system may satisfy the conditions for surge-free operation at part speed and be capable of surge at higher rotative speed.

The rotating stall phenomenon has such important ramifications in practice that questions concerning the advantages of eliminating it and the possibility of elimination arise immediately. As far as performance alone is concerned, it is not clear that a uniform stall, the alternative to rotating stall, would have any advantage whatever. Certainly there would be much gained by elimination of the stall as far as blade vibration and surging are concerned. No practical method of preventing rotating stall has been devised so far, and in fact the problems of doing so appear formidable. The introduction of some asymmetry in the annulus geometry could stop the rotating cells, but the wavelengths of the stall patterns are so large that asymmetries of sufficient magnitude to be effective would certainly compromise design-point operation excessively. There is evidence that the partial stalls that usually occur at the casing can be shifted to the hub by moderate increase of solidity toward the rotor tips, and this may have advantages if the stator blade supporting structure is more rigid than that of the rotor. Aside from this limited means of control, the designer has had to accept the rotating stall and learn to live with it.

Stationary asymmetric disturbances, for instance partial or complete blocking of a sector of the compressor annulus at the inlet, are the forced oscillation counterparts of the self-induced propagating stall disturbances. If the velocity disturbances introduced into the inlet flow are small perturbations, a linearized analysis [*21*] of the type developed for small disturbance propagating stall can be expected to be valid. However, just as for rotating stall, the velocity amplitudes that are of practical importance are usually so large that the linearized theory is not valid. A few detailed experimental investigations of partial blocking of a compressor annulus have been made, see for instance [*21*], but the number of such studies is still too small to justify statements as general as those possible for rotating stall.

F,6. Other Problems. In the previous articles it has been assumed that the effects of compressibility are slight. This assumption is reasonably correct up to inlet Mach numbers of 0.7 or 0.8 (see Sec. E and [*2*]).

For many compressor designs $M = 0.7$ or 0.8 is assumed to be the limiting design Mach number, but for some high duty compressors tip Mach numbers at the design point have been as high as 0.9. Below Mach numbers of 0.7 to 0.8, the characteristics of an axial compressor stage can normally be represented by a single curve of pressure rise, efficiency, or work done, plotted against the flow coefficient. As the Mach number increases, the stage begins to show a loss of efficiency. The loss is generally gradual because the incidence of the critical Mach number usually varies with the radius on both rotor and stator blades. As the Mach number is increased at air inlet angles below $\cos^{-1} o/s$, a choking condition is reached so that the mass flow is limited. This effect is gradual over the radius but eventually results in choking at all radii and limiting the flow that the stage can pass.

In gas turbines for aircraft, there is a strong incentive to increase the mass flow through a compressor, thereby increasing the Mach number relative to the blades and decreasing the hub-to-tip ratio in the first stage. The flow at which choking of the stage occurs depends on the average characteristics of the blades, from hub to tip. On the other hand, the loss of efficiency due to compressibility depends on the weighted effect of losses at different radii. In the free vortex stage, the highest Mach numbers occur at the tip of the rotor and the hub of the stator, as may be seen from the typical velocity triangles shown in Fig. F,4a. To avoid the incidence of such high Mach numbers, designers began in 1944 to depart from free vortex designs and to adopt designs that tended to give constant reaction at all radii. The use of inlet guide vanes was also influenced by the need to reduce the Mach number. Counterrotation of the flow by the guide vanes, which was originally intended to give increased work, was found to increase Mach numbers relative to the rotor; this was thought to be undesirable. Some compressors have been designed in which the Mach number relative to the blades has been transonic. The performance of these compressors has been satisfactory. The theoretical basis for their design has been somewhat elementary but, recently, McCune [*23*] has given a theoretical analysis of the transonic compressor problem. Apart from its effect at the inlet stages of a compressor, compressibility also affects the performance of the outlet stages, particularly at speeds well below the design speed. As will be shown later, the over-all performance of a multistage compressor is materially affected by the choking of its last stages.

In compressors of small hub-to-tip ratio, the effects of radial flow may be analyzed by the methods described in Sec. C and by methods depending on the actuator disk concept outlined in Art. 4 of this section. Even if a stage is designed for negligible radial flows, such flows will develop when operating off the design point. The effect becomes interesting near the stall. Louis and Horlock [*24*] have shown that, in a conventionally designed free vortex stage, the blade section at the root may stall first. The effect of the blockage at the stall is to force the flow radially outward to the mean and tip sections. Across these sections, therefore, the axial velocity may increase markedly owing to the radial flow, and the pressure rise across them then falls, owing to the increase in axial velocity, although the sections do not stall. The section that stalls first is determined by the variation of design loading with radius and by the type of vortex flow selected for the stage.

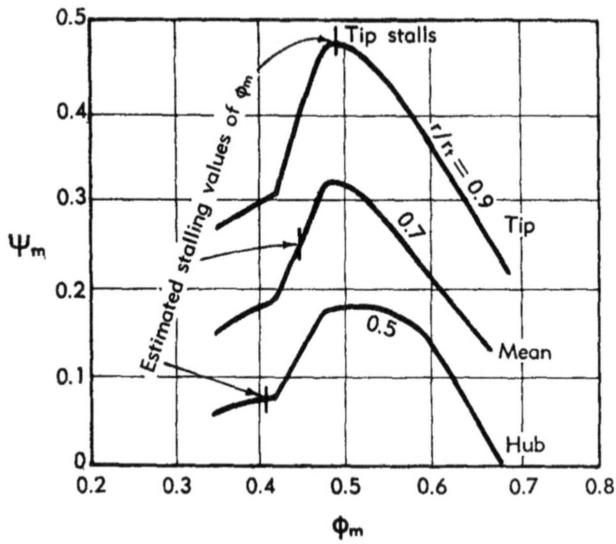

Fig. F,6a. Pressure rise at different radii for constant α_2 stage comprising guide vanes and rotor [24]

$$\Psi_m = \frac{\Delta p}{\frac{1}{2}\rho U_m^2} \qquad \phi_m = \frac{\text{average inlet axial velocity}}{\text{rotor speed at mean radius } (U_m)}$$

Some experimental results for a rotor following a row of inlet guide vanes are shown in Fig. F,6a (from [24]). At the design point the absolute outlet angle from the rotor was invariant with radius. With this design the tip section actually stalls, but the pressure rises at the mean and hub sections begin to droop at the flow coefficient at which the tip stalls, even though it is well above their stalling flow coefficients. The effect of the radial flow away from a stalled section provides one explanation for the inaccuracy of calculations of stage characteristics based on the prediction of the performance of the mean section. Generally the stage stalls sooner than such calculations predict because of the earlier stall of either tip or hub sections.

In Art. 5 of this section and in Sec. D, some of the phenomena associated with the complicated flow on the hub and casing boundary layers are described. These boundary layer flows reduce the effective area of the annulus and alter the work and pressure rise in the stage. In the design of a compressor it is customary to separate the flow into three regions: the casing or tip boundary layer, the mainstream region, and the hub boundary layer. Definition of the boundaries between the regions may be difficult, particularly if there are high losses or large velocity gradients in the mainstream region. In some schemes for design, blockage factors are estimated so that the annulus area can be adjusted to allow for the boundary layers. Values suggested for the blockage factors range between 0.98 and 1.0 for the early stages and 0.96 for the later stages in a compressor [*2*]. Corrections to allow for variations of work and losses in the boundary layers are also necessary and may amount to a few per cent. If a compressor is designed with the assumption that the thickness of the wall boundary layers is negligible, the most marked effect, when the compressor is run, will be a reduction of work at the designed mass flow. This reduction is due to the increased axial velocities in the mainstream, caused by the presence of wall boundary layers. Howell [*1*] evolved a design method in which the wall boundary layer thickness was neglected but the loss of work was allowed for by an empirical work-done factor which varies from unity to 0.85 from the first to the last stages in the compressor. Associated with the use of the work-done factor were recommendations for values of blade incidence and deflection to be used in the design. Clearly the actual incidence will be less than the value used in the design calculations. Various systems of design that allow for these effects may contain a number of interdependent assumptions or estimations which make comparison of the details difficult.

An increase in wall boundary layer thickness is often observed in cascades without wall suction and in rotors and stators. This increase causes an increase in axial velocity in the mainstream flow through the cascade and, particularly for cascades of high stagger, the axial velocity increase may be high enough to inhibit the development of the pressure rise. This effect, often described as "wall stall," has led some designers to suggest a limit to the pressure rise, which at high staggers is lower than the limit set by truly two-dimensional stalling. De Haller [*25*], for instance, recommends that the designed outlet velocity should not be less than 0.72 times the inlet velocity. Horlock [*26*, Fig. 3.16] compares some of the suggested limits.

The prediction of the characteristics of a multistage compressor is made difficult by the effects of interactions between the stages, the growth of the hub and casing boundary layers, and the stalling of one or more stages. A first, rough approximation may be obtained by estimating the characteristics of each stage from the performance of the blade elements at the mean radius. One such approximation has been suggested by Mellor [*27*] and Horlock [*28*]. The method assumes that there is negligible variation in outlet angle with flow coefficient. Then Eq. 3-22 leads to the result

$$\frac{\Psi}{\Psi^*} = \frac{2}{\Psi^*} - \frac{\phi}{\phi^*}\left(\frac{2 - \Psi^*}{\Psi^*}\right) \tag{6-1}$$

where the asterisk denotes the values at the design point. The stage flow coefficient is related to the density ρ and volume flow Q by

$$Q\rho_1 = A_1\rho_1 V_{z_1} = A\rho V_z = A\rho\phi U \tag{6-2}$$

where subscript 1 refers to the first stage.

Then

$$\frac{\phi U}{\phi^* U^*} = \frac{V_z}{V_z^*} = \frac{Q\rho^*}{Q^*\rho} \tag{6-3}$$

where $\rho_1 = \rho_1^*$ is assumed.

It is now assumed that the compressor consists of a large number of similar stages of equal Ψ^* and of constant polytropic efficiency, so that

$$\frac{\rho}{\rho_1} = \left(\frac{T}{T_1}\right)^n \tag{6-4}$$

where

$$n = \left(\frac{\gamma}{\gamma - 1}\right)\eta_{st} \tag{6-5}$$

Eq. 6-1 may then be written

$$\frac{\Psi}{\Psi^*} = \left(\frac{U^*}{U}\right)^2 \frac{dT}{dT^*} = \frac{2}{\Psi^*} - \left(\frac{2 - \Psi^*}{\Psi^*}\right)\frac{U^*}{U}\frac{Q}{Q^*}\left(\frac{T^*}{T}\right)^n \tag{6-6}$$

If $T/T^* = 1 + \epsilon$, where ϵ is small this equation may be reduced to the form

$$\frac{d\epsilon T^*}{dT^*} = A + B\epsilon \tag{6-7}$$

where

$$A = \frac{2 - \Psi^*}{\Psi^*}\left[\left(\frac{U}{U^*}\right)^2 - \frac{UQ}{U^*Q^*}\right] + \left(\frac{U}{U^*}\right)^2 - 1 \tag{6-8}$$

and

$$B = n\,\frac{2 - \Psi^*}{\Psi^*}\,\frac{UQ}{U^*Q^*} \tag{6-9}$$

The solution, in a form slightly different from that given by Horlock [*26*], is

$$\frac{T_2}{T_1} = \frac{T_2^*}{T_1} + \frac{A}{B - 1}\left[\left(\frac{T_2^*}{T_1}\right)^B - \frac{T_2^*}{T_1}\right] \tag{6-10}$$

where $T_1 = T_1^*$ and subscript 2 refers to the compressor outlet conditions.

Of the various deductions that can be made from Eq. 6-10 perhaps the most useful is that, when $UQ = U^*Q^*$, the temperature rise varies as U^2. This result is often used in calculations of the approximate off-design performance of gas turbines (see XI,K).

A simplified analysis of the type given above is of very limited value because the stage efficiency of compressors varies appreciably with the flow coefficient. Furthermore, stage characteristics derived from the flow at the mean radius may lead to errors in the prediction of the stall, owing

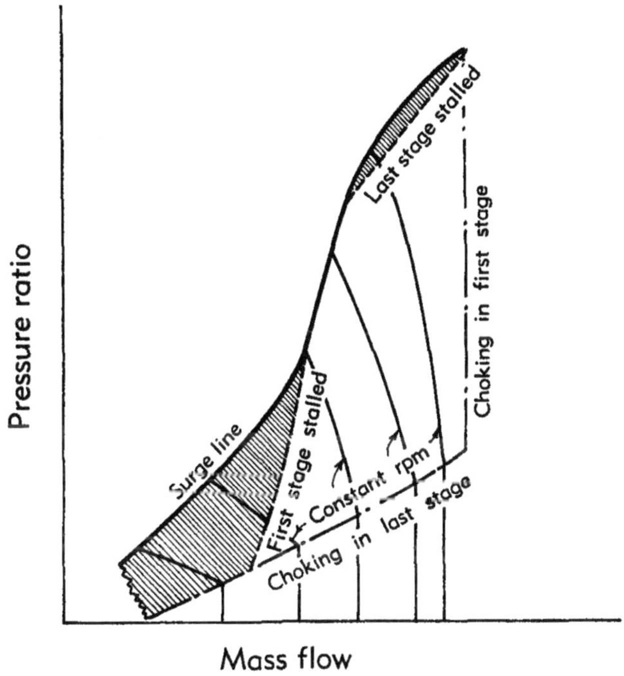

Fig. F,6b. Diagrammatic representation of the characteristics of a typical axial compressor.

to the effect of the earlier stalling at hub or tip, which has already been discussed. This disadvantage may be overcome by using the measured characteristics of the stage, or similar stages, as suggested in the NACA report [2], which describes the "stage-stacking" method of computing compressor characteristics. The stage-stacking method suffers in turn from the effects of neighboring stages which produce nonuniform flows at inlet to the stage and which may induce radial flows at both inlet and outlet from the stage, particularly when the rows of blades are closely packed. The effect of these nonuniform entry conditions is to alter the stage stalling point, an effect that is most marked when the previous stage is stalled. It is also observed that a stage that stalls progressively when tested alone may exhibit an abrupt stall when operated in a multistage compressor, particularly if it is one of the later stages.

Fig. F,6b shows diagrammatically the characteristics of a typical multistage compressor obtained by the stage-stacking method. There is a region of stall-free operation, the size of which decreases rapidly as the design pressure ratio is increased. As the flow is increased at constant rotational speed, the flow becomes limited by the choking of the last stage. The flow through the compressor is also limited eventually by choking of the first stage. The compressor will operate satisfactorily in the cross-hatched regions of the diagram when one or more stages are stalled. At high speeds, however, the stalling of the last or later stages tends to be abrupt and may cause complete compressor stall. The diagram in Fig. F,6b is simplified because it has been found (see [*2*]) that multiple-valued performance curves are obtained if discontinuities in the characteristics of stages near the front of the compressor are obtained when the front stage stalls. The operating point of the compressor may then depend on the manner in which it is approached. The kink in the compressor surge line often obtained in high-pressure-ratio compressors appears to be a result of the transition from operation with the front stages stalled to operation with them unstalled. It has been found that this kink tends to be less marked in compressors with stages of low values of Ψ^*.

F,7. Cited References.

1. Howell, A. R. Design of axial compressors. *Proc. Inst. Mech Engrs. 153*, 1945. Reprinted by *Am. Soc. Mech. Engrs.*, 452–462 (1947).
2. Aerodynamic design of axial-flow compressors. *NACA Research Mem. 56B0 3a, 3b, and 3c*, 1956.
3. Squire, H. B., and Winter, K. G. Secondary flow in a cascade of airfoils in a nonuniform stream. *J. Aeronaut. Sci. 18*, 271–277 (1951).
4. Hawthorne, W. R. Secondary circulation in fluid flow. *Proc. Roy. Soc. A206*, 374–387 (1951).
5. Armstrong, W. D. The secondary flow in a cascade of turbine blades. *Brit. Aeronaut. Research Council Reports and Memoranda 2979* (1957).
6. Briceland, R. H. *Secondary Flow in Cascades. PhD Thesis*, Calif. Inst. Technol., 1959.
7. Hawthorne, W. R., and Armstrong, W. D. Shear flow through a cascade. *Aeronaut. Quart. 7*, 247 (1956).
8. Mager, A. Three-dimensional laminar boundary layer with small cross flow. *J. Aeronaut. Sci. 21*, 335–345 (1954).
9. Sears, W. R. The boundary layer of yawed cylinders. *J. Aeronaut. Sci. 15*, 49–52 (1948).
10. Herzig, H. Z., Hansen, A. G., and Costello, G. R. A visualization study of secondary flow in cascades. *NACA Rept. 1163*, 1954.
11. Williams, A. D. *The Effect of Tip Clearance Flows on Performance of Axial Flow Compressors. A. E. Thesis*, Calif. Inst. Technol., 1960.
12. Rains, D. A. Tip clearance flows in axial turbomachines. Report to the Office of Naval Research, *Calif. Inst. Technol.*, June 1954.
13. Cheshire, L. J. The design and development of centrifugal compressors for aircraft gas turbines. *Proc. Inst. Mech. Engrs. (London) 153*, 426–440 (1945).

14. Iura, T., and Rannie, W. D. Experimental investigations of propagating stall in axial-flow compressors. *Trans. Am. Soc. Mech. Engrs. 76*, 463–471 (1954).
15. Emmons, H. W., Pearson, C. E., and Grant, H. P. Compressor surge and stall propagation. *Am. Soc. Mech. Engrs.*, New York, Dec. 1953.
16. Huppert, M. C., and Benser, W. A. Some stall and surge pheomena in axial-flow compressors. *J. Aeronaut. Sci. 20*, 835–845 (1953).
17. Sears, W. R. Rotating stall in axial compressors. *Z. angew. Math. u. Phys. 6*, 429–455 (1955).
18. Marble, F. E. Propagation of stall in a compressor blade row. *J. Aeronaut. Sci. 22*, 541–554 (1955).
19. Stenning, A. H., Kriebel, A. R., and Montgomery, S. R. Stall propagation in axial flow compressors. *NACA Tech. Note 3580*, 1956.
20. Benenson, D. *Stall Propagation in Axial Compressors. PhD Thesis*, Calif. Inst. Technol., 1957.
21. Rannie, W. D., and Marble, F. E. Unsteady flows in axial turbomachines. Communications aux *Journées Internationales de Sciences Aéronautiques*, Paris, May 1957.
22. Bodden, C. A. Vector velocity fluctuations of propagating stall in axial flow compressors. *M.E. Thesis*, Calif. Inst. Technol., 1961.
23. McCune, J. E. The transonic flow field of an axial compressor blade row. *J. Aero-Space Sci. 25*, 616 (1958).
24. Horlock, J. H., and Louis, J. F. A graphical method of predicting the off design performance of a compressor stage. *Brit. Aeronaut. Research Council, Current Paper 321*, 1957.
25. de Haller, P. Das Verhalten von Tragflügelgittem in Axialverdichtern und im Windkanal. *Brenst. Wàrmekr. 5*, 333 (1953).
26. Horlock, J. H. *Axial Flow Compressors; Fluid Mechanics and Thermodynamics.* Butterworths, London, 1958.
27. Mellor, G. *The Aerodynamic Performance of Axial Compressor Cascades with Application to Machine Design. Sc.D. Thesis*, Mass. Inst. Technol. Gas Turbine Lab. Rept. 38, (1957).
28. Horlock, J. H. A rapid calculation method for the off design performance of compressors and turbines. *Mass Inst. Technol. Gas Turbine Lab. Rept. 40*, 1957.

SECTION G

THE SUPERSONIC COMPRESSOR

CHAPTER 1. EARLY WORK ON SUPERSONIC COMPRESSORS

JOHN R. ERWIN

G,1. Introduction. Two desired features of aircraft engines are that the weight and frontal areas have the minimum value obtainable for a given net thrust and fuel economy. For military applications, an engine having a minimum number of accurately made parts is preferable if no significant sacrifice in performance or reliability is involved. Centrifugal compressors fulfill the requirements of light weight and minimum number of parts; however, the relatively small flow capacity for a given frontal area and the comparatively low efficiency presently obtainable in high compression ratio stages are negative features that subtract seriously from the advantages. The advantages of the axial flow compressor are the high flow capacity for given diameter and the relatively high efficiency. The major disadvantage of present axial compressors is the low pressure ratio produced by a single stage and the resultant requirement of many stages and many blades to obtain the over-all pressure ratio necessary. The mixed flow compressor is a compromise between the centrifugal and the axial flow compressors. The future use of the mixed flow compressor in aircraft gas turbines will probably be confined to smaller engines either as a primary unit or as a second compression stage. As yet, however, the aerodynamic design requirements for efficient, high flow, high pressure ratio, mixed flow compressors have not been determined to an extent permitting wide application. A need exists for a compact, high flow, efficient, high pressure ratio compressor.

The pressure ratios obtainable from one stage and, to a lesser extent, the flow capacity of presently used compressors has been limited by the apparent necessity of avoiding supersonic velocities relative to the blading. There is ample evidence that with conventional compressors, axial or radial, the efficiency decreases rapidly as the Mach number relative to the blades or vanes exceeds unity. The increased losses are believed to be due largely to the presence of shock waves originating near each blade and the flow separation attending the intersection of the shock waves with the blade surface. The total pressure loss suffered by the flow in passing through a single normal shock at a Mach number below about 1.4 is entirely acceptable. However, the losses due to separation, particularly if subsonic diffusion is attempted downstream of the shock, result in a severe increase in entropy. The problem of using supersonic velocities relative to compressor blade rows can be resolved by the reduction or elimination of strong compression waves and the separation of the flow from the aerodynamic surfaces.

For supersonic velocities relative to the blades, sharp-edged profiles are probably necessary to eliminate strong upstream waves. In the Busemann biplane [*1*], extended wave patterns are prevented by producing waves from the internal surfaces only (Fig. G,1a). Kantrowitz, in applying this principle to the design of the forward portion of the

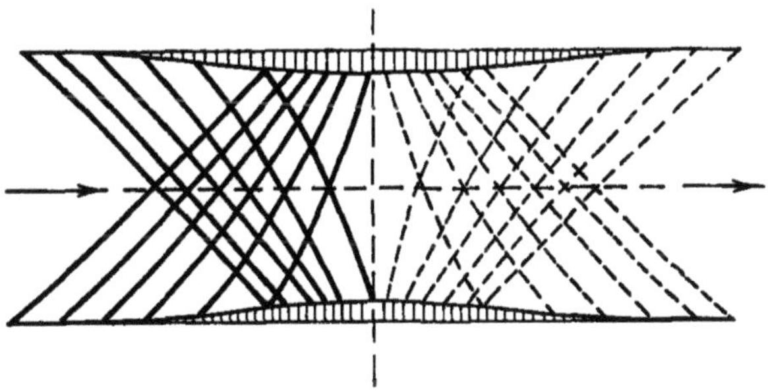

Fig. G,1a. Busemann biplane. Full lines: compression waves. Broken lines: expansion waves.

blade passages (hereafter called the entrance region) of supersonic compressor blades, was able to prevent the occurrence of extended wave systems upstream of supersonic compressor blade rows [*2*]. High pressure recovery and efficiency can be obtained with supersonic diffusers [*3,4*]. Thus no fundamental reason for supersonic compressors to be inefficient was apparent. The high pressure ratios potentially available in a single stage and the high flow capabilities of supersonic compressors have stimulated interest in their development.

The first active consideration of supersonic compressors appears to have taken place in Germany during 1935. Weise of the D.V.L., Berlin, and Encke and Betz of the A.V.A., Gottingen, realized the potentialities of employing supersonic flows in axial compressors about the same time. Encke's compressor was the first built but it was destroyed during early testing. No report was issued describing this machine, but Davidson [*5*] notes that the blading was of triangular section. The sections were located so that the shock wave from the leading edge of one blade intersected the adjacent blade at the vertex of the triangle. It was intended that the flow would become subsonic near the minimum section thus formed and diffuse in the expanding passage downstream of the throat. Insufficient data were recorded to determine whether these flow conditions could be achieved before the compressor was destroyed.

Weise proposed several types of axial flow supersonic compressors [6]. The first type tested employed guide vanes to prewhirl the flow entering the rotor in the direction of rotation. The flow entered the rotor at subsonic velocity and was turned about 90° without a net change in static pressure at the mean diameter (Fig. G,1b). Due to the large addition of

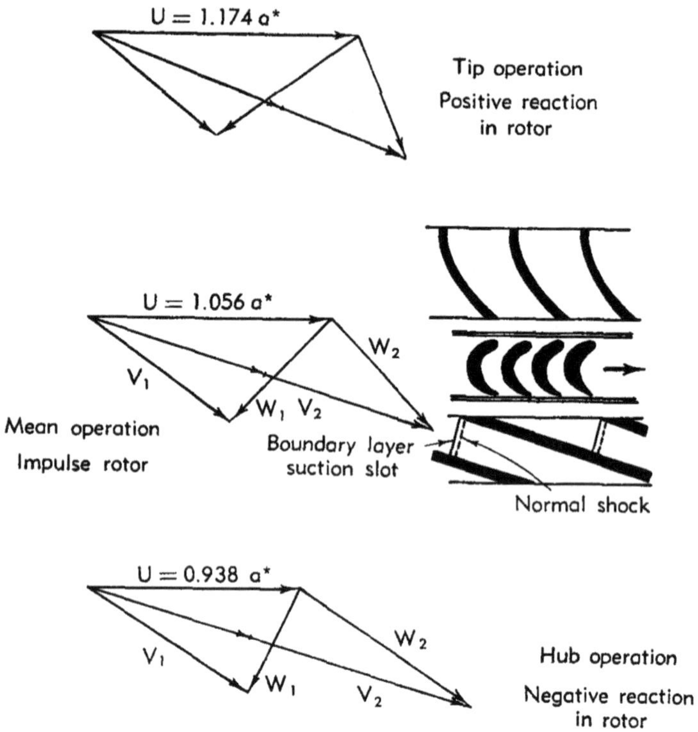

Fig. G,1b. Weise Type I compressor details. a^* is sonic velocity before intake.

tangential velocity imparted by the impulse rotor, the Mach number of the flow into the stators was over 1.5. Weise's intention was to maintain a normal shock at the stator entrance. Slots were provided in each passage to remove the casing boundary layers and to prevent flow separation. The stator sections were straight. Expansion of the passage area was obtained by divergence of the outer casing (Fig. G,1c).

Unfortunately, this compressor yielded poor performance in tests. At design speed, the maximum pressure ratio obtained was less than 1.4 compared to the design value of about 2. The adiabatic efficiency was very low (Fig. G,1d). These results were believed to be due to separation of the flow from the inner casing downstream of the rotor. To study this problem further, the stator blades were removed and were replaced by a cylindrical passage. Surveys of the flow at several stations downstream from the rotor showed that the flow separated from the case under these

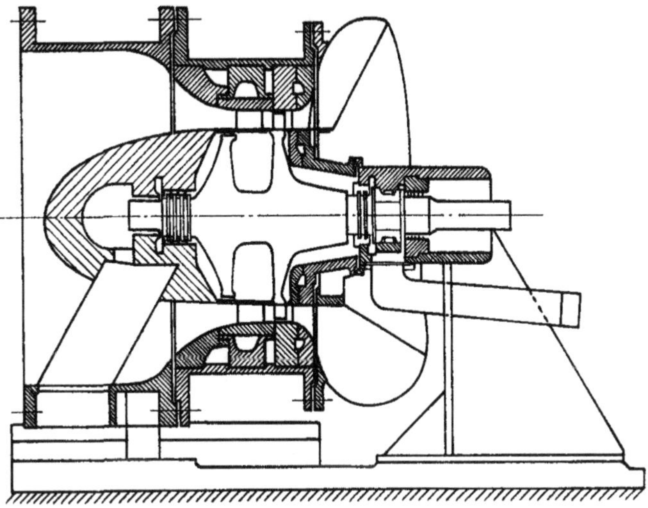

Fig. G,1c. Section through the D.V.L. test rig.

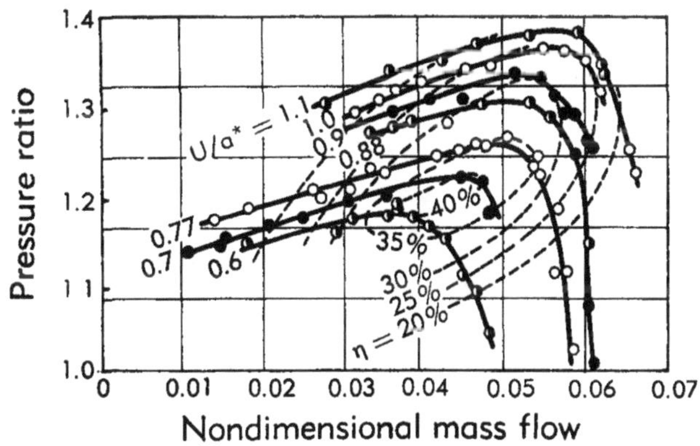

Fig. G,1d. Weise Type I compressor performance.

conditions also. No reasonable explanation for the observed behavior of the flow was discovered.

Weise considered a different stator design for this compressor (Fig. G,1e). In this configuration, all of the compression was to occur in a normal shock at the stator entrance. The stator blades were not intended to diffuse the flow but merely to return the flow direction to that leaving the guide vanes. To maintain constant pressure through the stators and to reduce the tendency of the flow to separate from the inner casing, this surface was to curve outward in the region of the blades. This modification was not built, however.

The second supersonic compressor built by Weise was quite different in principle from the first. In the second compressor, the rotor was intended to produce a high static pressure rise with little turning of the flow.

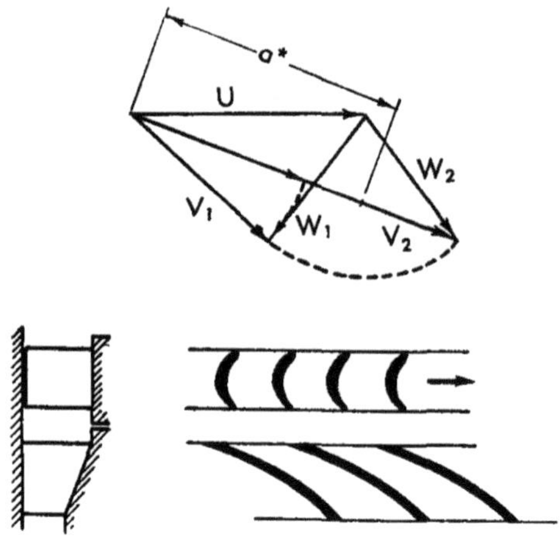

Fig. G,1e. Weise Type Ia compressor details. a^* is sonic velocity before intake.

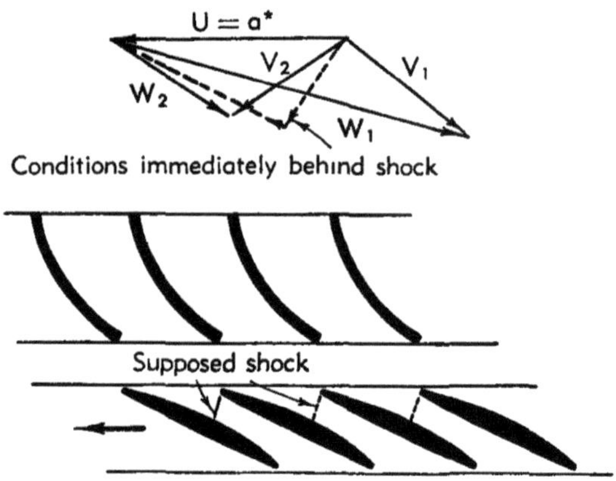

Fig. G,1f. Weise Type II compressor details. a^* is sonic velocity before intake.

Although the same inlet guide vanes were used and the rotational speed was slightly lower, supersonic velocities into the rotor were obtained by designing the rotor to turn in the opposite direction from the first compressor. The flow was to enter the rotor with a Mach number of about 1.55, undergo a normal shock near the leading edge of each blade, and exit with subsonic velocity. The blading was similar to typical subsonic compressor blading but had sharp leading edges and higher chord-spacing ratio. The passage between blades was designed so that the minimum area occurred at the entrance. The drawing (Fig. G,1f) neglects the wave pattern that must exist upstream of the rotor with the curved sections shown. This compressor was destroyed during the first high speed test

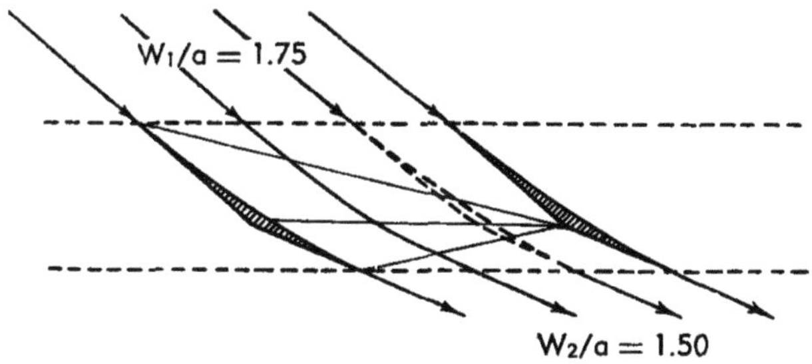

Fig. G,1g. Supersonic compressor blades without shock waves. The Prandtl-Meyer compression waves are cancelled within the passage.

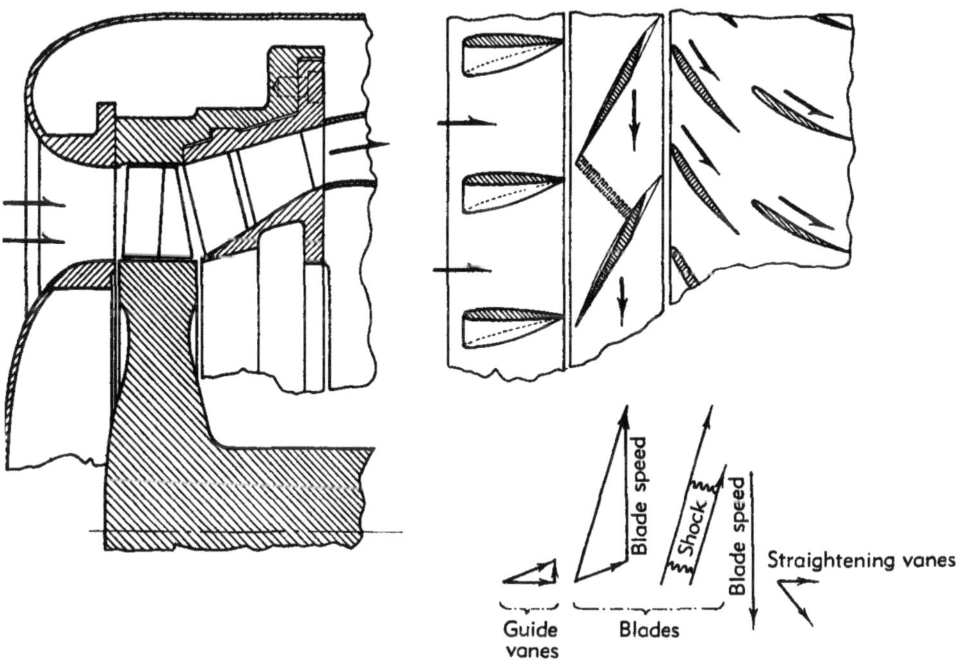

Fig. G,1h. Supersonic compressor patented by Redding.

run before design conditions could be established. As the machine was not rebuilt, no information as to the performance of this compressor is available.

A third type of supersonic compressor considered by Weise is illustrated in Fig. G,1g. The inlet velocity is presumed to be axial and supersonic so that compression waves from the leading edge region propagate downstream. This type of compressor can produce only limited pressure ratios but may have application because of a unique feature: a supersonic axial velocity can be converted to a subsonic axial velocity while the flow

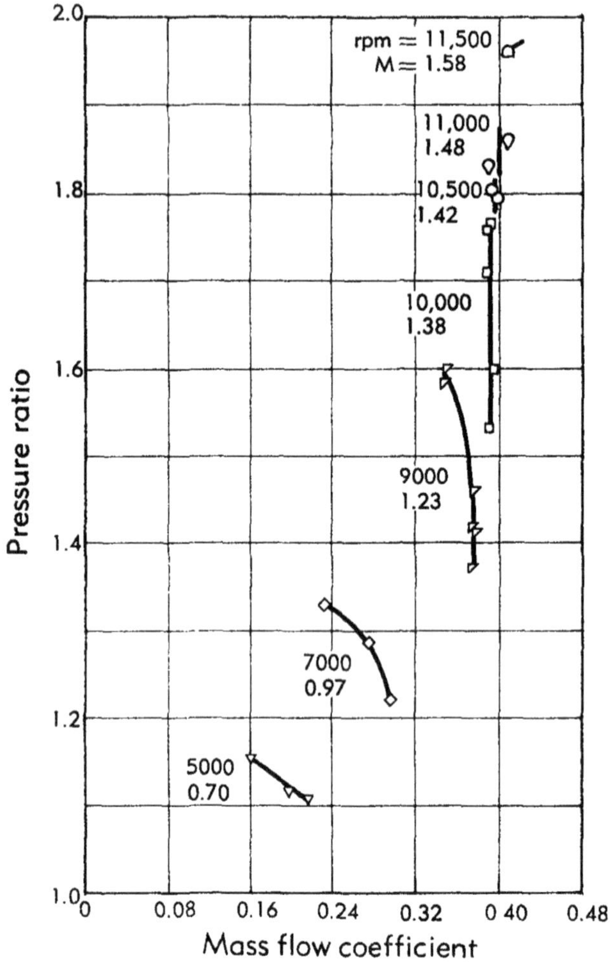

Fig. G,1i. Variation of pressure ratio with mass flow coefficient and compressor Mach number. These data were obtained from surveys behind the rotor and therefore do not include any stator losses.

in the rotor remains supersonic. By suitable design of the blading, such a rotor could be either self-driven, power-producing, or power-absorbing [7].

In the United States, investigation of supersonic compressors was begun independently by Redding [8] at the Westinghouse Electric Company and by Kantrowitz at the NACA Langley Laboratory in 1941.

Both studies proceeded along remarkably similar lines, the studies consisting first of theoretical considerations of possible vector diagrams and selection of similar designs; secondly, of experimental investigations of small circular supersonic diffusers [*9*] and then tests of rotors designed for similar performance (Fig. G,1h and Plates G,1a and G,1b). Early tests in both laboratories were discouraging; at Westinghouse, supersonic compressor investigations were discontinued for the time in favor of the immediately applicable, subsonic axial flow compressor.

The performance of the rotor tested at Langley [*10*] is shown in Fig. G,1i and G,1j. The characteristic behavior of supersonic compressors in operating at constant flow over wide ranges of pressure ratios at higher

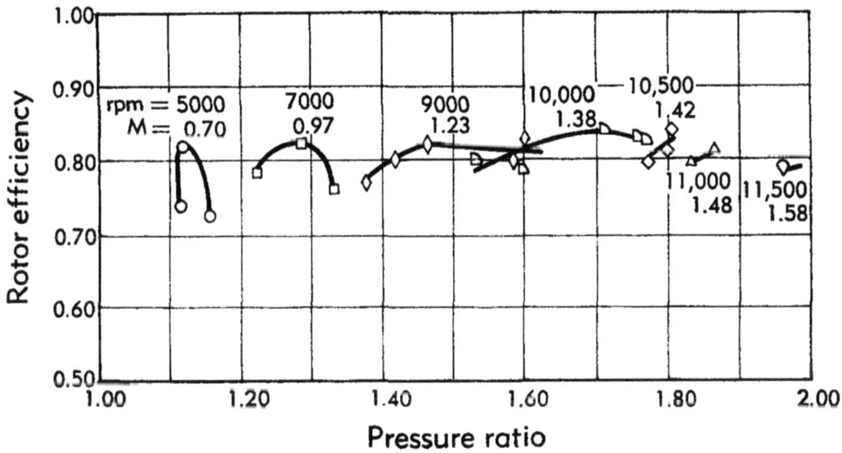

Fig. G,1j. Efficiency variation with pressure ratio and compressor Mach number. Data are from surveys in Freon-12 behind the rotor and do not include stator losses.

speeds is evident. The pressure ratio obtained at the design rotational speed (10,500 rpm in Freon-12) was 1.8 and the maximum rotor efficiency was about 81 per cent, considerably less than the design pressure ratio 2.9 and the assumed rotor efficiency of 90 per cent. A rather surprising result obtained was that the peak efficiency occurred in the transonic speed range. This result led to an interest in compressors operating in the transonic speed range. Although a number of supersonic compressors of the reaction type have been constructed, no major improvements in pressure ratio or efficiency have been obtained.

Most of the difficulties of the reaction-type supersonic compressor are believed to be associated with the outward flow of the blade and hub surface boundary layers and particularly of fluid from regions of separation toward the rotor blade tip. Thus the tip section, which ordinarily must produce the highest static pressure rise, is required to do so in the presence of an accumulation of low energy flow. It would seem reasonable that, by transferring the difficult task of producing the high static pressure rise to the stator, the rotor performance would be improved considerably and, further, higher work inputs would become possible. Transferring the static pressure recovery from the rotor to the stator makes possible the use of variable geometry, boundary layer control, higher solidities or multiple blade rows which would hardly be practical with a rotor. Weise's first compressor had a subsonic impulse rotor but the performance was poor. Investigation of impulse compressors began in 1945 at Langley Field and efficient rotor performance was obtained in low speed tests [*11,12*]. The supersonic impulse rotor shown in Plate G,1c was designed at Langley Field in 1948.

G,2. Isentropic Flow through Blade Rows. For visualizing possible two-dimensional isentropic flows through blade rows, the Mach vector diagram [*7*] is useful. On this diagram (Fig. G,2a), lines of constant Mach number are plotted as circles about the origin *O*. Lines of constant flow direction measured from the axis of rotation of the blade row are shown radiating from the origin. For any given cascade (one frame of reference), contours of constant mass flow or through-flow coefficient, $\lambda = A^*/A$ (where A^* is the area of the flow at $M = 1$ and A is the annulus area of the blade row), can then be drawn. The constant λ contours illustrate

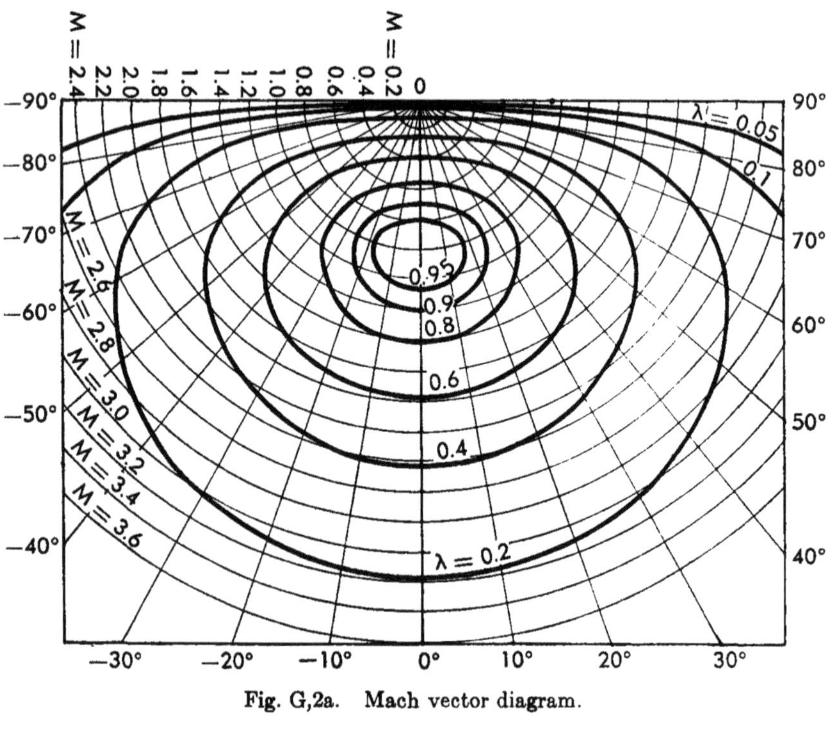

Fig. G,2a. Mach vector diagram.

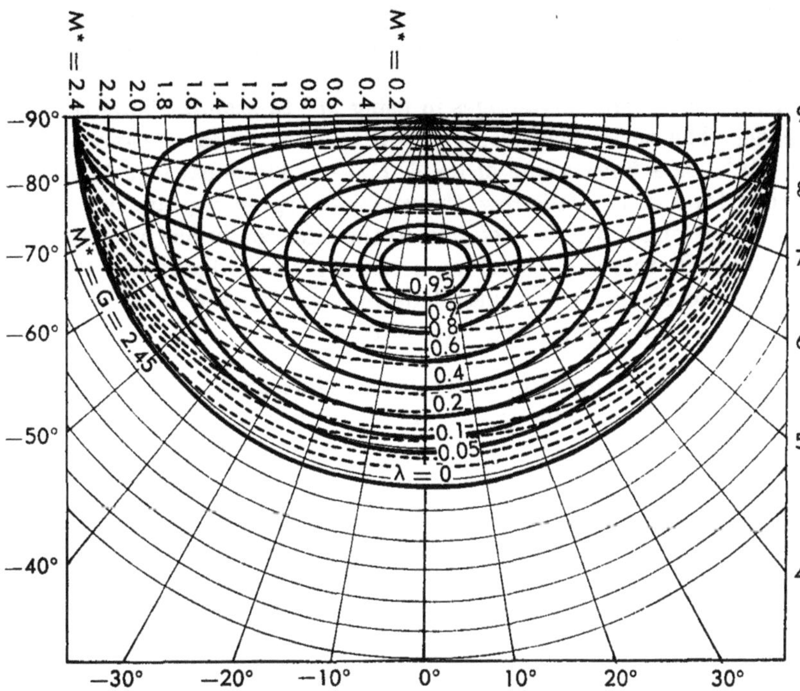

Fig. G,2b. Reduced velocity vector diagram.

possible flows through blade rows of constant annulus area. For example, consider a cascade having flow entering at a Mach number of 1.6 in a direction 60° from the axis of rotation. For this condition, $\lambda = 0.4$. If the flow is diffused and discharged in the same direction, the discharge Mach number for isentropic compression would be about 0.57. In a real flow, the growth of the boundary layers and the total pressure losses due to finite shocks restrict the flow, so that the value of λ and the Mach number of the main stream leaving the blade row are higher. The value of λ will also change if the physical flow area is changed. Within the cascade, the finite thickness of the blades increases the value of the flow coefficient.

The maximum tangential velocity that a given flow can attain is of interest to compressor and turbine designers. In [7], the maximum tangential velocity is shown to occur when the Mach number of the axial component of the flow is unity.

Another method of presenting possible flows in compressor or turbine blade rows is the "reduced velocity" vector diagram. This diagram (Fig. G,2b) is constructed in a manner similar to the Mach vector representation, except that values of constant $M^* = V/a^*$ (where a^* is the velocity of sound at $M = 1$) instead of $M = V/a$ are plotted as circles about the origin. Contours of the constant mass flow coefficient λ can be superimposed on this diagram to illustrate the possible conditions a given flow can obtain.

There are several advantages in using M^* to examine flows through cascades. The entire field from $M = 0$ to $M = \infty$ can be represented since M^* has a finite value of $[(\gamma + 1)/(\gamma - 1)]^{\frac{1}{2}}$ equal to 2.45 for air. Further, for a given blade row, the rotational speed appears as a vector of the same length whether referred to the inlet conditions or to the exit conditions. On this plot, therefore, a velocity vector diagram of the flow entering and leaving a rotating blade row can be used to estimate the conditions in stationary coordinates directly.

G,3. High-Turning Blade Sections. Two-dimensional blade sections for high-turning supersonic compressors can be constructed graphi-

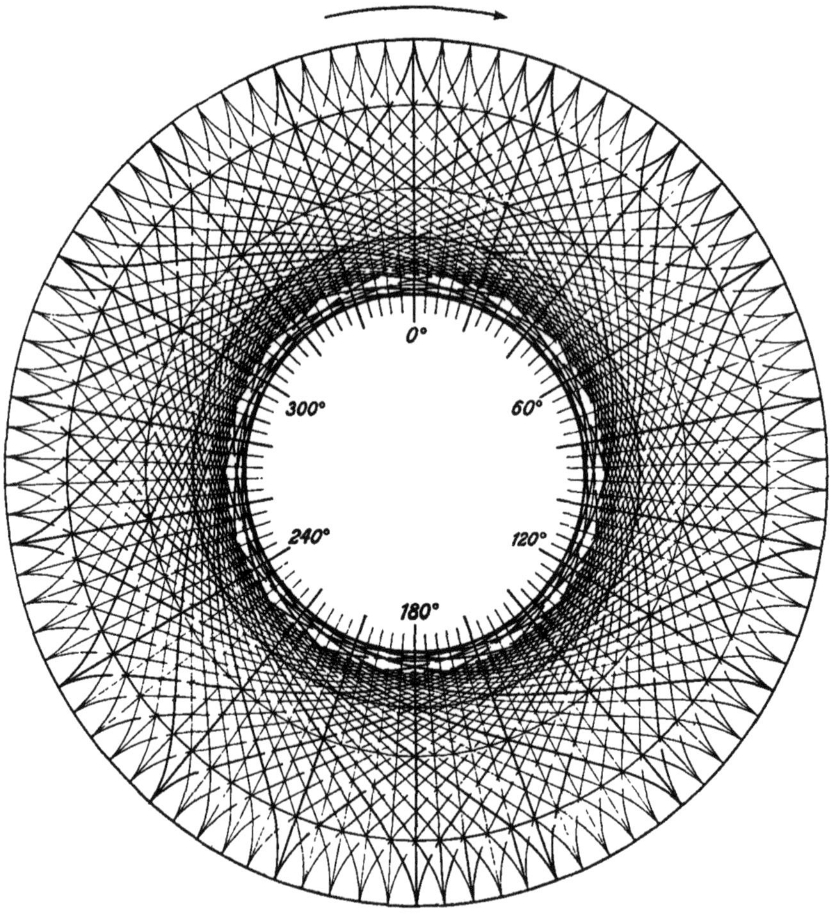

Fig. G,3a. Characteristic line network for supersonic vortex flow according to Busemann. $\gamma = 1.40$.

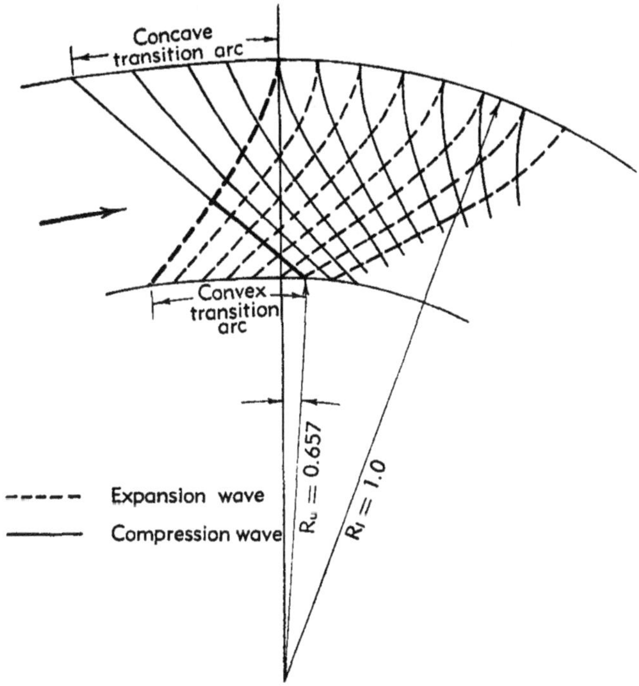

Fig. G,3b. Construction of transition arcs.

cally by the method of characteristics. Sections so designed for different conditions do not necessarily bear any physical relationship to each other. If the passages between blades are designed so that supersonic vortex flow exists [*9*], a number of advantages accrue: (1) sections designed for different conditions are related so that the properties of a series can be determined by a minimum of theoretical or experimental effort; (2) the desired turning of the flow can be achieved with a minimum of blade surface area if maximum and minimum surface Mach numbers are specified; (3) the major portions of the curved surfaces are either straight lines or circular arcs so that machining is simplified and fair blade surfaces are more easily obtained and checked; and (4) the graphical construction of the sections is quick and simple.

The classical compressible vortex flow in which the streamlines are concentric circles and in which the velocity is inversely proportional to the radius has been shown by Busemann in an unpublished paper to yield an analytic solution for the Mach lines in the supersonic portion. If the solution is solved in terms of a nondimensional velocity ratio it can then be tabulated or diagrammed for general use without the necessity of recalculating for each specific case.

The general vortex equation VR = const, where V is velocity and R is radius, can be rewritten $M^*R^* = 1.0$ with $M^* = V/a^*$ and $R^* =$

R/r^*. Here a^* is the sonic velocity at unit Mach number and r^* is the radius of the sonic velocity streamline in the vortex field. Then

$$\phi = \pm \frac{1}{2} \left\{ \sqrt{\frac{\gamma + 1}{\gamma - 1}} \sin^{-1} \left[\frac{\gamma - 1}{R^{*2}} - \gamma \right] + \sin^{-1} [(\gamma + 1)R^{*2} - \gamma] \right\} + \text{const}$$

where ϕ is the angular position of the radius vector. Figure G,3a shows the characteristic network for supersonic vortex flow according to Busemann for $\gamma = 1.40$. Using this diagram, the Mach wave pattern in the vortex flow region of any high-turning passage can be traced.

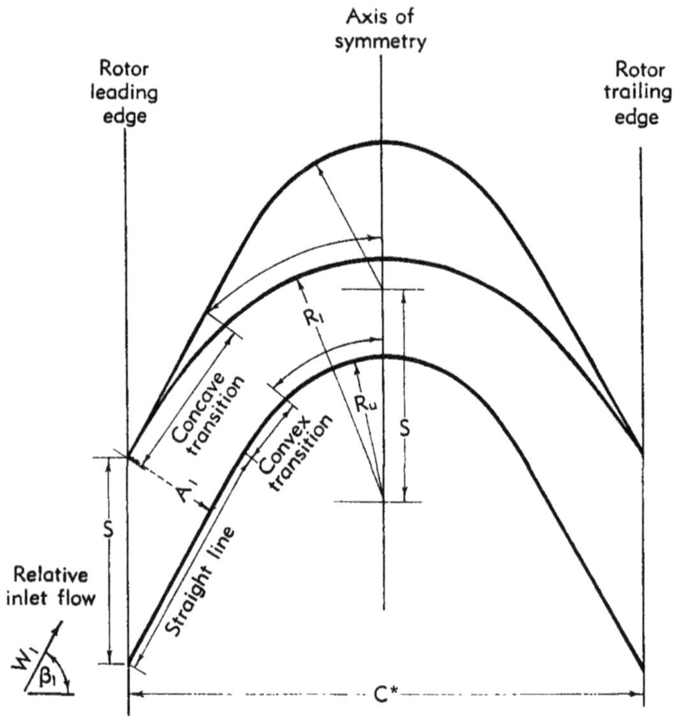

Fig. G,3c. Construction of typical symmetrical blade section. From [5].

The uniform relative inflow into a blade is converted to a vortex flow by means of transition arcs on the convex and concave surfaces which can be simply constructed by means of nonintersecting characteristic lines (Fig. G,3b). At the exit the process is reversed. The graphical design of a blade passage for given inlet, outlet, and surface Mach number conditions can be readily accomplished. The surface velocities of the passage determine the radius of one circular arc with respect to the other. The concave transition arc must generate compression waves to reduce the inlet velocity to that prescribed for the concave circular surface while, at the convex surface, expansion waves are generated. The transition sections must be so shaped that the Mach waves fair into the vortex wave pattern, only the leading and trailing portions of which need to be drawn. At the exit, the transition arc surfaces are created by shaping the contour so as to cancel the incident waves and return the flow to uniform parallel flow. To form a closed profile the fore and aft transition arcs on the suction surface are continued in a straight line parallel to the flow direction to the leading and trailing edges of the blade (Fig. G,3c). A schlieren photograph of flow through blade sections designed by this method is presented in Plate G,3. Disturbances visible in the photograph appear to follow a Mach line (dotted) of the characteristic network very closely.

G,4. Summary. Considerable interest has been shown in supersonic compressors for more than two decades. Although the test results of these compressors have generally been discouraging, research on this type of machine has been continued because of the design simplicity and reduced weight which appear to be available once the difficult problems are solved.

At the present time, the design of supersonic compressors is based largely on experience with supersonic diffusers, stationary cascades of blades, and previous supersonic compressors. Two-dimensional blade sections can be designed readily, using graphical characteristic methods. The diffusion rates that can be obtained efficiently with supersonic flow or with subsonic flows following normal shocks in the presence of strong radial pressure gradients due to centrifugal effects are not well established. Studies of laminar and turbulent boundary layers in the vicinity of a strong oblique shock intersecting and reflecting from the surface have aided the visualization of the problems. A satisfactory theoretical treatment of the three-dimensional flow, including the boundary layers, through supersonic compressor blading appears to await more thorough experimental results with cascades and compressors of various types.

CHAPTER 2. AERODYNAMIC PROPERTIES OF SUPERSONIC COMPRESSORS

ANTONIO FERRI

G,5. Introduction. A compressor is usually defined as a "supersonic" compressor when the flow enters the rotor at a velocity which is everywhere supersonic with respect to a coordinate system rotating with the rotor. When the velocity of the flow relative to the rotor, at the entrance of the rotor, is only partly supersonic, and thus in some region of the flow is subsonic, then the compressor is called a "transonic" compressor.

According to this definition only the velocity relative to the rotor must be supersonic; the absolute velocity of the flow can be supersonic or subsonic. Supersonic compressors have operating characteristics which depend on the value of the axial component of the velocity of the flow approaching the rotor (component normal to the plane parallel to the disk of the rotor). Therefore, in the following articles these subdivisions are introduced: supersonic compressors operating with subsonic axial

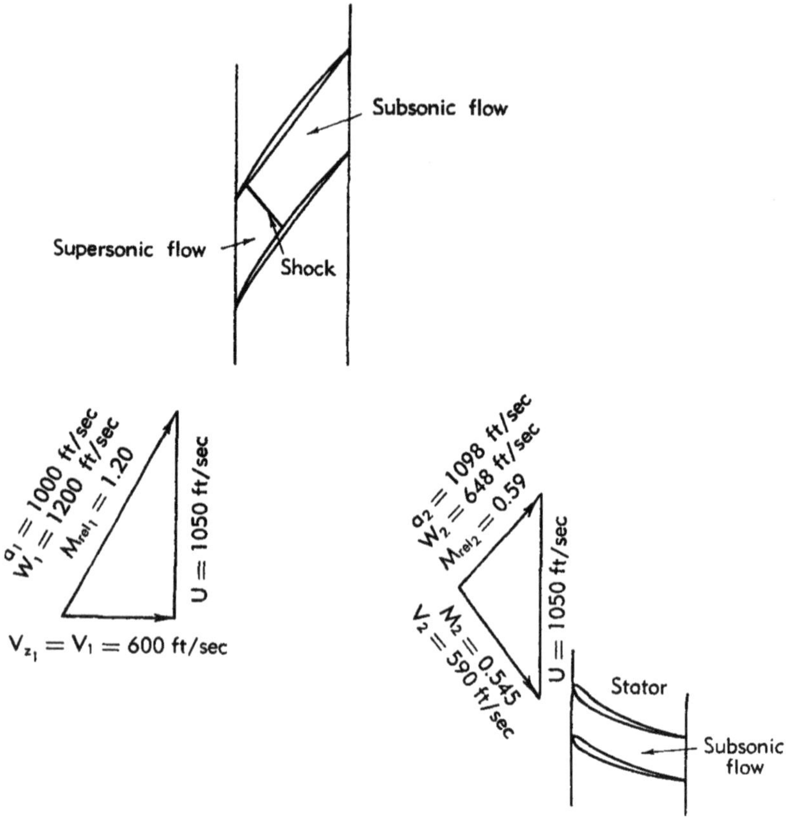

Fig. G,5a. Velocity diagram for shock in rotor compressor.

velocity and supersonic compressors operating with supersonic axial velocity at the entrance of the rotor.

Other subdivisions are often introduced depending on the value of the velocity leaving the rotor. If the velocity leaving the rotor in a relative coordinate system is subsonic, then transition from supersonic to subsonic flow must occur in the rotor. In this case the rotor is usually called shock in the rotor compressor, because it is assumed that the transition from supersonic to subsonic flow will occur in the rotor passages through a shock. When the exit velocity in relative and in absolute motion is supersonic, the transition from supersonic flow to subsonic flow must occur in the stator; then the compressor is called shock in the stator compressor. The possibility exists that the flow leaving the rotor of a supersonic compressor is subsonic in relative motion but supersonic in absolute motion; then transition from supersonic to subsonic flow occurs both in the rotor and in the stator. It is possible also that the velocity at the exit of the rotor in relative coordinates may be supersonic while the absolute value of the velocity is subsonic. Then the transition could occur without a strong shock in the rotor or in the stator. Typical velocity diagrams for compressors with shock in the rotor, shock in the stator, shock in the rotor and shock in the stator, and with no transition are

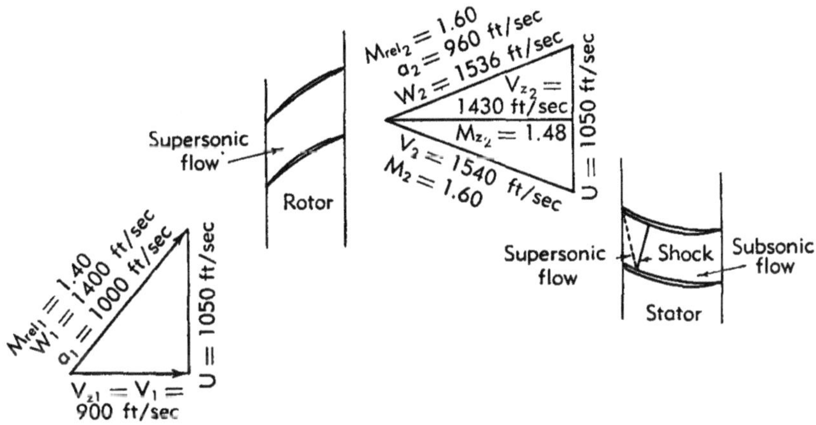

Fig. G,5b. Velocity diagram for shock in stator compressor.

given in Fig. G,5a, G,5b, G,5c, and G,5d respectively. The symbols in these figures are defined as follows:

V = the velocity of the flow in a stationary coordinate system
V_z = the axial component of the flow velocity
U = the rotational velocity of the rotor
W = the velocity of the flow relative to the rotor

The subscript 1 indicates properties at the entrance of the rotor. The subscript 2 indicates properties of the flow at the exit of the rotor. In the passage of Fig. G,5d a contraction in the annulus is assumed along the passage. The contraction decreases the relative velocity from 1400 ft/sec to 1250 ft/sec.

In the following articles the flow fields at the entrance and at the exit of the rotor passage are analyzed for different types of supersonic compressors, by assuming first that three-dimensional effects are small and can be neglected. In Art. 3 some of the important three-dimensional effects are qualitatively discussed, while Art. 4 briefly outlines possible performances of the different types of supersonic compressors.

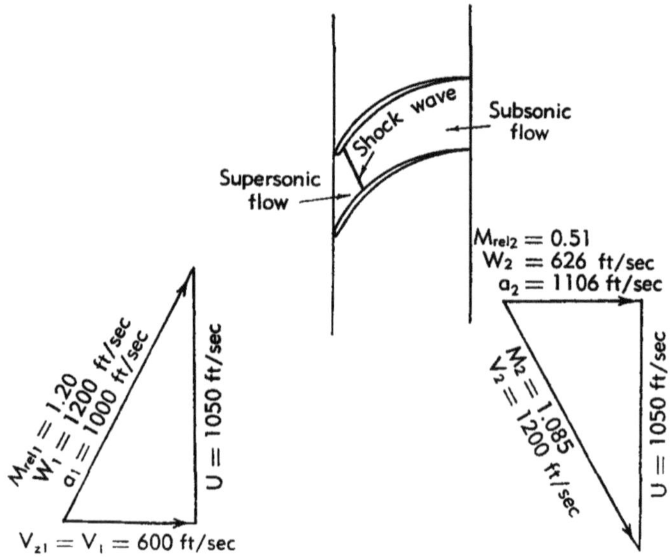

Fig. G,5c. Velocity diagram for shock in rotor and stator compressor.

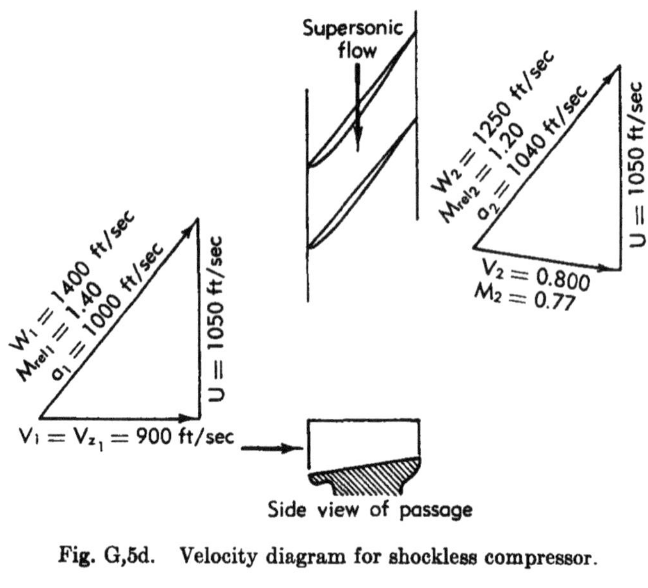

Fig. G,5d. Velocity diagram for shockless compressor.

G,6. Flow Field at the Entrance of a Supersonic Cascade with Subsonic Axial Velocity.

Blades of zero thickness and zero camber. Consider a rotor having blades of zero thickness and without camber, and assume that the radial effects are small so that the velocity component in the radial direction can be neglected. The flow field in such a rotor can be analyzed by representing the rotor as a two-dimensional cascade having an infinite number of blades with zero thickness and no camber, and having the same spacing as the rotor blades at the radial station at the middle of the blade.

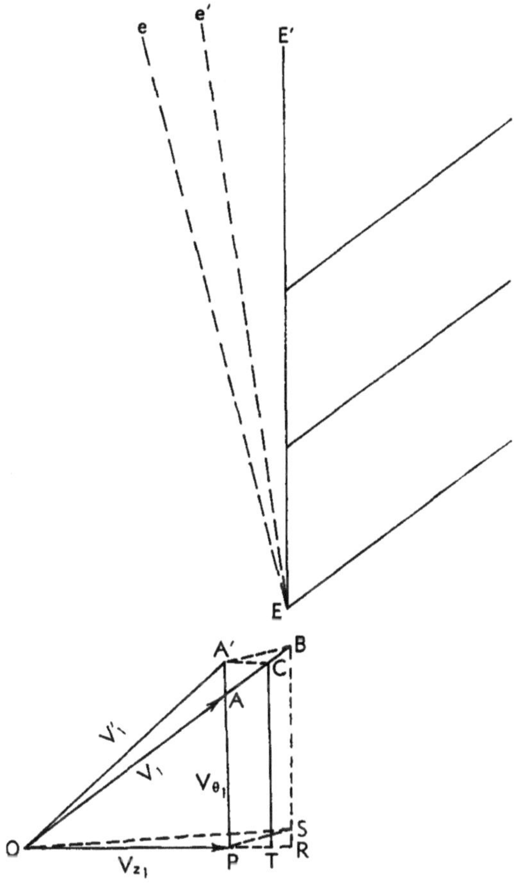

Fig. G,6a. Semi-infinite supersonic cascade.

Assume that the axial velocity V_{z_1} of the flow entering the rotor is subsonic, while the absolute velocity V and the rotational velocity U of the rotor are such that the velocity relative to the blade W_1 is supersonic.

In order to analyze the flow in such a cascade let us consider at first a cascade of semi-infinite type, as shown in Fig. G,6a, where the blade at E is the first blade of the cascade and an infinite number of blades follow the blade in the direction of the rotation of the rotor. The cascade is placed in a supersonic flow field of velocity V_1. The inclination of the entrance plane of the cascade EE' to the undisturbed velocity V_1 is such that the velocity component V_{z_1} normal to the plane EE' is equal to the axial velocity of the flow entering the rotor, while the tangential component V_{θ_1} parallel to the plane EE' is equal to the rotational velocity of the rotor plus the tangential component of the absolute velocity of the flow entering the rotor.

When the axial component V_{z_1} of the velocity, relative to the blade, is subsonic, waves produced at the leading edge of the blades are less inclined with respect to the entering velocity than the entrance of the cascade EE', because the velocity component normal to the wave must at least be sonic. Therefore, these waves move ahead of the cascade and change the conditions of the entering stream.

Consider at first the flow field in the region of the entrance of the cascade and assume that the as-yet-unspecified flow conditions downstream of the cascade are such that supersonic flow can be established in the passages between the blades of the cascade. Then the entering volume flow is determined only by the conditions at the entrance of the cascade.

When the subsonic axial velocity component V_{z_1} and the tangential velocity component V_{θ_1} are such that the velocity V_1 relative to the cascade is parallel to the blades (vector OA of Fig. G,6a), no waves are produced at the entrance of the cascade and therefore no waves are transmitted upstream in the flow. If the tangential velocity V_{θ_1} increases with respect to this value while V_{z_1} remains constant, the new direction of the velocity OA' relative to the cascade becomes inclined with respect to the blade direction, and expansion waves e, e' are produced at the leading edge of the first blade at E (Fig. G,6a). Because of the subsonic value of the axial component, the expansion waves produced at the first blade accelerate the flow in front of the other passages, and because behind the last waves e' the flow again becomes parallel to the blades, no other waves are produced at the leading edge of all the other blades following blade a.

The equilibrium conditions behind the waves e, e' correspond to a value of the entering relative velocity having a direction parallel to the blade or in the direction of OA; therefore the expansion waves produce an increase of velocity from OA' to OB. The point B, from two-dimensional supersonic flow considerations, is on the epicycloid $A'B$, corresponding to the velocity OA'. The waves, moving upstream in the flow, change the value of the axial velocity from OP to OR and the value of the tangential velocity from PA to RB.

Consider now a supersonic rotor obtained by wrapping the cascade of Fig. G,6a around a cylinder. If the velocity relative to the rotor is in the direction of the vector OA, the flow conditions at the entrance of the rotor are such that no waves are produced at the entrance of the rotor. Now if the direction of the velocity V_1 relative to the rotor changes abruptly because, for example, the rotational velocity U changes from the value PA to the value PA' (Fig. G,6b), then waves similar to the waves ee' of Fig. G,6a are generated simultaneously at the leading edges of all blades.

Plate G,1a. Supersonic compressor rotor developed by A. R. Kantrowitz.

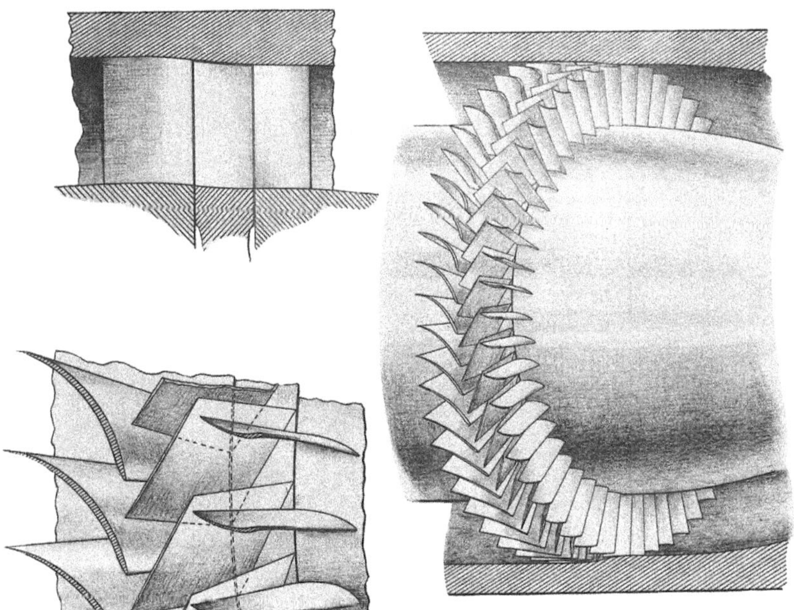

Plate G,1b. Supersonic compressor developed by Kantrowitz.

Plate G,1c. Impulse supersonic compressor rotor.

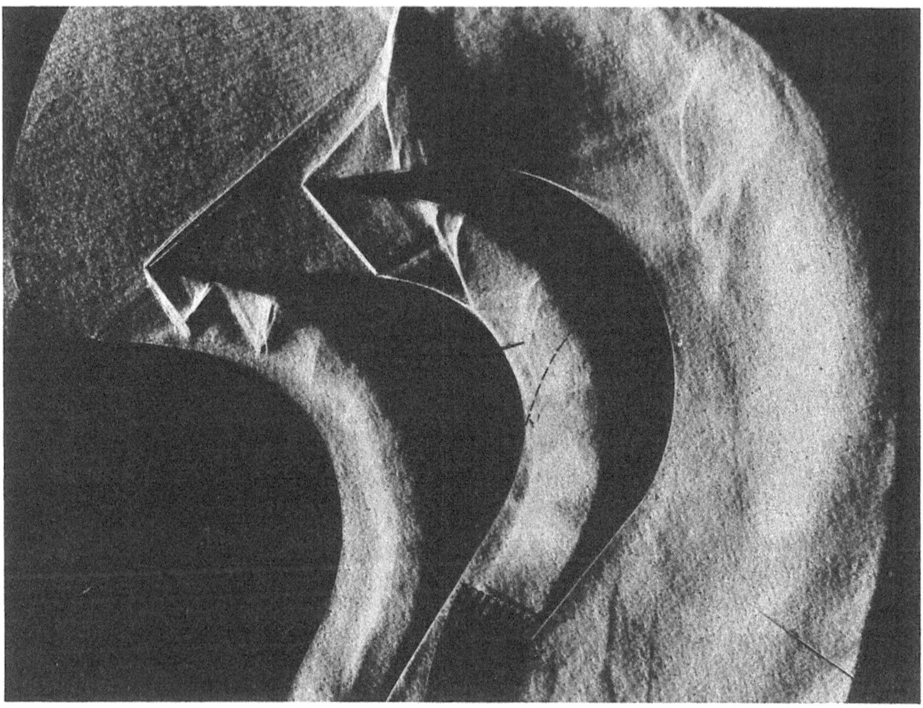

Plate G,3. Schlieren photograph of the flow in a blade passage designed for supersonic vortex flow.

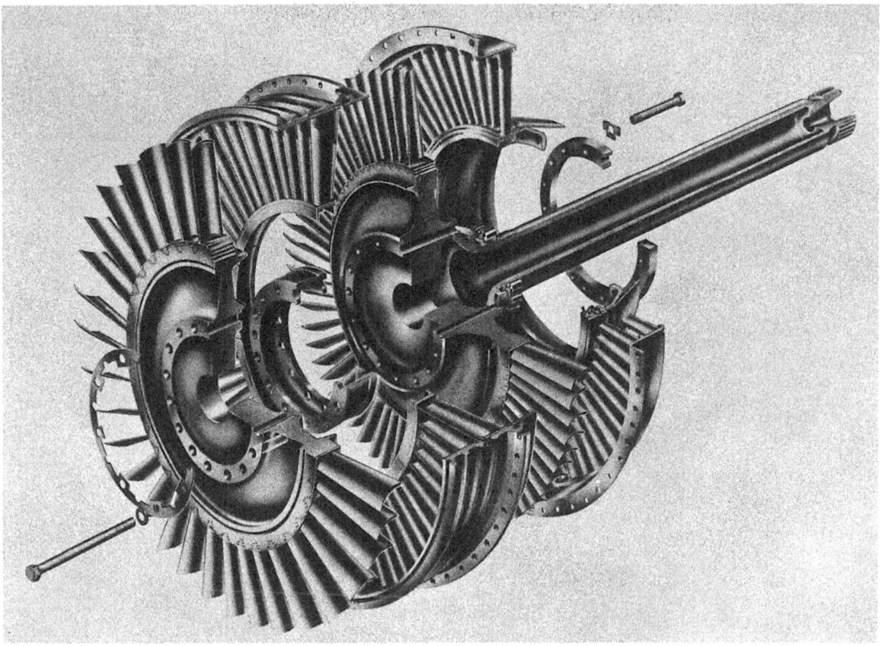

Plate H,3. Exploded view of two-stage turbine element from Westinghouse J34 turbojet. Gas flows through turbine from right to left.

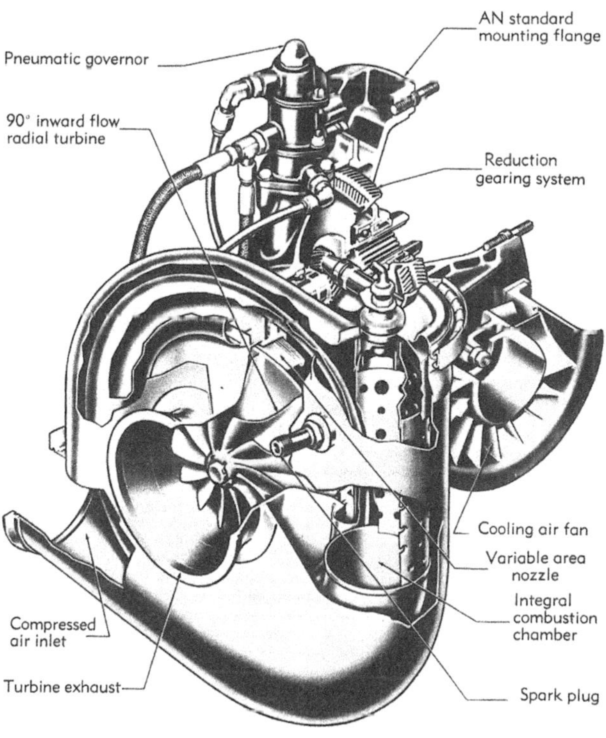

Plate I,13a. Gas turbine motor with centripetal turbine. (The Garrett Corporation.)

Plate I,13b. Three radial turbine wheels used in turbochargers of comparable characteristics.

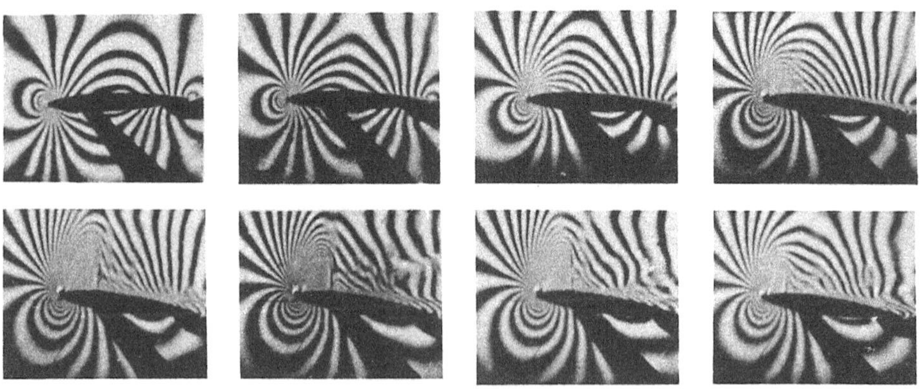

Plate K,3. Unsteady density field around a fluttering profile. $M = 0.6$; $\nu = 11$ cycles/sec.

The production of such waves occurs through a nonsteady process and at a time t_1, slightly after the time of the change of the velocity U the family of waves produced by each blade travel only at a small distance in front of the blade. The qualitative form of the wave pattern is shown in Fig. G,6b. Each family of waves moves upstream of the blade and interferes with the family of waves produced by the other blades. The wave

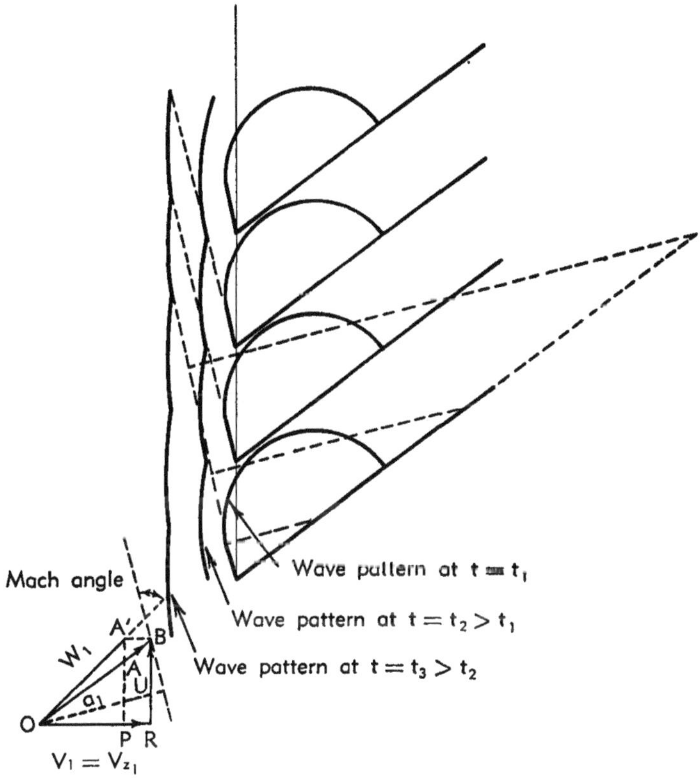

Fig. G,6b. Wave pattern in front of supersonic rotor.

resulting from the interference of all the families of waves moves upstream of the compressor in a direction normal to the axis of the compressor and, therefore, changes only the axial velocity [*13*].

Waves are produced at the leading edge of each blade unless the direction of the velocity relative to the blade is tangent to the blade; therefore, the steady conditions in front of the compressor when the nonstationary waves have moved far away in front of the rotor corresponds to the velocity diagram *ORB*, where *OR* is the new axial velocity corresponding to the new rotational velocity *BR* and the vector *OB* is tangent to the blade. The increase in axial velocity corresponds to *PR*.

If the tangential velocity U decreases with respect to the original value, compression waves (shock waves) are produced at the leading edge of every blade of the rotor. Each compression wave moves upstream and interferes with the compression waves produced by the other blades. The nonsteady wave pattern is again similar to the wave pattern shown in Fig. G,6b. The compression wave resulting from the interference of all the compression waves produced by each blade moves upstream and decreases the value of the axial velocity.

The equilibrium condition in the absence of waves moving upstream from the rotor for a given value of the tangential component U again corresponds to a value of the axial velocity V_{z_1} such that the velocity W_1 entering the rotor is parallel to the direction of the entrance of the blade. Therefore, in an axial supersonic compressor or turbine having blades as those considered in Fig. G,6b and with a known blade angle, the value of the axial flow component at the rotor entrance and the direction of the entering flow can be determined directly from the value of the rotational component.

It is important to keep in mind that such indications have been obtained by assuming two-dimensional flow in front of the compressor and straight blades without thickness; in the following discussion it is shown that both parameters can affect the results of such an analysis.

Blades having finite thickness and camber. Consider now a semi-infinite cascade having blades with finite but small thickness, such that an attached shock is possible at the leading edge of the blade for the existing Mach number and direction of the incoming flow. If the design of the entrance of the passage between two blades is such that no curvature exists at the trailing surface AB of the passage in the region of the entrance (Fig. G,6c), then the flow field at the entrance is similar to that for blades of zero thickness and the considerations presented in the preceding discussion still apply. In this case the thickness can be considered to be obtained by deflecting the leading surface of the blade from AB to AC. A shock wave AD is produced which is contained inside the passage and does not affect the flow in front of the cascade.

Consider now a blade shape as shown in Fig. G,6d with curved upper surface in the region of the leading edge of the blade. The velocity relative to the blade at the leading edge is supersonic and is tangent to the blade at the leading edge A; therefore the flow along the trailing surface AC of the blade undergoes a deviation which produces waves. Some of the waves produced at the trailing surface near the leading edge move upstream because of the subsonic character of the axial component of the stream and change the flow field in front of the other passages of the cascade, while other waves are confined inside the passage. The waves generated downstream of the point B of the blade a, defined as the point which produces the wave meeting the leading edge of the next blade b, do not affect the flow entering the other passages, while the waves produced along AB move upstream of all the passages. The region AB can be defined as the entrance region of the blade.

Consider at first a semi-infinite cascade as shown in Fig. G,6d, where the curvature of the trailing surface of the blades is such that expansion waves are produced along the entrance region AB. The expansion waves produced by the curvature along the surface AB expand and turn the

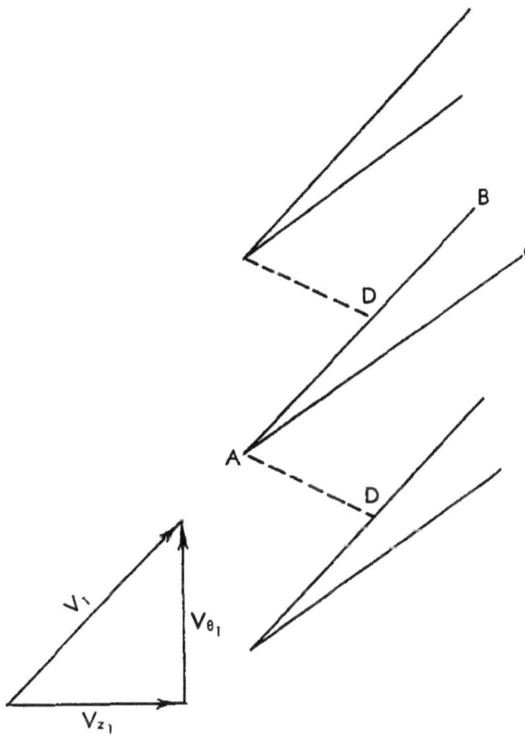

Fig. G,6c. Supersonic cascade having no curvature at entrance.

flow in front of the blade b, so that the flow in front of the blade b is expanded and turned from the free stream direction OA to a direction OB parallel to the tangent at B to the blade a. The blade b is geometrically identical to blade a and the tangent to A' is parallel to the tangent at A; therefore a shock wave is produced at A' followed by expansion waves. If the deviation from A to B is small, the flow field can be analyzed by neglecting waves reflected from the shock, because the intensity of the shock is small and entropy gradients are not important.

The shock at A' produces a deviation equal to the angle between the tangent at B and to the tangent at A, and interferes with some of the waves produced along AB and with some of the waves produced along

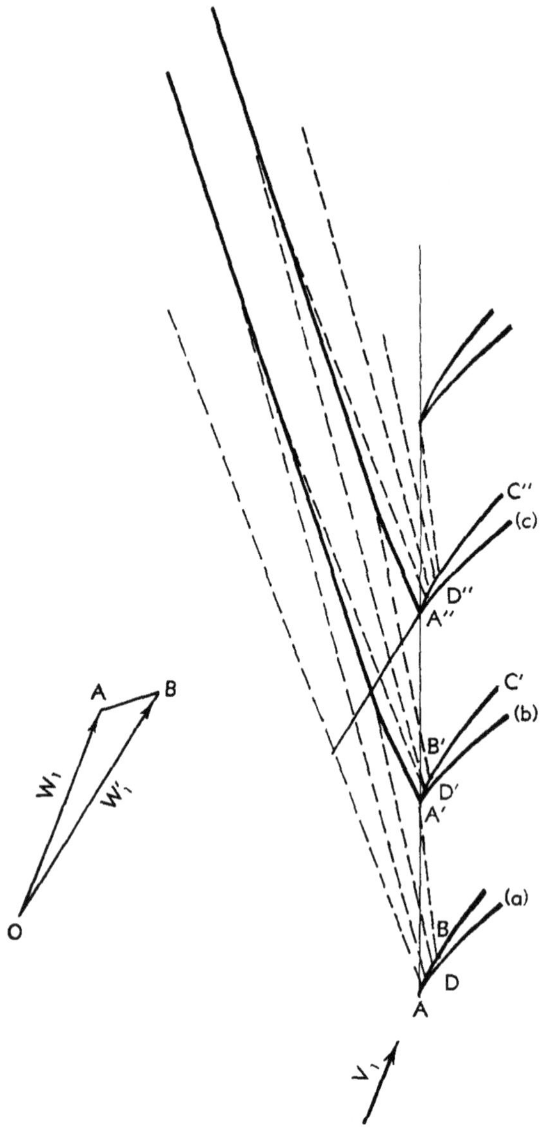

Fig. G,6d. Supersonic cascade having curved entrance.

A'B'. The strength of the interacting waves in front of the shock plus the strength of the interacting waves behind the shock must be such that the shock wave disappears at infinity; therefore, the deviation of expansion across the two groups of waves must be equal and opposite in sign to the deviation across the shock at *A'* and hence equal to the deviation from *A* to *B*. Call *D* the point on blade *a* where the first wave that interacts with the shock produced at *A'* is generated and *D'* the point on blade *b* where the last expansion wave behind the shock that interacts with the shock from A'. Point D' is at the same position on blade b that point D is on blade a, because the expansion from A to D must be equal to the expansion from A' to D'. If the blade c is considered, all the waves produced between A' and D' on blade b and A'' and D'' on blade c interfere with the shock produced at A'', and hence the flow on all the blades of the cascade is similar to the flow on b.

The waves from A to D move in front of the cascade and change the free stream velocity from a direction parallel to A to the direction parallel to D. These waves are the only waves that reach infinity and that are not canceled out by the following shocks. Therefore the flow direction and intensity behind these waves corresponds to the undisturbed flow for an infinite cascade having the same geometry.

The position of the point D which determines the direction of the undisturbed flow depends on the free stream Mach number and on the geometry of the entrance region. These in turn determine the mechanism of interaction between expansion and shock waves (see [*13*]).

For any entrance blade design to a given position of the point B along the trailing surface of the blade, there corresponds a given value of the Mach number at B defined by the condition that the Mach wave from B must pass through the leading edge A'. From the position of B and the Mach number at B the wave pattern along AB can be designed; similarly for the wave pattern along $A'B'$. The shock wave shape from A' and the points D and D' which produce the waves corresponding to free stream direction can be obtained. Then the velocity diagram for the cascade corresponding to steady conditions or to no waves moving to infinity can be determined. This condition is obtained by taking the direction of the relative velocity W_1' tangent to the blade at D. The direction and intensity of the free stream velocity vector W_1' for the case of an infinite cascade corresponding to any given point B which bounds the entrance region can also be determined from the pressure distribution at the entrance region of the blade.

Consider the cascade of Fig. G,6e. The entrance region AB corresponding to the chosen point B on the blades produces waves that move in front of the cascade, while the wave from B by definition meets the leading edge of the following blade. Construct now a new cascade having the entrance region of the blades $EACBD$ where the line BD is a straight line tangent to the original blade at B, the line EA is parallel to BD, $E'D$ is parallel and equal in length to the Mach line $A'B$, and ED is identical to AB. The entrance region ABC is equivalent to the entrance region $EACBD$, because the pressure distribution in EA or $E'A'$ is equal to the pressure distribution on DB or $D'B'$; therefore the flow properties between $E'D$ and $A'B$ are constant and the pressure along ED is equal to the pressure distribution along AB, while the pressure along AC is unaffected. Then the flow in the cascade can be considered in two parts, one corresponding to the region of the blade EAD in front of the line AA', and the second corresponding to the region behind the line AA', which does not affect the entering conditions because no waves are produced along DB.

Assume that the flow at infinity of the cascade is uniform and neglect waves reflected at the shock or by the entropy gradients; these waves

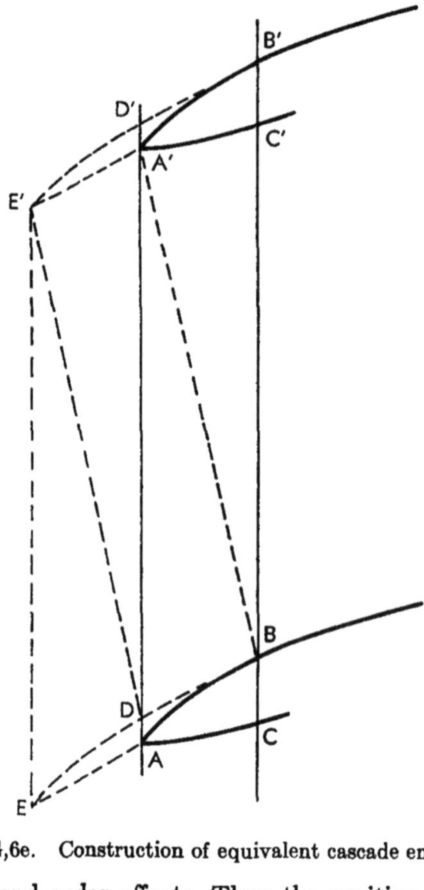

Fig. G,6e. Construction of equivalent cascade entrance.

produce only second order effects. Then the position of point B determines the inclination of the Mach wave AB and therefore the Mach number at B. The pressure distribution along the surface ED can be obtained from the shape of the blade and the pressure on EA is the same as the pressure at B; therefore the force acting on the blade EAD can be determined. In an infinite cascade the number of the passages is equal to the number of the blades, so that the force produced by each blade EAD acts on the mass flow of one passage, which is known because the flow along $A'B$ is known. This force changes the direction of the flow from the free stream direction at infinity to the direction $A'B$, and therefore the direction and intensity of the velocity of the flow at infinity can be obtained from continuity and conservation of impulse equations.

In all these considerations, both compression and expansion waves have been considered isentropic. However, extended in the flow in front of the cascade the shock waves produce an increase of entropy and vorticity in the flow, and therefore produce reflected waves. This variation affects the position of point B along the blade surface; in the usual case, however, the difference is small. The increase in entropy is produced by the compression waves at the expense of the power transmitted by the compressor to the flow. It affects the efficiency of the compressor, and at the same time decreases the entering volume flow by decreasing the value of the axial velocity in front of the compressor.

When a rotor is considered, the shock waves extend in front of the entrance of the rotor and the vortices produced by the intersection between expansion and shock waves are of helicoidal type.

When the tangential velocity of the rotor U increases (from RP to $R''P''$ of Fig. G,6f), the Mach number at the entrance of the rotor increases, and for the type of curved entrance region shown in Fig. G,6f the point B which defines the entrance region moves from B to a position B' closer to the leading edge of the blade. Also the point D dividing the region of the blade, which affects the shock from the leading edge A, from the region of the blade which affects the shock from the leading edge E, moves ahead to a position D'. This occurs because the expansion along AB' is smaller than the expansion along AB and therefore the direction of the free stream velocity vector also changes from OP tangent to the blade at D to OP' tangent to the blade at D'. Because of the sign of the direction change obtained for the type of blade shown in Fig. G,6f for a given increase of tangential velocity, the axial velocity changes less than for the blade with straight entrance tangent at D (OR' in place of OR''). The variation of the entering volume flow due to the variation of rotational velocity is smaller for the blade considered than for the blade having a straight entrance.

Consider now a cascade of blades having a blade shape as shown in Fig. G,6g. Compression waves are produced at the upper surface of the blade a along the entrance region AB; they move outside of the passage in front of the next blade b. At the leading edge of blade b expansion waves are produced, followed by compression waves of the same strength produced along the upper surface EF. The wave pattern is similar to the wave pattern considered in Fig. G,6d, but inverted; expansion waves are produced at the leading edge followed by compression waves produced along the entrance region. The compression waves produced along AB between A and D interact and cancel the expansion waves produced at the leading edge A, while the compression waves produced along DB

interact and cancel the expansion waves produced at the leading edge E of the following blade. In order to represent the flow in an infinite cascade with a semi-infinite cascade test, the free stream velocity in front of the first blade a must have a direction and Mach number such that, at the leading edge A, the correct amount of expansion waves is produced to interact and cancel the part of the compression waves produced

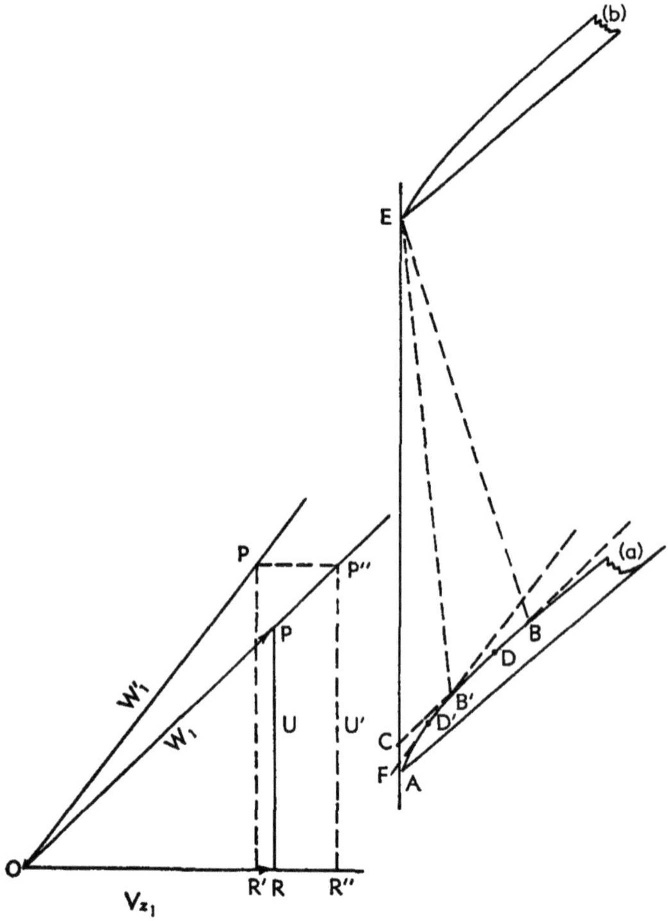

Fig. G,6f. Effect of variation of rotor speed for convex entrance curvature.

between A and D in every blade. The free stream direction must be parallel to the tangent at D, defined as the point that divides the part of the entrance region affecting the expansion waves from A, from the part of the entrance region affecting the expansion waves from E. Also in this case when the rotational velocity of the rotor (tangential velocity of the cascade) changes, the direction of the flow at infinity in relative coordinates changes because point D moves along the surface of the blade. However, in this case when the rotational velocity increases the axial velocity increases in larger proportion than for straight blades. If the tangential velocity component U changes from RP to $R'P'$, the Mach number in relative coordinates increases and point D moves to D'. The

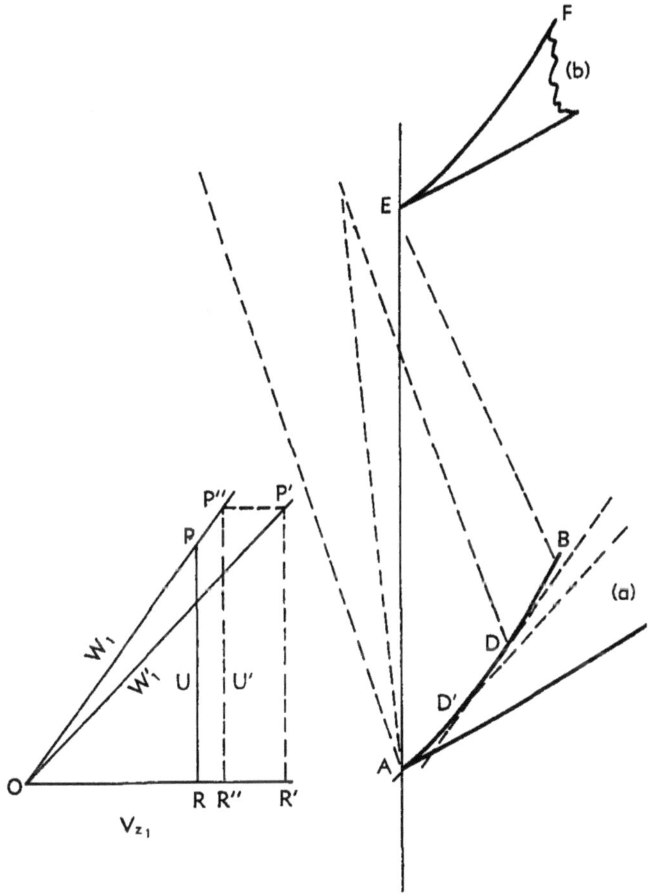

Fig. G,6g. Effect of variation of rotor speed for concave entrance curvature.

relative velocity changes direction from OP to OP' and the new value of the axial velocity component corresponds to OR', which is larger than OR'' corresponding to the same variation of rotational velocity for a blade having a straight entrance region tangent at D.

Blades producing detached shock waves. In a supersonic rotor cascade, conditions can be found where supersonic flow cannot exist at the leading edge of the blades. This can occur with blades having round leading edges or with leading edge wedge angles greater than the maximum deviation possible for the incoming stream Mach number. Both possibilities have practical importance, especially for low rotational velocities when the entering relative Mach number is close to unity.

Consider at first a cascade having blades with sharp leading edges and a straight entrance region (Fig. G,6h). Assume that the Mach number of

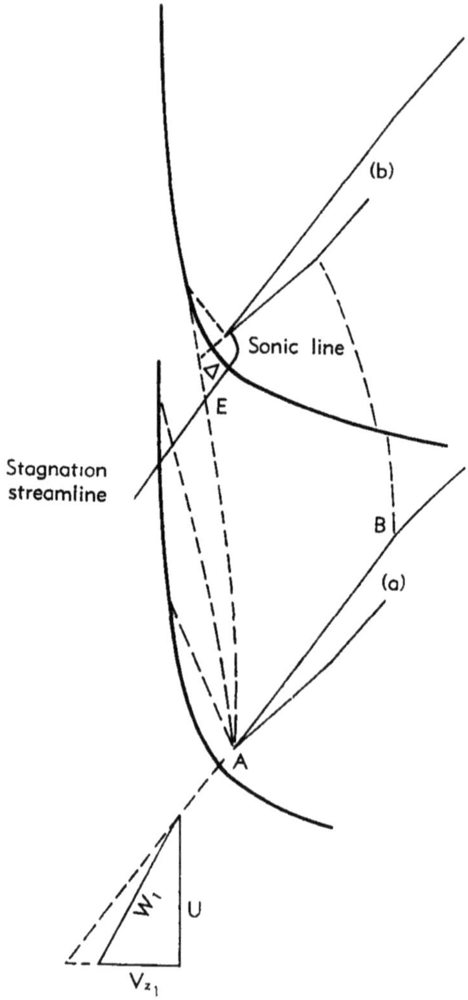

Fig. G,6h. Compressor blades having detached shock pattern.

the flow relative to the blade is small and the blade has a wedge angle that produces a detached shock at the leading edge of the blade. The detached shock moves in front of the passage and a local subsonic region exists behind the shock near the leading edge of the blade.

In order to obtain steady conditions in the flow in front of the cascade, the shock wave at each leading edge must be neutralized by expansion waves which must also move in front of the passage; therefore when a detached shock exists in front of each blade, the incoming flow for steady conditions cannot be parallel to the trailing surface of the blade AB, because in this case a stagnation point would exist at the leading edge of each of the blades followed by a subsonic stream. The velocity would become sonic only at a point B inside the passage where the blade starts to curve (see VI,H), and therefore no expansion waves could be transmitted in front of the passage to neutralize the shock, as is required for steady coditions.

If the incoming flow is inclined with respect to AB, then expansion waves are produced at A downstream of the shock, because the stagnation point occurs at the leading surface of the blade and the Mach number along AB becomes larger than 1, while the expansion waves produced at A move in front of the cascade. Therefore, in the cascade considered, the direction of the entering stream for steady condition must be such that the blade has an angle of attack.

The analysis of the flow field at the entrance region is difficult and can be obtained only in some cases with the help of experimental investigations. Also in the case of detached shocks the expansion waves produced at the leading edge A can be divided in two parts: the first part interferes with the detached shock produced by the blade a while the other part interferes with the shock produced by the trailing blade b; the wave that divides the two parts corresponds to the free stream conditions in front of the infinite cascade. If the expansion waves interact with the following shock in a region which corresponds to supersonic flow behind the shock, then the localized subsonic region at the leading edge of each blade is not affected by the presence of the other blades and the same considerations presented before can be applied. The flow field at the leading edge of the blades can be obtained from the theoretical work on isolated wings having a sharp leading edge and a detached shock (see VI,H) or from experimental data on an isolated wing under the same conditions. If the expansion waves produced by blade a at A interfere with the subsonic region of the detached shock produced by blade b, then a test or analysis of a complete cascade configuration is required. The same is also required when the intensity of the shock is such that reflected waves produced at the interaction of expansion waves and shock are important and affect the subsonic region of the trailing blade.

G,7. Mechanism of Steady Flow at the Entrance of a Cascade for Supersonic Axial Velocity.

Blades of zero thickness and zero camber. Consider a supersonic infinite cascade having blades of zero thickness and zero camber as shown in Fig. G,7a. Assume that the flow in front of the cascade is steady and that the axial component of the velocity OB is subsonic. For steady conditions no waves can travel in front of the infinite cascade; therefore the velocity OA relative to the blades in front of the passages must be parallel to the blades. If the tangential component of the velocity AB increases, then for steady conditions the axial component of the velocity OB must also increase because the direction of the entering velocity for steady conditions is determined by the direction of the entrance of the blades and cannot change. This situation prevails until the axial component becomes sonic. For the condition of sonic axial velocity OB, if infinitesimal waves are produced at the leading edges of the blades they are contained inside the cascade and disturbances can be transmitted upstream of the cascade

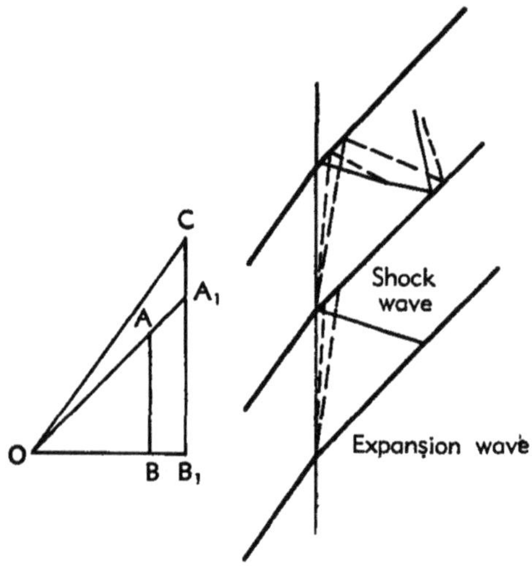

Fig. G,7a. Cascade having zero thickness and camber with entering axial velocity supersonic.

only by means of shock waves of finite strength. Assume now that the tangential component to the velocity increases from B_1A_1 corresponding to a relative velocity OA_1 parallel to the blades to a value B_1C. Expansion waves and a shock wave are produced on opposite sides of the leading edge of each blade, but are contained inside the passage (Fig. G,7a); the entering velocity OC is inclined with respect to the blade direction, but the axial component does not change. No waves are transmitted upstream of the cascade, and therefore the conditions considered represent a steady state. The maximum value of the axial velocity that can be obtained in front of a cascade having such a blade entrance by increasing the tangential component of the velocity of the compressor is equal to sonic. When sonic velocity is reached, small disturbances produced downstream cannot increase the velocity upstream.

Consider now a change of the axial component of the velocity for a given value of the tangential component of the velocity relative to the blades. When the axial velocity OB (Fig. G,7a) is subsonic, a change of the axial velocity without a change of tangential velocity produces a change of direction of the velocity OA relative to the blades. Then nonstationary waves are produced at the leading edge of the blades. These move upstream and change the axial velocity at the entrance of the cascade. Therefore, when steady flow is again established in front of the cascade, the axial component is again equal to OB. However, the situation is different if the axial velocity in front of the compression is sonic; a wave pattern as shown in Fig. G,7b for the velocity diagram OB_1C then exists.

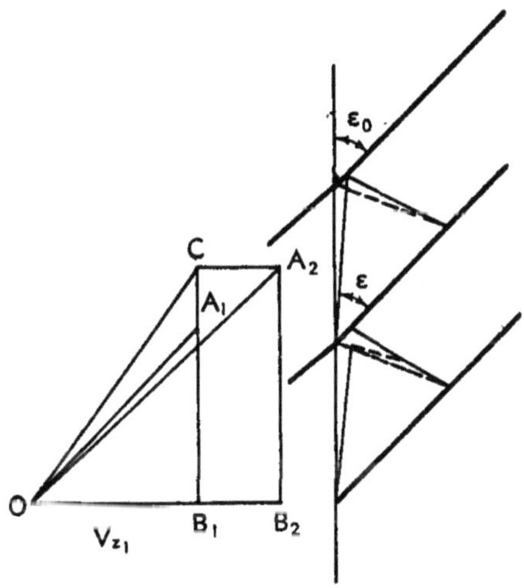

Fig. G,7b. Cascade having zero thickness and camber with entering axial velocity sonic.

An increase of the axial component from OB_1 to OB_2 produces a variation of relative velocity from OC to OA_2 (Fig. G,7b) and thus a small deviation of compression at the leading edge of the blades which are parallel to OA_1. This increase of axial velocity is possible without producing nonstationary waves in front of the cascade, because the Mach number corresponding to OA_2 is larger than the Mach number corresponding to OA_1 and therefore a small shock wave is still contained inside the passage. In this case the cascade can have entering velocity with an axial supersonic component.

The same considerations apply to a supersonic compressor. By increasing the rotational speed of the rotor, the axial velocity is increased. If the blades have a straight entrance region, the velocity relative to the blades does not change direction; therefore the value of the axial velocity is determined, cannot be changed independently of the rotational velocity, and is independent of the flight conditions. This is true until the rotational speed is such that the axial velocity becomes sonic. Then an increase of rotational speed cannot change the axial velocity but changes the direction of the relative velocity. For this condition it is possible to change the value of the axial velocity component by changing the axial velocity, for example, by changing the flight speed. The axial velocity can change from the sonic value to a supersonic value which produces a shock at the leading edge of the blades tangent to the entrance of the cascade (angle ϵ of the shock equal to ϵ_0 in Fig. G,7b). Therefore, for a given rotational speed of the rotor higher than B_1A_1, corresponding to OB_1 and equal to sonic, it is possible to have different values of the mass flow entering the compressor for given rotational speed and given values of the static pressure and temperature of the flow.

Blades having curved entrance region. When the entrance region of the blade is curved and the axial velocity component is subsonic, expansion and compression waves exist in front of the cascade, which for steady conditions, cancel each other in front of the cascade. Consider the blades as shown in Fig. G,6f and G,6g. When the tangential velocity increases from OR to OR', the relative velocity changes from OP to OP' and becomes parallel to the tangent to the blade at a new point B' which is closer to the leading edge A than point B. For the condition of W_1 parallel to the tangent to the leading edge of the blade, the axial component V_{z_1} is sonic and point B' is at A. For this condition the curvature of the upper surface of the blade does not affect the flow in front of the cascade, and the entrance region AB disappears. Therefore, the same considerations previously applied to straight blades are valid when the tangential velocity is larger than the velocity corresponding to sonic axial velocity and when the axial velocity is supersonic.

G,8. Mechanism of Starting.

Single passage. In the preceding sections the flow conditions at the entrance of the cascade have been discussed as functions only of the design of the entrance region of the blades. It has been assumed that the downstream conditions do not interfere with the upstream flow. Here the effect of the geometry of the blade downstream of the entrance region and of the exit conditions on the entering flow properties are discussed.

Consider first a single passage as shown in Fig. G,8a. Assume that supersonic flow exists in the converging part of the diffuser. Then the flow inside the passage behaves as the flow inside a supersonic convergent-divergent diffuser discussed in detail in VII,E. The transition from supersonic to subsonic velocity can occur, for steady conditions, only in the divergent part of the passage; for example, at a section 3 between the throat 2 and the exit 4 of the passage or downstream outside the passage. The position of the transition region 3 depends on the value of the static pressure at the discharge section 4 and can be changed by changing flow properties at the end of the passage; for example, by increasing the static pressure at the discharge as in a throttling process.

By changing the position of the shock 3, the static pressure, the stagnation pressure, and the volume flow passing at 4 change. However, the mass flow and the volume flow of the stream tube at free stream conditions are not changed; therefore, so long as the throttling is such that region 3 is downstream of the throat, the throttling process does not influence the conditions at the entrance and the entering mass flow.

If the back pressure is increased, the shock 3 moves toward section 2, becomes weaker, and the pressure recovery of the diffuser increases. If

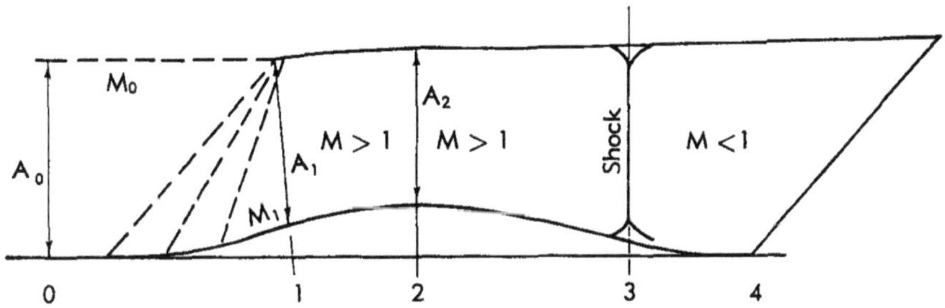

Fig. G,8a. Single blade passage with shock swallowed.

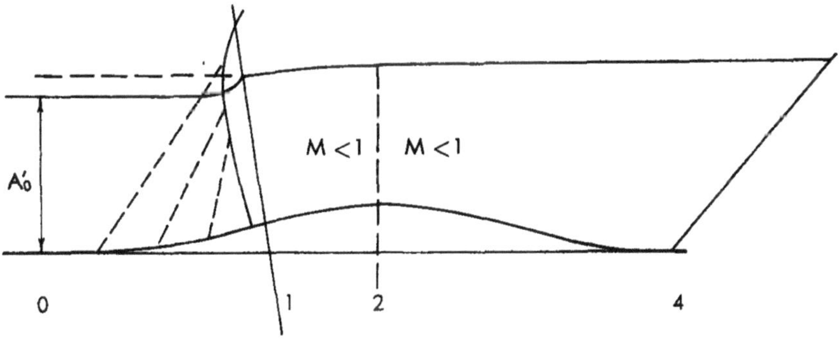

Fig. G,8b. Single blade passage with shock not swallowed.

the throttling downstream is still increased, the shock moves from the region of the throat to the front of the passage (Fig. G,8b) because no stable configuration can exist with the shock in the convergent region. With the shock in front of the passage the flow between sections 1 and 4 is all subsonic, and the mass flow entering the passage is affected by the throttling process; therefore, it can be changed by changing the conditions downstream of the passage.

If the back pressure at 4 is again decreased, the strong shock wave, which establishes the transition from supersonic to subsonic flow, moves again inside the passage provided the area contraction ratio A_1/A_2 is such that the diffuser can start. If the contraction ratio A_1/A_2 is too large the strong shock remains in front of the diffuser. However, when the back pressure decreases, sonic velocity is reached at section 2 and the flow downstream of section 2 becomes supersonic; a transition region 3 again exists in the diverging part. In this case, the passage is "choked" and downstream conditions again do not affect the flow field at the entrance; the entering mass flow is again independent of the downstream pressure conditions.

The possibility of the passage starting or not starting at the incoming stream Mach number M_0 is discussed in detail in VII,E and can be briefly presented as follows: The minimum value of A_2 for given values of A_0 and M_0 corresponds to $M_2 = 1$. If the transformation is not isentropic, the relation between A_0 and A_2 for given values of M_0 and M_2 is given for one-dimensional flow by

$$\left(\frac{A_2}{A_0}\right)_{is} \frac{p_0^0}{p_2^0} = \frac{A_2}{A_0} \tag{8-1}$$

where $(A_2/A_0)_{is}$ represents the ratio for isentropic transformation from M_0 to $M_2 = 1$, and p_0^0/p_2^0 is the ratio between the stagnation pressures at 0 and 2; this ratio is larger than 1 because of the increase of entropy in the flow between sections 0 and 2.

The transformations between stations 0 and 1 in front of the shock and inside the diffuser between 1 and 2 occur with small increases in entropy and are, therefore, nearly isentropic. All the losses can be considered in the first approximation to occur across the strong shock at station 3 (Fig. G,8a) or across the shock in front of station 1 (Fig. G,8b). Then when the flow is supersonic in the diffuser as in Fig. G,8a, the ratio A_2/A_0 can be close to the isentropic value, while in order to satisfy the continuity law for the condition of Fig. G,8b the ratio A_2/A_0' at the same Mach number M_0 must be smaller than A_0 of Fig. G,8a.

In order to have the possibility of pushing the strong shock inside the passage, it is required that when the shock is at station 1, $A_0 = A_0'$; therefore, if A_2 corresponds to the value determined from isentropic transformation for M_2 close to 1, the diffuser cannot start unless the value of free stream Mach number M_0 is increased.

The starting Mach number is a function of the contraction ratio A_2/A_0 and of the Mach number M_1 at the entrance of the channel (and not of M_0). The difference between the isentropic contraction ratio and the starting contraction ratio increases when M_1 increases; therefore, the Mach number M_2 for supersonic flow between stations 1 and 2 increases when, for a given M_0, M_1 increases, while the losses due to the increase of entropy in the passage also increase.

Infinite cascade. Similar considerations can be made when in the place of a single passage a cascade is considered.

Consider first a cascade having diverging passages (without internal contraction). In this case, supersonic flow can be established inside each passage for any supersonic Mach number in front of the cascade, for a condition downstream of the cascade corresponding to no-throttling.

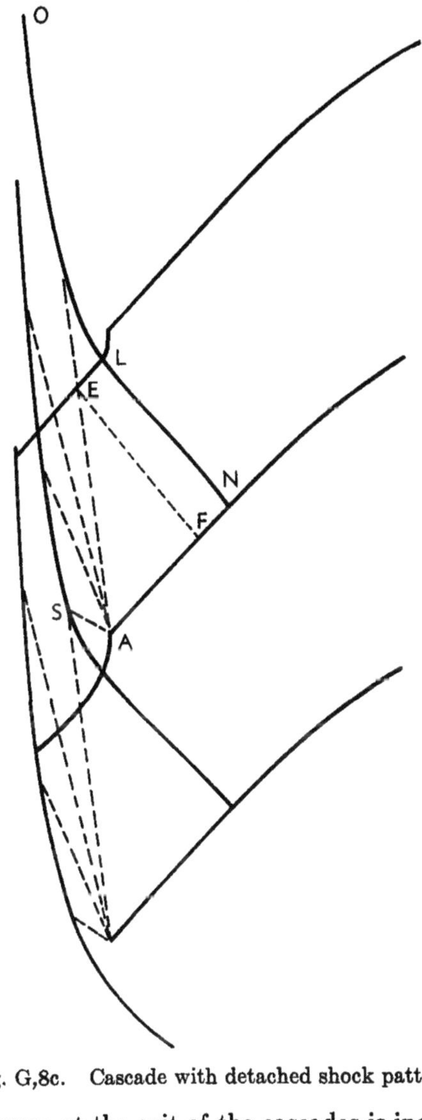

Fig. G,8c. Cascade with detached shock pattern.

When the pressure at the exit of the cascades is increased, the region of transition from supersonic to subsonic velocity moves upstream until it reaches the entrance of the cascade (Fig. G,8c). For this condition, detached shocks exist in front of each blade and compression waves move upstream. These waves decrease the value of the axial entrance velocity component corresponding to the free stream condition, so that the value of the mass flow entering each passage decreases. The compression waves which move upstream change the direction of the entering velocity; therefore the blades are at an angle of attack, and the equilibrium conditions for each value of the throttling are similar to those described in connection with Fig. G,6h. In this case, however, the flow inside the passage is all subsonic. The detached shock existing in front of each blade is canceled by expansion waves produced at the leading edge of the preceding blade; no waves extend to infinity. Under some conditions local separation can occur on the trailing surface of the strong shock and interferes with the boundary layer on that surface of the preceding blade; then instability of the flow field can occur.

Each blade produces first a shock wave and then a fan of expansion waves. Therefore the free stream tube entering the passage for choked conditions is less than the free stream tube which would enter the passage in the absence of detached shocks. The continuity law can thus be satisfied. The expansion waves produced at the leading edge of the blades and in front of the passage must cancel the detached shock for steady conditions.

Consider now the case of a cascade having some internal contraction. In this case, when the tangential velocity is increased from low values to larger values, the Mach number in front of the cascade increases. For some tangential velocity, the velocity at the throat of the passage becomes sonic, while the velocity in the converging region is still subsonic. The free stream velocity at infinity in front of the cascade is supersonic; therefore, the transition from supersonic to subsonic flow must occur in front of the passage (Fig. G,8d). The strong shock in front of each passage produces a deviation of the flow in front of the next blade; therefore, each blade has an angle of attack. Expansion waves are produced at the leading edge of each blade which interfere with the preceding and with the following shocks, and for equilibrium conditions expansion waves and shocks cancel completely at infinity. These waves persist as the tangential velocity increases. Until the contraction ratio A_1/A_2 corresponds to the value given by the criteria outlined for single passage, however, the intensity of the waves decreases when the free stream Mach number increases. A decrease of the static pressure at the exit of the cascade increases the supersonic expansion in the divergent part of the passage, but the flow in the converging part remains subsonic as for the case of the single passage.

When the flow is not started, less mass flow enters the cascade than for started conditions and the pressure recovery across the cascade decreases; therefore, the associated compressor under choked conditions has a lower efficiency than under started conditions.

Starting Mach number depends on the contraction ratio of the passages and on the blade shape at the entrance. In a supersonic rotor the passages can be started by increasing the rotational speed of the rotor.

When the passage has started, it is possible to decrease the rotational speed of the rotor below the value corresponding to starting conditions, still maintaining supersonic flow inside the passage as for the case of a convergent-divergent diffuser (see VII,E). For these conditions the pressure recovery of each passage is higher than for the case of an unstarted compressor and the mass flow entering the compressor is larger.

The influence of the shape of the blade entrance region on the starting conditions can be illustrated from the following considerations:

Consider the blade design shown in Fig. G,8d. The entrance region is straight and the entrance section A_1 is larger than the minimum section

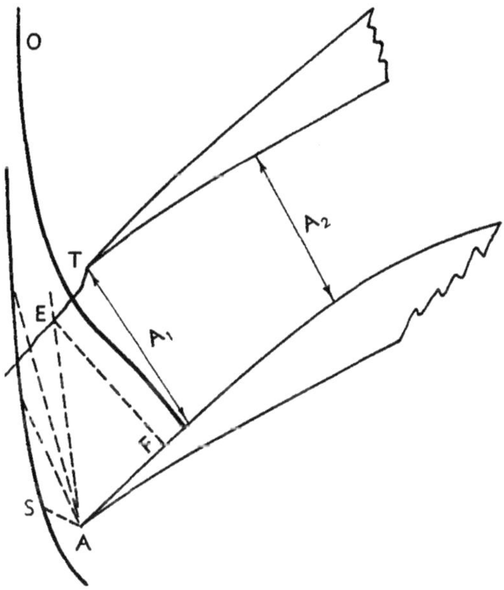

Fig. G,8d. Cascade with detached shock pattern having no entrance compressions.

A_2. Assume that the internal contraction is such that for the stream velocity considered the passage is choked. Then the wave pattern in front of the cascade will be as shown in Fig. G,8d. If separation is avoided along *AF* the relative velocity in the region *AEF* is in a rough approximation parallel to the surface *AF*.

Consider now the blade of Fig. G,8e, obtained from the blade of Fig. G,8d, by curving the surface from *AF* to *AF'*, in such a manner that compression waves are produced along *AF'* and that these compression waves are contained inside the passage in the absence of normal shock in front of the passage. The compressions reduce the Mach number in front of the strong shock; thus they increase the pressure recovery p_2^0/p_0^0 which enters in Eq. 8-1, decrease the starting Mach number, and permit a larger area contraction which corresponds to a larger pressure recovery for a given starting Mach number. The opposite result is obtained if expansion waves are introduced along *AF*.

These considerations are of particular importance when the starting Mach number is large and the axial component is close to sonic or is supersonic. In this case, the amount of compression which can be produced in front of the strong shock along *AF′* is large. For started conditions, the compression waves can be reflected at the upper surface of the passage and hence the advantages of passages having external compression can be effectively utilized.

The flow field in front of the unstarted cascade is similar to the flow field in front of a single diffuser and differs only in the mechanism of

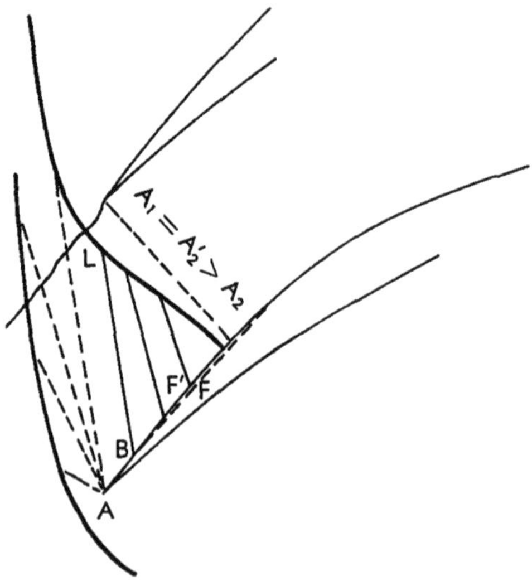

Fig. G,8e. Cascade with detached shock pattern having entrance compressions.

cancellation by means of expansion waves of the strong shock produced at the entrance. It is known that convergent-divergent diffusers can encounter phenomena of instability called "buzz" when the flow at the minimum section is subsonic (see VII,E). These phenomena of instability are due to differences of pressure recovery in different regions of the stream entering the passage and to interference between shocks or strong pressure gradients and the boundary layer. These phenomena can be expected to occur also in supersonic cascades or in supersonic compressors and can lead to stalling in the rotor.

G,9. Mechanism of Steady Flow at the Exit of a Cascade for Supersonic Discharge Velocity.

Equilibrium conditions determined by static pressure. In this article the equilibrium conditions at the exit of a cascade are discussed for the condition of supersonic exit velocity such as exists in shock in the stator compressors. It is assumed at first that the equilibrium conditions can be obtained by fixing only the value of the static pressure behind the compressor and not both static pressure and direction of the stream in the downstream flow. This assumption corresponds to requiring the conditions downstream of the cascade to be effected by only the axial component of the exit velocity while the tangential component of the exit velocity is determined only by the cascade design. However, when the axial component of the discharge axial velocity is subsonic, an effect of the downstream conditions on the tangential velocity is possible. This case is discussed later.

Consider a cascade having converging and diverging, or all diverging, passage. The velocity diagram at the exit of the cascade depends upon

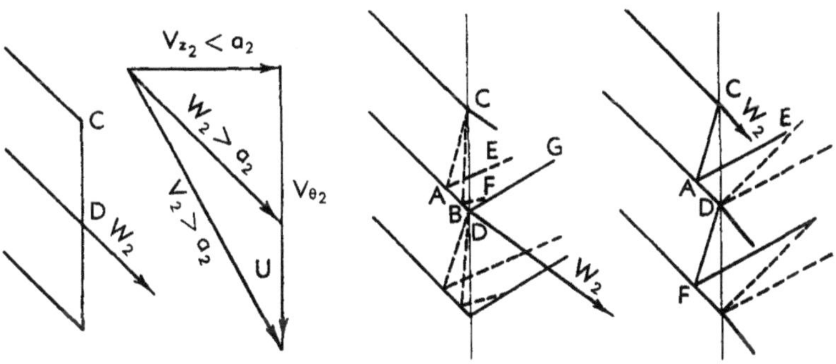

Fig. G,9a. Velocity diagrams at rotor exit.

the conditions at the exit. If the static pressure at the exit of the cascade is sufficiently low, the stream leaving the cascade has supersonic velocity, while for larger values of the back pressure, the stream leaving the cascade is subsonic and transition from supersonic to subsonic velocity occurs inside the passage. By changing the back pressure, the value of the velocity at the exit and its axial component can be gradually changed from a maximum value which corresponds to sonic or supersonic axial velocity, depending upon the cascade design to zero. During the throttling process a wave pattern is produced in the flow at the exit of the rotor which can be described by means of two-dimensional considerations [*14*].

Assume that the blades near the exit region are straight and have zero thickness and that the flow at the exit of the passage is uniform (Fig. G,9a). Assume that the exit relative velocity W_2 is supersonic and parallel to blades at the exit while the axial velocity component for the blade design considered is subsonic $V_{z_2} < a_2$ (a_2 speed of sound at the exit). The absolute exit velocity V_2 is also supersonic.

If the static pressure at the exit of the cascade is lowered with respect to the value considered in Fig. G,9a, there exists at the exit of each passage an expansion which tends to increase the velocity inside the passage. Because the velocity inside the passage is supersonic but the axial component is subsonic, the expansion is propagated inside the passage from the trailing edge of each blade by means of expansion waves (Fig. G,9a, left and center). The region AD of the lower blade of each passage is affected by the expansion induced at the trailing edge of the upper blade of the passage, and along AD the expansion waves are reflected.

The amount of expansion occurring between wave AC and the exit CD depends upon the difference of pressure required at C. However, when the axial component of the velocity at the exit CD becomes sonic, the expansion wave from C is along CD; therefore, a further decrease of the downstream pressure does not affect the velocity at the exit of the rotor but changes only the conditions downstream of the rotor. Therefore, for the design considered, the maximum value of the axial component of the flow at the exit CD is equal to sonic velocity. At the trailing edge D of each blade, the direction of the stream and the pressure cannot have discontinuous values; therefore, in order to obtain equilibrium conditions at D, a wave DG exists at the trailing edge of the upper surface of the blade. This wave has opposite sign and the same intensity as the expansion waves CAD produced at the lower surface of each blade.

The shock wave moves downstream in the flow in nearly the same direction of the waves AE and BF produced by the reflection at the lower blade of the passage of the expansion waves from C and tends to neutralize these waves. The compression wave DG produces an increase of entropy in the flow downstream of the cascade.

If the pressure at the exit of the cascade is increased with respect to the condition considered in Fig. G,9a, left, a shock wave CA is produced at the trailing edge C (Fig. G,9a, right), which is reflected at the lower surface as another shock wave AE. At the trailing edge D of each blade, on the opposite surface from the shock wave DF, an expansion fan is produced in order to obtain equilibrium of pressure and direction of flow at the trailing edge. The expansion fan tends to neutralize the reflected shock wave EA.

The intensity of the shock wave AC of Fig. G,9a, right, increases with increasing of the downstream pressure until the reflection at A of the shock wave AC is not possible. Then if separation is avoided at A, the reflection of the wave AC becomes a Mach reflection. The shocks CA and AE form a lambda shock with a leg normal to the surface AD. The flow behind the shock is subsonic and becomes again sonic at D by expansion.

When the reflection of the shock wave at A produces separation, the separation travels upstream of the reflection point producing compression waves in front of A. The shock CA is then reflected at the separated region and the reflected waves at some distance from the surface AD have the same intensity as in the absence of separation.

For larger values of the back pressure, the flow behind the shock CA becomes all subsonic, the strong shock from C becomes normal to the surface AD and then moves gradually upstream in the passage as the back pressure increases. For this condition the discharge velocity W_2 is subsonic and the considerations discussed in Art. 8 apply.

In practical cases the flow at the exit of the passages is not uniform because the blades have finite thickness and the passages are short; then the condition of no waves downstream of the passage as shown in Fig. G,9a, left, is not possible. However, in this case the equilibrium conditions at the exit of each passage can still be determined from considerations similar to those presented previously. For each value of the static pressure at the trailing edge of each blade D of the cascade, the intensity of the waves CAB of Fig. G,9a, center, or CA of Fig. G,9a, right, can be determined. From the velocity distribution in front of wave CA the flow field between CA and CD can be determined. Then the intensity of the wave DG in Fig. G,9a, center, or of the waves DE in Fig. G,9a, right, can be obtained from the consideration that at D the pressure and flow direction are known.

Consider now a cascade design for which the condition presented in Fig. G,9a, left, corresponds to supersonic axial velocity ($V_{z_2} > a_2$). Then the downstream conditions can change the flow at the exit of the passage only by means of shock waves. The conditions corresponding to Fig. G,9a, center, are not possible because expansion waves from C move downstream of the passage. However, a wave pattern as shown in Fig. G,9a, right, is still physically possible.

In this case, an increase of the back pressure with respect to the values corresponding to Fig. G,9a, left, tends to produce a shock that at first is normal to the direction of the axial flow component. This shock moves toward the exit section of the cascade and then becomes attached at the exit CD of the cascade (Fig. G,9a, right). For this condition, the flow at the exit of the cascade has a subsonic axial component. A further increase of the back pressure pushes the shock inside the passage as for the case discussed before.

Interference between waves produced by different stages. In the preceding discussion, the flow fields at the entrance or exit of a cascade have been analyzed under the assumption that the flow at infinity in front of the cascade was uniform or that the flow at the exit of the cascade also for the case of supersonic exit velocity was a function only of the design of the cascade considered and of the discharge static pressure. However, in practical applications, other rotor or stator stages, preceding or following the element of the compressor considered, can change the upstream or the downstream flow. Thus waves which move downstream or upstream in the flow and which enter the following or the preceding stage can be produced. These waves interfere with and change the wave pattern discussed previously.

Interference between different elements, such as rotor and stator, as guide vanes and rotor of a given stage or as stator and rotor of two different stages exist in any practical machine. The analysis of this interference is involved.

One aspect of such interference can be qualitatively illustrated by assuming in first approximation that the flow is two-dimensional in each cylinder parallel to the axis of the machine and by neglecting the boundary layer effects. Consider two elements of a machine, one of which is in movement with respect to the other and assume that the velocity relative or absolute leaving one of the elements and entering the other is supersonic.

When waves are produced at the exit region of the first element, the waves travel downstream and are partially reflected at the entrance region of the second element. If the axial component of the flow is subsonic, the reflected waves travel upstream. Therefore, in this case, a train of waves moves from one element to the other. The intensity of these waves depends on the distance in the axial direction between the two elements, because waves of opposite sign and of the same family as those produced by a cascade, tend to neutralize each other and to become zero at an infinite distance. However, because of the relative motion existing between the two elements, the intensity and sign of the waves depends also on the relative position of the two elements in the tangential direction. This relative position changes with time, and therefore the flow conditions at the exit of one element and at the entrance of the other, also change with time. Pulsating conditions in the flow between the two elements can be established for particular geometrical conditions and for particular values of the relative velocity.

Consider, for example, a rotor having supersonic exit relative velocity W_2 with subsonic axial component V_{z_2} (Fig. G,9b). Assume for simplicity that the blades of the rotor and of the stator have zero thickness. Call W_2' and V_2' the velocity vectors in rotating and absolute coordinates of the flow inside the rotor passage at the exit of the rotor and in front of any wave AB produced at the trailing edge of the rotor blades (Fig. G,9b); call W_2 and V_2 the corresponding velocity vectors of the stream at the exit of the rotor downstream of the trailing edge of the blades, and therefore downstream of any wave AB existing inside the passage and produced at the trailing edge. The angle between the vector W_2' and W_2 is due to the deviation produced by the wave BA. The waves AD and the expansion fan bounded by BE and BF produce equal and opposite deviation.

If the value of the tangential velocity U is such that the resultant vector V_2' is not parallel to the stator blades, then the waves AB, AD

and AE, AF are produced at the exit of the rotor such that the velocity vector V_2 is parallel to the stator blades. The reason for the existence of this wave pattern is that, as shown in Art. 5, the stator cannot accept for steady conditions a supersonic flow field with subsonic axial velocity

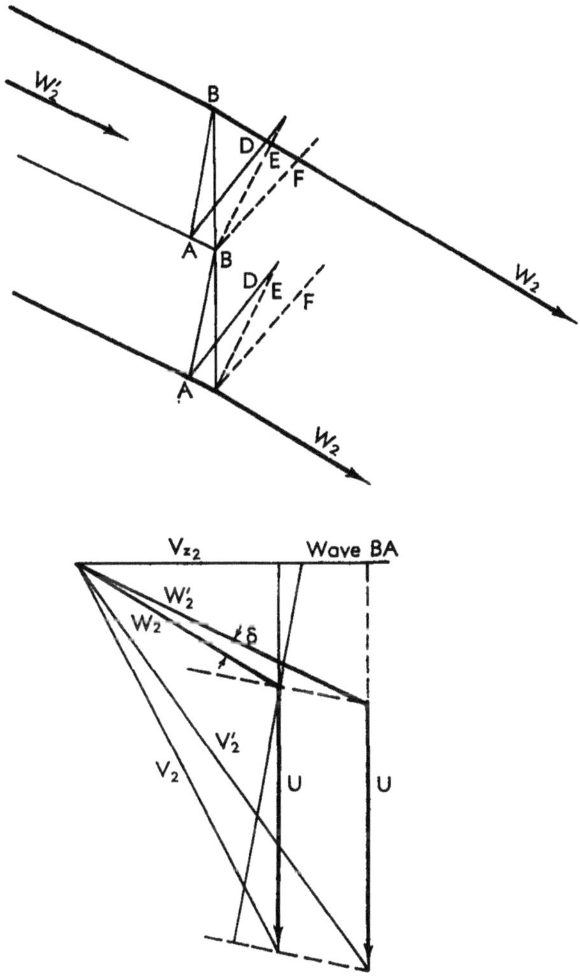

Fig. G,9b. Rotor exit with supersonic relative velocity and subsonic axial velocity.

at an angle of attack with the blade entrance region; when the incoming flow is at an angle of attack, the stator produces a nonsteady family of waves that moves upstream. These waves reach the rotor exit, become steady waves at the rotor exit and change the exit velocity from V_2' to V_2. If the rotor and stator were far apart from each other, the equilibrium conditions would correspond to a wave pattern at the exit of the rotor as shown in Fig. G,9b. The compression wave AD of Fig. G,9b would be

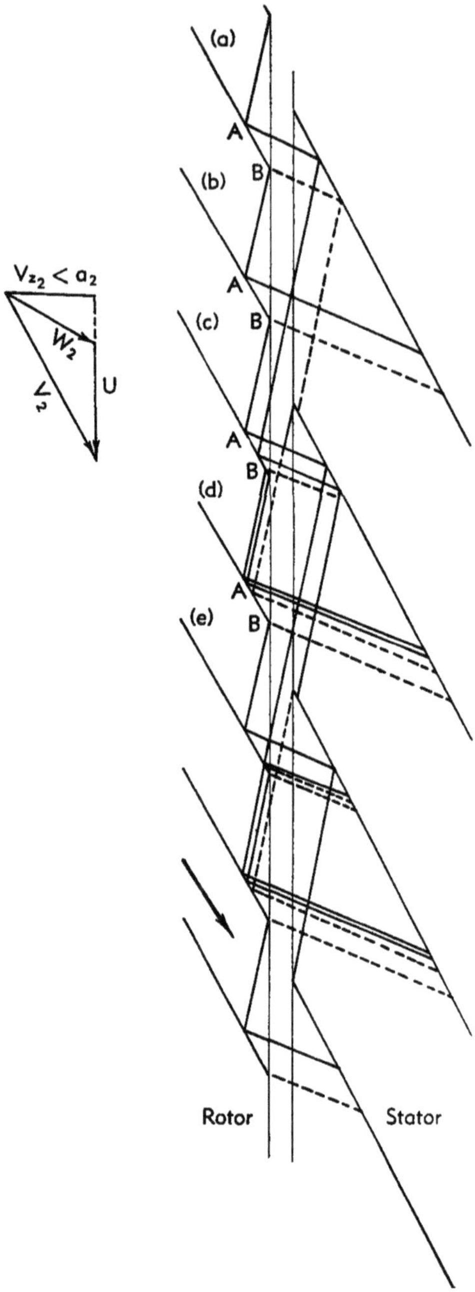

Fig. G,9c. Rotor wave pattern with respect to the stator at a fixed rotor position.

canceled by the following expansion waves *EB*, *FB* and no waves would exist in the flow in front of the stator. However, in the case of a small distance between rotor and stator, some of the downstream waves are reflected by the stator before being neutralized.

In Fig. G,9c, the wave pattern for a given position of the rotor, with respect to the stator, is presented in stator coordinates. In this representation the trajectory of the particles in absolute coordinates is considered and therefore the lines *a*, *b*, *c*, and *e* do not represent the shape of the rotor blades but are parallel to the absolute velocity vector V_2. The equilibrium conditions without interference would correspond to the conditions at blade *a*. A shock wave is produced inside the passage which is reflected at *A*. The region *AB* is the exit region and at the trailing edge *B* expansion waves are produced which tend to cancel the compression wave reflected at *A*. However, if stator blades are present, expansion waves and compression waves can be reflected before they are canceled and can interfere with the other passages. Because of the interference of the stator blades at some blade position (blades *c* and *d*), the compression waves reflected along *AB* have twice the intensity of the compression waves reflected along *AB* at the position *a*. Then the expansions and compressions produced at the trailing edges are also twice as strong. The increased pressure gradient along *AB* can produce large variations in the flow conditions at the exit of the passage with the possibility of pulsation in the flow.

If the axial component of the velocity is supersonic, the flow conditions at the exit of the rotor are steady because the waves produced by the stator cannot move upstream; variations occur only at the stator entrance. Therefore this type of interference is not possible.

If the velocity inside the rotor is subsonic in rotor coordinates, and supersonic in stator coordinates, the waves *BA* of Fig. G,9b are transient and travel upstream in the passage and are absorbed at the region of transition from supersonic to subsonic flow which in this case is pulsating.

G,10. Three-Dimensional Effects. Supersonic compressors can be designed to have an axial velocity component very close to sonic. With other conditions fixed, they can compress more mass flow per unit frontal area of the compressor annulus than subsonic or transonic compressors. However, this advantage is small when transonic or high velocity subsonic compressors are considered. Therefore, in order to be competitive from the point of view of mass flow per compressor frontal area, the supersonic compressor must be designed with a large annular area; that is, with long compressor blades, resulting in a root-to-tip diameter ratio of the same order as the subsonic or transonic compressors. For small values of the ratio of the root-to-tip diameters the change of velocity diagram along the blade is very large due to the large change in rotational velocity.

The two-dimensional considerations presented before must be modified in order to take into account large three-dimensional effects.

The three-dimensional effects are important both for the inviscid and for the viscous phenomena and have a primary influence on the performances of supersonic compressors. Their analysis is very complex and at present incomplete. A few of the known effects on the inviscid flow are briefly outlined here.

Three-dimensional effects at the entrance of the compressor. Consider a supersonic rotor having blades with straight leading edges and uniform axial velocity at the entrance of the compressor. In order to have constant entrance axial velocity along the blade height, the blade entrance sections must be rotated from root to tip of the blade according to the change of the rotational speed, Fig. G,10a (sections a_1 and b_1). The Mach number of the velocity relative to each blade section increases, moving from the hub to the tip of the blades, while the static pressure in front of the compressor is constant. This type of velocity diagram does not introduce any special difficulty in the design of a shock in the stator compressor. In this case the work is produced in the compressor by large turning in the passage, without a large reduction of relative speed in the rotor, and therefore without an appreciable static pressure rise. The same amount of work at the tip and hub of the compressor can be obtained by introducing a different amount of turning along the blade height. However, such a velocity diagram can introduce complications for the design of a shock in the rotor compressor where transition from supersonic to subsonic velocity occurs in the rotor passage. This transition is efficient only if the Mach number in front of the shock is reduced to the minimum value possible. Then the passage must be contracted as much as possible according to the starting considerations. Now if the design is based on two-dimensional considerations, then a larger contraction ratio of the passage is required at the tip, where the Mach number is larger, while from geometrical and structural considerations a larger contraction of the passage near the hub and a smaller contraction near the tip of the blades are required, in order to obtain a convenient thickness distribution along the blade. This difficulty can be solved only by introducing three-dimensional effects inside the rotor passage, for example, by contracting the annulus in radial direction (which is discussed later) or by introducing three-dimensional effects in front of the passage by twisting the blade leading edge in a convenient way.

Assume, for example, that the cross sections of the blades at different radii considered before in Fig. G,10a and indicated as section a_1 and b_1 are twisted as shown in Fig. G,10a by sections a_2 and b_2. This twist is such that the value of the entering axial velocity decreases at the tip of the blades and increases at the hub of the blades. Because of this change the difference in relative Mach number between hub section and tip section of the blade in the second design is smaller than in the first design. The static pressure at the entrance of the compressor is not uniform because the absolute velocity is not uniform. The streamlines are not parallel to the axis of the compressor. In the second design, the axial velocity at the hub can at most be sonic and the velocity at the tip must be somewhat

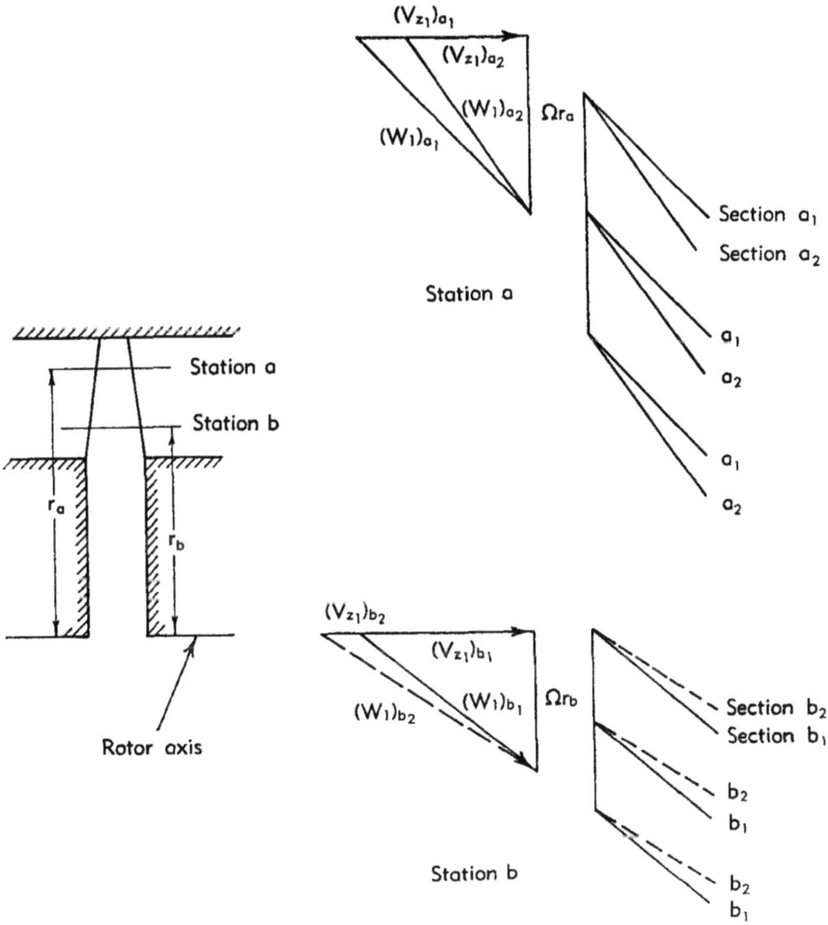

Fig. G,10a. Three-dimensional effects at blade entrance.

below sonic; therefore, in the second design a decrease in mass flow with respect to the maximum possible for the same frontal area must be accepted.

Similar three-dimensional effects can be introduced by using blades having the leading edge swept with respect to the radial direction.

Three-dimensional effects inside the passage. The three-dimensional effects inside the passage can be produced by several different sources:

the entrance conditions can be nonuniform; static pressure gradients can exist at the entrance of the rotor, due to the direction and intensity of the relative velocity at the entrance along the blade and to the existence of a radial velocity component. Centrifugal effects produce static pres-

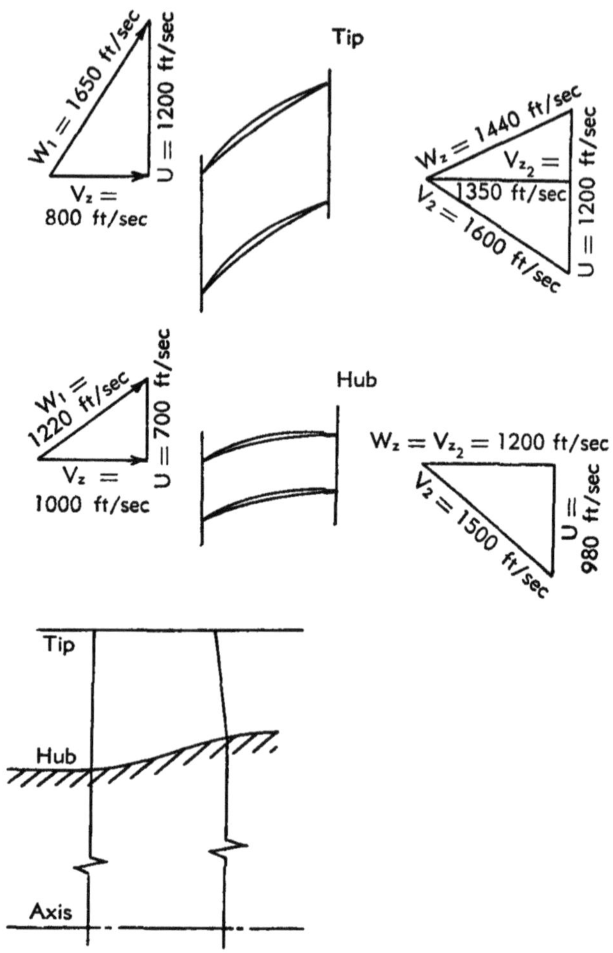

Fig. G,10b. Three-dimensional effects in blade passage.

sure gradients inside the passage which usually results in radial velocity components.

In practical compressor design, the passage must be highly three-dimensional. Large variations of the height of the annulus are usually required in order to permit the independent choice of the entrance and the exit velocity diagram while the blades have sharp trailing edges. Consider, for example, the velocity diagram shown in Fig. G,10b. The axial velocity at the tip is equal to 800 ft/sec and at the hub is equal to 1000 ft/sec, while at the exit the axial velocity at the hub is equal to 1200 ft/sec and at the tip is equal to 1350 ft/sec. The flow in relative coordinates has been slightly decelerated along the passage and the static pressure has increased, and at the same time the axial component at the exit of the rotor is larger than at the entrance. The annular area at the exit must be smaller than the annular area at the entrance in order to satisfy the continuity law. The contraction of the annulus in Fig. G,10b is produced at the root of the blades. In this way the rotational velocity at the hub exit is larger than at the entrance.

Three-dimensional effects are important and can be analyzed with some approximation in the absence of large viscous effects for the supersonic part of the flow. Experimental evidence exists which indicates that these effects can influence sensibly the performance of the machine. For the supersonic part of the flow field two different types of analyses have been performed [*15,16,17,18,19*]. The rotor has been represented as a rotor having an infinite number of blades. With this approximation the flow properties inside the passage become a function only of two spacial coordinates x and r. The x axis corresponds to the axis of the compressor: r is the coordinate in the radial direction. Distributed body forces are introduced in order to represent the load on the blades. In this approximation the variation of flow properties from point to point on a cylindrical surface is neglected.

The second method is based on the linearized characteristics method described in VI,G. The flow is considered three-dimensional but the three-dimensional effects are assumed small. The flow is considered as a basic nonlinear two-dimensional flow on each cylindrical cross section of the compressor plus a linear flow; the linear flow which is superposed on the basic two-dimensional flow takes into account the three-dimensional effects. This method permits the shape of the blade to be considered. Both the methods can be used for stationary or rotating elements. The method of [*17*] can be extended to the case where the radial and axial variations are considered large and the variation in tangential direction can be considered small and need be determined only in first approximation.

An application of the first method to a compressor design has been discussed in detail in [*19*]. An application of the second method has been presented in [*20*]. A typical comparison between experimental and theoretical data obtained with the method of [*19*] is presented in Fig. G,10c and G,10d (from [*20*]). In Fig. G,10c the shape of the passage is shown while in Fig. G,10d a comparison between the Mach number distribution along the walls obtained from the two-dimensional analysis, from the three-dimensional analysis, and from measurements is given for a station at the middle region of the passage.

Three-dimensional effects are more difficult to analyze when transition from supersonic to subsonic flow occurs in the passage, because it is necessary to determine accurately the shape of the shock which is sensitive to three-dimensional effects and to the presence of the boundary layer. Efficient transition from supersonic to subsonic flow in compressor passages is required in order to have efficient compressors. The values of pressure recovery measured in isolated passages of axially symmetric or two-dimensional-type diffusers having uniform inlet flow are sufficiently high to permit high efficiency in supersonic compressors to be expected, when it is assumed that the same values of pressure recovery can be realized in machines where large three-dimensional effects are

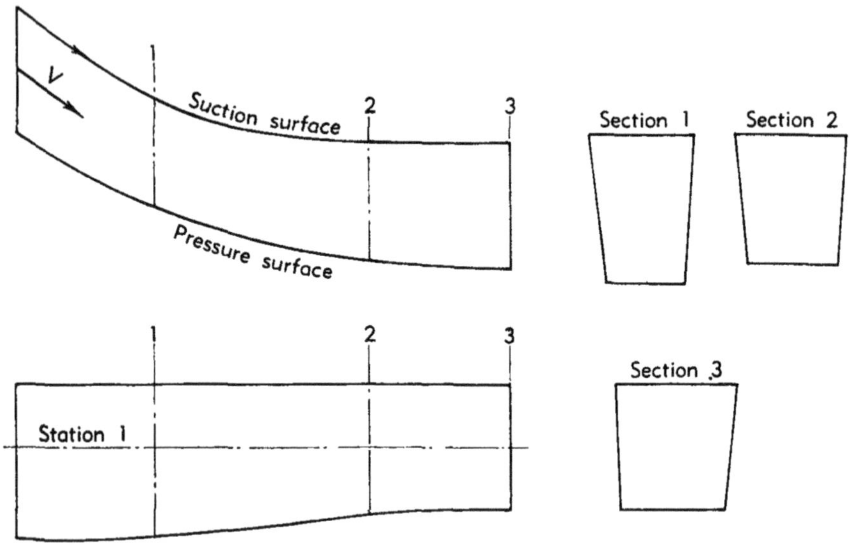

Fig. G,10c. Typical three-dimensional blade passage.

present. It is also possible that an understanding of three-dimensional effects and of boundary layer phenomena could permit the realization in compressor passages of pressure recoveries higher than those measured in isolated passages having essentially one-dimensional flow. However, the three-dimensional effects can also produce large losses of pressure recovery.

Consider, for example, a convergent-divergent passage stator for a shock in the stator compressor. Assume that the velocity and Mach number and total enthalpy at the rotor discharge are such that they are much larger at the tip than at the root of the passage. The variation of static pressure along the passage is determined by three-dimensional effects. If the shape of the passage is not carefully chosen, it may be impossible to decelerate such a flow efficiently because the pressure rise along the passage is dictated by the equilibrium between both the high and the low total enthalpy parts of the stream. Unless large centrifugal effects are introduced, for example, by curving the annulus of the passage, the static pressure cannot change a large amount in the radial direction from the hub to the tip of the passage. Therefore it is not possible to decelerate the high enthalpy region of the stream to very low Mach numbers. This deceleration would produce a static pressure rise which cannot be achieved

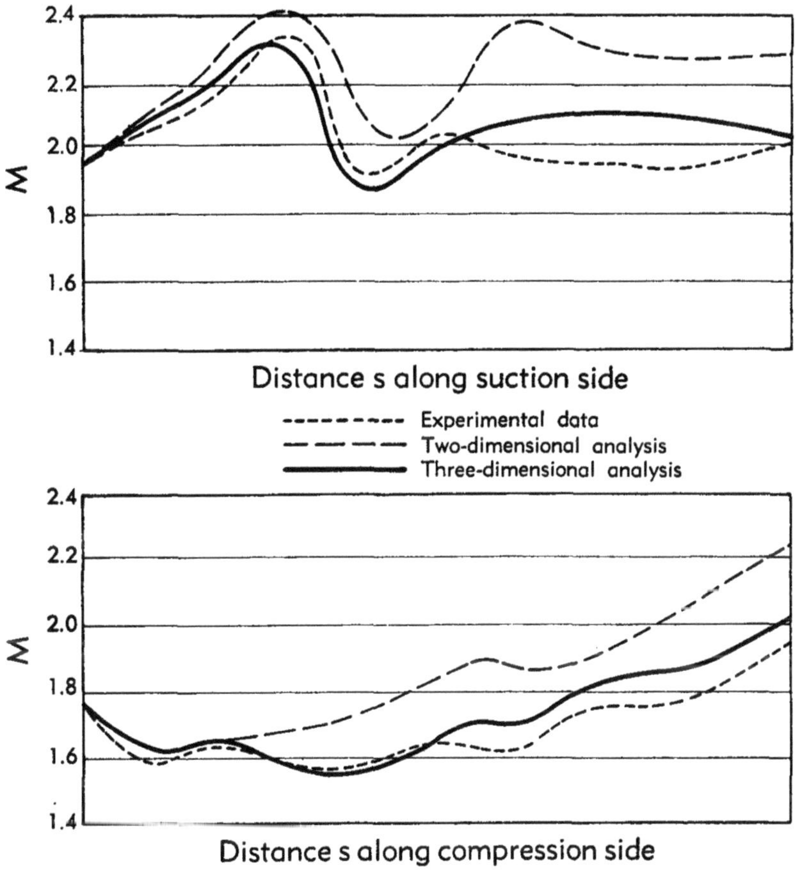

Fig. G,10d. Comparison of theoretical and experimental results for passage shown in Fig. G,10c.

in the part of the stream having lower total enthalpy unless large mixing takes place during the deceleration. Such a rise in the static pressure may be larger than the stagnation pressure of the low total enthalpy stream.

Three-dimensional considerations and the correct use of centrifugal forces can eliminate this difficulty and can improve significantly the compressor performance. Similar considerations can be made for the shape of the shock. In this case the equilibrium of the static pressure in front of and behind the shock determines the actual shock shape and position.

G,11. Performance of Supersonic Compressors.

Performance at the design point. Consider a rotor passage of a supersonic compressor and assume that the flow field in the passage can be analyzed to a first approximation by considering only the flow field at an average section of the passage. Assume that the passage is a convergent-divergent one, that changes the direction of the relative velocity from the entrance to the exit of the passage by an angle τ. The passage has a contraction ratio between the minimum section and a section normal to the flow direction at the entrance equal to A_{min}/A_1 and an expansion ratio between the minimum section and the section normal to the flow direction at the exit equal to A_{min}/A_2 (Fig. G,11a). The performance of such a rotor is a function of the entrance velocity diagram and of the exit pressure.

At a given rotational speed of the rotor the work done by the rotor is a function of the static pressure at the exit. If the pressure at the exit of the rotor is sufficiently low the flow at the exit of the rotor is supersonic; a gradual increase in the back pressure moves the transition from supersonic to subsonic flow gradually upstream toward the minimum section. When the shock moves from outside to inside the passage the velocity diagram at the exit of the rotor changes and the adiabatic efficiency of the rotor also changes. When the shock is inside the rotor the maximum efficiency corresponds to the condition of the shock near the throat of the convergent-divergent passage.

The amount of work performed by the rotor is a function of the exit Mach number and of the turning angle of the passage. It can decrease or increase when the static pressure at the discharge is increased, depending on the value of the turning angle τ. If the flow at the exit of the rotor is turned in the direction of the axis only slightly, then the turning angle τ is smaller than the angle β_1 between the relative entering velocity W_1 and the axis; the angle β_2 between the relative exit velocity W_2 and the axis has the same sign as β_1. Then a decrease of the exit relative velocity corresponding to an increase of the value of the discharge static pressure produces an increase of the work performed (Fig. G,11a) because the variations ΔV_θ of the tangential component of the relative velocity increases when W_2 decreases. In this case a compressor with subsonic exit velocity (Fig. G,11a, bottom) produces more work than when the exit velocity is supersonic (Fig. G,11a, top). The reverse is true if the value of the turning angle τ is larger than the angle β_1 between M_{rel_1} and the axis (Fig. G,11b) and β_2 is negative. A decrease of the exit velocity or an increase of the discharge static pressure produces a reduction of the amount of work produced. This characteristic could produce instability in the case of an actual engine.

The adiabatic efficiency of the rotor is a function of the pressure recovery of the passage and of the amount of work performed. In order to obtain high values of the compression ratio, it is convenient to produce large turning angles in the rotor; however, it is difficult to produce large turning in a passage having transition from supersonic flow to subsonic flow in the rotor. The turning of the flow produces large pressure gradients along the surface of the passage, which interfere with the shock

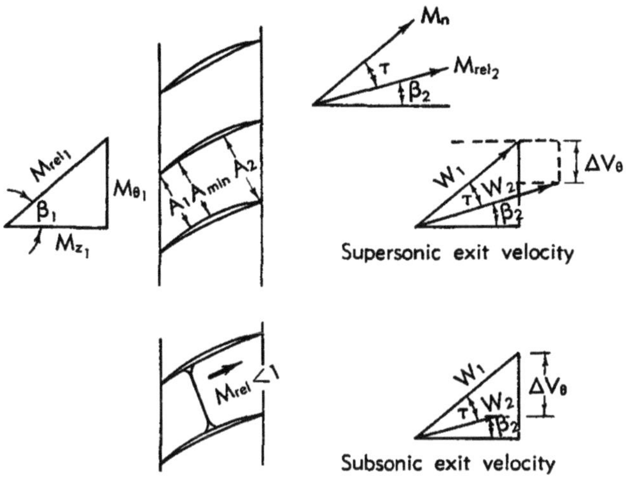

Fig. G,11a. Velocity diagrams for compressor with $\tau < \beta_1$.

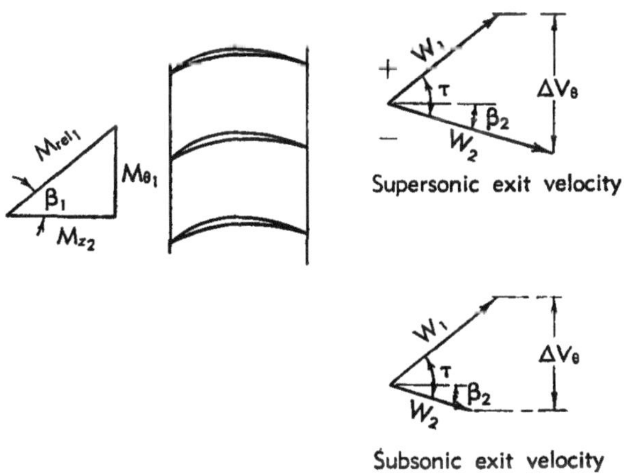

Fig. G,11b. Velocity diagrams for compressor with $\tau > \beta_1$.

inside the passage and tend to produce local separation of the boundary layer. Therefore, in order to obtain high values of the compression ratio it has been considered practical to turn the flow in the rotor without producing a large static pressure rise and then to decelerate the flow in the stator (shock in the stator compressor). In this type of machine the adiabatic efficiency of the rotor is very high because the turning can be performed with high pressure recovery (see e.g. [14]). However, large losses are encountered in the stator where the flow must be turned to the axial direction and must be decelerated to subsonic speed. However, the stator can in principle be designed with boundary layer control and variable geometry and hence can produce efficient deceleration from supersonic to subsonic flow.

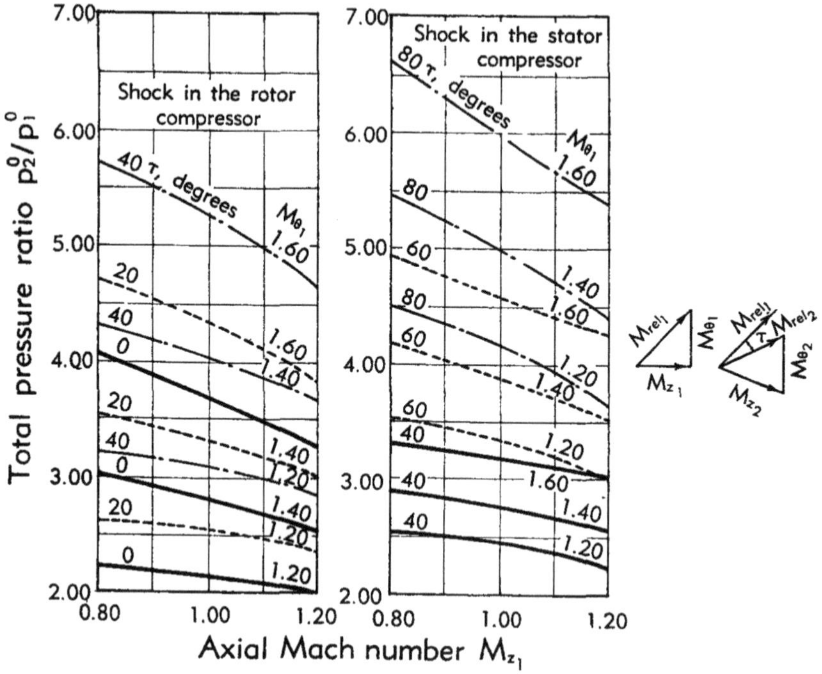

Fig. G,11c. Comparison of the total pressure ratio characteristics of shock in rotor and shock in stator compressors.

In the absence of systematic experimental data, the possible performances of supersonic compressors with shock in the rotor or shock in the stator have been determined analytically and are presented in Fig. G,11c to Fig. G,11i. Similar data can be found in [14,21]. Experimental data for shock in the rotor compressors are presented in [13,22]. The analysis has been performed with the following assumptions: For the case of the shock in the rotor compressor the stagnation pressure recovery of the rotor passage in relative coordinates has been assumed equal to the pressure recovery across a normal shock for the value of the relative Mach number at the entrance of the passage. The stagnation pressure recovery of the following stator has been assumed constant and equal to 0.95. For the case of shock in the stator compressor the stagnation pressure recovery in the rotor passage in relative coordinates has been assumed constant and equal to 0.95, while the stagnation pressure recovery in the stator has been assumed equal to stagnation pressure recovery across a normal shock for the Mach number entering the stator. In Fig. G,11c the single-stage pressure ratio p_2^0/p_1^0 for shock in the rotor and shock in the stator compressors is presented as a function of the axial entering Mach number M_{z_1}. The turning angle τ and the rotational Mach number M_{θ_1} at a section considered representative of the compressor are included as parameters. In the data of Fig. G,11c the rotor passage for the shock in the

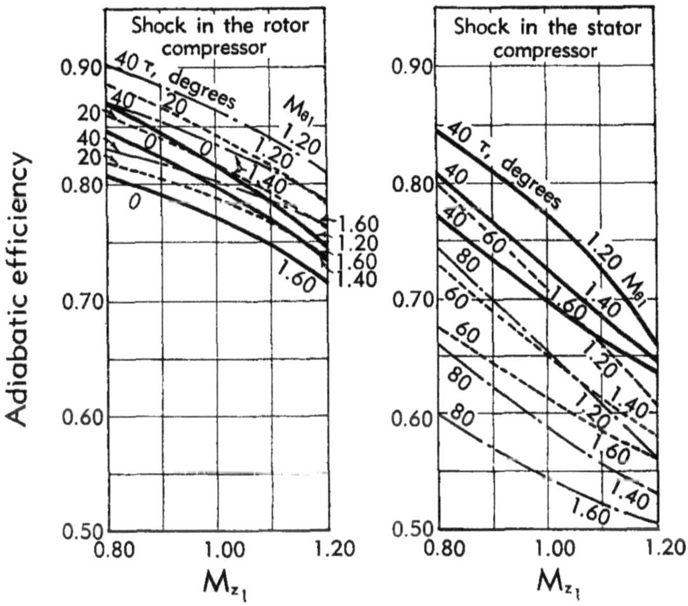

Fig. G,11d. Comparison of the adiabatic efficiency characteristics of shock in rotor and shock in stator compressors.

rotor compressor has been assumed to have an area ratio A_2/A_1 equal to 1.20 and for the shock in the stator compressor the value of 1. (The quantity A_2/A_1 is defined in Fig. G,11a.) Three values of the tangential velocity of the rotor at the representative section of the rotor have been considered; these values correspond to $M_{\theta_1} = 1.2$, 1.4, and 1.6. Three values of the turning angle τ as defined in Fig. G,11a have been considered. The values of τ chosen for the shock in the rotor compressor are $\tau = 0°$, 20°, 40° while for the shock in the stator compressor they are $\tau = 40°$, 60°, 80°. Because of the large turning angles considered for the shock in the rotor compressor, the flow at the exit of such a compressor is turned beyond axial in a large part of the curves (β_2 of Fig. G,11a is negative). In Fig. G,11d, there is given the corresponding adiabatic efficiency η_{ad} which is defined as the ratio between the increase in total enthalpy required for an adiabatic compression for the same pressure ratio and the actual total enthalpy rise. In Fig. G,11e the value of the exit Mach number entering the stator (in fixed coordinates) M_2 is given. In Fig. G,11f the angles β_3 between the axis and the velocity vector entering the stator are given. The angles are considered negative if the flow is turned beyond the axis (see Fig. G,11b).

In order to give some indication of the effect of the expansion ratio and of the effect of pressure recovery in the shock in the rotor compressor

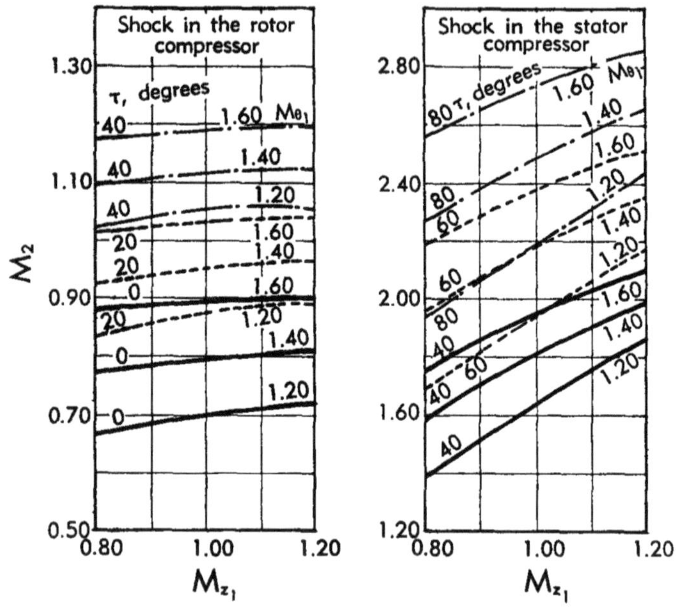

Fig. G,11e. Comparison of rotor exit Mach numbers for shock in rotor and shock in stator compressors.

in Fig. G,11g, the compression ratio p_2^0/p_1^0, the adiabatic efficiency η_{ad}, the exit Mach number M_2, and the exit angle β_3 are presented as functions of the turning angle τ. As shown in these figures the pressure recovery of the rotor stage has an important effect on the compression ratio and the adiabatic efficiency of the compressor. In Fig. G,11h, the effect of different values of the ratio A_2/A_1, the pressure recovery in the rotor, and of the pressure recovery in the stator are considered for a shock in the stator compressor. It can be seen that a large gain in compression ratio and in adiabatic efficiency can be obtained by improving the stator pressure recovery.

Compressor map for supersonic compressors having subsonic axial velocities. For a supersonic compressor the relation between the entering mass flow and the rotational speed is determined by the blade design, and the entering mass flow remains constant provided that the flow relative to the rotor remains supersonic. Therefore the compressor map of a supersonic compressor when the flow relative to the rotor is completely supersonic has a well-defined shape. Compressor maps are usually obtained by plotting the stagnation pressure ratio through the compressor p_2^0/p_1^0 (in absolute coordinates) as a function of the corrected weight flow $w\sqrt{\Theta_1}/\delta_1$ for given values of the corrected rotational speed $N/\sqrt{\Theta_1}$ where w is the entering weight flow per unit area, Θ_1 is the ratio of total stagnation temperature at the entrance of the compressor to a

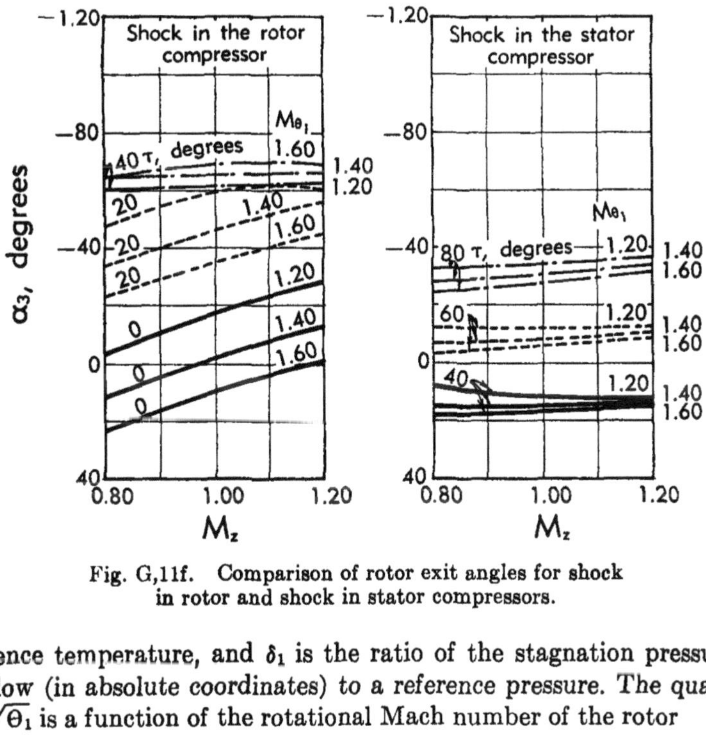

Fig. G,11f. Comparison of rotor exit angles for shock in rotor and shock in stator compressors.

reference temperature, and δ_1 is the ratio of the stagnation pressure of the flow (in absolute coordinates) to a reference pressure. The quantity $N/\sqrt{\Theta_1}$ is a function of the rotational Mach number of the rotor

$$M_{\theta_1} = \frac{\pi ND}{60\sqrt{\gamma \mathcal{R} T_1}} = \frac{N}{\sqrt{\Theta_1}}\frac{\pi D}{60\sqrt{\gamma\mathcal{R}}}\sqrt{\frac{T_1^0}{T_1}T_{\text{ref}}}$$

or

$$\frac{N}{\sqrt{\Theta_1}} = \text{const}\, M_{\theta_1} f(M_{\theta_1})$$

where T_1^0/T_1 is the ratio of stagnation-to-static temperature which is a function of M_{θ_1} and D is the diameter of the section considered. The

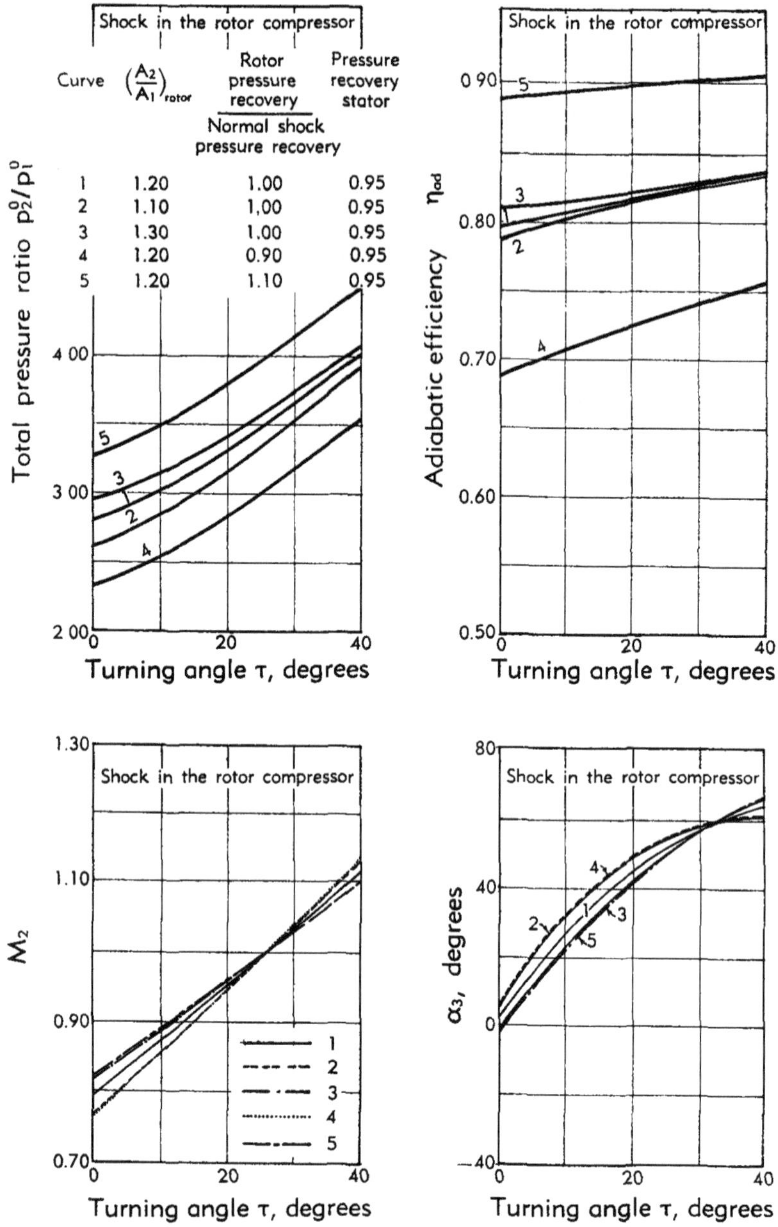

Fig. G,11g. Effects of rotor area ratio and rotor pressure recovery on the performance of shock in rotor compressors. (Axial Mach number $M_2 = 1.0$; tangential Mach number $M_\theta = 1.4$.)

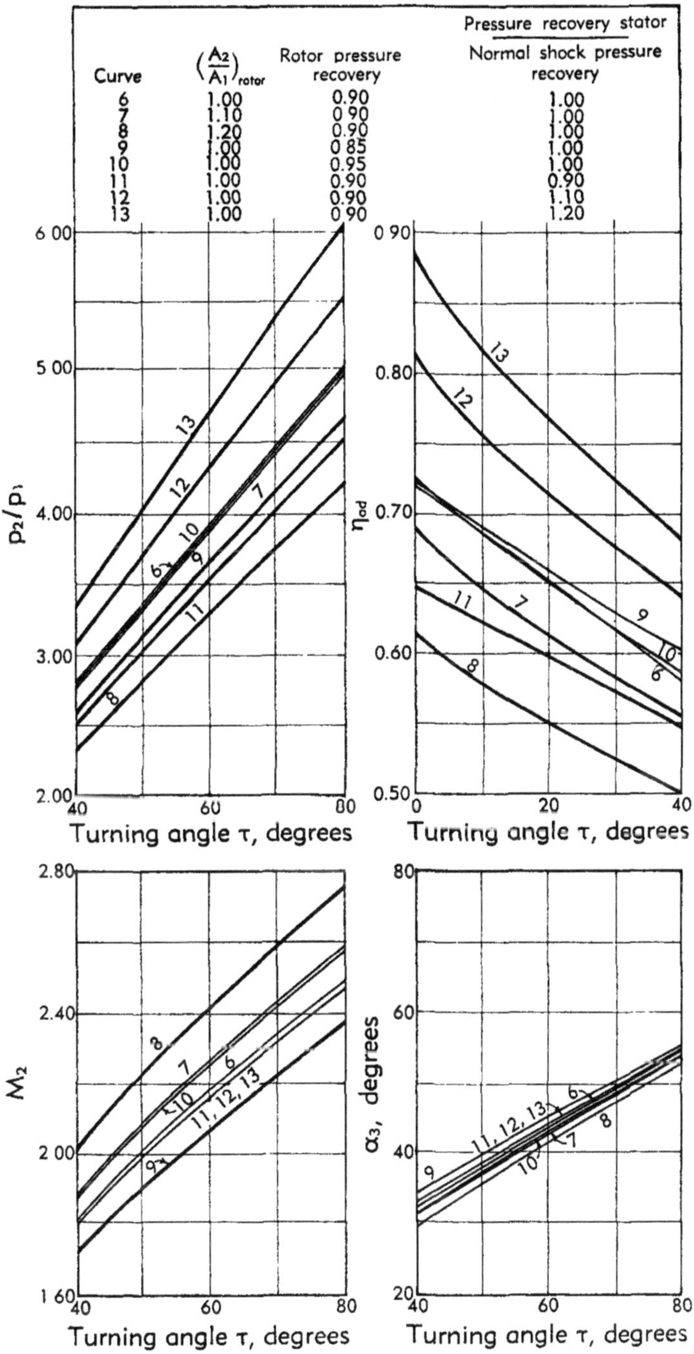

Fig. G,11h. Effects of rotor area ratio, rotor pressure recovery, and stator pressure recovery on the performance of shock in stator compressors. (Axial Mach number $M_z = 1.0$; tangential Mach number $M_\theta = 1.4$.)

quantity $w\sqrt{\Theta_1}/\delta_1$ is a function only of the axial entrance Mach number,

$$w = V_z\rho_1 g = \frac{M_z p_1}{\sqrt{T_1}}\sqrt{\frac{\gamma}{\mathcal{R}}}\,g = \frac{M_z p_1}{\sqrt{T_1^0}}\frac{p_1}{p^0}\sqrt{\frac{T_1^0}{T_1}}\sqrt{\frac{\gamma}{\mathcal{R}}}\,g$$

or

$$\frac{w\sqrt{\Theta_1}}{\delta_1} = \text{const } M_{z_1}f(M_{z_1})$$

Now for a given blade entrance region and a given rotational speed the entrance axial Mach number in a supersonic compressor is constant; so that, provided the flow at the entrance is supersonic the variation of $w\sqrt{\Theta_1}/\delta_1$ for a given value of the parameter $N/\sqrt{\Theta_1}$ is zero. A variation of N changes the velocity diagram at the entrance and hence changes the value of M_{z_1}, while a change of the exit static pressure changes the velocity diagram at the exit and therefore the compression ratio and the adiabatic efficiency. Thus a supersonic compressor map, where the value of the compression ratio is plotted as a function of the parameter $w\sqrt{\Theta_1}/\delta_1$ for any given value of the parameter, corresponds to a vertical line. An experimental compressor map for a supersonic compressor obtained from [*22*] is shown in Fig. G,11i. Here the pressure compression ratio P_2^0/P_1^0 as a function of the mass flow coefficient defined as $w/gA\rho^*a^*$ is presented for different values of M. (ρ and a are the initial density and speed of sound at the entrance.) The tests have been performed in Freon 12.

Compressor map for supersonic compressors having supersonic axial velocities. In supersonic compressors the possibility exists that the axial velocity at the entrance is supersonic. This can be obtained, for example, when an airplane flies at supersonic speed. For these conditions the velocity diagram at the entrance of the compressor is not uniquely defined by the rotational speed of the compressor and, therefore, by the parameter $N/\sqrt{\Theta_1}$ as for the case when the axial velocity is subsonic. Thus with a supersonic axial velocity the compressor map is different from the compressor map corresponding to the subsonic axial velocity. A given value of the parameter $N/\sqrt{\Theta_1}$ does not define a single value of the parameter $w\sqrt{\Theta_1}/\delta_1$. The value of $w\sqrt{\Theta_1}/\delta_1$ and of the compression ratio p_2^0/p_1^0 can be altered by changing the value of the axial velocity component, even if the parameter $N/\sqrt{\Theta_1}$ and the rotational velocity remain constant.

Consider, for example, Fig. G,11j. The compressor passage is designed in such a way that at the design rotational speed the axial velocity is sonic. If the flight velocity is subsonic, the compressor at the design rotational speed produces sonic velocity at the entrance and the velocity diagram is as shown in Fig. G,11j, left. The relative velocity vector is tangent to the entrance region of the blade. However, the compressor can accept a supersonic incoming velocity. An increase of the axial velocity at first will reduce the expansions at the leading edge if they are present; when all the expansions are eliminated by a further increase in axial velocity, a shock is produced at the leading edge of the blade (Fig. G,11j, right). Because of the supersonic axial component the shock

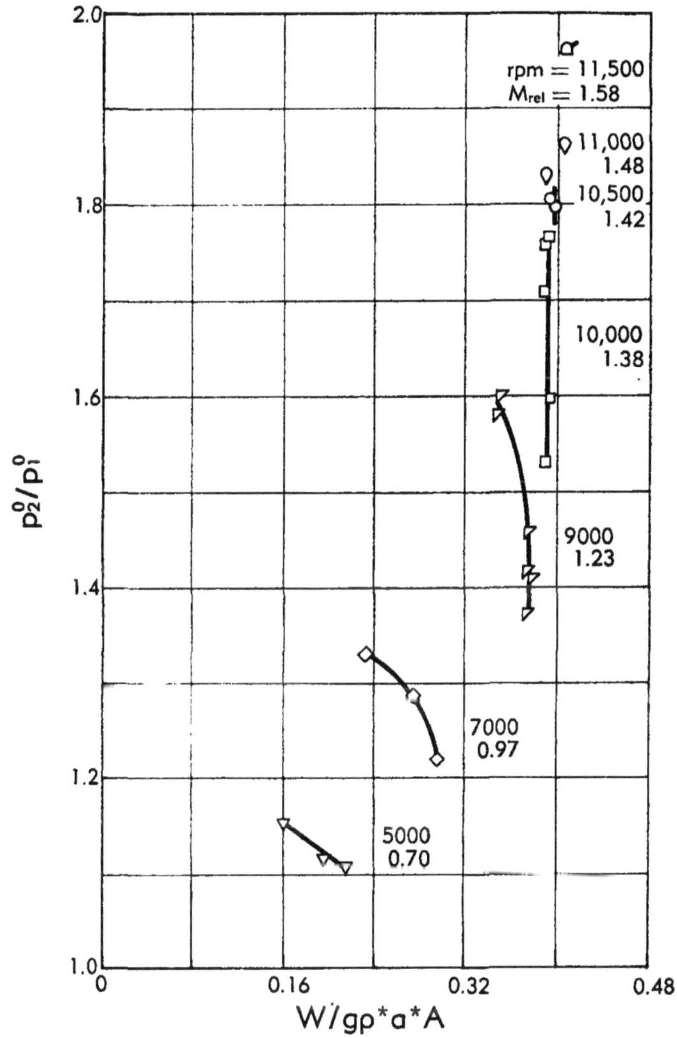

Fig. G,11i. Experimental supersonic compressor performance from [22].

goes inside the passage. The maximum axial component that still produces steady conditions corresponds to the condition of shock parallel to the entrance of the compressor. The values of $w\sqrt{\Theta_1}/\delta_1$ and p_2^0/p_1^0 corresponding to a given value of $N/\sqrt{\Theta_1}$ change when the axial velocity becomes supersonic, and all the possible stable solutions of p_2^0/p_1^0 as a function of $w\sqrt{\Theta_1}/\delta_1$ for a given value of $N/\sqrt{\Theta_1}$ lie not along a single line, as for the case of the subsonic axial velocity, but in a region of the diagram.

A typical compressor map for a given value of $N/\sqrt{\Theta_1}$ is represented in Fig. G,11k. The compressor considered is a shock in the rotor compressor and has a passage as shown in Fig. G,11j. The rotational Mach number is $M_{\theta_1} = 1.4$ and the axial Mach number for the design conditions is $M_{z_1} = 1$. The passage is assumed to have a pressure recovery for the design condition equal to the pressure recovery of a convergent-divergent inlet with all internal compression for the design entering Mach number and a corresponding contraction ratio. For conditions of supersonic axial velocity, the losses across the inclined shock are added to the

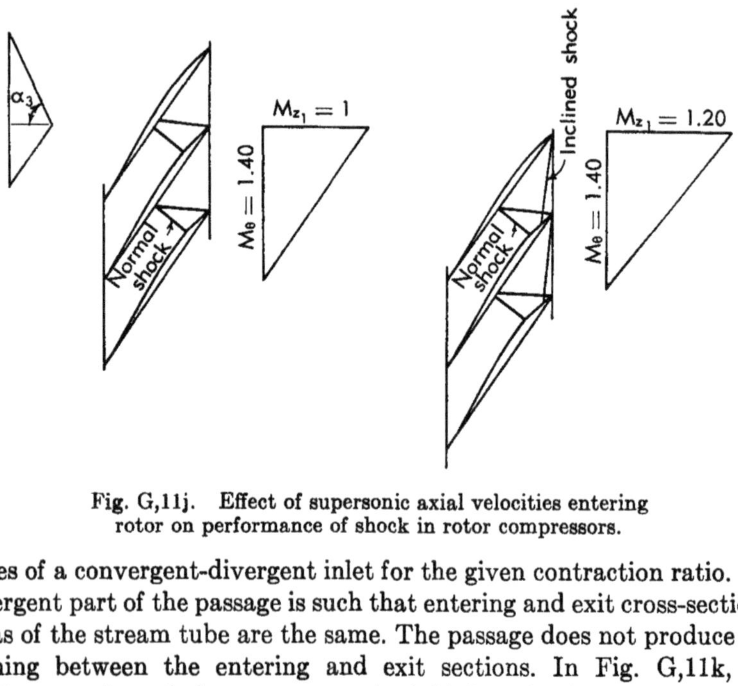

Fig. G,11j. Effect of supersonic axial velocities entering rotor on performance of shock in rotor compressors.

losses of a convergent-divergent inlet for the given contraction ratio. The divergent part of the passage is such that entering and exit cross-sectional areas of the stream tube are the same. The passage does not produce any turning between the entering and exit sections. In Fig. G,11k, the maximum value of the pressure ratio $(p_2^0/p_1^0)_{max}$ is given as a function of M_{z_1} for a value of $M_{\theta_1} = 1.40$ and, therefore, for a constant value of $N/\sqrt{\Theta_1}$. Each point between the abscissa and the curve $(p_2^0/p_1^0)_{max}$ is a possible operating point. The optimum adiabatic efficiency and the value of the angle β_3 of the exit velocity with respect to the axis are also given. The data of Fig. G,11k show that for a given value of $N/\sqrt{\Theta_1}$ the value of $w\sqrt{\Theta_1}/\delta_1$ can be increased above the value corresponding to $w\sqrt{\Theta_1}/\delta_1$ for $M_z = 1$, when the axial velocity is supersonic. This property of supersonic compressors can be of interest for engines designed for supersonic airplanes.

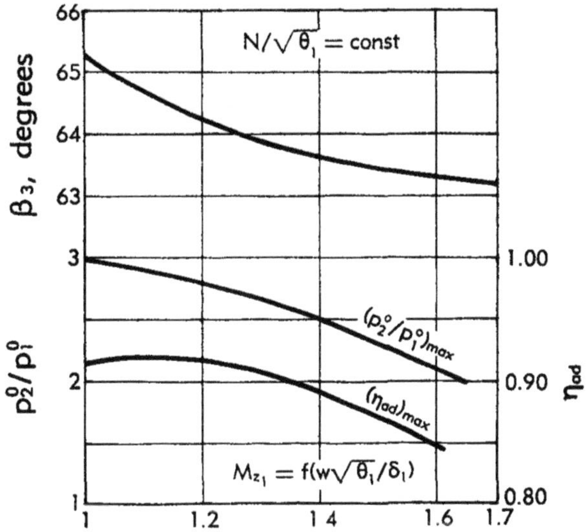

Fig. G,11k. Performance of a rotor having a supersonic axial entrance velocity, at a given rotational speed, as a function of the axial Mach number.

G,12. Cited References.

1. Busemann, A. *Atti del V Convegno Volta, accad. nazl. Lincei, Rome,* 1935.
2. Kantrowitz, A. The supersonic axial-flow compressor. *NACA Rept. 974,* 1950.
3. Kantrowitz, A., and Donaldson, C. duP. Preliminary investigation of supersonic diffusers. *NACA Wartime Rept. L713,* 1945.
4. Neumann, E. P., and Lustwerk, F. Supersonic diffusers for wind tunnels. *Mass. Inst. Technol. Guided Missiles Program, Meteor Rept. 13,* Nov. 1947.
5. Davidson, I. M. Some data pertaining to the supersonic axial-flow compressor. *Brit. Aeronaut. Research Council Repts. and Mem. 2554,* 1951.
6. Weise, A. Supersonic axial compressor. *Ministry of Supply Transl. Rept. GDC 10/6116 T(H),* Oct. 1943.
7. Buhler, R. D. *Supersonic Flow through Cascades, with Application to Diffusers. Thesis,* Calif. Inst. Technol., 1948.
8. Redding, A. H. Superacoustic compressor. *U.S. Patent 2,435,236,* 1948.
9. Boxer, E., Sterrett, J., and Wlodarski, J. Application of supersonic vortex flow theory to the design of supersonic impulse compressor or turbine blade sections. *NACA Research Mem. L52B06,* 1952.
10. Erwin, J. R., Wright, L. C., and Kantrowitz, A. Investigation of an experimental supersonic axial-flow compressor. *NACA Research Mem. L6J01b,* Oct. 1946.
11. Erwin, J. R., and Schulze, W. M. Investigation of an impulse axial-flow compressor. *NACA Research Mem. L9J05a,* 1950.
12. Schulze, W. M., Erwin, J. R., and Westphal, W. R. Investigation of an impulse axial-flow compressor rotor over a range of blade angles. *NACA Research Mem. L50F27a,* 1950.
13. Kantrowitz, A. The supersonic axial-flow compressor. *NACA Rept. 974,* 1950.
14. Ferri, A. Preliminary analysis of axial-flow compressors having supersonic velocity at the entrance of the stator. *NACA Research Mem. L9G06,* 1949.
15. Wu, C.-H. A general through-flow theory of fluid flow with subsonic or supersonic velocity in turbomachines of arbitray hub and casing shapes. *NACA Tech. Note 2302,* 1951.
16. Wu, C.-H. A general theory of three-dimensional flow in subsonic and super-

sonic turbomachines of axial-, radial-, and mixed-flow types. *NACA Tech. Note 2604*, 1952.
17. Goldstein, A. W. Axisymmetric supersonic flow in rotating impellers. *NACA Rept. 1083*, 1952.
18. Costilow, E. L. Application of a characteristic blade to blade solution to flow in a supersonic rotor with varying stream-filament thickness. *NACA Tech. Note 2992*, 1953.
19. Vaglio-Laurin, R. Three-dimensional effects in supersonic compressors. Part 2: The analytical method and its application for rotating and fixed coordinate systems. *Polytech. Inst. Brooklyn Aerodynamics Lab. Rept. 234*, 1953.
20. La Rocca, A. Three-dimensional effects in supersonic compressors. Part 3: The design of two stator passages and comparison between theory and experiment. *Polytech. Inst. Brooklyn Aerodynamics Lab. Rept. 271*, 1954.
21. Wright, L. C. Investigation to determine contraction ratio for supersonic-compressor rotor. *NACA Research Mem. E7L23*, 1948.
22. Erwin, J. R., Wright, L. C., and Kantrowitz, A. Investigation of an experimental supersonic axial-flow compressor. *NACA Research Mem. L6J01b*, 1946.

SECTION H

AERODYNAMIC DESIGN OF AXIAL FLOW TURBINES

E. DUNCOMBE

CHAPTER 1. FUNDAMENTAL PRINCIPLES

H,1. Introduction. The availability of energy associated with a compressed fluid or gas is commonly observed both as a natural occurrence (geysers, waterfalls, springs) and in many man-made tools (bellows, waterwheels, rockets). Turbines may be defined as those devices that make use of this energy by continuous rotary motion. However, it was only in the nineteenth century that successful efforts were made to obtain pressure energy by conversion from heat energy through the steam boiler. At that time the turbine in the form of the windmill and waterwheel was well known and it was thus quite natural to combine it with the steam boiler, producing the forerunner of the modern steam turbine plant. The high degree of technical development accomplished is well exemplified by Stodola's classical treatise on the subject of steam turbines [*1*].

The gas turbine calls for a single-phase gas cycle rather than a two-phase, liquid-vapor cycle, but since the turbine element is very similar its development is largely based on earlier steam and water experience. After the advent of flight and the development of wing theory the very considerable body of knowledge accumulated on the lift and drag of wings could be brought to bear on problems associated with blade rows. Boundary layer theory, potential flow analysis, and data on high speed flow could then be applied to the field of turbomachinery. It would be misleading, however, to convey the impression that the field is not still fraught with a number of unknowns. Problems in nonviscous flow are largely mathematical in nature in that the equations of flow are known, and can be solved, at least in theory, by means of elaborate numerical techniques. On the other hand, problems relating to viscous effects (losses in general) are of a different order since the laws governing the mechanisms are still being shaped. In cases where local Mach numbers exceed unity, shock-boundary layer interaction effects present added unknowns. In the field of turbomachinery, unsteady flow conditions are always present as an added complication, since each blade row experiences unsteady inlet conditions as it moves relative to a tangentially nonuniform pattern of wakes from the preceding row. These effects are usually ignored in turbine design and, in fact, there is experimental evidence that the row spacing must be very small before they become appreciable.

In general, turbine designs proceed by the application of fundamental physical laws where possible and otherwise by reference to a logical range of tests from which the necessary data can be extracted.

The operation of a turbine can be explained in terms of fundamental mechanical, aerodynamic, and thermodynamic principles, and therefore the remainder of Chap. 1 is concerned with the application of these principles to the study of axial flow gas turbines. Convenient parameters are considered which define the work capacity or govern the efficiency of the machine, and relations between them are deduced. Chap. 2 is concerned with experimental data on blade performance, chiefly as affected by the passage configuration and height. Performance is considered in the light of both pressure loss and control of gas direction. In Chap. 3 the presence of a three-dimensional field of flow is recognized. This chapter deals in particular with radial variations in flow properties, though a more general analysis is also introduced. Various phases of turbine design are considered in Chap. 4. The final design will consist of compromises between the various factors, which affect efficiency, such as blade shape, blade speed, the number of stages, etc., and those affecting weight and degree of reliability. The interplay of these factors is treated in Chap. 4. The off-design performance of single and multistage turbines is also treated here.

H,2. Thermodynamic Properties of a Turbine.

Stagnation conditions. The paragraphs under this heading are not intended to contain a complete treatment of the subject but to provide the reader with an introduction to some concepts and equations which will be widely used thereafter.

Although the gas constituting the working fluid of the turbine element is usually composed of a variety of combustor exhaust products in addition to nitrogen and some unburned oxygen, it is often treated as a unique gas, the values of the specific heats c_p and c_v being either unique functions of temperature or, more approximately, having some average value. This practice is justified by the comparatively narrow range of temperature and fuel-air ratios over which the turbine is required to operate. The use of fuels other than hydrocarbons will thus necessitate some revision of this method of treatment. If the working fluid can be treated as a perfect gas the following equation of state applies [*2*, p. 94]

$$\frac{p}{\rho} = \mathcal{R}T \tag{2-1}$$

where p, ρ, and T are absolute pressure, density, and temperature, and $\mathcal{R}$ is an experimentally determined constant.

The enthalpy h can now be defined as

$$\int dh = \int c_v dT + \frac{1}{J}\int d\left(\frac{p}{\rho}\right) = \int c_v dT + \frac{1}{J}\int d(\mathcal{R}T) \tag{2-2}$$

Using the identity $\mathcal{R}/J = c_p - c_v$ [*2*, p. 99], Eq. 2-2 may also be written

$$\int dh = \int c_p dT = \bar{c}_p \int dT \tag{2-3}$$

The bar sign over c_p, indicating an average, will henceforth be omitted as understood.

Tabulated values of h, c_p, c_v, and T are available both for pure air and combustion product mixtures [*3*].

The assumption of the perfect gas law (Eq. 2-1) leads to some simple relations between the gas state, its velocity, and heat supplied or work done. Under conditions of steady flow from state 1 to state 2, the combined energy input from external sources, either in the form of heat or work per pound of gas, must equal the internal energy increase plus the kinetic energy increase plus the net work done by the gas [*2*, p. 35].

Heat + work supplied per pound of gas

$$= \left[c_{v_2}T_2 + \frac{1}{J}P_2\left(\frac{1}{\rho_2}\right) + \frac{1}{J}\frac{V_2^2}{2g}\right] - \left[c_{v_1}T_1 + \frac{1}{J}P_1\left(\frac{1}{\rho_1}\right) + \frac{1}{J}\frac{V_1^2}{2g}\right] \tag{2-4}$$

Eq. 2-4 becomes, using Eq. 2-2,

$$\text{Energy input per pound} = \left(h_2 + \frac{V_2^2}{2gJ}\right) - \left(h_1 + \frac{V_1^2}{2gJ}\right) \tag{2-5}$$

Let us now define

$$h + \frac{V^2}{2gJ} = h^0 \tag{2-6}$$

The term h^0 is known as stagnation enthalpy. It will be seen (Eq. 2-5) to be the enthalpy of the gas when it is brought to zero velocity without energy exchange from the surroundings. The stagnation temperature T^0 is then defined similarly to Eq. 2-2 or 2-3,

$$\int dh^0 = \int c_p dT^0 = \bar{c}_p \int dT^0 \tag{2-7}$$

If the gas is also brought to rest isentropically, the isentropic law applies as follows (*2*, p. 101):

$$\frac{p^0}{p} = \left(\frac{T^0}{T}\right)^{\frac{\gamma}{\gamma-1}} \tag{2-8}$$

$$\frac{\rho^0}{\rho} = \left(\frac{T^0}{T}\right)^{\frac{1}{\gamma-1}} \tag{2-9}$$

Eq. 2-8 and 2-9 define the stagnation pressure and density p^0 and ρ^0, respectively, where $\gamma = c_p/c_v$.

If Eq. 2-5 and 2-6 are combined, it follows that

$$\text{Energy input per pound of gas} = h_2^0 - h_1^0 \tag{2-10}$$

Thus, under conditions of steady flow, the net energy input is equal to the increase in stagnation enthalpy.

Since Eq. 2-6 contains a velocity term which must be measured relative to some frame of reference, the stagnation quantities, like energy, are relative rather than absolute. The stagnation relationships relative to the moving blades are thus defined by substituting h^{00} for h and W for V in Eq. 2-6 through 2-10, for example,

$$h + \frac{W^2}{2gJ} = h^{00} \tag{2-11}$$

Definition of efficiency. Without considering the detailed fluid mechanics of operation it is possible to consider a turbine merely as a device or "black box" through which there is a change of state and extraction of work. The inlet and outlet gas states can then be analyzed in terms of what is ideally possible and what is practically achievable.

The change of gas state across a turbine is illustrated in Fig. H,2a. For steady processes, the work done by the gas, in the absence of heat transfer from the surroundings, between the state points 4 and 5, is given by $h_4^0 - h_5^0$. The corresponding pressure drop is $p_4^0 - p_5^0$. If this pressure drop occurred isentropically, the work would be $h_4^0 - h_{5_{is}}^0$. The efficiency of energy conversion can then be defined as the ratio of actual to ideal work done by expansion from p_4^0 to p_5^0,

$$\eta_{is} = \frac{\text{actual enthalpy drop}}{\text{isentropic enthalpy drop}} = \frac{h_4^0 - h_5^0}{h_4^0 - h_{5_{is}}^0} \tag{2-12}$$

Efficiency defined in this manner is commonly called the "isentropic efficiency."

Other definitions of efficiency may be employed. In cases where the leaving velocity from the turbine is not recovered, for instance in stationary power plants, it may be more meaningful to substitute $h_4^0 - h_{5_{is}}$ for $h_4^0 - h_{5_{is}}^0$ in the denominator of Eq. 2-12.

Reheat factor. A turbine consisting of N stages each of efficiency $\eta_{N_{is}}$ is found to have an over-all efficiency η_{is}, somewhat higher than $\eta_{N_{is}}$ (*2*, p. 163). The ratio of these two efficiencies is known as the "reheat factor," R, where

$$R = \frac{\eta_{is}}{\eta_{N_{is}}} \tag{2-13}$$

An expression for the reheat factor R is most easily developed when N becomes infinitely large and the work per stage becomes infinitely small.

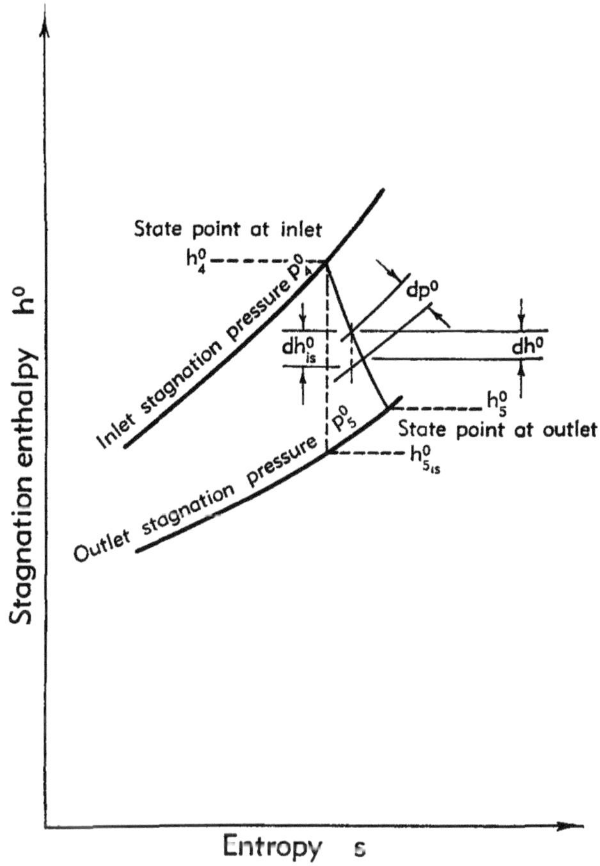

Fig. H,2a. Change of gas state through a turbine.

Such a small stage is illustrated in Fig. H,2a, dh^0 and dh^0_{is} being the actual and ideal enthalpy drops across the pressure differential dp^0.

Eq. 2-12 can then be written in differential form:

$$\text{Limit value of } \eta_{is} = \frac{dh^0}{dh^0_{is}} = \eta_p \tag{2-14}$$

The efficiency η_p is then sometimes known as the "small stage" or "polytropic" efficiency. Eq. 2-14 leads to a relationship between end points somewhat different from that given by Eq. 2-12. Application of Eq. 2-7 and the isentropic law [*2*, p. 101] leads to the relation

$$\frac{dp^0}{dT^0} = \frac{1}{\eta_p}\frac{\gamma}{\gamma - 1}\frac{p^0}{T^0}$$

or, on integration between the limits 4 and 5,

$$\frac{T^0_4}{T^0_5} = \left(\frac{p^0_4}{p^0_5}\right)^{\frac{\gamma-1}{\gamma}\eta_p} \tag{2-15}$$

The ratio η_{is}/η_p can be called the polytropic reheat factor R_p. It is related to the turbine pressure ratio in accordance with Eq. 2-12, 2-15, and 2-3, as follows:

$$R_p = \frac{\eta_{is}}{\eta_p} = \frac{\left[1 - \left(\frac{p_5^0}{p_4^0}\right)^{\frac{\gamma-1}{\gamma}\eta_p}\right]}{\left[1 - \left(\frac{p_5^0}{p_4^0}\right)^{\frac{\gamma-1}{\gamma}}\right]\eta_p} \tag{2-16}$$

This relationship is shown in Fig. H,2b.

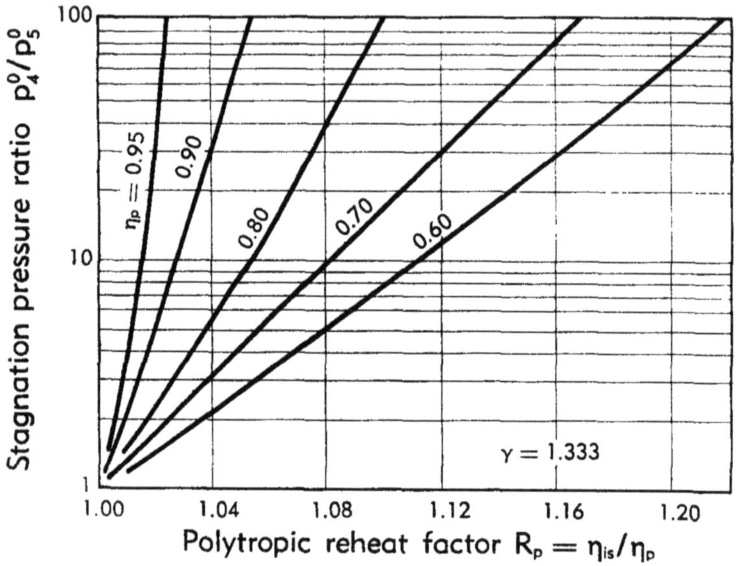

Fig. H,2b. Relationship between isentropic efficiency, polytropic efficiency, and turbine pressure ratio.

The efficiency of an N-stage turbine, η_{is}, may now be related to the efficiency of each separate stage, $\eta_{N_{is}}$, through their separate polytropic reheat factors,

$$\frac{R_p}{R_{N_p}} = \frac{\eta_{is}/\eta_p}{\eta_{N_{is}}/\eta_p} = \frac{\eta_{is}}{\eta_{N_{is}}} \tag{2-17}$$

The significance of this relationship is that a multistage turbine, all stages having the same efficiency, will have an ever-increasing over-all isentropic efficiency as the number of stages increases. Therefore, if efficiency is to be used as a criterion of excellence, comparisons on the basis of $\eta_{N_{is}}$ are more valid. However, the isentropic efficiency η_{is} is often more convenient to use in calculations.

H,3. Fluid Mechanics of a Turbine Stage.

Velocity triangles. The windmill, the waterwheel, and other early rotating devices which are made to extract energy from a fluid stream can all be classified as turbines. The axial flow turbine from a turbojet engine is illustrated in Plate H,3 in an "exploded" form to show the component parts with greater clarity. It bears considerable resemblance to its forerunners. The number of stages depends on the blade speed and over-all pressure ratio and usually varies from one to four for aviation gas turbines. Each stage consists of a row of stationary blades or nozzles which serve to accelerate the gas and give it a swirl component. A following row of rotating blades serves to absorb this swirl energy. These blades may also accelerate the gas relative to themselves, in which case they are said to have "reaction," which will be defined more formally later.

The tangentially averaged gas conditions after the nozzles and blades can be represented graphically by velocity triangles, illustrated for a single stage in Fig. H,3a. Each triangle represents the vector relationship between velocities relative to the preceding and succeeding blade row at some chosen axial station and radius. If the annulus area varies between the blades the triangles only represent conditions at one axial position, the flow adjacent to the preceding or succeeding blades being modified. Tangential velocities shown in Fig. H,3a are positive where they are in the same direction as the blade speed U. This will normally mean positive nozzle outlet swirls and angles but negative blade relative outlet swirls and angles. In order to treat nozzle and blade relative conditions with uniform terminology, some authors label $\alpha_{4.5}$ and β_5 with signs reversed to those shown in Fig. H,3a. This produces no confusion here, since conditions are always assumed to be relative to a rotating row, being the most general case. However, any general qualitative statement on relative blade outlet angles or swirl velocities such as "large," "small," "increasing," etc. in subsequent articles refer to the absolute magnitude of these quantities without regard to sign. For a single row, the gas angles are referred to simply as β for the inlet angle and α for the exit angle. Inlet and outlet values have the suffixes $_i$ and $_o$ respectively.

Expressions for torque and work. The laws of motion give a mathematical relation between the change of gas conditions in passing through the blade row and the work output. Let a stream tube passing through the blade row be defined as in Fig. H,3b. The blading is here assumed to experience tangentially averaged conditions at inlet and outlet for any given radial positions. For an elementary stream tube of the type illustrated, the blading can therefore be replaced by a hypothetical force field or, as a more physical concept, by an infinite set of infinitely thin blades. The equilibrium of pressure, internal forces, and external forces on any

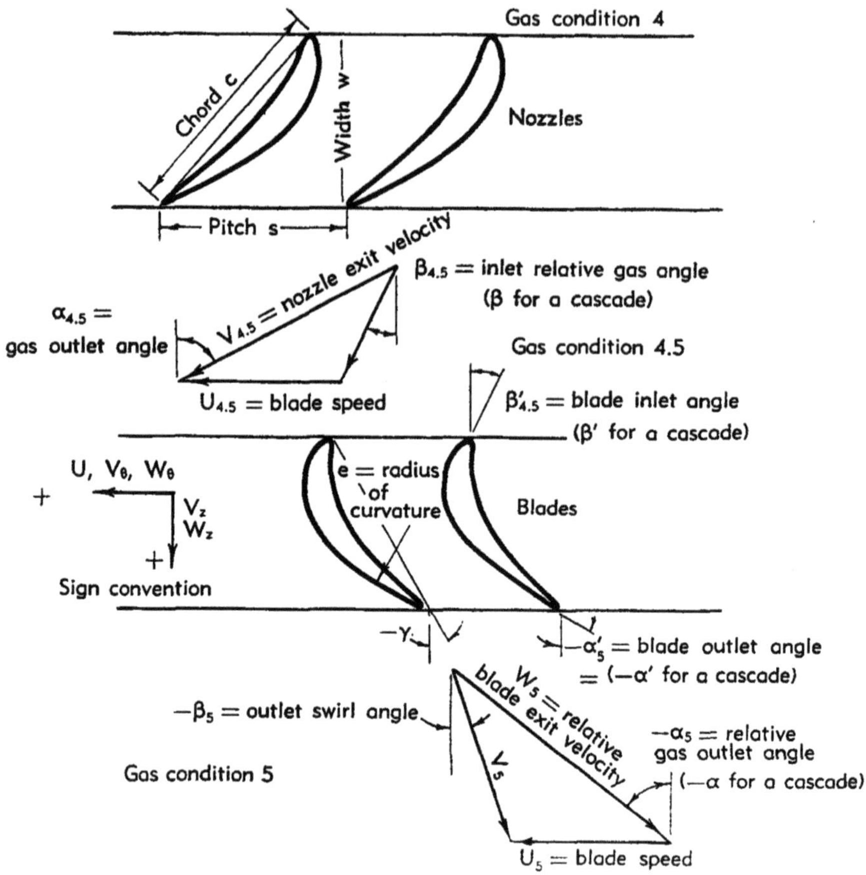

Fig. H,3a. Definition of interstage gas conditions by velocity triangles.

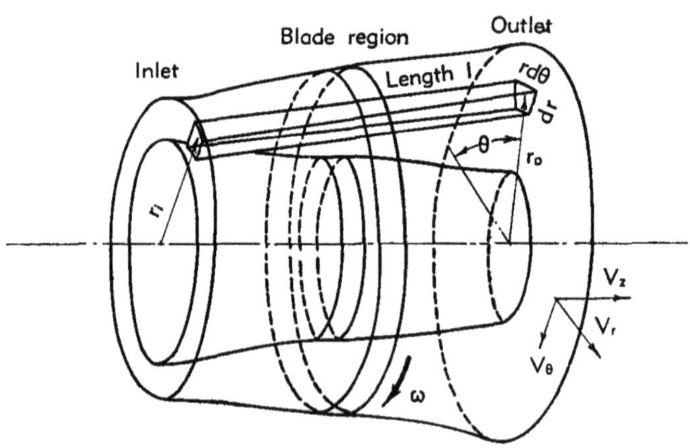

Fig. H,3b. An elementary stream tube.

length of stream tube follows from Newton's second law of motion, leading to the following vector equation [*4*, pp. 73, 74]:

$$-\int \nabla p dv + \int \rho \mathbf{F} dv - \frac{\partial}{\partial t}\int \rho \mathbf{V} dv = \int (\mathbf{n}\cdot\mathbf{V})\rho\mathbf{V} dA \qquad (3\text{-}1)$$

where $\mathbf{n}$ is the inward-directed unit vector normal to the surface, dv and dA are volume and surface elements respectively, $\mathbf{F}$ is the body force per unit mass, and $\mathbf{V}$ is the velocity vector. The symbols p and ρ have their usual meanings of pressure and density. Viscous force terms are assumed negligible and have been omitted from Eq. 3-1. This equation is valid for any complete surface, under steady or unsteady conditions, whether portions of it are stream tube boundaries or not. Now, in the case of turbomachinery, all the shaft movement is in a tangential direction and therefore the only work-producing forces are also in a tangential direction. The shaft torque is given by the sum of the moments of the body forces $\mathbf{F}$ about the axis. The following simplifications can now be made.

1. All unsteady terms are neglected, i.e. partial derivatives with respect to time are zero.
2. Terms involving pressure on the sides (ldr) of the stream tube cancel out due to symmetry and those on the ends ($rd\theta dr$) or top and bottom ($rd\theta l$) have no components about the axis.
3. In the case of a stream tube, the weight flow components $(\mathbf{n}\cdot\mathbf{V})$ normal to the surface will only exist at the two ends and, from continuity, they will be equal and opposite at these ends.

The right-hand side of Eq. 3-1 can then be written

$$\int (\mathbf{n}\cdot\mathbf{V})\rho\mathbf{V} dA = (\rho V_z V_\theta A)_i - (\rho V_z V_\theta A)_o$$
$$= dQ(V_{\theta_i} - V_{\theta_o}) \qquad (3\text{-}2)$$

where dQ is defined by the continuity equation

$$dQ = \rho_i V_{z_i} A_i = \rho_o V_{z_o} A_o$$

Taking moments of the tangential force terms now left in Eq. 3-1,

$$\text{Torque on a stream tube} = \int \rho(F_\theta r) dv = dQ(V_{\theta_o} r_o - V_{\theta_i} r_i) \qquad (3\text{-}3)$$

Now by the definition of work and from Eq. 2-5 and 2-6,

$$\text{Work done on stream tube per unit time} = \omega \times \text{torque}$$

$$\text{Work done by stream tube per pound of gas} = \frac{-\omega \times \text{torque}}{dQ} \qquad (3\text{-}4)$$

$$\text{Work done per pound (Eq. 2-10)} = (h_i^0 - h^0)_o gJ \qquad (3\text{-}5)$$

Using Eq. 3-3, 3-4, and 3-5,

$$\begin{aligned} gJ(h_i^0 - h_o^0) &= -\omega(V_{\theta_o}r_o - V_{\theta_i}r_i) \\ &= \frac{U_i}{r_i}V_{\theta_i}r_i - \frac{U_o}{r_o}V_{\theta_o}r_o \\ &= U_iV_{\theta_i} - U_oV_{\theta_o} \end{aligned} \tag{3-6}$$

Eq. 3-6 is thus the expression for the work per pound of gas, or total enthalpy drop along any stream tube on passing through a blade row. If V_{θ_o} is negative, as shown in Fig. H,3a, the last two terms will be positive numbers and will both contribute to the enthalpy drop across the rotor. As a special case, when no blading exists between o and i, Eq. 3-3 leads to the law of conservation of angular momentum:

If $F_\theta = 0$ then

$$r_iV_{\theta_i} = r_oV_{\theta_o} \tag{3-7}$$

Eq. 3-6 can be written in a variety of different forms by using the geometrical relationships between V_z, V_θ, W_z, W_θ, and V described by Fig. H,3a. An important expression for the change in relative stagnation enthalpy h^{00} across the rotor can be developed as follows. The change in relative stagnation enthalpy can be written (Eq. 2-11) as

$$h_i^{00} - h_o^{00} = \left(h_i + \frac{W_i^2}{2gJ}\right) - \left(h_o + \frac{W_o^2}{2gJ}\right) \tag{3-8}$$

The squared velocity terms can be expanded into their components (Fig. H,3a)

$$\begin{aligned} h_i^{00} - h_o^{00} &= \left(h_i + \frac{V_{r_i}^2 + V_{z_i}^2 + W_{\theta_i}^2}{2gJ}\right) - \left(h_o + \frac{V_{r_o}^2 + V_{z_o}^2 + W_{\theta_o}^2}{2gJ}\right) \\ &= \left[h_i + \frac{V_{r_i}^2 + V_{z_i}^2 + (V_{\theta_i} + U_i)^2}{2gJ}\right] \\ &\qquad - \left[h_o + \frac{V_{r_o}^2 + V_{z_o}^2 + (V_{\theta_o} - U_o)^2}{2gJ}\right] \\ &= \left(h_i^0 - \frac{V_{\theta_i}U_i}{gJ} + \frac{U_i^2}{2gJ}\right) - \left(h_o^0 - \frac{V_{\theta_o}U_o}{gJ} + \frac{U_o^2}{2gJ}\right) \end{aligned} \tag{3-9}$$

Substituting Eq. 3-6 in Eq. 3-9,

$$h_i^{00} - h_o^{00} = \frac{U_i^2 - U_o^2}{2gJ} \tag{3-10}$$

Since U is defined as ωr it is clear from Eq. 3-10 that $h_i^{00} - h_o^{00}$ will be zero if the streamline leaves the rotor at the same radius at which it enters it.

Work coefficient. The work per pound of gas delivered to the rotating blading has been shown (Eq. 3-6), for each streamline, to be represented by the difference in the product of $r\omega$ and V_θ before and after the blading. Work achieved by large changes in swirl and low blade speed penalizes aerodynamic performance. On the other hand, the same work achieved by high blade speed and small changes in swirl causes increased rotational stresses. A useful criterion of aerodynamic severity of design could therefore be defined as:

$$\text{Work coefficient} = \phi$$

$$= \frac{\text{Work per pound of gas}}{\frac{1}{2}(\text{blade speed})^2} = \frac{(h_4^0 - h_5^0)gJ}{\frac{1}{2}U^2} \qquad (3\text{-}11)$$

The parameter ϕ, if expressed in consistent units, is dimensionless. It represents the division of responsibility, as it were, between blade speed and change of tangential velocity in forming their product which defines the work per pound of gas. In comparing machines it is necessary to specify the radial station at which the quantities in Eq. 3-11 are measured. The practical expression for the work coefficient in the case of a multistage machine with N stages would be

$$\phi_{\mathrm{m}} = \frac{h_{4_{\mathrm{m}}}^0 - h_{5_{\mathrm{m}}}^0}{U_{\mathrm{m}}^2}\frac{2gJ}{N} \qquad (3\text{-}12)$$

where the suffix m indicates that values are measured at some mean value of radius.

The difference between the design of stages with low and high values of work coefficient ϕ is illustrated in Fig. H,3c. The velocity triangles at the nozzle and rotor outlets are here superposed to indicate the relation between the change in swirl across the rotor, $V_{\theta_{4.5}} - V_{\theta_5}$, and the gas angles. Blade velocities $U_{4.5}$ and U_5 will be equal ($U = r\omega$) if no radial streamline movement takes place. The necessary gas deflections can be seen to increase as ϕ is increased, a factor which tends to make such designs less efficient.

Fig. H,3d shows the trend of turbine efficiency at its design operating condition as a function of the work coefficient. A zone is shown rather than a single line since this parameter is not the only criterion of design severity, being also influenced by blade shapes and operating Mach numbers (Art. 5, 6, and 7). Nevertheless, the work coefficient is usually the largest single factor affecting design severity. Specific designs have an efficiency at off-design operating points, shown by the chain-dotted line in Fig. H,3d, which is lower than the locus of design points. This decrease is caused chiefly by excessive gas incidences under these conditions.

Expressions equivalent to the work coefficient ϕ as defined in Eq.

3-11 and 3-12 are often encountered. One such expression is the isentropic velocity ratio ν_{is}, defined as

$$\nu_{is} = \frac{\text{blade speed}}{\text{velocity equivalent of isentropic work}} = \frac{U}{\sqrt{2gJ(h_4^0 - h_{5_{is}}^0)\frac{1}{N}}} \tag{3-13}$$

There is an obvious algebraic relation between the work coefficient ϕ (Eq. 3-11) and the isentropic velocity ratio ν_{is} (Eq. 3-13). The expected

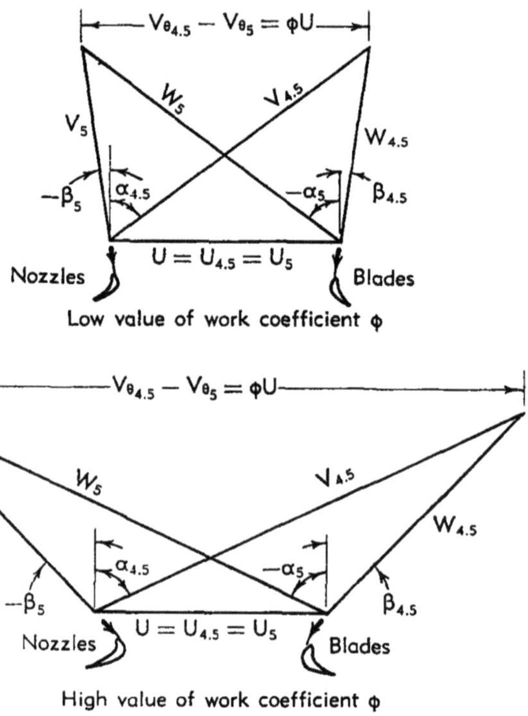

Fig. H,3c. Effect of work coefficient ϕ on gas velocities and angles.

efficiency in Fig. H,3d has therefore been shown as a function of both ν_{is_m} and ϕ_m.

Degree of reaction. By the term "reaction" the turbine designer refers to the amount of acceleration that occurs across the moving blade row. The degree of reaction can be defined formally as

$$D = \frac{\text{enthalpy drop across rotating blades}}{\text{sum of enthalpy drops across nozzles and blades}} \tag{3-14}$$

The reaction is zero by this definition if the velocity relative to the moving blades at inlet and outlet is constant and no change in relative stagnation enthalpy h^{00} occurs (Eq. 3-8). The latter condition is fulfilled if no radial streamline shifts occur (Eq. 3-9). In this case the outlet pressure is slightly less than the inlet value due to pressure losses. Alternatively, if the zero reaction condition is defined by zero pressure change across the

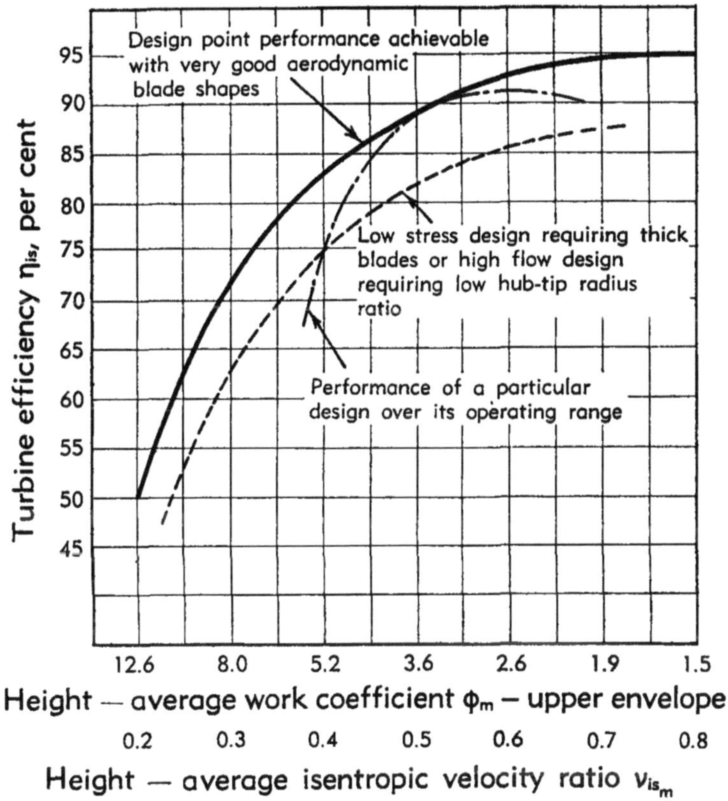

Fig. H,3d. Attainable efficiency as a function of work coefficient ϕ or isentropic velocity ratio ν_{is}.

rotor, there is a slight decrease in relative velocity, pressure losses being just offset by gas deceleration. Definitions of reaction other than by Eq. 6-1 are sometimes employed, since many are simpler for measurement or computation. One such definition is

$$D = \frac{W^2_{\theta_s} - W^2_{\theta_{4.s}}}{V^2_{\theta_{4.s}} - V^2_{\theta_s} + W^2_{\theta_s} - W^2_{\theta_{4.s}}} \tag{3-15}$$

This definition is the same as that given by Eq. 3-14 if the axial velocities are constant and the gas inlet tangential velocity to the nozzles is equal to the stage outlet tangential velocity V_{θ_s}.

Velocity triangles for values of reaction D, equal to zero (known as "impulse" condition) and unity, are shown in Fig. H,3e. As in Fig. H,3c the triangles corresponding to the axial positions 4.5 and 5 have been superposed and values of U at these two stations have been assumed equal. This is usually nearly the case. The 100 per cent reaction case applies strictly only when the stage is part of a multistage unit and the stage outlet velocity V_5 is equal to the stage inlet velocity V_4 in magnitude and direction. In the case of a single stage, where V_4 is axial as shown, the reaction is somewhat smaller, though trends are similar.

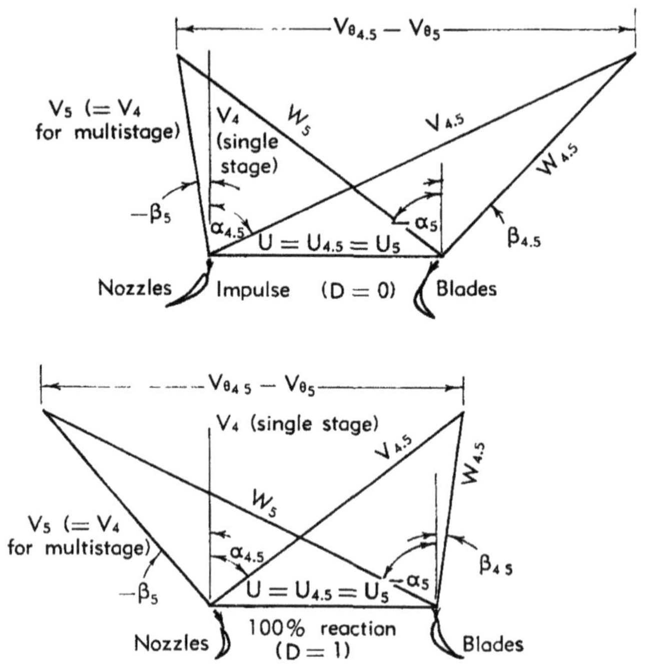

Fig. H,3e. Velocity triangles for impulse and reaction stages having the same blade speed and work.

Although it might be thought at first sight that 50 per cent reaction or "symmetric" stages are the best design choice, in view of the fact that nozzles and blade angles are then identical, other considerations also apply. Impulse stages, for example, give less leaving swirl from the stage and therefore eliminate the need for further swirl vanes when they form the final stage. The temperature reached by impulse blades is also lower due to the decreased relative stagnation temperature in accordance with the following relation (Eq. 2-6, 2-7, and 2-11),

$$T^{00}_{4.5} = T^{0}_{4.5} + \frac{W^2_{4.5}}{2gJc_p} - \frac{V^2_{4.5}}{2gJc_p} \tag{3-16}$$

This is mechanically advantageous.

If the design is of the free vortex type (Art. 10) the degree of reaction varies with the radius, being low at the hub and high at the tip. Constant reaction designs necessitate nonuniform axial velocity and radial streamline shifts which detract from the accuracy of design predictions.

Characteristic diagram. The effect of blade-outlet-angle choice on turbine work capacity may be examined by deriving a relation between ϕ, V_z/U, and the gas outlet angles. This relationship is much simplified and yet still provides the necessary insight if it is assumed that no radial streamline movement takes place. This means that U ($= \omega r$) is constant for any particular streamline. Eq. 3-6 thus becomes, using the nomenclature of Fig. H,3a,

$$gJ(h_4^0 - h_5^0) = U(V_{\theta_{4.5}} - V_{\theta_5}) \tag{3-17}$$

The peripheral velocities V_θ can now be expressed in terms of the axial velocity V_z and the gas angles,

$$V_{\theta_{4.5}} = V_{z_{4.5}} \tan \alpha_{4.5} \tag{3-18}$$

$$\begin{aligned} V_{\theta_5} &= W_{\theta_5} + U \\ &= V_{z_5} \tan \alpha_5 + U \end{aligned} \tag{3-19}$$

Substituting Eq. 3-18 and 3-19 into Eq. 3-17 and dividing by U^2,

$$\frac{\phi}{2} = gJ\,\frac{h_4^0 - h_5^0}{U^2} = \frac{V_{z_{4.5}}}{U}(\tan \alpha_{4.5} - \mu \tan \alpha_5) - 1 \tag{3-20}$$

where the axial velocity ratio μ is defined by

$$\mu = \frac{V_{z_5}}{V_{z_{4.5}}} \tag{3-21}$$

and is usually in the neighborhood of unity. Eq. 3-20 indicates a linear relationship between the work $h_4^0 - h_5^0$ and the axial velocity V_z if U, $\alpha_{4.5}$, and α_5 are constant.

A relationship similar to Eq. 3-20 can be derived, using the inlet angles $\beta_{4.5}$ and β_5 instead of outlet angles $\alpha_{4.5}$ and α_5, as follows,

$$gJ\left(\frac{h_4^0 - h_5^0}{U^2}\right) = \frac{\phi}{2} = \frac{V_{z_{4.5}}}{U}(\tan \beta_{4.5} - \mu \tan \beta_5) + 1 \tag{3-22}$$

Eq. 3-20 and 3-22 are represented graphically in Fig. H,3f. This figure must be considered as the locus of design points of various turbines rather than the performance of given turbines over their operating range, since at other than design points compressibility effects alter the axial velocity ratio μ. Also the gas outlet angles $\alpha_{4.5}$ and α_5 do not remain constant over a large range of operating conditions (see Art. 5).

Eq. 13-20 and 13-22 and Fig. H,3f put into numerical form the relationships indicated qualitatively by Fig. H,3c and H,3e. The latter figure

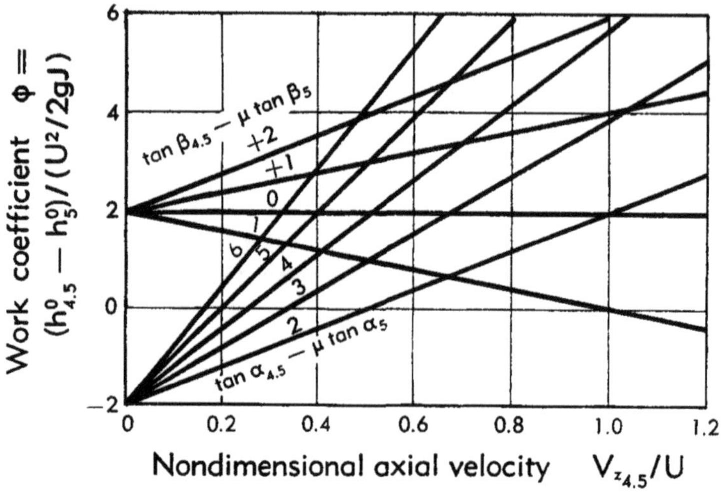

Fig. H,3f. Design work coefficient ϕ as a function of velocity coefficient $V_{z_{4.5}}/U$ and gas angles.

illustrates that angles $\beta_{4.5}$ and $\alpha_{4.5}$ can be decreased only at the expense of increases in α_5 and β_5 for the same level of work, and the former figure illustrates the increase in values of all the gas angles with increase in the work coefficient ϕ.

H,4. Relation between Blade Losses and Turbine Efficiency.

Blade losses. Fluid friction gives rise to a drag on the blade surfaces which, from the laws of motion, must be balanced by a momentum defect (see, for instance, Eq. 3-1). This concept is brought out most clearly in conventional pipe friction analysis. Blade frictional loss is usually defined either in terms of stagnation pressure loss or in terms of the amount of kinetic energy present at the exit from the row, compared with the maximum or ideal value at the prevailing outlet pressure. In view of the nonuniformity of the outlet flow (existence of wakes), some circumferentially averaged value must be taken. In the case of a moving blade row, the stagnation pressure reduction due to absorption of energy by the blades must obviously be excluded as a loss. It has been shown that, in axial flow machines, no energy is added relative to the moving blades (Eq. 3-10). Consequently, if all quantities are taken relative to the moving blades, both stator and rotor losses can be analyzed from stationary cascade data.

Let the gas states and corresponding relative stagnation values at inlet to, and outlet from, a blade row be represented on the enthalpy-entropy diagram of Fig. H,4a as conditions i and o. After expansion to the outlet pressure isentropically, the outlet relative velocity energy would be represented by $h^{00}_{o} - h_{o_{is}}$ and the loss in energy is $h_o - h_{o_{is}}$.

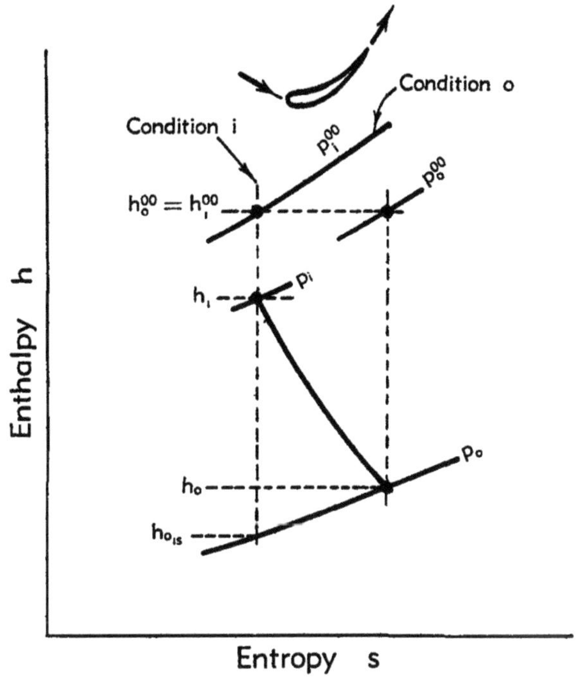

Fig. H,4a. Inlet conditions to, and outlet conditions from, a moving blade row.

The loss coefficient can therefore be defined as

$$\lambda = \frac{h_o - h_{o_{is}}}{h_o^{00} - h_o} \tag{4-1}$$

The alternative method of defining loss is in terms of the reduction in stagnation pressure divided by outlet dynamic head, as follows:

$$Y = \frac{p_i^{00} - p_o^{00}}{p_o^{00} - p_o} \tag{4-2}$$

The symbol Y without a suffix will hereafter signify an average taken over both blade height and blade width. The term Y_r denotes a pitchwise average at a specific value of r or blade height and $Y_{r\theta}$ denotes a local value for specific values of radius r and angle position θ. The terms λ and Y can be shown to be approximately equal numerically for, by the following transformation, Eq. 4-2 gives

$$Y = \frac{(p_i^{00}/p_o) - (p_o^{00}/p_o)}{(p_o^{00}/p_o) - 1}$$

Using the isentropic expansion law,

$$Y = \frac{(T_i^{00}/T_{o_{is}})^{\frac{\gamma}{\gamma-1}} - (T_i^{00}/T_o)^{\frac{\gamma}{\gamma-1}}}{(T_i^{00}/T_o)^{\frac{\gamma}{\gamma-1}} - 1}$$

$$= \frac{\left(1 + \frac{T_i^{00} - T_{o_{is}}}{T_{o_{is}}}\right)^{\frac{\gamma}{\gamma-1}} - \left(1 + \frac{T_i^{00} - T_o}{T_o}\right)^{\frac{\gamma}{\gamma-1}}}{\left(1 + \frac{T_i^{00} - T_o}{T_o}\right)^{\frac{\gamma}{\gamma-1}} - 1} \tag{4-3}$$

First terms of binominal expansion can be used for the above powers since temperature differences are generally small compared with temperature levels. Eq. 4-3 thus becomes

$$Y = \frac{\left(1 + \frac{\gamma}{\gamma - 1}\frac{T_i^{00} - T_{o_{is}}}{T_{o_{is}}}\right) - \left(1 + \frac{\gamma}{\gamma - 1}\frac{T_i^{00} - T_o}{T_o}\right)}{1 + \frac{\gamma}{\gamma - 1}\frac{T_i^{00} - T_o}{T_o} - 1}$$

$$= \frac{T_o - T_{o_{is}}}{T_i^{00} - T_o}\frac{T_i^{00}}{T_{o_{is}}} \tag{4-4}$$

Now, within the previous assumption that the ratio of stagnation temperature to actual temperature is nearly unity, specific heat c_p may be taken as constant and thus Eq. 4-4 becomes (Eq. 4-1)

$$Y = \frac{h_o - h_{o_{is}}}{h_i^{00} - h_o}\frac{T_i^{00}}{T_{o_{is}}}$$

$$= \lambda\left(\frac{p_i^{00}}{p_o}\right)^{\frac{\gamma-1}{\gamma}} \tag{4-5}$$

In most turbojet applications the pressure ratio p_i^{00}/p_o is less than 2 and the term $(p_i^{00}/p_o)^{\frac{\gamma-1}{\gamma}}$ is therefore less than 1.2 ($\gamma = \frac{4}{3}$). The loss coefficients Y and λ are thus interchangeable for most engineering purposes.

Turbine efficiency. The relationship between the over-all efficiency of a stage and the blade loss coefficients may now be derived. Gas state conditions at the three axial stations 4, 4.5, and 5 are defined in Fig. H,4b. In accordance with Eq. 2-12 the isentropic efficiency can be expressed as $(h_4^0 - h_5^0)/[h_4^0 - (h_{5_{is}}^0)_4]$. The suffixes $_{5_{is}4}$, etc. indicate here, and subsequently, the gas state corresponding to an isobar from condition 5 and a constant entropy line from condition 4, etc. (See Fig. H,4b.) The re-

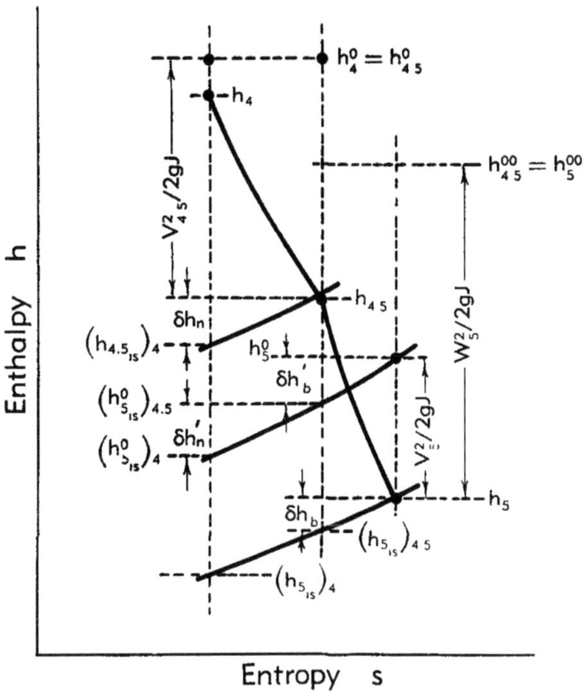

Fig. H,4b. Changes of gas state in a turbine stage.

lationship between the isentropic work and actual work is now as follows,

$$h_4^0 - (h_{5_{is}}^0)_4 = h_4^0 - h_5^0 + \delta h_n' + \delta h_b' \tag{4-6}$$

The enthalpy increments $\delta h_n'$, $\delta h_b'$ are related to the nozzle and blade loss increments δh_n, δh_b as follows,

$$\frac{\delta h_b'}{\delta h_b} = \frac{T_5^0 - (T_{5_{is}}^0)_{4.5}}{T_5 - (T_{5_{is}})_{4.5}} = \frac{T_5^0}{T_5}\,\frac{1 - \dfrac{(T_{5_{is}}^0)_{4.5}}{T_5^0}}{1 - \dfrac{(T_{5_{is}})_{4.5}}{T_5}} \tag{4-7}$$

Since all isentropic temperature ratios across the isobars p_5^0 and p_5 are equal we can write

$$\frac{T_5^0}{T_5} = \frac{(T_{5_{is}}^0)_{4.5}}{(T_{5_{is}})_{4.5}} \quad \text{or} \quad \frac{(T_{5_{is}}^0)_{4.5}}{T_5^0} = \frac{(T_{5_{is}})_{4.5}}{T_5}$$

Eq. 4-7 then reduces to

$$\frac{\delta h_b'}{\delta h_b} = \frac{T_5^0}{T_5} \tag{4-8}$$

Similar analysis leads to

$$\frac{\delta h'_n}{\delta h_n} = \frac{(T^0_{5_{is}})_{4.5}}{T_{4.5}} \tag{4-9}$$

Combining Eq. 4-6, 4-8, and 4-9 into the expression for isentropic efficiency (Eq. 2-12),

$$\eta_{is} = \cfrac{1}{1 + \cfrac{\cfrac{(T^0_{5_{is}})_{4.5}}{T_{4.5}}\,\delta h_n + \cfrac{T^0_5}{T_5}\,\delta h_b}{h^0_4 - h^0_5}} \tag{4-10}$$

Now, using the definitions of λ_n, λ_b (Eq. 4-1) and ϕ (Eq. 3-11),

$$\eta_{is} = \cfrac{1}{1 + \cfrac{\left(\cfrac{V_{4.5}}{U}\right)^2 \cfrac{(T^0_{5_{is}})_{4.5}}{T_{4.5}}\,\lambda_n + \left(\cfrac{W_5}{U}\right)^2 \cfrac{T^0_5}{T_5}\,\lambda_b}{\phi}} \tag{4-11}$$

$$= \cfrac{1}{1 + \cfrac{\left(\cfrac{V_{z_{4.5}}}{U}\right)^2 \left[\sec^2 \alpha_{4.5} \cfrac{(T^0_{5_{is}})_{4.5}}{T_{4.5}}\,\lambda_n + \mu^2 \sec^2 \alpha_5 \cfrac{T^0_5}{T_5}\,\lambda_b\right]}{\phi}} \tag{4-12}$$

The above expression for efficiency has been derived along similar lines by Ainley [*5*]. By combining Eq. 4-12 and 3-20 this expression can also be written purely in terms of gas angles and velocity coefficient V_z/U,

$$\eta_{is} = \cfrac{1}{1 + \cfrac{1}{2}\cfrac{V_{z_{4.5}}}{U}\,\cfrac{\cfrac{(T^0_{5_{is}})_{4.5}}{T_{4.5}}(\sec^2 \alpha_{4.5})\lambda_n + \mu^2 \cfrac{T^0_5}{T_5}(\sec^2 \alpha_5)\lambda_b}{\tan \alpha_{4.5} - \mu \tan \alpha_5 - \cfrac{U}{V_{z_{4.5}}}}} \tag{4-13}$$

The contributions of loss coefficients λ_n and λ_b, gas angles, and velocity ratio μ to turbine inefficiency can be examined with the aid of Eq. 4-13. The temperature ratio terms are always in the neighborhood of unity but tend to weight the blade loss at high outlet Mach numbers and the nozzle loss when $T_{4.5}$ is low, i.e. for low reaction stages (Eq. 3-14). A low velocity coefficient $V_{z_{4.5}}/U$ is seen to be desirable for a given work coefficient ϕ, but this causes increases in the angles $\alpha_{4.5}$ and α_5, thus tending to restore the balance. The values of λ_n and λ_b would also increase, but they also depend greatly on blade shape, pitching, and height (Art. 5, 6, 7, and 8) and therefore questions concerning the choice of triangles for optimum efficiency will be deferred until later (Art. 13).

CHAPTER 2. EXPERIMENTAL DATA ON PERFORMANCE

H,5. Effect of Profile Shape on Blade Performance.

Evaluation procedure. Blade profile loss may be considered as that portion of the over-all loss which would be measured by tests on the profile in the absence of any disturbance effects due to the presence of walls at the blade extremities. The corresponding loss coefficient is designated as Y_p. Stationary cascade tunnels are the most commonly used tools for investigation of these losses (Sec. E) although certain difficulties may arise in their interpretation. As indicated in Art. 4, the gas passing through a stationary row of blades experiences some average loss in stagnation pressure caused by the blade wakes. If the blades are moving, a relative stagnation pressure can still be defined in terms of the pressure and relative velocity (Art. 2), but radial streamline movements cause increases in this relative stagnation pressure (Eq. 3-10). The pressure loss may then be defined as the difference between the ideal frictionless relative stagnation pressure at the exit and the actual value. It is clear that large radial streamline shifts, such as take place in a centrifugal impeller, cause the relative conditions to a rotating surface to be completely different from conditions if the rotor is stationary [*6*]. Nevertheless, in the case of axial flow machinery it is usual to apply data from stationary blade tests to analyze rotating blade performance, the justification being that energy addition relative to the rotating row is small.

The gas outlet angle from the blading is also an important variable. Fig. H,3f and Eq. 3-20 show that the work capacity is closely related to these angles, and in order to obtain gas velocities and directions as predicted for a given amount of work their accurate prediction is desirable. In the case of turbines, the gas outlet angle is usually correlated with the throat opening o, indicated in Fig. H,5a, rather than the true blade outlet angle α' as defined by the tangent to the camber line at the outlet.

The following approximate relationship between blade geometry and gas outlet angle is often used as a guidance rule.

$$|\alpha_o| \cong \left|\cos^{-1}\frac{o}{s}\right| \tag{5-1}$$

This relationship can be physically interpreted as an adjustment of stream tube area at the exit to the same value as the blade throat area. The actual gas outlet angle is conveniently recorded as a deviation angle δ from this standard condition.

$$\delta = \left|\cos^{-1}\frac{o}{s}\right| - |\alpha| \tag{5-2}$$

A full set of cascade data for given blade geometry should depict the loss coefficient Y and the gas outlet angle α as a function of incidence i, Mach number, and Reynolds number. However, unless variable density cascade tunnels are employed, the two latter quantities vary together over a test. Fortunately, in many tests the regions of high Mach numbers cover Reynolds number ranges to which the blades are not sensitive.

Typical data on impulse blading ($\beta' = -\alpha'$) due to Todd [*7*] are indicated in Fig. H,5b. Data on reaction blading or nozzles ($\beta' = o$) due to Andrews and Schofield [*8*] are indicated in Fig. H,5c. These figures show

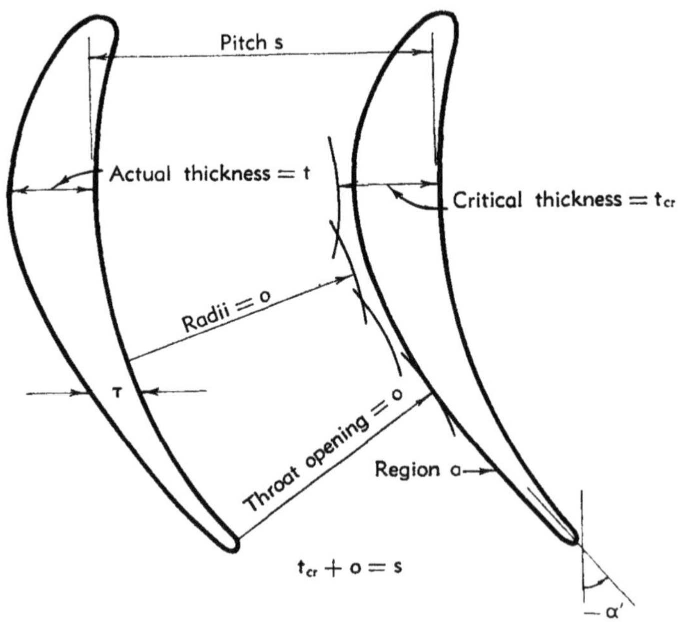

Fig. H,5a. Relation between throat opening, thickness, and pitch.

the order of magnitude of operating incidence range, sensitivity to Mach number, and values of deviation. Specific values of these quantities will depend greatly on blade form and pitching. Losses at high Mach numbers are particularly sensitive to blade trailing edge form (Art. 6). Reaction blades in general have a larger operating range of incidence than impulse blades [*9*], but this generalization may be obscured in actual instances by differences in operating Mach number, thickness, pitch, etc.

Relationship between profile losses and pressure distribution. Although profile losses can be related to physical blade properties, such as blade pitch, thickness, and throat opening, and to gas incidence, they can also be related more fundamentally to the form of the velocity and pressure distribution around the blade. This distribution controls the nature of the boundary layers which form the blade wake. Such distribution can be obtained analytically [*10,11*], experimentally [*12*], or by analogue techniques [*11,13*]. Typical distributions are shown in Fig. H,5d, showing the

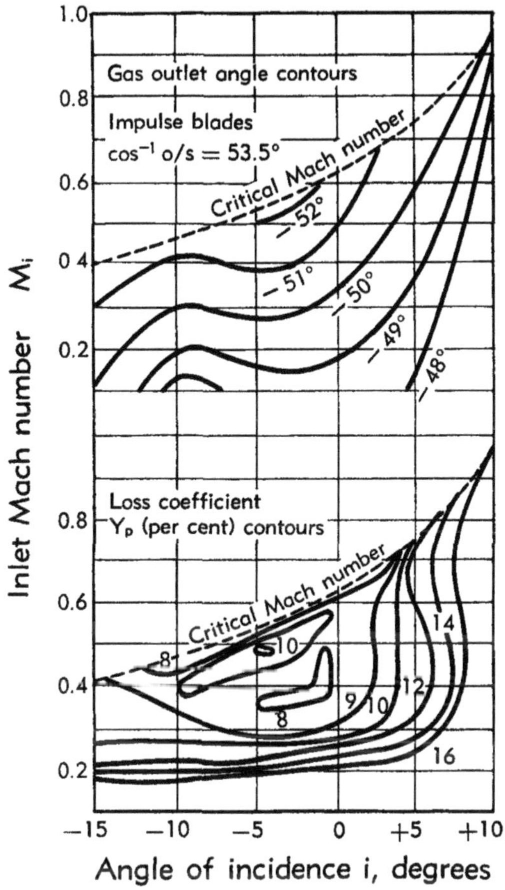

Fig. H,5b. Effects of relative inlet Mach number and angle of incidence $i = \beta - \beta'$ [7].

effect of changes in blade patch. The surface pressure is plotted here as the nondimensional coefficient,

$$C_p = \frac{p - p_o}{p_i^{00} - p_o} \tag{5-3}$$

The pressure coefficient, C_p, is a function of the blade width at which the pressure reading p is taken. The cascade data of Dunavant and Erwin [*12*] may be cited as an example of the correlation between blade losses and pressure distribution. The higher losses are generally associated with highly negative values of C_p on the suction surface, which lead to adverse pressure gradients and boundary layer thickening over the trailing edge.

Boundary layer calculations, based on previously determined pressure distributions for blade cascades, have been carried out by Schlichting [*14*] and compared with experimentally observed losses. These are shown in Fig. H,5e, indicating that losses can be predicted with reasonable accuracy by these means. Other boundary layer calculation techniques have been developed by Tetervin [*15*], Dowlen [*16*], MacGregor [*17*], and others.

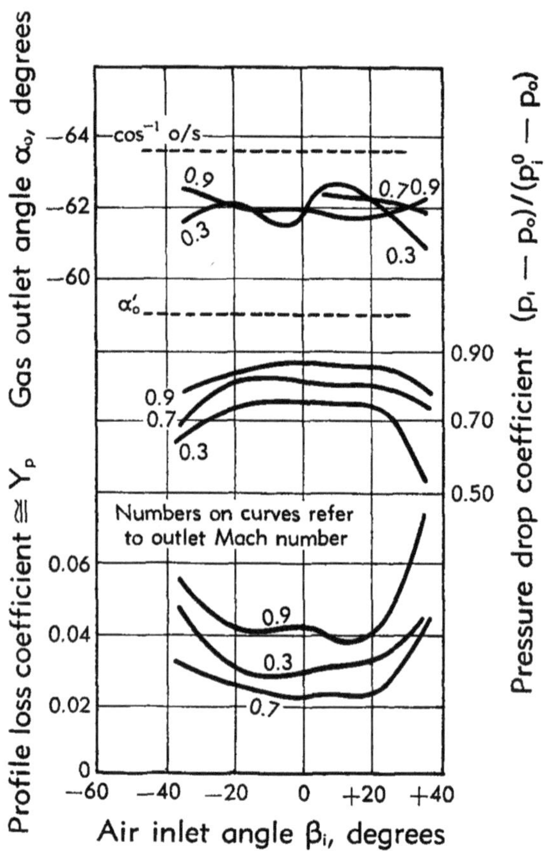

Fig. H,5c. Cascade data on turbine nozzles [*8*].

Some experimental data giving the effect of blade geometry on profile loss and deviation as a function of Reynolds and Mach numbers are given below. The large variety of blading that is possible renders this type of presentation somewhat approximate, but trends can certainly be discerned, leading to better initial decisions and less development effort.

Effect of blade pitch and thickness. Since losses are related to surface diffusion, it follows that there should be an optimum pitching for each value of thickness-to-chord ratio t/c. Lower pitching than this optimum

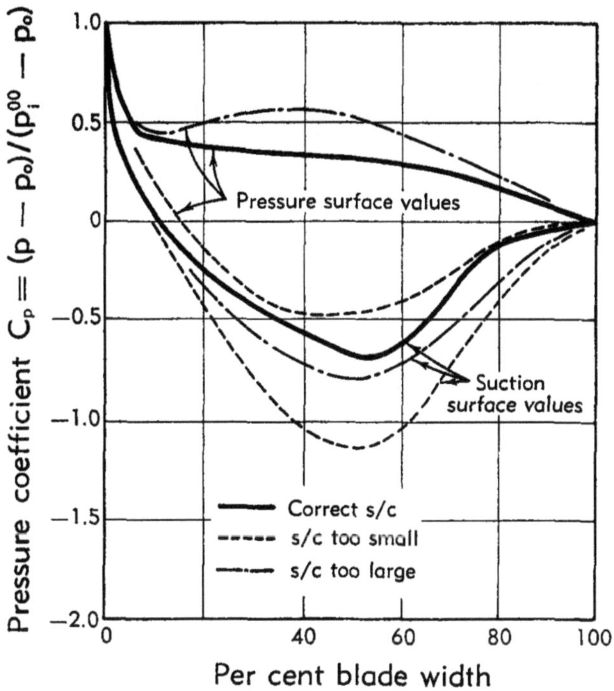

Fig. H,5d. Variation of blade surface pressure distribution with pitch-to-chord ratio s/c.

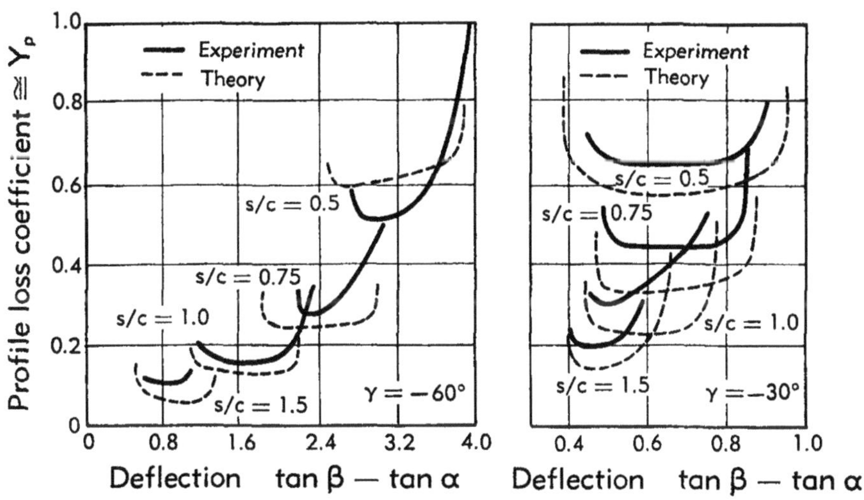

Fig. H,5e. Comparison between predicted and observed profile losses [14].

drives the operating condition to an extreme where $s/c = t/c$ and the blades touch. On the other hand, higher pitching also leads to extremely low suction surface pressure, since each blade must sustain a large pressure difference across it to provide the same total force on the fluid. The above two alternatives have been illustrated in Fig. H,5d. Comprehensive data on these effects, at low Mach number, have been reported by

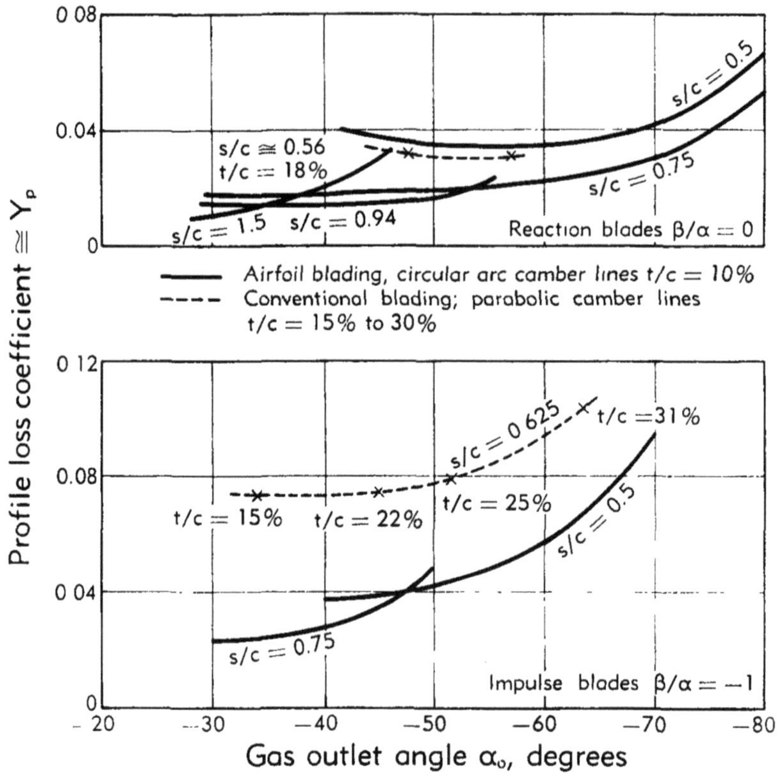

Fig. H,5f. Effect of blade pitch, thickness, and gas angles on loss at low Mach numbers [9].

Reeman [18] and also by Ainley [9]. Ainley's summary is reproduced in Fig. H,5f. These tests covered a range of pitch-to-chord ratio s/c and turning angles $(\beta - \alpha)$, the lines shown being envelopes for optimum incidence. The continuous lines refer to airfoil section blading with a thickness-to-chord ratio t/c of 10 per cent. Broken lines refer to more conventional turbine sections of greater thickness as indicated. The minimum possible loss coefficient increases with gas outlet angle, but much more rapidly in the case of impulse blades than reaction blades. Optimum pitching also decreases with the gas outlet angle. Blade thickness is seen to have a great influence on loss level but very little data is available regarding its effect on optimum pitching. Designers often employ a practical guidance rule that the passage opening should diminish smoothly from inlet to throat as in Fig. H,5a. If the passage opening is uniform the following critical relation applies,

$$\left(\frac{t}{c}\right)_{\text{cr}} = \left(\frac{s}{c}\right)_{\text{cr}}\left(1 - \frac{o}{s}\right) \tag{5-4}$$

This relationship is useful in the case of preliminary stress calculations to arrive at some estimate of the maximum section area over which the centrifugal and bending loads must be distributed (Art. 12).

H,6. Influence of Trailing Edge Form.

Trailing edge thickness effects. Finite trailing edge thickness may be expected to give rise to pressure losses in much the same manner as a sudden enlargement of a pipe through which fluid is flowing. If the velocity, pressure distribution, and gas direction just before the gas leaves the blades are known, the gas condition far downstream where the flow is uniform can be calculated from considerations of continuity, energy, and momentum. The downstream unknown quantities are velocity, direction, density, and pressure. In the two-dimensional case, momentum considerations (Eq. 3-1) provide two relations. Continuity and conservation of energy provide the remaining two conditions. The downstream conditions can be found simply if the flow is assumed incompressible. Let suffix b refer to conditions immediately at the blade outlet and suffix o refer to (uniform) conditions far downstream (Fig. H,6a). Two-dimensional flow is assumed here ($r \to \infty$) and the pitchwise direction is taken as y. The momentum relations (Eq. 3-1) applied normal to the pitch and along it are

$$\int_0^s \rho W_b \cos \alpha_b W_b \cos \alpha_b dy + \int_0^s p_b dy = \rho W_o^2(\cos^2 \alpha_o)s + p_o s \tag{6-1}$$

$$\int_0^s \rho W_b \cos \alpha_b W_b \sin \alpha_b dy = \rho W_o^2(\cos \alpha_o \sin \alpha_o)s \tag{6-2}$$

Continuity of flow gives

$$\int_0^s \rho W_b \cos \alpha_b dy = \rho W_o(\cos \alpha_o)s \tag{6-3}$$

Eq. 6-1, 6-2, and 6-3 can be solved to obtain W_o, α_o, and p_o. The outlet gas angle is obtainable from Eq. 6-2 and 6-3 only and is given by

$$\tan \alpha_o = \frac{\int_0^s W_b^2 \sin \alpha_b \cos \alpha_b d\left(\frac{y}{s}\right)}{\left[\int_0^s W_b \cos \alpha_b d\left(\frac{y}{s}\right)\right]^2} \tag{6-4}$$

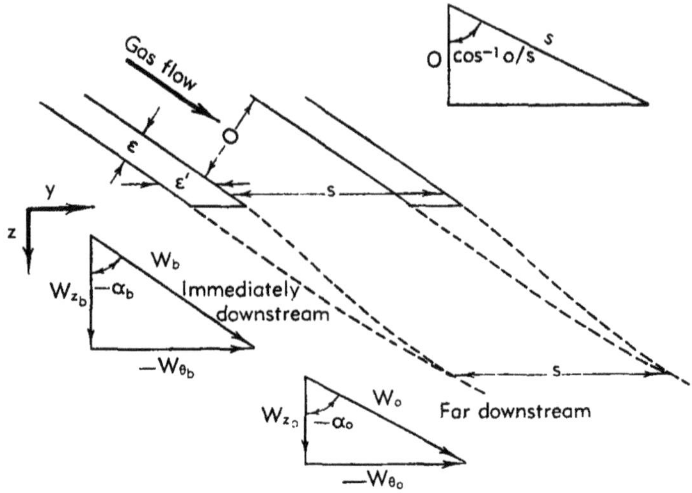

Fig. H,6a. Effect of trailing edge thickness on conditions far downstream.

Some reasonably accurate conclusions regarding the value of α_o can be reached if the assumption is made that W_b and α_b are uniform except in the blade trailing edge region where W is zero over a pitchwise distance of ϵ'. Eq. 6-4 then becomes

$$\tan \alpha_o = \frac{(\tan \alpha_b)}{\left(1 - \frac{\epsilon'}{s}\right)} \tag{6-5}$$

Reference to Fig. H,8a also gives

$$\tan \alpha_b = \frac{\sqrt{(s - \epsilon')^2 - o^2}}{o} \tag{6-6}$$

Combining Eq. 6-5 and 6-6, the outlet angle can be written,

$$\tan \alpha_o = \frac{\left[\left(1 - \frac{\epsilon'}{s}\right)^2 - \left(\frac{o}{s}\right)^2\right]^{\frac{1}{2}}}{\left(1 - \frac{\epsilon'}{s}\right)\frac{o}{s}} \tag{6-7}$$

The deviation angle δ, defined by Eq. 5-2, then becomes

$$\delta = \cos^{-1}\frac{o}{s} - \tan^{-1}\frac{\left[\left(1 - \frac{\epsilon'}{s}\right)^2 - \left(\frac{o}{s}\right)^2\right]^{\frac{1}{2}}}{\left(1 - \frac{\epsilon'}{s}\right)\frac{o}{s}} \tag{6-8}$$

The relationship given by Eq. 6-8 is shown in Fig. H,6b. Deviations are often greater than these values because the surface boundary layers pre-

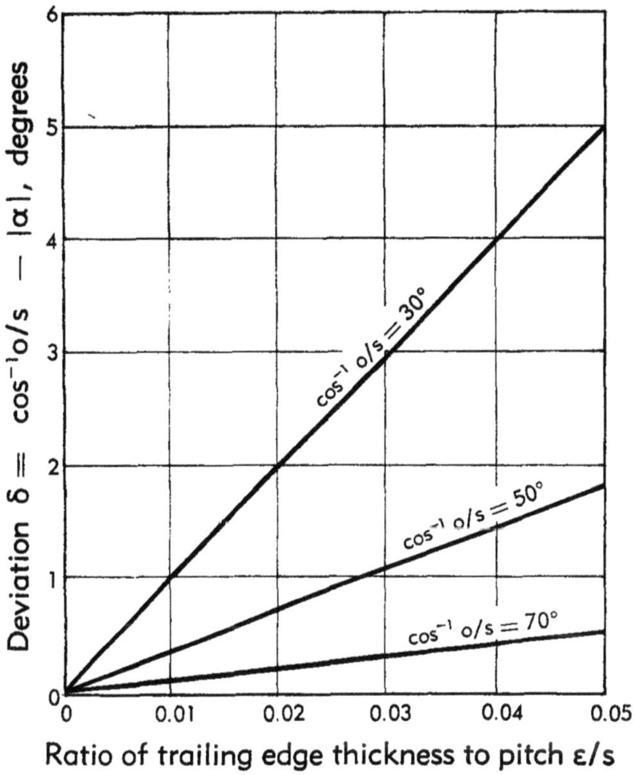

Fig. H,6b. Calculated effect of trailing edge thickness on outlet deviation angle—incompressible flow.

sent a blockage to the flow which increases the effective trailing edge thickness of the blading.

Trailing edge thickness losses can be deduced in a manner similar to the foregoing analysis, leading to the relation

$$Y_\epsilon = \frac{(p_b + \frac{1}{2}\rho W_b^2) - (p_o + \frac{1}{2}\rho W_o^2)}{\frac{1}{2}\rho W_o^2}$$

$$\cong (\cos^2 \alpha_b)\left(\frac{\epsilon'}{s}\right)^2 = \left(\frac{\epsilon}{s}\right)^2 \qquad (6\text{-}9)$$

An example of the use of Eq. 6-9 to determine loss increment due to trailing edge thickness is shown in Fig. H,6c. The experimentally determined value of a known profile loss coefficient Y_p can be related, through Eq. 6-9, to an equivalent value of ϵ/s. An increase in blade thickness can be associated with a new value of profile loss coefficient.

More detailed analyses of trailing edge thickness effects have been carried out by Reeman and Simonis [*19*], Goldstein [*20*], and Stewart [*21*].

Trailing edge curvature. Some effects of suction surface form in the region downstream of the throat on blade performance have been published by Kraft [*22*], who carried out reaction tests on blade packs using steam as the fluid. Similar data have been reported by Stephenson and Naylor using air [*23*]. Curvature after the throat tended to decrease the efficiency substantially at sonic or supersonic outlet velocities, though a

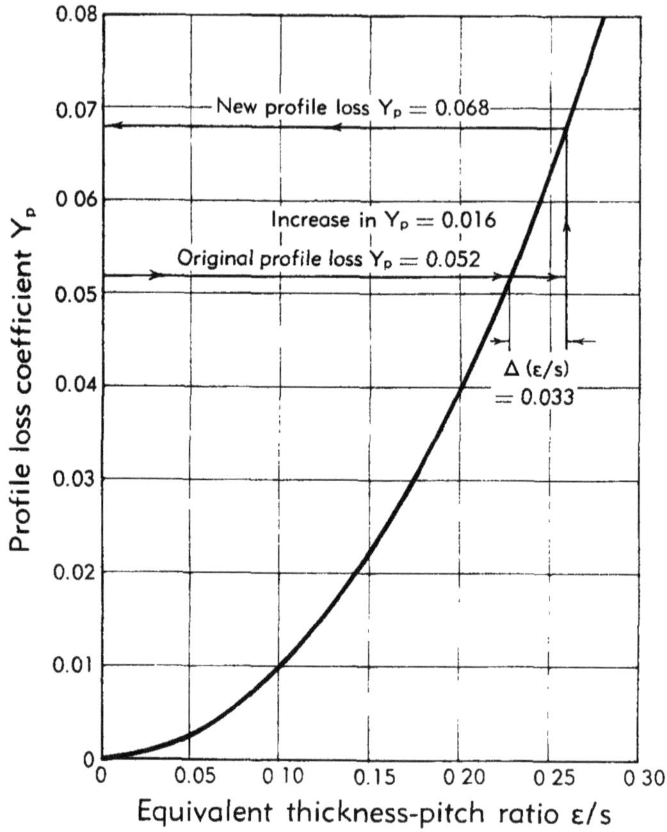

Fig. H,6c. Calculated effect of trailing edge thickness on profile loss—incompressible flow.

slight increase (perhaps 1 per cent in efficiency) was observed at low subsonic outlet velocities. Correlation of a variety of tests has been given by Ainley [*24*], resulting in a loss relationship shown in Fig. H,6d, which supports that of the other authors. The value of suction surface radius of curvature (Fig. H,3a) is the mean from throat to trailing edge in cases where variations take place along the surface.

The disadvantage of suction surface trailing edge curvature at supersonic exit conditions is probably connected with the blade pressure dis-

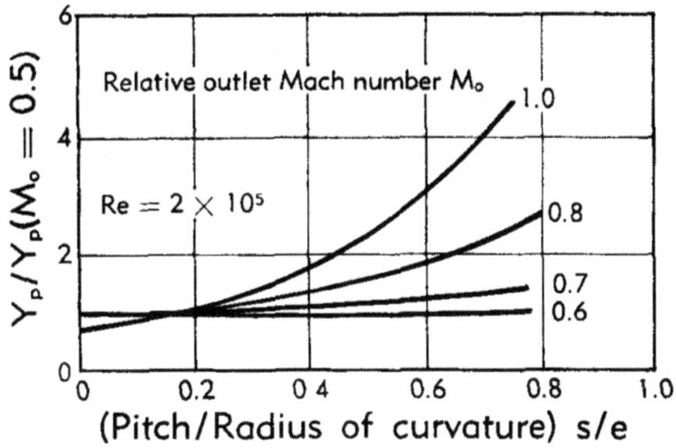

Fig. H,6d. **Effect of trailing edge curvature on profile loss [24].**

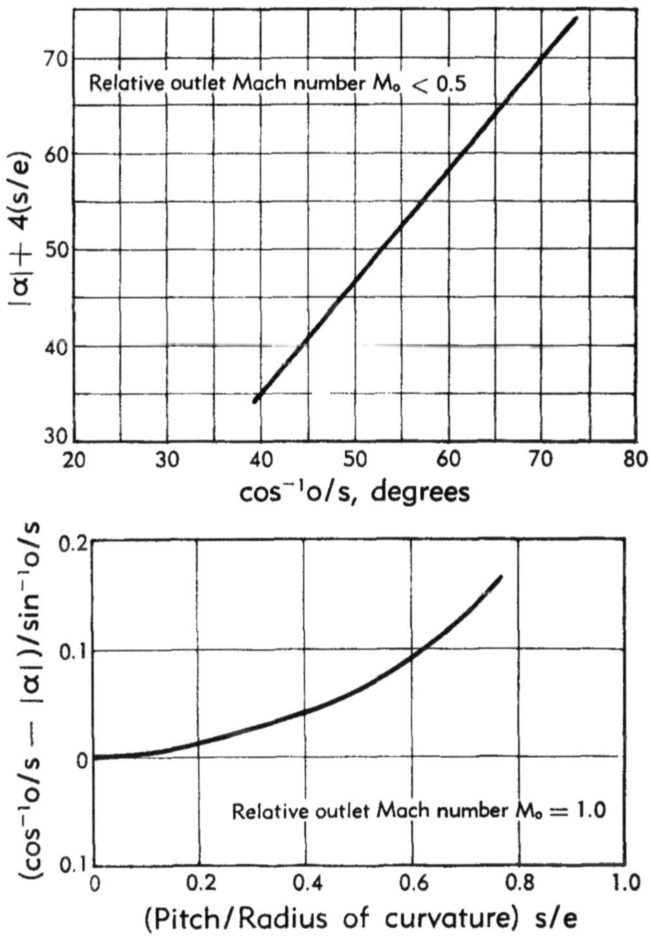

Fig. H,6e. **Effect of trailing edge curvature on deviation [24].**

tribution over the suction surface. Curvature of this surface after the throat tends to shift the point of peak suction (maximum velocity) to a position near the trailing edge. This may decrease losses for subsonic outlet conditions by preserving a greater length of surface with a laminar boundary layer. However, under supercritical conditions, more severe shock-boundary-layer separation phenomena reverse this trend. If the suction surface is to resemble in any manner an ideal convergent-divergent supersonic expansion nozzle, the curvature should in fact be reversed; the radius e (Fig. H,3a) should be negative for exit Mach numbers greater than unity.

In view of the foregoing remarks it might be expected that the influence of the trailing edge curvature on the gas outlet angle is also strongly dependent on the outlet Mach number. Deviation rules, based on available experimental data [*23*], have been suggested by Ainley [*24*] for low subsonic and sonic exit velocities. These are shown in Fig. H,6e. At other relative outlet Mach numbers some judicious interpolation is necessary.

H,7. Supercritical Blade Exit Conditions. If the exit static pressure behind a blade row, stationary or rotating, is progressively reduced, conditions will exist that are somewhat similar to those occurring in a simple convergent-divergent nozzle [*25*, p. 229]. The implications of indefinitely lowering the outlet pressure from a turbine stage can be considered first on a one-dimensional basis. In a multistage machine, where velocity triangles are repeated for each stage and relative outlet velocities are sonic or slightly supersonic, the immediate result of a decrease in outlet static or total pressure is to increase W_5 (Fig. H,3a) for the last rotating row. If this row is choked, i.e. the Mach number relative to the blades at their throats is unity, then the conditions upstream will be unaffected, however much the outlet pressure is reduced. Even if the row is not choked, the Mach number corresponding to W_5 will probably be nearly unity whereas that corresponding to $W_{4.5}$ will be well in the subsonic range. A very large increase in W_5 is then obtained with a small increase in $W_{4.5}$, and the referred flow relative to the blades, $Q\sqrt{\theta_5^{00}}/\delta_5^{00}$, will remain at a maximum, nearly constant figure. This results from standard one-dimensional compressible flow relations in a duct [*3*, pp. 139, 140]. The effect is illustrated in Fig. H,7a, where broken lines refer to the conditions for reduced outlet pressure. The slight change in gas outlet angle will be explained later. The presence of compressibility thus tends to isolate the earlier stages from the final stages (refer also to Art. 14). The highest proportion of change in work, corresponding to change in outlet pressure, is taken up by increased velocities from the final rows. Similar reasoning indicates that the referred flow through the turbine $Q\sqrt{\theta_4^0}/\delta_4^0$ is set predominantly by the first nozzle row if the Mach number at outlet from this row is in the neighborhood of unity. If changes in value of opening *o* (Fig. H,5a) take place in the later rows, effects on the earlier rows rapidly diminish. All these effects can be investigated quantitatively by a fairly simple row-by-row analysis procedure at the mean diameter [*26,27,28*]. This is often representative enough to analyze effectively the results of restaggered blading or off-design operating conditions.

Returning to two-dimensional considerations, accommodation of a blade row to progressively increasing pressure ratios has been investigated by Hauser, Plohr, and Sonder [*29,30*]. Effects are illustrated in Fig. H,7b and H,7c. The shock waves and expansion conditions were observed with schlieren and shadowgraph apparatus using straight cascades of

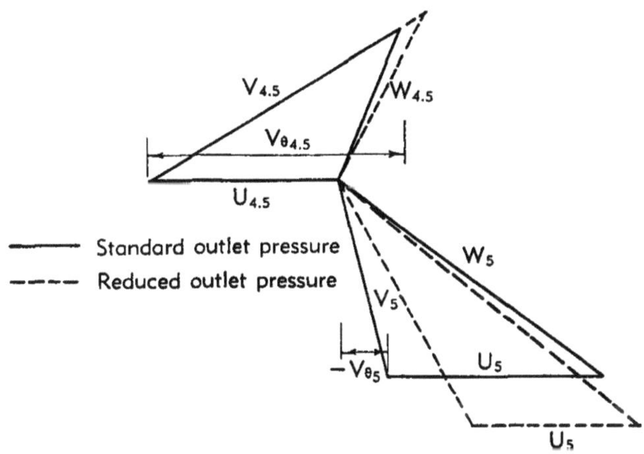

Fig. H,7a. Effect of reduction in outlet pressure from blading that is nearly choked.

blades. Fig. H,7b gives the regime where supercritical conditions are just established. A supersonic expansion fan radiates from the pressure surface of the trailing edge. In this case the expansion is followed by a shock, shown by the letter "a" in Fig. H,7b, emanating from the suction surface. Shocks "b" and "c" emanating from the wakes on the pressure and suction surfaces were also observed. Still lower outlet pressures produce a flow regime shown in Fig. H,7c. Shocks "a" and "b" are now completely outside the blade passage so that an uninterrupted expansion over the suction surface from the throat to the trailing edge takes place. Typical streamlines (broken lines) are illustrated in Fig. H,7b and H,7c and indicate gas deflections in the axial direction which can be interpreted as a deviation (Eq. 5-2).

If the pressure distribution over the blade surface is known as a function of blade width, the area of this diagram will then represent the tangential force on each blade. A representative plot is shown in Fig. H,7d. For supercritical conditions an increased area and associated tangential force are produced by the decreased pressure on the suction surface. This force can be related to the average increase in value of W_{θ_0} (Fig. H,7a). It will be a maximum when the supersonic expansion fan after the blade throat (Fig. H,7b and H,7c) first reaches the extremity of its adjacent blade. This condition can be referred to as that of "limiting loading." Turbine efficiency deteriorates rapidly at pressure ratios exceeding

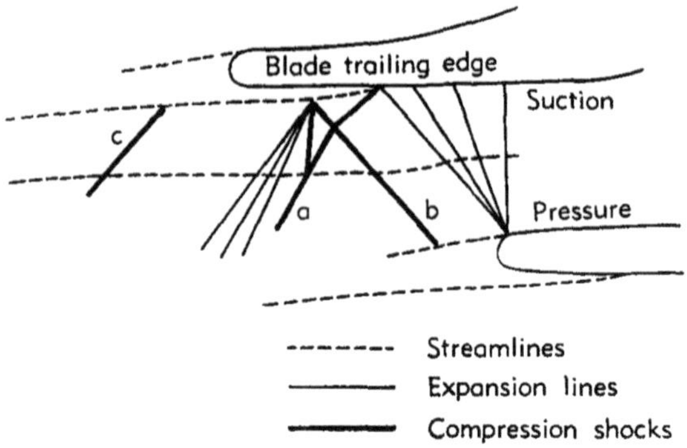

Fig. H,7b. Shock and streamline configuration for an expansion ratio p_o/p_i^{00} of 0.42 [29].

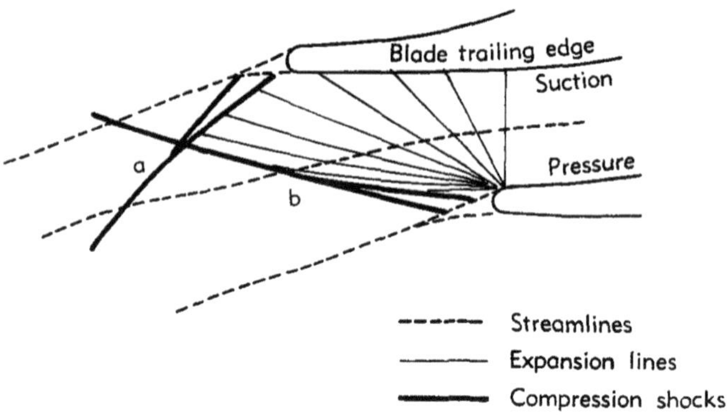

Fig. H,7c. Shock and streamline configuration for an expansion ratio p_o/p_i^{00} of 0.23 [29].

the limiting loading value [22,31], since no increase in work per pound of gas is possible.

For design purposes it is necessary to represent the two-dimensional regime shown in Fig. H,7b and H,7c by equivalent one-dimensional values which represent averages felt by blade rows or components downstream. The aforementioned authors [29,30] have carried out a calculation procedure, which checks very well with experiment even for blades with surface curvature downstream of the throat, as follows:

1. Until W_{θ_0} reaches limiting loading value, it is assumed that W_{θ_0} and W_{z_0} increase together in an isentropic manner, so that $\rho_0 W_{z_0}$ is maintained constant.
2. When W_{θ_0} has reached limiting loading value, it is assumed that it remains constant but ρ_0 and W_{z_0} change, still maintaining their product constant. This is a nonisentropic process.

This method of calculation requires advance knowledge of the breakpoint between assumptions 1 and 2, which can be obtained by plotting the flow using the method of characteristics (VI,G) as shown in Fig. H,7e. A relation between the blade outlet angle and the outlet tangential

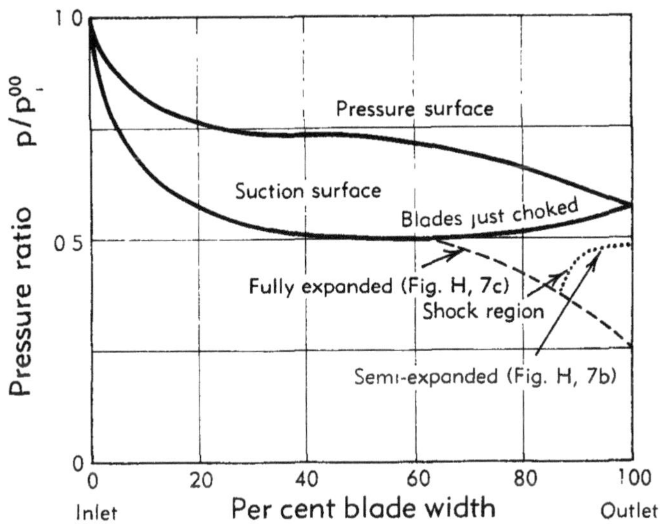

Fig. H,7d. Variation of blade surface pressure distribution with outlet pressure.

velocity, deduced in this manner for straight-backed blades, is reproduced in Fig. H,7f. Velocities are given nondimensionally as (W_θ/W_{cr}), where suffix $_{cr}$ denotes the value attained if the gas is brought isentropically to the sonic condition.

A comparison, made by Hauser, Plohr, and Sonder [*29*] between theoretical and observed outlet conditions, averaged in a pitchwise direction, for the cascade in question is reproduced in Fig. H,7g and H,7h. The ratio $(W/W_{cr})_{o,is}$ may be regarded as equivalent to the pressure ratio p_o/p_i^{00}, being related to it as follows [*29*].

$$\left(\frac{W}{W_{cr}}\right)_{o,is} = \left\{\frac{\gamma+1}{\gamma-1}\left[1-\left(\frac{p_o}{p_i^{00}}\right)^{\frac{\gamma-1}{\gamma}}\right]\right\}^{\frac{1}{2}} \tag{7-1}$$

The term p_i^{00} relates to conditions relative to the rotating blades. Also shown by broken lines on Fig. H,7g and H,7h are estimates of outlet

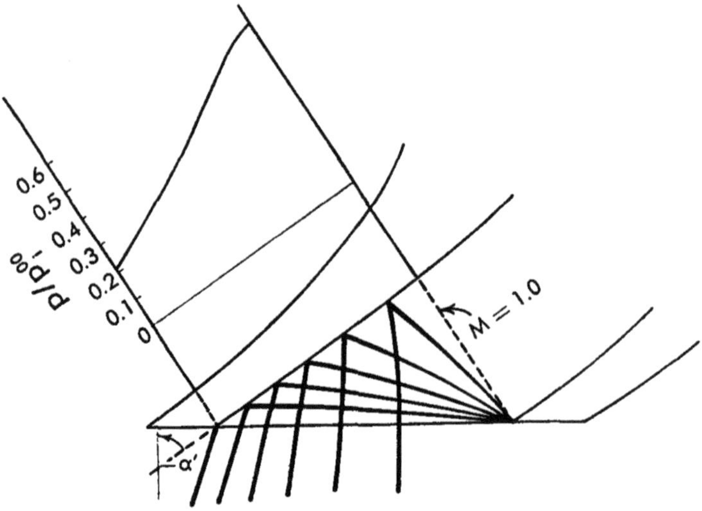

Fig. H,7e. Characteristic diagram and static pressure distribution under conditions of limiting loading [30].

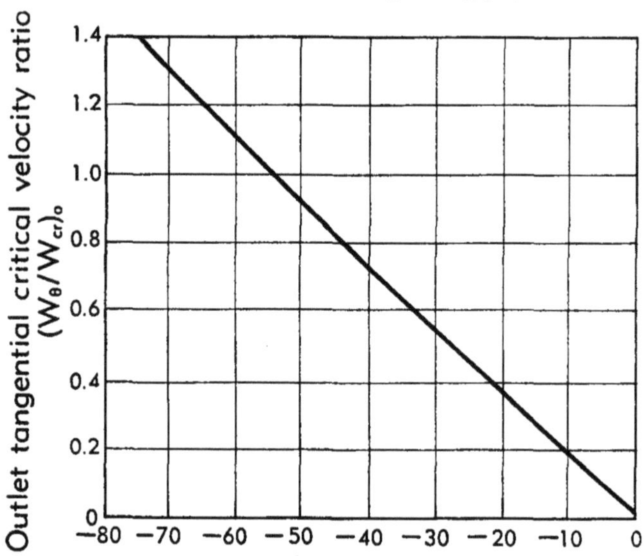

Fig. H,7f. Tangential outlet velocity under limiting loading conditions calculated by method of characteristics [30].

velocities and angles obtained by analytical procedures (1) and (2) above. The sharp discontinuity in the curve of tangential velocity W_{θ_0} (Fig. H,7g) is explained by the limiting loading condition, verifying that no further change in outlet swirl is possible after point A has been reached. These figures also show that it was not possible to obtain axial Mach

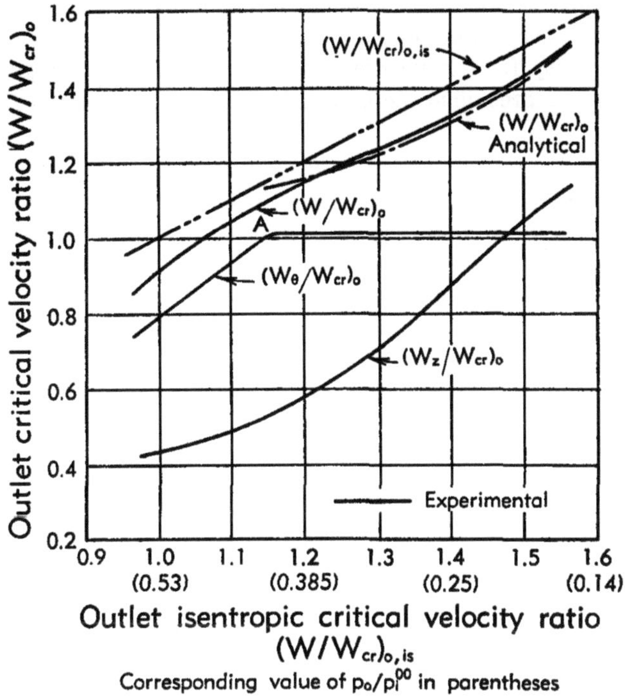

Fig. H,7g. **Variation of outlet velocities under supercritical conditions [29].**

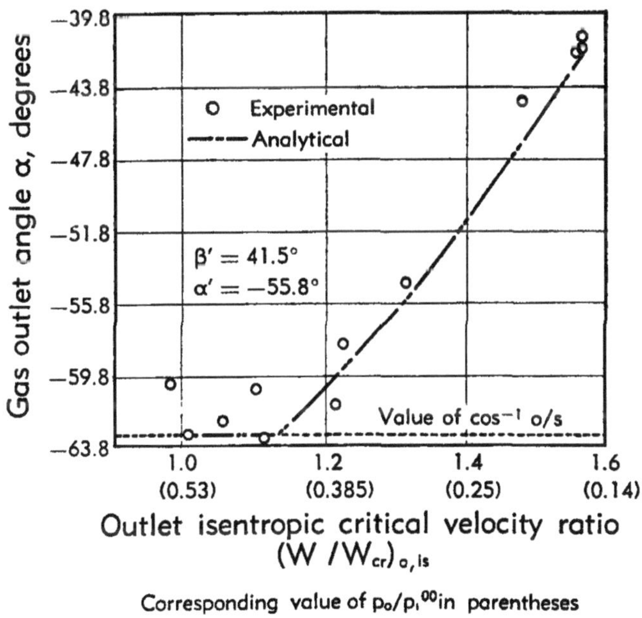

Fig. H,7h. **Gas deflection resulting from supercritical expansion [29].**

numbers higher than a certain value corresponding to $(W_z/W_{cr})_o = 1.15$, however much the outlet pressure was reduced. This value would theoretically be unity for turbines or cascades with a wall configuration at the outlet which precluded increase in the flow area normal to the axial direction.

H,8. Secondary Flow Losses. Secondary flow losses, or contributions to loss in turbine efficiency caused by secondary flow, are usually of about the same magnitude as profile losses. The flow takes the form of circulatory or eddy flow between adjacent blades. This pattern may be regarded as composed of several distinguishable effects indicated below.

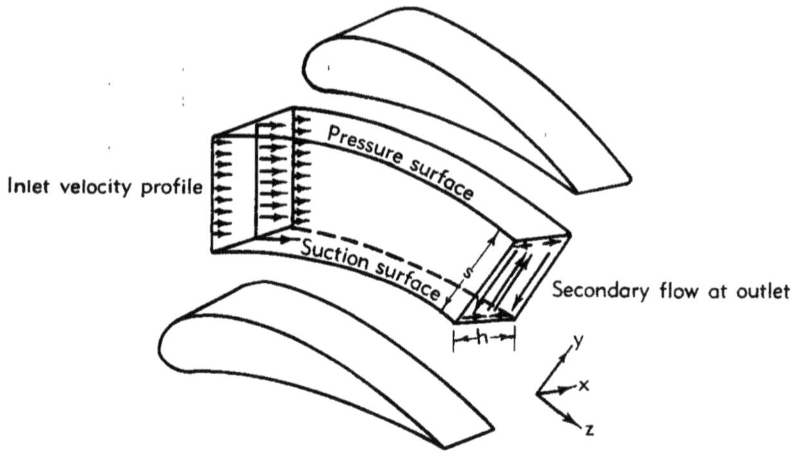

Fig. H,8a. Circumferential secondary flow mechanism.

Tangential momentum transport (end effects). This is the predominant effect. It can be described physically, although only qualitatively, as follows. The pressure differential between the pressure and suction surfaces of adjacent blades in the central portion of the annulus is in equilibrium with the centrifugal force field caused by streamline curvature as the gas turns through the blade deflection angle. At the blade ends, where gas velocities are very small, this pressure differential causes a flow of gas from pressure to suction surfaces. The effect is illustrated by Fig. H,8a. A measured three-dimensional loss pattern taken from experiments by Rohlik, Kofskey, Allen, and Herzig [*32*] is shown in Fig. H,8b.

The generation of tangential secondary flow may also be explained [*33*] by following the pitchwise vortex filaments which result from a heightwise axial velocity profile. Fig. H,8c illustrates a vortex filament A_1A_1 traveling successively to positions A_2A_2, A_3A_3, etc., finally reaching position A_5A_5. The portion adjacent to the suction (convex) surface is washed further downstream than the portion adjacent to the pressure (concave) surface, because there it moves on a small radius and also because fluid velocities on the suction surfaces are greater. In its final position the vortex filaments have components in the fluid direction, which generate the flow patterns shown in Fig. H,8a.

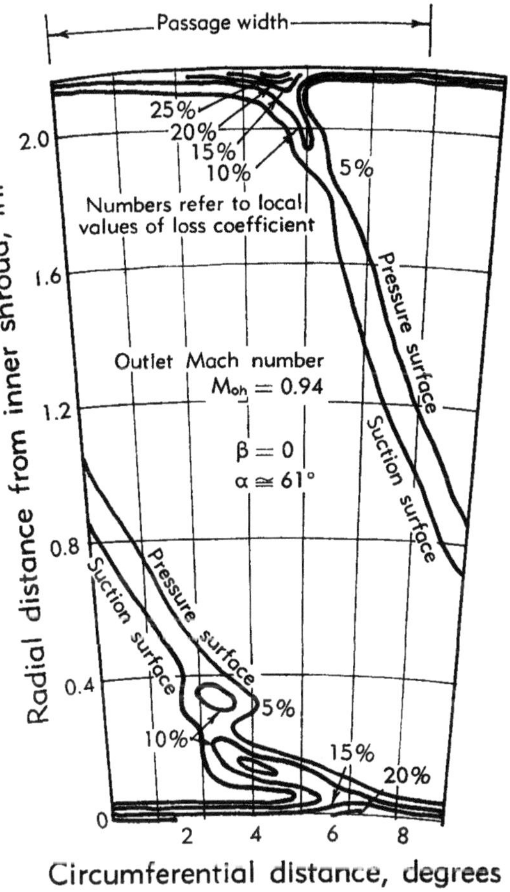

Fig. H,8b. Three-dimensional loss contours for a stationary blade. Subsonic outlet Mach number [*40*].

Hawthorne [*34*] has derived an expression for the change in vorticity component along a streamline in the absence of dissipation and in an incompressible medium as

$$\left(\frac{\zeta}{V}\right)_o - \left(\frac{\zeta}{V}\right)_i = 2\int_i^o \left|\nabla\left(\frac{p^0}{\rho}\right)\right| \sin\chi \frac{d\Omega}{V^2} \tag{8-1}$$

where χ is the angle between the normal to the plane of constant total pressure and the principal normal to a streamline, and Ω is the angular direction of the streamline. The regions immediately adjacent to the blade surfaces must be excluded from this analysis by reason of predominant viscosity effects. For straight cascades with a heightwise velocity gradient (dW_1/dx) at the entry, Eq. 8-1 leads to the expression

$$\zeta_0 - \zeta_1 \cong -2 \frac{dW_1}{dx} (\beta - \alpha) \tag{8-2}$$

This equation was first developed independently by Squire and Winter [35].

Evaluation of the losses associated with tangential secondary flow involves the solution of an induced velocity field due to a distributed

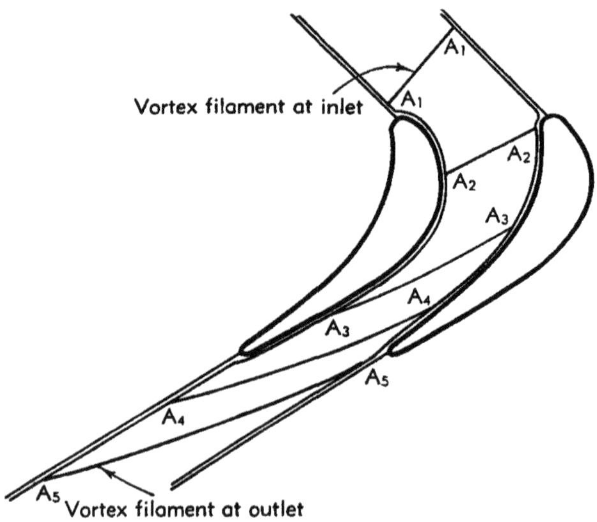

Fig. H,8c. Progress of a vortex filament through a blade passage.

vorticity, requiring, in general, numerical methods [36]. Quantitative prediction of the value of induced losses is also hampered by the approximate nature of Eq. 8-2 for large secondary flows, an unknown value of dW_1/dx in actual turbine blading, and by complicated wall boundary layers produced by this type of flow. Even if the heightwise velocity gradient is zero at the entrance to a blade row, a gradient builds up on the annulus walls inside the row [36]. Finally, the energy associated with secondary flow appears in the next row in a fluctuating manner with an attendant increase in effective Reynolds numbers, which effect may, in fact, decrease the losses (Art. 9). An experimental approach is therefore necessary.

Experimental data on the effect of blade height on the local loss coefficient Y_r, due to New [37], are reproduced in Fig. H,8d. A significant feature of this pattern is the comparative freedom of the center portions of the blade from end wall interference at the higher aspect ratios. This suggests that the secondary loss might be inversely proportional to the blade height. Let a secondary loss coefficient be defined as

$$Y_s = Y - Y_p \tag{8-3}$$

where Y_p is the loss coefficient obtained from separate two-dimensional tests and Y is the mean loss over the whole blade height for the aspect ratio in question. The suggested relationship between Y_s and blade height is thus

$$Y_s \sim \frac{1}{h} \tag{8-4}$$

The validity of Eq. 8-4 can be verified quantitatively by plotting the

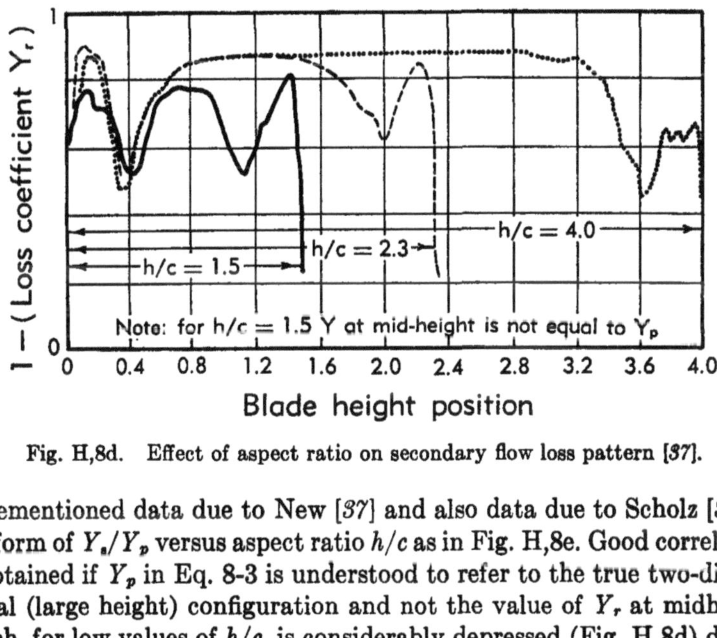

Fig. H,8d. Effect of aspect ratio on secondary flow loss pattern [*37*].

aforementioned data due to New [*37*] and also data due to Scholz [*38*] in the form of Y_s/Y_p versus aspect ratio h/c as in Fig. H,8e. Good correlation is obtained if Y_p in Eq. 8-3 is understood to refer to the true two-dimensional (large height) configuration and not the value of Y_r at midheight which, for low values of h/c, is considerably depressed (Fig. H,8d) due to merging of the secondary flow zones.

The effects of varying gas conditions and geometry other than blade height have been investigated by Ainley and Mathieson [*24*] and Carter and Cohen [*39*]. (Refer to E,3.) Ainley has postulated [*24*] that in an actual turbine the value of Y_s is related most closely to the mean acceleration of gas in passing through the row, since this controls the build-up of the wall boundary layers which constitute the term dW_i/dx in Eq. 8-2. The mean acceleration can be expressed as a ratio $A_{\text{ann}_o}/A_{\text{ann}_i}$ of the two annulus areas. As a result of Carter and Cohen's analytical work, he has

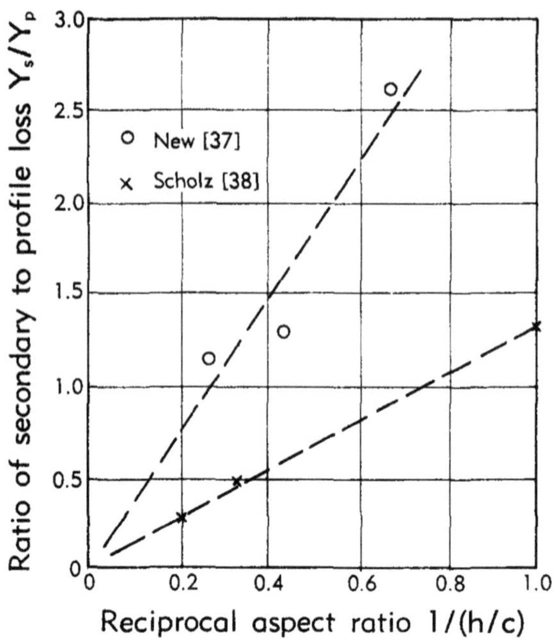

Fig. H,8e. Effect of aspect ratio on secondary loss coefficient [37,38].

also suggested that the values of Y_s should be related by their associated drag and lift coefficients as follows:

$$C_{D_s} \sim \frac{C_L^2}{s/c} \tag{8-5}$$

where C_{D_s}, C_L are defined by the following equations:

$$C_L = 2\frac{s}{c}(\tan\beta - \tan\alpha)\cos\alpha_\infty \tag{8-6}$$

$$C_{D_s} = \frac{s}{c}Y_s\frac{\cos^3\alpha_\infty}{\cos^2\alpha} \tag{8-7}$$

$$\alpha_\infty = \tan^{-1}\frac{\tan\beta + \tan\alpha}{2} \tag{8-8}$$

A practical correlation was therefore achieved by plotting the parameter $(C_{D_s}/C_L^2)(s/c)$ against an acceleration and blade height parameter $(A_{\text{ann}_o}/A_{\text{ann}_i})^2/(1 + \lambda_h)$. This gives recognition to the part played by acceleration of the flow in maintaining small and stable boundary layers on the annulus walls. The term $\lambda_h(= r_h/r_t)$ is introduced to account for the increased values of work coefficient ϕ at the blade hub, due to reduced values of U which tend to produce increased losses in these regions (Fig. H,3d). The opposing contributions of effects illustrated in Fig. H,3d (hub loading) and H,8e (merging of secondary flow vortices) were thus not separated out. Correlation data is reproduced in Fig. H,8f.

Radial flow movements. Other flow movements in the form of shed vorticity or eddying motion have been observed. Although little numerical data on corresponding losses are available, their role in modifying boundary layer phenomena on the blade surfaces and in causing radial interchange of mass is of interest and importance.

A radial flow movement inside the blade wakes and in thickened profile boundary layers has been observed by Allen, Kofskey, and Chamness

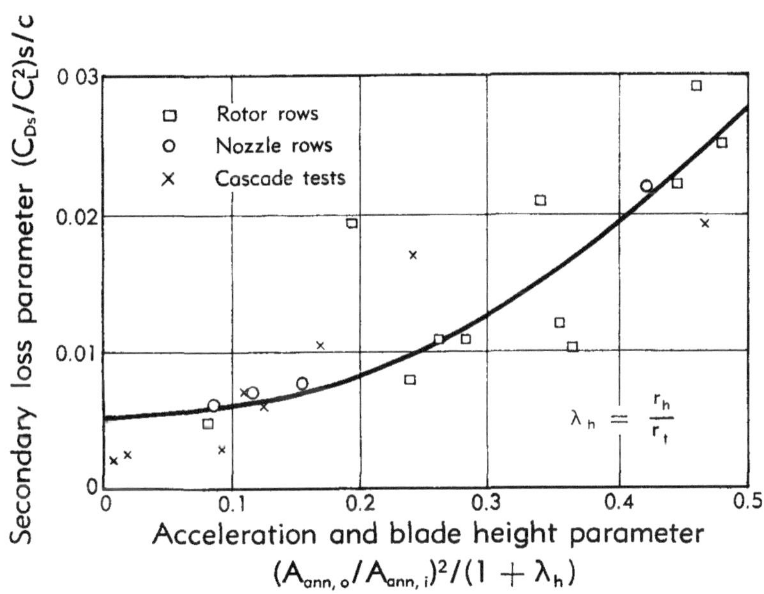

Fig. H,8f. Correlation of secondary loss data due to Ainley [*24*].

[*40*] at high outlet Mach numbers. This is attributable to the change in (tangential) velocity in the boundary region. Since the radial pressure gradient adjusts itself to the general tangential velocity pattern outside the boundary layer, adjustment inside the boundary layers must take place by radial flow. This flow can be expected to be very much accentuated by the presence of a thickened boundary layer caused by inefficient blades. The transport of stagnant air from the tip to hub regions was in fact found to be greatly in evidence when supercritical pressure ratios existed across the blading. However, radial flows have also been observed at intermediate and low gas speeds [*41*]. Fig. H,8b and H,8g present these comparisons, identical stationary blading being used for each test.

In the case of rotating blading, the radial mass shift in the boundary layer is directed outward over the exit portions of the blade surface where the tangential velocity of the blade is greater than the tangential velocity of the fluid; otherwise the flow is directed inward as for stationary blades.

Shed circulation. Blades operate with varying circulation up their height even if the axial velocity is uniform with height, unless free vortex designs are used (Art. 10). There is a circulatory flow pattern at the exit

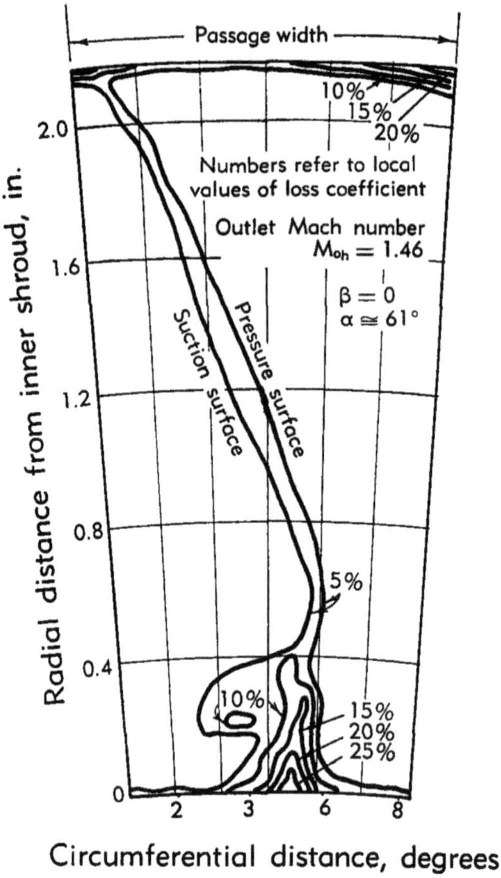

Fig. H,8g. Three-dimensional loss contours for a stationary blade. Supersonic outlet Mach number [*40*].

due to the shedding of this circulation. This combines with the tangential secondary flow pattern to produce some net effect. Fig. H,8h illustrates combined flows calculated and experimentally verified by Ehrich [*36*].

Reduced axial velocities in the hub and tip regions at the outlet from a row reduce or even reverse the turning, and hence reduce the circulation, for these regions in the following row. This can be seen by examining typical velocity triangles (Fig. H,3a). Exactly similar effects are imposed on stationary nozzles by the relative motion of preceding blades

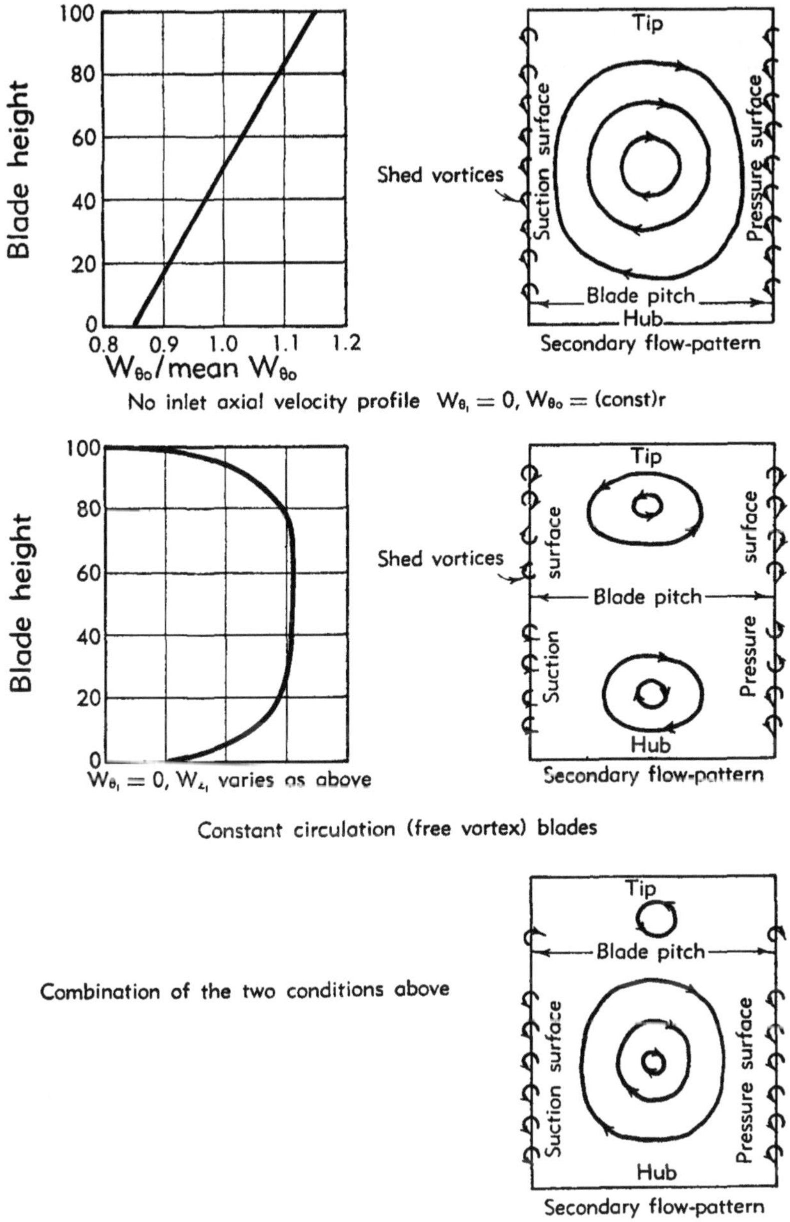

Fig. H,8h. Secondary flow pattern resulting from end effects combined with varying blade circulation [*36*].

(for compressors, the blades suffer increased incidence at their extremities, giving the opposite effect). Some shed circulation can therefore be expected even when the blading is designed for constant circulation.

H,9. Other Factors Influencing Performance.

Windage losses. Rotating disks and shafts give rise to a power loss associated with air friction. Experimental data have been summarized by Stodola (*1*, p. 197), also by Theodorsen and Regier [*42*]. The drag coefficients are not strong functions of Reynolds numbers for turbulent flow. The latter authors give the following experimental data:

$$\text{Disks} \qquad c_d = 0.053(Re_d)^{-\frac{1}{5}} \tag{9-1}$$

$$\text{Shafts} \quad \frac{1}{\sqrt{c_{sh}}} = -0.6 + 4.07 \log (Re_{sh} \sqrt{c_{sh}}) \tag{9-2}$$

where c_d or c_{sh} is the local value of the drag force per unit wetted area divided by $\frac{1}{2}\rho U^2$, ρ is the gas density, and U is the speed of the disk or shaft at the point in question. The Reynolds number Re is given as

$$Re_{sh} \text{ or } Re_d = \frac{\rho\omega r^2}{\mu} \tag{9-3}$$

where ρ, ω, r, and μ are the local values of the density, the angular velocity, the radius, and the coefficient of viscosity. Eq. 9-1 and 9-2 can be integrated to give the drag horsepower. For a disk (Eq. 9-1) the expression is

$$\text{hp} = \frac{0.146}{550} (Re_d)_r^{-\frac{1}{5}} \frac{1}{2} \rho r_r^5 \omega^3 \tag{9-4}$$

where all values are now taken at the disk rim, ρ being in slugs/ft^3, r_r, the rim radius, in ft, and ω in sec^{-1}.

Losses due to unsteady flow. The mechanism of loss associated with unsteady flow can be attributed [*43*] to the shedding of vortices from each blade as the inlet conditions vary with time. These effects might therefore be expected to diminish as a blade row is separated from the disturbing influence of the wakes from the row immediately upstream. Data published by Wu [*44*] are shown in Fig. H,9a, where over-all turbine efficiency is plotted as a function of axial clearance between blade rows. For normal loading about 0.15 inches appeared to be the critical distance, which is outside the range of usual designs. Unsteady effects are further considered in Sec. K.

Cooling air flows. Air is sometimes deliberately introduced into the stream after it has cooled certain highly stressed portions. Such air is shown in Fig. H,9b as introduced into the tips of hollow first-stage nozzles and discharged again in the hub region. Effects on performance depend on the method of air ejection into the main stream. It is reasonable to assume a loss of work equivalent to the cooling air mass flow multiplied by the enthalpy drop associated with the lost pressure. This ignores secondary disturbances in the main stream caused by such ejection, and,

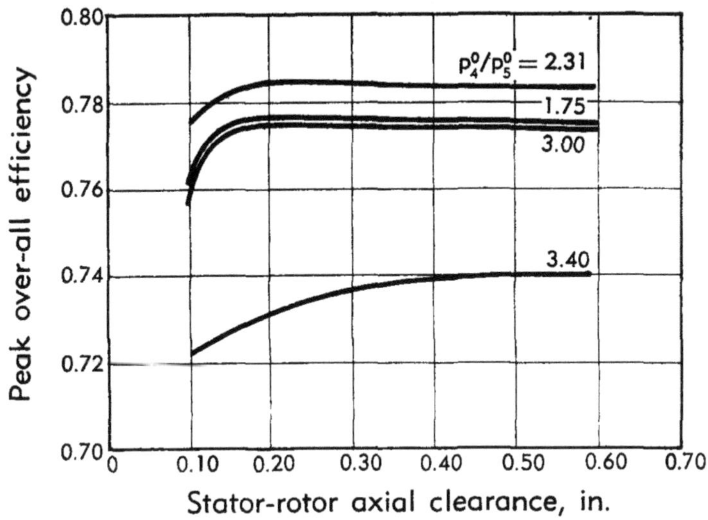

Fig. H,9a. Turbine peak efficiency as a function of axial clearance [*44*].

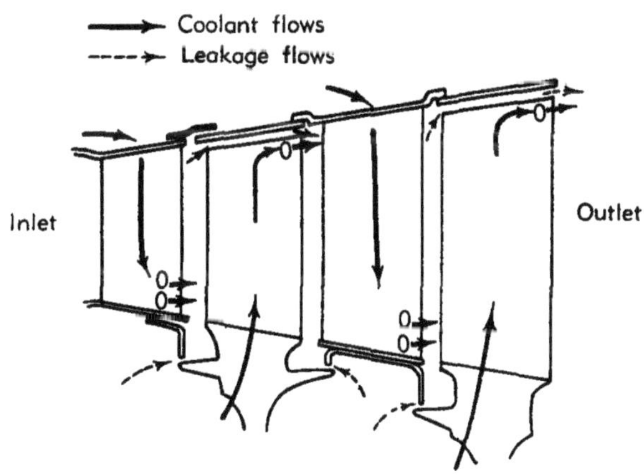

Fig. H,9b. Leakage and cooling flows.

of course, loss of heat into the blade, which must be investigated separately [*45*,*46*,*47*,*48*].

Leaving losses. It is generally the purpose of the turbine designer so to define the turbine component that the engine has maximum over-all efficiency. It is often possible to improve the turbine efficiency alone at the expense of high leaving Mach numbers or exit swirl which penalizes the performance of components downstream. Some cognizance of this effect must be taken, such as by including the loss due to a standard diffuser or exhaust nozzle element with which the turbine is to be used.

Tip leakage and scrubbing effects. These effects have been discussed by Carter [*49*]. They are illustrated schematically in Fig. H,9c. The effects have also been observed by Herzig, Hanson, and Costello [*50*] with smoke-tracing techniques. The relative motion of blades with respect to walls causes a roll-up of gas which, for a turbine, reinforces the secondary

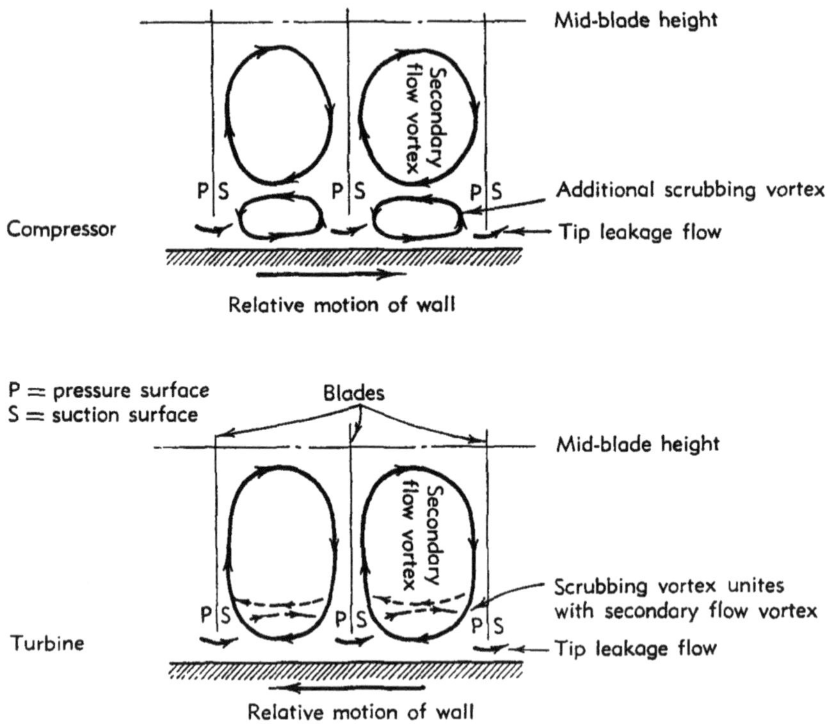

Fig. H,9c. Flows caused by scrubbing action and tip clearance [*49*].

flow vortex but opposes the leakage flow over blade tips. In the case of compressor blading, both leakage flow and scrubbing effects oppose the secondary flow. Analytical evaluation of associated drops in turbine efficiency is inherently somewhat intractable due to viscous effects. Consequently turbine designers usually carry out tests to estimate leakage effects for particular designs. For unshrouded blades the per cent efficiency has been observed to drop between one and two points for an average leakage area per row which is one per cent of the annulus area. The figure for shrouded blading depends on the number of seals employed.

Reynolds number effects. It is not strictly possible to generalize on Reynolds number effects except for geometrically similar blade configurations. In addition, these effects, though valid for the turbulence level at which the tests were undertaken, may be modified under operating conditions by nonuniformity of pressure and temperature at the combustor exit and by the wakes from previous blade rows and secondary flow patterns. Even for a given turbine, Reynolds number data obtained by complete scaling or density changes must be used with caution to predict effects of changing the blade chords only (at constant values of s/c) and not blade height, or vice versa, since this would presume that neither the profile nor secondary losses are dependent on the unchanged dimension.

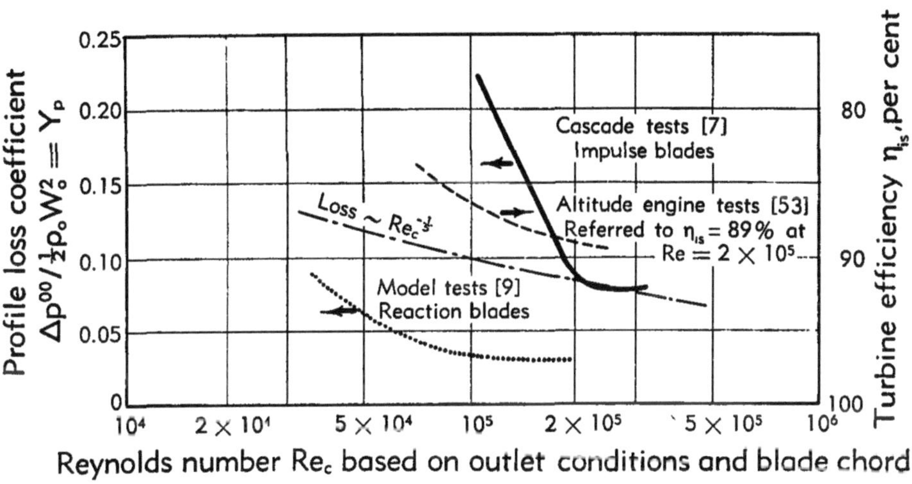

Fig. H,9d. Effect of Reynolds number on performance of turbine blading.

In practice certain trends are nevertheless found to hold for a wide variety of turbine blades. Typical cascade data on Reynolds number effects [7] for impulse blades are indicated by the solid line in Fig. H,9d, where the profile loss coefficient is expressed as a function of the Reynolds number based on the blade chord and outlet conditions.

$$Re_c = \frac{\rho_o V_o c}{\mu_o} \tag{9-5}$$

Cascade tests by Bridle [*51*] and rotating tests by Ainley [*9*] indicate that reaction blading is less sensitive to Reynolds number effects than impulse blading. Typical results are indicated by the dotted line in Fig. H,9d. These show a critical blade Reynolds number in the region of 2×10^5.

Several authors [*26,52*] have found correlations by using a Reynolds number dependence of the form

$$\text{Loss} \sim (Re)^{-\frac{1}{5}} \tag{9-6}$$

This relationship is shown by the dot-dash line on Fig. H,9d for a basic value of Y_p equal to 10 per cent at Reynolds number Re_c of 10^5. Altitude engine tests [53], indicated by the broken line on Fig. H,9d, show similar trends. Correlation of a number of compressor and turbine Reynolds number tests has been undertaken by Davis, Kottas, and Moody [54] using a machine Reynolds number defined as

$$Re_U = \frac{\rho U_t(2r_t)}{\mu} \tag{9-7}$$

This Reynolds number is at least an order of magnitude higher than that defined by Eq. 9-5. Average figures for the drop in efficiency from that at $Re_U = 10^7$, taken from the above data, are given in Table H,9. When allowance is made for the difference between Re_c and Re_U, these data compare reasonably well with trends shown in Fig. H,9d.

Table H,9.

Reynolds numbers		Total percentage points efficiency drop
Re_U (Eq. 9-6)	Re_c (Eq. 9-5) (Approx.)	
10^7	0.5×10^6	0
10^6	0.5×10^5	5
10^5	0.5×10^4	15
10^4	0.5×10^3	35

CHAPTER 3. THREE-DIMENSIONAL EFFECTS

H,10. Radial Equilibrium. The term "radial equilibrium" is employed by the turbine designer with reference to those fluid flow equations which apply in a radial direction. For frictionless steady flow the equation of motion in direction r can be shown to be [4, p. 94, Ex. 8 and 9],

$$F_r - \frac{g}{\rho}\frac{\partial p}{\partial r} = V_r\frac{\partial V_r}{\partial r} + \frac{1}{r}V_\theta\frac{\partial V_r}{\partial \theta} + V_z\frac{\partial V_r}{\partial z} - \frac{V_\theta^2}{r} \tag{10-1}$$

where V_r, V_θ, and V_z represent velocity components in the radial, circumferential and axial directions, and F_r is the radial body force per unit mass of fluid. At an inter-row position the body force F_r will be zero; also, the assumption of axial symmetry implies that $\partial V_r/\partial\theta$ will be zero. If V_r is small (generally true for axial flow machines) the term $V_r\partial V_r/\partial r$ can be neglected compared to $V_z\partial V_r/\partial r$. These assumptions introduced in Eq. 10-1 lead to

$$\frac{g}{\rho}\frac{\partial p}{\partial r} = \frac{V_\theta^2}{r} - V_z\frac{\partial V_r}{\partial z} \tag{10-2}$$

The third term in Eq. 10-2 may be expressed in a more significant manner in terms of r_m, the radius of curvature of the "meridional streamlines" or projections in the "meridian plane," defined by θ = const. The following equation results,

$$\frac{g}{\rho}\frac{\partial p}{\partial r} \cong \frac{V_\theta^2}{r} - \frac{V_z^2}{r_m} \tag{10-3}$$

The approximate equivalence of the third terms in Eq. 10-2 and 10-3 can be deduced by expanding as follows:

$$\begin{aligned} V_z\frac{\partial V_r}{\partial z} &= V_z\frac{\partial}{\partial z}\left(V_z\frac{dr}{dz}\right) \\ &= V_z\left(V_z\frac{d^2r}{dz^2} + \frac{\partial V_z}{\partial z}\frac{dr}{dz}\right) \end{aligned} \tag{10-4}$$

For axial flow machines, $\partial V_z/\partial z$ and dr/dz will be small. If their product is neglected compared with $d^2r/\partial z^2$, Eq. 10-4 can be written

$$\begin{aligned} V_z\frac{\partial V_r}{\partial z} &= V_z^2\frac{d^2r}{dz^2} \\ &\cong \frac{V_z^2}{r_m} \end{aligned} \tag{10-5}$$

This establishes Eq. 10-3 from Eq. 10-2.

A physical interpretation of Eq. 10-3 is that the radial pressure gradient is caused by two easily conceived centrifugal force fields. The first is caused by conventional rotational motion about the axis of the machine and the second arises when the meridional streamlines are moving along curved paths. The term "simple radial equilibrium" can be applied to cases where the second term on the right-hand side of Eq. 10-3 is absent or neglected and the term "modified radial equilibrium" to cases where both terms are taken into account. If conditions sufficiently far downstream of a blade row are considered and the annulus is of a fixed cylindrical shape, the terms involving V_r vanish and simple radial equilibrium then applies exactly.

The use of both simple and modified radial equilibrium to determine inter-row gas conditions has been thoroughly explored by Wu [*55*]. If the circumferential velocity V_θ is prescribed either graphically or analytically, as in the "design" case, for all values of radius r at a particular inter-row station, Eq. 10-3 can be used, ignoring the terms V_z^2/r_m initially, to connect the pressure at two adjacent radial positions. This can then be used to determine the required value of V_z at the second radial position if the first is known. In this way consistent values of V_z for all radii are found. This is essentially a numerical integration process. The choice of V_z at the first radial position, e.g. the hub, is arbitrary and can be made to produce a set of V_z values which integrate out to some desired mass flow. If this process is performed at more than one inter-row position, the initial (hub) value of V_z must be chosen so that the total mass flow is the same at all inter-row positions. The inter-row gas states at all radii having been determined using simple radial equilibrium, it is then possible to plot streamline paths and determine some approximation to the third term, V_z^2/r_m, in Eq. 10-3, which can then be used to further refine the calculations. "Performance" solutions proceed in a similar manner by prescribing gas angles in place of tangential velocities.

The more general solution of axially symmetric flow conditions, which will involve solution of Eq. 10-1, requires more involved though similar techniques. Such solutions are considered in Art. 11. The advent of high speed computing machines has obviated much of the need for closed analytical solutions which are tractable to hand calculation.

Free vortex design. Turbine designers often proceed on the basis that some mean stagnation temperature $\bar{T}^0$ applies and the variation of tangential velocity V_θ with radius r is prescribed so that their product is constant at inter-row positions, i.e.

$$V_\theta r = \text{const} \tag{10-6}$$

This condition is known as a "free vortex" pattern, since the relation given in Eq. 10-6 applies to a two-dimensional natural vortex [*4*, p. 175]. This type of flow pattern will be seen to produce constant work at all radii if Eq. 3-6 and 10-6 are combined. The circulation around each blade will also be constant, resulting in the absence of shed vorticity. This results from the fact that V_θ varies inversely with r whereas blade pitch s varies directly with r and their product is therefore constant. Hence, for each blade, the circulation

$$\begin{aligned}\Gamma &= (V_{\theta i} - V_{\theta o})s \\ &= (rV_{\theta i} - rV_{\theta o})\frac{s}{r} = \text{const}\end{aligned} \tag{10-7}$$

It will now be proved directly that Eq. 10-6 leads to constant axial velocity if the following assumptions are made:

1. Stagnation temperature T^0 is constant with radius.
2. The term V_z^2/r_m is neglected in Eq. 10-3.
3. Stagnation pressure p^0 is constant at all radii.

Condition 2 will be true if there are no pronounced streamline shifts. This is consistent with a flow regime established where V_z (or, better,

ρV_z) is constant at all radii for all inter-row stations. Also, conditions 2 and 3 lead to constant entropy [2, pp. 104, 105].

The proof of constant V_z then proceeds as follows. Simple radial equilibrium (Eq. 10-3), neglecting meridional streamline curvature, leads to

$$\frac{g}{\rho}\frac{dp}{dr} = \frac{V_\theta^2}{r} \tag{10-8}$$

The relation between temperature and stagnation temperature is (Eq. 2-3, 2-6, and 2-7)

$$c_pT + \frac{V_z^2}{2gJ} + \frac{V_\theta^2}{2gJ} = c_pT^0 = \text{const} \tag{10-9}$$

Constant stagnation pressure p^0 and temperature T^0 [2, p. 104] for a perfect gas leads to

$$\frac{ds}{dr} = \frac{c_p}{T}\frac{dT}{dr} - \frac{\mathcal{R}}{p}\frac{1}{J}\frac{dp}{dr} = \frac{c_p}{T^0}\frac{dT^0}{dr} + \frac{\mathcal{R}}{p^0}\frac{1}{J}\frac{dp^0}{dr} = 0 \tag{10-10}$$

where s is the entropy of the gas at radius r. Variables p and T can be eliminated from the above three equations as follows. Differentiating Eq. 10-9 for r,

$$c_p\frac{dT}{dr} + \frac{V_z}{gJ}\frac{dV_z}{dr} + \frac{V_\theta}{gJ}\frac{dV_\theta}{dr} = c_p\frac{dT^0}{dr} = 0 \tag{10-11}$$

The term dp/dr can be eliminated from Eq. 10-8 and 10-10, which yields

$$\frac{\mathcal{R}}{p}\frac{\rho}{gJ}\frac{V_\theta^2}{r} = c_p\frac{1}{T}\frac{dT}{dr}$$

or, if the perfect gas law $p/\rho = \mathcal{R}T$ is assumed,

$$\frac{V_\theta^2}{rgJ} = c_p\frac{dT}{dr} \tag{10-12}$$

The temperature differential dT/dr can now be eliminated by combining Eq. 10-11 and 10-12, which yields

$$-\frac{1}{2}\frac{d}{dr}(V_z^2) = \frac{V_\theta}{r}\frac{d}{dr}(V_\theta r) \tag{10-13}$$

The introduction of Eq. 10-6 into Eq. 10-13 then leads to $d(V_z^2)/dr = 0$, i.e. constant V_z with r.

Apart from nonuniformity in stagnation temperature and pressure, T^0 and p^0, compressibility of the gas introduces errors into the calculations leading to Eq. 10-13 by violating the validity of assumption 2. For incompressible flow, with zero flare in the annulus area and constant

V_z, meridional streamlines are all parallel to each other and to the axis. For compressible flow, the radial position of a streamline, as a function of axial distance z, is that value of r for which

$$\int_{r=r_h}^{r=r} \rho V_z 2\pi r dr = \text{const} \tag{10-14}$$

The streamline radial position, r, is large where ρ is low and vice versa. This results in a fluctuating streamline path shown in Fig. H,10a. A strict solution, even of the free vortex flow pattern (rV_θ = const), entropy and total temperature being also constant with radius, then involves modified radial equilibrium (Eq. 10-3), using values of r from streamline plots

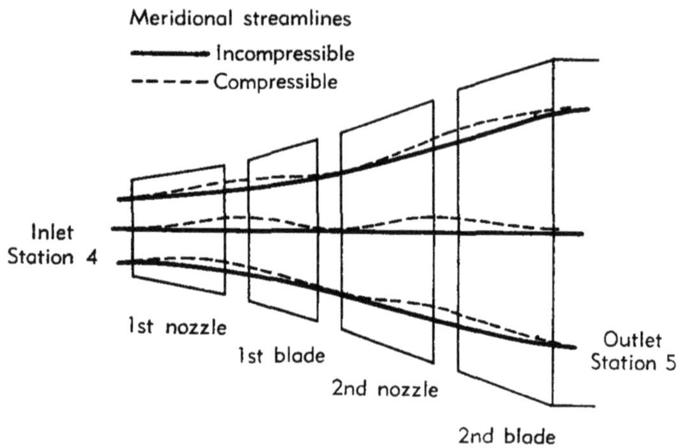

Fig. H,10a. Distortion of meridional streamlines due to density changes for a free vortex design.

such as in Fig. H,10a. However, it is found in practice that, for free vortex designs, blade angles arrived at by the more rigorous process usually differ by no more than a degree or two from the value obtained by assuming V_z to be constant.

Free vortex designs are often further simplified by calculating V_z on the assumption that ρV_z or the "specific mass flow" is constant with the radius. This technique leads to a direct determination of V_z at all radii from the known flow required and eliminates a trial-and-error technique to determine the level of V_z which will match the total required flow. Here again the justification is that of sufficient accuracy, bearing in mind the other larger disturbing influences, such as secondary flow patterns, unsteady effects, and radial distribution of stagnation pressure and temperature. Typical velocity triangles and blades for a single-stage free vortex (Eq. 10-6) design, with constant specific mass flow ρV_z, are shown in Fig. H,10b.

Forced vortex designs. In the case of this type of design the free vortex condition (Eq. 10-6) does not apply. A disadvantage of free vortex design is the high degree of twist in the rotating blades to accommodate the large changes in inlet and outlet angles (Fig. H,10b). Other variations of V_θ with r can give rise to a more uniform nozzle or blade section. The

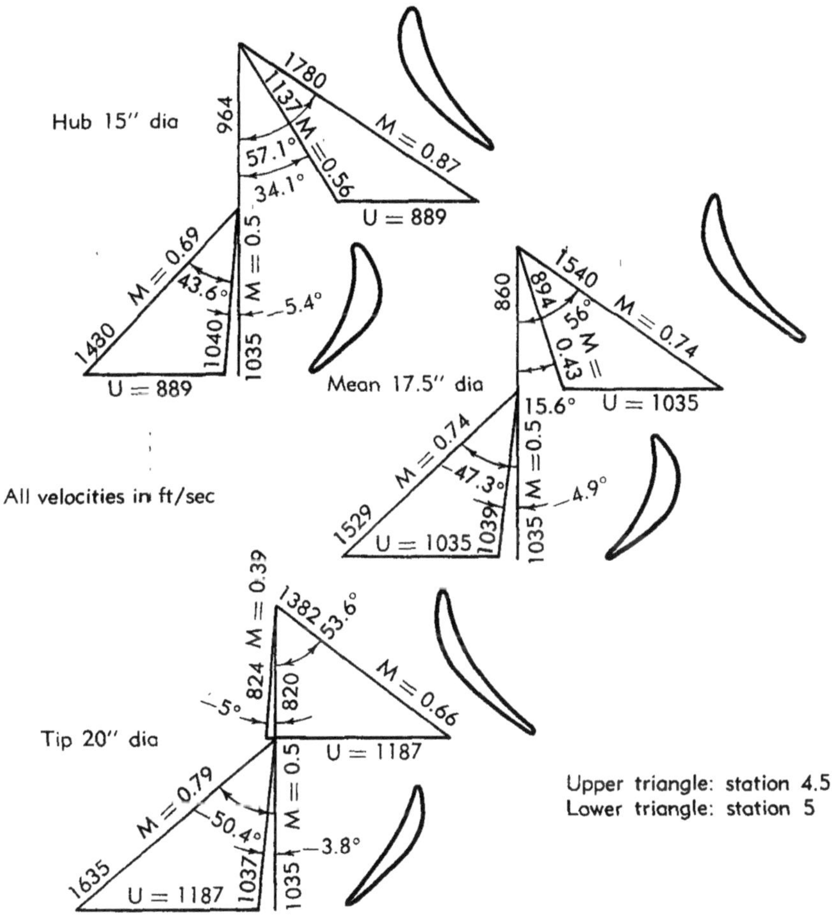

Fig. H,10b. Typical velocity triangles for free vortex stage with constant specific mass flow.

design for a blade section that is constant with radius has been investigated by Slivka and Silvern [*56*]. Once the gas direction on each side of the blade is set, the actual velocities fix themselves to satisfy radial equilibrium and therefore the variation of work with radius cannot be set in an independent manner. A typical radial variation of work is shown in Fig. H,10c.

A variety of nozzle and blade flow angles can be designed into a forced vortex turbine, still maintaining constant or any desired variation of work with radius. This is achieved by choosing a desired V_θ distribution in front of the moving blade, the distribution behind the blade then being fixed by virtue of Eq. 3-6. Some iteration may be required for desired accuracy, since the radial streamline positions at the exit from a blade row must be calculated from Eq. 10-8 (or, better, Eq. 10-3) and 10-14, but the V_θ distribution required for the solution of Eq. 10-8 depends (Eq. 3-6) on these positions.

Axial velocities that are dictated by radial equilibrium equations impose a severe limitation on the range over which V_θ can be chosen for a given design. A study has been undertaken by Pinnes [57] of the consequences of a relationship of V_θ with r of the general form

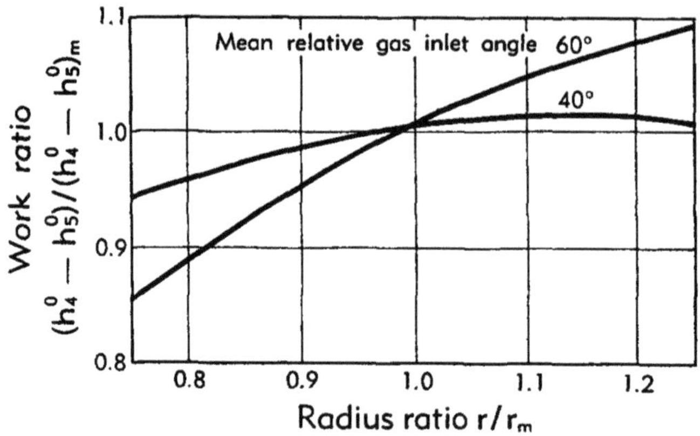

Fig. H,10c. Variation of work per lb of gas with blade height for constant section rotor blades [56].

$$V_\theta = ar^n \tag{10-15}$$

where a is a constant of proportionality. Total temperature and pressure were assumed uniform with radius. When $n = -1$, Eq. 10-15 reduces to the free vortex condition (Eq. 10-6) which leads, under idealized conditions, to constant axial velocity (Eq. 10-8, 10-9, 10-10, 10-11, 10-12, and 10-13). However, Eq. 10-13 serves to indicate the effect of using a more generalized distribution of V_θ with radius such as in Eq. 10-15. The contribution of the term $d/dr(V_\theta r)$ in Eq. 10-13 is positive if $n > -1$ and negative if $n < -1$. Consequently V_z^2 decreases more with r as n increases and vice versa. A cut-off point is therefore reached, in the former case with increasing r and in the latter case with decreasing r, at which V_z^2 becomes negative, leading to no real solution. Before this point is reached, designs will be impracticable due to excessive gas angles. Increase in the value of n at the nozzle exit may thus increase the enthalpy drop at the blade tips (Eq. 3-6) but reduce the flow in this region. Fig. H,10d, due to Pinnes [*57*], shows the limiting radius ratios r_h/r_t as a function of gas angle and value of n in Eq. 10-15.

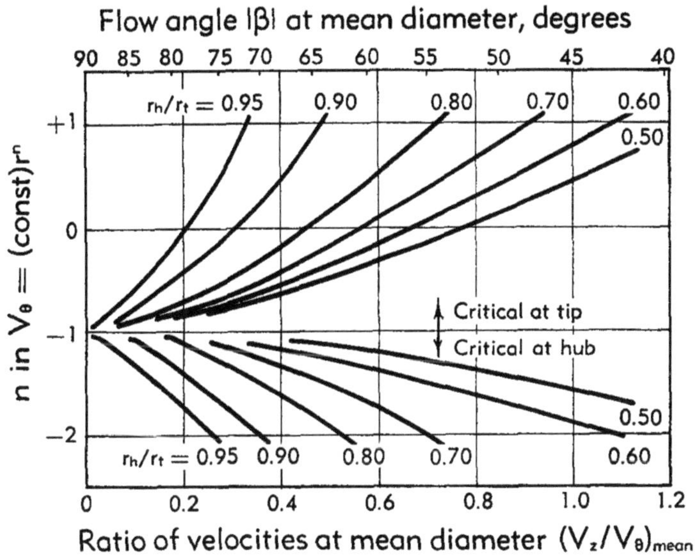

Fig. H,10d. Limiting conditions for balanced flow [*57*].

H,11. General Three-Dimensional Analysis. Any serious attempt to examine blade performance in the light of pressure and velocity distributions around the blading necessitates a re-examination of the accuracy of two-dimensional studies at specific inter-row axial stations. This examination is most necessary when Mach numbers are high, since radial shifts of flow are mandatory if the blades are choked at some radial positions and not at others. This type of effect is illustrated in Fig. H,11. If the flow conditions are calculated before and after the blade rows (Art. 10) and it is proposed to obtain estimates of the blade surface pressure distributions, a simple approach would be to assume a conical or some other simple type of streamline path through the blading. This is shown in Fig. H,11 as a linear variation of flow per unit blade height with axial distance z indicated by the solid lines, the true variation being indicated by the broken lines. These variations correspond to very significant changes in the pressure coefficient C_p as shown at the bottom of Fig. H,11.

Solution of the three-dimensional flow equations in turbomachinery is very tedious without the aid of modern high speed computing machinery. Approximate solutions have been proposed by various authors [*58,59*], using what might be called a quasi three-dimensional approach. In this technique the axially symmetric solution is first obtained. It is then assumed that this solution provides valid answers for the mean flow condition through the blades. The velocities on the blade surfaces can

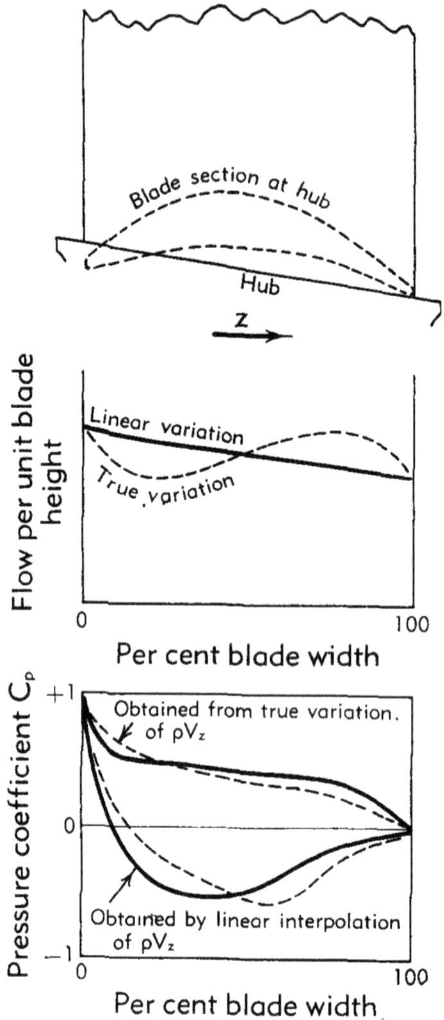

Fig. H,11. Changes in blade pressure distribution due to meridional streamline shifts in blade regions.

then be deduced as perturbation effects from the mean conditions. The development of secondary flows can only be determined analytically by replacing the two-dimensional solutions with a full three-dimensional solution.

Axially symmetric solution. The physical picture of a blade row consists of an infinite number of infinitely thin blades which can be treated mathematically as a force field. A simple transformation of the nonviscous, isentropic equations of motion [*4*, p. 74] to a coordinate system rotating with the blade [*61*, p. 103] yields the following relation for steady flow:

$$\mathbf{F} - \frac{1}{\rho}\nabla p = (\mathbf{W}\cdot\nabla)\mathbf{W} + 2\boldsymbol{\omega}\times\mathbf{W} + \boldsymbol{\omega}\times(\boldsymbol{\omega}\times\mathbf{r}) \tag{11-1}$$

In the absence of body forces **F**, that is, taking the blades as constituting boundaries rather than force fields, Eq. 11-1 yields three scalar equations but four unknowns. These latter are the three components of velocity **W** and the pressure p. The density ρ need not be considered as an unknown since it is calculable in terms of the pressure if entropy is constant or known [*2*, p. 19]. Entropy gains associated with actual machines render Eq. 11-1 invalid because of the omission of friction terms. Nevertheless it is often the practice, dictated by the necessity of getting a reasonably representative value of ρ, to include some practical entropy increase for establishing density levels, and yet use the perfect fluid equation (Eq. 11-1). The fourth unknown in Eq. 11-1 is covered by the law of mass conservation [*4*, p. 68],

$$\nabla\cdot(\rho\mathbf{W}) = 0 \tag{11-2}$$

The momentum equation (Eq. 11-1) can be written in cylindrical polar coordinates and expanded to three scalar equations [*4*, p. 95, Ex. 9],

$$W_r\frac{\partial W_z}{\partial r} + \frac{W_\theta}{r}\frac{\partial W_z}{\partial\theta} + W_z\frac{\partial W_z}{\partial z} = F_z - \frac{1}{\rho}\frac{\partial p}{\partial z} \tag{11-3}$$

$$W_r\frac{\partial W_r}{\partial r} + \frac{W_\theta}{r}\frac{\partial W_r}{\partial\theta} + W_z\frac{\partial W_r}{\partial z} - \frac{W_\theta^2}{r} - \omega^2 r - 2\omega W_\theta = F_r - \frac{1}{\rho}\frac{\partial p}{\partial r} \tag{11-4}$$

$$W_r\frac{\partial W_\theta}{\partial r} + \frac{W_\theta}{r}\frac{\partial W_\theta}{\partial\theta} + W_z\frac{\partial W_\theta}{\partial z} + \frac{W_rW_\theta}{r} + 2\omega W_r = F_\theta - \frac{1}{\rho}\frac{\partial p}{r\partial\theta} \tag{11-5}$$

These equations are reduced as follows if variations in tangential direction are omitted (axial symmetry).

$$W_r\frac{\partial W_z}{\partial r} + W_z\frac{\partial W_z}{\partial z} = F_z - \frac{1}{\rho}\frac{\partial p}{\partial z} \tag{11-6}$$

$$W_r\frac{\partial W_r}{\partial r} + W_z\frac{\partial W_r}{\partial z} - \frac{W_\theta^2}{r} - \omega^2 r - 2\omega W_\theta = F_r - \frac{1}{\rho}\frac{\partial p}{\partial r} \tag{11-7}$$

$$W_r\frac{\partial W_\theta}{\partial r} + W_z\frac{\partial W_\theta}{\partial z} + \frac{W_rW_\theta}{r} + 2\omega W_r = F_\theta \tag{11-8}$$

The equation of continuity (Eq. 11-2) provides the following modified relationship to account for finite blade thickness:

$$\nabla \cdot (\rho \mathbf{W} d) = 0 \tag{11-9}$$

where d is a blockage factor defined as

$$d = \frac{s - \tau}{s} \tag{11-10}$$

s being the pitch distance from blade to blade at a particular value of r and z, τ being the corresponding blade thickness (Fig. H,5a).

In the case of a performance solution the blade geometry is given. The body forces **F** and the velocity **W** = **V** − **U** relative to the blade are normal to the blade surface and along this surface respectively. This introduces further relationships,

$$\frac{F_\theta}{u} = \frac{F_r}{v} \tag{11-11}$$

$$\frac{F_r}{v} = \frac{F_z}{w} \tag{11-12}$$

$$F_\theta W_\theta + F_r W_r + F_z W_z = 0 \tag{11-13}$$

where u, v, and w are the direction cosines of the normal to the blade surface. If the density is considered calculable from the pressure and entropy, assumed constant or known as above, the seven equations (Eq. 11-6, 11-7, 11-8, 11-9, 11-11, 11-12, and 11-13) are sufficient for a solution of the seven unknowns p, **F**, and **W**. Outside the blade regions, between the rows, forces **F** no longer exist and there are therefore four unknowns, p and **W**, with four equations (11-6, 11-7, 11-8, and 11-9) for solution.

Prior to the advent of high speed digital computers, general axially symmetric solutions as indicated above would not have been practicable. Detailed procedures for exact solution have been worked out by Wu [*55*].

Blade surface velocity. The ultimate purpose of the solution of quasi three-dimensional flow equations is the determination of boundary layer behavior, which is in turn a function of surface velocities. The blade-to-blade solution is an approximation in that the general character or flow change in a radial direction is taken from the axially symmetric solution, assumed to be unchanged by the presence of actual blades.

If attention is focused on a blade-to-blade region in the θ, z domain, the momentum equations are represented by Eq. 11-3 and 11-5, where the body forces in the fluid region are now zero, since they are replaced by the discontinuous forces exercised by discrete blades. Some assumptions must be made to recognize the axially symmetric solution as the average condition pertaining to the conditions between the blades. A reasonable assumption would be that W_r and all derivatives in the radial direction are uniform over the θ, z domain and possess their axially symmetric values. Eq. 11-3 and 11-5 then reduce to

$$\frac{W_\theta}{r}\frac{\partial W_z}{\partial \theta} + W_z \frac{\partial W_z}{\partial z} + a_1 = -\frac{1}{\rho}\frac{\partial p}{\partial z} \qquad (11\text{-}13)$$

$$\frac{W_\theta}{r}\frac{\partial W_\theta}{\partial r} + W_z \frac{\partial W_\theta}{\partial z} + \frac{a_2}{r} W_\theta + a_3 = -\frac{1}{\rho}\frac{\partial p}{r\partial \theta} \qquad (11\text{-}14)$$

where a_1, a_2, and a_3 are functions of z derived from the axially symmetric solution. The unknowns are now p, W_θ, and W_z and the necessary relationships are given by Eq. 11-2, 11-13, and 11-14. The density ρ again need not be regarded as a variable since it is directly related to the pressure [*2*, p. 19] if some estimate of entropy increases in the flow direction is made.

Detailed computational techniques to obtain blade surface velocities and pressures from axially symmetric solutions have been worked out by Wu and Brown [*61*], also Wu, Brown, and Prian [*62*].

More approximate three-dimensional calculation procedures, suitable for hand calculation, have been developed by Hamrick, Ginsburg, and Osborn [*63*], Huppert and MacGregor [*10*], and Stewart [*64*].

CHAPTER 4. TURBINE DESIGN

H,12. Mechanical Considerations. The foregoing articles will enable the designer to obtain some insight into the probable blade losses and turbine efficiency of a given design. However, the aerodynamic configurations chosen may lead to excessive stresses in the materials for a given weight, or excessive weight for a given stress level. Allowable blade speed, height, and width are usually dictated by considerations of mechanical stress. The availability of materials superior to those currently available will cause modifications in aerodynamic design. The use of blade-cooling devices to decrease material temperatures provides this increase in strength even in present-day materials and therefore leads to similarly modified aerodynamic designs. Stresses in rotating blades caused by dynamic effects of rotation are known as centrifugal stresses whereas those in both stationary nozzles and blades due to gas forces are called gas bending stresses.

Centrifugal blade stresses. The designer will wish to know the effect of the following variables on centrifugal stress at some radius r:

1. Shaft rpm, n
2. Blade density, ρ_b
3. Blade height, $(r_t - r)$

4. Radius at tip, r_t
5. Ratio of section areas at r and r_t, A_{bt}/A_b
6. The manner in which A_b varies, e.g. the exponent f in the relation

$$\frac{A_b - A_{bt}}{A_{bh} - A_{bt}} = \left(\frac{r_t - r}{r_t - r_h}\right)^f \tag{12-1}$$

The following analysis applies rigorously only to unshrouded rotating blades. For shrouded blades, the force exerted by the shroud must be determined separately and distributed over the section at any required radius to form an additional increment of stress. The relationship between

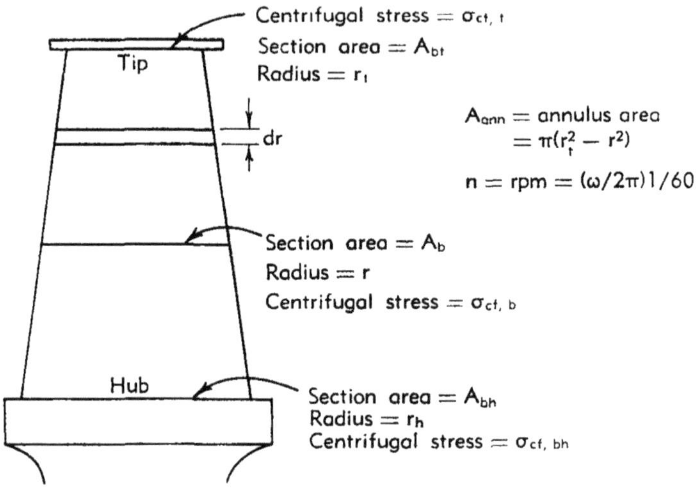

Fig. H,12a. Rotor blade under centrifugal forces.

centrifugal stress in the blade and the above variables (Eq. 1–6) can be deduced by integrating the total centrifugal force over any section A_b at radius r (see Fig. H,12a),

$$\text{Total force over } A_b = \sigma_{cf_b} A_b = \int_{r=r}^{r=r_t} (r\omega^2)\rho_b A_b dr \tag{12-2}$$

By using Eq. 12-1, the section area A_b can be expressed in terms of the radius r in Eq. 12-2, which leads to

$$\frac{\sigma_{cf_b} A_b}{A_b \omega^2} = \int_r^{r_t} \left[\left(\frac{r_t - r}{r_t - r_h}\right)^f (A_b - A_{bt}) + A_{bt}\right] r dr \tag{12-3}$$

where ω is the angular velocity $(2\pi n/60)$. Eq. 12-3 can then be integrated to obtain

$$\frac{\sigma_{cf_b}}{\rho_b \omega^2 (r_t^2 - r^2)} = \frac{1}{2}\frac{A_{bt}}{A_b} + \frac{1 - A_{bt}/A_b}{1 + \lambda}\left(\frac{1}{f+1} - \frac{1-\lambda}{f+2}\right) \tag{12-4}$$

Eq. 12-4 can also be written

$$\frac{\sigma_{cf_b}}{\rho_b(n/60)^2 A_{ann}} = 2\pi \frac{A_{bt}}{A_b} + \frac{4\pi}{1+\lambda}\left(1 - \frac{A_{bt}}{A_b}\right)\left(\frac{1}{f+1} - \frac{1-\lambda}{f+2}\right) \qquad (12\text{-}5)$$

Eq. 12-5 can be written

$$\frac{\sigma_{cf_b}}{\rho_b(n/60)^2 A_{ann}} = 2\pi T \qquad (12\text{-}6)$$

where the taper factor T accounts for the combined effects of the area ratio A_{bt}/A_b, the hub-to-tip radius ratio λ, and the exponent f, and will

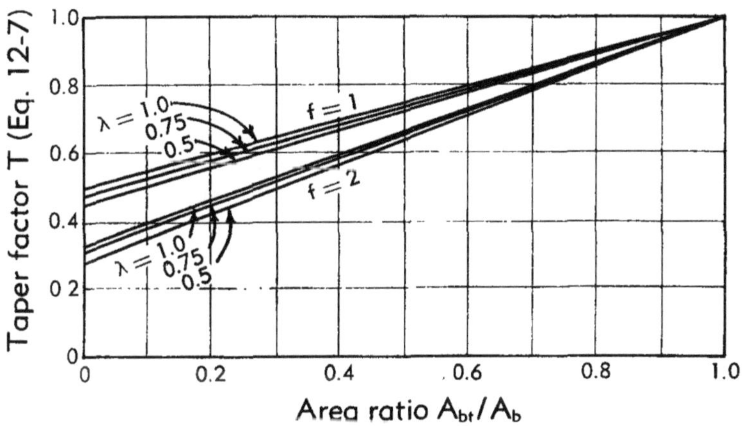

Fig. H,12b. Effect of blade thickness on centrifugal stress.

be unity if A_{bt}/A_b is unity. The expression for taper factor T is therefore

$$T = \frac{A_{bt}}{A_b} + \frac{2}{1+\lambda}\left(1 - \frac{A_{bt}}{A_b}\right)\left(\frac{1}{f+1} - \frac{1-\lambda}{f+2}\right) \qquad (12\text{-}7)$$

This relationship is shown in Fig. H,12b. The rather weak effect of the hub-to-tip radius ratio λ, the strong effect of A_{bt}/A_b, and the intermediate effect of the exponent f can be seen.

A given maximum value of centrifugal stress σ_{cf_b} dictates a maximum hub value of the term $\rho_b n^2 A_{ann}$ for a given blade section area ratio, A_{bt}/A_b, and value of f. If this value is exceeded, some compromise solution must be reached. The options at the choice of the designer are:

1. Reduce rpm, n. This increases the work coefficient ϕ (Eq. 3-11) and therefore reduces the efficiency (Fig. H,3d) unless the weight penalty of additional stages is taken. Moreover, the compressor, which also rotates at rpm n, suffers similarly.
2. Reduce the annulus area A_{ann}. This increases the axial Mach numbers and ultimately leads to high shock losses in the blading.

3. Reduce A_{bt}/A_b. Where this is already small, little advantage is gained in going to the limit $A_{bt}/A_b = 0$ (see Fig. H,12b).
4. Increase f in Eq. 12-1. This is only effective when the ratio A_{bt}/A_b is small (Fig. H,12b). In that case it will ultimately lead to knifelike blade tips which are subject to fatigue failure. They may also be unsatisfactory aerodynamically, since the thin leading edges do not allow a large incidence range. Unduly thin blade tips can be avoided only by increasing the blade section area A_b at the point in question. A limit is reached here when the passage areas through the hub regions of the blades become highly blocked and pressure losses increase sharply.
5. Maintain centrifugal stress at its present level and reduce bending stress instead. This involves increased blade width and consequent increased turbine weight as it will be seen later.

An optimum design requires as its ingredients those proportions of the compromises 1–5 which penalize the performance least. The choice is complex and emphasizes the need for more precise knowledge of the performance penalties which these compromises entail.

Gas bending stresses. The width of each blade row is determined largely by the allowable level of stress due to gas bending forces. The intensity of the latter will be a function of radius and are due to:

1. Static pressure drop across the nozzles or rotors.
2. A momentum term, $\rho_o V_{z_o}^2 - \rho_i V_{z_i}^2$.
3. Tangential forces due to change in swirl velocity across the blades.

The reaction force on the blade can be expressed more precisely as a function of fluid states in the form of Eq. 3-1, if applied to the flow between two cylindrical surfaces generated by r and $r + dr$. This yields the following relation if unsteady terms are ignored.

$$-\int \rho \mathbf{F} dv = \text{force of total fluid volume on blades}$$

$$= \int \mathbf{n} p dA + \int (\mathbf{n} \cdot \mathbf{V}) \rho \mathbf{V} dA \qquad (12\text{-}8)$$

The right-hand side of Eq. 12-8 yields force items 1, 2, and 3 above as the major contribution to axial and tangential forces. Only if large radial flows take place will additional contributions from the last term of Eq. 12-8 take place. These can be regarded effectively as transferring forces from one radius to another. Exact calculation requires exact knowledge of the whole fluid flow pattern.

Unfortunately the gas bending stresses as determined by the time-averaged bending forces and section shape do not lead directly to an allowable blade chord or width. Any flexibility in blade-to-disk fixation causes the blade to set itself in a slightly nonradial direction so that centrifugal forces then exert a moment to cancel out the bending forces. Complete cancellation would reduce the effects of gas-bending forces to a cyclic loading caused by unsteady effects, such as wakes and secondary flow patterns upstream. The harmonic content of this loading varies with turbine design and its effects vary with natural frequencies of the blade and its damping qualities. Considerable experience, coupled with experimental verification, is therefore required to fix the minimum allowable chord for any blade row. If blades are shrouded, the system will be stiffer but the shrouds will impose additional centrifugal stresses on the blades. It may nevertheless be advantageous to exchange some reduction of

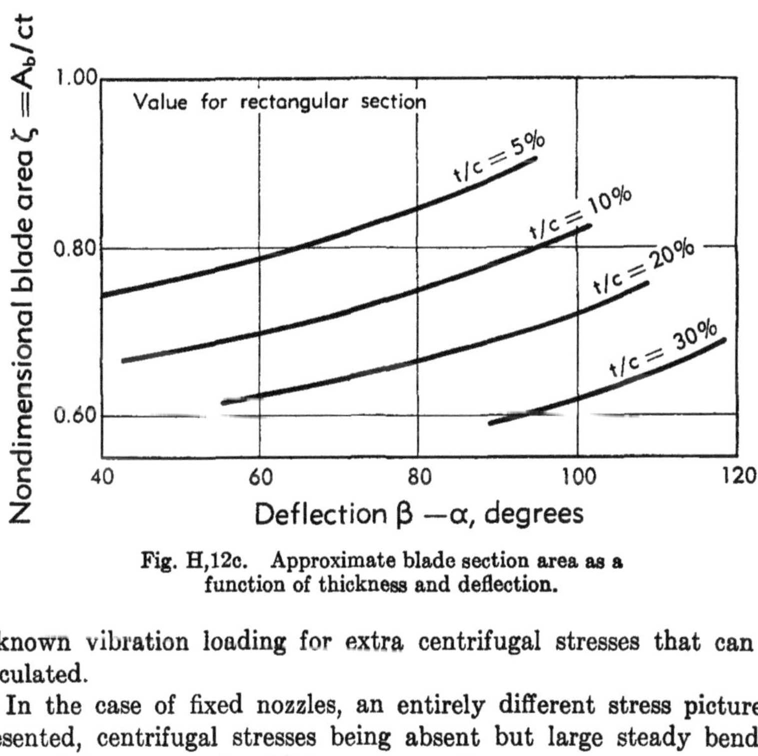

Fig. H,12c. Approximate blade section area as a function of thickness and deflection.

unknown vibration loading for extra centrifugal stresses that can be calculated.

In the case of fixed nozzles, an entirely different stress picture is presented, centrifugal stresses being absent but large steady bending stresses being present in addition to unsteady stresses.

For general design studies a statistical relationship between blade section properties and angles is useful. Fig. H,12c and H,12d show such relations for nondimensional area of cross section $\zeta = A_b/ct$ and section modulus $\xi = (A_b k_b^2)/(bt^2c)$, where c is the blade chord, t is the blade thickness, k_b is the radius of gyration of the blade section about the principle axis, and b is the distance of the furthest blade fiber from this axis. Some variation naturally exists between blades of the same value of t/c and deflection, depending on camber shape and thickness form, and

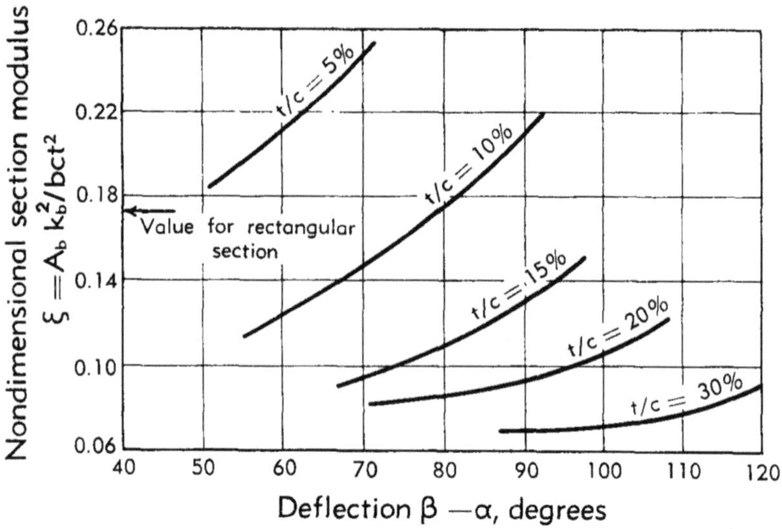

Fig. H,12d. Approximate blade section modulus as a function of thickness and deflection.

therefore it must be emphasized that Fig. H,12c and H,12d are for survey purposes rather than mechanical design.

Standard relations between applied forces and stress σ_{be} for a rectangular bar of thickness t, width c, and height h indicate approximate relations for a blade fixed at one end as follows:

σ_{be} varies directly as: (a) Total bending force L on blade row.
(b) Blade height $h = r_t(1 - \lambda_h)$.

σ_{be} varies inversely as: (c) Number of blades $N_b = \dfrac{2\pi r_h}{s_h}$
(d) Section modulus $= \xi c t^2$

An example of the use of these relations would be to determine the variation of chord at the hub, c_h, with allowable bending stress σ_{be_h}, height h, gas load L, and value of s/c, t/c at the hub.

$$\sigma_{be_h} \sim \frac{L r_t (1 - \lambda_h) s_h}{r_h c_h t_h^2 \xi_h}$$

$$\sim L \left(\frac{s}{c}\right)_h \left(\frac{c}{t}\right)_h \frac{1 - \lambda_h}{\lambda_h} \frac{1}{\xi_h} \left(\frac{1}{c_h}\right)^2$$

Consequently,

$$c_h \sim \left[\frac{L}{\sigma_{be_h}} \left(\frac{s}{c}\right)_h \left(\frac{c}{t}\right)_h^2 \frac{1}{\xi_h} \frac{1 - \lambda_h}{\lambda_h}\right]^{1/2} \tag{12-9}$$

Eq. 17-2 will be used later for weight estimation.

Weight estimation. The considerable influence of turbine aerodynamic design on weight is well brought out by a brief speculation on probable design features if mechanical properties constitute no barrier. Blade speeds would then be increased, giving increased work (Eq. 3-6) if velocity triangles (Fig. H,3f) remained similar, compressibility considerations introducing the first barrier. Blade widths would be a minimum consistent with Reynolds number effects (Art. 9). Much higher turbine inlet temperatures could be employed, which generally increase the engine output and sometimes also decrease fuel consumption [*65*].

Factors affecting disk and blade weights are considered separately although it may be seen that many of these are common. It must be

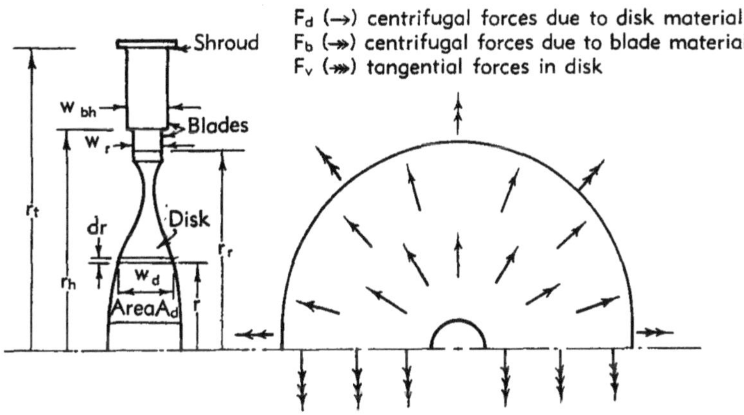

Fig. H,12e. Equilibrium of external and internal forces on rotating disk.

emphasized here that the relations which follow are often only approximate and therefore will not serve for detailed stress or weight calculations, particularly in extreme cases. They are rather intended for preliminary design purposes, so that the necessary processes of adjustment between considerations of aerodynamics, stress, and weight can be facilitated. A method of disk weight analysis alternate to that given below has been derived by LaValle and Huppert [*66*] on the basis of a disk thickness which varies exponentially with the radius.

Disk weight analysis. For the purposes of this analysis a sufficiently accurate criterion of stress intensity in a rotating disk is that of average tangential stress. This is the stress that would be experienced over the area A_d (Fig. H,12e) due to the integrated centrifugal forces arising from both disk and blades. The total radial centrifugal force exerted on the disk material is

$$F_d = \int r\omega^2 dM$$

where dM is an element of mass and ω is the angular velocity. This summation can be written as

$$F_d = \left(\frac{2\pi}{60}\right)^2 n^2\rho_d \int w_d r 2\pi r dr \tag{12-10}$$

where w_d is the disk width at any radius r and ρ_d is the density of the disk material.

Writing $\int w_d r^2 dr = A_d k_d^2$, Eq. 12-10 becomes

$$F_d = 2\pi \left(\frac{2\pi}{60}\right)^2 n^2\rho_d A_d k_d^2 \tag{12-11}$$

To assess the disk stresses, the centrifugal force due to the rotating disk must be added to the similar force exerted by the blades. The latter can be written as

$$F_b = \sigma_{cf_{bh}} A_{bh} N_b \tag{12-12}$$

where A_{bh} is the area of cross section of a blade at the hub.

In the case of shrouded blades, $\sigma_{cf_{bh}}$ cannot be determined in a simple fashion such as by Eq. 12-5. Nevertheless $\sigma_{cf_{bh}}$ is set at some maximum figure and it is this quantity which is assumed to be carried through the subsequent analysis. In cases where the hub section is connected to the root by a shank ($r_r \neq r_h$), the value of $\sigma_{cf_{bh}}$ is increased by some factor. Eq. 12-4, 12-5, and 12-6 show that this factor is of the order

$$\frac{\left(1 - \lambda_h^2 \frac{r_r^2}{r_h^2}\right)}{(1 - \lambda_h^2)}$$

The total radial centrifugal force is thus

$$F = F_d + F_b = 2\pi \left(\frac{2\pi}{60}\right)^2 n^2\rho_d A_d k_d^2 + \sigma_{cf_{bh}} A_{bh} N_b \frac{1 - \lambda_h^2 \frac{r_r^2}{r_h^2}}{1 - \lambda_h^2} \tag{12-13}$$

The average tangential stress in the disk may be defined as (Fig. H,12e)

$$\bar{\sigma}_d = \frac{F_v}{2A_d} = \frac{F_d + F_b}{2A_d}\frac{1}{\pi} \tag{12-14}$$

where F_v is the numerical sum of the vertically resolved components of the radial force $F/2$ acting on each half of the disk. Combining Eq. 12-13 and 12-14,

$$\bar{\sigma}_d = \frac{2\pi \left(\frac{2\pi}{60}\right)^2 n^2\rho_d A_d k_d^2 + \sigma_{cf_{bh}} A_{bh} N_b \frac{1 - \lambda_h^2 \frac{r_r^2}{r_h^2}}{1 - \lambda_h^2}}{2\pi A_d} \tag{12-15}$$

Eq. 12-15 may now be rearranged to determine the disk area A_d if the other quantities are known,

$$A_d = \frac{\sigma_{cf_{bh}} A_{bh} N_b \dfrac{1 - \lambda_h^2 \dfrac{r_r^2}{r_h^2}}{1 - \lambda_h^2}}{2\pi \left[\bar{\sigma}_d - \left(\dfrac{2\pi}{60}\right)^2 n^2 \rho_d k_d^2 \right]} \tag{12-16}$$

The possibility that the denominator in this expression for A_d can be zero ($A_d = \infty$) is a logical outcome of the possible equality of the centrifugal stress term $(2\pi/60)^2 n^2 \rho_d k_d^2$ and the allowable stress term $\bar{\sigma}_d$. The disk area A_d would then necessarily be infinite in order to carry the additional centrifugal load due to the blading. In practice, A_d and consequently the disk weight increase very rapidly with blade loading if the disk is only just self-supporting. This requires a reduction in rotational speed. The total disk weight can be written as

$$M_d = \rho_d \int w_d 2\pi r dr = 2\pi \rho_d \bar{r}_d A_d \tag{12-17}$$

where $\bar{r}_d$ is the distance of the centroid of A_d from the axis. Eq. 12-16 and 12-17 can be combined to give

$$M_d = \frac{\rho_d \bar{r}_d \sigma_{cf_{bh}} A_{bh} N_b \dfrac{1 - \lambda_h^2 \dfrac{r_r^2}{r_h^2}}{1 - \lambda_h^2}}{\bar{\sigma}_d - \left(\dfrac{2\pi}{60}\right)^2 n^2 \rho_d k_d^2} \tag{12-18}$$

Some of the terms in Eq. 12-18 are more conveniently transformed. The number of blades may be expressed in terms of the hub pitch-to-chord ratio, since this ratio is an important ingredient of good performance (Fig. H,3g):

$$N_b = \frac{\dfrac{2\pi r_h}{c_{bh}}}{\left(\dfrac{s}{c}\right)_{bh}} \tag{12-19}$$

The blade area at the hub A_{bh} may be expressed in nondimensional terms as $(A/ct)_{bh}$ (Fig. H,12c):

$$A_{bh} = \left(\frac{A}{ct}\right)_{bh} \left(\frac{c}{t}\right)_h t_h^2 = \zeta_h \left(\frac{c}{t}\right)_h t_h^2 \tag{12-20}$$

The blade width and chord may be related approximately to the stagger angle γ (Fig. H,3a) as

$$\cos \gamma_m \cong \left(\frac{w}{c}\right)_{bm} \cong \cos \alpha_{\infty m} \tag{12-21}$$

where $\alpha_{\infty m}$ is defined by Eq. 8-8. The angle γ is used here as a convenient approximation to define a direction perpendicular to the average line of action of the total force on a blade. If no losses take place this angle should properly be α_∞ (see Sec. A). The blade chord at hub c_h is a function of allowable bending stress, the form of which has already been developed (Eq. 12-9). Since the tangential load on the blades is related to the work by Eq. 3-5, the blade force L can be expressed in the following manner:

$$\begin{aligned} L &\sim \frac{Q(h_4^0 - h_5^0)}{U_m \cos \gamma_m} \\ &\sim Q\phi_m \frac{n}{2} r_t(1 + \lambda_h) \frac{1}{\cos \gamma_m} \end{aligned} \tag{12-22}$$

where Q is the total mass flow rate through the turbine and the mean work coefficient ϕ_m is defined by Eq. 3-11. Inserting Eq. 12-19, 12-20, 12-9, and 12-22 into Eq. 12-18,

$$\frac{M_d}{\rho_d r_t^3} = a_1 \frac{\left(\frac{Q}{A_t}\right)^{1/2} \phi_m^{1/2} (nr_t)^{1/2} \frac{\bar{r}_d}{r_r} \frac{\sigma_{cf_{bh}}}{(\sigma_{be_{bh}})^{1/2}}}{\frac{144}{g}\frac{\bar{\sigma}_d}{\lambda_h^2} - \left(\frac{2\pi}{60}\right)^2 (nr_t)^2 \rho_d \left(\frac{k_d}{r_r}\right)^2 \left(\frac{r_r}{r_h}\right)^2} \left[\frac{\zeta}{\left(\frac{s}{c}\right)^{1/2} \xi^{1/2}}\right]_{bh} \frac{1}{\cos^{1/2} \gamma_m} \left[\frac{\left(1 - \lambda_h^2 \frac{r_r^2}{r_h^2}\right)\frac{r_r}{r_h}}{\lambda_h^{1/2}(1 - \lambda_h^2)^{1/2}}\right] \tag{12-23}$$

when (Q/A_t) is the flow rate per unit frontal area and a_1 is a constant. The constants 144 and g are added if $\bar{\sigma}_d$ is expressed, as usual, in psi.

The thickness-to-chord ratio t/c does not appear explicitly in Eq. 12-22, which is attributable to opposing effects of centrifugal forces (Eq. 12-12, 12-19, and 12-20) and gas bending forces (Eq. 12-9). Intuitively it can be seen that, for a given hub centrifugal force, an increase in blade thickness tends to produce higher rim loading, F_b, but also leads to reduced bending stresses, thus permitting smaller blade chords, restoring the status quo. In practice, blade thickness-to-chord ratio has an effect through changed values of ζ and ξ (Fig. H,12c and H,12d). Also a decrease in t/c makes the blade less stiff, probably increasing the tendency to vibrate and thus increasing the constant of proportionality in Eq. 12-9. The constant is an ingredient of the constant a_1 in Eq. 12-23.

The fact that σ_{cf} occurs in the numerator of Eq. 12-23 rather than the denominator, as does σ_{be}, implies an increase of disk weight with allowable centrifugal stress. This is due to the higher disk loading that this increase entails. The presence of the ratio $\sigma_{cf_{bh}}/(\sigma_{be_{bh}})^{\frac{1}{2}}$ indicates that increase of bending stress at the expense of centrifugal stress is favorable to weight reduction. However, for a given value of the hub-to-tip ratio λ_h, this enforces thicker blade sections at the hub and thinner sections at the tip (Eq. 12-6 and Fig. H,12b), resulting in decreased aerodynamic performance. In addition, an increase in the bending stresses increases the vibratory portion of the loading, which is least amenable to exact calculation.

Disk weight is a complicated function of λ_h. Eq. 12-23 shows a zero weight at the extremes of $h = 0$ and $h = 1$. In the former case this is due to the vanishing disk size and in the latter case to the absence of blades altogether.

The first and second moment ratios $\bar{r}_d/r_r$ and k_d/r_r in Eq. 12-23 emphasize the need for placement of disk material as near the axis as possible. For weight comparison purposes these can be assumed to remain constant from one design to the next.

Blade weight. If the blades were replaced by solid material of width equal to the average value for the disk, the entire rotor weight might be expected to have the value M_d/λ_h^2. Due to partial solidity of the blading, the rotor weight is often very nearly equal to M_d/λ_h. For optimization purposes a more precise investigation of factors influencing blade weight can proceed as follows.

The weight per blade varies linearly with the area of the cross section, the height, and the density. The weight of a complete blade row can then be written

$$M_b \cong \frac{A_{bh}}{2} \rho_b (r_t - r_r) N_b \tag{12-24}$$

This assumes that the section area varies linearly with radius and is substantially zero at the tip. The values of A_{bh} and N_b can now be expressed in terms of allowable pitching (Eq. 12-19) and allowable chord (Eq. 12-9). Using also Eq. 12-22, the dimensionless weight of a blade row becomes

$$\frac{M_b}{\rho_b r_t^3} \sim \zeta_{bh}\lambda_h \left(1 - \lambda_h \frac{r_r}{r_h}\right) \frac{1}{(s/c)_{bh}} \left[\frac{Q}{A_t} \frac{\phi_m(nr_t)}{\cos\gamma_m} \frac{1 - \lambda_h^2}{\lambda_h} \frac{1}{\xi} \frac{1}{\sigma_{be_{bh}}} \left(\frac{s}{c}\right)_{bh}\right]^{\frac{1}{2}}$$

$$= a_2 \left(\frac{Q}{A_t} \phi_m n r_t\right)^{\frac{1}{2}} \left[\frac{\zeta}{(\xi s/c)^{\frac{1}{2}}}\right]_{bh} \frac{1}{\cos^{\frac{1}{2}}\gamma_m}$$

$$\left[\lambda_h^{\frac{1}{2}}\left(1 - \lambda_h \frac{r_r}{r_h}\right)(1 - \lambda_h^2)^{\frac{1}{2}}\right]\left(\frac{1}{\sigma_{be_{bh}}}\right)^{\frac{1}{2}} \tag{12-25}$$

where a_2 is a constant.

Eq. 12-25 may be used in this form for studying influences on the weight of a stator row.[1] It can be combined with Eq. 12-23 for study

[1] Eq. 12-25 uses Eq. 12-22 to obtain the torque on the blade row. This is strictly accurate only for the rotor. It is accurate for the stator if the swirl velocities entering and leaving the stage are equal.

Table H,12. **Effect of turbine design variables on disk and blade weight.**

Standard values of design variables For these values M_d = 25.97 lb, M_b = 23.45 lb, $M_d + M_b$ = 49.42 lb			Changed values			
			Variable	M_b, lb	M_d, lb	$M_b + M_d$, lb
Q	Flow	100 lb/sec	110 lb/sec	24.62	27.2	51.82
r_t	Tip radius	1 ft	1.1 ft	33.43	38.1	71.53
ϕ_m	Work coefficient at mid-blade height	0.55	0.50	22.38	24.7	47.08
n	Shaft speed	10,000 rpm	11,000 rpm	24.62	28.9	53.52
$(s/c)_h$	Pitch/chord ratio at hub	0.6	0.55	25.57	28.23	53.80
$\beta_{h4.5}$	Gas inlet angle at hub	40°	45°	23.87	26.35	50.22
α_{h5}	Gas outlet angle at hub	−50°	−55°	24.64	27.21	51.85
γ_m	Stagger at mid-blade height	−10°	−30°	25.28	27.96	53.24
$(t/c)_h$	Thickness/chord ratio at hub	0.2315	0.18	23.53	25.99	49.52
r_r	Disk rim radius	0.57 ft	0.6 ft	21.75	26.72	48.47
r_h	Hub radius	0.6 ft	0.55 ft	20.69	27.24	47.93
$\sigma_{cf_{bh}}$	Blade centrifugal force stress at hub	25,000 psi	20,000 psi	23.45	20.67	44.12
$\sigma_{be_{bh}}$	Blade bending stress at hub	10,000 psi	15,000	19.10	21.75	40.85
$\bar{\sigma}_d$	Average disk tangential stress	60,000 psi	65,000	23.45	23.33	46.78
ρ_d	Disk density	0.288 lb/in.^{3}	0.368 lb/in.^{3}	23.45	36.08	59.53
ρ_b	Blade density	0.311 lb/in.^{3}	0.368 lb/in.^{3}	27.75	25.97	53.72
$\bar{r}_d/r_r$	1st moment ratio of disk	0.55	0.6	23.45	28.26	51.71
k_d/r_r	2nd moment ratio of disk	0.60	0.65	23.45	27.36	50.81
a_1	Const of proportionality—disk	0.49	—	—	—	—
a_2	Const of proportionalty—blades	0.17	—	—	—	—

of the rotor weight as a whole, as follows,

$$\frac{M_{\text{rotor}}}{\rho_d r_t^3} = \left(\frac{Q}{A_t}\phi_m n r_t\right)^{\frac{1}{2}} \left[\frac{\zeta}{(\xi s/c)^{\frac{1}{2}}}\right]_{bh} \frac{1}{\cos^{\frac{1}{2}}\gamma_m} \left[\lambda_h^{\frac{1}{2}}\left(1 - \lambda_h \frac{r_r}{r_h}\right)(1-\lambda_h^2)^{\frac{1}{2}}\right]$$
$$\times \left(\frac{1}{\sigma_{be_{bh}}}\right)^{\frac{1}{2}} \left[a_1 \frac{\frac{1}{\lambda_h}\left(\frac{1+\lambda_h \frac{r_r}{r_h}}{1-\lambda_h^2}\right)\frac{r_r}{r_h}\frac{\bar{r}_d}{r_r}\sigma_{cf_{bh}}}{\frac{144}{g}\frac{\bar{\sigma}_d}{\lambda_h^2} - \left(\frac{2\pi}{60}\right)^2 (nr_t)^2 \rho_d \left(\frac{k_d}{r_r}\right)^2 \left(\frac{r_r}{r_h}\right)^2} + a_2 \frac{\rho_d}{\rho_b}\right] \quad (12\text{-}26)$$

In assessing the influence of aerodynamic parameters contained in Eq. 12-26 on rotor weight, the contents of the first four terms in the product influence both the disk and blades equally and remarks relative to disk weight (Eq. 12-23) therefore also apply to the complete rotor weight. The contents of the last product term reveal those items which influence the disk or blade weights only, i.e. those multiplied by the factors a_1 and a_2. These factors are best determined by examination of a number of designs similar to that under investigation.

Although Eq. 12-25 and 12-26 are useful for weight assessments in a preliminary way, the weights of actual turbines are influenced considerably by detailed design which may modify the constants a_1 and a_2 or the first and second moment ratios $\bar{r}_d/r_r$ and k_d/r_r. Also the turbine casing should rightly be part of the turbine weight, as should any coupling between the disks of multistage turbines for driving purposes. Nevertheless the relationships given supply the aerodynamic designer with some indication of factors governing weight and enable fairly accurate comparisons to be made between two similar designs.

Some variations in disk and blade weight derived from Eq. 12-23, 12-25, and 12-26 are indicated in Table H,12 in the form of a representative weight for standard conditions and revised weights for changes in the variables, one at a time.

H,13. Aerodynamic Design Methods. Aerodynamic turbine design procedure is often divided into two distinct phases, the first stage being the calculation of velocity triangles defining the gas conditions between the blading and the second stage being the design of the blading itself. This procedure is only fruitful if the first phase is carried out with the implications of the second phase and stress considerations in mind. Experience must be the guiding factor for choice of the particular combination of gas angles and velocities that leads to the desired combination of stress, weight, and efficiency. These estimates may require modification after the blade shapes are studied.

The probable effect of velocity triangle choice on blade losses and turbine efficiency can be estimated if a general relationship between gas angles α and β, pitch-to-chord ratio s/c, and loss coefficient Y can be found. Such a relationship has been estimated by Ainley [5] on the basis of a variety of tests such as those indicated in Fig. H,5f. These relationships are necessarily approximate in that Mach number effects, secondary flow losses, and blade thickness variations are not included. They are shown in Fig. H,13a and H,13b, the first figure relating to gas inlet angles β of zero (reaction) and the second for angles β equal to $-\alpha$ (impulse). Other values of β can be obtained by interpolation. Each pair of gas inlet

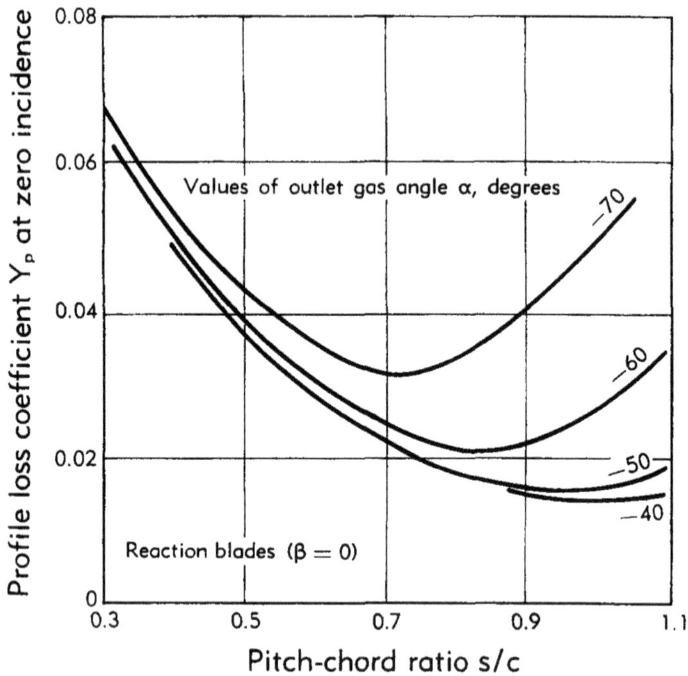

Fig. H,13a. Profile loss coefficients for reaction blades [5].

and outlet angles thus gives a minimum value of profile loss coefficient Y_p, for a particular value of pitch-to-chord ratio s/c, and these minima may be inserted into relationships for turbine efficiency, such as Eq. 4-12. Data obtained in this way are shown in Fig. H,13c. They indicate that, other things being equal, the best efficiencies for a given value of work coefficient ϕ are obtained at values of blade outlet angle α_5 in a definite range. The desirability of choosing a value of V_z/U for a given value of ϕ which is neither too high nor too low can be appreciated qualitatively. Very high values of V_z/U cause high friction losses, which are proportional to dynamic head, whereas very small values give rise to high gas angles α and β (Fig. H,3a) and therefore high losses (Fig. H,13a and H,13b).

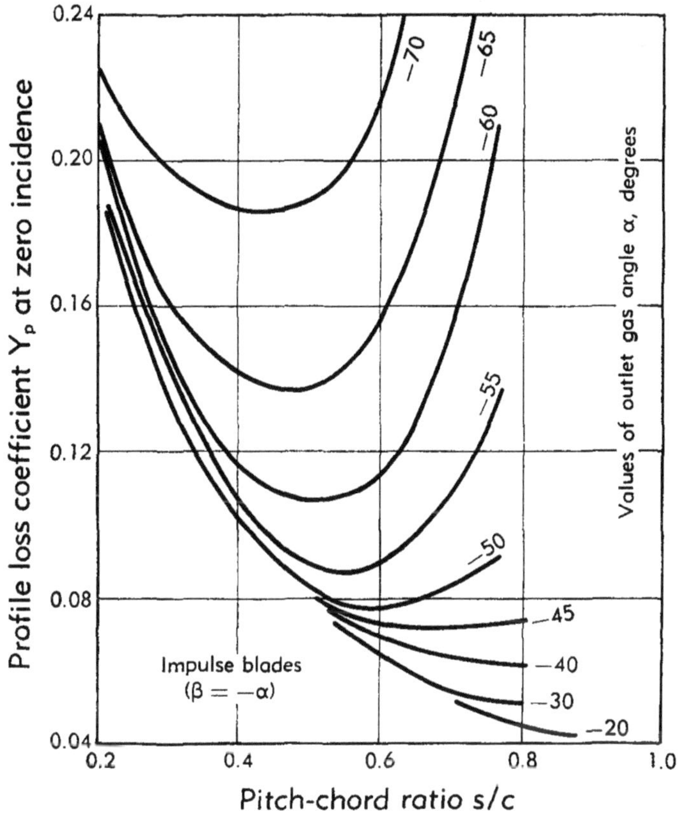

Fig. H,13b. Profile loss coefficients for impulse blades [*6*].

The detailed procedure by which the best possible set of velocity triangles is attained varies with the extent to which the duty quantities, such as blade speed, pressure ratio, and gas flow are fixed by over-all engine design. The general approach nevertheless is apparent from a typical example which follows:

Quantities that are fixed (turbine duty)

Shaft rpm	n
Inlet referred flow	$\dfrac{Q\sqrt{\theta_4^0}}{\delta_4^0}$
Desired work per pound of gas	$h_4^0 - h_5^0$
Number of stages	N
Maximum tip radius	$(r_t)_{max}$
Vortex type	e.g. value of n in Eq. 10-15

The choice of interstage gas conditions is best made by the selection of hub conditions, where the design is most critical owing to the minimum

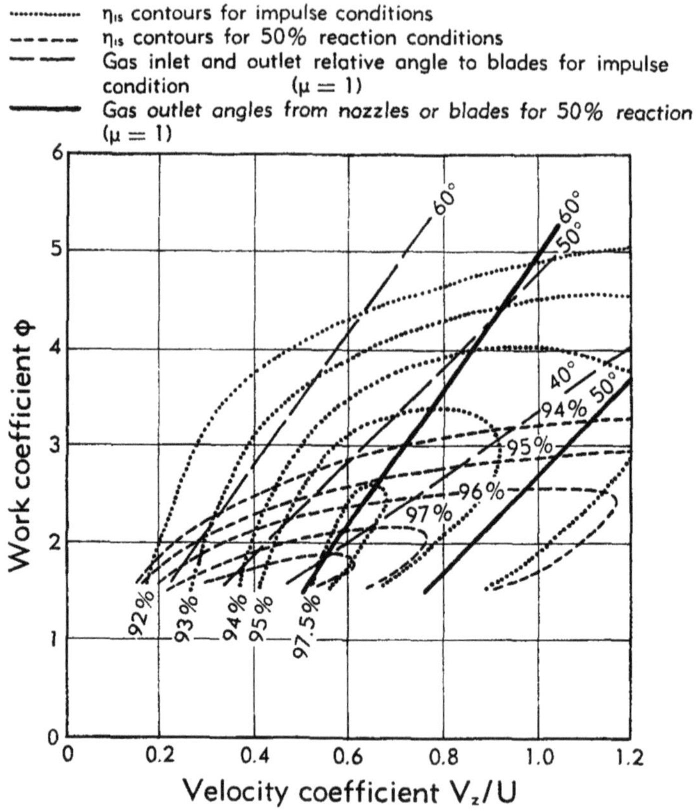

Fig. H,13c. Stage efficiency as a function of reaction, work, and flow parameters.

blade speed at that point. It is sometimes the practice to design for a referred flow slightly higher than the actual value at which the turbine is expected to operate. This is justified by the blockage presented by boundary layers which causes increased axial velocities in midstream positions. Some important design quantities that must be selected to give a turbine the highest efficiency compatible with its life and weight are as follows:

Design criteria—Hub values.	*Severe*	*Mild*
a. Exit Mach number M_5	0.65	0.40
b. Degree of reaction D for each stage	0	35 per cent
c. Gas deflection through rotating blades	120°	80°
d. Axial velocity coefficient $V_{z4.5}/U$	3	½
e. Mach number relative to blading at outlet $W_5/\sqrt{g\gamma \mathcal{R} T_5}$	1.2	0.6
f. Hub centrifugal stress severity criterion $n^2 A_{ann,h}$ (rpm² ft²) (See Eq. 12-6) First stage	250 × 10⁶	150 × 10⁶
Second stage	300 × 10⁶	250 × 10⁶
g. Exit swirl angle β_5 (Fig. H,3a): with outlet guide vanes:	−30°	−15°
without outlet guide vanes:	−20°	0°

The values given under "severe" and "mild" columns represent current values for designs that are highly loaded and lightly loaded (either aerodynamically or mechanically) respectively. The former are pertinent to a very light turbine for a low-life engine, due sacrifices being made for maximum power per unit weight. The latter column represents a highly efficient and durable but rather heavy multistage turbine, perhaps for a long range power plant. The distribution of work between the stages must be adjusted to produce a balanced design, that is, with none of the quantities (a through g) abnormally high at the expense of any of the others. The shape of the annulus must also be considered. It is advisable to introduce annulus flare so that density reductions are offset by larger annulus areas. If the tip diameter is held constant and all radius changes are brought about at the hub, use will be made of the maximum possible blade speed but the gas may suffer excessively high changes of direction in the meridian plane, in the hub region. This position can be alleviated by introducing some of the flare at the outer annulus, but the mean work coefficient ϕ_m (Eq. 3-12) will be higher.

The progress of a design can now be outlined as follows:

1. Decide on an outlet Mach number and swirl angle. For a given duty the exit area is then fixed and the hub diameter can be found.
2. Fix an annulus shape, i.e. hub and tip radii at all inter-row stations, a work distribution, and also a degree of reaction at the hub for all stages except the last, where it is already fixed by the outlet conditions.
3. The hub tangential velocities at all inter-row positions can now be determined (Eq. 3-6 and 3-15). A value of n in Eq. 10-15 must be fixed to give values for V_θ at all radii. For free vortex designs, $n = -1$. The value of n cannot be chosen arbitrarily, both before and after the rotor, if constant work at all radii is required.
4. Axial velocities must now be determined. Their precise determination, even assuming axial symmetry, is a lengthy process (Art. 10 and 11) due to streamline shifts (Fig. H,10a) to accommodate the radial force balance. For free vortex designs the assumption of constant ρV_z with radius is usually sufficiently accurate (Art. 10).
5. The inter-row conditions are now fixed and it only remains to examine them in the light of the design criteria (a through g) which largely control the efficiency and stress level. By systematic examination and adjustment of these quantities for each stage, it is possible to achieve hub and tip radii, hub reactions, work distribution, and exit conditions so that all quantities (a) through (g) are roughly in the same position relative to "severe" and "mild" values. A balanced design is then achieved.

The indicated range of values to be given to design criteria (a) through (f), though representative of modern practice, is still somewhat a matter of debate. Turbine efficiency is logically related most directly to the velocity distribution over the blade surface (Art. 5) and to secondary flow considerations (Art. 8). Criteria (a) through (g) are related to performance in a more indirect manner by test experience.

Typical velocity triangles for a single-stage turbine have been given in Fig. H,10b. This turbine is designed for the following duty:

Inlet stagnation temperature T_4^0	2000°R
Rotor rpm n	13,500
Total flow Q at 73.5 psi inlet pressure	62.7 lb/sec
Work per pound of gas $h_4^0 - h_5^0$	56.4 Btu/lb
Stagnation pressure ratio p_4^0/p_5^0 at 95 per cent for η_{is}	1 57
Hub diameter (const), $2r_h$	15 in.
Tip diameter (const), $2r_t$	20 in.

Values of design criteria (a) through (g) for the hub condition are as follows:

a. Exit Mach number M_5	0.5
b. Degree of reaction (Eq. 3-14)	24 per cent
c. Deflection through rotating blade	77.7°
d. Axial velocity coefficient $V_{a4.5}/U$	1.08
e. Mach number relative to blading at exit	0.69
f. Stress criterion n^2A_{ann} (rpm² ft²)	176×10^6
g. Exit swirl angle β_5	−5.4°

Charts and graphical aids for rapid determination of velocity triangles for given values of some of the variables (a) through (g) have been developed by Cavicchi and English [*67*] and by Reeman and Simonis [*68*]. Fig. H,13d shows a typical design chart due to the latter authors whereby all relevant data for the hub condition are given in terms of two variables, such as the turbine enthalpy drop and the blade speed, for a given value of exit swirl angle and degree of reaction.

Blading design. Final velocity diagrams can now lead to the design of blade shapes, which are a compromise between aerodynamic and mechanical considerations. The former are best suited by thin, closely spaced blading, whereas the latter call for thick blades, which are therefore of necessity pitched farther apart to avoid excessive blockage (Eq. 5-4).

General factors influencing the choice of blade shape have already been indicated in Art. 5, 6, and 7. There are very clear advantages in keeping the blading thin (Fig. H,5f), the trailing edge thickness small (Fig. H,6c), and the operating Mach numbers low (Fig. H,6d). In the case of limiting designs, where all of these quantities are increasing, an exact three-dimensional study (Art. 11) of the pressure distributions around the blades is desirable. The blading performance can then be predicted by boundary layer calculations [*14*] or by correlation with surface pressure distributions [*12*]. There are good indications [*22,23*] that straight-backed blading is most suitable for relative exit Mach numbers

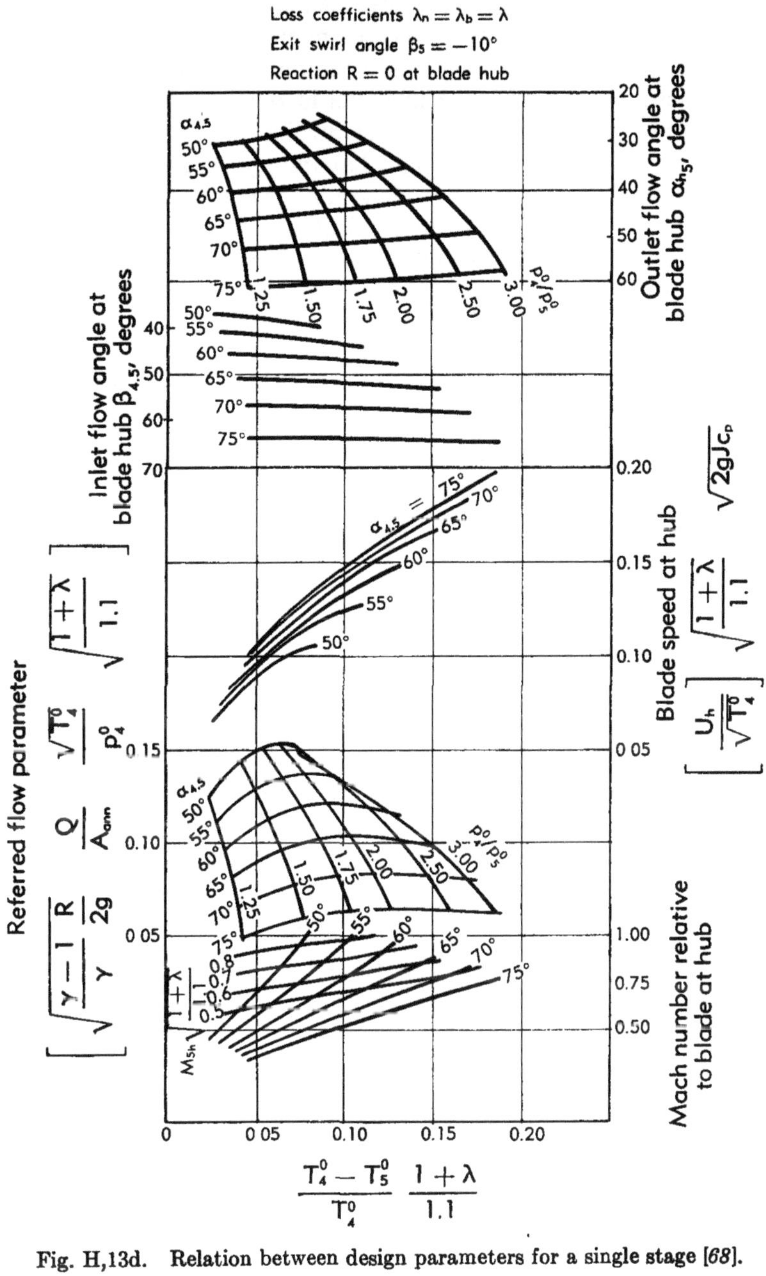

Fig. H,13d. Relation between design parameters for a single stage [68].

of the order of unity or higher, whereas some trailing edge curvature is desirable for lower relative exit Mach numbers.

The possibility of obtaining high subsonic velocities relative to the blade surfaces without severe reductions in turbine efficiency has been demonstrated by a series of transonic turbine tests *[69,70,71,72,73,74]* conducted at the NACA Lewis Flight Laboratory. Approximate methods due to Stewart *[64]* were used to determine the three-dimensional flow pattern over the blading, and, in particular, as a tool to minimize the pressure rise over the pressure and suction surfaces of the blading. In this series of tests it was found that turbine efficiency correlated well with a total diffusion parameter defined as $D_s + D_p$, where

$$D_p = \frac{\text{(inlet relative velocity)-(minimum surface relative velocity)}}{\text{inlet relative velocity}}$$

$$D_s = \frac{\text{(maximum surface relative velocity)-(outlet relative velocity)}}{\text{maximum surface relative velocity}}$$

The total diffusion parameter $D_s + D_p$ can be interpreted with the aid of Fig. H,5d as representing the sum of the increases in pressure on blade pressure and suction surfaces, which will trigger boundary layer growth and will therefore be major causative agents for turbine inefficiency. An accurate determination of the exact meridional streamline pattern is necessary in order to determine the velocity and pressure distributions over blading, since, under transonic conditions, velocity is very sensitive to stream tube area.

Pitching criteria. Optimum pitching criteria developed by Ainley have already been indicated in Fig. H,13a and H,13b. Other pitching criteria due to various authors are available. Criteria based on a limiting value of the blade lift coefficient have been given by Howell *[75]* and by Zweifel *[76]*. Howell's criterion is of the form

$$C_{L_0} = 1.125B \tag{13-1}$$

where

$$B = \frac{\left(6\frac{s}{c} - 1\right)}{5\frac{s}{c}} \tag{13-2}$$

Here C_{L_0} is the blade lift coefficient based on outlet velocity (cf. Eq. 8-6) and defined as

$$C_{L_0} = 2\frac{s}{c}(\tan\beta - \tan\alpha)\frac{\cos^2\alpha}{\cos\alpha_\infty} \tag{13-3}$$

where α_∞ is the vector mean gas angle already defined in Eq. 8-8. The blockage term B is introduced with the thought that, for blades of finite thickness, neighboring blades will touch when $t/c = s/c$ (Fig. H,5a) and therefore the allowable lift coefficient under these extreme conditions must become vanishingly small. The blockage factor B, in Eq. 13-2 becomes unity when s/c is equal to one and zero when s/c equals $\frac{1}{8}$.

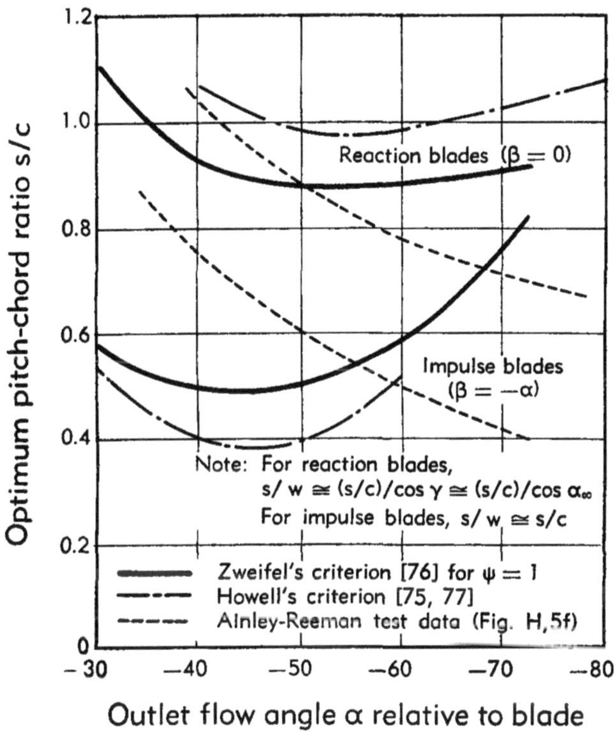

Fig. H,13e. Optimum pitch rules for turbine blading.

Zweifel has defined a loading coefficient ψ, which is substantially the same as Howell's definition of C_{L_0} (Eq. 13-3), namely

$$\psi = 2\frac{s}{w}\cos^2\alpha\left[-\tan\alpha + \frac{r_i}{r_o}\frac{W_{z_i}}{W_{z_o}}\tan\beta + \frac{V_o}{W_{z_o}}\left(\frac{r_i^2}{r_o^2} - 1\right)\right] \tag{13-4}$$

Values at inlet and outlet are taken along the same streamline. If there is a negligible streamline shift across the row, Eq. 13-4 becomes

$$\psi = 2\frac{s}{w}\cos^2\alpha\left(-\tan\alpha + \frac{W_{z_i}}{W_{z_o}}\tan\beta\right) \tag{13-5}$$

In comparing the pitching criteria of Eq. 13-1, 13-2, and 13-3 with those of Eq. 13-5, these will be identical for $1.125B = \psi$, $W_{z_i} = W_{z_o}$, and $W = c\cos\alpha_\infty$. The latter is a geometrical relation nearly true for most blades. The coefficient ψ is usually chosen to have a value in the range 0.75 to 1.2, thicker blades having the lower values. Eq. 13-4 then fixes the required pitch-to-width ratio s/w if the gas inlet and outlet angles are known.

Values of s/c obtained by applying Howell's formula (Eq. 13-1) are compared with values using Zweifel's formula (Eq. 13-5) and with Ainley's relationship (Fig. H,13a and H,13b) in Fig. H,13e.[2] The fairly large discrepancies do not in practice cause concern since variations in

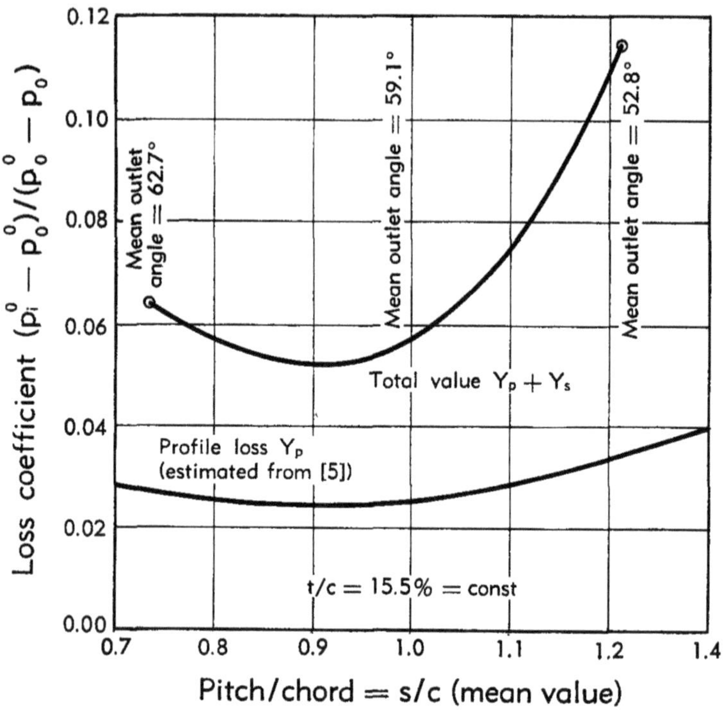

Fig. H,13f. Effect of pitching on losses, for a nozzle [*79*].

blade thickness and form give rise to discrepancies of this order. Moreover, in the usual operating range of angles ($\alpha = -50°$ to $-60°$) agreement is good.

Other methods of fixing the desired value of s/c or s/w, using boundary layer separation criteria, have been developed by Goldstein and Mager [*78*].

It has been customary to set blade-pitching on the basis of two-dimensional loss data mentioned above and to ignore secondary flow considerations. This is justified only if the ratio Y_s/Y_p remains constant for a given cascade as the pitch is changed. There are some indications that

[2] The sign for the term tan α in Eq. 13-3, 13-4, and 13-5 is in accordance with the sign convention of Fig. H,3a for a rotating blade. This sign is reversed if α_5 is considered positive as drawn in this figure.

this is true. Johnston has carried out tests on a nozzle for a range of pitching [*79*], comparing the pitching for minimum over-all loss with that predicted from data [*5*] on profile losses. The upper curve in Fig. H,13f shows the experimentally determined over-all loss $Y_s + Y_p$, while the lower curve indicates the expected variation of Y_p from the two-dimensional cascade data. Very good agreement between optimum s/c based on Y_p and that based on $Y_p + Y_s$ is obtained.

H,14. Performance Characteristics.

Analyses of test data. The measured performance of a given turbine is conveniently expressed in terms of efficiency η_{is} and flow $Q\sqrt{\theta_4^0}/\delta_4^0$ as

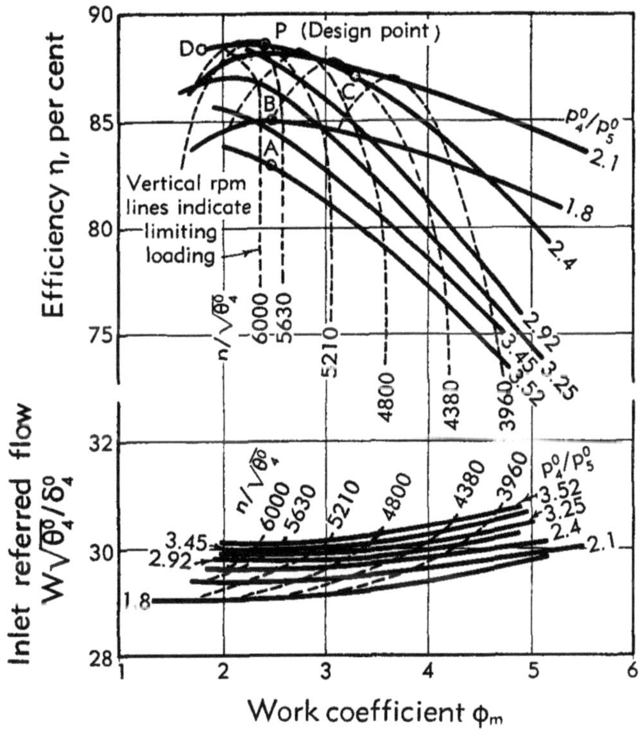

Fig. H,14a. Typical turbine characteristics obtained from test data.

functions of the work coefficient ϕ_m and the over-all pressure ratio p_4^0/p_5^0. A choice of variables alternative to p_4^0/p_5^0 and η_{is}, such as $n/\sqrt{\theta_4^0}$, $(h_4^0 - h_5^0)/\theta_4^0$ can be made, these being directly derivable from each other through Eq. 3-12 and 3-13. The inclusion of the parameter ϕ_m as a variable is significant in that this single parameter will suffice to define the turbine operating point for incompressible flow, as it does for instance in

Fig. H,3f and H,13c. The second parameter, p_4^0/p_5^0, exerts various modifications to the incompressible operating mode, being usually large enough to give rise to significantly high Mach numbers in the turbine.

A typical experimentally determined relationship between η_{i}, or $Q\sqrt{\theta_4^0}/\delta_4^0$, ϕ_m and p_4^0/p_5^0 is shown in Fig. H,14a. Lines of constant referred rpm $n/\sqrt{\theta_4^0}$ are also shown. Other important data often required are the mean outlet Mach number and the outlet swirl angle. Referred flow is nearly constant for most operating conditions because the Mach numbers in the blade throats are generally high. Large changes in delivered work are produced by large changes in these Mach numbers, which correspond, however, to comparatively small changes in flow [*3*, pp. 139–140].

As discussed in Art. 7, continuous reduction in back pressure will load up a blade only to the point where the supersonic expansion fan completely envelops the suction surface (Fig. H,7b and H,7c). Further increases in the pressure ratio then manifest themselves as decreases in efficiency with constant work. This effect can be seen in Fig. H,14a as vertical lines of constant rpm when over-all pressure ratios become very high.

Off-design conditions which lead to reduced efficiency are indicated in Fig. H,14a as points *A*, *B*, *C*, and *D*. These can be categorized as shown in Table H,14. Changes are given relative to the design point *P* in Fig. H,14a.

Table H,14.

Point in Fig. H,14a	Work $h_4^0 - h_5^0$	Pressure ratio p_4^0/p_5^0	Work coefficient ϕ_m	Wheel speed
A	Increased	Increased	Design	Increased
B	Decreased	Decreased	Design	Decreased
C	Decreased slightly	Design value	Increased	Decreased
D	Decreased slightly	Design value	Decreased	Increased

	Blade performance		
	Incidence	Outlet Mach number	Outlet swirl
A	Normal	Decreased	More negative
B	Normal	Increased	More positive
C	Positive	Normal	More negative
D	Negative	Normal	More positive

Velocity triangles corresponding to the off-design conditions *A*, *B*, *C*, and *D* are compared in Fig. H,14b with the design triangles. All comparisons are made by scaling the wheel speed to the same dimension, since this best indicates angle relationships. A comparison of velocities requires rescaling the diagrams.

The performance curves shown in Fig. H,14a can also be deduced theoretically [*28*] provided the blade element pressure losses as a function

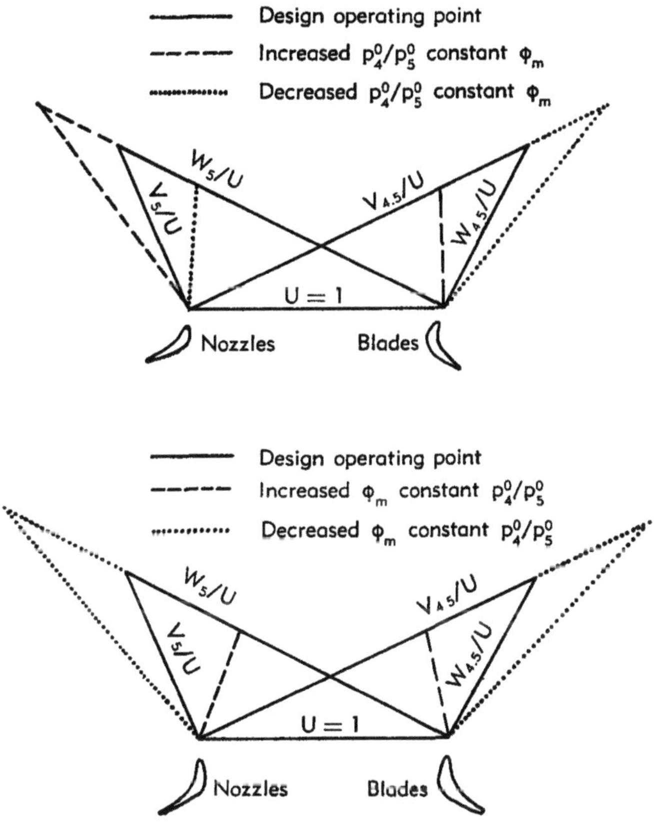

Fig. H,14b. Changes in operating conditions caused by off-design values of ϕ_m and p_4^0/p_5^0.

of their operating conditions are known. In the presence of compressibility effects, simple relationships such as Eq. 3-20, 3-22, and Fig. H,3f are no longer sufficiently valid. However, a row-by-row technique can be employed [*26,27*] whereby the relative outlet conditions from each row are deduced from the relative inlet conditions by regarding the blade passage as a simple converging duct. In this way the work, pressure drop, etc. can be determined in respect to any number of rows for prescribed values of inlet mass flow and rotational speed. If the blading varies substantially in form from hub to tip, a simple calculation based on mid-height will no longer be accurate. Performance calculations then require three-dimensional procedures (Art. 10 and 11).

Multistage turbines. Matching. If the performance characteristics of the separate stages of a multistage turbine are known, the combined characteristic can be worked out by assuming the input conditions from one stage as the output conditions from the previous stage. Stages which are choked, or nearly so, i.e. where referred flow is changing slowly with stagnation pressure ratio across the stage, do not redistribute their work evenly when the over-all pressure ratio is reduced at a given referred rotational speed (rpm) $n/\sqrt{\theta_4^0}$. This can be demonstrated by considering three such stages, say 1, 2, and 3, each designed for a certain pressure drop. The behavior of each stage is illustrated in Fig. H,14c, being a unique line for a given referred rpm $n/\sqrt{\theta^0}$ and very nearly so for different values of referred rpm. Let the design points be A_1, A_2, and A_3. A very small alteration in pressure ratio and flow of the first stage to some point B_1 will lead to a greater divergence from design for the next stage to some point B_2, as can be seen from the following identity:

$$\frac{Q\sqrt{T_2^0}}{p_2^0} = \frac{Q\sqrt{T_1^0}}{p_1}\frac{p_1^0}{p_2^0}\sqrt{\frac{T_2^0}{T_1^0}} \tag{14-1}$$

Using the polytropic relation between the pressure ratio and the temperature ratio (Eq. 2-15), Eq. 14-1 can be written

$$\frac{Q\sqrt{T_2^0}}{p_2^0} = \frac{Q\sqrt{T_1^0}}{p_1^0}\left(\frac{p_1^0}{p_2^0}\right)^{1-\frac{\gamma-1}{\gamma}\eta_p} \tag{14-2}$$

Therefore

$$\frac{\left(\frac{Q\sqrt{\theta_2^0}}{\delta_2^0}\right)_B}{\left(\frac{Q\sqrt{\theta_2^0}}{\delta_2^0}\right)_A} = \frac{\left(\frac{Q\sqrt{\theta_1^0}}{\delta_1^0}\right)_B}{\left(\frac{Q\sqrt{\theta_1^0}}{\delta_1^0}\right)_A}\left[\frac{\left(\frac{p_1^0}{p_2^0}\right)_B}{\left(\frac{p_1^0}{p_2^0}\right)_A}\right]^{1-\frac{\gamma-1}{\gamma}\eta_p} \tag{14-3}$$

This causes a still greater divergence from design operating point in respect to stage 3 to an operating point B_3. An increased pressure drop across the three-stage unit is thus accommodated almost entirely by the last stage. Similar reasoning shows that decreased pressure ratios cause the unit to rematch at points C_1, C_2, and C_3. The operating line of the whole unit is shown in the lower half of Fig. H,14c, where

$$\frac{p_1^0}{p_4^0} = \frac{p_1^0}{p_2^0}\frac{p_2^0}{p_3^0}\frac{p_3^0}{p_4^0} \tag{14-4}$$

Fig. H,14c indicates that the referred flow of the multistage turbine as a whole remains very near the design value for stage 1, even though this stage has a higher flow capacity if operated alone. Eq. 14-2 shows that the full capacity of this machine will only be used if p_1^0/p_2^0 is decreased. This tends to occur at the lower values of rpm, as borne out by the experimental data shown in Fig. H,14a.

The division of the pressure ratio among the three stages is indicated in Fig. H,14d. The absorption of most of the changes in over-all shaft

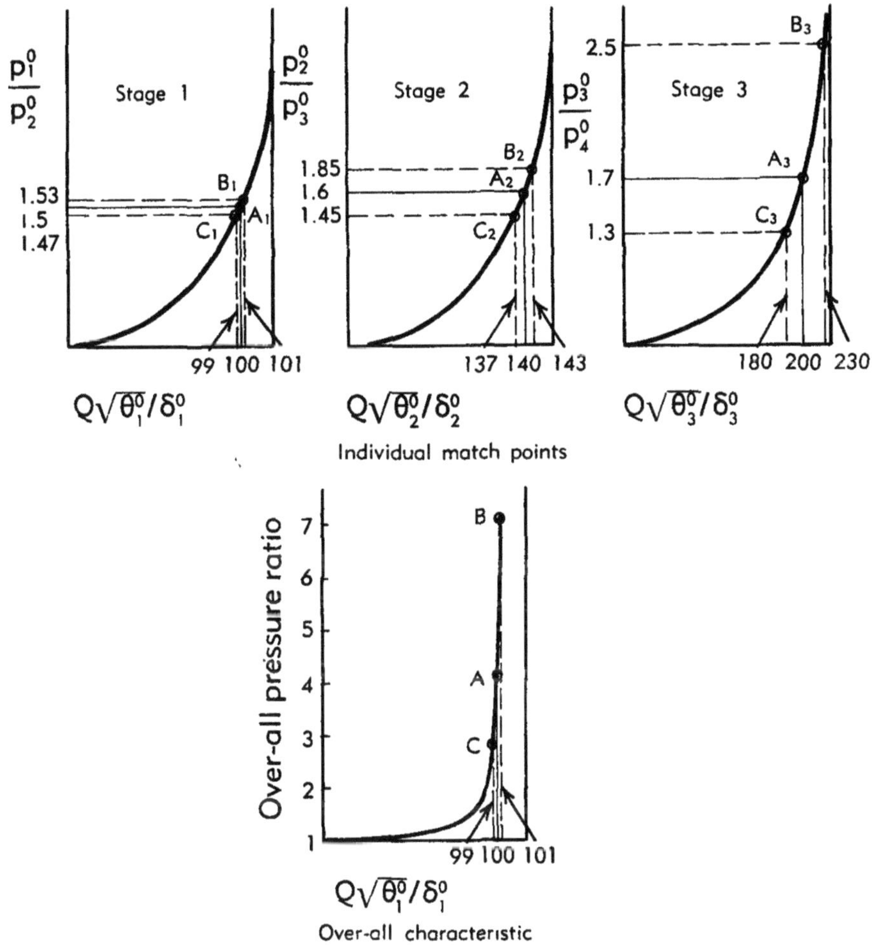

Fig. H,14c. Stage matching for a multistage turbine.

work by the later stages is a phenomenon which must be taken into account in design. Any greatly increased pressure ratios that take place in the final stage lead to supersonic flow relative to the rotating blades, eventually leading to limiting loading conditions and very low efficiencies (Art. 7 and Fig. H,14a).

Very similar considerations apply if the blading is analyzed on a row-by-row rather than stage-by-stage basis [*27,28*]. Curves similar to

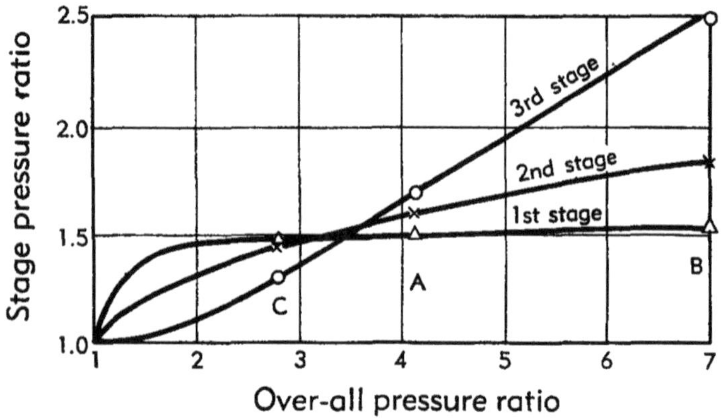

Fig. H,14d. Off-design operation of a multistage turbine.

those of Fig. H,14c will then depict the relation between the relative pressure drop p_1^{00}/p_2 and the referred flow $Q\sqrt{\theta_1^{00}}/\delta_1^{00}$ for each blade row.

H,15. Cited References

1. Stodola, A. *Steam and Gas Turbines*. McGraw-Hill, 1927. Reprinted by Peter Smith, New York, 1945.
2. Keenan, J. H. *Thermodynamics*. Wiley, 1946.
3. Keenan, J. H., and Kaye, J. *Gas Tables*. Wiley, 1948.
4. Milne-Thomson, L. M. *Theoretical Hydrodynamics*, 2nd ed. Macmillan, 1950.
5. Ainley, D. G. An approximate method for estimation of the design point efficiency of axial flow turbines. *Brit. Aeronaut. Research Council Current Paper 30*, 1950.
6. Stanitz, J. D., and Ellis, G. O. Flow surfaces in rotating axial-flow passages. *NACA Tech. Note 2834*, 1952.
7. Todd, K. W. Practical aspects of cascade wind tunnel research. *Proc. Inst. Mech. Eng. 157*, 482–497 (1947).
8. Andrews, S. J., and Schofield, N. W. An experimental investigation of a thick-aerofoil nozzle cascade. *Brit. Aeronaut. Research Council Repts. and Mem. 2883*, 1956.
9. Ainley, D. G. Performance of axial-flow turbines. *Proc. Inst. Mech. Eng. 159*, 230–244 (1948). Reprinted in U. S. by *Am. Soc. Mech. Engrs.* Apr. 1949.
10. Huppert, M. C., and MacGregor, C. Comparison between predicted and observed performance of gas turbine stator blades designed for free-vortex flow. *NACA Tech. Note 1810*, 1949.
11. Poritsky, H., Sells, B. E., and Danforth, C. E. Graphical, mechanical and electrical aids for compressible fluid flow. *J. Appl. Mech. 17*, 37–46 (1950).
12. Dunavant, J. C., and Erwin, J. R. Investigation of a related series of turbine-blade profiles in cascade. *NACA Tech. Note 3802*, 1956.
13. Hargest, T. J. The theoretical pressure distribution around some conventional turbine blades in cascade. *Brit. Aeronaut. Research Council Repts. and Mem. 2765*, 1953.
14. Schlichting, H. Some problems of cascade flow. *Proc. Conference on High-Speed Aeronautics, Polytech. Inst. Brooklyn*, Jan. 20–22, 1955.
15. Tetervin, N. A method for the rapid estimation of turbulent boundary layer thickness for calculating profile drag. *NACA Advance Confid. Rept. L4G14*, 1944.
16. Dowlen, E. M. A shortened method for the calculation of aerofoil profile drag. *J. Roy. Aeronaut. Soc. 56*, 109–116 (1952).

17. MacGregor, C. A. Two-dimensional losses in turbine blades. *J. Aeronaut. Sci. 19*, 404–408, 432 (1952).
18. Reeman, J. Performance of cascades of aerofoils at positive stagger. *Power Jets* (now *Natl. Gas Turbine Establishment*) *Mem. M1203*, 1946.
19. Reeman, J., and Simonis, E. A. The effect of trailing edge thickness on blade loss. *Roy. Air Establishment Tech. Note Eng. 116*, 1943.
20. Goldstein, A. W. Analysis of performance of jet engine from characteristics of components. I: Aerodynamic and matching characteristics of the turbine component determined with cold air. *NACA Rept. 878*, 1947.
21. Stewart, W. L. Analysis of two-dimensional compressible-flow loss characteristics downstream of turbomachine blade rows in terms of basic boundary-layer characteristics. *NACA Tech. Note 3515*, 1955.
22. Kraft, H. Reaction tests of turbine nozzles for subsonic velocities. *Trans. Am. Soc. Mech. Engrs. 71*, 781–787 (1949).
23. Stephenson, J. M., and Naylor, V. D. High speed test on a turbine cascade of aerofoil blading having a maximum profile thickness of 10% chord. *Natl. Gas Turbine Establishment, England, Mem. M82*, 1950.
24. Ainley, D. G., and Mathieson, G. C. R. An examination of the flow and pressure losses in blade rows of axial flow turbines. *Brit. Aeronaut. Research Council Repts. and Mem. 2891*, 1955.
25. Durand, W. F., ed. *Aerodynamic Theory*, Vol. 3, Div. H. Springer, Berlin, 1935.
26. Ainley, D. G., and Mathieson, G. C. R. A method of performance estimation for axial-flow turbines. *Brit. Aeronaut. Research Council Repts. and Mem. 2974*, 1957.
27. Duncombe, E. A method of estimating optimum turbine operating conditions for a range of nozzle and blade angles. *Natl. Aeronaut. Establishment, Canada, Rept. 17*, 1952.
28. Kochendorfer, F. D., and Nettles, J. C. An analytical method of estimating turbine performance. *NACA Rept. 930*, 1949.
29. Hauser, C. H., Plohr, H. W., and Sonder, G. Study of flow conditions and deflection angle at exit of two-dimensional cascade of turbine rotor blades at critical and supercritical pressure ratios. *NACA Research Mem. E9K25*, 1950.
30. Hauser, C. H., and Plohr, H. W. Two-dimensional cascade investigation of the maximum exit tangential velocity component and other flow conditions at the exit of several turbine blade designs at supercritical pressure ratios. *NACA Research Mem. E51F12*, 1951.
31. Keenan, J. H. Reaction tests of turbine nozzles for supersonic velocities. *Trans. Am. Soc. Mech. Engrs. 71*, 773–780 (1949).
32. Rohlik, H. E., Kofskey, M. G., Allen, H. W., and Herzig, H. Z. Secondary flows and boundary-layer accumulations in turbine nozzles. *NACA Rept. 1168*, 1954.
33. Preston, J. H. A simple approach to the theory of secondary flows. *Aeronaut. Quart. 5, Pt. 3*, 218–234 (1954).
34. Hawthorne, W. R. Secondary circulation in fluid flow. *Proc. Roy. Soc. London A206*, 374–387 (1951).
35. Squire, H. B., and Winter, K. G. The secondary flow in a cascade of airfoils in a nonuniform stream. *J. Aeronaut. Sci. 18*, 271–277 (1951).
36. Ehrich, F. F. Secondary flows in cascades of twisted blades. *J. Aeronaut. Sci. 22*, 51–60 (1955).
37. New, W. R. An investigation of energy losses in steam-turbine elements by impact-traverse static test with air at subacoustic velocities. *Trans. Am. Soc. Mech. Engrs. 62*, 489–502 (1940).
38. Scholz, N. Secondary flow losses in turbine cascades. *J. Aeronaut. Sci. 21*, 707–708 (1954).
39. Carter, A. D. S., and Cohen, E. M. Preliminary investigation into the three dimensional flow through a cascade of aerofoils. *Brit. Aeronaut. Research Council Repts. and Mem. 2339*, 1946.
40. Allen, H. W., Kofskey, M. G., and Chamness, R. E. Experimental investigation of loss in an annular cascade of turbine-nozzle blades of free vortex design. *NACA Tech. Note 2871*, 1953.

41. Kofskey, M. G., and Allen, H. W. Smoke study of nozzle secondary flows in a low-speed turbine. *NACA Tech. Note 3260*, 1954.
42. Theodorsen, T., and Regier, A. Experiments on revolving disks, cylinders and streamline rods at high speeds. *NACA Rept. 793*, 1944.
43. Stanitz, J. D., and Ellis, G. O. Two-dimensional flow on general surfaces of revolution in turbomachines. *NACA Tech. Note 2654*, 1952.
44. Wu, C-H. Survey of available information on internal flow losses through axial turbo-machines. *NACA Research Mem. RME 50J13*, 1951.
45. Hawthorne, W. R. The thermodynamics of cooled turbines, Part 1. *Trans. Am. Soc. Mech. Engrs. 78*, 1765–1779 (1956).
46. Hawthorne, W. R. The thermodynamics of cooled turbines, Part 2. *Trans. Am. Soc. Mech. Engrs. 78*, 1781–1786 (1956).
47. Rohsenow, W. M. Effect of turbine-blade cooling on efficiency of a simple gas-turbine power plant. *Trans. Am. Soc. Mech. Engrs. 78*, 1787–1794 (1956).
48. Burke, J. C., Buteau, B. L., and Rohsenow, W. M. Analysis of the effect of blade cooling on gas-turbine performance. *Trans. Am. Soc. Mech. Engrs. 78*, 1795–1806 (1956).
49. Carter, A. D. S. Three-dimensional flow theories for axial compressors and turbines. *Proc. Inst. Mech. Eng. 159*, 255–268 (1948). Reprinted in U.S. by *Am. Soc. Mech. Engrs.*, Apr. 1949.
50. Herzig, H. Z., Hanson, A. G., and Costello, G. R. A visualization study of secondary flows in cascades. *NACA Rept. 1163*, 1954.
51. Bridle, E. A. Some high speed tests on turbine cascades. *Brit. Aeronaut. Research Council Repts. and Mem. 2697*, 1949.
52. Wong, R. Y., and Monroe, D. E. Investigation of a 4.5 inch-mean-diameter two-stage axial-flow turbine suitable for auxiliary power drives. *NASA Mem. 4-6-59E*, 1959.
53. Urban, L. A. An approach to the analysis of altitude effects on the performance of axial flow turbojet engines. *Westinghouse AGT Rept. A-1750*, 1954.
54. Davis, H., Kottas, H., and Moody, A. M. G. The influence of Reynolds number on the performance of turbomachinery. *Trans. Am. Soc. Mech. Engrs. 73*, 499–509 (1951).
55. Wu, C-H. A general through-flow theory of fluid flow with subsonic or supersonic velocity in turbomachines of arbitrary hub and casing shapes. *NACA Tech. Note 2302*, 1951.
56. Slivka, W. R., and Silvern, D. H. Analytical evaluation of aerodynamic characteristics of turbines with nontwisted rotor blades. *NACA Tech. Note 2365*, 1951.
57. Pinnes, R. W. The theoretical possibilities for balanced flow in compressor and turbine design. *Am. Soc. Mech. Engrs. Paper 50A66*, 1951.
58. Stanitz, J. D., and Shreldrake, L. J. Application of a channel design method to high-solidity cascades and tests of an impulse cascade with 90° of turning. *NACA Tech. Note 2652*, 1952.
59. Wu, C-H. General theory of three-dimensional flow in subsonic and supersonic turbo-machines of axial-, radial-, and mixed-flow types. *NACA Tech. Note 2604*, 1952.
60. Page, L. *Introduction to Theoretical Physics*, 2nd ed. Van Nostrand, 1951.
61. Wu, C-H., and Brown, C. A. Method of analysis of compressible flow past arbitrary turbo-machine blades on general surface of revolution. *NACA Tech. Note 2407*, 1951.
62. Wu, C-H., Brown, C. A., and Prian, V. D. An approximate method of determining the subsonic flow in an arbitrary stream filament of revolution cut by arbitrary turbo-machine blades. *NACA Tech. Note 2702*, 1952.
63. Hamrick, J. T., Ginsburg, A., and Osborn, W. M. Method of analysis for compressible flow through mixed-flow centrifugal impellers of arbitrary design. *NACA Rept. 1082*, 1952.
64. Stewart, W. L. Analytical investigation of flow through high-speed mixed-flow turbine. *NACA Research Mem. E5H06*, 1951.
65. Mallinson, D. H. General performance calculations for gas turbine engines. *Brit. Aeronaut. Research Council Repts. and Mem. 2684*, 1946.

66. LaValle, V. L., and Huppert, M. C. Effects of several design variables on turbine-wheel weight. *NACA Tech. Note 1814*, 1949.
67. Cavicchi, R. H., and English, R. E. A rapid method for use in design of turbines within specified aerodynamic limits. *NACA Tech. Note 2905*, 1953.
68. Simonis, E. O., and Reeman, J. Gas turbine design based on free vortex flow. *Brit. Aeronaut. Research Council Repts. and Mem. 2541*, 1944.
69. Stewart, W. L., Wong, R. Y., and Evans, D. G. Design and experimental investigation of transonic turbine with slight negative reaction across rotor hub. *NACA Research Mem. E53 L29a*, 1954.
70. Wong, R. Y., Monroe, D. E., and Wintuckey, W. T. Investigation of effect of increased diffusion of rotor-blade suction-surface velocity on performance of transonic turbine. *NACA Research Mem. E54 F03*, 1954.
71. Whitney, W. J., Monroe, D. E., and Wong, R. Y. Investigation of transonic turbine designed for zero diffusion of suction surface velocity, *NACA Research Mem. E54 F23*, 1954.
72. Whitney, W. J., Wong, R. Y., and Monroe, D. E. Investigation of a transonic turbine designed for a maximum rotor-blade suction-surface relative Mach number of 1.57, *NACA Research Mem. E54 G27*, 1954.
73. Miser, J. W., Stewart, W. L., and Monroe, D. E. Effect of high rotor pressure-surface diffusion on performance of a transonic turbine, *NACA Research Mem. E55 H29a*, 1955.
74. Whitney, W. J., Buckner, H. A., and Monroe, D. E. Effect of nozzle secondary flows on turbine performance as indicated by exit surveys of a rotor. *NACA Research Mem. E54 B03*, 1954.
75. Howell, A. R. The aerodynamics of the gas turbine. *J. Roy. Aeronaut. Soc. 70*, 329–348 (1948).
76. Zweifel, O. The spacing of turbo-machine blading especially with large angular deflection. *Brown Boveri Rev. 32*, 436–444 (1945).
77. Howell, A. R., and Carter, A. D. S. Fluid flow through cascades of aerofoils. *Sixth Intern. Congress Appl. Mech.*, Paris, 1946; *Natl. Gas Turbine Establishment, England, Rept. R6*, 1946.
78. Goldstein, A. W., and Mager, A. Attainable circulation about airfoils in cascade. *NACA Rept. 953*, 1950.
79. Johnston, I. H. An analysis of the air flow through the nozzle blades of a single stage turbine. *Brit. Aeronaut. Research Council Current Paper 131*, 1953.

SECTION I

THE RADIAL TURBINE

WERNER T. VON DER NUELL

In the field of water turbines it has long been common practice to select the turbine wheel configuration dependent on the characteristic power supply factors (flow rate per time unit and pressure head), i.e. dependent on specific speed considerations. The Pelton turbine with jet nozzles and the typical buckets for very low speed, the Francis type with a radial or semiradial wheel and, often, variable nozzles for medium speed, and the Kaplan type with a propeller-like wheel for highest specific speed values are the significant representatives. Single-stage units are the rule.

With steam and gas turbines, the picture is different. The majority of such turbines are of the axial flow type and multistaging, partial admission, and other means have been used to arrive at practical solutions. Around the turn of the century radial steam turbines [*1*, Fig. 551, 741, 743] were introduced, but their utilization remained very limited.

In the last two decades intensified development of gas turbines has revived the search for new design approaches in an effort to simplify and otherwise improve turbine construction and performance. A thorough re-evaluation of the merits of radial turbines[1] for compressible fluids has occurred and configurations very similar to radial pumps for gases and air have thus begun to become of practical significance. This type of radial turbine (both centripetal and centrifugal) will be dealt with in the following, mainly for single-stage units. Although the principles discussed can easily be applied to multistage arrangements, only very limited experience with multistage units of these types is available.

For reasons of simplicity and space, restriction to the use of the so-called one-dimensional flow theory [*2*] must be accepted here. This simplification, while of stronger influence in the design of compressors for higher pressure ratios, is rather satisfactory for radial turbines with predominantly accelerated flow, as practical results have shown.

[1] Birmann selected the diagonal flow type, such as that used in De Laval turbochargers; von Ohain used a Francis-type wheel in the first Heinkel aircraft gas turbine. The author concentrated on developments with the simplest turbine wheel, the 90° type, now in use and in production in very large numbers.

CHAPTER 1. IDEAL FLOW THROUGH A RADIAL TURBINE

I,1. Velocity Triangles. In a "radial turbine" the direction of the flow of the working substance is predominantly radial, either inwardly (centripetal), i.e. toward the axis of rotation, or outwardly (centrifugal) through a rotating cascade of blades arranged around the center in a circular fashion.

Fig. I,1 illustrates a radial turbine blade system of the rotor blades proper and a surrounding stationary nozzle ring with vanes. Moving with

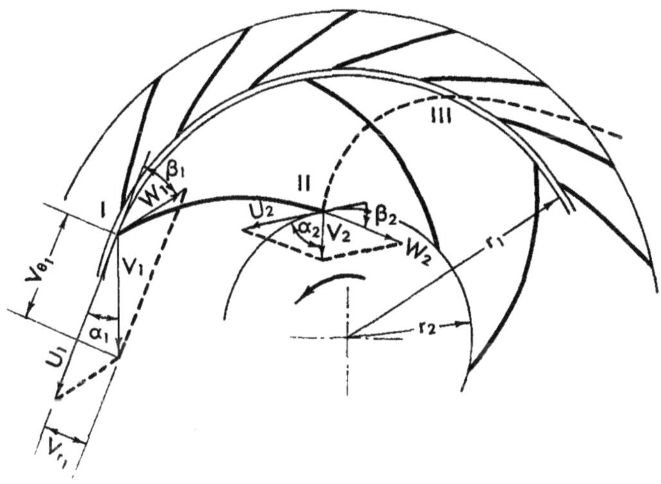

Fig. I,1. Blading of a radial turbine.

the wheel, one would observe the velocity of the flow relative to the rotating blades, i.e. the relative velocity W; whereas, while at rest in relation to the rotating wheel, one observes the absolute velocity V resulting from the vectorial addition of W and the peripheral velocity U as shown in the so-called velocity triangles. The blades of the rotor deflect the flow from the approaching to the leaving direction, thus producing a torque.

Assuming the flow filaments to be congruent and proceeding parallel to the surrounding surfaces, the relative path of a gas particle is given by the blade shape I-II, while the absolute path has the shape III-II.

I,2. Specific Blade Work. The amount of work theoretically extracted by the turbine wheel from each pound of gas, H_t (ft lb/lb), represents the difference of the energy contents at the entrance to, and the exit from, the blading and is called "head." This head (change in enthalpy

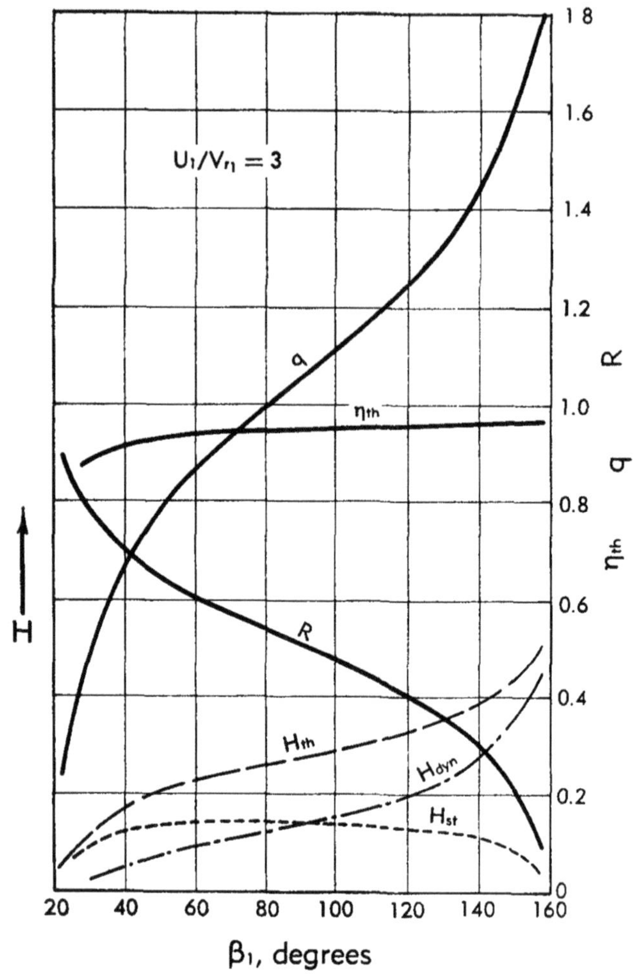

Fig. I,2a. Calculated theoretical characteristic values for radial turbines as functions of the inlet blade angle.

or "specific theoretical blade work") can easily be determined through the "moment of momentum" consideration [2] to be

$$gH_t = U_1V_{\theta_1} - U_2V_{\theta_2} \tag{2-1}$$

with V_θ being the circumferential component of the absolute velocity of the flow. Naturally, H_t can be divided into a portion H_s, representing the change in static pressures, and a portion $H_d = \Delta V^2/2g$, representing the change in absolute velocities or dynamic head, that is

$$H_t = H_s + H_d \tag{2-2}$$

If the flow at the exit from the turbine blading is to leave the wheel along the shortest possible path, the absolute velocity must be perpendicular to the peripheral velocity, i.e. $\alpha_2 = 90°$, which means $V_{\theta_2} = 0$, and thus,

$$gH_t = U_1 V_{\theta_1} = qU_1^2 \tag{2-3}$$

(Different outlet conditions, resulting in positive or negative V_{θ_2} values,

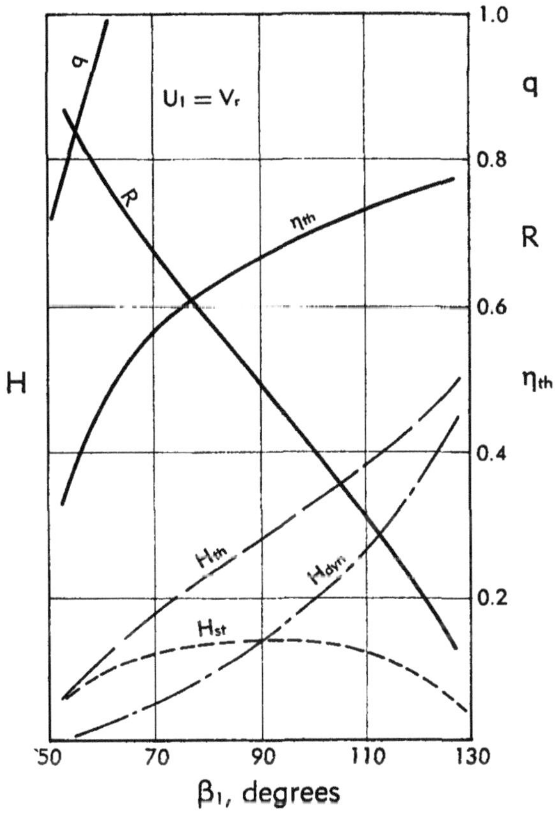

Fig. I,2b. Calculated theoretical characteristic values for radial turbines as functions of the inlet blade angle.

obviously decrease or increase the H_t value and must be accounted for accordingly.)

The blade angle β_1 theoretically can have any value $0° < \beta_1 < 180°$. Its influence can be demonstrated through changing Eq. 2-3 by means of $V_{\theta_1} = U_1 - W_1 \cos \beta_1$ into

$$gH_t = U_1(U_1 - W_1 \cos \beta_1) \tag{2-4}$$

demonstrating that, for a given value U_1 of the tip speed, H_t increases with increasing β_1. Fig. I,2a and I,2b illustrate this in principle for two different ratios of tip speed and meridional velocity, as well as the corresponding changes in the static and dynamic heads as functions of the blade angle β_1. (For better comparison, Fig. I,2a and I,2b are based on identical velocity triangles at the blade exit. Fig I,2c, shows the type of blading.)

I,3. Turbine Wheel Reaction. The reaction of a turbine wheel is defined as the ratio:

$$R = \frac{\text{static head of wheel}}{\text{total head}}$$

which also means that $R = 0$ if the static pressures at the blade inlet and outlet are equal. In line with the above,

$$R = \frac{H_s}{H_t} = 1 - \frac{H_d}{H_t} \tag{3-1}$$

With $\alpha_2 = 90°$ and, for simple demonstration, assuming $V_{r_1} = V_{r_2} = V_2$ it can be written that

$$2gH_d = V_1^2 - V_2^2 = V_1^2 - V_{r_1}^2 = V_{\theta_1}^2$$

resulting in

$$R = 1 - \frac{V_{\theta_1}}{2U_1} \tag{3-2}$$

It is necessary and common practice to differentiate between turbines with different degrees of reaction. For the special case of $\beta_1 = 90°$, $\alpha_2 = 90°$, there is $R = \frac{1}{2}$; while with $V_{\theta_1} = 2U_1$ there is $R = 0$, characteristic for the so-called impulse-type turbine. Fig. I,2a and I,2b show clearly[2] that small β_1 values result in a higher degree of reaction than do larger β_1 values which in turn, however, give higher H_t values for a given tip speed. The decrease of R with increasing β_1 results from the fact that the discharge velocity from the nozzle vanes increases simultaneously, thus requiring an increasing portion of the available head.

For centripetal turbines the magnitude of the influences discussed can be demonstrated easily by showing three typical wheel blading arrangements, Fig. I,3a [3], for given equal values of the head, the through flow, and the number of revolutions per unit time. The latter condition explains the difference in diameter which is proportional to the tip speed required.

Attention is invited to the important fact that even though the blading systems illustrated assume radial inward flow, the general equations have no such restriction as long as inlet and outlet conditions are used properly, as shown for a centrifugal turbine in Fig. I,3b.

[2] Each and every point of the curves represents operation at the design condition because one (and only one) set of velocity triangles applies to each point.

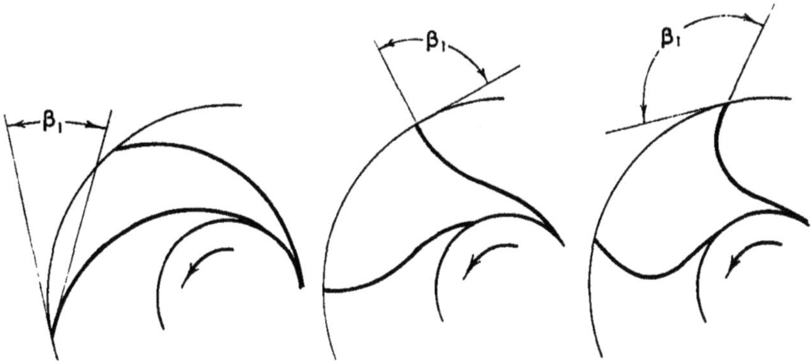

Fig. I,2c. Centripetal turbine wheels with $\alpha_2 = 90°$ and equal outlet angles.

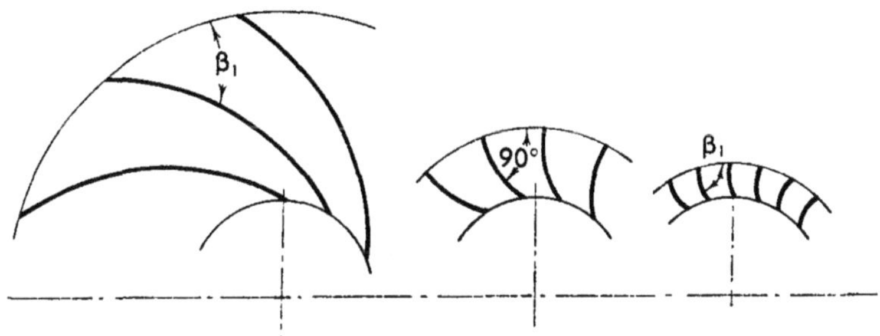

Fig. I,3a. Three turbine wheels for equal head, through flows, and number of revolutions with equal blade outlet angle.

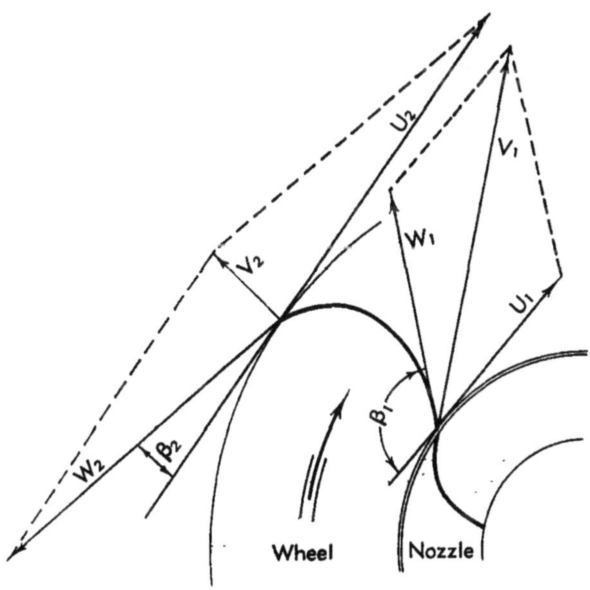

Fig. I,3b. Velocity triangles for centrifugal turbine.

I,4. Theoretical Efficiency. In addition to the energy transfer matters discussed so far, it is necessary to study the influence of the various configurations on the efficiency. It must be realized that as long as the exit velocity V_2 has, of necessity, a finite value, some of the energy available to the turbine wheel has not been converted into mechanical work. In order to even theoretically extract the specific blade work H_t from the gas, the head available to the turbine must be greater by a head corresponding to the exit loss. Thus the theoretical efficiency, by definition, ignoring all other losses (friction, leakage, etc.), can be calculated from

$$\eta_t = \frac{2gH_t}{2gH_t + V_2^2} = \frac{1}{1 + (V_2^2/2gH_t)} \tag{4-1}$$

and, compatible with the assumptions, has been included in Fig. I,2a and I,2b.

CHAPTER 2. REAL FLOW THROUGH A RADIAL TURBINE

I,5. Deflection and Friction. The actual conditions of flow through a turbine, as contrasted to the assumptions made above, can be accounted for in some of the equations discussed previously. Basically, the actual deviation of the flow angles from the blade angles requires a larger deflection of the flow (or larger turning angle) than results from the "theoretical" consideration [*1,2,3,4*]. If this angle increase is not provided for, the energy output effected will be smaller than the value calculated. This influence can be held relatively small for centripetal turbines with diameter ratios $D_1/D_2 > 1.5$ and a reasonably great number of blades.

The friction losses resulting from viscous flow can be accounted for through the introduction of loss coefficients, e.g. by a method similar to the classical method used for steam turbines for decades [*1*]. This method is simple and has shown satisfactory agreement with test data [*5*]. One of the physical effects of viscosity, flow separation, occurs especially in a decelerated flow and upon drastic changes in direction or magnitude of the flow velocity. Some dead area filled with eddies results from such separation as well as from the blade thickness. This phenomenon causes a decrease in effective flow area, i.e. an increase in meridional velocity which counteracts the influence of the lack of deflection. Thus it becomes understandable why, in many cases, simplified methods [*5,6*] are quite acceptable as a design basis. When the "cascade solidity" becomes small a more correct calculation of turning angles or flow deviations may be necessary [*2*].

I,6. Turbine Head, Efficiency, and Other Characteristics. For this presentation the turbine is assumed not to have provisions for gas cooling while expanding, so that the reference equation for the change of state of the gas is that of the adiabate. For a pressure p_{in} and a gas temperature T_{in} at the turbine inlet and a pressure p_{ex} at its exit,

$$H_{ad} = \frac{\gamma}{\gamma - 1} \mathcal{R} T_{in} \left[1 - \left(\frac{p_{ex}}{p_{in}} \right)^{\frac{\gamma-1}{\gamma}} \right] = \frac{V_0^2}{2g} \qquad (6\text{-}1)$$

represents the energy available per pound of gas. V_0 is introduced as

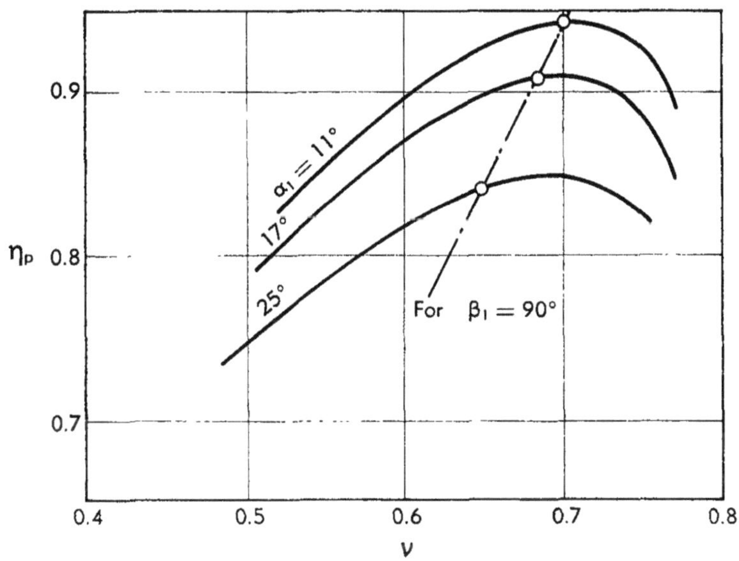

Fig. I,6. Calculated influence of nozzle angle on peripheral efficiency [6].

head velocity. The gas velocity at the nozzle throat, with φ_1 accounting for friction loss and with a reaction R, is

$$V_1 = \varphi_1 \sqrt{2gH_{ad}(1 - R)} = \varphi_1 V_0 \sqrt{1 - R} \qquad (6\text{-}2)$$

A *peripheral* or blading efficiency, i.e. the efficiency not accounting for disk friction and mechanical and leakage losses, can be defined as

$$\eta_p = \frac{\text{effective specific blade work}}{\text{adiabatic head}} = \frac{2(U_1 V_{\theta_1} - U_2 V_{\theta_2})}{V_0^2} \qquad (6\text{-}3)$$

Introducing as the characteristic velocity ratio

$$\nu = \frac{U_1}{V_0} \qquad (6\text{-}4)$$

and

$$V_{r_2} = jV_{r_1} \qquad (6\text{-}5)$$

Fig. I,6 shows calculated values [6] of η_p as a function of the velocity ratio ν, with the nozzle angle α_1 as parameter. Although this calculative approach is too much simplified, the results show the trends clearly and are essentially in agreement with applicable test data.

In addition to the friction in all flow channels, there are other inner losses influencing the energy transfer, such as fluid friction on the disk and/or the shroud of the wheel and internal leakage or bypass losses, because a certain amount of the gas, such as direct heat loss to the outside, does not participate in the normal energy transfer. These losses influence the temperature of the gas and find expression in the inner efficiency:

$$\eta_{i,ad} = \frac{\Delta t_i}{\Delta t_{ad}} \tag{6-6}$$

This value can obviously be measured via temperature and pressure surveys at a turbine in operation. External leakage and mechanical losses can be determined and accounted for with the commonly accepted definitions. Also, the polytropic efficiency [2, p. 442], helpful in compressor calculations, can be useful when comparing turbines. The over-all or total efficiency, of course, is

$$\eta_{tot} = \frac{\text{measured power output}}{\text{power input}} = \frac{550P}{wH_{ad}}\left[\frac{\text{hp}}{\text{lb/sec} \times \text{ft lb/lb}}\right] \tag{6-7}$$

where P is the power out and w is the weight flow through the turbine.

With $\alpha_2 = 90°$, i.e. $V_{\theta_2} = 0$, the general turbine equation can be written as

$$gH_t = U_1 V_1 \cos \alpha_1$$

With reference to Eq. 4-1 one can write

$$V_2^2 + 2gH_t = \xi 2gH_{ad}$$

which with

$$V_2 = \zeta \sqrt{2gH_{ad}} = \zeta V_0 \tag{6-8}$$

leads to a "velocity ratio"

$$\frac{U_1}{V_0} = \frac{\xi - \zeta^2}{\varphi_1} \frac{1}{2 \cos \alpha_1 \sqrt{1 - R}} = \nu \tag{6-9}$$

A numerical evaluation of more average values for centripetal turbines with β_1 approximating 90° results in values for ν from 0.68 to 0.73 for the best *inner* efficiency. The head coefficient q_{ad}, defined by

$$q_{ad} U_1^2 = gH_{ad} = \frac{V_0^2}{2} \tag{6-10}$$

and the velocity ratio are obviously related by

$$q_{ad} = \frac{1}{2\nu^2} \tag{6-11}$$

For best *over-all* efficiency, it is generally advantageous to design for a ν value somewhat smaller than calculated, in the manner shown.

I,7. Calculation of the Main Dimensions. Assuming the normal case of prescribed inlet and outlet pressures and inlet gas temperature,

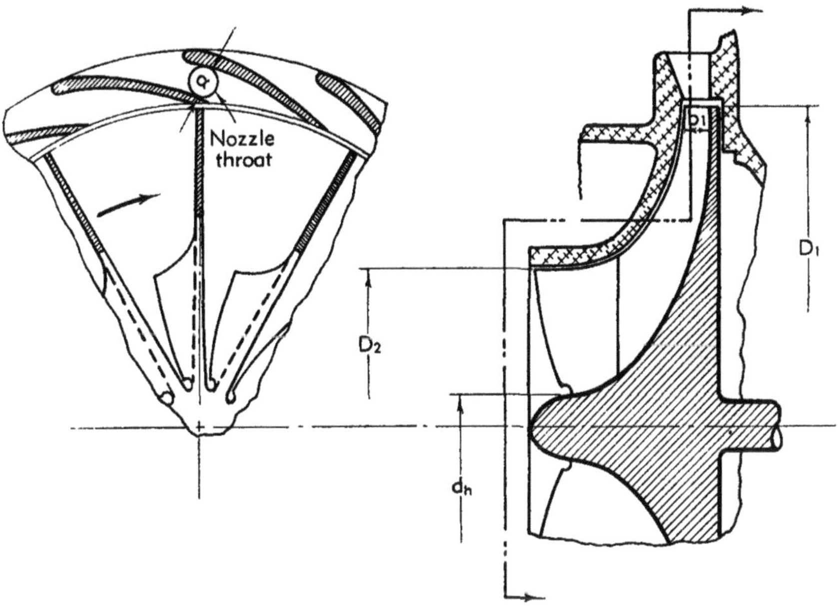

Fig. I,7a. Centripetal turbine with 90° blading.

Eq. 6-1 serves to determine H_{ad}. The discharge volume flow results from Eq. 6-7:

$$Q_{ex} = w \frac{\mathcal{R}T_{ex}}{p_{ex}} = \frac{550P\mathcal{R}T_{ex}}{H_{ad}\eta_{tot}p_{ex}} \tag{7-1a}$$

wherein the gas temperature at the wheel exit,

$$T_{ex} = T_1 - \eta_{i,ad}\,\Delta t_{ad}$$

with

$$\eta_{tot} < \eta_{i,ad}$$

The selection of a tolerable discharge velocity, i.e. ζ (Eq. 6-8), establishes the wheel discharge area (Fig. I,7a)

$$A_2 = \frac{Q_{ex}}{V_{r_2}} = \frac{\pi}{4}(D_2^2 - d_h^2) \tag{7-1b}$$

The wheel inlet diameter is $D_1 = mD_2$ with m values dependent on the type of the wheel blading. For 90° wheels (Fig. I,7a), diameter ratios in the range from $m = 1.4$ to 2.2 are quite common with both larger and smaller values possible in extreme conditions and dependent upon turbine requirements. The tip speed required for a given H_{ad} depends on the choice of blading and the efficiency-velocity ratio relationship. If D_1, ν, and the wheel rotational speed n are not compatible, different assumptions and/or blading are indicated, or a multistage or a multiflow arrangement may be required. No general method for the calculation of the most favorable number of blades as functions of turning angle, diameter ratio, etc. has been established sufficiently firmly. Large numbers of blades (Fig. I,7b) obviously cause higher flow area restrictions and more friction, whereas very low numbers permit greater deviation from the energy

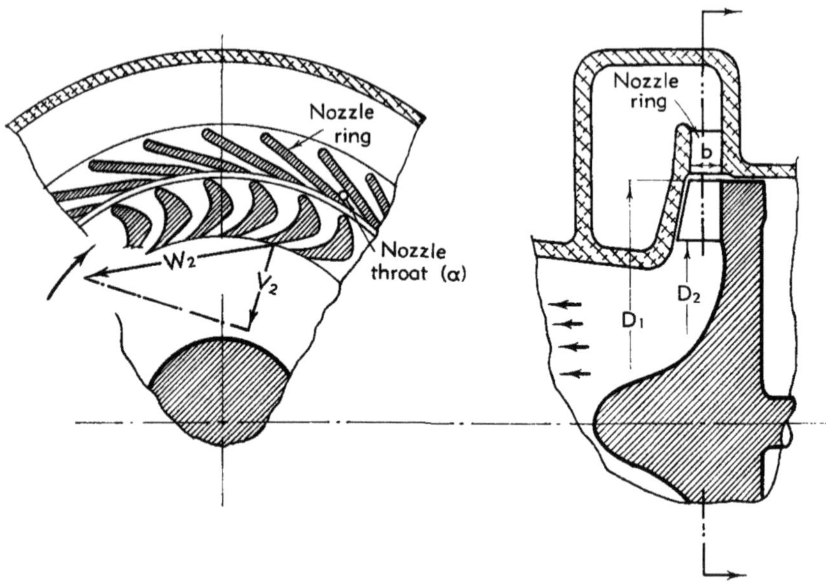

Fig. I,7b. Centripetal turbine with cantilever-type blading.

transfer calculated on the basis of the one-dimensional approach. With 90° wheels of from 5 to about 15-in. diameter, good efficiencies have been obtained with blade numbers between 10 and 20.

With respect to the blade angles at the wheel outlet the desired discharge velocities and the circumferential velocities determine these angles as functions of the radii. For discharge conditions characterized by $\alpha_2 =$ 90° and V_2 being constant along the radius, the blade angle β_2 varies according to

$$r \tan \beta_2 = \text{const} \tag{7-2}$$

The throat area $A_{th} = \sum ab$ (Fig. I,7a and I,7b) needed for the mass flow rate w is

$$A_{th} = \frac{w\sqrt{\mathcal{R}T_{in}}}{\mu\psi p_{in}} \tag{7-3}$$

with μ = coefficient of contraction, dependent greatly on nozzle configuration, and the value ψ [1; 7, Fig. 25] defined as

$$\psi = \sqrt{2g \frac{\gamma}{\gamma - 1} \left[\left(\frac{p_{no}}{p_{ni}} \right)^{\frac{2}{\gamma}} - \left(\frac{p_{no}}{p_{ni}} \right)^{\frac{\gamma+1}{\gamma}} \right]} \tag{7-4}$$

with p_{no}/p_{ni} representing (p nozzle out)/(p nozzle in) and resulting from the introduction of the critical gas velocity into the continuity equation.

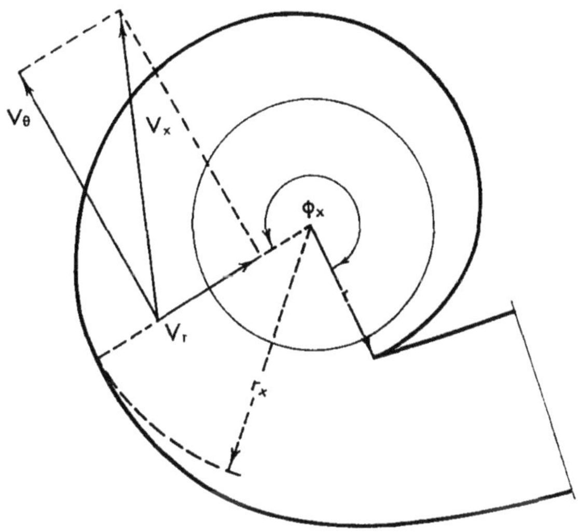

Fig. I,7c. Volute-type turbine scroll.

The pressure ratio across the nozzle ring results from

$$H_{ad,no} = (1 - R)H_{ad} = \frac{\gamma}{\gamma - 1} \mathcal{R} T_{in} \left[1 - \left(\frac{p_{no}}{p_{ni}} \right)^{\frac{\gamma-1}{\gamma}} \right] \tag{7-5}$$

If the turbine inlet casing is of the volute type (Fig. I,7c), the dimensioning of its cross-sectional area can be based on the flow with constant angular momentum

$$rV_\theta = \text{const} = K \tag{7-6}$$

At a station (Fig. I,7c and I,7d) defined by φ_x and r_x the flow rate is

$$Q(\varphi_x, r_x) = \int_{r_0}^{r_x} V_\theta dA = rV_\theta \int_{r_0}^{r_x} b \frac{dr}{r} \tag{7-7}$$

For the common single inlet volute with $\varphi = 360°$, the flow area at each station can be calculated from

$$\varphi \text{ (degrees)} = \frac{360 r V_\theta}{Q} \int_{r_0}^{r} b \frac{dr}{r}$$

after b as a function of r has been selected for a design.

By proper modification, compressibility and flow friction can be taken into account. This complication may be indicated for high Mach number flow conditions or very rough volute walls.

The inlet angle α_{in} to the nozzle ring, Fig. I,7e, should be adapted to the volute out-flow conditions in order to avoid losses through excessive deflection. With V_r calculated by means of the continuity equation and V_θ from the angular momentum equation, the flow angle (Fig. I,1) results from $\tan \alpha = V_r/V_\theta$.

Properly matching the nozzle area, the shape of the volute and the angles of the nozzle vanes complete the design of the flow areas in the turbine inlet housing.

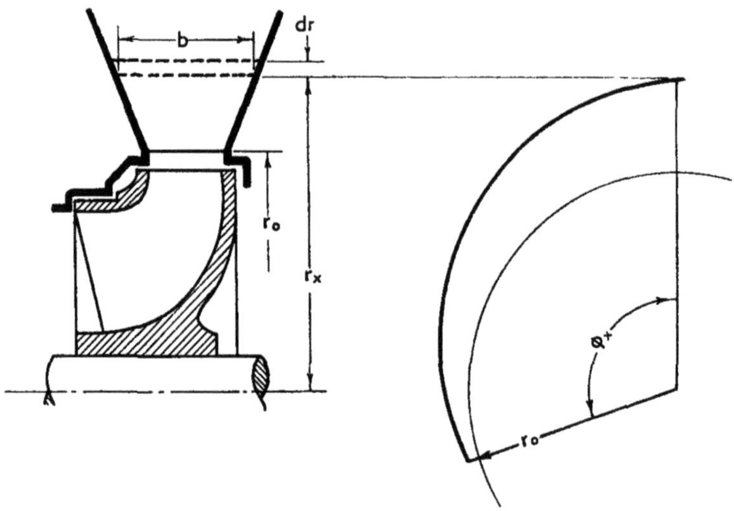

Fig. I,7d. Volute-type turbine scroll.

Special attention is required with respect to the critical pressure ratio and the critical velocity whenever dealing with turbomachinery for gases. At high pressure ratios, the jet deflection (Prandtl-Meyer effect [*4*]) at the nozzle exit should be taken into account. Humidity effects may require special attention in refrigeration turbines. Otherwise, common nozzle design knowledge [*1*] is, of course, applicable. However, the spacing or radial distance from the trailing edge of the nozzle ring to the inlet edge of the wheel blades is more or less a matter of free choice. Although tests regarding the influence of this dimension are known to have been conducted, no results are available for publication. It has been recognized, though, that rather large radial distances, say up to one eighth of the wheel diameter, may show little, if any, loss in over-all turbine efficiency. Naturally, this greatly depends on the flow pattern leaving the nozzle ring and the design of the flow passages. Studies regarding the "sink flow" should be helpful. Also, mechanical considerations, such as blade excitation, erosion, etc., may influence the spacing.

Now, one may ask whether nozzle vanes must be used. Compressors and pumps with vaneless diffusers and volute-type scrolls as diffusing

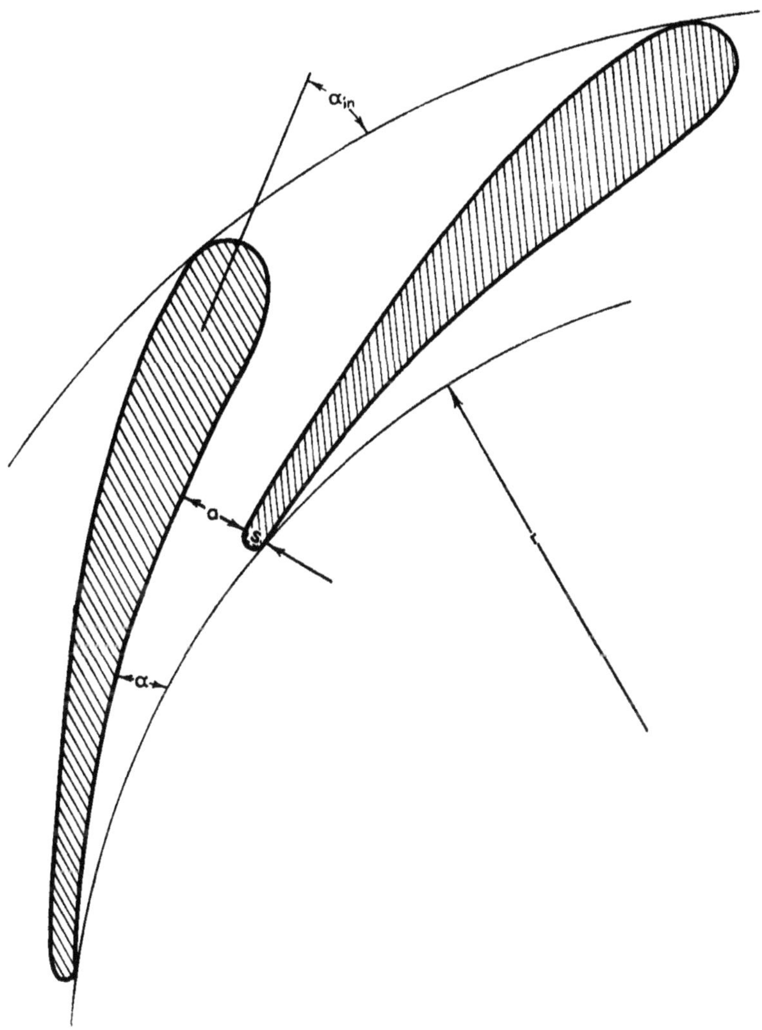

Fig. I,7e. Nozzle vanes for centripetal turbine.

means are well known. They offer certain advantages such as a flat head vs. volume curve and less tendency for pulsation. A reversal of the flow in such compressor configuration represents a turbine without nozzle vanes. The calculation of such a turbine offers no special difficulties. For simplicity, the gas is considered to be a perfect, nonviscous gas and the side walls of the vaneless nozzle ring are assumed to be perpendicular to the axis of rotation and parallel to each other with an axial distance b. The thermodynamic and flow conditions to be observed are

$$pv^\gamma = \text{const (isentrope)}$$
$$rV_\theta = K = \text{const (free vortex)}$$
$$\rho AV = \text{const (continuity)}$$
$$2g\Delta H_{ad} = \Delta c^2 = 2gc_p\Delta TJ \text{ (constant energy)}$$

With x and y representing two radii in the vaneless ring, and with V_r and V_θ for the meridional and the tangential components of the absolute velocity,

$$p_x v_x^\gamma = p_y v_y^\gamma$$
$$r_x V_{\theta_x} = r_y V_{\theta_y}$$
$$b_x r_x V_{r_x} \rho_x = b_y r_y V_{r_y} \rho_y$$
$$V_{r_x}^2 + V_{\theta_x}^2 - (V_{r_y}^2 + V_{\theta_y}^2) = 2gJc_p(T_y - T_x)$$

Combining and rearranging leads to

$$\frac{V_{r_x}^2 - V_{r_y}^2}{2g} + \frac{V_{\theta_y}^2}{2g}\left[\left(\frac{r_y}{r_x}\right)^2 - 1\right] + \frac{\gamma p_y v_y}{\gamma - 1}\left[\left(\frac{r_y V_{r_y}}{r_x V_{r_x}}\right)^{\gamma-1} - 1\right] = 0 \quad (7\text{-}8)$$

permitting the calculation of the gas velocity at any radius. With $V_r = f(r)$ and $rV_\theta = K$, there is $\tan \alpha = rf(r)/K$. For a centri-angle Φ included between the radii to two points representing the beginning and the end of a flow filament between subject radii, there is, per geometry, $\tan \alpha = dr/rd\Phi$ or transformed $d\Phi = Kdr/r^2f(r)$. Thus the path of a gas particle in a vaneless ring with parallel side walls is described. Considering a sink flow (centripetal turbine) of a compressible substance, one finds that the fluid leaves the ring at an angle steeper than that with which it entered the ring. Accounting for viscosity the deviation from a logarithmic spiral [3], representing the "ideal" sink flow, decreases somewhat.

I,8. Design of Blades. Curved blades in the radial portion of a wheel are sometimes designed as a single arc (Fig. I,8a). Describing the sum of the two blade angles from radius MB, connecting A and B, extending to the intersection at C, drawing a perpendicular line bisecting AC, the crosspoint O_1 represents the center for a single-arc blade with β_1 at the inlet and β_2 at the outlet [3]. If, however, certain functions between blade thickness, relative velocity, and radius are prescribed, the blade shape can be calculated. Combining (see Fig. I,8b and I,8c) $\sin \beta = V_r/W$, $zt = \pi D$ and $\tau = t/(t - \sigma)$ with the continuity equation $V_r\pi Db = \tau Q$,

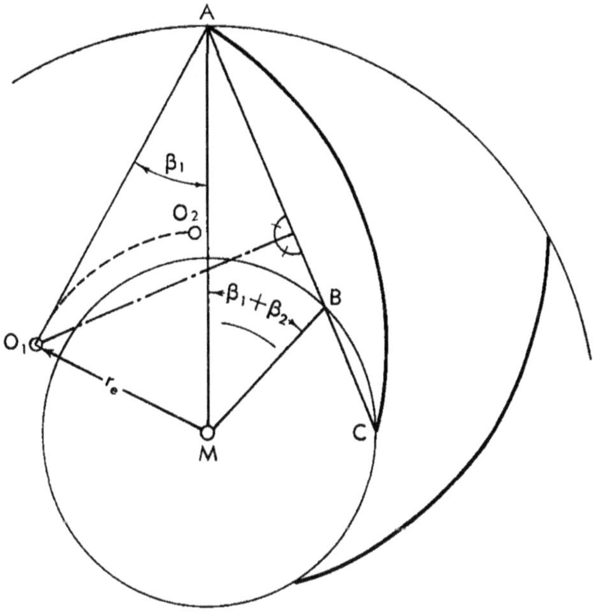

Fig. I,8a. Single-arc blade.

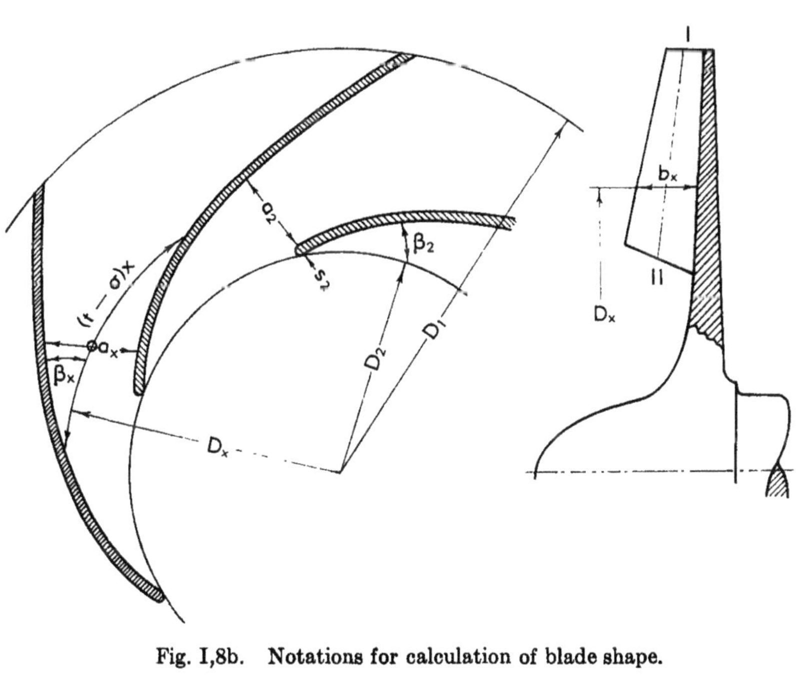

Fig. I,8b. Notations for calculation of blade shape.

there results after rearrangement

$$\sin \beta = \frac{s}{\sigma} = \frac{\tau Q}{\pi b D W}$$

It can also be written that $\tan \beta = CB/AB = dr/rd\varphi$ (Fig. I,8d). The length of the blade measured in the central angle φ is obtained by integration between the inner and outer diameters via

$$\varphi \text{ (degrees)} = \frac{180}{\pi} \int_{r_1}^{r_2} \frac{dr}{r \tan \beta}$$

For the numerical evaluation, a graphical integration can easily be used.

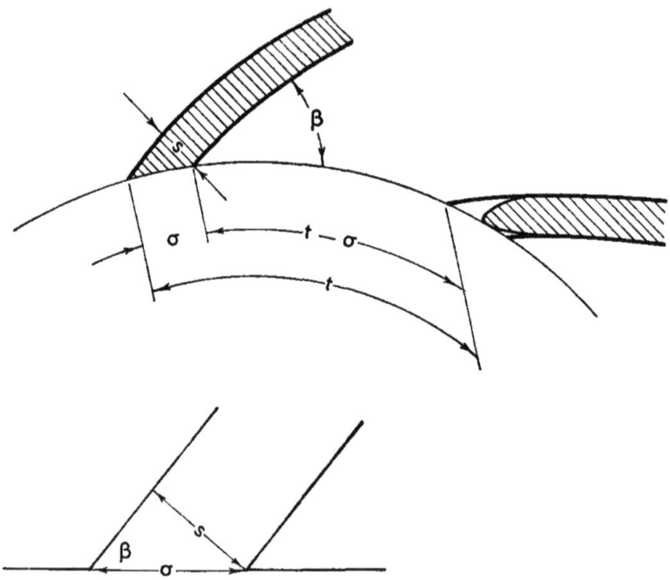

Fig. I,8c. Area reduction through blade thickness.

The design of exducers for turbine wheels is very similar to the methods used for "inducers" of impellers of centrifugal compressors. For stress reasons, cross sections of exducer blades may show a taper comparable to axial blades. Cantilever-type blading (Fig. I,7b) is designed with methods known, for example, from Ljungstroem turbines [*1*].

Where radial wheels are required to operate at high tip speeds and temperatures, semishrouded wheels or the "star type" normally have to be used because of the stress problems connected with front shrouds. Efficiency-wise, fully shrouded impellers may be superior to semishrouded and unshrouded wheels, but a fully satisfactory method to calculate stresses in radial wheels has not been published. In applications with great temperature gradients along the radius of the wheel, significant thermal stresses are superimposed over the mechanical stresses. The resulting stress analysis is equally difficult and time-consuming, but very essential even if only based on approximate methods. With highly loaded wheels, the temperature change along the disk can become very great, thus possibly leading to compressive stresses where, with no thermal gradient, tensile stresses would be the criterion.

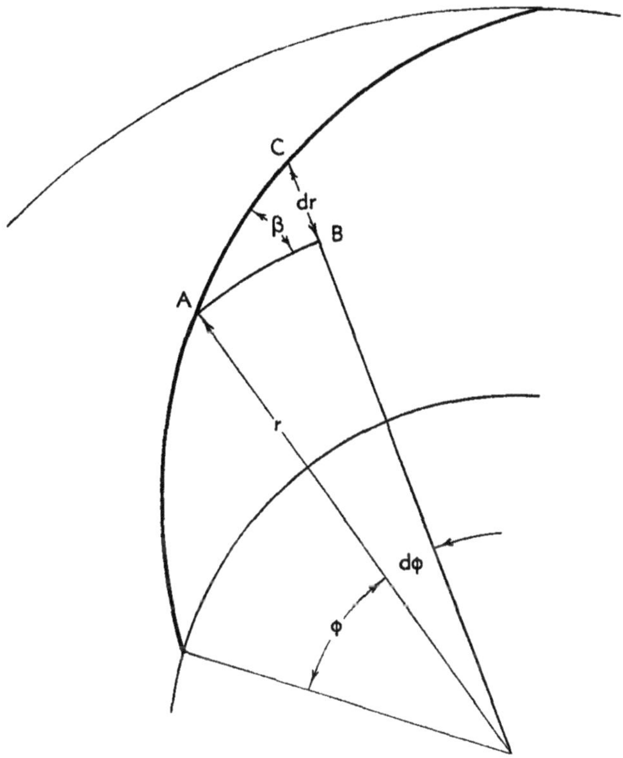

Fig. I,8d. Notations for blade calculation.

I,9. Adaptation of Power Output and Energy Supply. In publications around the turn of the century, it was stressed that a very important factor in favor of radial-type turbines was the simplicity with which the most economical load adaptation over rather wide ranges could be accomplished.

Instead of controlling the output by throttling or by partial admission, the *variable area nozzle* with individually pivoted nozzle vanes, which has been widely accepted for use in water turbines, has been demonstrated to be equally favorable for centripetal turbines for compressible fluids as long as the special requirements of compressible fluids with very high, i.e. up to sonic, velocity are carefully observed in the turbine design. For purposes of demonstration, Fig. I,9a shows the first very simple arrange-

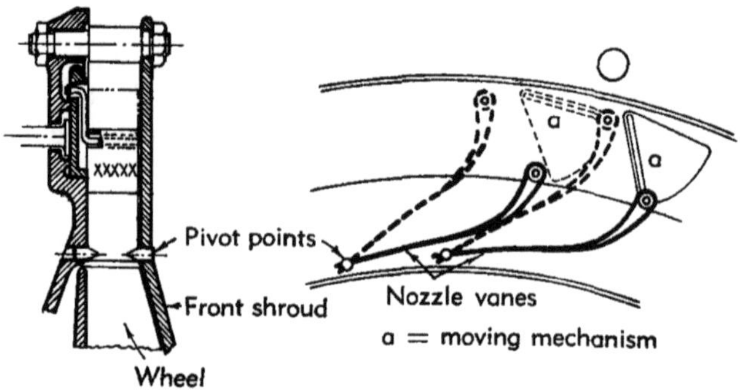

Fig. I,9a. Variable area nozzle with pivoted nozzle vanes for small, simple centripetal gas turbine.

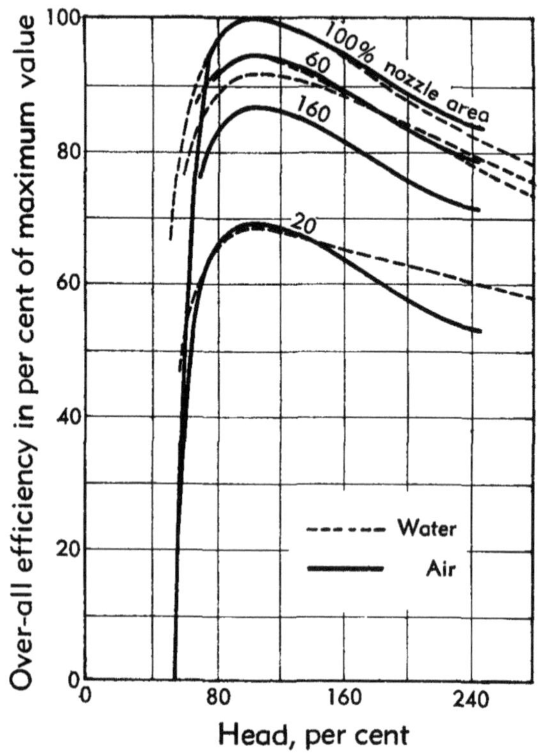

Fig. I,9b. Comparison of test results observed with centripetal turbines for water and air [8].

ment known to have been used up to high temperatures in a small gas turbine over twelve years ago. Test results shown in Fig. I,9b and I,9c [8] can be considered representative of performance characteristics easily obtainable with centripetal turbines (90° turbine wheel with exducer) and fairly simple nozzle arrangements.

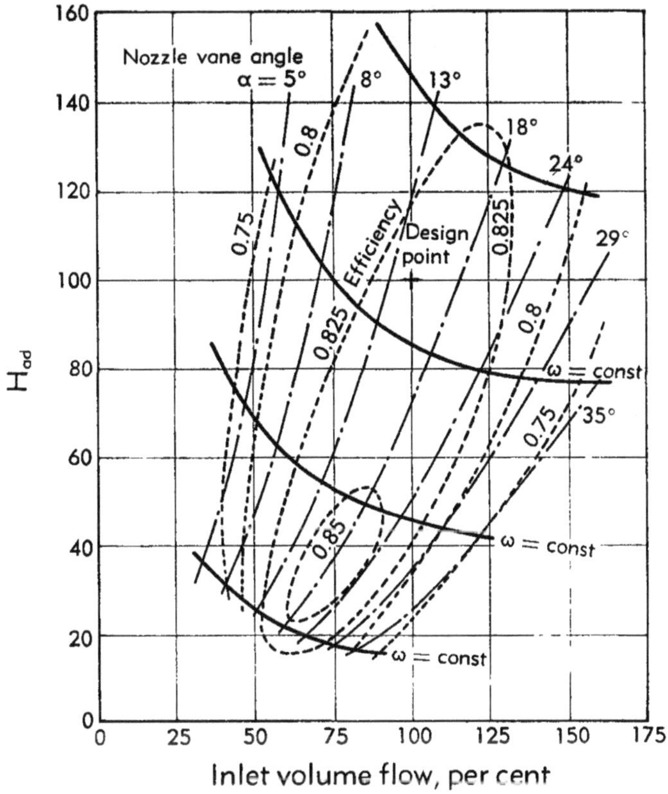

Fig. I,9c. Performance characteristics of centripetal air turbine with variable area nozzle.

I,10. Multistage and Multiflow Arrangements. As is the case with almost all types of turbomachinery, head or enthalpy drop which can be used in a single rotating blade system is proportional to the square of the angular velocity. The determining stresses resulting from centrifugal forces are equally dependent on the square of the angular velocity. This means that the stresses limit the possible head per blading row. A general statement with respect to the maximum possible head per stage can not be made because the very wide range of velocity ratio vs. efficiency may or may not be utilized, dependent on the turbine requirements. Moreover, the strength of materials available is continuously being improved. Limitations can also result from gas dynamical problems such as sonic velocity.

Whenever a single-stage machine cannot handle the head supplied, *multistage* arrangements are indicated. Fig. I,10a shows in principle a simple multistage centripetal turbine arrangement consisting of single-stage elements combined in one housing on one shaft. The two-stage arrangement shown in Fig. I,10b combines a centripetal first and centrifugal second stage, offering certain attractive features with respect to compactness and the design of housing and bearings. Another combination of two stages, both of the centripetal type (Fig. I,10c) appears

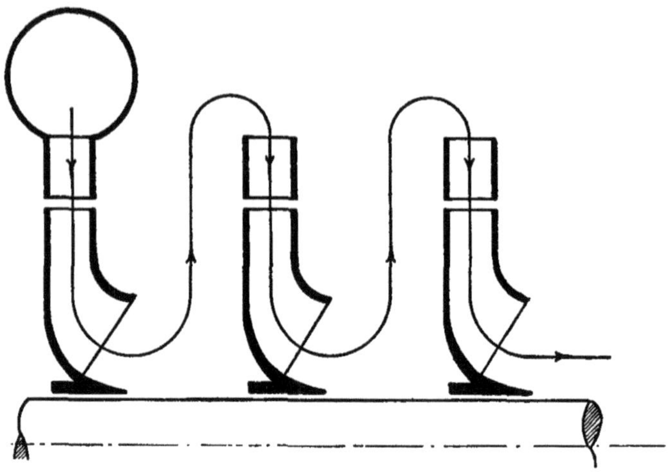

Fig. I,10a. Multistage arrangement.

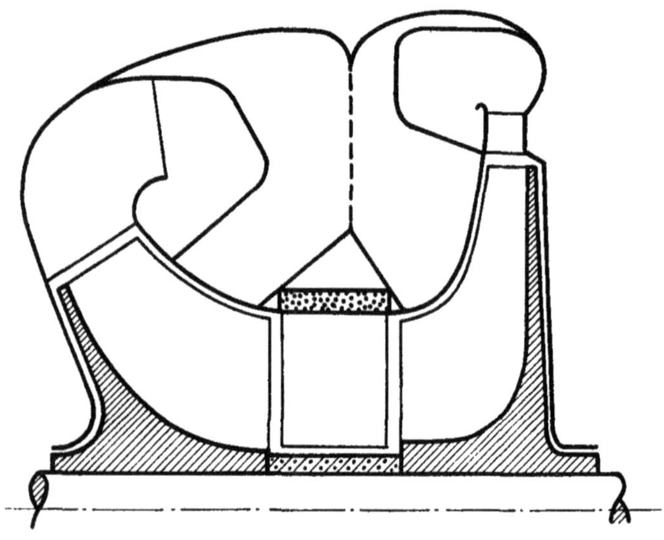

Fig. I,10b. Two-stage turbine with centripetal and centrifugal stage.

feasible, for example, for gas turbines with a free power turbine. In such a unit, one stage would drive the compressor of the gas generator, whereas the other one delivers shaft power.

Similarly, the *multiflow* arrangement, Fig. I,10d, can be used when the through flow required exceeds the flow capacity of a single-wheel outlet. Blading rows similar to Fig. I,7b have been used successfully since the turn of the century in the Eyermann turbine, a simple multistage radial turbine with outward flow. The blading consists of alternate rows of stator vanes (nozzle rings) and rotating blade rows. It is not difficult to

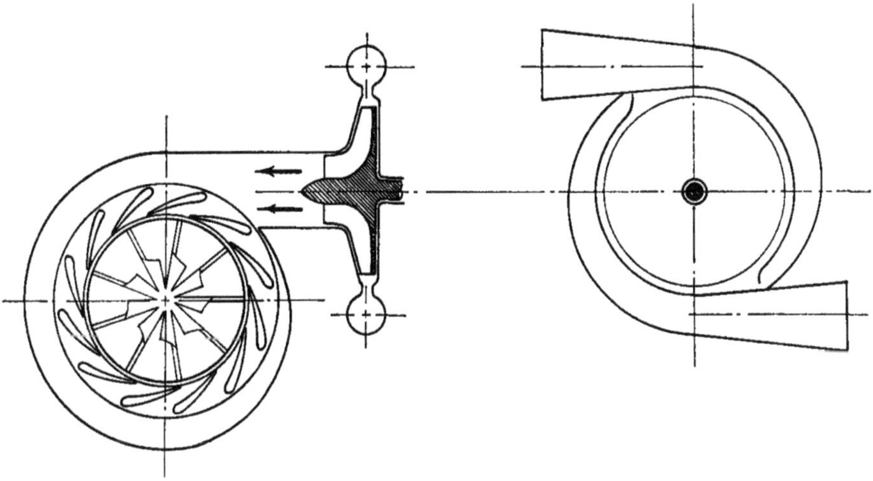

Fig. I,10c. Two-stage turbine arrangement.

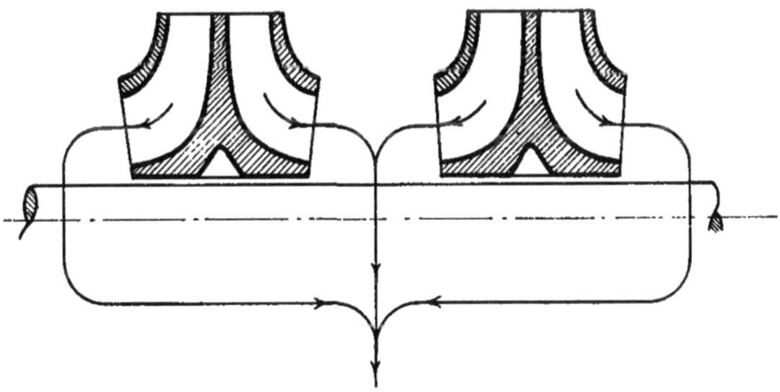

Fig. I,10d. Multiflow arrangement.

use congruent bladings for both nozzles and wheels. Combinations of impulse and reaction stages have been used since about 1900, and the advantages of using multistage arrangements permitting low rotational speeds were recognized early. The next natural step was the Ljungstrom turbine [*1*] which differs from the Eyermann turbine in that there are no stators; both rows of blades rotate in opposite directions, thus producing a higher head drop in a given space for the same angular velocity.

I,11. Brief Comparison between Single-Stage Axial and Radial Turbines. One of the commonly known methods of determining the suitability of a radial or an axial turbine wheel (or compressor impeller) for a given problem statement is to determine the specific speed characteristic which, for a certain rotational speed, gives best efficiencies for the given pressure ratio and inlet temperature of the working medium. This specific speed characteristic has been presented many times in turbomachinery literature [*2*,*3*,*9*]. Another way of illustrating the typical characteristics of the axial vs. the radial wheel configuration is to compare

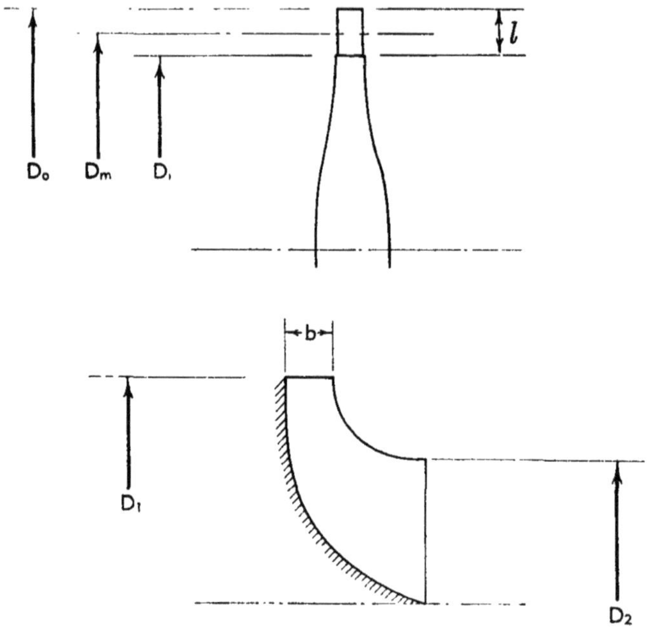

Fig. I,11a. General dimensions for axial- and radial-type turbines.

characteristic through-flow areas. With the relationship of the hub ratio $h = D_i/D_o$ of the axial wheel (Fig. I,11a) and the diameter ratio m of the radial wheel and with $y = A_i/A_o$ representing the ratio of inlet to outlet area of the radial wheel and assuming equal outer diameters, i.e. $D_o = D_1$, through simple arithmetic one finds

$$D_o^2(1 - h^2) = y\left(\frac{D_1}{m}\right)^2$$

which, when simplified, leads to

$$m = \sqrt{\frac{y}{1 - h^2}}$$

The evaluation of this equation, Fig. I,11b, shows that an axial impeller of a hub ratio of 0.7 and with the conditions outlined above corresponds to a diameter ratio of about 1.4 for a radial wheel with $y = 1$. Assuming that a diameter ratio $m = 1.4$ is considered the low reasonable limit for a radial wheel, Fig. I,11b shows that, for example, an axial wheel with a hub ratio of 0.6 cannot be replaced by a radial wheel with a y value of about 1. Whereas a hub ratio of, say, 0.9 is not considered particularly desirable in an axial wheel because of the bypass or leakage losses, over

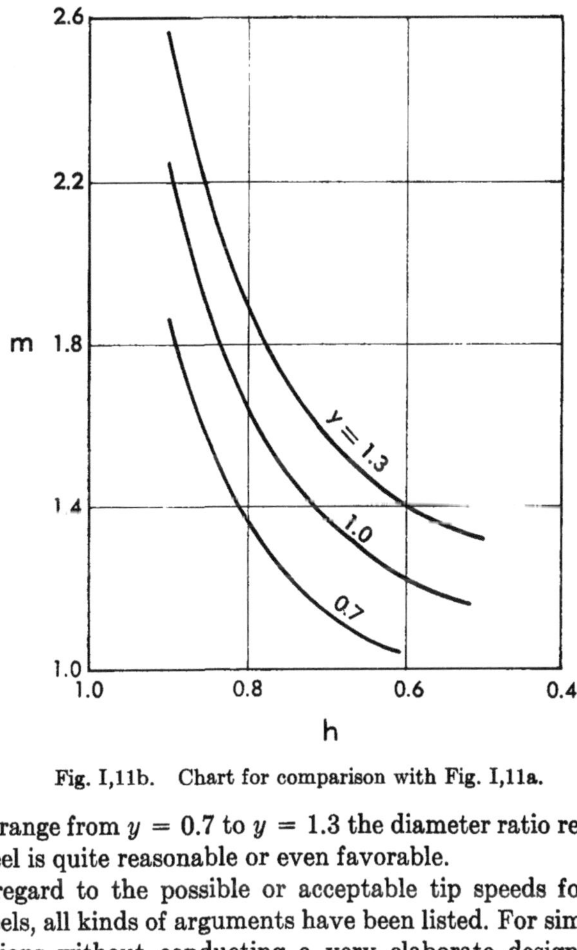

Fig. I,11b. Chart for comparison with Fig. I,11a.

the entire range from $y = 0.7$ to $y = 1.3$ the diameter ratio required for a radial wheel is quite reasonable or even favorable.

With regard to the possible or acceptable tip speeds for axial and radial wheels, all kinds of arguments have been listed. For simple or basic considerations without conducting a very elaborate design and stress study and under otherwise comparable conditions (safety, temperature, material), a 90° radial wheel lends itself to somewhat higher tip speeds than an axial wheel. If an axial wheel machined integrally in one piece with all blades is assumed for the comparison, there is little reason to consider one or the other superior regarding maximum permissible tip speeds.

I,12. Performance Data. As pointed out elsewhere [8], the efficiency characteristics of centripetal turbines for compressible fluids are quite similar to those known for water turbines, which is demonstrated by a comparison of test data obtained on two such turbines (Fig. I,9b).

The range of efficiency versus head coefficient obtained with numerous types of centripetal turbines with wheel Mach numbers around 0.75 (defined as the ratio of the tip speed and sonic velocity of the gas under inlet conditions) is shown in Fig. I,12a based on test data from various units with small wheels (15-in. diameter and less). Single-stage centrifugal compressors (outlet blade angles between 70 and 90°) when operated as air turbines with reversed directions of flow and rotation have shown similar results. As must be expected, the maximum efficiency moves toward

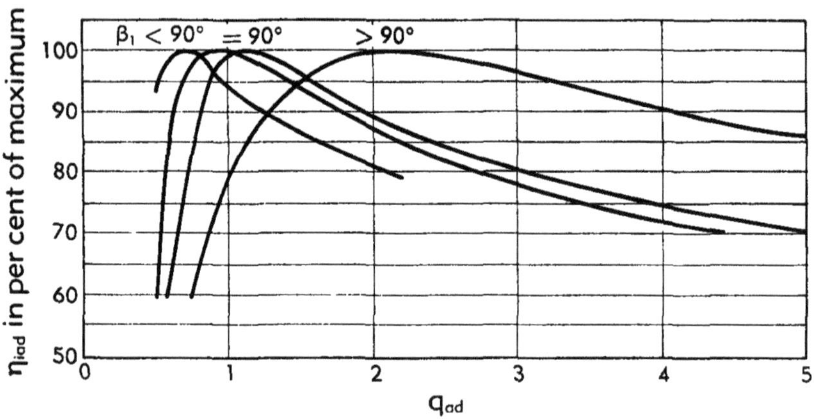

Fig. I,12a. Efficiency vs. head coefficient curves for several centripetal turbines with $\beta_1 \lesseqqgtr 90°$.

smaller q_{ad} values for turbine wheels with small β_1 (Fig. I,1), whereas the cantilever-type blading (Fig. I,7b) produces its range of maximum efficiency at higher q_{ad} values.

Typical test results obtained with 90°-type centripetal turbines, 4 to 8-in. wheel diameter, with pressure ratios up to about 4:1 and inlet temperatures up to around 1300° (Fig. I,12b), and over a range of head coefficients from $q_{ad} = 0.5$ to $q_{ad} = 2.5$ show turbine efficiencies above 0.65 and up to about 0.88 for the best velocity ratio to be quite common. These test data do not represent the highest values known to have been obtained. The band indicated over the entire range at the upper limit represents better nozzle configurations tested and at the lower limit the simpler nozzle configurations tested.

By calculation alone, the degree of reaction of a radial turbine can be approximated [5,6] in good agreement with test results (Fig. I,12c) over a wide range of head coefficients and various turbine configurations. The essentials for the calculation of the degree of reaction are fairly simple and, generally, the results are satisfactory for dimensioning a turbine design.

In order to give some first indication about the influence of face clearance, i.e. clearance between the blade edges and the stationary shroud of

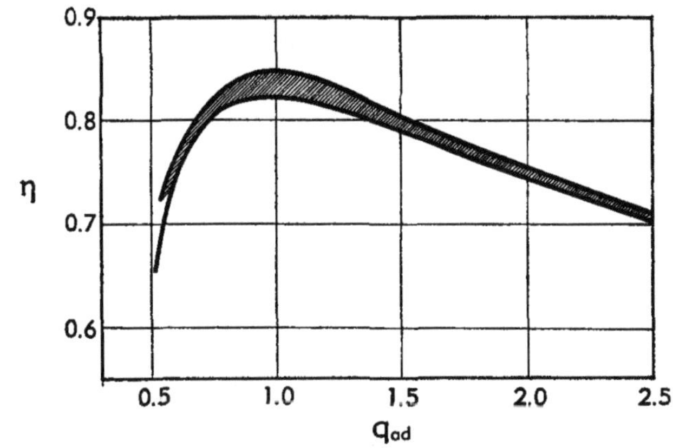

Fig. I,12b. Measured efficiencies of small centripetal turbines of the 90° type.

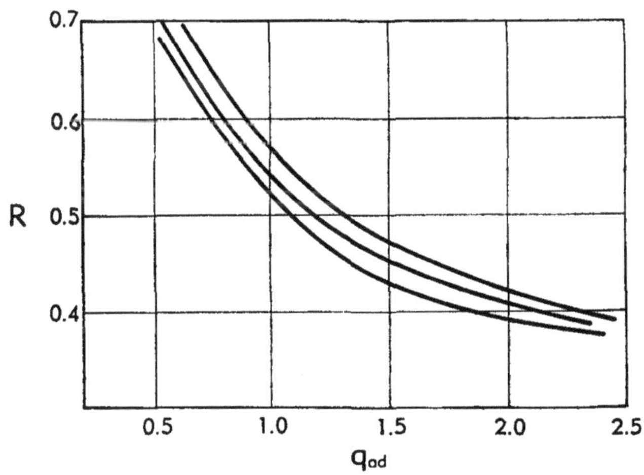

Fig. I,12c. Measured values of degree of reaction of small 90° centripetal-type turbines.

the housing, Fig. I,12d shows over a wide head coefficient range how much the efficiency changed in a very simple vaneless centripetal turbine with a semishrouded wheel when increasing this clearance s from a value $D_1/s = 300$ to one half of this ratio of outer wheel diameter D_1 to face clearance s. Whereas in the range of maximum efficiency the drop in efficiency shown might be objectionable, its influence almost disappears if operational conditions require the turbine to work at higher head coefficients.

When choosing the most favorable blade configuration of a radial turbine for the given energy supply and output conditions, inner adiabatic efficiencies well above 80 per cent (and approaching 90 per cent) have been obtained consistently with wheel tip Mach numbers up to approximately unity and down to very small wheel dimensions, e.g. 3 in. and less. Such performance and the simplicity and the ruggedness explain why the simple 90°-type centripetal turbine has gained so fast in practical importance.

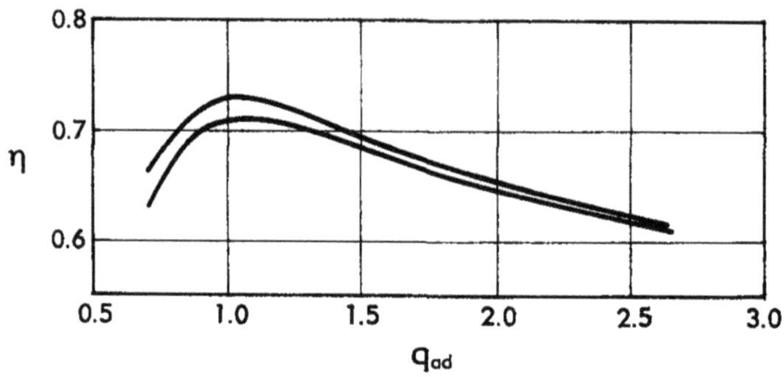

Fig. I,12d. Influence of face clearance on efficiency of small 90° centripetal turbines.

I,13. Examples of Centripetal 90-Degree-Type Turbines in Production. Data published and presently known to the author indicate that radial turbine wheels of the types described and used in production turbines range up to 20 in. in diameter. It has been stated that the radial turbine only lends itself to small sizes. Obviously this is not so. Certain desirable manufacturing techniques for big radial wheels may not have been developed fully. Just as axial turbines with wheel diameters as small as 1 inch and as large as, say 15 feet, are being used, there is no absolute size limit with the radial wheel configuration as long as the problem statement and its relationship with specific speed characteristics lead to a favorable wheel configuration of the radial type. Little, if any, real practical experience exists with multistage arrangements although some appear quite feasible.

Typical examples for gas turbines with radial turbines produced and used in large numbers for secondary aircraft power have been published [*7,8,9,10,11*]. More recently, developments have been indicated with a centrifugal compressor impeller and a centripetal turbine wheel manufactured in one piece, Fig. I,13, both for auxiliary gas turbine power units and for exhaust-gas turbochargers.

The possibility of power control through variable area nozzles is being utilized in centripetal turbine units for air and gas (see Plate I,13a). Although such developments have already resulted in practical products, further simplifications appear promising and feasible.

While the utilization of the simple 90°-type centripetal turbine is rapidly increasing, varied configurations of the wheel itself, comparable to the centrifugal compressor development [9], can be observed (Plate I,13b). The explanation of why such different shapes came into being is

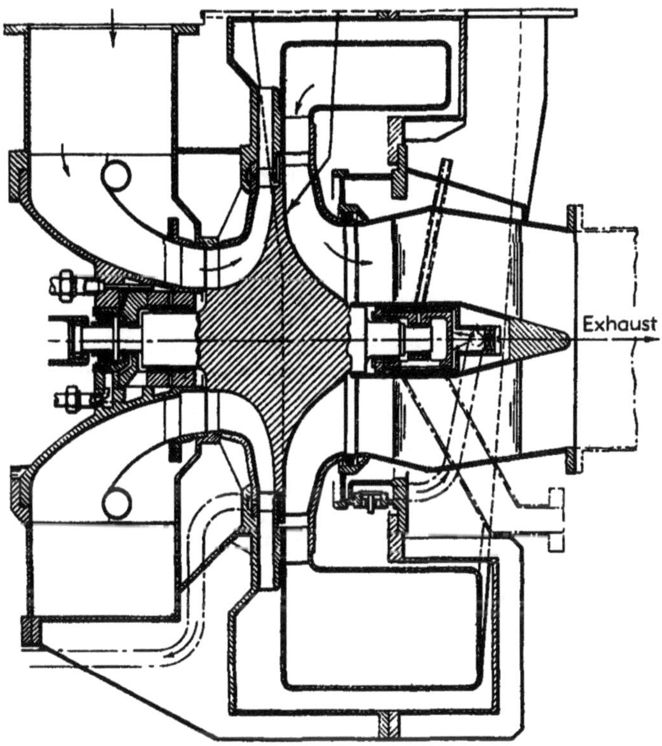

Fig. I,13. 200-horsepower gas turbine with centripetal turbine and centrifugal compressor. (W. H. Allen, Sons, and Co., Ltd.)

to be found in stress, weight, and especially manufacturing problems. The progress in the precision casting of heat-resistant alloys of satisfactory ductility gives the designer great freedom in the blade-shaping. Further advancements in the stress analysis, in the aerodynamical knowledge about the centripetal flow, and in precision-forging techniques can be expected to lead to new shapes of radial wheels differing more and more from the common-style centrifugal compressor impeller.

Whereas, to some degree, the emphasis may have been on radial wheels with straight 90° blading in the radial part, backward and/or forward curved blading is quite feasible for certain applications. Naturally the fact of the curvature in itself can impose stress problems and manufacturing problems; however, where precision-cast materials have sufficient strength for the tip speeds required, radial wheels with curved blading can have significant merits. Notable examples are some superchargers for aircraft cabin pressurization and air conditioning (The Garrett Corporation). In this system, the first compressor stage which is mechanically driven by the main power plant of the aircraft, as well as the second compressor stage and the turbine which drives it, all have the same radial-type rotor and the same housing, except that the diffuser vaning for the compressors is different from the nozzle vaning of the turbine. In all three applications of this radial configuration of small size, the unit efficiencies are above 80 per cent.[3]

I,14. Cited References and Bibliography.

Cited References

1. Stodola, A. *Steam and Gas Turbines.* (Transl. by L. S. Loewenstein.) McGraw-Hill, 1927.
2. Wislicenus, G. F. *Fluid Mechanics of Turbomachinery.* McGraw-Hill, 1947.
3. Pfleiderer, C. *Stroemungsmaschinen.* Springer, Berlin, 1952.
4. Prandtl, L. *Fuehrer durch die Stroemungslehre.* Vieweg, Braunschweig, 1944.
5. Baljé, O. E. A contribution to the problem of designing radial turbomachines. *Trans. Am. Soc. Mech. Engrs. 74,* 451 (1952).
6. von der Nuell, W. T. The radial turbine. *ATI Tech. Data Digest 12,* (*5*), 1947.
7. von der Nuell, W. T. Superchargers and their comparative performance. *SAE Quart. Trans. 6,* (*4*), 1952.
8. Wood, H. J. Characteristics and limitations of expansion turbine for auxiliary power purposes in turbine-propelled aircraft. *SAE Quart. Trans. 6,* 443 (1952).
9. von der Nuell, W. T. Single-stage radial turbines for gaseous substances with high rotative and low specific speed. *Trans. Am. Soc. Mech. Engrs. 74,* 419 (1952).
10. Wosika, L. R. Radial-flow compressors and turbines for the simple small gas turbine. *Trans. Am. Soc. Mech. Engrs. 74,* 1337 (1952).
11. Anon. A versatile range of small gas turbines. *The Oil Engine and Gas Turbine 24,* (*278*), 237 (1956).

Bibliography

Ainley, D. G. Some notes concerning inward radial flow turbines. *Natl. Gas Turbine Establishment, England, Mem. M136,* Nov. 1951.

Anon. Marine auxiliary and emergency gas turbines. *The Engineer 197,* 709 (1954).

Birmann, R. The elastic-fluid centripetal turbine for high specific outputs. *Trans. Am. Soc. Mech. Engrs. 76,* 173 (1954).

Cox, H. R., consulting ed., and twenty-four specialist contributors. *Gas Turbine Principles and Practice.* Van Nostrand, 1955.

Gelpke, V. *Turbinen und Turbinenanlagen.* Springer, Berlin, 1906.

Hunsaker, J. C., and Rightmire, B. G. *Engineering Applications of Fluid Mechanics.* McGraw-Hill, 1947.

Keenan, J. H. *Thermodynamics.* Wiley, 1941.

[3] The manuscript for this section was revised in 1953.

SECTION J

THE CENTRIFUGAL COMPRESSOR

EDWARD S. TAYLOR

J,1. Introduction. Elementary analysis of the centrifugal compressor explains many of its characteristics and has the advantage of being easily comprehended. The usual elementary analysis involves a one-dimensional approach, that is, all fluid properties are assumed to be constant over certain flow cross sections. These cross sections are conveniently taken before and after the rotating wheel, as well as at the inlet and at the discharge of the entire machine. In the case of machines where guide vanes are used ahead of the impeller, the flow at this point is no longer one-dimensional, but is easily handled by an extension of the analysis.

J,2. The Energy Equation. Application of the first law of thermodynamics to a fluid flowing steadily through a compressor gives the result

$$P = m\left[Jh_2 + \frac{V_2^2}{2g} - \left(Jh_1 + \frac{V_1^2}{2g}\right) + JQ\right] \tag{2-1}$$

where P is the power supplied to the compressor, m the mass of fluid entering the compressor in unit time, h_1 the enthalpy per unit mass of fluid entering, h_2 the enthalpy per unit mass of fluid leaving, V_1 the velocity of fluid entering, V_2 the velocity of fluid leaving, Q the heat lost from the fluid between the points of measurement of the entrance and exit conditions per unit mass of fluid flowing, g a dimensional constant numerically equal to the acceleration produced by the action of a unit force on a unit mass, and J is the mechanical equivalent of heat.

If it is assumed that the fluid is a perfect gas with constant specific heat c_p,

$$P = m[Jc_p(T_2^0 - T_1^0) + JQ] \tag{2-2}$$

where T_2^0 and T_1^0 are respectively the stagnation temperatures at the exit and the entrance. Stagnation temperature is defined by the relation

$$T^0 = T + \frac{V^2}{2gJc_p}$$

In accordance with the one-dimensional simplification, it is assumed that the stagnation temperature at the entrance (and at the exit) is uniform across the cross section where it is measured.

In high capacity compressors such as those used in turbojet engines, the quantity Q is negligible compared to $c_p(T_2^0 - T_1^0)$, resulting in a further simplification of Eq. 2-2.

$$P = mJc_p(T_2^0 - T_1^0) \tag{2-3}$$

J,3. The Momentum Equation. Consider the flow through a rotating impeller as indicated in Fig. J,3a. It is assumed that the flow at cross sections a and b is axially symmetric.

Under this assumption it is possible to derive Euler's turbine equa-

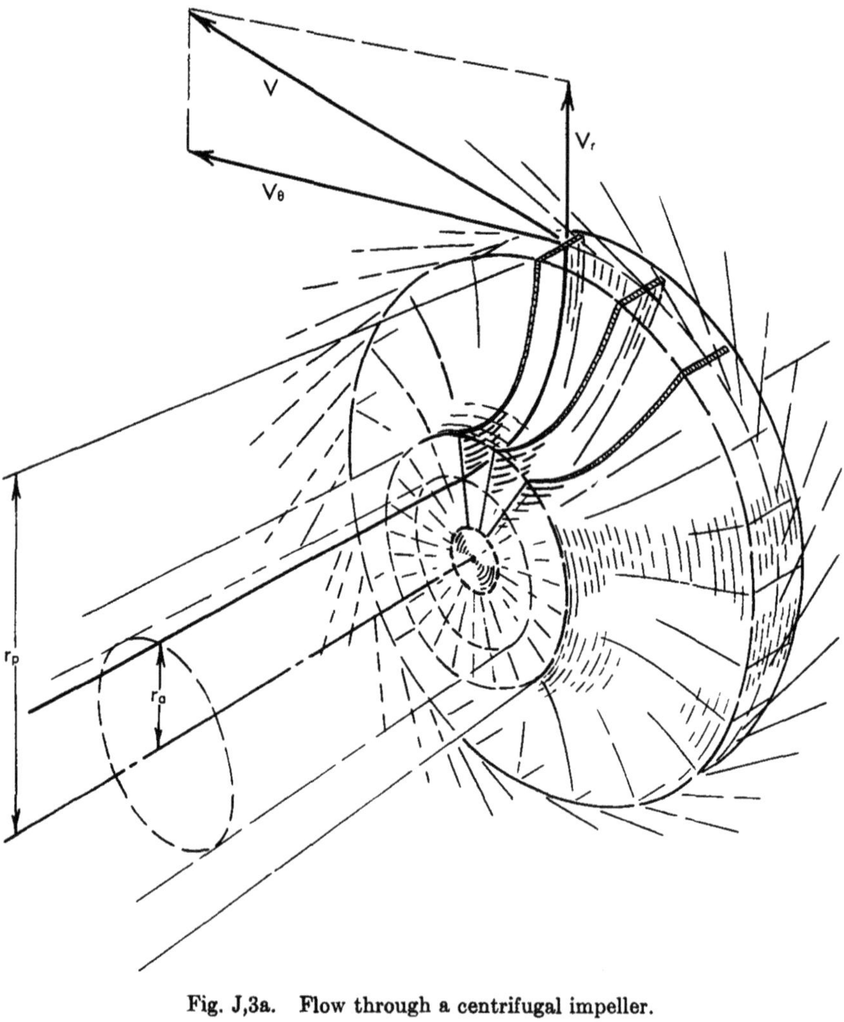

Fig. J,3a. Flow through a centrifugal impeller.

tion in the following form from the application of Newton's second law to a thin annular stream tube passing through the impeller:

$$\delta\tau = \delta m(r_b V_{\theta_b} - r_a V_{\theta_a})/g \tag{3-1}$$

where $\delta\tau$ is the external torque applied to the fluid flowing in the stream tube between sections a and b, δm the mass flowing per unit time through the stream tube, r_b the radius at section b, V_{θ_b} the tangential component of the fluid velocity at section b, r_a the radius of the stream tube at section a, V_{θ_a} the tangential component of fluid velocity, and at section a.

If it is further assumed that the velocity distribution at section a is such that $r_a V_{\theta_a}$ is independent of radius (free vortex distribution) so that $r_a V_{\theta_a} = r_1 V_{\theta_1}$ and that the tangential component of velocity at section b is uniform over a cylindrical cross section of constant radius r_2, it is possible to integrate Eq. 3-1 to obtain the external torque acting on the fluid

$$\tau = m(r_2 V_{\theta_2} - r_1 V_{\theta_1})/g \tag{3-2}$$

where m is the mass flow per unit time passing through the wheel, r_1 is the radius of the inlet pipe, and V_{θ_1} is the tangential velocity of the fluid at radius r_1, section a.

Assuming that the tangential friction between the fluid and the housing is negligible, the external torque applied to the wheel shaft is equal to the torque τ applied to the fluid by the wheel. Thus the power P applied to the wheel shaft is given by

$$P = m\omega(r_2 V_{\theta_2} - r_1 V_{\theta_1})/g \tag{3-3}$$

where ω is the angular velocity of the wheel.

If Eq. 3-3 is combined with Eq. 2-3 the following relation is obtained

$$\frac{T_2^0 - T_1^0}{T_1^0} = \frac{U_2^2}{gJc_pT_1^0}\left[\frac{V_{\theta_2}}{U_2} - \left(\frac{r_1}{r_2}\right)^2 \frac{V_{\theta_1}}{U_1}\right] \tag{3-4}$$

where $U_1 = r_1\omega$ is the tangential velocity of the wheel at radius r_1.

Note that the velocity of sound a_1^0 corresponding to the inlet stagnation temperature is given by

$$a_1^{02} = (\gamma - 1)gJc_pT_1^0 \tag{3-5}$$

where γ is the ratio of specific heats. Then Eq. 3-4 becomes

$$\frac{T_2^0 - T_1^0}{T_1^0} = (\gamma - 1)\left(\frac{U_2}{a_1^0}\right)^2\left[\frac{V_{\theta_2}}{U_2} - \left(\frac{r_1}{r_2}\right)^2 \frac{V_{\theta_1}}{U_1}\right] \tag{3-6}$$

If the inlet circulation is constant with the radius, there will be an increase in tangential velocity of the fluid toward the center of the wheel, while the wheel velocity increases from the center outward. It is clear that at any operating condition there exists a radius r_j where $U_1 = V_{\theta_1}$.

Eq. 3-6 may thus be written

$$\frac{T_2^0 - T_1^0}{T_1^0} = (\gamma - 1)\Pi_m^2 \left[\frac{V_{\theta_2}}{U_2} - \left(\frac{r_j}{r_2}\right)^2\right] \tag{3-7}$$

Where the quantity Π_m, called the "Mach index" (sometimes called the compressor Mach number), is defined as the tip speed divided by the velocity of sound at the inlet stagnation condition,

$$\Pi_m \equiv \frac{U_2}{a_1^0}$$

Note that r_j is not a geometric property of the design, but may depend upon the operating condition.

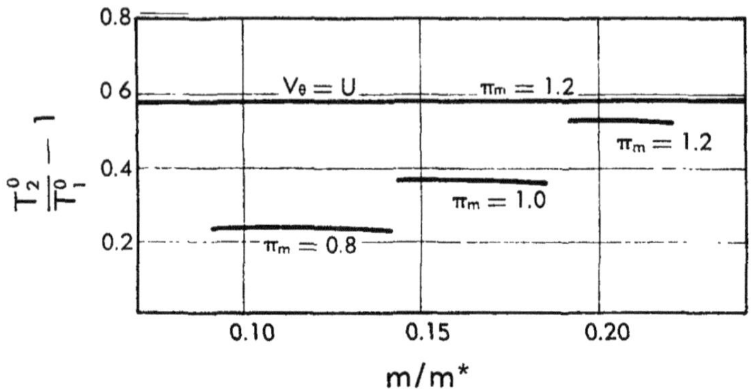

Fig. J,3b. Temperature rise in a centrifugal compressor with 16 radial impeller blades.

If there is no inlet circulation, $r_j = 0$. Under these conditions we should expect that the stagnation temperature rise through a given impeller would depend only on the rotative speed, provided that V_{θ_2}/U_2 does not vary with operating conditions. With radial-bladed impellers, both analysis and experiment indicate that V_{θ_2}/U_2 is a characteristic of the design and very nearly invariant with operating conditions.

Fig. J,3b shows the temperature rise through an impeller plotted against the mass rate of flow through the machine. The abscissa scale of this plot is a convenient dimensionless ratio to represent mass flow. The quantity m^* is defined as the mass that would flow at sonic velocity through an orifice having an area of πr_2^2 with stagnation temperature and stagnation pressure the same as corresponding values at the inlet to the compressor.

It will be noted that the actual temperature rise ratio is about 0.9 of the temperature rise ratio found by assuming $V_{\theta_2} = U_2$. In tests of compressors having radial impeller blades it is invariably found that $V_{\theta_2} < U_2$. The ratio V_{θ_2}/U_2 is called the *slip factor*. A number of efforts have been made to evaluate the slip factor on the basis of the solution of the equations of flow of a perfect fluid around the rotor blades. A review of this work may be found in [*1*]. The simplest of these efforts is the approximation of Stodola [*2*], which gives for an impeller with straight radial blades

$$\zeta = 1 - \frac{\pi}{Z} \tag{3-8}$$

where ζ is the slip factor and Z is the number of blades. This approximation gives values which in general are lower than experimentally observed values.

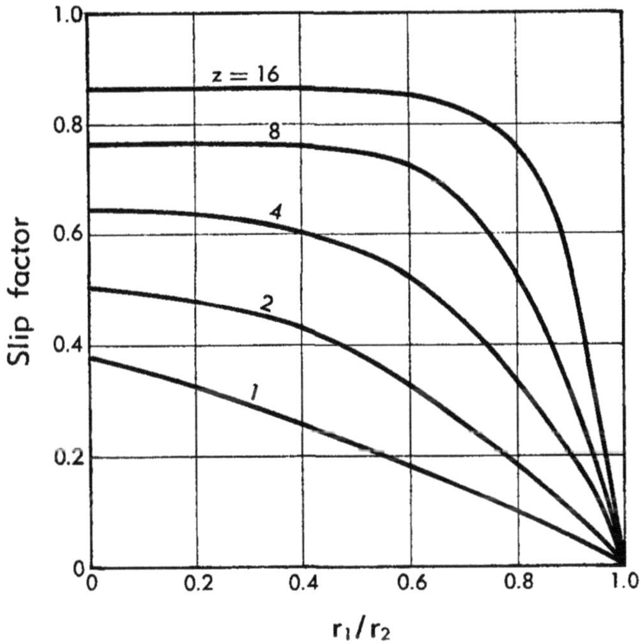

Fig. J,3c. Slip factors for radial blade compressors [*3*].

A more accurate analysis based on the flow of a perfect fluid around blades having the form of logarithmic spirals is due to Busemann [*3*]. This derivation assumes an impeller of uniform axial depth having blades extending from r_1 to r_2.

The results of Busemann's analysis for straight radial blades are shown in Fig. J,3c. At a radius ratio $r_1/r_2 = 0$ the Busemann values of slip factor are somewhat greater than those given by the Stodola equation but still slightly lower than are generally observed in practice. Stanitz and Ellis [*4*], extrapolating from two computations of compressible flow in radial-bladed impellers by the relaxation method, suggest the use of $\zeta = 1 - (1.98/z)$. A comparison of the three values is given in Table J,3.

Table J,3.

Number of blades Z	Slip factor ζ		
	Stodola	Busemann*	Stanitz
16	0.80	0.86	0.88
18	0.83	0.87	0.89
20	0.84	0.88	0.90**
22	0.86	0.89	0.91
26	0.88	0.91	0.92
30	0.90	0.92	0.93**

* Values extrapolated to $r_1/r_2 = 0$.
** Values from relaxation computation.

Busemann's analysis and Stanitz's computations both indicate that the slip factor for impellers with radial blades is independent of the mass rate of flow.

J,4. Effect of Inlet Circulation and Nonradial Blades. If the inlet circulation is produced by stationary blades (inlet guide vanes)

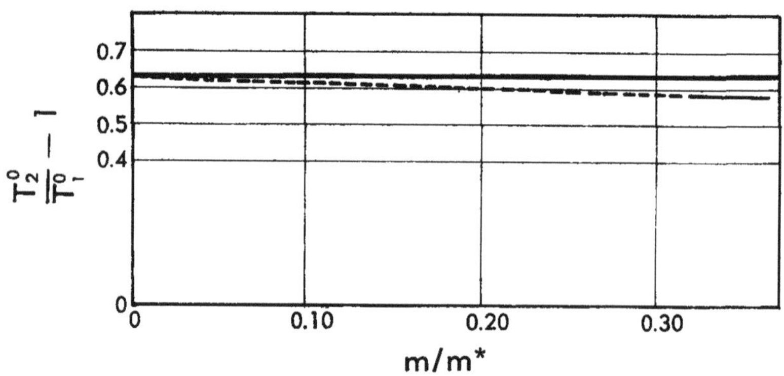

Fig. J,4a. Temperature rise in a centrifugal compressor having inlet circulation. $r_j = 0.275$; r_2 at $m/m^* = 0.36$.

ahead of the impeller, the tangential component of the absolute inlet velocity V_{θ_1} varies with the inlet flow. If it is assumed that the velocity through the inlet guide vanes is low so that the flow may be assumed to be incompressible, then V_θ is proportional to m and the relation between $(T_2^0 - T_1^0)/T_1^0$ and m/m^* is a linear one starting from the value at $m/m^* = 0$ which would be given without inlet circulation. Fig. J,4a shows the temperature rise ratio of an impeller where $r_j = 0.275r_2$ at $m/m^* = 0.36$.

A similar characteristic results from the use of impeller blades which are not radial at the impeller exit. If such blades are inclined backward with respect to the direction of rotation, an increase in mass flow decreases the value of V_{θ_2}. It may be shown that here also the decrease in temperature rise is very nearly proportional to mass flow. Forward-curving blades result in an increase in V_{θ_2} with mass flow and a rising

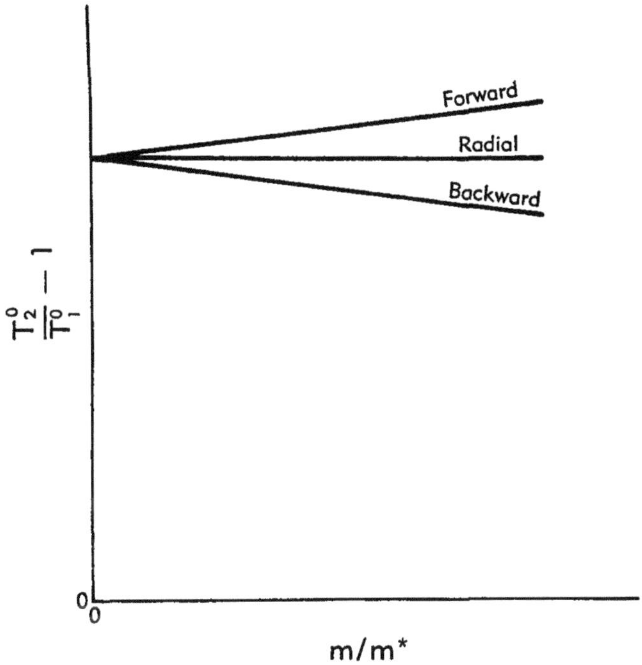

Fig. J,4b. **Temperature rise in centrifugal compressors having forward-leaning, backward-leaning, and radial impeller blades.**

temperature characteristic. Typical characteristics for forward-curved, backward-curved, and radial blades are shown in Fig. J,4b.

J,5. Effect of Disk Friction. Ippen [*5*], extending an analysis by von Kármán, proposed the following expression for the fluid friction on a thin disk running in a casing:

$$\tau_f = \frac{0.0418}{(Re_d)^{\frac{1}{5}}} \frac{\rho r_2^3 U_2^2}{2} \tag{5-1}$$

where τ_f is the friction torque on one side of a disk, Re_d the disk Reynolds number $= \omega r_2^2 \rho/\mu$, ρ the density of the fluid, r_2 the outer radius of the disk, U_2 the ring speed of the disk, ω is the disk angular velocity, and μ the viscosity of the fluid.

As indicated in [5], Eq. 5-1 checks reasonably well with experimental results.

Assuming that the friction on the casing is equivalent to that on both sides of a thin disk of diameter equal to the impeller diameter, Eq. 3-6 may be corrected for disk friction as follows:

$$\frac{T_2^0 - T_1^0}{T_1^0} = (\gamma - 1)\left(\frac{U_2}{a_1^0}\right)^2 \left\{\frac{V_{\theta_2}}{U_2} - \left(\frac{r_1}{r_2}\right)^2 \frac{V_{\theta_1}}{U_1} + \left[\frac{0.0133}{(Re_d)^{\frac{1}{5}}}\right]\left[\left(\frac{\gamma + 1}{2}\right)^{\frac{\gamma+1}{2\gamma-2}} \frac{m^*}{m} \frac{U_2}{a_1^0} \frac{\rho}{\rho_1^0}\right]\right\} \tag{5-2}$$

In order to evaluate this equation it is necessary to determine ρ/ρ_1^0 and Re_d in terms of known quantities. Since the friction term is small compared to other terms in the equation it is unnecessary to determine its value with great precision. Even less precision is satisfactory in the evaluation of the disk Reynolds number Re_d.

Using for ρ the arithmetic average of the inlet stagnation density and the density which would result from reversible adiabatic compression with the temperature rise given by Eq. 3-6, and using a value of viscosity corresponding to the discharge temperature, the author has found that for several single-entry high performance centrifugal machines, the temperature rise due to disk friction is between one and two per cent of the temperature rise given by Eq. 3-6 throughout the entire operating range (with normal atmospheric inlet conditions). The proportion of disk friction is little affected by the disk Reynolds number over the range encountered in normal operation. Evidently the proportion of disk friction will be approximately halved in a double-entry machine.

It is interesting to note that the conditions under which high output centrifugal machines work are such as to reduce the proportion of disk friction loss. Small flow, backward-curved blades, and low Reynolds numbers all increase the disk friction compared to the total power input.

J,6. Pressure Ratio. Except for the small effect of disk friction, the ratio of discharge to inlet stagnation temperature is *independent of the losses* in a centrifugal compressor having radial blades. The ratio of discharge to inlet total pressure on the other hand must depend upon the losses as well as upon the temperature ratio. These losses are in general of two kinds: (1) those due directly to friction forces on the surfaces of the rotating wheel and the stationary housing, and (2) those due to secondary flow resulting in local velocities in undesired directions. Of these two, the first appears to be relatively unimportant. The second type of loss can become large when separation of the flow occurs somewhere in the machine. The problem of separation is discussed later in this section where the various component parts of the machine are considered separately. For present purposes it is sufficient to consider the over-all losses. These are usually evaluated in terms of an "efficiency." If the machine were without losses, that is, reversible, the rise in total pressure would be uniquely determined by the temperature ratio through the isentropic relation

$$\frac{p_2^0}{p_1^0} = \left(\frac{T_2^0}{T_1^0}\right)^{\frac{\gamma}{\gamma-1}}$$

where p_2^0 is the discharge total pressure and p_1^0 is the inlet total pressure.

It is customary to define the efficiency of a compressor as the ratio of the temperature rise produced by a reversible machine working between the inlet conditions of total pressure and stagnation temperature and the final total pressure or

$$\eta = \frac{\left(\frac{p_2^0}{p_1^0}\right)^{\frac{\gamma-1}{\gamma}} - 1}{\frac{T_2^0}{T_1^0} - 1} \tag{6-1}$$

It is clear then that if the stagnation temperature ratio is known or can be estimated from Eq. 6-1, the total pressure ratio can be determined if the efficiency can be estimated. It has thus far not been found possible to improve on the rather unsatisfactory procedure of estimating the over-all efficiency on the basis of experience with similar machines.

J,7. Components. For further study, the centrifugal compressor will be divided into the following components: the inducer, the impeller, and the diffuser.

The inducer. The inducer is that portion of the wheel or impeller near the entrance which serves to produce a solid body rotation of the fluid which is necessary to match the flow in the impeller. Inducers are sometimes made as a part of the impeller and sometimes made as a separate part fastened to the wheel. While it is often difficult to distinguish a division between the inducer and the remainder of the wheel, since the two parts function together as a unit, the distinction is convenient for analytic purposes. It is assumed here that there is no change in radius of any streamline passing through the inducer. With this assumption, we may write Euler's turbine equation for an annular stream tube or radius r passing through the inducer as follows:

$$\delta\tau = \delta m(r_b V_{\theta_b} - r_a V_{\theta_a}) \tag{7-1}$$

where sections a and b are taken before and after the inducer.

In order to match the tangential velocity required in the impeller,

$$\delta P = \delta m U^2\left(1 - \frac{V_{\theta_a}}{U}\right) \tag{7-2}$$

It will be noted that the work done in the inducer at a given radius per pound of fluid is identical to the work done by a centrifugal impeller with radial blades (and no slip) having the same inlet conditions and having an outer radius equal to the radius under consideration. Without inlet circulation ($V_{\theta_a} = 0$), the work done at radius r in the inducer is $(r/r_2)^2$ times the work of the impeller neglecting slip. For centrifugal machines such as those used in high speed aircraft,

$$0.40 > \left(\frac{r_1}{r_2}\right)^2 > 0.25$$

where r_1 is the outer diameter of the inducer. Since the rise in stagnation temperature of such machines may be 300°F, the rise in the inducer may be of the order of 100°F. This is truly an extraordinary temperature rise to expect from a single subsonic axial flow stage. Coupled with the fact that the relative Mach number at the outer radius of the inducer is usually high, it is not surprising to find that inducers often operate at least partially stalled, and sometimes with reversed flow at the outer radius [*6,7*].

A somewhat different approach to the same problem results from considering the flow through the outer radius of the inducer equivalent to the flow through a two-dimensional cascade.

If the cascade be considered incompressible and frictionless the lift coefficient will be given by

$$C_L = \frac{2}{\sigma}\frac{\tan\alpha}{\sqrt{1 + \frac{1}{4}\tan^2\alpha}} \tag{7-3}$$

where σ = blade chord/blade spacing and α = the angle between the velocity vector at inlet and the perpendicular to the plane of the cascade.

The ratio of lift coefficient to solidity is plotted in Fig. J,7a. It will be seen that with the usual turning angle of the inducer (70° to 50° at the tip) the required solidity to give a reasonable lift coefficient is high. Furthermore, the angle of turn required ($= \alpha$) is also very much larger than is considered good practice in axial flow machines. In addition the relative Mach number at the inducer tip is necessarily high in order to achieve high capacity. The net result is that the inducer is one of the critical points in the design of centrifugal compressors [*8*].

It might be mentioned here that in at least two respects the conditions in an inducer are more favorable than in a two-dimensional cascade. The fact that the streamlines in an actual inducer do not lie on cylindrical surfaces is slightly favorable at the inducer tip. Probably more important is the fact that each inducer blade is joined to an impeller blade and it is therefore not necessary that the pressure on the two sides of the inducer blade be equalized where the flow leaves the inducer (Kutta condition). Tests of inducers as separate components [*9,10*] have neglected this important consideration. In order that such tests of inducers should more nearly approximate actual conditions of use, the inducer should be followed by a relatively long section of straight radial blades.

It is sometimes stated that the work done by the inducer is a small fraction of the work of the entire impeller and that therefore the efficiency of the inducer is not of great consequence. On the other hand, the result of stalling in the inducer is reflected in the flow through the entire machine. In particular the diffuser cannot be expected to operate at high

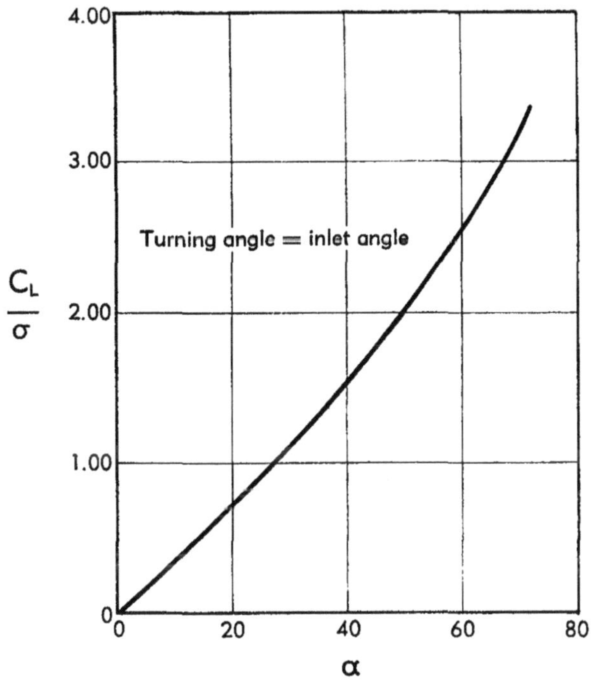

Fig. J,7a. Lift coefficient for incompressible cascade with axial discharge. Turning angle = inlet angle.

efficiency with the irregular velocity distribution at the diffuser entrance which results from a stalled inducer. Experimental results justify this point of view.

Relative Mach number at the inducer tip. If the velocity into the inducer is assumed to be axial and uniform, the velocity triangle at the outer radius of the inducer gives the following relation:

$$W_1^2 = V_1^2 + U_1^2 \tag{7-4}$$

where W_1 is the velocity relative to the wheel, V_1 the absolute entrance velocity, and U_1 the tangential velocity of the wheel at the inducer outer radius.

The following dimensionless relation can be derived from Eq. 7-4 by introducing the velocity of sound:

$$\frac{r_1}{r_2}\frac{U_2}{a_1^0} = \sqrt{\frac{M_{1_r}^2 - M_1^2}{1 + \frac{\gamma - 1}{2} M_1^2}} \tag{7-5}$$

where U_2 is the tip speed of the wheel, a_1^0 the velocity of sound at inlet stagnation temperature, r_1 the outer radius of the inducer, r_2 the outer

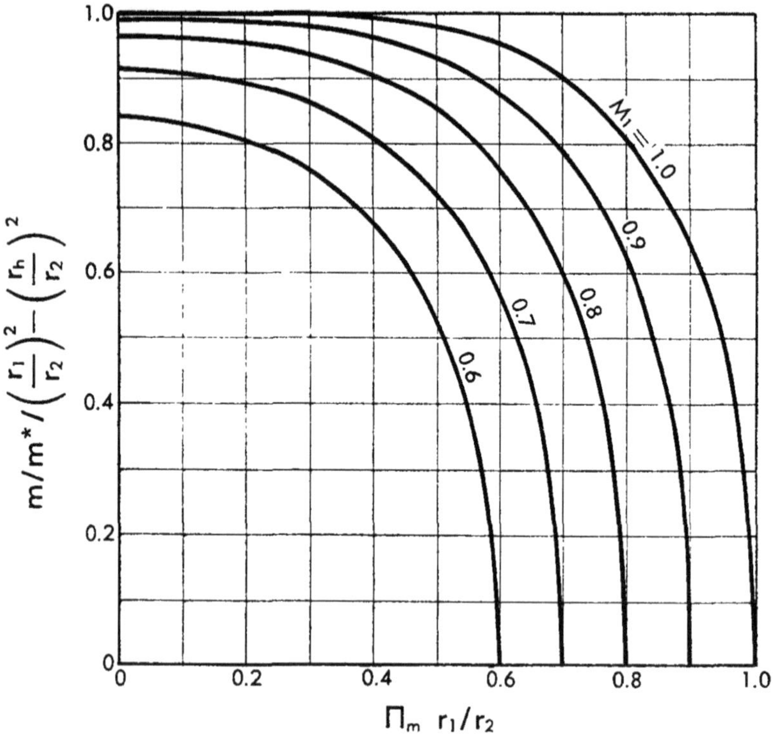

Fig. J,7b. Air capacity of axial inlet compressor.

radius of the wheel, M_{1_r} the relative Mach number at the inducer outer radius, M_1 the absolute Mach number at the inlet, and γ the ratio of specific heats.

The mass rate of flow m into the impeller is given by the continuity equation

$$m = \pi(r_1^2 - r_h^2)\rho_1 V_1 \tag{7-6}$$

where r_h is the inner (hub) radius of the inducer and ρ is the density at the inlet. Now m^* is defined by

$$m^* = \pi r_2^2 \rho^* c^*$$

where ρ^* is the density resulting from isentropic expansion from inlet conditions to the velocity of sound, and c^* is the velocity resulting from isentropic expansion from inlet conditions to the velocity of sound.

Eq. 7-6 may be written in dimensionless form as

$$\frac{\frac{m}{m^*}}{\left(\frac{r_1}{r_2}\right)^2 - \left(\frac{r_h}{r_2}\right)^2} = M_1 \left(\frac{\frac{\gamma + 1}{2}}{1 + \frac{\gamma - 1}{2} M_1^2} \right)^{\frac{\gamma+1}{2\gamma-2}} \tag{7-7}$$

Fig. J,7b showing the relation between mass flow, radius ratio, Mach

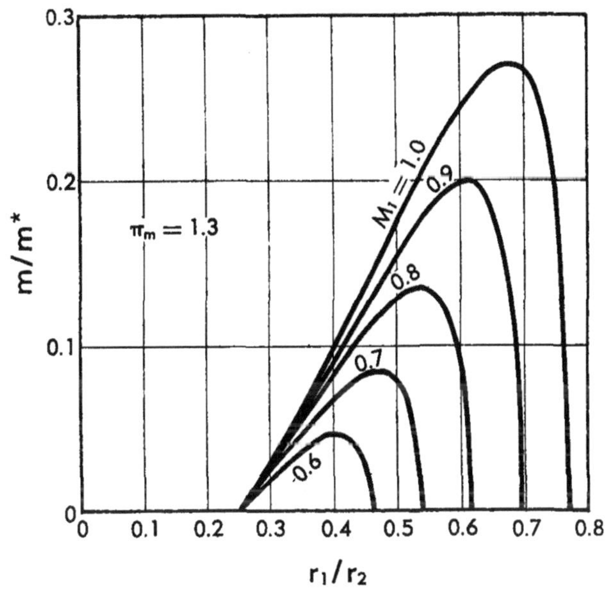

Fig. J,7c. Air capacity vs. inlet radius. $r_h/r_2 = 0.25$; $\Pi_m = 1.3$.

index, and relative inlet Mach number has been constructed from Eq. 7-5 and 7-6.

If values for U_2/a_1^0 and for r_h/r_2 are assumed, a plot similar to Fig. J,7b may be constructed from the data of Fig. J,7c. This plot shows the variation of mass flow resulting from a change in r_1/r_2 and indicates that for a given value of the relative Mach number there is a radius ratio giving maximum mass flow.[1] Conversely, with a given mass flow, there is a radius ratio which gives a minimum value of relative Mach number. In centrifugal compressors for application to high speed aircraft, the relative Mach number is high, and it is therefore important that this quantity be minimized. If inlet guide vanes are provided, it is possible to give

[1] This analysis is essentially that given by Hawthorne in [11].

initial rotation to the air stream to reduce the relative Mach number at the inlet. Analysis of the problem is algebraically complicated and is not attempted here. However, it is clear that even with circulation of the entering stream, limitation of flow and an inlet diameter for minimum relative Mach number will exist.

The amount of inlet circulation which can be used is limited by conditions at the hub of the inducer. If the inlet circulation Γ is such that

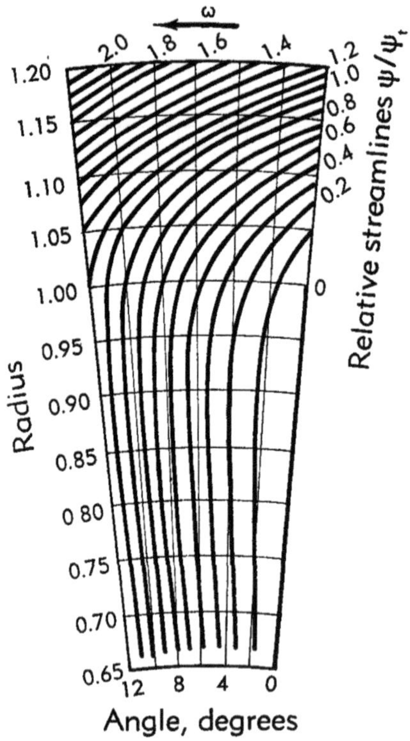

Fig. J,7d. Relative streamlines for flow through a centrifugal impeller [4].

$V_{\theta h} = U_h$, then the work done by the inducer at the hub is zero. Furthermore it is clear from Eq. 3-7 that inlet circulation reduces the work done by the impeller at a constant tip speed. A larger circulation results in negative work done at the hub of the inducer, that is, this part acts as a turbine. No doubt the inlet circulation can be made somewhat larger in certain cases, notably those in which the ratio r_h/r_1 is large.

If, for example, we assume 20 radial blades in the impeller and $r_j = 0.25r_2$, the work is reduced by the introduction of inlet circulation in the ratio $(0.90 - 0.06)/0.90 = 0.93$ and to do the same work the impeller tip speed must be increased approximately by 3.5 per cent. However, the improved inducer efficiency resulting from (1) decreased relative

Mach number, (2) decreased angle of turn, and (3) decreased lift coefficient can easily offset the decrease in work done, and the resulting pressure ratio obtained at the same tip speed is not necessarily less than that without inlet circulation.

The impeller. The problem of incompressible two-dimensional potential flow through a rotating circular lattice of blades resembling a centrifugal impeller has been considered by many writers [*3,12,13,14,15,16,17*].

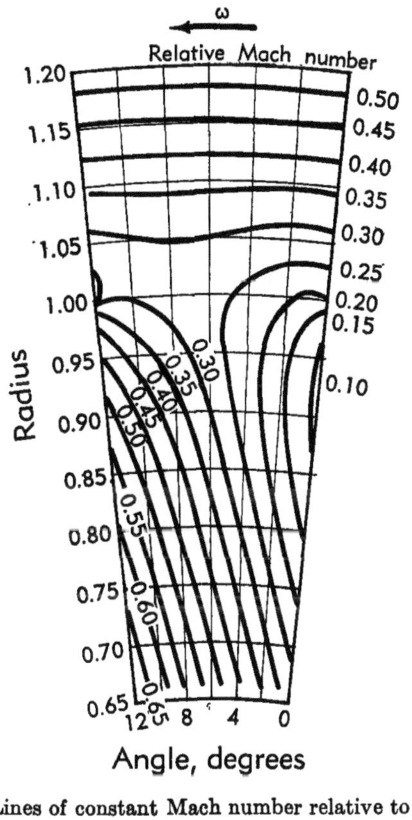

Fig. J,7e. Lines of constant Mach number relative to impeller [*4*].

By the use of relaxation methods, Stanitz [*4,13,18*] has computed a number of cases for compressible flow through an impeller having radial and logarithmic spiral blades. The slip factors previously mentioned are the result of such consideration.

Fig. J,7d and J,7e taken from [*4*] show, respectively, streamlines and lines of constant relative Mach number obtained by this method. One of the conclusions from [*4*] is the possibility of radial inward flow near the pressure face of the impeller blade at small flow rates and high rotative speeds. The following simplified analysis is sufficient to demonstrate this possibility.

The flow in an impeller having radial vanes can be approximated by assuming that the relative flow in the blade passages is radial and that the absolute flow is irrotational. This latter consideration gives

$$\frac{dV_r}{d\vartheta} = 2r\omega \tag{7-8}$$

where V_r is the radial velocity, ϑ is the angle with respect to an impeller blade, and r is the radius under consideration.

This relation shows that the variation in radial velocity along a circumference is linear with angular position. It is clear that the Kutta condition at the outer radius of the impeller blades cannot be satisfied with this simplification since the velocity on opposite sides of a blade differs by $4\pi r\omega/Z$ at all points, and thus cannot be equal at the blade tips as required by the Kutta condition. The approximation therefore cannot be expected to apply near the outer radius of the wheel. The approximation does check well, however, with the relaxation analysis up to 80 per cent of the outer radius of the blades for the cases considered by Stanitz. No doubt, with larger numbers of blades, the approximation is good to larger radii.

Eq. 7-8 reveals that the *variation* of radial velocity with angle is independent of mass flow. Since the average value of the radial velocity depends upon the mass flow, it is evident that as the mass flow is reduced, the radial velocity at the pressure side of the blade must go to zero, and with further reduction in flow the radial velocity must be negative. The resulting flow has a permanent eddy, always containing the same fluid. It is questionable whether such a flow can be stable. It seems more likely that eddies of this nature would be shed downstream and reformed periodically. In any case, there is no doubt that such a flow configuration is undesirable. Fortunately, there is no great problem in making impeller passages of sufficiently small cross section so that the average radial velocity will be sufficient to prevent negative velocities at any point.

It is interesting to note that while increasing the number of impeller blades produces a more uniform radial velocity and a decreased tendency to reverse flow, the variation of radial velocity with angle is independent of the number of blades. This consideration is of interest in the speculation as to whether an impeller with an infinite number of blades can be said to produce a one-dimensional flow pattern.

In the conventional axial flow compressor, the fluid is caused to flow between blades in such a manner that the relative discharge is always more nearly axial than the relative inlet to any blade row (except the first row of entrance guide vanes). This means that there is a diffusion process—a decrease in relative velocity with a corresponding increase in pressure in each blade row. In fact in the axial flow machine the over-all pressure rise is entirely due to the sum of these pressure rises accompanying decreases in relative velocity. In the centrifugal machine, on the other hand, only a part of the pressure rise is due to diffusion. Another part, of approximately equal magnitude, is due to what may be called centrifugal action, that is, direct compression of the fluid due to the fact that fluid enters at a small radius, is whirled, and then discharged at a larger radius. This fact has an important bearing on the potentialities of the centrifugal machine. The diffusion process is limited by the ability of the shear forces to carry the boundary layer downstream against the "adverse pressure gradient." If these forces are insufficient, the flow direction in the boundary layer reverses and the high losses associated with separation appear. It is thus necessary to design diffusing passages in such a manner as to give shear forces an opportunity to act on the boundary layer, that is to spread the pressure rise over a considerable distance. Since at least a part of the losses is associated with shear forces, it is not strange to find that in practice the diffusion process is never as efficient as its inverse, the expansion process in a nozzle.

The process of compression by centrifugal action differs from that by diffusion. The centrifugal compression is not due to a relative retardation of the stream; furthermore, the forces due to centripetal acceleration act on the boundary layer and the main stream alike, and hence do not decelerate the boundary layer, but merely produce a pressure gradient in addition to that produced by the change in relative velocity. Except insofar as the density of the boundary layer may differ from that of the main stream, it may be stated that *the pressure rise due to centrifugal action has no effect on boundary layer separation.* The significance of this fact is that it appears possible to achieve high efficiency in that portion of the compression process involving centrifugal action. For this reason the centrifugal compressor may have potentialities of a higher efficiency than the axial flow machine.

Secondary flow in impeller passages. Consideration of the boundary layer on the back wall of a centrifugal impeller leads to the conclusion that the boundary layer on this wall assumes a higher tangential component of velocity than the main flow and can easily have a higher tangential velocity than the wheel itself. This is a fortunate circumstance since it indicates that the work done on this boundary layer is greater than the work done on the main stream, which tends to improve the distribution of kinetic energy entering the diffuser. In order to verify the forward motion of the boundary layer, it is necessary merely to note that the circumferential pressure gradient in the passage between the leading and trailing sides of radial impeller blades can be looked upon as the result of the Coriolis acceleration ($2\omega V_r$). Since the radial velocity in the boundary layer is less than that of the free stream, the boundary layer is deflected forward under the influence of the pressure gradient.

If the inducer is stalled at the tip, the resulting low energy fluid will likewise be deflected in the direction of rotation, giving rise to secondary circulation. Depending on how far this effect has progressed, the low energy air may be discharged near the front wall, near the trailing side of the blades, or near the back wall of the impeller. Any secondary circulation of course greatly affects the performance of the diffuser.

The diffuser. Fluid leaves the impeller of a centrifugal compressor at a velocity which is much higher than is desirable for use. For this reason centrifugal machines are invariably provided with some form of diffuser to slow down the flow. It is desirable that this process be as nearly reversible as possible. If axial symmetry of the flow leaving the impeller is assumed, the flow at this point is in the nature of a source-vortex combination. This type of flow is particularly interesting since, especially in the supersonic region, it exhibits unusual characteristics (see [*19,20,21*]).

For steady frictionless compressible flow in a source-vortex the following equations may be written:

Continuity.

$$m = \rho(2\pi r h) V \sin \alpha \tag{7-9}$$

where m is the mass flow crossing the radius r per unit time, ρ is the density at the radius r, h is the axial dimension of the passage, V is the velocity, and α is the angle of the velocity vector with the tangential direction.

If r^* is used to denote the radius at which the Mach number is unity and the asterisk is used to distinguish other conditions at this radius, and if h is assumed constant,

$$\rho r V \sin \alpha = \rho^* r^* V^* \sin \alpha^* \tag{7-10}$$

Momentum. The conservation of angular momentum requires that

$$r V \cos \alpha = r^* V^* \cos \alpha^* \tag{7-11}$$

From Eq. 7-10 and 7-11 the following may be derived

$$\frac{\rho}{\rho^*} = \frac{\tan \alpha^*}{\tan \alpha} = \left(\frac{\frac{\gamma + 1}{2}}{1 + \frac{\gamma - 1}{2} M^2} \right)^{\frac{1}{\gamma - 1}} \tag{7-12}$$

$$\frac{r^* \sin \alpha^*}{r \sin \alpha} = \frac{\rho V}{\rho^* V^*} = M \left(\frac{\frac{\gamma + 1}{2}}{1 + \frac{\gamma - 1}{2} M^2} \right)^{\frac{\gamma + 1}{2\gamma - 2}} \tag{7-13}$$

From Eq. 7-12 and 7-13 the relation between r/r^* and α may be determined for any assumed value of α^*. Fig. J,7f shows curves of these relationships. It will be noted that there is a minimum value of r/r^* below which no flow is possible. This limit line for the flow occurs when $M \sin \alpha = 1$, that is when the radial component of the Mach number is unity. At larger values of r/r^* two flows are possible: a flow in which the velocity increases with increasing radius and one in which the velocity decreases with increasing radius. The latter type of flow is of interest in connection with centrifugal machines since it provides diffusion without vanes. Since the radial component of the Mach number is always less than unity with this type of flow, no shock can exist without violating the condition of axial symmetry. It thus appears possible with a vaneless

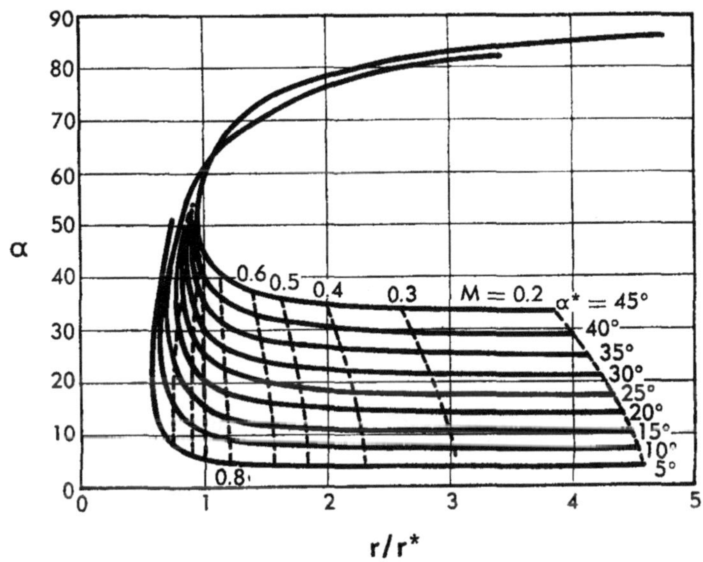

Fig. J,7f. Relation of angle of flow to radius for a compressible source-vortex.

diffuser to pass from supersonic to subsonic velocity without shock. Furthermore, disturbances will be propagated upstream when the radial component of the Mach number is less than unity; thus the entire flow in a vaneless diffuser will change with a change in back pressure even though the velocity is supersonic.

For aircraft purposes it is usually not feasible to use a vaneless diffuser because of the large outer radius required. For example, in a typical jet engine compressor the Mach number at the exit of the impeller is 1.10 and the absolute velocity at this point is 15° from the tangential direction. Starting from these conditions, Fig. J,7f may be used to determine the radius of a vaneless diffuser to reduce the Mach number to 0.3. In this case the outer radius would be more than three times the radius of the impeller. A considerable saving in space can be achieved by using diffuser vanes to reduce the tangential velocity of the fluid leaving the impeller more rapidly than can be accomplished in a free vortex flow. It is clear that in order to accomplish this end it is necessary that the diffuser vanes exert a torque on the fluid in a direction opposite to its rotation in order to reduce its angular momentum.

A vaned diffuser is subject to limitations as to the angle of flow which it can accept without flow separation and consequent losses. For this reason it is pertinent to inquire into the change in angle of incidence with changes in mass flow.

For a radial-bladed impeller, the tangential component of velocity leaving the impeller is equal to the product of the slip factor ζ and the wheel tip speed U. If the angle of the flow with the tangent is α then the radial component of velocity leaving the impeller is $\zeta U \tan \alpha$. The mass flow is equal to

$$m = \rho A \zeta U \tan \alpha \tag{7-14}$$

where A is the area of the annulus at the wheel discharge and ρ is the corresponding density.

We are interested in the change in α for a given percentage change in mass flow or

$$\frac{d\alpha}{d(\ln m)} = \frac{d\alpha}{d[\ln (\rho A \zeta U \tan \alpha)]} \tag{7-15}$$

or since A, ζ, and U are constant

$$\frac{d(\ln m)}{d\alpha} = \frac{d(\ln \rho)}{d\alpha} + \frac{d(\ln \tan \alpha)}{d\alpha} \tag{7-16}$$

If it is assumed that the change in ρ with a change in mass flow is small, then

$$\frac{d(\ln m)}{d\alpha} = \frac{d(\ln \tan \alpha)}{d\alpha} = \frac{2}{\sin 2\alpha} \tag{7-17}$$

and

$$\frac{d\alpha}{d(\ln m)} = \frac{1}{2} \sin 2\alpha \tag{7-18}$$

Thus the change in the angle of incidence of the diffuser blades for a given change in the proportion of mass flow is a maximum for $\alpha = 45°$.

Assuming that all diffusers accept a given variation in angle of incidence without stall, then the maximum variation in mass flow for unstalled operation will be achieved if the machine is designed to have a small angle at the entrance to the diffuser. Note that this condition is diametrically opposed to the requirement of sufficiently high radial velocity to insure the absence of an eddy on the driving face of the impeller.

Design of a vaned diffuser. The essential problem of the vaned diffuser is to arrange the blades (or passages) in such a manner that the necessarily adverse pressure gradient will not result in flow separation. An essentially one-dimensional approach to this problem is given by Campbell [*8*]. This sort of analysis has given good results when supplemented by a large background of experience. A two-dimensional approach has been proposed by Giraud [*22*].

In view of the large effects due to wall friction, as noted by Giraud [*20*], it seems unlikely that any approach to the problem based on potential flow is likely to give more than a rough indication of the actual flow. Under these circumstances it seems probable that diffuser design will remain for some time essentially a trial-and-error process.

J,8. Stability. One of the difficulties, experienced in dynamic compressors as opposed to volumetric types, is that operation is often found

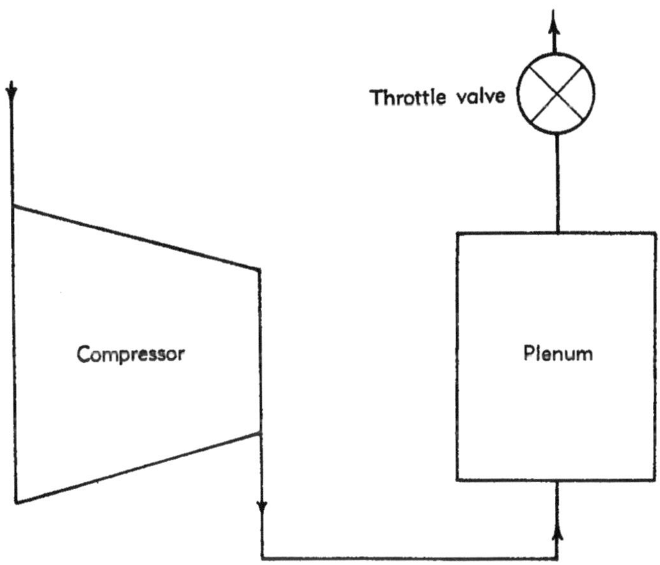

Fig. J,8a. Diagrammatic compressor test setup.

to be unstable when the flow is reduced (holding constant rotative speed) below a certain value. An analysis of this problem, which may help to explain this phenomenon follows.

Fig. J,8a is a diagrammatic sketch of a compressor test setup. The throttle valve is necessary to establish the desired pressure ratio across the compressor, and the plenum provides a convenient spot to measure pressures and temperatures. In some setups the plenum is replaced by a relatively long pipe operating with a low gas velocity. Also, the throttle valve and plenum are sometimes attached on the inlet side of the compressor. None of these possible variations appreciably affects the analysis which follows.

Analysis of the actual arrangement would be too complex to be useful, if it were possible, and simplification is in order. For this purpose we make the following assumptions:

1. The density at any point within the compressor is a function of position only, at any running condition of the compressor. With this assumption, at any instant the mass flow out of the compressor is equal to the mass flow into the compressor.
2. The process in the plenum is assumed to be isothermal.
3. The characteristic pressure versus flow relation of the compressor is assumed to be unaffected by transient phenomena, that is the compressor is assumed to be in a quasi-steady state of operation at all times.
4. Friction forces are assumed to be independent of time.

With these assumptions it is possible to make a diagram of the simplified system as in Fig. J,8b. The static pressure flow characteristic of the

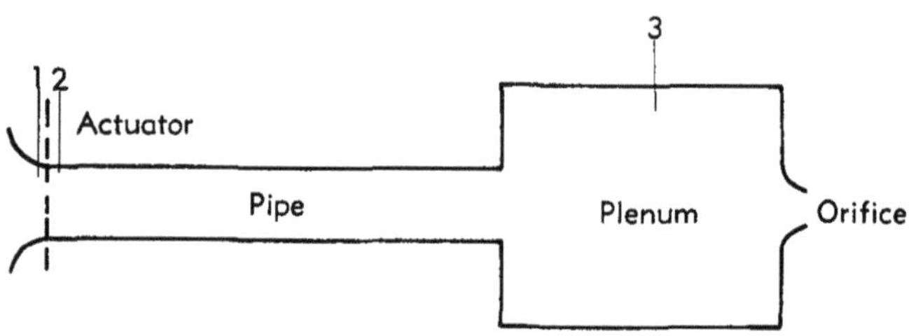

Fig. J,8b. Analogue of compressor test setup.

compressor is assumed to be supplied by an actuator between sections 1 and 2. The compressor is replaced by this actuator and a straight pipe in which the flow is incompressible and which is proportioned so that the rate of change of mass flow through the pipe, produced by a pressure difference at the ends, is the same as the rate of change of mass flow through the compressor for the same value of pressure difference (i.e. the difference between the discharge pressure of the compressor and the pressure for equilibrium).

It may be shown that the pipe thus dynamically simulates the compressor if

$$\left(\frac{l}{A\rho}\right)_{\text{pipe}} = \int_l \left(\frac{dl}{A\rho}\right)_{\text{compressor}} \tag{8-1}$$

where l is the length taken in the flow direction (for no rotation of the compressor wheel), A is the area of cross section perpendicular to l, and ρ is the density at any point.

The equation of motion for the air in the pipe may be written as

$$\left[p_1\left(\frac{p_2}{p_1}\right) - p_3\right] A = \frac{\rho A l}{g_0}\frac{du}{dt} \tag{8-2}$$

where p_1, p_2, and p_3 are pressures measured as shown in Fig. J,8b. A is the cross-sectional area of the pipe, ρ the density in the pipe, l the length of the pipe, and u the velocity in the pipe (assumed uniform throughout the pipe).

Since $u = m_c/\rho A$ where m_c is the mass flow through the pipe (compressor), Eq. 8-2 may be written in dimensionless form

$$\frac{p_2}{p_1} - \frac{p_3}{p_1} = T\frac{d}{dt}\left(\frac{m_c}{m^*}\right) \tag{8-3}$$

where the characteristic time T of the compressor is given by

$$T = \gamma\left(\frac{2}{\gamma + 1}\right)^{\frac{\gamma+1}{2\gamma-2}}\frac{\pi r_2^2}{A}\frac{l}{a_1^0} \tag{8-4}$$

Assuming for simplicity that the process in the plenum is isothermal, the rate of change of pressure in the plenum may be written in dimensionless form as

$$\frac{d}{dt}\left(\frac{p_3}{p_1}\right) = F\left(\frac{m_c}{m^*} - \frac{m_o}{m^*}\right) \tag{8-5}$$

where m_o is the mass flow per unit time in the orifice and the characteristic frequency F of the plenum is given by

$$F = \frac{T_2^0}{T_1^0}\left(\frac{2}{\gamma + 1}\right)^{\frac{\gamma+1}{2\gamma-2}}\frac{\pi r_2^2}{\mathcal{V}^{2/3}}\frac{a_1^0}{\mathcal{V}^{1/3}} \tag{8-6}$$

where $\mathcal{V}$ is the volume of the plenum.

There are two further relationships expressing, respectively, the characteristic of the compressor

$$\frac{p_2}{p_1} = f_1\left(\frac{m_c}{m^*}\right) \tag{8-7}$$

and the characteristic of the orifice

$$\frac{p_3}{p_1} = f_2\left(\frac{m_o}{m^*}\right) \tag{8-8}$$

The flow through the orifice is assumed to respond instantaneously to changes in pressure.

Fig. J,8c is a plot of f_2 for a converging nozzle of unity coefficient. Fig. J,8d is a typical plot of f_1, the compressor characteristic.

These two functions are in general complicated, and when substituted in Eq. 8-3 and 8-5 result in nonlinear equations which cannot be solved analytically, since Eq. 8-6 represents an experimental function.

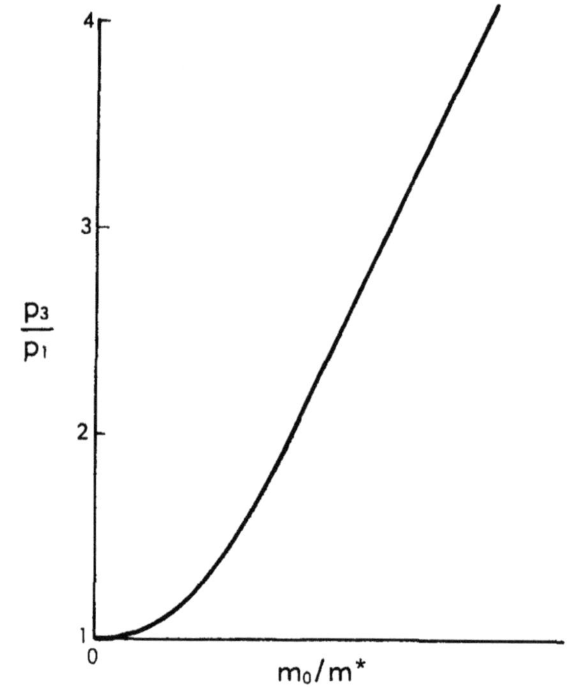

Fig. J,8c. Orifice characteristic.

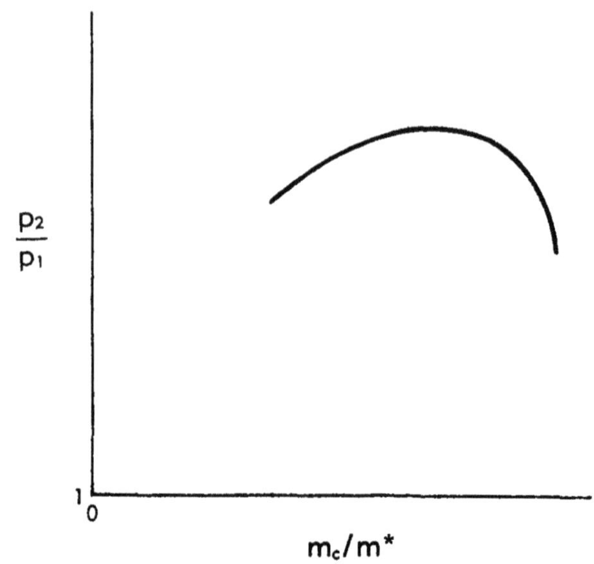

Fig. J,8d. Static compressor characteristic.

For purposes of studying stability to small disturbances, it suffices to replace Eq. 8-6 and 8-7 with linear relationships near the equilibrium point, provided that in the neighborhood of this point the two functions are continuous and have continuous first derivatives. The equilibrium point is defined by the relations

$$\left(\frac{p_2}{p_1}\right)_{\text{equil}} = \left(\frac{p_3}{p_1}\right)_{\text{equil}} \tag{8-9}$$

and

$$\left(\frac{m_o}{m^*}\right)_{\text{equil}} = \left(\frac{m_c}{m^*}\right)_{\text{equil}} \tag{8-10}$$

With these assumptions, Eq. 8-7 and 8-8 may be written

$$\xi_1 = X\eta_1 \tag{8-11}$$

$$\xi_2 = Y\eta_2 \tag{8-12}$$

where

$$\xi_1 = \frac{p_2}{p_1} - \left(\frac{p_2}{p_1}\right)_{\text{equil}} \tag{8-13}$$

$$\xi_2 = \frac{p_3}{p_1} - \left(\frac{p_3}{p_1}\right)_{\text{equil}} \tag{8-14}$$

$$\eta_1 = \frac{m_c}{m^*} - \left(\frac{m_c}{m^*}\right)_{\text{equil}} \tag{8-15}$$

$$\eta_2 = \frac{m_o}{m^*} - \left(\frac{m_o}{m^*}\right)_{\text{equil}} \tag{8-16}$$

$$X = f_1^1\left(\frac{m_c}{m^*}\right) \text{ at } \left(\frac{m_c}{m^*}\right)_{\text{equil}} \tag{8-17}$$

$$Y = f_2^1\left(\frac{m_o}{m^*}\right) \text{ at } \left(\frac{m_o}{m^*}\right)_{\text{equil}} \tag{8-18}$$

It is seen that X and Y are the slopes of the characteristic curves of the compressor and the orifice at the equilibrium point.

Using the new variables ξ and η, Eq. 8-3 and 8-5 may be rewritten

$$\xi_1 - \xi_2 = T\frac{d}{dt}(\eta_1) \tag{8-19}$$

$$\frac{d}{dt}(\xi_2) = F(\eta_1 - \eta_2) \tag{8-20}$$

By the substitution of values from Eq. 8-11 and 8-12, Eq. 8-19 and 8-22 finally become

$$X\eta_1 - Y\eta_2 = T\frac{d}{dt}(\eta_1) \tag{8-21}$$

$$Y\frac{d}{dt}(\eta_2) = F(\eta_1 - \eta_2) \tag{8-22}$$

A solution of Eq. 8-21 and 8-22 is

$$\eta_1 = C_1 e^{bt}$$

$$\eta_2 = C_2 e^{bt}$$

where C_1 and C_2 are arbitrary constants and b is given by the quadratic

$$b^2 + \left(\frac{F}{Y} - \frac{X}{T}\right) b + \frac{F}{T} - \frac{F}{T}\frac{X}{Y} = 0 \tag{8-23}$$

of which the roots are

$$b = -\frac{1}{2}\left(\frac{F}{Y} - \frac{X}{T}\right) \pm \sqrt{\frac{1}{4}\left(\frac{F}{Y} - \frac{X}{T}\right)^2 + \frac{F}{T}\left(\frac{X}{Y} - 1\right)} \tag{8-24}$$

It is clear that b may be real or complex, and that the criterion for stability is that the *real part* of b must not be positive for either root.

If b is complex, the variation in ξ is oscillatory and stability is indicated when, and only when,

$$\frac{F}{Y} - \frac{X}{T} > 0 \tag{8-25}$$

If b is real, stability is indicated when, and only when, both

$$\frac{F}{Y} - \frac{X}{T} > 0$$

and the radical is not larger in absolute value than the first term in Eq. 8-24, that is, when

$$\frac{F}{T}\left(\frac{X}{Y} - 1\right) < 0 \tag{8-26}$$

It may be noted that Y, F, and T are always positive quantities so that Eq. 8-25 can be written as

$$X < \frac{FT}{Y} > 0 \tag{8-27}$$

and Eq. 8-26 gives

$$X < Y \tag{8-28}$$

Therefore, stability is assured if X is negative, i.e. a compressor with a falling pressure flow characteristic is always stable. A rising characteristic is also stable if the slope X is sufficiently small. How large the slope may be without encountering instability depends on the characteristics of the circuit into which the compressor is connected. The smaller the volume of the plenum, the larger the permissible slope of the compressor characteristic and, also, the smaller the slope of the orifice characteristic may be, with the exception that if $X > Y$ instability always results.

Two special cases are of interest. If the plenum is made very large,

F becomes very small and, for a plenum of infinite size, stability can only be assured if $X < 0$.

On the other hand, if the volume of the plenum is reduced to zero the case is reduced to the equivalent of that involving an incompressible fluid, and stability is assured if $X < Y$. Of course in this case the assumption that the density within the compressor is invariant with time is inappropriate.

In view of the very broad assumptions made in this analysis, qualitative agreement with experiment is perhaps all that can be expected. However, it may be worth noting that if the controlling orifice is choked, $X = 1$ and if the plenum is of such size that $FT > 1$, the criterion $X > Y$ will be sufficient to assure stability. It appears that with a centrifugal compressor and a plenum of reasonable volume, say $\mathcal{V} \leqq (2r_2)^3$, the product FT will indeed exceed unity, and stability might be expected to depend only on the inequality $X > Y$.

Comparison with observations. In testing compressors, it is well known that stable operation sometimes occurs with a rising pressure flow characteristic. The analysis presented gives at least a qualitative notion of the condition required for this possibility, namely a not-too-large plenum between the controlling orifice and the machine.

On the other hand, in centrifugal compressors, it is often noted that the onset of instability is at or very near the flow for maximum pressure ratio. This is especially true of machines having vaned diffusers when operated at a high pressure ratio. With machines having vaneless diffusers, it is usual to find stable operation with a rising characteristic.

It should be pointed out that it is not possible to discover experimentally the shape of the *unstable* part of the characteristic by the ordinary test procedure. It may be inferred, however, that instability in centrifugal machines is associated with violent stall, probably in the vaned diffuser, and that the characteristic in the region of stall is either discontinuous or at least very steeply rising.

It should be noted here that, in any compressor, there are a number of passages operating in parallel with the same pressure difference across the various passages. If each passage is assumed to have a pressure flow characteristic similar to that of the complete compressor, which is assumed to be a rising and falling characteristic having two values of flow corresponding to a given pressure ratio as shown in Fig. J,8d, it is then possible that the end condition of pressure and flow will be satisfied if some of the passages are operating on the rising and some on the falling part of the characteristic. Under these circumstances, instability could result from flow in those passages which are on the rising part of their characteristic, even though the characteristic of the entire machine might be such as to indicate stable operation.

In any dynamic machine operating with a high relative Mach number at the entrance to a set of blades, stall may be expected when the angle of attack of the blades becomes only slightly larger than the design angle. For this reason stall may be expected to occur in such machines at a flow only slightly less than the design flow, or very close to the flow for maximum efficiency. When the relative Mach number is sufficiently low so that shock waves are absent, the angle of attack required to produce stall is of course greater, and a larger range of stable flow may be expected. Since, in the centrifugal machine, high pressure ratio is always associated with high Mach number at the diffuser entrance, and since stall in the diffuser makes such a great difference in the over-all pressure ratio of the machine, it is not surprising to find that instability in a centrifugal machine with a vaned diffuser occurs very near the point of maximum efficiency when operating at high pressure ratios, while at low pressure ratios there is a more comfortable margin between the flow for best efficiency and the flow at which instability occurs.

Since the axial-entry, radial-vaned centrifugal compressor has an approximately constant temperature rise for all flows in the operating region, it is inevitable that maximum efficiency and maximum pressure will be at nearly the same flow. In this case the onset of instability is likely to occur very close to the flow for maximum pressure when operating at high pressure ratios. Such reasoning may account for the observation of stable regions of operation where the pressure-flow characteristics have a positive slope at low speeds and low pressure ratios, as well as the absence of such regions at high pressure ratios. Furthermore, a stable region of operation in the region of positive characteristic slope is invariably observed in machines having vaneless diffusers which presumably do not stall under any conditions. It would seem that a desirable margin of safety between the flow for maximum efficiency and the flow which will produce stall can be achieved either by the use of a vaneless diffuser or by operation at low relative Mach number. Either of these solutions has serious disadvantages. There is also the possibility that with a sufficiently high supersonic velocity entering the diffuser a design could be worked out which would help to solve this problem.

It should not be assumed that stall somewhere in the machine will always produce instability. Stall will only produce instability when it produces a sufficiently large positive slope of the characteristic of the entire machine. Stable operation can therefore be expected in the neighborhood of stall if either (1) the stall is in only a portion of the machine (as one stage of a multistage axial compressor) or (2) if the stall is not sudden but progressive, as might be the case if stall started at the tip of a blade and worked inward toward the root as flow was decreased.

J,9. Performance. Fig. J,9 shows a typical curve of performance of a centrifugal compressor. As is characteristic of some highly developed machines, the efficiency holds up well at high pressure ratios and choking is very abrupt. Also the flow range between surge and choke is quite small. An interesting analytic possibility presents itself if the relation between the minimum diffuser area and the area πr_2^2 is known. If it is assumed that

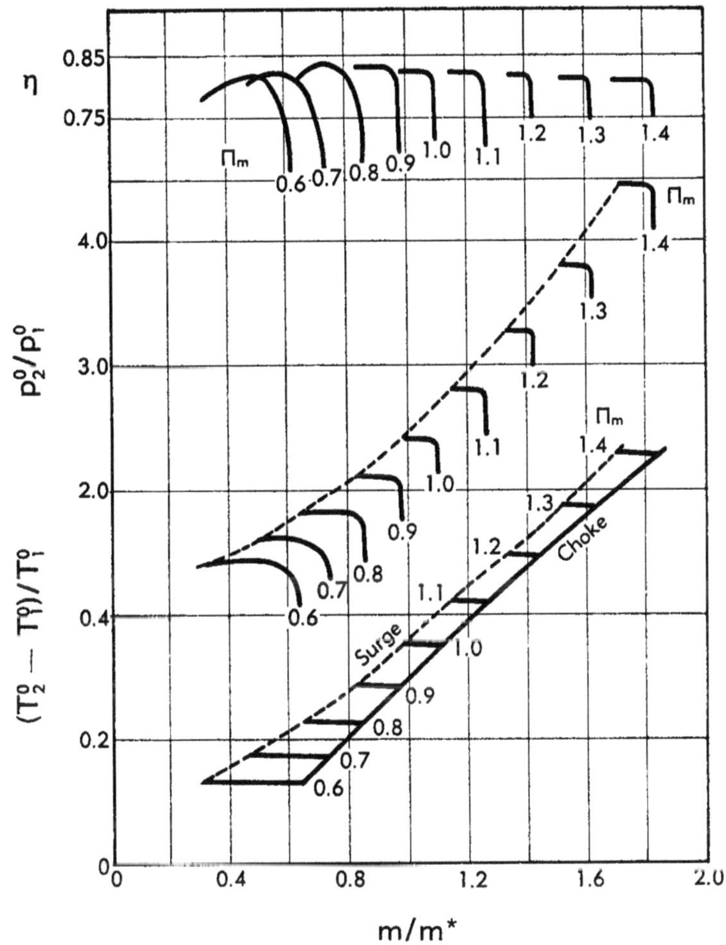

Fig. J,9. Typical performance of a centrifugal compressor.

the flow through the diffuser minimum area is at a uniform Mach number of unity at the choking condition, and since the mass flow and stagnation temperature at this point are known, the stagnation pressure may be computed.

If it is further assumed that the stagnation pressure entering the diffuser is unchanged for a small change in mass flow, the efficiency of the impeller and of the diffuser may be separately evaluated for a point at slightly less flow than the choking flow. Application of this technique to one of modern design has shown that for this particular case:

1. The diffuser efficiency is nearly constant over a range of Mach index from 0.8 to 1.3.
2. The impeller efficiency is very high at a Mach index of 0.9 and falls off at higher values.

With the value of stagnation pressure ahead of the diffuser available from the above analysis, and assuming one-dimensional flow and an appropriate slip factor, it is possible to estimate the Mach number and the angle of the absolute velocity leaving the impeller.

Effect of Reynolds number. While there have been a number of attempts to isolate the effect of Reynolds number in centrifugal compressors [*5*,*23*], no very convincing data are known to the author. The problem is complicated by the fact that if the Reynolds number is altered by altering the pressure level, a considerable change in geometry due to elastic distortion of the casing is possible. A number of small compressors have been built with high efficiency but never quite as high as larger compressors. A part of the difference is undoubtedly due to the fact that the urge to obtain high efficiency is much greater in the case of large machines. It is therefore very difficult to make good estimates of the effect of Reynolds number at the present state of the art.

J,10. Limitations of the Centrifugal Compressor. A primary limitation on the use of the centrifugal compressor has been its low efficiency when operating at high pressure ratio. It has recently been found possible by careful design to achieve reasonably high efficiency at pressure ratios of four and over, and there seems reason to believe that at least as good efficiency should be attainable at still greater speeds and pressure ratios. The Mach number of the flow leaving an impeller operating at a pressure ratio of 4.0 is approximately 1.1. It is, of course, particularly difficult to handle fluid flowing at a Mach number so near unity, and it seems entirely possible that the problem of diffusion would be easier for a somewhat higher Mach number.

It appears characteristic of all dynamic compressors that the range of mass flow between surge and choke becomes smaller as the machine is designed for higher pressure ratio. There is reason to believe that in the case of centrifugal machines this is associated with diffuser entrance Mach numbers near unity. A Mach number of unity occurs at a pressure ratio of approximately 3.0 for a radial bladed impeller. While it is possible that increased range can be obtained at higher Mach numbers by means of a properly designed supersonic diffuser, the limitation of range may be the factor that ultimately determines the maximum pressure ratio of practical machines.

The tip speed required in a centrifugal compressor to produce a pressure ratio of 4.0 with sea level atmospheric temperature is nearly as high as can be tolerated from the point of view of centrifugal stress in the impeller. Any attempt to increase the pressure ratio of a centrifugal machine must reckon with this stress problem. A material with a higher strength-to-weight ratio than aluminum alloy or steel would make the problem of higher tip speeds easier. In this connection it is interesting to note that the highest wheel temperature (at the rim of the wheel) can be no more than the arithmetic mean of the inlet and discharge stagnation temperatures in a radial-bladed machine. This temperature is moderate, and in single-stage machines it has been found possible to use aluminum alloy wheels without serious sacrifice in strength due to high temperature.

Forward-inclined impeller blades make possible higher pressure ratios without increase in tip speed. However, departure from radial blades involves bending due to centrifugal loads and consequently increased stresses. In addition, it appears likely that forward-curving blades result in increased difficulty with surging due to the characteristically higher slope of the pressure-volume characteristic which they produce. In spite of these theoretical drawbacks, it would be interesting to see what could be done with a machine with forward-curving impeller blades.

A large number of blades is assumed so that slip can be neglected, and it is further assumed that the entrance to the impeller is axial with a velocity $V_1 = W_2$, where W_2 is the relative discharge velocity. With these assumptions the fluid experiences deceleration relative to the wheel only in the inducer section. A stream filament which entered the center of the wheel would experience no diffusion, while a filament entering at a larger radius would experience diffusion in the inducer only. The pressure rise in the wheel is therefore essentially due to centrifugal action under this assumption.

Consider the stream filament which enters the center of the wheel.

$$T^0_{2r} - T^0_1 = T^0_2 - T^0_1 - \frac{V_2^2 - W_2^2}{2gJc_p}$$

where T^0_{2r} is the stagnation temperature which would be measured by a thermometer moving with the wheel at the outer radius of the wheel.

It can be shown that

$$\frac{T^0_{2r} - T^0_1}{T^0_2 - T^0_1} = \frac{1}{2}\frac{U_2}{V_{\theta_2}}$$

Since the velocity relative to the wheel is constant, $T^0_{2r} - T^0_1 = T_2 - T_1$, and $T^0_{2r} - T^0_1$ is the temperature rise occurring in the wheel; by our assumption all of this rise is due to centrifugal action.

It may be concluded that the fraction of the temperature rise occurring in the wheel by centrifugal action depends upon the ratio V_{θ_2}/U_2.

For the case of radial vanes, approximately one half of the temperature rise occurs in the wheel and one half in the diffuser. A backward-leaning impeller blade exit results in a larger fraction of the temperature rise occurring in the wheel.

On the other hand, forward-leaning impeller blades result in a larger proportion of the temperature rise in the diffuser. Thus the diffusion problem becomes more acute with forward-leaning blades, and the advantages of centrifugal compression are less in this case.

Some of the greatest recent advances in centrifugal machines have been in air-handling capacity. While advances in this respect are still possible, they can be achieved only by one or more of the following changes:

1. Increased relative inlet Mach number.
2. Increased inlet circulation in the direction of impeller rotation.
3. Increased ratio of inlet diameter to outside diameter.

There are obvious disadvantages to any of these changes, so that any additional increase in the capacity of centrifugal machines is likely to be at the expense of either efficiency or pressure ratio.

It is of course clear that a double-entry impeller has approximately double the air-handling capacity of a single-sided impeller of the same diameter.

J,11. Cited References and Bibliography.

Cited References

1. Wislicenus, G. F. *Fluid Mechanics of Turbomachinery.* McGraw-Hill, 1947.
2. Stodola, A. *Steam and Gas Turbines*, Vol. 1 and 2. McGraw-Hill, 1927.
3. Busemann, A. Das Förderhühenverhältuis radialer Kreiselpumpen mit Logarithmish-spiraligen Shaufeln. *Z. angew. Math. u. Mech. 8*, (*5*), 372 (1928).
4. Stanitz, J. D., and Ellis, G. O. Two-dimensional compressible flow in centrifugal compressors with straight blades. *NACA Tech. Note 1932*, 1949.
5. Ippen, A. T. The influence of viscosity on centrifugal pump performance. *Annual Meeting Am. Soc. Mech. Engrs.* New York, Nov. 1945.
6. Benser, W. A., and Moses, J. J. An investigation of the backflow phenomenon in centrifugal superchargers. *NACA Advance Restricted Rept. E5E16a*, June 1945.
7. Spannhake, W. *Centrifugal Pumps, Turbines, and Propellers.* Technology Press, 1934.
8. Campbell, K., and Talbert, J. E. Some advantages and limitations of centrifugal and axial aircraft compressors. *SAE Quart. Trans. 53*, (*10*), 1945.
9. Ritter, W. K., Ginsburg, A., and Beede, W. Performance comparison of two deep inducers as separate components and in combination with an impeller. *NACA Wartime Rept. E-9*, Oct. 1945.
10. Ritter, W. K., and Johnsen, I. A. Preliminary investigation of deep inducers as separate supercharger components. *NACA Wartime Rept. E-13*, Nov. 1945.
11. Hawthorne, W. R. Factors affecting the design of jet turbines. *Quart. Trans. SAE 54*, (*7*), 347–357 (1946).
12. Betz, A., and Flugge-Lotz, I. Design of centrifugal impeller blades. *NACA Tech. Mem. 902*, 1939.
13. Ellis, G. O., and Stanitz, J. D. Two dimensional compressible flow in centrifugal compressors with logarithmic-spiral blades. *NACA Tech. Note 2255*, 1951.

14. Hamrick, J. T., et al. Method of analysis for compressible flow through mixed-flow centrifugal impellers of arbitrary design. *NACA Tech. Note 2165*, 1950.
15. Lin, C. C. On the subsonic flow through circular and straight lattices of airfoils. *J. Math. and Phys. 28*, (*2*), 1949.
16. Muller, W. *Mathematische Stromungslehre.* Springer, Berlin, 1928.
17. Sorensen, E. Potential flow through centrifugal pumps and turbines. *NACA Tech. Mem. 973*, 1941.
18. Stanitz, J. D. Two dimensional, compressible flow in conical, mixed-flow compressors. *NACA Tech. Note 1744*, 1948.
19. Cohen, H. Compressibility effect on the flow of an ideal fluid in a radial passage. *Roy. Aircraft Establishment Rept. Aero E3940*, 1942.
20. Giraud, F. L. Theoretical and experimental investigations on supersonic free-vortex flow. *Mass. Inst. Technol. Gas Turbine Lab. Rept.*, June 1951. (Unpublished.)
21. Ringleb, F. Exakte Losungen der Differentialgleichungen einer Adiabatischen Gasstromung. *Z. angew Math. u. Mech. 20*, 185–198 (1940).
22. Giraud, F. L. Determination of solutions of the hydrodynamical equation as an initial value problem. *Mass. Inst. Technol. Gas Turbine Lab. Rept.*, 1950. (Unpublished.)
23. Anderson, R. J., Ritter, W. K., and Parsons, S. R. Apparent effect of inlet temperature on adiabatic efficiency of centrifugal compressors. *NACA Tech. Note 1490*, 1947.

Bibliography

Barina, F. J. Comparative performance of two vaneless diffusers designed with different rates of passage curvature for mixed-flow impellers. *NACA Tech. Note 1490*, 1947.

Beede, W. L., Kovach, K., and Creagh, J. W. R. Performance of J-33-A-21, turbojet-engine compressor, K-overall performance characteristics at equivalent impeller speeds from 6000 to 13,400 rpm. *NACA Research Mem. SE8C15*, Mar. 1948.

Brooke, G. V. Surging of centrifugal superchargers. *Brit. Aeronaut. Research Council Repts. and Mem. 1503*, 1932.

Brooke, G. V. Tests on the effect of fuel evaporation on the performance of a centrifugal supercharger. *Brit. Aeronaut. Research Council Repts. and Mem. 1574*, 1934.

Bullock, R. O., Wilcox, W. W., and Moses, J. J. Experimental and theoretical studies of surging in continuous-flow compressors. *NACA Tech. Note 1213*, 1947.

Capon, R. S., and Brooke, G. V. The application of dimensional analysis to air compressors with special reference to the variation of engine performance with inlet conditions. *Brit. Aeronaut. Research Council Repts. and Mem. 1336*, 1930.

Cheshire, L. J. Design and development of centrifugal compressors for aircraft gas turbines. *Proc. Inst. Mech. Engrs. London 153*, 1945.

Church, A. H. *Centrifugal Pumps and Blowers.* Wiley, 1944.

Dugan, P. D., Mahoney, J. J., and Benser, W. A. Effect of Mach number on performance of an axial flow compressor rotor-blade row. *NACA Research Mem. E8D29*, Sept. 1948.

Eckert, B. A collection of compressor test results. *Navy Dept. Buships 338*, (*22*), 1946.

Eckert, B. Calculation and design of flow machines in aircraft power plants. *Navy Dept. Buships 338*, (*12*), 1946.

Eckert, B. Supercharger for aviation engine FKFS 9-900VI (D engine). *Navy Dept. Buships 338*, (*16*), 1946.

Eckert, B. Superchargers for automotive engines. *Navy Dept. Buships 338*, (*17*), 1946.

Ellerbrock, H. J., and Goldstein, A. W. Effect of humidity on several supercharger parameters. *NACA Restricted Rept. 33*, Nov. 1941.

Fischer, K. Investigation of flow in a centrifugal pump. *NACA Tech. Mem. 1089*, 1946.

Fischer, K. Untersuchung der Stromung in Einer Zentrifugal Pumpe. *Mitt. Hyd. Inst. der T. H. Munchen 4*, 1931.

Frietsch, D. E. *Ver Deut. Ing. Berlin Forschungsheft 384*, 1937.

Ginsburg, A., Creagh, J. W. R., and Ritter, W. K. Performance investigations of a large centrifugal compressor from an experimental turbojet engine. *NACA Research Mem. E8H13*, Oct. 1948.

Ginsburg, A., Johnsen, I. A., and Redlitz, A. C. Determination of centrifugal-compressor performance on basis of static-pressure measurements in vaneless diffuser. *NACA Tech. Note 1880*, 1949.

Goldstein, A. W., and Wager, A. Attainable circulation of airfoils in cascade. *NACA Tech. Note 1941*, 1949.

Information on unconventional compressors. *Navy Dept. Buships 338*, (*9*), 1946. (Written by group of German engineers.)

Kearton, W. J. *Turbo Blowers and Compressors*. Isaac Pitman & Sons, Ltd., 1931.

King, J. A., and Klein, H. Performance characteristics of a Junkers Jumo 211F engine supercharger with a DVL fully shrouded impeller and scroll diffuser. *NACA Advance Restricted Rept. E5B03*, Feb. 1945.

Kranz, H. Stromung in Spiralgehausen. *Ver. deut. Ing. Berlin Forschungsheft 370*, 1935.

Ku, P. M., and Wang, C. H. A study of the centrifugal supercharger. *NACA* Advance confidential report, June 1940. (Declassified.)

Laskin, E. B., and Kofskey, M. G. Increase in the stable operating range of a mixed flow compressor by means of a surge inhibitor. *NACA Research Mem. L6J01B*, June 1947.

Laskin, E. B., and Kofskey, M. G. Performance of a mixed flow impeller in combination with a semi-vaneless diffuser. *NACA Research Mem. E7C05a*, Apr. 1947.

Legalise, P. Superchargers. *NACA Tech. Mem. 631*, 1931.

Liang, S. P. Performance analysis of centrifugal compressors. *J. Aeronaut. Sci. 16*, (7), 1949.

NACA Subcommittee on Supercharger Compressors. Standard method of graphical presentation of centrifugal compressor performance. *NACA Advance Restricted Rept. E5F13a*, Aug. 1945.

Osborne, W. C., and Morelli, D. A. Head and flow observation on a high efficiency free centrifugal-pump impeller. *Trans. Am. Soc. Mech. Engrs. 72*, 999 (1950).

Patterson, G. N. Modern diffuser design. *Aircraft Eng. 10*, 267 (1938).

Pearson, H., and Bowmer, T. Surging of axial compressors. *Aeronaut. Quart. 1*, 195 (1949).

Seewald, F. The work at the D. V. L. *J. Roy. Aeronaut. Soc. 43*, 508 (1939).

Sheets, H. E. Non-dimensional compressor performance for a range of Mach numbers and molecular weights. *Trans. Am. Soc. Mech. Engrs. 74*, 93 (1952).

Sheets, H. E. The flow through centrifugal compressors and pumps. *Trans. Am. Soc. Mech. Engrs. 72*, 1009–1015 (1950).

Stanitz, J. D. Two-dimensional compressible flow in turbo-machines with conic flow surfaces. *NACA Rept. 935*, 1949.

Stanitz, J. D., and Prian, V. D. A rapid approximate method for determining velocity distribution on impeller blades of centrifugal compressors. *NACA Tech. Note 2421*, 1951.

Stepanoff, A. J. *Centrifugal and Axial Flow Pumps, Theory, Design, and Application*. Wiley, 1948.

Taylor, G. I. Some cases of compressible flows. *Brit. Aeronaut. Research Council Repts. and Mem. 1382*, Feb. 1930.

von der Nuell, W. The design of airplane-engine superchargers. *NACA Tech. Mem. 902*, 1939.

von der Nuell, W. The maximum delivery pressure of single-stage radial superchargers for aircraft engines. *NACA Tech. Mem. 949*, 1940.

von der Nuell, W., and Garve, A. Leistung und Wirkungsgrad bei Flugmotorenladern. *Jahrbuch deut. Luftfahrtforschung*, 1937.

Whittle, F. The turbo-compressor and the supercharging of aero-engines. *J. Roy. Aeronaut. Soc., 35*, 1047 (1931).

SECTION K

BLADING INTERACTION EFFECTS IN TURBINES

H. KRAFT

K,1. Introduction. Gas turbine design practice developed quite naturally from the art and science of the design of steam turbines. As the steam turbine grew from its relatively crude beginnings to the efficient machine of today the conceptions of its operating process had to be refined from the original, one-dimensional picture, corrected by suitable coefficients, to a simultaneous consideration of two, and later three, dimensions.

As soon as good pictures of the two-dimensional flow pattern through turbine elements such as stationary nozzles and rotating "buckets" or "vanes" were available, it was rather obvious that the fourth important basic variable, time, also needed to be considered. These patterns show clearly that the deflection of the flow produced by the element in question is never confined solely to the flow passage. Rather severe distortions of the pattern exist for a good distance both upstream and downstream. Fig. K,1a shows the density contours, obtained by an interferometer [*1*], of the stationary cascade representing the nozzle of an impulse turbine, and Fig. K,1b gives the theoretically computed lines[1] of constant pressure in the flow from the exit of a high deflection cascade.

There are many valid reasons for designing the stationary and the rotating flow elements to perform at a close axial distance from each other. It follows that interaction between these two flow elements must exist and that this mutual interference must be a function of time.

The same conclusion can be arrived at by entirely practical considerations. By far the vast majority of turbine elements perform at subsonic velocities at which it is possible to find the flow performance of the elements by impact tube traverse. This method of "cascade testing" is wel known and needs no further explanation. If such a test is carried through with the necessary care, there is little reason to doubt that the performance thus found is that of the cascade in question under the given upstream and downstream boundary conditions. In general these are such as to insure an evenly distributed upstream and downstream pressure.

[1] By a method developed for the General Electric Company by Th. von Kármán.

Thus it is possible to find, by stationary testing, the flow losses of cascades representing both the stationary and the rotating element of a turbine. The two can also be tested working together as a running turbine stage, a test of which can be carried through with a high degree of

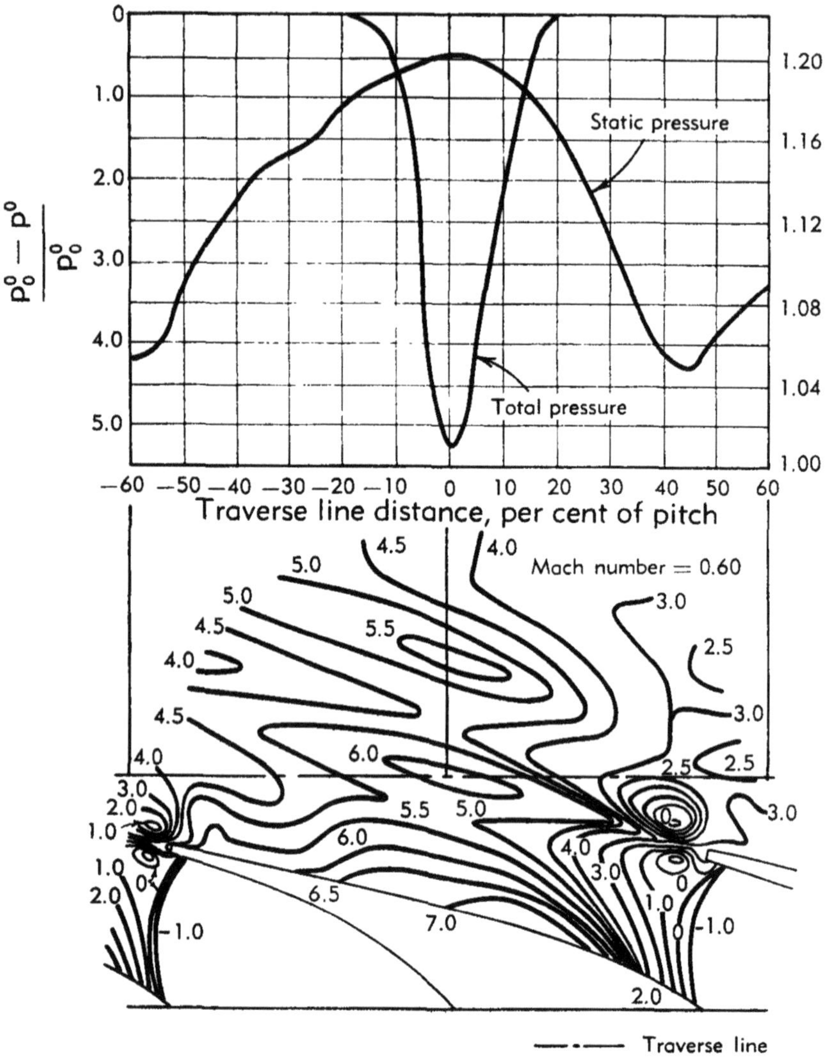

Fig. K,1a. Density field at exit of turbine nozzle.

accuracy. This test will permit the researcher to eliminate, by measurement and calculation, all losses extraneous to those represented by the two cascades. It is then generally found that the flow losses in the running stage exceed those of the combined cascades by an appreciable

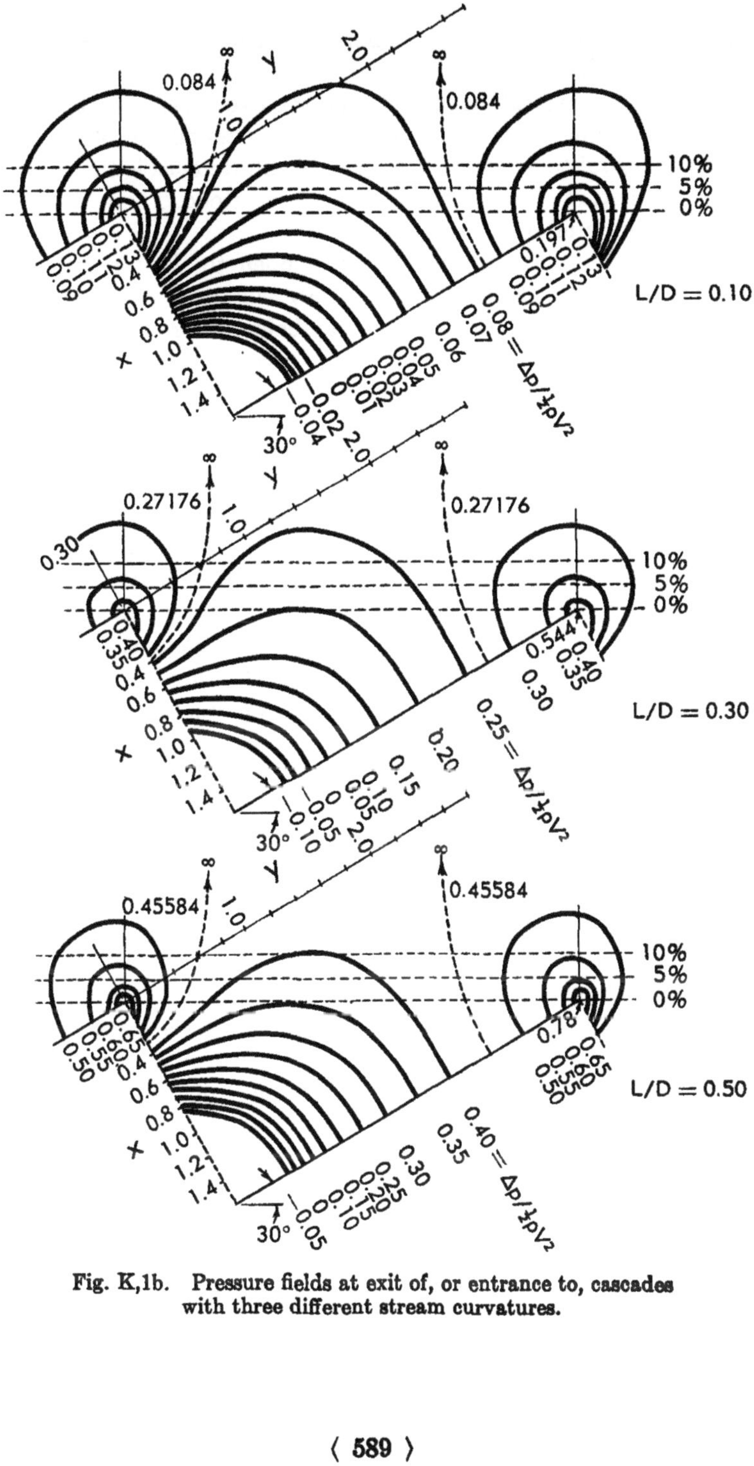

Fig. K,1b. Pressure fields at exit of, or entrance to, cascades with three different stream curvatures.

amount, which ranges between one to three per cent of the stage energy. The picture is not entirely clear—there are some cases where little discrepancy is found.

Since the principal difference between the two types of tests is due to the fact that in the stage test one cascade is passing the other at high speed, it is natural to conclude that mutual interference is the cause of a flow loss, the nature of which is as yet not understood. There is a scarcity of published material on this phenomenon.

In such a situation the only recourse open to the researcher is to examine whatever evidence there is, draw whatever conclusions seem justified, and make suggestions which may help the formation of a theory that correctly describes the phenomenon. Design procedures to minimize this loss will then easily follow.

K,2. Mechanism of Loss. An energy loss in a turbine is primarily a loss in kinetic energy. It is well known that such a loss must result from some viscosity action somewhere. The only exception is the kinetic energy which leaves a given stage. This is accounted for as a loss for the stage under consideration but should be available, at least theoretically, for a succeeding stage.

Fluid friction originates in the boundary layer. It must be concluded that the response of the boundary layers of the running stage to time changes of the flow pattern is the key to an understanding of the mechanism of the loss.

The boundary layer flow may be either laminar or turbulent. If, at Reynolds numbers where a stable turbulent pattern may exist, a laminar boundary layer is maintained by proper contour design, a lower viscous drag will be achieved. If, on the other hand, such a laminar boundary layer is artificially "turbularized" an increased drag must be expected and there is some evidence that such boundary layer changes actually occur in turbines.

Of greater importance is the actual separation of the boundary layer from the surface. It is known that in diffusers, where the static pressure is expected to rise, such a separation can all but destroy the expected pressure rise, with a disastrous effect on the efficiency of the process.

Two conditions must exist to induce a boundary layer to separate from the wall. The pressure in the main stream outside the boundary layer must actually rise, and the boundary layer flow itself must have lost so much of its kinetic energy that the remainder, together with the viscous drag of the main stream, cannot maintain equilibrium with the rising pressure.

Euler's equation relates changes of velocity (both with space and with time) to changes in pressure. If a flow boundary undergoes changes with time and the law of continuity prevails, there are velocity changes due to the contour change and the passage of time. Both contribute to the changes in pressure as shown in Eq. 2-1. It is an integration for one case of Euler's equation attributed to Kelvin,

$$\rho \frac{\partial \phi}{\partial t} + \frac{\rho V^2}{2} + p = \text{const} \tag{2-1}$$

It is not difficult to visualize a boundary layer separation whenever the pressure increases with time at a sufficiently high rate.

Consider the case of a running turbine. A row of moving blades passes a row of stationary profiles. Each of the two rows deflects the flow opposite to the other or, in aerodynamic language, the circulation of the rotating cascade is opposite to that of the stationary row. A periodic mutual interference must be expected which can but mean periodic pressure changes along the profiles. The most obvious phenomenon to be expected is the shedding of periodic vortices if the conditions for separation prevail. This must be expected nearest the place where the interference is highest, i.e. at the exit of the stationary element or at the entrance to the rotating element. That such a rapid shedding of vortices is possible has been demonstrated in the case of many acoustic phenomena.

The questions to be discussed are:

1. Can there be analogical phenomena in turbines involving much higher amounts of energy than those responsible for the acoustic phenomenon?
2. Will such a periodic shedding of vortices occur as a forced phenomenon, not at its own "natural" frequency?
3. Can such a flow pattern occur at frequencies sufficiently variable with time or with such a large number of harmonics that the resulting loud noise will not favor a particular frequency, or a tone, but will be "white," as is the case in most applications?
4. What velocity patterns may be expected?

There is not enough experimental evidence available to answer any of those questions; nevertheless a discussion of the various angles of the phenomenon may be useful in estimating its importance in the design of turbines.

Ideal, incompressible fluid flow. The simplest kind of fluid flow is that of an ideal, incompressible fluid flowing two-dimensionally. Furthermore it may be assumed that no rotation is present, except that represented by the circulation around the profiles. In other words the flow is potential. Fig. K,2 represents the boundaries for the case of four moving profiles extending for the same distance along the cascade axis as do three stationary profiles. The flow pattern repeats itself in this case every three stationary blade spacings. Within this distance it is distorted by the relative position of stationary to rotating blades. Such periodicity, if it does not rigorously exist, can at least be approximated with technical accuracy for all possible relative blade spacings. The total of the circulations around the three stationary elements must equal the circulation around the four corresponding moving elements if no change in tangential component should appear between the entrance and the exit of the stage as a whole. A study of the possible flow pattern in the vicinity of the

Fig. K,2. Example of cascades of a turbine stage. 4 buckets for 3 nozzles.

trailing edge of the stationary and leading edge of the moving profile reveals the following important points:

1. By adjusting the circulation around a given profile, one of the two stagnation points can move to a desired location which may be either a leading or trailing edge, but not both.
2. If a profile is designed to have stagnation points at the two edges in an otherwise undisturbed stream, then the stagnation points must leave this favored position in a stream distorted by another profile.
3. Within the condition that the sum of the circulations of the succeeding profiles of one period is constant, the circulation around the single profiles may vary with relative profile position, i.e. with time, to move either stagnation point to a more probable position.
4. Within the condition that the sum of the flow quantities through the succeeding passages of one period is constant, the flow through the single passages must vary as the circulation varies to assure a constant velocity vector at upstream and downstream infinity.
5. Thus it should be possible by proper adjustment along the cascade axes of both flow and circulation to keep one stagnation point at a position of the profile which may seem to be that preferred by the flow of the actual viscous fluid. Therefore the pressure distribution around the profiles depends to a large extent on the choice the fluid makes in locating the stagnation points.

Therefore, a good prognostication of the boundary layer behavior along a profile has to rely greatly on the ability of the designer to "outguess" the flowing medium. Such guesses, in turn, must conform with the physical phenomenon establishing circulation.

The physical process that establishes the lift on a wing profile is well known. As soon as boundary layers build up, the downstream stagnation point becomes unstable. The more energetic flow of the boundary layer on the suction side pushes that from the pressure side into the interior of the fluid. A "starting vortex" is formed and the stagnation point moves to the trailing edge. The rotation of the starting vortex equals that of the circulation which is established.

An airplane profile is, in general, a large structure. Its trailing edge can be made rather well pointed. Thus the final location of the downstream stagnation point approaches rather closely the geometrical condition which is known as the Joukowski condition. There is little doubt as to its location. In turbine elements the trailing edge is much thicker in per cent of the thickness of the profile and generally it is far from being sharp. While it is physically small it is rather blunt-edged as far as the flow is concerned. Here the Joukowski condition is a far less definite statement than it is in the case of the airfoil. This factor introduces an element of uncertainty with respect to the actual existing circulation of a turbine cascade, much more so because the flow field in the vicinity of the trailing edge of a high deflection cascade is highly distorted.

The theory of vortex motion rules that in a frictionless fluid the amount of rotation present must remain unchanged with time. Boundary layer theory explains the generation of rotation within the fluid by a separation of the boundary layer. It must be concluded that *if* the circulation of a profile changes, rotation is insinuated into the flowing medium which is kinetic energy which has not been used in the stage. It is certainly an energy loss for the stage in question, although it is probably not a total loss as far as succeeding stages are concerned.

However, it may well be asked whether the circulation actually does change. As explained above, a certain transition process is necessary until a starting vortex is formed, a process which must require some finite time. In flow which is well below sonic velocity, it may safely be assumed that the pressure changes and the resulting changes in velocity will establish themselves instantaneously. The viscous actions within the boundary layers must find a mechanism to escape the boundary layer to such an extent that a change in circulation can result. It is difficult to assume that such a process can take place without a time lag. Some students of the problem even assert that, for very high frequencies, circulation cannot change at all.

In a steam turbine as built today, the blade-passing frequency is between 5000 and 20,000 cycles per second. In from $\frac{1}{5000}$ to 1/20,000 of a second the flow pattern must go through a complete cycle of shedding the rotation of the starting vortex. This is a hard assumption to make especially in view of the fact that the phenomenon is not symmetrical to the trailing edge. It probably is easier to decrease rather than increase circulation since the latter requires a change in flow pattern on the pressure side of the profile.

It should be realized that the usual assumption of infinite velocity of the flow around a sharp corner is modified by the presence of the boundary layer. The character of such a flow is rather that of a physical vortex with a rotational vortex core (the boundary layer) and outside of it a potential vortex field. Such a conception makes the assumption of zero change in circulation much more plausible than the mere potential pattern.

In the absence of reliable data it may be assumed that the true solution represents a compromise. It is as difficult to assume that circulation will not change at all as it is to imagine that the full vortex shedding phenomenon can take place in a very short time.

The establishment of a reliable theory proceeds best by using both assumptions and comparing their results with known flow phenomena.

Viscosity. So far in this discussion viscosity entered the picture only insofar as the flow pattern of the frictionless fluid affects the boundary layer and, with it, the circulation. However, there is also a *direct* effect of viscosity on the phenomenon. The boundary layers of the upstream cascade unite at the trailing edge, establishing a wake which periodically impinges on the downstream profiles. The flow through the downstream cascade consists of regions of potential flow and of flow carrying rotation. The wake decays with the distance from its trailing edge, i.e. it gains in width and loses in velocity gradient. The total velocity head of the wake is below that of its surroundings; hence it will definitely change the circulation of the downstream profile if time permits.

The entrance contours of the downstream, moving, cascade are usually designed to furnish the most favorable entrance flow pattern at a given relative velocity between the entering fluid and the rotating profiles. A large, slow wake flowing between high velocity fluid will thus impinge sidewise into the convex wall of the profile. This phenomenon can be readily observed in a water tank. It may be assumed that under the right conditions such periodic impingement may even be beneficial. At the entrance of the profile the wake momentarily moves the stagnation point to the convex wall. This action superimposes on the normal swing of the stagnation point and creates a hydrodynamic phenomenon that should challenge any physicist in search of difficult problems. It is natural to assume that a close study of the effect of the wake will be possible only after a good theory of the ideal flow is available.

Compressibility. The preceding discussion assumed the fluid to be incompressible. One consequence of this simplification is the assumption of an infinitely high sound velocity or a Mach number of zero. While a strictly incompressible theory assumes that every change in flow pattern occurs instantaneously over the whole plane, a compressible theory must consider the spreading of changes of the pattern at a finite velocity. Now, there is not only a distortion due to position but also due to the passage of time. While in the incompressible case there is strict periodicity of the phenomenon, i.e. the flow patterns repeat themselves for the same stator to rotor position, this can only rarely be so in the compressible case. As the Mach numbers increase toward unity, the flowing medium lags only a little behind the changing flow pattern, and wave reflection and interferences will play a role. The problem is very complicated. Approximate solutions of specific cases may be obtained in the near future by some of the large scale electronic digital computers.

Three-dimensional flow. It is generally well realized that two-dimensional considerations furnish only a good approximation to the working process of an axial turbine. Flow components in the third dimension contribute materially to the formation of the energy losses. Within the boundary layer along the end walls of a turning passage there is a parasitic flow pattern closely related to that of the induced drag of the airfoil of finite length. This flow contributes very materially to the energy loss of cascades of small profile length (radial height), a phenomenon which is a direct function of the circulation around the profile. Periodic changes in circulation of the profile can be expected to increase this type of loss. The same can be expected of radial flow components which are either due to profiles notably inclined to the radius or with a circulation varying along the radius. In the former case it is obvious that if the leading edge of the rotor and the trailing edge of the stator are mutually inclined a radial sweep of the interference pattern must be expected, which may be either detrimental or beneficial. In the case of radially variable circulation any time changes of circulation, if they occur, must affect the radial pattern. In a strictly axial machine, radial patterns do not in general contribute to the working process. More light will be shed on these complicated phenomena once a fair picture is obtained of the important two-dimensional phenomenon.

The outline given of this problem so far is, by its very nature, vague and general, and clearly indicates that it is a large and difficult one. It is fitting to ask if there is experimental indication that periodic phenomena of magnitude actually do exist.

K,3. Experimental Evidence. It has already been mentioned that there is a "no man's land" of unexplored and unexplained flow losses in turbine stages. However, the strongest argument that strong periodic forces are at work appears in the record of fatigue failures of turbine blading. Both stationary and rotating members are involved although the failures of rotating blades are, of course, in the vast majority. Every turbine designer is careful to hold the working stresses of turbine blading well below what would normally be considered a very safe stress of the material being used. This is a clear indication of a periodic change of the force exerted by the working fluid on the blade. The change can be either in direction or in magnitude, or both. The former may take place without a time change of blade circulation, the latter involves circulation changes.

A good description of periodic forces acting on the blades of a centrifugal impeller is given in [2]. Here a pump impeller was visually observed by a moving picture camera rotating with the impeller. The flow through the impeller was varied to produce a number of approach conditions different from smooth entrance. The pictures clearly showed a periodic formation of vortices. The reaction on one blade was measured and appeared to be of a sinusoidal nature. It must be remarked that this phenomenon was self-excited, i.e. the periodicity was due entirely to the

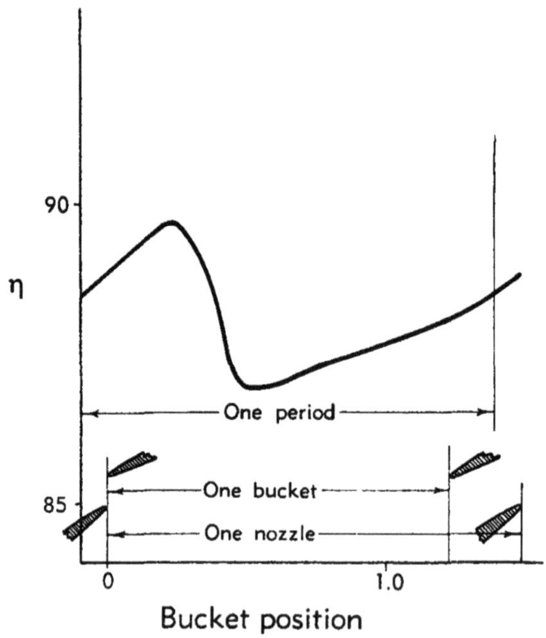

Fig. K,3a. Steady state efficiency variation with bucket to nozzle position. Approximately 3 buckets per 2 nozzles.

flow pattern itself. However, it must be considered that in the turbine interference phenomenon one of the most important effects is also the periodic variation of the effective approach angle. The experiments were made at a Reynolds number too low to be of great interest in turbine design; and the Reynolds number will have an important effect on the viscosity effects originating from the changing flow patterns.

The writer is familiar with a number of experiments on a single-stage air turbine. Three rotating cascades (buckets) of identical contour but varying through a range in scale of three to one were tested behind the same stationary cascade. Thus the effect of one important variable, the ratio of the number of stationary to that of rotating profiles was varied in this ratio. The results were rather inconclusive insofar as no appreciable variation in over-all performance was measured. It must be realized, however, that such an experiment can never be made by changing merely one variable. In order to make this test, not only the ratio of stator to rotor blades was changed, but also the Reynolds number of the rotor passages and their important aspect ratio. The fact that the over-all performance did not change much would indicate that there must have been

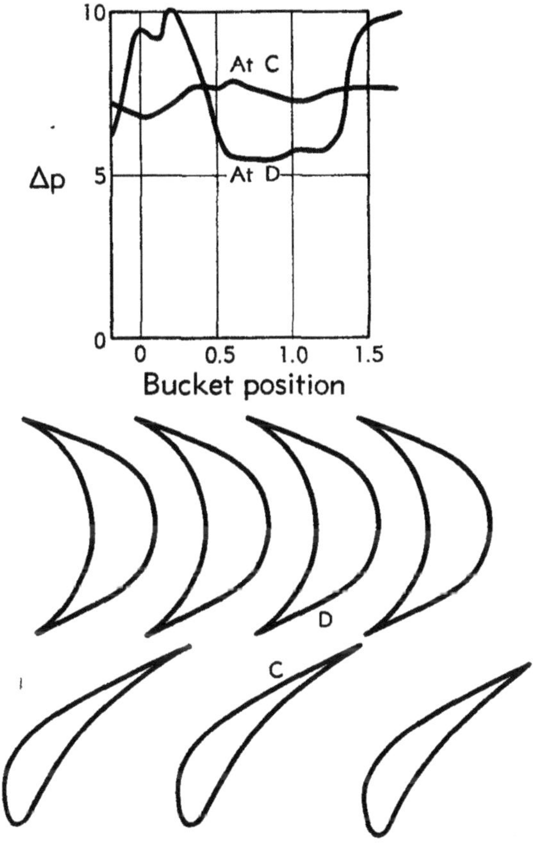

Fig. K,3b. Steady state pressure variation with bucket to nozzle position. Approximately 3 buckets per 2 nozzles.

some change in the loss pattern which counteracted the substantial effects which the latter two variables were expected to have. This short discussion may serve to illustrate the great difficulty the turbine designer faces when he tries to dissect the turbine working process into partial, interconnected phenomena.

Somewhat more interesting evidence was found by means of cascade testing. Rotor cascades of different scale were tested behind a stator cascade of correct angle at a large number of relative trailing to leading edge positions, thus freezing, so to speak, the motion into a large number of steady state patterns. Impact tube traverses and pressure measurements were made. They both gave conclusive evidence that the flow pattern undergoes large variations with relative position. Fig. K,3a, K,3b, K,3c, and K,3d give representative samples of these results.

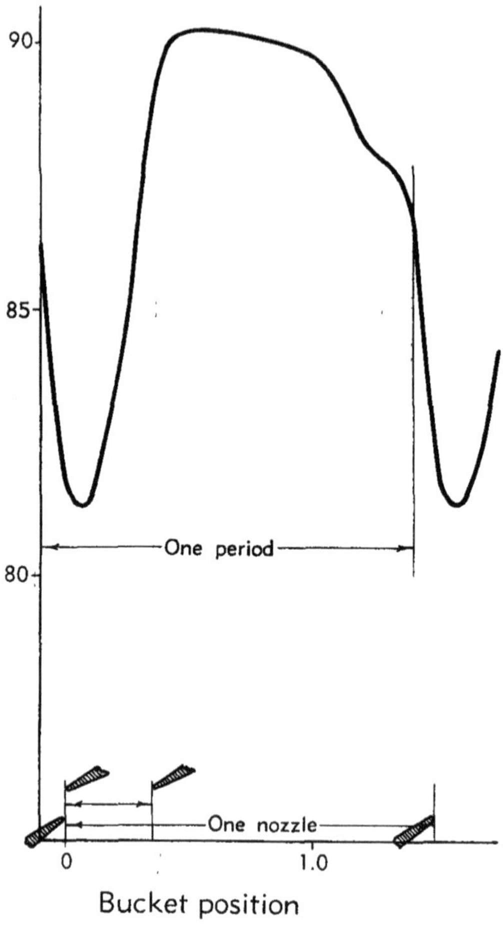

Fig. K,3c. Steady state efficiency variation with bucket to nozzle position. Approximately 5 buckets per nozzle.

A further indication of the effect of flow patterns changing with time is shown in Plate K,3 and Fig. K,3e which are taken from experiments made at Volkenrode, Germany, by Zobel [3]. Interferometer pictures were made of an airfoil oscillating in a wind tunnel, which clearly show the effects of the unsteadiness of the flow. Specifically it is of interest to see the large effect on the boundary layer on the suction side. The interferometer stripes are lines of constant air density. It is apparent that here, in the case of low frequencies, the circulation does change.

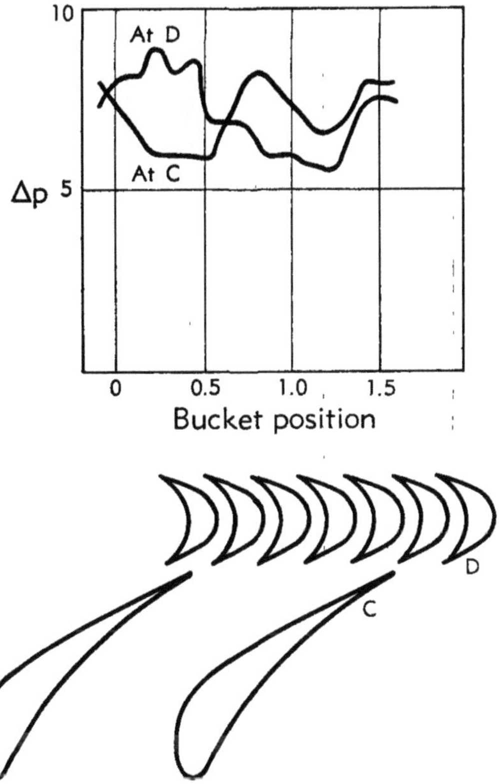

Fig. K,3d. Steady state static pressure variation with bucket to nozzle position. Approximately 5 buckets per nozzle.

K,4. Theory. There are several theories for the flutter of airplane wings, but very few for the running turbine stage. A pioneering approach was made by Keller in his dissertation [*4*].

This theory assumes a periodic change of the circulation of the rotating profiles. As described above, this change must be accompanied by a periodic rotation (starting vortices) in the flow leaving the profiles. The kinetic energy of this rotational flow pattern is taken as the loss due to the blade interference. Since this theory does not concern itself directly with the potential flow pattern involved in the interference phenomenon, the author of this section began a study that was aimed at computing such flow patterns.

The flow under investigation is the flow relative to the moving bucket cascade, i.e. the flow as charted by an observer traveling with the buckets; it is assumed to be incompressible, i.e. the Mach number is zero. Such a flow pattern is obtained if a constant velocity vector equal to that of the bucket speed is subtracted from the actual flow pattern; thus an absolute

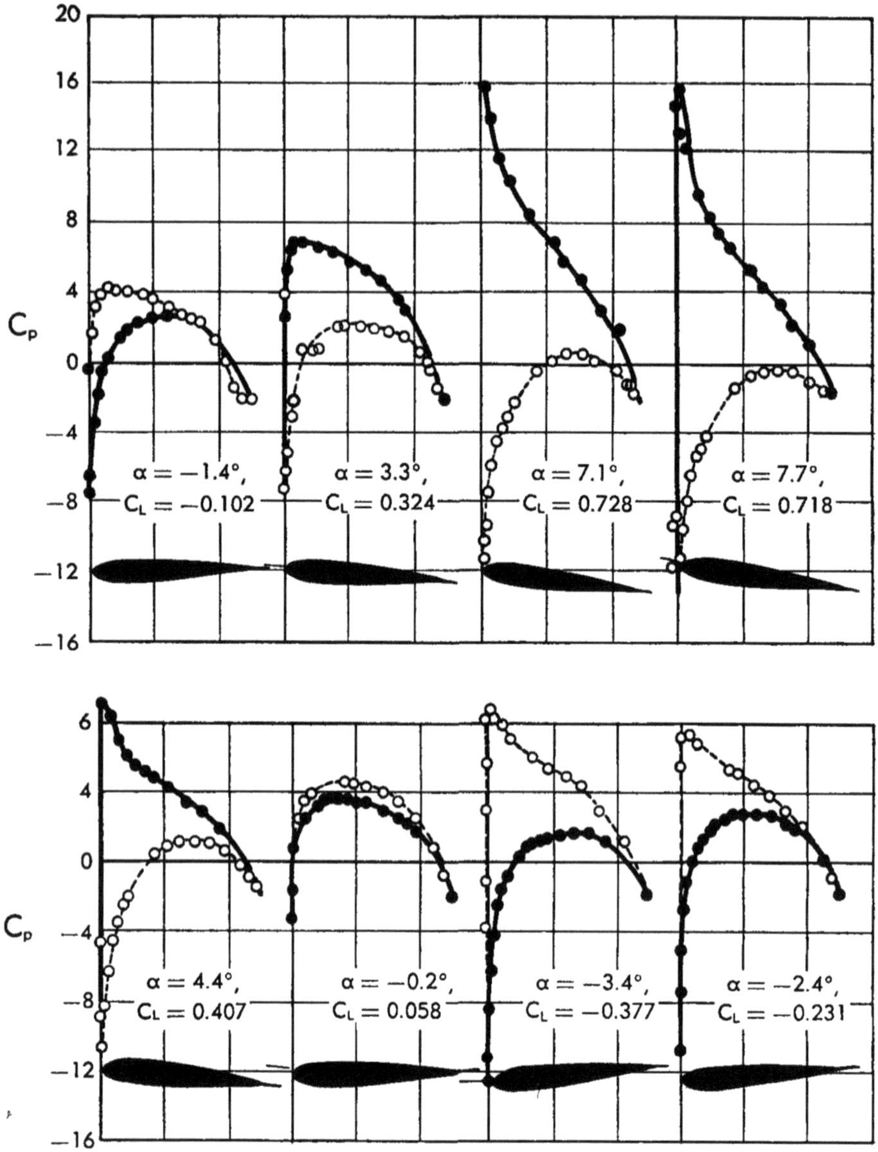

Fig. K,3e. Unsteady pressure distribution on a fluttering airfoil.

axial approach flow to the stationary nozzle cascade appears, relatively, as composed of oblique parallel streamlines.

To simplify the computation the nozzle cascade is represented by a row of single vortices, the effect of which is studied as they pass the bucket entrance. Some distance downstream from the vortex centers, the velocity distribution due to these vortices does not materially differ from

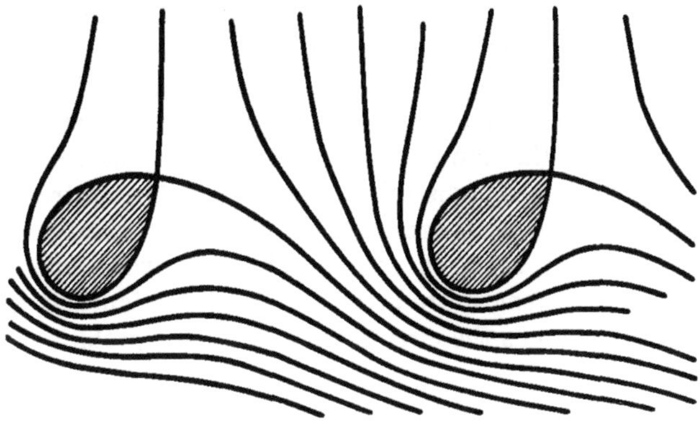

Fig. K,4a. Equivalent absolute nozzle flow furnished by vortices.

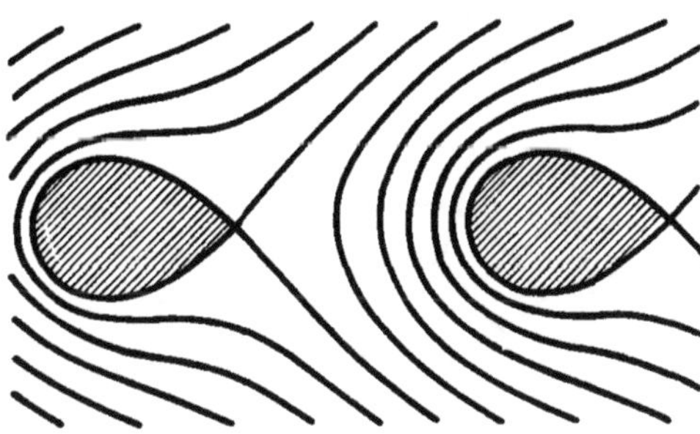

Fig. K,4b. Equivalent relative nozzle flow furnished by vortices.

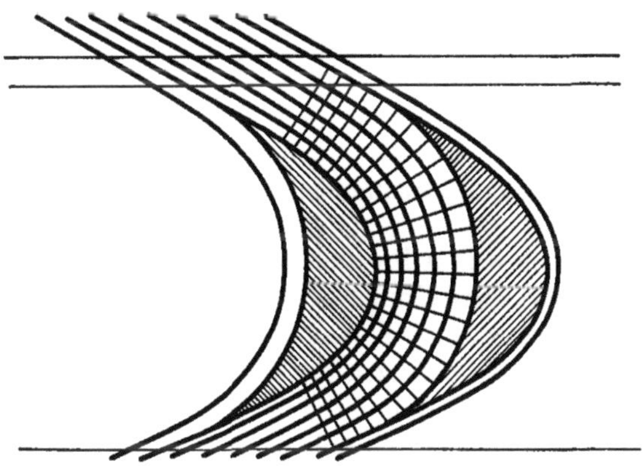

Fig. K,4c. Relative bucket cascade flow for undisturbed approach.

that measured behind an actual nozzle cascade. Fig. K,4a and K,4b explain this simplification. The moving cascade is the row of buckets, computed for parallel approach, shown in Fig. K,4c. Its pattern is transformed by conformal transformation into flow within a circle, the latter representing the bucket contour. (See Fig. K,4d.)

The center is upstream infinity, while downstream infinity is located very close to the circle. For this reason the flow approaching downstream infinity is shown in Fig. K,4e enlarged approximately 10^6 times. This situation illustrates the fact that, in a bucket cascade, upstream conditions cannot affect the downstream pattern to any noticeable degree. In the case of the straight undisturbed cascade flow the upstream infinity pattern consists of a vortex source, while downstream infinity is a vortex sink. Source and sink strength equal each other while upstream and downstream vortex strength are given by the deflection diagram of the cascade; their sum equals the circulation around the profile.

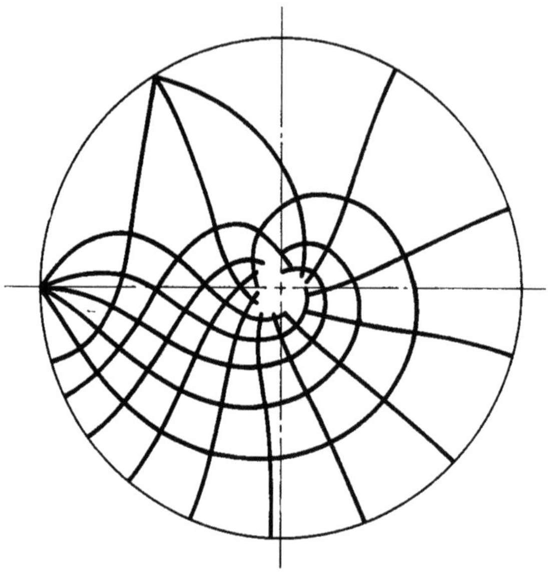

Fig. K,4d. Conformal transformation of flow shown in Fig. K,4c.

If the upstream flow is redirected by a vortex located in front of the bucket cascade this vortex must contain twice the upstream vortex strength of the previous case while at upstream infinity (center of circle) is a vortex of the original strength but with opposite rotation. To maintain the unit circle as a streamline the vortex of double strength must be reflected on the circle. By an additional simple transformation this pattern can be transformed into another circle representing a number of adjoining flow passages with the bucket contours located in sequence on the circle as shown in Fig. K,4f. A vortex path in front of the bucket cascade and parallel to its axis appears as a closed curve surrounding the center. Along this path one or more equally spaced vortices can be thought to move in front of the images of the buckets. The effect of their motion on the velocities along the bucket contours is computed.

To do this, sequences of flow patterns corresponding to the passage of one rotating profile past one stationary passage were calculated. This is equivalent to an infinitely slow rotation of the bucket picture circle around its center and thus around the stationary vortices. The pressure

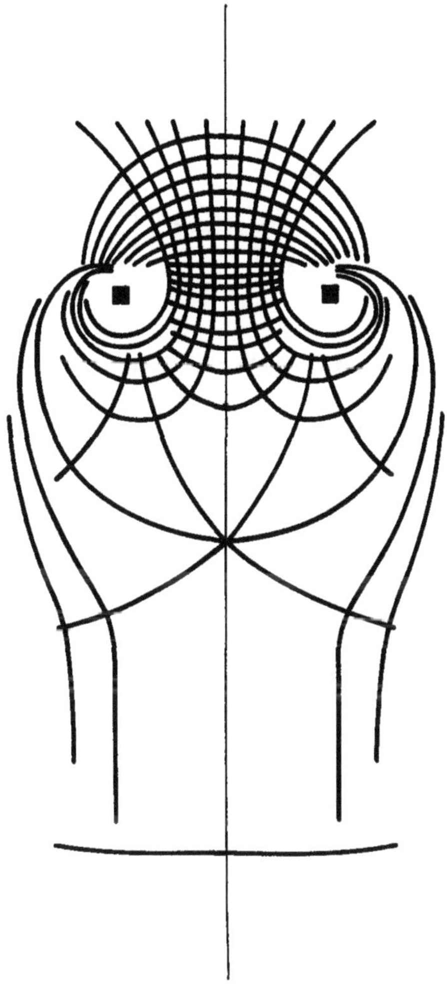

Fig. K,4e. Enlarged downstream section of picture circle shown in Fig. K,4d.

along the rotating profile was then found from Kelvin's equation. The pressure changes thus arrived at are very high, especially in the case where many rotating profiles pass few stationary blades.

The representation of the stationary nozzle profile by a single vortex undoubtedly exaggerates the flow pattern. On the other hand, the computation permits the stationary "profile" to adapt its shape to the flow

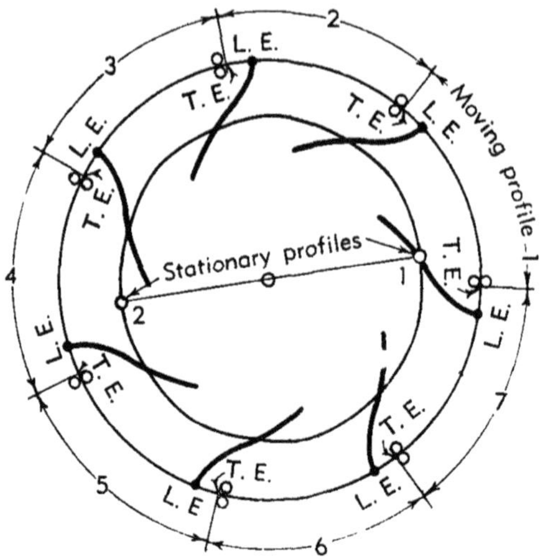

Fig. K,4f. Picture circle flow for 7 buckets and 2 nozzles.

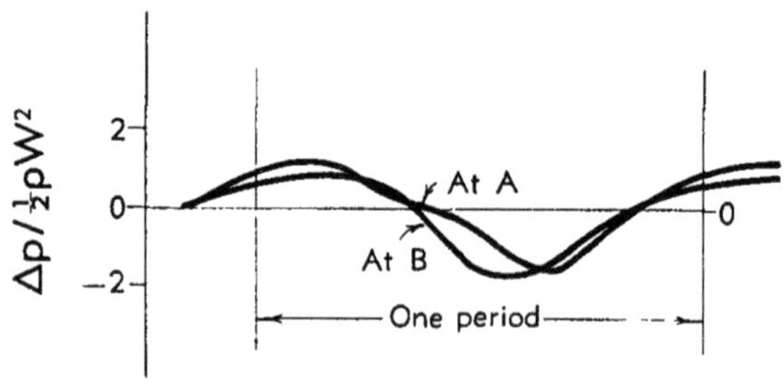

Fig. K,4g. Pressure wave at two positions at convex entrance. Ratio of nozzle profiles to bucket profiles, 1:1.

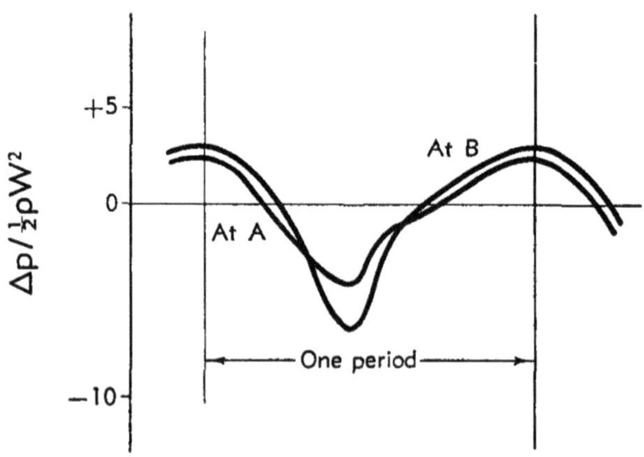

Fig. K,4h. Pressure wave at two positions at convex entrance. Ratio of nozzle profiles to bucket profiles, 2:3.

pattern while an actual profile will not adapt its shape. This factor must counteract the exaggerating effect of the single vortex. Samples of the results thus obtained appear in Fig. K,4g, K,4h, K,4i, K,4j, K,4k, K,4l, and K,4m. They are derived for the assumption that the circulation does not change during the period.

A and *B* refer to two positions on the convex bucket wall, *A* being immediately after the leading edge and *B* at the entrance throat. *S* marks

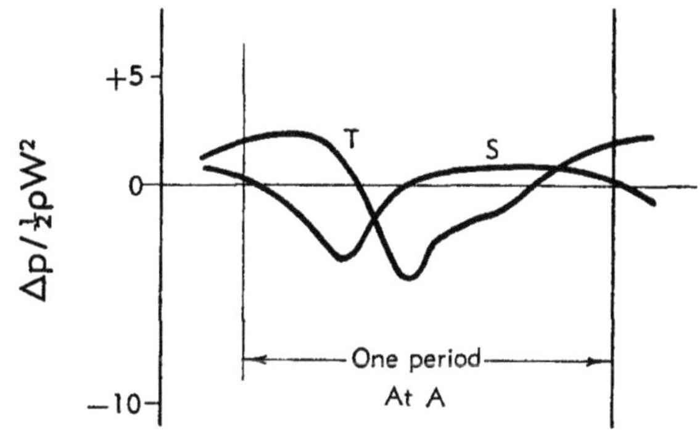

Fig. K,4i. Pressure wave with space and time near convex leading edge. Ratio of nozzle profiles to bucket profiles, 2:3.

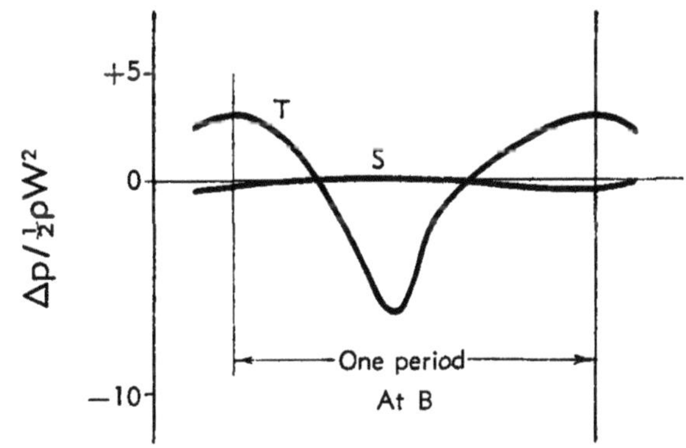

Fig. K,4j. Pressure wave with space and time near convex throat wall. Ratio of nozzle profiles to bucket profiles, 2:3.

the steady state pressure characteristic obtained from the flow pattern while *T* marks the additive time transient $\partial\phi/\partial t$.

It is interesting to note that the curve for the 1:1 spacing ratio is nearly sinusoidal which agrees with the experimental evidence that such turbine stages emit an intolerable siren tone. It is also remarkable that, while the "steady state" variation attenuates from the entrance edge to the throat, the time transient does not. Thus the bucket, over almost the whole entrance, is subjected to an extremely severe pressure pulse. The good agreement in character with the measured curves shown in Fig.

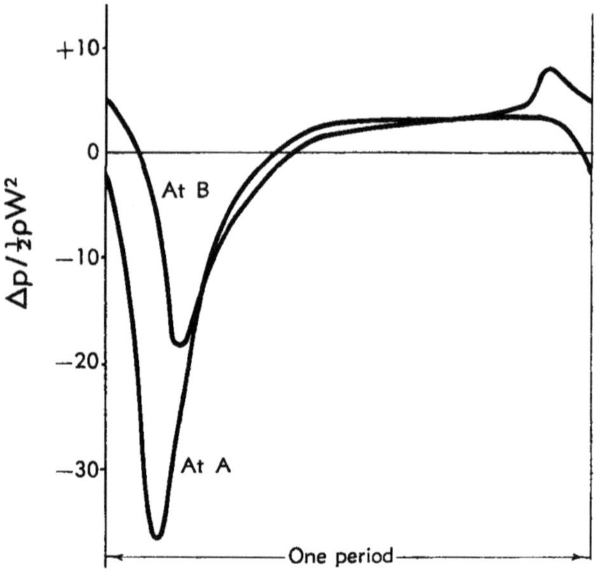

Fig. K,4k. Pressure wave at two positions at convex entrance. Ratio of nozzle profiles to bucket profiles, 1:4.

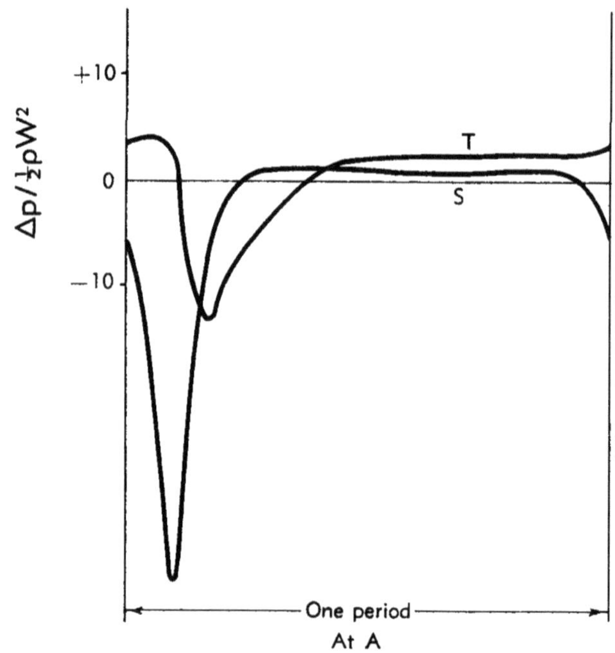

Fig. K,4l. Pressure wave with space and time near convex leading edge. Ratio of nozzle profiles to bucket profiles, 1:4.

K,3a, K,3b, K,3c, and K,3d should also be noted. The latter are steady state measurements of flow with viscosity and actual nozzle profiles while the former apply to the ideal fluid using an idealized nozzle.

The theory can and will be refined and perfected for actual rigid stationary blades.

Compressors. Practically all of the foregoing discussions of turbine phenomena apply equally well to the axial compressor.

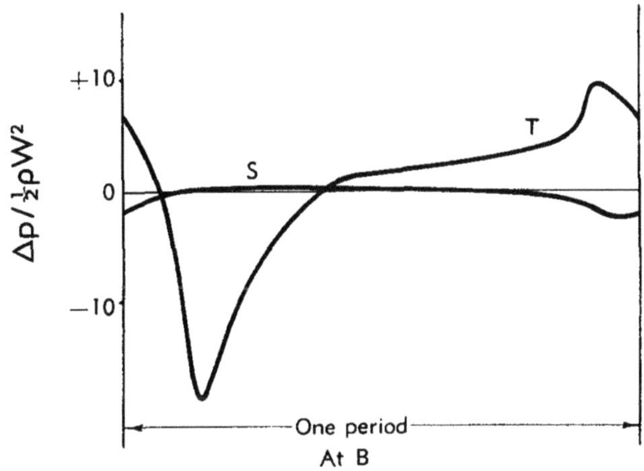

Fig. K,4m. Pressure wave with space and time near convex throat wall. Ratio of nozzle profiles to bucket profiles, 1:4.

K,5. Conclusions. The nature of the meager evidence so far available makes most conclusions conditional.

It can be taken for granted that there are large transient pressure effects in turbine stages.

The computed magnitudes are necessarily modified by compressibility and viscosity. The first will spread the harsh incompressible pattern; the latter can modify it by jet separation and vortex formation.

The amount of separation to be expected is in doubt. At high frequencies the fluid may be quite able to circumnavigate a sharp edge.

It is often proposed to resort to a variable or even a random spacing of nozzles. The present state of knowledge of this problem makes such a complication appear to be of doubtful value.

The most important variable next to the axial edge-to-edge clearance is the relative spacing of stator and rotor profiles. The shape of the profile is also of importance, as can readily be deduced if it is imagined that the picture circle used in the calculation may represent, by stint of different transformation functions, quite different profiles.

Boundary layers in rapid time transients should be studied both in theory and experiment.

K,6. Cited References.

1. Faulders, C. R., and Cole, E. G. Interferometer study of flow through a steam turbine nozzle cascade. *Mass. Inst. Technol. Gas Turbine Lab. Rept.* Dec. 15, 1950.
2. Frietsch, D. E. Wirbelbildung und Kraeftewirkung an umlaufenden Kreiselradschaufeln. *Ver. deut. Ing. Berlin Forschungsheft 384*, 1937.
3. Zobel, T. Fortschritte in der optischen Stroemungsmessung. *Mitt. deut. Akad. Luftfahrtforschung 5008/44g*, 1944.
4. Keller, C. *Axialgeblaese vom Standpunkt der Tragfluegeltheorie. Dissertation*, Eidgenössische Technische Hochschule, Zurich, 1934.

INDEX

Milton Keynes UK
Ingram Content Group UK Ltd.
UKHW020226300324
440331UK00006B/667